MONOGRAPHS ON THE PHYSICS AND CHEMISTRY OF MATERIALS

General Editors

C. E. H. BAWN, H. FRÖHLICH,
P. B. HIRSCH, N. F. MOTT

NEUTRON DIFFRACTION

BY

G. E. BACON

UNIVERSITY OF SHEFFIELD

THIRD EDITION

CLARENDON PRESS · OXFORD
1975

Oxford University Press, Ely House, London W.1

GLASGOW NEW YORK TORONTO MELBOURNE WELLINGTON
CAPE TOWN IBADAN NAIROBI DAR ES SALAAM LUSAKA ADDIS ABABA
DELHI BOMBAY CALCUTTA MADRAS KARACHI LAHORE DACCA
KUALA LUMPUR SINGAPORE HONG KONG TOKYO

ISBN 19 851353 4

FIRST EDITION 1955
SECOND EDITION 1962
THIRD EDITION 1975

PRINTED IN GREAT BRITAIN
J. W. ARROWSMITH LTD,
BRISTOL, ENGLAND

PREFACE TO THE THIRD EDITION

IT IS NOW twenty years since this title was first published and twelve years since the appearance of the second edition. It has become increasingly unrealistic to think of neutron diffraction as no more than a recent extension of the X-ray technique, for it is now in fact half as old as its familiar predecessor, and new fields of application continue to be developed. With improvements in experimental methods, made possible by increases in neutron flux and encouraged by developments in computer programming and data analysis, it has become very much simpler to distinguish and measure separately the different types of neutron scattering which may occur. This is now leading to very rapid progress in fields such as liquids, defects, polymers, biological materials, and many others. In trying to indicate broadly these diverse applications we have concentrated our attention on the principles which are involved, retaining the essential division of the book into two parts, devoted predominantly to fundamentals and to applications. Inevitably the choice of examples from the literature, to illustrate the applications, will prove to be a personal one. It is fashionable to judge the coming of age of a physical technique as the point at which it becomes widely used by chemists, and with neutrons this stage has certainly now been reached.

Nevertheless, the number of users of neutron beams remains quite small, and they form a readily identifiable group. This is because a nuclear reactor with a high flux is needed and for the most recent applications a *very* high flux is required, and at present there are only three suitable reactors in the world. Inevitably, therefore, many readers of this book will not have carried out neutron experiments or may be starting them for the first time. Accordingly it has been judged worthwhile retaining in this edition descriptions of some of the early techniques and measurements, where these are felt to illustrate best the important principles of the subject, although these measurements may now appear to be very unsophisticated in an age of intensively pre-programmed experiments. At the same time I have aimed to present an up-to-date view of the subject, selecting examples from both early and recent work to illustrate the applications as widely as possible.

We have tried to use a consistent set of symbols and units, but this has not always been practicable, bearing in mind a wish to keep the symbols employed by the original authors of papers. Readers of the literature of magnetic scattering and inelastic scattering, in particular, will have found a good deal of variety among the symbols. In the face of this we have generally retained those symbols which we used in the earlier editions.

It is again a great pleasure to acknowledge the many individuals and institutions who have granted permission to use illustrations from their published papers and their photographic records. In each case the sources are indicated in the text and in the captions to the figures. Particular thanks are due to the Brookhaven National Laboratory and Dr. L. M. Corliss for Fig. 6, to the United Kingdom Atomic Energy Authority, Dr. B. T. M. Willis, and Mr. R. F. Dyer for Figs 7, 8, 11, 14, 59, 65, 66, and 186, to the Institut Laue–Langevin for Figs 5 and 91, to Dr. A. L. Rodgers at Aldermaston for Fig. 15, to the Editors of the *Physical Review* and the *Journal of Chemical Physics* in recognition of the large number of new illustrations which come from these journals, and to the International Atomic Energy Agency for the use of many illustrations from their series of conference proceedings devoted to neutron scattering. Drs W. M. Lomer, M. W. Stringfellow, and J. S. Plant are thanked for enlightenment on various points.

Finally it is a personal pleasure to note that so many of the original researchers in neutron diffraction are still actively producing papers, and to recall that many of their successors have commented that they gained their first introduction to the subject from earlier editions of this book.

Edale G.E.B.
April, 1975

FROM THE

PREFACE TO THE SECOND EDITION

DURING the seven years since the first edition of this book was published, neutron diffraction has developed from a research topic to an accepted method of studying solids and liquids. Apparatus for extracting and using beams of neutrons has been set up in most of the many countries all over the world where nuclear reactors have now been installed. At the same time, the appearance of reactors giving higher neutron fluxes has increased considerably the scope and accuracy of the work which can be carried out.

With the large increase in the number of published papers which has followed this growth of the subject, it has become quite impossible to discuss all of them. Our aim has been to devote most space to those which make clear the fundamentals of the subject or indicate most plainly the types and range of applications. The main fields of application remain in magnetism and the location of light atoms, and in each case structural studies are developing increasingly along conventional crystallographic lines, using single crystals. At the same time the availability of more intense neutron beams has increased greatly the attractiveness of studies of inelastic scattering, both from the point of view of magnetism and of lattice dynamics, and a chapter is now devoted to an outline of this work.

Goring G.E.B.
January, 1962

FROM THE

PREFACE TO THE FIRST EDITION

THE aim of this monograph is to give a comprehensive account of the principles, practice, and achievements of neutron diffraction—a technique which has been developed during the last ten years as intense beams of thermal neutrons have become available from atomic piles. The book is written from the point of view of one who sees this technique principally as a method of investigating experimentally the atomic and molecular structure of matter, both as the complement and the extension of the well-established method of study by the diffraction of X-rays. It is hoped that it will be of particular interest to those who are concerned with the study of the solid state, whether from some aspect of fundamental research or of more technological application, and especially to those who would use neutron diffraction methods if the means of doing so were more readily and widely available.

The book is divided into two parts. The first six chapters deal with the physical principles of the subject, the experimental methods, and the determination of the fundamental data essential for structural investigations. The remaining seven chapters deal with the applications of neutron diffraction, with the aim of both giving a detailed account of the studies made so far and also indicating the many and varied fields in which contributions to our knowledge are being made.

Goring G.E.B.
August, 1954

CONTENTS

1. INTRODUCTION 1
1.1. The wave nature of the neutron 1
1.2. Neutron beams from nuclear reactors 2
1.3. Monochromatization 5
1.4. Neutron diffractometers 7
1.5. Neutrons and X-rays 19

2. THE SCATTERING OF NEUTRONS BY ATOMS 21
2.1. Introduction 21
2.2. Neutron scattering by a single nucleus 24
2.3. Assemblies of nuclei in solids 35
2.4. Experimental values of the scattering constants of elements and isotopes 38
2.5. The integrated scattering by polycrystalline substances 50
2.6. The effect of finite nuclear mass: inelastic scattering 54
2.7. X-ray and neutron scattering amplitudes 63

3. DIFFRACTION BY CRYSTALS 65
3.1. Primary and secondary extinction 65
3.2. The coefficients of true absorption for X-rays and neutrons 71
3.3. Ideally-imperfect crystals: plane-parallel slabs 75
3.4. Experimental measurements with single crystals 78
3.5. Measurement of Q values for a zone of reflections 80
3.6. Zachariasen's calculation 84
3.7. Further development of Zachariasen's calculation 85
3.8. Correction for thermal diffuse scattering 90
3.9. The use of powdered crystals 92

4. EXPERIMENTAL TECHNIQUES FOR DIFFRACTION MEASUREMENTS 94
4.1. Production of the monochromatic beam 94
4.2. The choice of monochromator 97
4.3. Analysis of the monochromatic beam: parallel and antiparallel positions 101
4.4. The optics of collimating systems 103
4.5. The powder diffraction method 111
4.6. Single-crystal measurements 124

4.7. Profile refinement in powder diffraction 130
4.8. Measurements at low and high temperatures 133
4.9. Use of 'cold' and 'hot' sources 138
4.10. Techniques using 'white' radiation: time-of-flight crystallography 140

5. FUNDAMENTAL MEASUREMENTS OF SCATTERING AMPLITUDES 155
5.1. General determination of the sign of $\bar{b}$ 155
5.2. General methods of determining the magnitude of $\bar{b}$ 162
5.3. Subsidiary methods of determining $\bar{b}$ 167
5.4. The gravity mirror refractometer: Christiansen filters 170
5.5. *Pendellösung* fringes 178
5.6. Complex scattering amplitudes 182

6. THE PRINCIPLES OF MAGNETIC SCATTERING 189
6.1. Introduction 189
6.2. Paramagnetic scattering 191
6.3. Spin and orbital coupling 198
6.4. Scattering by antiferromagnetic and ferromagnetic materials 202
6.5. Use of unpolarized neutrons 206
6.6. Polarized-neutron beams 213
6.7. Scattering by rare-earth atoms 221

7. MAGNETIC FORM FACTORS 226
7.1. The iron group of transition metals 226
7.2. The 4d metals 241
7.3. The rare-earth elements 245
7.4. Uranium 250
7.5. Covalency effects in salts 251

8. OBSERVATION OF MAGNETIC SCATTERING 259
8.1. Paramagnetic materials 259
8.2. Ferromagnetism 260
8.3. Antiferromagnetism 260
8.4. Spiral-spin arrangements: helimagnetism 261
8.5. Antiphase domains and sinusoidal variations of moments 271
8.6. Magnetic symmetry 275

9. INELASTIC SCATTERING 283
9.1. Introduction 283
9.2. Energy and momentum relations 286

9.3. Experimental measurements and apparatus 288
9.4. Incoherent scattering 299
9.5. Magnetic scattering 303
9.6. Critical scattering 313

SUMMARY of principles and fields of application of neutron diffraction, with a table of comparative properties of X-rays and neutrons 319

10. DETERMINATION OF THE ATOMIC POSITIONS OF LIGHT ELEMENTS 323
10.1. Introduction 323
10.2. Studies with polycrystalline samples 324
10.3. The development of single-crystal methods 328
10.4. Application of Fourier-synthesis methods to single-crystal data 332

11. THE STUDY OF MOLECULAR STRUCTURE 343
11.1. Hydrates 343
11.2. Hydrogen bonds in aromatic compounds 349
11.3. Correlation of X-ray and neutron data 352
11.4. Biological materials 358
11.5. Thermal motion and bond-length determination 363
11.6. Study of inelastic incoherent scattering 369

12. STRUCTURAL INVESTIGATIONS INVOLVING DISTINCTION BETWEEN ATOMS OF NEIGHBOURING ATOMIC NUMBER 380
12.1. Alloys of the transition elements 380
12.2. Spinel structures 386
12.3. Inorganic cyanides 391
12.4. σ-phases 394

13. INVESTIGATIONS OF MAGNETIC MATERIALS I. METALS AND ALLOYS 395
13.1. The transition metals of the 3d group 395
13.2. Rare-earth metals 407
13.3. Rare-earth alloys 418
13.4. Transition-metal alloys 419
13.5. Disorder and defects in ferromagnetic alloys 432

14. INVESTIGATIONS OF MAGNETIC MATERIALS II. OXIDES, SULPHIDES, HALIDES, ETC. 444
14.1. Oxides, sulphides of types MX, MX_2 444
14.2. Transition-group halides MX_2 and related compounds 462
14.3. Haematite and ilmenite structures 470
14.4. Perovskite structures 477
14.5. Manganites $RMnO_3$: non-collinear structures 487
14.6. Iron-group trifluorides 496
14.7. Magnetite, ferrites, and chromites 499
14.8. The sulphates RSO_4 and $CrVO_4$ 506

15. SOME TOPICS IN NEUTRON POLARIZATION 512
15.1. Introduction 512
15.2. Single transmission effect 514
15.3. Double transmission effect 517
15.4. Measurement of domain size 519
15.5. Magnetic refraction at domain boundaries 521
15.6. Reflection by magnetized mirrors 523
15.7. Polarization of the mirror-reflected beam 526
15.8. Polarization effects in Bragg-reflected beams 527
15.9. Polarization analysis 530
15.10. Polarizing filters 541

16. STUDY OF LIQUIDS, GLASSES, AND GASES 544
16.1. Introduction 544
16.2. Space–time correlation functions 546
16.3. Neutron scattering patterns 548
16.4. Experimental measurements for simple liquids 552
16.5. Binary alloys 561
16.6. Molecular liquids 563
16.7. Atomic motion 567
16.8. Structure of glasses 573
16.9. Diffraction by gases 578

17. DEFECTS AND SMALL-ANGLE SCATTERING 581
17.1. Use of long-wavelength neutrons 581
17.2. Transmission measurements of defects 582
17.3. Angular variation of scattering 586
17.4. Non-stoichiometric compounds 588
17.5. Small-angle scattering 591

LIST OF SYMBOLS 603

REFERENCES 605

AUTHOR INDEX 621

SUBJECT INDEX 628

FIGS 5, 6, 7, 8, 14, 64, 91, and 186 appear as separate plates between pp. 10–11.

1
INTRODUCTION

1.1. The wave nature of the neutron

ALTHOUGH it is only since 1945, with the advent of nuclear reactors, that neutron diffraction has become a subject of some importance in the study of solids and liquids, nevertheless it was shown as long ago as 1936 that neutrons could be diffracted. Elsasser (1936) first suggested that the motion of neutrons would be determined by

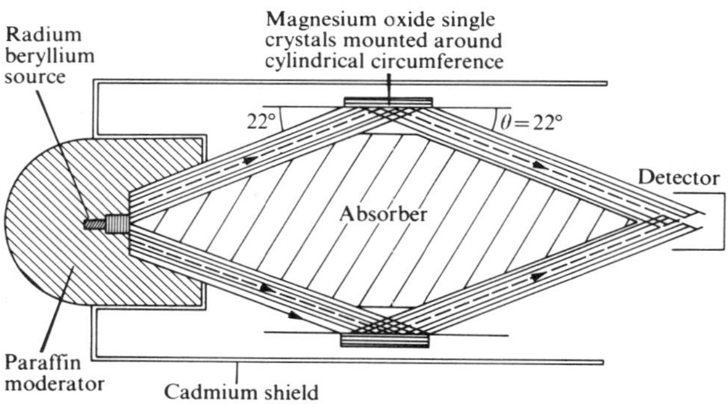

FIG. 1. Mitchell and Powers's apparatus for demonstrating the diffraction of neutrons. (After Mitchell and Powers 1936.)

wave mechanics, so that they would be diffracted by crystalline materials, and shortly afterwards an experimental demonstration of this was given both by Halban and Preiswerk (1936) and by Mitchell and Powers (1936). These experiments were done using a radium–beryllium neutron source, the neutrons from which were by no means of constant velocity, or 'monochromatic', and, while sufficient to demonstrate that diffraction occurred, they were not of course sufficient to provide any quantitative data.

Fig. 1 shows, diagrammatically, a section through the apparatus used by Mitchell and Powers. A radium–beryllium source was surrounded by paraffin to provide a source of thermal neutrons, the peak of whose velocity distribution was at 1·6 Å. Sixteen large crystals of magnesium oxide with their faces parallel to the (100) plane were mounted around the circumference of the cylindrical apparatus, the geometry being such that they were inclined to the neutron beam at 22°, the expected angle of diffraction for the (200) reflection. A large decrease in the number of neutrons received by the counter was noted when the crystals were rotated from this position.

1.2. Neutron beams from nuclear reactors

With the development of nuclear reactors, neutrons have been made available in sufficient numbers to allow collimation into beams and segregation in energy within a fairly narrow band. Thus there has grown up a technique of neutron diffraction on rather similar lines, but on a different scale, to the familiar technique of X-ray diffraction which developed from von Laue's classic discovery of 1912 that crystals would diffract X-rays. The first apparatus of this sort, initially termed a 'neutron spectrometer' but now usually described by the more correct name of diffractometer, was built at the Argonne National Laboratory in the U.S.A. in 1945 (Zinn 1947). Since then many different instruments have been described in the literature, and a recent summary of some of the current apparatus and techniques has been given by Stirling (1973).

From a practical point of view the immense technological investment which is necessary to provide and maintain a nuclear reactor is a disadvantage, and during the last few years several groups of researchers have used neutron beams from linear electron accelerators for diffraction studies (Moore, Kaspar, and Manzel 1968; Kimura, Sugawara, Oyamada, Tomiyoshi, Suzuki, Watanabe, and Takeda 1969).

If neutrons or any other radiation with wave properties are to be of use in investigating the arrangement of atoms in solids, it is necessary for their velocity to be such that they have a wavelength of the same order as the distance of separation of these atoms. This requirement accounts for the very general use of the X-radiation of copper of wavelength 1·54 Å by the X-ray crystallographer. In the case of neutrons the equivalent wavelength λ is given, according to wave theory, by the equation

$$\lambda = h/mv, \tag{1.1}$$

where h is Planck's constant and m and v are the mass and velocity respectively of the neutrons. If we consider neutrons which have made a large number of collisions with atoms in a reactor at temperature T before being allowed to escape from the reactor, then they will have a root-mean-square velocity v appropriate to that temperature, and given by the equation

$$\tfrac{1}{2}mv^2 = \tfrac{3}{2}kT, \tag{1.2}$$

where k is Boltzmann's constant. From eqns (1.1) and (1.2) it follows that

$$\lambda^2 = h^2/3mkT, \tag{1.3}$$

from which it can be shown that the wavelengths corresponding to the root-mean-square velocities of neutrons in equilibrium at temperatures of 0 °C and 100 °C are 1·55 Å and 1·33 Å respectively. It is a fortunate circumstance that these wavelengths are of just the magnitude desired for the investigations of atomic arrangements, since it will be neutrons having temperatures of this order which can be most conveniently obtained from a reactor. These neutrons will have been slowed down by making many collisions with a moderator of graphite or heavy water and will tend to come into thermal equilibrium at the reactor temperature. They will have a distribution of velocities which follows a Maxwellian curve appropriate to a temperature of the order of 100 °C. By inserting a collimator into the face of the reactor a beam of these neutrons can be extracted. The nature of the wavelength distribution in such a collimated beam has been given by Bacon and Thewlis (1949), assuming that the neutrons in the reactor are in equilibrium at temperature T. It is shown that if $\nu_\lambda\,\mathrm{d}\lambda$ is the number of neutrons emerging per second with wavelengths between λ and $\lambda+\mathrm{d}\lambda$, then

$$\nu_\lambda = \frac{2\mathcal{N}_1}{\lambda}\left(\frac{E}{kT}\right)^2 \mathrm{e}^{-E/kT}, \tag{1.4}$$

where $\mathcal{N}_1$ is the total number of neutrons of all velocities emerging per second and E is the energy of a neutron of wavelength λ. The nature of the spread of wavelengths in the beams obtained in practice is illustrated in Fig. 2(a). This curve shows the variation of counting rate, or number of neutrons counted per minute, in the beam diffracted by a single crystal of calcium fluoride set, successively, to reflect various wavelengths. The precise shape of this

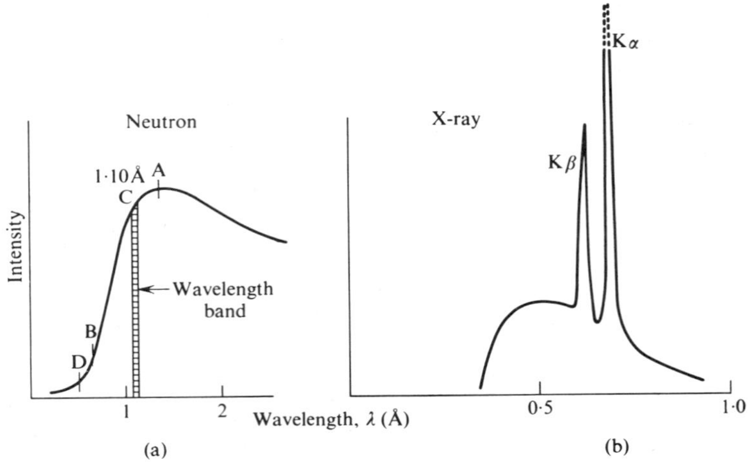

FIG. 2. The intensity versus wavelength distribution (a) for the neutron beam emerging from a reactor, indicating the band of wavelength selected by a monochromator, is contrasted with the distribution (b) from an X-ray tube which gives intense lines of 'characteristic' radiation.

curve will not be identical with that giving the distribution of wavelengths in the incident beam from the reactor, since it will also depend, in particular, on the variation of the effective reflectivity of the crystal with wavelength, which will be discussed later. However, the curve is sufficient to indicate the general nature of the neutron spectrum and, more especially, it will indicate the value of the wavelength which gives maximum counting rate at the detector. It is clear that the spectrum is 'white', and contains nothing to correspond to the characteristic lines in the spectrum from an X-ray tube which is contrasted in Fig. 2(b).

We emphasize that the velocity v which is given by eqn (1.2) is the root-mean-square velocity for neutrons in equilibrium at temperature T. Alternatively, if we wish to specify the velocity v_{p} corresponding to the peak of the distribution curve then it will be given by

$$\tfrac{1}{2}mv_{\mathrm{p}}^2 = kT.$$

On the other hand, if we measure the distribution of velocities among the neutrons which emerge from a collimator we can show that the peak of this curve is given by $\tfrac{1}{2}mv^2 = \tfrac{3}{2}kT$.

Finally, as a further useful way of characterizing the emerging neutrons, we may plot the spectrum in terms of wavelength using a function $\phi(\lambda)$ which is defined by saying that $\phi(\lambda)\,d\lambda$ is the number of neutrons emerging from the collimator per second with wavelengths between λ and $\lambda+d\lambda$. It can be shown that

$$\phi(\lambda) = \frac{\text{constant}}{\lambda^5} \exp(-h^2/2mkT\lambda^2),$$

and the peak of this curve occurs at a wavelength

$$\lambda = h/\sqrt{(5mkT)} \tag{1.5}$$

1.3. Monochromatization

Fig. 3 illustrates, diagrammatically, the experimental arrangement for obtaining a wavelength distribution curve. The neutron beam from the reactor impinges on a crystal which is amply large enough to receive the whole of the beam, and the counting rate is recorded for various values of θ. At any particular angle neutrons will be reflected into the counter if they lie within a small band of wavelength centred about a value λ, satisfying the Bragg equation

$$\lambda = 2d \sin\theta, \tag{1.6}$$

where d is the interplanar spacing of planes parallel to the crystal surface.

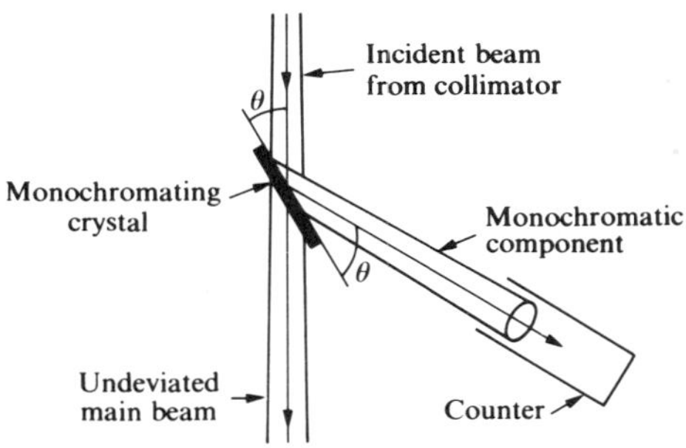

FIG. 3. Measurement of neutron spectrum.

The foregoing considerations of the arrangement indicated in Fig. 3 lead to the simplest method of producing a monochromatic beam of neutrons which, though alternative methods will be discussed later, is commonly an essential preliminary for the examination of crystal structures. By suitable choice of the angle θ, a beam of neutrons of any desired wavelength can be separated out by the crystal and is then available for diffraction by a specimen placed in the position occupied by the counter in the Figure. At first sight the angle chosen would seem to be, for preference, that at which neutrons at the peak A of the spectrum distribution curve in Fig. 2(a) are reflected. In fact, however, the presence of second-order reflections modifies this conclusion. The calcium fluoride crystal used was cut with its surface parallel to the (111) planes; thus, when oriented to reflect the peak wavelength A, at about 1·5 Å, it would also be in the correct position to give a (222) reflection for a wavelength of a half this value, namely 0·75 Å, corresponding to B. From the shape of the curve it is clear that there would be a strong component of this wavelength, giving adulteration of the beam and confusion in interpreting the subsequent diffraction patterns. It is preferable, therefore, to choose a wavelength such as C on the short-wavelength side of the peak intensity, for in this case the second-order wavelength D is at a position well down the counting-rate curve. In fact the intensity of the second-order component in the diffracted beam will depend not only on the shape of the incident neutron spectrum but also on the relative values of the crystal structure factors of the (111) and (222) reflections. This is a subject which will be discussed further; it can be stated here, however, that in practice the second-order contamination of the beam must not exceed 1 per cent. It is particularly important to get the second-order component as small as possible when weak reflections are being sought from antiferromagnetic materials. This is illustrated in Fig. 4 which shows the result of an analysis of the reflected beam by a second crystal. The latter was also a calcium fluoride crystal but, in this case, was arranged to reflect from the (220) planes. The reflection of the main wavelength component, for which $\lambda = 1{\cdot}16$ Å, has a peak intensity of about 4000 counts min^{-1}, whereas the reflection of the component having a wavelength of half this value gives an intensity of only 30 counts min^{-1}. This diffraction peak, which is barely discernible in the Figure, corresponds to less than 1 per cent of the main component. We shall return in Chapter 4 to a discussion of suitable crystals for use as monochromators.

In comparing the above method of producing a 'monochromatic' neutron beam with the familiar use of a monochromator for produc-

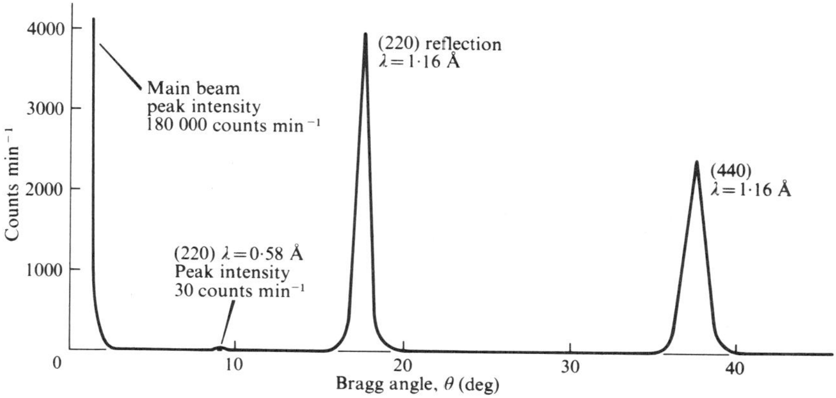

FIG. 4. Analysis of beam from monochromator.

ing crystal-reflected X-rays, an important difference must be emphasized. In the latter case the reflected radiation comprises solely a single line or close doublet of the X-ray spectrum, for example, the molybdenum Kα line, and the monochromator separates this line from the general range of wavelengths which constitute the background of white radiation in Fig. 2(b). In the neutron case there is no equivalent of the line spectrum and the monochromator merely selects a band of wavelengths as indicated in Fig. 2(a). The width of this band of wavelengths is determined largely by the angle of divergence in the horizontal plane of the incident beam from the collimator. In practice this angle has to be relatively large, amounting to $\pm\frac{1}{2}°$ in many cases, in order to give adequate counting rates, with the result that the wavelength band is of the order of 0·05 Å in width. This width is one of the factors which limits the resolution which can be obtained in the diffraction patterns, particularly at the larger values of the Bragg angle.

1.4. Neutron diffractometers

It will be clear that, in practice, the production of a monochromatic beam of neutrons, as briefly described above, has similarities to the production of crystal-reflected X-rays, apart from certain differences which have been emphasized. However, it is when the physical sizes of the two sets of apparatus are considered that it is realized that the details of the two diffraction techniques will be quite different. The intensity of the neutron beam which can be obtained from a nuclear reactor is greater by some orders of magnitude than that from a radium–beryllium source. Nevertheless, in comparison with the number of quanta in the X-ray beams from orthodox diffraction tubes, it is effectively a very weak beam. Consequently, a beam of large cross-section has to be used, collimation can only be relatively poor, specimen dimensions are much greater, and in place of the usual X-ray camera, which may be about 10 cm in diameter, a neutron diffractometer may measure 1 m. It is indeed fortunate that the absorption coefficients of most materials for neutrons are low, otherwise it would not be possible to use the large samples which are essential in order to give adequately large diffracted intensities.

The actual design of the diffractometer and collimator will depend very much on the magnitude of the flux available from the reactor and the kind of sample which it is intended to examine. During the first 10 years or so of neutron diffraction studies the maximum flux available from most reactors was about 2×10^{12} neutrons $cm^{-2} s^{-1}$, but by 1960 most laboratories were using more advanced reactors giving fluxes of about 5×10^{13} neutrons $cm^{-2} s^{-1}$. Consequently the intensities of the neutron beams had been increased by 20 or more times. Since then there have been further increases of flux as reactor technology has developed, and a few reactors are now available with fluxes greater than 10^{15}. At the present time, (1974), the most powerful reactors which are available for experimental work with neutron beams are at the Brookhaven National Laboratory, the Oak Ridge National Laboratory in the U.S.A., and the Institut Laue–Langevin at Grenoble, France. The latter, with a power of 57 MW, is used jointly by scientists in France, Germany, and the United Kingdom and a general account has been given by Ageron (1972). Needless to say, such reactors are immensely expensive, not only to design and construct but also to operate and maintain. For practical reasons no universal design of diffractometer has yet emerged, and although most instruments are very similar in principle they are generally 'tailor-made' to suit local conditions at the reactor face. Particularly is this so with the large instruments which are used for recording the diffraction patterns of powders and which are often only accommodated with difficulty in the congested environment set by neighbouring experimental apparatus. Figs. 5, 6, 7, and 8 (between pp. 10–11) will give the reader a general impression of some of these instruments and of the surroundings in which they are used.

The general principles of these diffractometers are indicated in Fig. 9 which represents the original scheme used by Wollan and Shull at the Oak Ridge Laboratory. A beam of neutrons from a collimating tube in the reactor falls on to a monochromating crystal

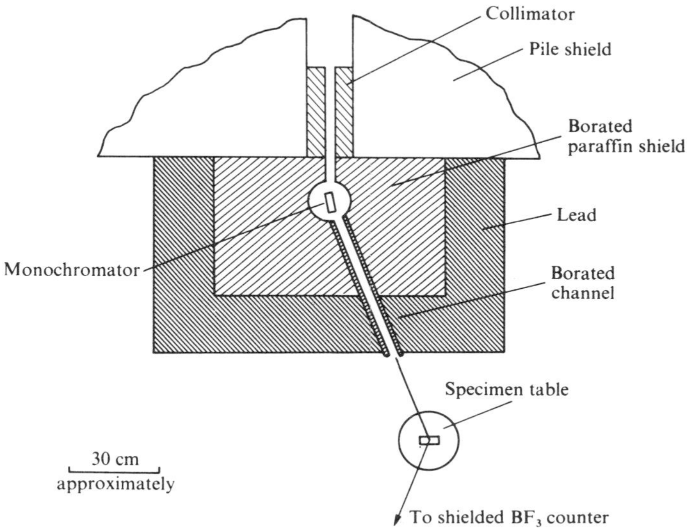

FIG. 9. Experimental layout at Oak Ridge National Laboratory. (Wollan and Shull 1948.)

which is surrounded by a massive shield attached to the reactor face. Shielding of this sort is essential for protecting both the operator and the detecting apparatus from the main beam of neutrons and γ-rays which emerges from the collimator. Only about 1 per cent of this mixed radiation, namely, thermal neutrons having wavelengths within a narrow band, is reflected by the monochromator and used in the subsequent neutron diffraction measurements. Practically the whole of the beam is undeviated by the monochromator, and unless steps were taken to prevent this it would undergo scattering by air molecules and any solid matter in its path. Many of the scattered neutrons would enter the counter and give a large background count against which diffracted neutron beams would have to be distinguished. In practice the peak counting rates from weak powder reflections may only be a few tens of counts per minute, so that it is essential to reduce the background count as far as possible. In the layout of Fig. 9 the fast neutrons are slowed down by collision with the hydrogen atoms in the borated paraffin shield and they are then subsequently absorbed by the boron which has a high absorption coefficient for slow neutrons. On the other hand, the γ-rays are absorbed by the lead screen. Depleted uranium, i.e. uranium from which all the ^{235}U has been extracted, is often used as a very effective shielding material for γ-rays and paraffin wax, as a slowing-down material for fast neutrons, is often replaced by slabs of dense impregnated wood. The latter can very readily be cut and shaped to provide shielding of the required form. An effective system of shielding is important not only to reduce the background of scattered neutrons which would be picked up by the counter, but also to ensure that the general level of radiation in the neighbourhood is sufficiently low from health and safety points of view. Any deficiency in the shield would set a limit on the distance within the reactor that the effective source may be positioned and hence would reduce the intensity available from this source.

Normally the neutron beam emerging from the reactor will have dimensions of the order of a few centimetres square, in order to provide adequate diffracted intensities but without requiring the diffracting specimen and counter to be of impracticable size. Some details of the factors determining the optimum dimensions of the various components will be given later. For the present introduction to the main features of the apparatus it will only be remarked that the collimator may be just a simple hole through the pile shielding, or may be of rather more elaborate construction. Fig. 10 shows a sketch of a practical collimator used in a graphite-moderated reactor. Here the total aperture measures 4 cm vertically and 3 cm horizontally, and the external dimensions of the collimator are such that it can be fitted into an experimental hole in the reactor face which would otherwise be filled with graphite reflector blocks. In order to limit the horizontal divergence of the beam, to increase the angular resolution, the aperture is divided into three sections of approximately 1 cm in width by two vertical steel slats of about $\frac{1}{2}$ mm in thickness. With a collimator 1·5 m in length this yields a beam with a horizontal divergence of about $\pm\frac{3}{8}^{\circ}$. The absorption of steel for neutrons is not high but any neutrons which the steel is required to absorb will be passing through it at very small angles, up to one degree away from the surface, so that their path length

FIG. 5. General view of the high-flux reactor at the Institut Laue–Langevin, Grenoble, during the installation of the neutron scattering apparatus. (Courtesy Institut Laue–Langevin)

FIG. 6. A close-up view of a powder diffractometer at the high-flux reactor of the Brookhaven National Laboratory, U.S.A. The sample under observation is held in a liquid helium cryostat within a magnetic field. (Courtesy Dr. L. M. Corliss)

FIG. 7. A view of the 'Panda' powder diffractometer at the PLUTO reactor of the Atomic Energy Research Establishment, Harwell. (Courtesy U.K.A.E.A.)

FIG. 8. A further view of the 'Panda' diffractometer at Harwell, emphasizing the massive circular shielding around the monochromator, which can be rotated to provide a number of alternative wavelengths. (Courtesy U.K.A.E.A.)

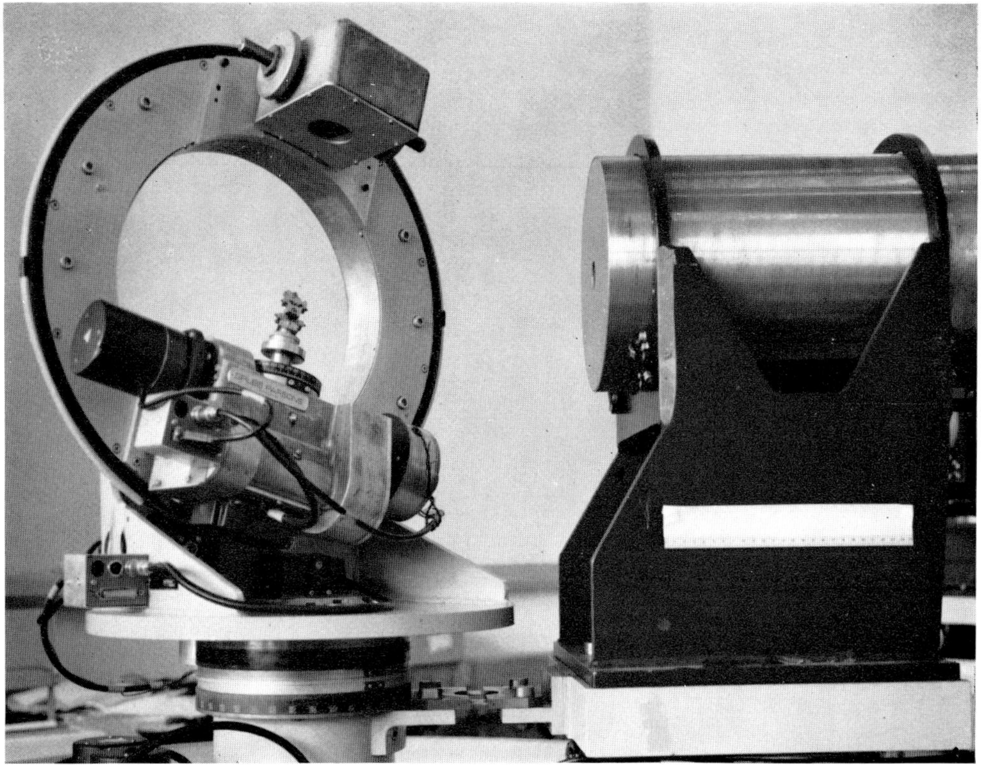

FIG. 14. The 'Mark VI' four-circle diffractometer at the A.E.R.E. Harwell, for three-dimensional structure analysis of single crystals.

FIG. 64. An automatic single-crystal diffractometer at the A.E.R.E. Harwell: the vertical circle which carries the crystal is illustrated in detail in Fig. 65.

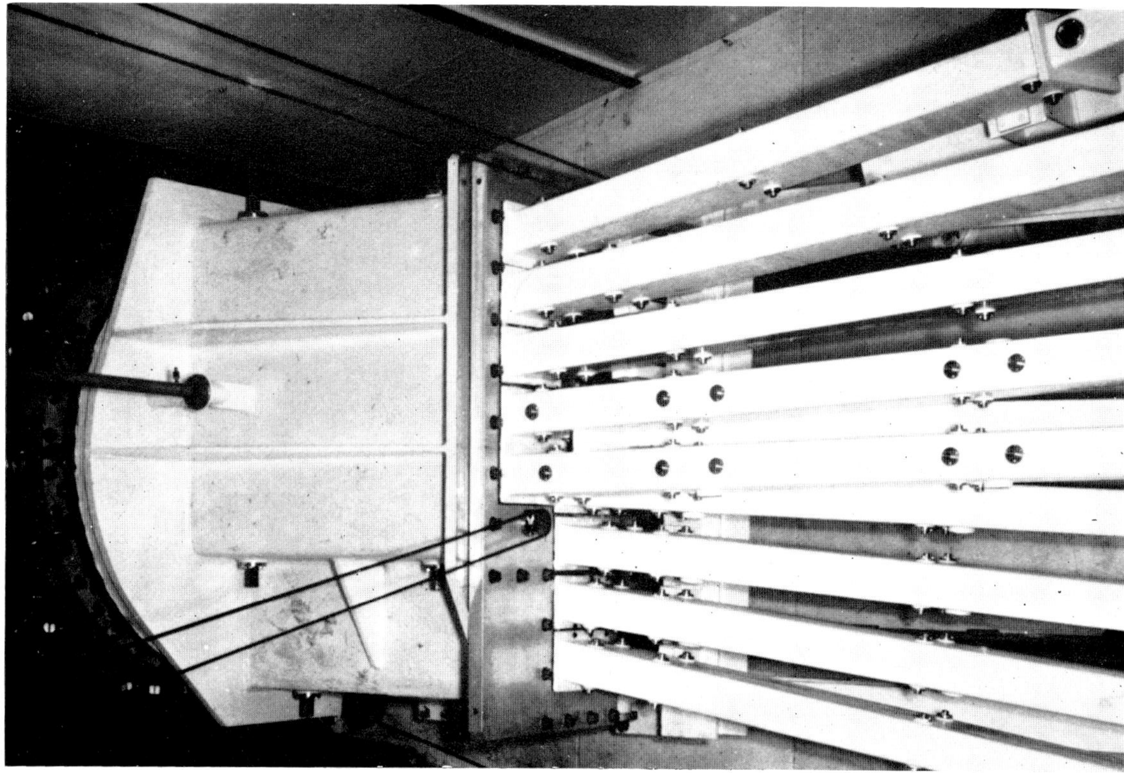

FIG. 91. The group of guide tubes leading from the thermal column of the reactor to the remote guide-hall at the high-flux reactor of the Institut Laue-Langevin at Grenoble.

FIG. 186. A triple-axis spectrometer for measurement of inelastic scattering at the A.E.R.E. Harwell; the large shielding drum in the right foreground houses the monochromating crystal.

(after Chumbley *et al.* 1968).

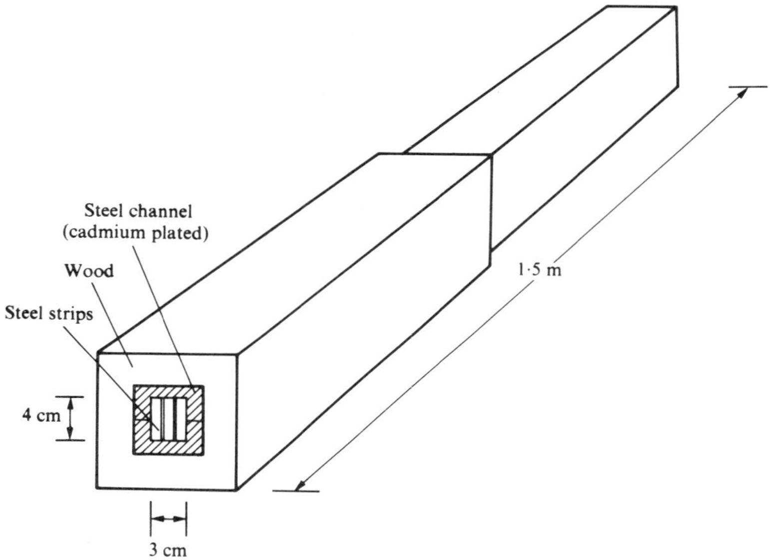

FIG. 10. Sketch of collimator.

in the steel is very considerable. It is not necessary, therefore, to coat the steel with a high neutron absorber such as cadmium. The three sections of the collimator are often called 'Soller slits'. When this particular collimator was used with a lead crystal as monochromator at the first Harwell reactor BEPO, the intensity of the neutron beam reflected from the monochromator was about 10^6 neutrons min^{-1}. This beam, which may be later defined by cadmium slits, is then available for diffraction by a specimen placed on the rotating table at the axis of the spectrometer.

When reactors giving higher neutron fluxes are used a more elaborate form of collimator will be required. Fig. 11 shows a collimator used with the Harwell reactor DIDO, where provision is made for a beam switch which is needed in order to enable adjustments to be made at the monochromator when the reactor is running. This is achieved by making two provisions. First, the collimator tube can be flooded with water which acts as a fast-neutron shield, and secondly, a lead-filled plug can be rotated across the collimator hole to provide a γ-ray shield. With this reactor it is also necessary to seal the collimator into the reactor hole, so

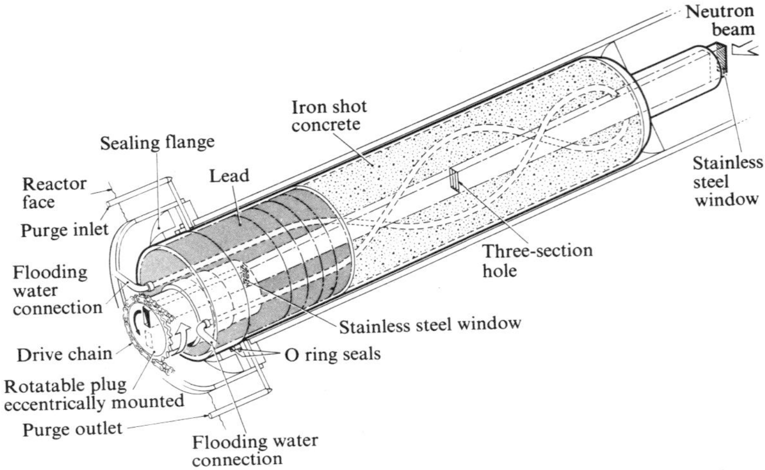

FIG. 11. Details of collimator in DIDO reactor, A.E.R.E. Harwell, with provision for closing aperture and flooding with water. (Courtesy of U.K.A.E.A.)

that any spaces may be purged, for example, with carbon dioxide to remove air, because the nitrogen in the air is converted into nitric acid during irradiation and may cause corrosion.

A general idea of the arrangement of the distribution of the neutron-beam tubes within the interior of a reactor may be gained from Fig. 12, which is a horizontal section through the reactor vessel of the HFBR at the Brookhaven National Laboratory. This vessel is filled with heavy water and its diameter at the height shown, at which the beam tubes emerge, is about 2 m. It is surrounded by a biological shield having a minimum thickness of 2·5 m and made of a mixture of heavy concrete and steel punchings.

The diffracted neutrons are received in a counter which can be rotated about the main axis of the diffractometer. In the majority of the experimental work carried out so far this has been a proportional counter filled with boron trifluoride enriched in the boron isotope of mass number 10. Such cylindrical counters were first described in detail by Fowler and Tunnicliffe (1950). Neutrons, being uncharged, do not produce ionization themselves, and the detection process depends on an absorption reaction from which the products produce ionization. To obtain an adequate counting efficiency it is essential to use the highly absorbing boron isotope ^{10}B. Even then it is necessary to use a counter about 60 cm in length in order to get a counting efficiency as high as 80 per cent. The diameter of the counter may need to be 5 cm in order to receive the diverging beam adequately. The neutrons react with the ^{10}B, which has a capture

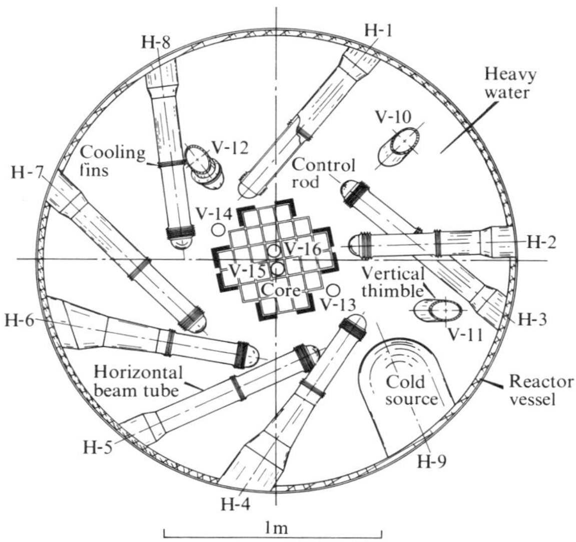

FIG. 12. Horizontal section through the reactor vessel of the high flux beam reactor at the Brookhaven National Laboratory indicating the arrangement of the horizontal beam holes H-1 to H-9 around the core of the reactor. The vertical thimbles V-10 to V-16 for the irradiation of samples are also shown. The large size of H-9 permits the provision of a 'cold source' (see p. 138) for the production of long-wavelength neutrons.

cross-section for thermal neutrons of about 2000 barns, to produce helium and lithium nuclei, according to the reaction

$$^{10}_{5}\mathrm{B} + ^{1}_{0}\mathrm{n} = ^{7}_{3}\mathrm{Li} + ^{4}_{2}\mathrm{He} + 2{\cdot}8\ \mathrm{MeV}.$$

The lithium and helium nuclei have a path length of about 5 mm in BF_3 gas at normal temperature and pressure, and together they produce about 7×10^4 ion pairs. When the counter is operated in the proportional region the electrons produced are accelerated to the anode and produce more electrons by collision with gas molecules. Thus each neutron entering the counter produces a voltage pulse. The pulses produced by γ-radiation are very much smaller and for an energy of 1 MeV the pulse-height is about 1 per cent of the maximum pulse-height due to neutrons. Thus any γ-radiation is well discriminated against. With some counters the terminating support for the central wire may give an unduly large absorption which is troublesome when small crystals and small beams are being used. It is then advisable to off-set the axis of the counter from the line of the beam, either by a slight twist or a parallel displacement. When a counter is surrounded by a sufficient thickness of shielding material, consisting in practice of perhaps 10 cm of borated paraffin or layers of paraffin and boron carbide, the completed counter assembly weighs about 100 kg.

The boron trifluoride counter remains by far the most widely used detector for this work, although several other types have been carefully investigated. Of these the most promising is the helium-3 counter (see Cocking and Webb 1965), in which the filling gas is the helium isotope of mass number 3, which can be produced by irradiating ^{6}Li in a nuclear reactor; the first product ^{3}H subsequently decays to ^{3}He.

The reaction in the ^{3}He counter is

$$^{3}_{2}He + ^{1}_{0}n = ^{1}_{1}H + ^{3}_{1}H + 0{\cdot}76 \text{ MeV},$$

for which the cross-section is about 3000 barns for neutrons of wave length 1 Å. The higher cross-section means that this counter is more efficient than a BF_3 counter of the same length and the same gas pressure, whereas the operating voltage is substantially lower. In fact voltages of 1500 are adequate, compared with 2500 for BF_3. On the other hand, the BF_3 counter is more robust and cheaper.

Because of the heavily shielded counter the conventional neutron spectrometer is not only large in dimensions but massive in construction. In some respects the heavy form of construction of these instruments is an advantage, because it means that they are sufficiently robust to carry the heavy magnets which are often needed in experiments with magnetic materials. Such an assembly can be seen in Fig. 6 which shows a powder diffractometer at the Brookhaven National Laboratory, U.S.A., carrying a low-temperature cryostat and a magnet.

Our description so far has related to general purpose instruments which would be suitable for work with both powdered polycrystalline samples and liquids or with single crystals. Their general dimensions have been decided largely by the low intensities of reflection which are given by powdered samples, which mean that a fairly large sample must be used and the detecting counter must have a high efficiency. If the reactor flux is about 2×10^{12} then a volume of powdered sample of about 5 cm^3 is needed; with a flux of 4×10^{13} the volume need not be more than $\frac{1}{4}$ cm^3 and proportionately for higher fluxes. Even in the latter case the sample diameter may be several millimetres when weak reflections are being studied, and it is necessary to space the counter some distance away from the sample if reasonably good angular resolution is to be obtained. With polycrystalline samples the various diffracted beams are always present, irrespective of any rotation of the sample, and the recording counter will respond to such a beam throughout its transit across it, that is over the angular range from A to B in Fig. 13. Hence the counter must not be too close to the sample if the angular

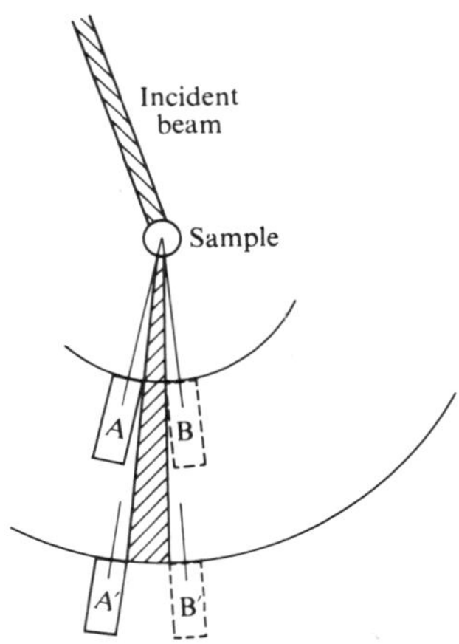

FIG. 13. Diagram to illustrate the improvement in angular resolution of a powder pattern when the sample-to-counter distance is increased.

resolution is not be be impaired unduly. With this large distance a rather massive construction is necessary to carry the heavy shielding around the counter. On the other hand, when single crystals are being used the angular resolution is no longer dependent on the aperture of the rotating counter and the latter can be placed much closer to the sample. Moreover, the single crystal's reflections are more intense, and a shorter counter, say 15 cm in length, will give sufficient sensitivity. Both of these factors mean that a much smaller apparatus is possible if its use is to be limited to studies of single crystals.

Fig. 14 (see between pp. 10–11) illustrates such a single-crystal instrument, and advantage is sometimes taken of the smaller size of this type of diffractometer by incorporating two such instruments at a single beam exit from the reactor. This can be done by mounting two separate monochromators at the reactor face, either in parallel to intercept different areas of the emerging beam or in series to extract different wavelengths, thus ensuring more economical use of the neutrons which the reactor provides. The general layout of the apparatus will depend very much on the nature of any experimental work which is being conducted on the neighbouring holes in the reactor face and, particularly, the usual requirements for heavy massive shielding which is essential to reduce the neutron background to a tolerable level.

In all the illustrations of apparatus which we have shown the detecting counter moves in a horizontal plane, rotating about a vertical axis. This is generally the simplest mechanical arrangement, particularly when it is necessary to mount the sample in a cryostat for observation at low temperature, but it is not the only possible arrangement and has the drawback that it is wasteful of floor space. Accordingly a number of designs have been produced in which the counter moves in a vertical plane, rotating about a horizontal axis. An early design of single-crystal diffractometer at the Oak Ridge National Laboratory was of this type, and Fig. 15 illustrates a recently-developed powder diffractometer of this kind at U.K.A.E.A., Aldermaston. The latter instrument has the further novelty of possessing nine detecting counters, all functioning simultaneously and thus covering a given sector of scattering angle 9 times as quickly. The underlying purpose of this is of course to reduce the time taken to complete a given experiment, bearing in mind that reactors are very costly to operate. The need for carrying out the measurements economically is increasingly great for modern high-flux reactors, in exactly the same way as large modern aircraft have to be employed intensively.

Similar factors to the above must also be taken into account when we consider the manner in which the diffractometers are operated and, particularly, the detailed way in which the diffraction pattern is recorded and measured. In 1945 the procedure was primitive but simple: the detecting counter, connected to a scaling unit, was set at a chosen angular position, and a count made of the number of neutrons arriving in, say, a time interval of 1 min. The counter

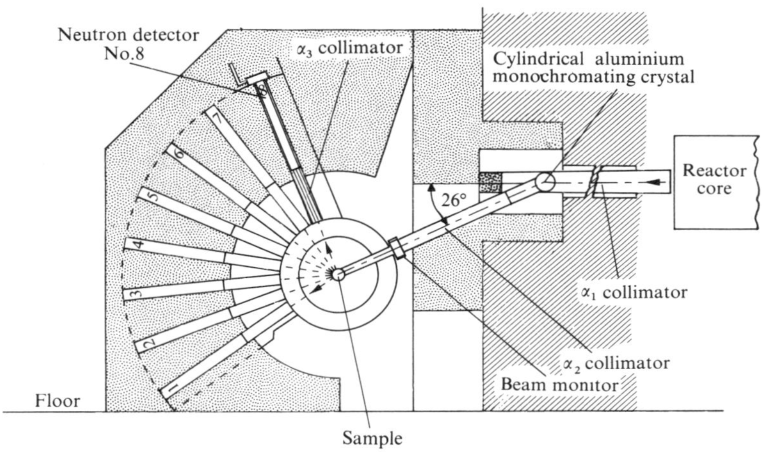

FIG. 15. Diagram of VANESSA powder diffractometer at A.W.R.E. Aldermaston, incorporating eight counters moving in a vertical plane. (Courtesy of A. L. Rodgers.)

was then moved to a new angular position, perhaps 0·2° different in azimuth, and a further count was made. In this way the shape of the diffraction pattern could be plotted out slowly and laboriously. Modern operation is pre-programmed to advance the counter automatically from point to point in the diffraction pattern, remaining at each point for a time which is appropriate to the intensity of the diffracted beam which is found there and taking account of any variations in intensity of the incident beam of neutrons during the period over which the measurements are made. At the same time, when desired, the programme may give instructions for the temperature of a sample to be raised, for a magnetic field to be applied or—systematically in the case of a single crystal—for the sample to be rotated to some new angular position. The instructions may be fed into the machine by, for example, magnetic or paper tape, and the measured data may be stored or recorded in a variety of ways, such as in tabular form or on tape or, by using automatic plotting tables, as a finished diffraction pattern. Often the progress of the experiment can be assessed from time to time when desired by calling for a display on a fluorescent screen of the portion of the diffraction pattern which has been accumulated at that stage of the experiment.

Apart from its dependence on the flux available from the reactor, the form and size of the diffracting sample may vary considerably with the particular problem under investigation. If a single crystal is available it can be mounted directly on a goniometer head on the centre table of the diffractometer, provision often being needed for raising or lowering the sample temperature. If the sample is in the form of polycrystalline powder it will be necessary to enclose it in a thin-walled container, generally of a metal like aluminium which has a low absorption coefficient for neutrons. Usually such a sample is in the form of a cylinder, with the axis vertical and uniformly bathed in the neutron beam, but occasionally it is convenient to use a parallel-sided box as a container. The choice of suitable material for containers and any supporting structures which may lie in the neutron beam becomes much more important, and restricted, with the small samples which can be used with high-flux reactors. Applications of special materials, designed to give very low neutron scattering, will be discussed in Chapter 4.

A further glance at Fig. 2(a) may suggest to the reader that the procedure which we have described, using a monochromator to select a narrow band of almost monochromatic neutrons of closely similar velocities, is a very uneconomical one. Only a very small proportion, about 1 per cent, of the neutrons which emerge from the collimator are utilized during the measurement of the diffraction pattern. The reason for persisting with this wasteful situation is that the task of interpreting the diffraction pattern, in terms of crystallographic structure and atomic and molecular motions, is vastly simplified if the neutrons which contribute to the pattern are all of a single, known wavelength. Nevertheless, a great deal of ingenuity has been applied in recent years to the task of trying to use a much greater proportion of the neutrons within the reactor spectrum. We shall describe some of these investigations in a later chapter, and it will be seen that the alternative methods have some intrinsic advantages of their own apart from ensuring more economical use of the nuclear reactors.

1.5. Neutrons and X-rays

The foregoing introduction will have provided a general background against which the subject matter of the following chapters can be considered. The details of technique which have been briefly mentioned above will be discussed more fully subsequently. In the next two chapters we shall proceed to a study of the process of neutron scattering, first by atoms and then by crystals. We shall draw particular attention to the ways in which neutron scattering differs, either qualitatively or quantitatively, from X-ray scattering, since it will be these differences which will give an advantage to the use of beams of neutrons, rather than X-rays or electrons, in particular cases. Neutrons have indeed proved to be one of the most powerful and versatile tools for studying the constitution of solids and liquids. The field of their applications has developed so much over the last ten years that we are bound to exercise some selectivity of topics for discussion in this book. The continued use of the title *Neutron diffraction* rather than the wider *Neutron scattering* is meant to indicate the way in which this selectivity is to be exercised. We shall be interested primarily in studying the regularity of arrangement which is the basis of the structure of condensed matter rather than with, say, defect structures and the study of lattice dynamics. However, we shall endeavour to deal in outline with these latter topics in order to indicate more fully the versatility of the neutron. It is important, however, at the same time to point out that there is no pretence that neutron diffraction techniques are, or are ever likely to be, a substitute for the classical methods using X-rays.

In certain cases, however, neutron diffraction techniques provide significant additional information where X-rays fail, and it will, of course, be examples of this kind which provide the particular achievements of the technique which we shall consider later in this book. Nevertheless, it cannot be too strongly emphasized that in all investigations of atomic arrangements it is essential to obtain first all the information which can be deduced from X-ray studies of the material under investigation. Usually we shall find that the general pattern of investigation which is adopted is very similar to that employed with X-rays. When single crystals are available they will be employed and the results interpreted by Fourier synthesis or 'least-squares' methods of analysis, with the assistance of electronic computers to reduce the time involved in computation to manageable proportions. In other cases the materials will only be available as polycrystalline fragments or powders, with the consequence, as for X-ray diffraction, that the interpretation of the results in terms of atomic structure may be ambiguous and has to be conducted on a 'trial-and-error' basis. However, we shall see that a process of 'profile refinement' has considerably extended the scope of the powder pattern with neutrons.

2

THE SCATTERING OF NEUTRONS BY ATOMS

2.1. Introduction

A CHARACTERISTIC of crystalline solids is their regular periodic atomic arrangement in three dimensions, and this regularity can be detected and detailed by examining the way in which solids scatter radiation. Diffracted beams are built up from the components of radiation scattered by the individual atoms. These diffracted beams will be our chief concern, although we shall see also that important information can be found by studying the accompanying background scattering. We have already seen in the previous chapter that the wavelengths of the monochromatic beams of thermal neutrons which we shall use for investigating the structures of solids are a little more than 1 Å. This value is not much less than that of the characteristic X-rays used for the corresponding studies by X-ray diffraction. We shall show, however, that the processes by which the two radiations are scattered by atoms are quite different.

In the case of X-ray scattering the fundamental scattering body is the electron which, by virtue of its charge, interacts with the incident X-radiation. It is well known (Compton and Allison 1935) that a free *electron* in an electromagnetic field of amplitude $\mathscr{A}$ gives at a distance r a scattered amplitude equal to

$$\mathscr{A}\frac{e^2}{mc^2}\frac{1}{r}\sin\chi,$$

where e, m are the electron charge and mass, c is the velocity of light, and χ is the angle between the direction of travel of the scattered beam and the direction of vibration of the incident radiation. In particular, for the polarized component whose electric intensity is at right angles to the plane of the incident and reflected beams, the amplitude scattered by the electron is

$$\mathscr{A}\frac{e^2}{mc^2}\frac{1}{r}.$$

The scattered wave from an *atom* will be built up from the contributions from the extra-nuclear electrons, which will be equal in number to Z, the atomic number. In the forward direction these will be in phase with one another, thus giving a resultant amplitude

$$Z\mathscr{A}\frac{e^2}{mc^2}\frac{1}{r},$$

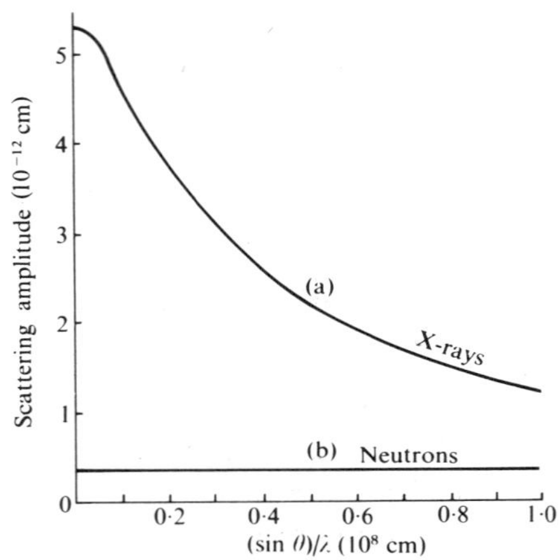

FIG. 16. X-ray and neutron scattering amplitudes for a potassium atom.

which is proportional to the atomic number. As the angle between the incident and scattered directions increases, this amplitude will fall off quite rapidly because the dimensions of the electronic cloud are comparable with the wavelength of the X-rays. For a given atom the rate of fall depends on $(\sin\theta)/\lambda$, where θ is the Bragg angle and 2θ is the angle of scattering, and is described by the well-known 'atomic scattering factor' or 'form factor' f_X. Fig. 16 shows, in curve (a), the variation of f_X with $(\sin\theta)/\lambda$ for the scattering of X-rays by potassium, using the data given in the *International tables for X-ray crystallography.*

In the case of neutron scattering the fundamental scattering body in most atoms is the nucleus and not the electron, except for magnetic materials where we shall find later in Chapter 6 that electronic scattering is also appreciable. The process of nuclear scattering will be studied in detail in the present chapter, but at this point we may simply state that, within a factor of 2 or 3, most atoms scatter neutrons equally well, in contrast to the rapid increase with atomic number of the X-ray scattering amplitude. For example, the neutron scattering amplitude of a lead atom is only about 50 per cent greater than that of a carbon atom, in contrast with a ratio of about 20:1 for the X-ray scattering amplitudes of the two atoms.

Not only does the neutron scattering amplitude show no regular or rapid increase with atomic number but it also shows no variation with the angle θ. This isotropic nature of the scattering is due to the fact that the dimensions of the nucleus, unlike those of the cloud of extranuclear electrons, are small in comparison with the wavelength of 1 Å. Consequently the 'form factor' for neutron scattering is a straight line in Fig. 16. Nor is there any angularly dependent polarization factor, such as occurs for X-rays and which depends on the direction of the incident electromagnetic vibration. However, we shall see in Chapter 6 that magnetic materials may in magnetic fields show polarization effects depending on the direction of neutron spin.

More exactly, there are a number of small interactions between neutrons and the electric charges in an atom, quite apart from the magnetic effect in which the neutron interacts with any atom which possesses a magnetic moment. However, these interactions are vastly smaller, by a factor of about 10^4, than the main interaction of the neutron with the nucleus. Nevertheless, they have been detected experimentally, and the results have been discussed in some detail by Shull (1967). The two most important contributions come from the Foldy (1952, 1958) effect and the Schwinger (1948) effect. The former is a relativistic interaction between the electric charges in the atom and the magnetic moment of the neutron, and it can be shown that the additional scattering length per electronic charge is

$$a_e = -2\pi\frac{e}{hc}\mu_{mag},$$

where μ_{mag} is the neutron moment, and this expression evaluates to

$$a_e = -1{\cdot}463 \times 10^{-16}\ \text{cm}.$$

When both the nuclear and electron charges are taken into account it follows that there is a net Foldy correction for an *atom* given by

$$a_{\mathrm{F}} = 1{\cdot}463Z(1-f_{\mathrm{X}})\times 10^{-16}\ \mathrm{cm}, \tag{2.1}$$

where Z is the atomic number and f_{X} is the form factor which is operative in X-ray scattering. The occurrence of this form factor means that there is a small asymmetry in the angular scattering of neutrons, which was measured experimentally by Hamermesh, Ringo, and Wattenberg (1952) and subsequently by Krohn and Ringo (1966). The observations are consistent with the above expressions. The Schwinger effect is of the same order of magnitude as the Foldy correction and arises through the relative motion of the magnetic moment of the neutron and the atomic electric field, as the neutron passes through the atom. It has been detected by Shull and Ferrier (1963) in observations with polarized neutrons and is further discussed by Shull (1967).

With the foregoing general remarks in mind we proceed to a detailed discussion of the scattering of neutrons by nuclei, atoms, and assemblies of atoms. We shall see that particular attention has to be paid to nuclear spin and to the existence of more than one isotope in many elements. These factors will be found to have very considerable influence on the intensities of both the Bragg reflections, i.e. the diffracted spectra characteristic of the three-dimensional crystal structure, and the background of scattered neutrons.

2.2. Neutron scattering by a single nucleus

If a plane wave of neutrons described by a wave-function

$$\Psi = \mathrm{e}^{\mathrm{i}\kappa z}, \tag{2.2}$$

where $\kappa = 2\pi/\lambda$ is the wave-number, is incident on a nucleus, the scattered wave will be spherically symmetrical of the form

$$\Psi = -(b/r)\,\mathrm{e}^{\mathrm{i}\kappa r}, \tag{2.3}$$

where r is the distance of the point of measurement from the origin at which the nucleus is considered to be rigidly fixed. The quantity b, which has the dimensions of length, is defined as the scattering length and is a complex quantity

$$b \equiv \alpha + \mathrm{i}\beta.$$

However, the imaginary component only becomes important for nuclei which have a high absorption coefficient, such as cadmium and boron which have been investigated experimentally. We shall discuss the example of cadmium in more detail in Chapter 5, but in the present discussion we shall treat the scattering length b as being completely real. The resultant neutron wave will be given by

$$\Psi = e^{i\kappa z} - (b/r)\, e^{i\kappa r}. \tag{2.4}$$

We define the scattering cross-section of the nucleus by

$$\begin{aligned}\sigma &= \frac{\text{outgoing current of scattered neutrons}}{\text{incident neutron flux}}\\ &= 4\pi r^2 v \frac{|(b/r)\, e^{i\kappa r}|^2}{v|e^{i\kappa z}|^2}\\ &= 4\pi b^2,\end{aligned} \tag{2.5}$$

where v is the neutron velocity.

The actual value of the scattering length of any particular nucleus will be determined by boundary conditions at the nuclear surface as discussed by Feshbach, Peaslee, and Weisskopf (1947). Although these conditions cannot be calculated in the present state of knowledge of nuclear structure, it is possible to express them in terms of certain quantities relating to the compound nucleus which is formed by addition of the incident neutron. It is the nature and position of the energy levels of this unstable compound nucleus which determine the cross-sections of the possible nuclear reactions, which may give rise to scattering or absorption of the incident neutrons.

The Breit–Wigner formula: potential and resonance scattering

Breit and Wigner (1936) gave a formula for the scattering and absorption cross-sections in terms of a single resonance energy level. This formula is usually referred to as the 'dispersion formula', owing to its analogy with the formula for the dispersion of light by atomic systems. Various workers have modified and extended this formula, and a fuller discussion can be found in *The theory of atomic collisions* by Mott and Massey (1949). For our present purpose it will be sufficient to give general consideration to the simple formula, according to which the scattering cross-section is given by

$$\sigma = \frac{4\pi}{\kappa^2}\left|\kappa\xi + \frac{\frac{1}{2}\Gamma_n^{(r)}}{(E-E_r)+\frac{1}{2}i(\Gamma_n^{(r)}+\Gamma_a^{(r)})}\right|^2, \tag{2.6}$$

where E is the energy of the incident neutron and E_r is the energy which the neutron must have to produce resonance in the compound nucleus, i.e. the excitation energy of the compound nucleus for resonance is $E_r + E_b$, where E_b is the binding energy of the neutron. $\Gamma_n^{(r)}$ and $\Gamma_a^{(r)}$ are the 'widths' of the resonance for re-emission of the neutron with its original energy and for absorption, respectively. It can be shown that the neutron width $\Gamma_n^{(r)}$ is proportional to the wave-number κ, so that the expression for σ reduces to

$$\sigma = 4\pi \left| \xi + \frac{\text{constant}}{(E - E_r) + \frac{1}{2}i(\Gamma_n^{(r)} + \Gamma_a^{(r)})} \right|^2. \tag{2.7}$$

The two terms in eqn (2.7) correspond to 'potential' and 'resonance' scattering respectively. The potential term ξ is always positive and is equal to the nuclear radius R, so that for potential scattering alone the cross-section σ would be equal to $4\pi R^2$. This is the value which would be expected, according to wave-mechanical theory (Mott and Massey 1949), for the scattering cross-section at low velocities of an impenetrable sphere of the same radius as the nucleus, the value being four times as great as πR^2, which is what would be expected from classical theory. At the highest velocities wave-mechanical theory calculates the cross-section to be $2\pi R^2$. The nuclear radius R is approximately equal to $1{\cdot}5 \times 10^{-13} A^{\frac{1}{3}}$ cm, where A is the mass number, the appearance of the factor $A^{\frac{1}{3}}$ meaning that nuclear matter is of constant density. This relationship is discussed by Bethe (1937). Consequently it follows that when there are no resonance levels sufficiently close in energy to be effective, the scattering amplitude b for thermal neutrons should increase as the cube root of the mass number of the nucleus; under these circumstances the quantity $(E - E_r)$ in the denominator of the resonance term in eqn (2.7) will be large, thus making the resonance term small. This relation is borne out quite well in practice for the heavier elements, as shown in Table 1, from a paper by Feshbach, Peaslee, and Weisskopf (1947), where the scattering cross-section is compared with the value of $4\pi R^2$.

The resonance term will become increasingly large as E_r approaches E, that is when the resonance level comes closer to thermal energy. $\Gamma_n^{(r)}$ and $\Gamma_a^{(r)}$ are essentially positive but $(E - E_r)$ may be positive or negative. Under certain conditions it is possible for the resonance term to be negative and also sufficiently large numerically

TABLE 1

Total elastic scattering cross-sections σ for thermal neutrons

Element	Atomic weight	$4\pi R^2$ (cm^2)	σ (cm^2)
Ti	47·9	$3{\cdot}6\times10^{-24}$	6×10^{-24}
Cr	52·0	3·8	3·8
Mn	54·9	3·8	2·2
Fe	55·8	4·0	11·7
Co	58·9	4·2	5
Ni	58·7	4·4	17
Cu	63·6	4·5	8
Zn	65·4	4·6	4·2
Ge	72·6	5·0	8·5
Se	78·9	5·3	10
Br	79·9	5·4	6
Sr	87·6	5·5	10
Cb	92·9	5·9	6·2
Mo	95·9	6·0	7·4
Ru	101·7	6·1	6
Pd	106·7	6·3	4·8
Ag	107·9	6·4	6·6
Cd	112·4	6·5	5·3
Sn	118·7	6·6	5
Sb	121·8	6·9	4·2
Te	127·6	7·0	5
Ba	137·4	7·5	8
Ta	180·9	9·0	7·2
W	183·9	9·2	5·7
Os	190·2	9·3	14·9
Pt	195·2	9·4	11·2
Hg	200·6	9·8	26·5
Tl	204·3	9·8	9·7
Pb	207·2	9·9	11·6
Bi	209·0	10·0	10·0

Based on Feshbach, Peaslee, and Weisskopf (1947) with some values of σ replaced by later determinations.

to outweigh the potential term, thus giving a resultant scattering amplitude b which is negative.

As an illustration we can calculate the variation of cross-section with neutron wavelength, according to the Breit–Wigner formula, for a particular example. Let us take $\Gamma_n^{(r)}$, $\Gamma_a^{(r)}$ as 10^{-3} eV and 0·1 eV respectively, a value of 1 eV for E_r, the resonance energy, and a value of 10^{-12} cm for the nuclear radius. A little re-arrangement of eqn (2.6) gives

$$\sigma = 4\pi\left|R+\frac{\frac{1}{2}\Gamma_{\mathrm{n}}^{(\mathrm{r})}/\kappa}{(E-E_{\mathrm{r}})+\frac{1}{2}\mathrm{i}\Gamma}\right|^2 \tag{2.8}$$

and, hence,

$$\sigma = 4\pi\left[\left\{R+\frac{\frac{1}{2}(E-E_{\mathrm{r}})\Gamma_{\mathrm{n}}^{(\mathrm{r})}/\kappa}{(E-E_{\mathrm{r}})^2+\frac{1}{4}\Gamma^2}\right\}^2+\frac{1}{16}\left\{\frac{\Gamma\,.\,\Gamma_{\mathrm{n}}^{(\mathrm{r})}/\kappa}{(E-E_{\mathrm{r}})^2+\frac{1}{4}\Gamma^2}\right\}^2\right], \tag{2.9}$$

where we have written Γ for the total width of the resonance, i.e. for the sum $\Gamma_{\mathrm{n}}^{(\mathrm{r})}+\Gamma_{\mathrm{a}}^{(\mathrm{r})}$. Since, as we have previously stated, $\Gamma_{\mathrm{n}}^{(\mathrm{r})}$ is proportional to κ it follows that the expression $\Gamma_{\mathrm{n}}^{(\mathrm{r})}/\kappa$ is a constant value and can be evaluated numerically at the outset of our calculation. We can then evaluate σ for various values of the incident neutron energy E, with the result shown in Fig. 17. As will be seen

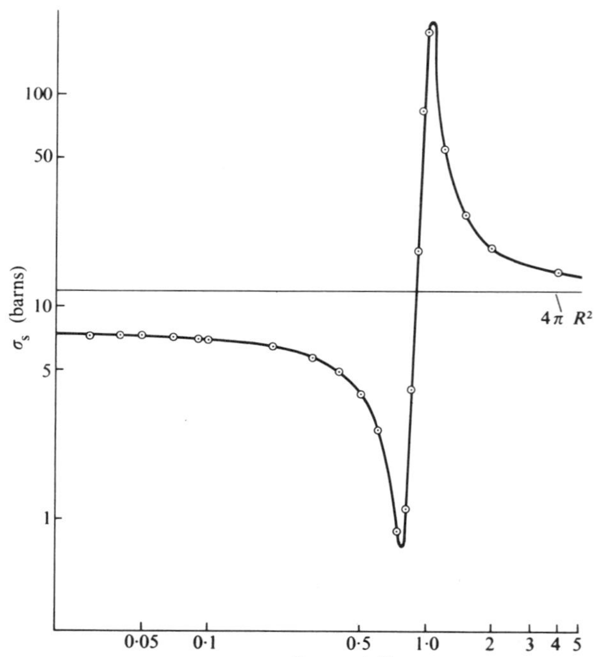

FIG. 17. Calculated variation of scattering cross-section σ_s with neutron energy for a nucleus of radius 10^{-12} cm and a single resonance at 1 eV.

from this Figure, the cross-section goes through a sharp minimum and maximum, reminiscent of the curve for anomalous optical dispersion. At high energies σ tends to the value $4\pi R^2$, and the scattering length is equal to the nuclear radius, but at low energies, i.e. for our thermal neutrons, the values of cross-sections and scattering lengths are substantially smaller than this. This can be deduced easily from eqn (2.9) by noting that for large values of $(E - E_r)$ the term in the second pair of braces will become negligible in comparison with that in the first pair, which for small values of E will then approach

$$\left(R + \frac{\frac{1}{2}\Gamma_{\mathrm{n}}^{(\mathrm{r})}/\kappa}{-E_{\mathrm{r}}}\right). \tag{2.10}$$

Thus the scattering length is less than the value of R by a constant value which depends on the details of the nuclear resonance. The data from the above calculation are plotted in a different manner in Fig. 18 for neutrons in the thermal region. It will be seen that the scattering length b varies by about 2 per cent over the neutron wavelength range 0·9–1·4 Å.

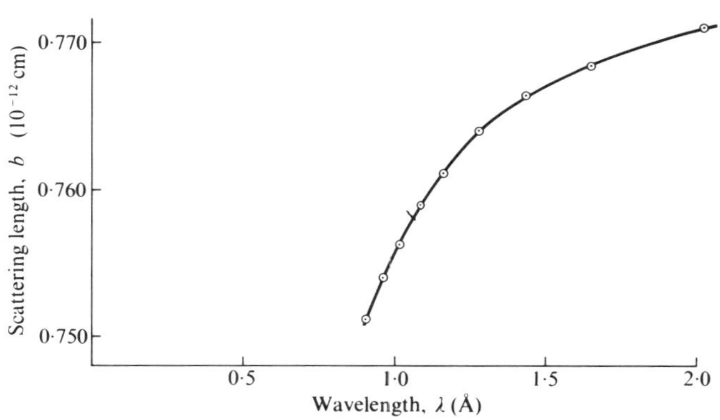

FIG. 18. Calculated variation of scattering length with neutron wavelength, for a nucleus of radius 10^{-12} cm and a single resonance at 1 eV.

The parameters which we used for the resonance in the calculation are quite close to those for the element rhodium which has a resonance at 1·26 eV. If the resonance energy is closer to thermal energy than this then the wavelength variation of b in the thermal

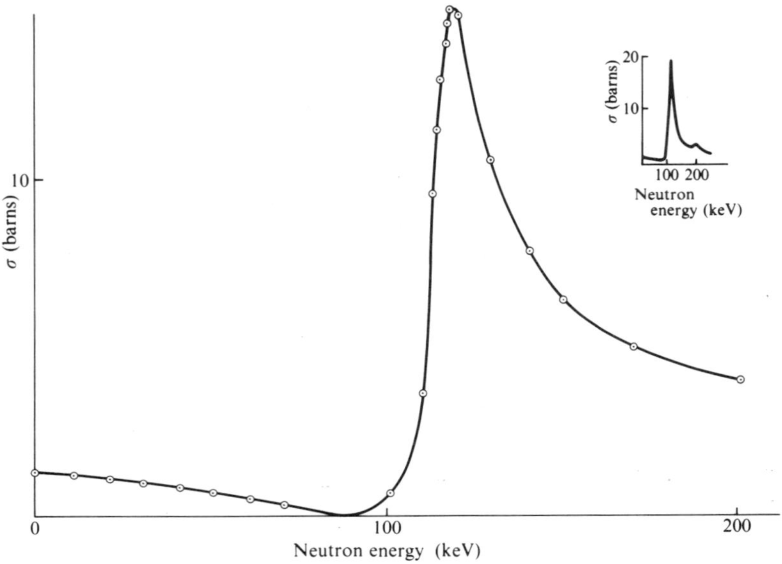

FIG. 19. Calculated variation of scattering cross-section with neutron energy for sulphur with (inset) the experimentally measured data.

region will be much greater and this is the case for the nucleus ^{113}Cd which has a resonance at 0·178 eV. We shall study this unusual case in more detail later on. On the other hand, when the resonance is more remote from thermal energies the variations of b with λ are much smaller, although the effective value at thermal energy may be well below that of the nuclear radius. As a further example we can calculate the cross-section and scattering length for sulphur ^{32}S for which $E_r = 111$ keV, $\Gamma_n^{(r)} = 18$ keV, and $\Gamma_a^{(r)}$ is negligibly small. In this case the nuclear radius is given by the formula $1{\cdot}5 \times \times 10^{-13}\,A^{\frac{1}{3}}$ as $0{\cdot}44 \times 10^{-12}$ cm and the quantity $\Gamma_n^{(r)}/2\kappa E_r$, which appears in eqn (2.10), is found to equal $0{\cdot}11 \times 10^{-12}$ cm. Thus the value of the scattering length at thermal energies should tend to $0{\cdot}33 \times 10^{-12}$ cm. The result of the calculation is shown in Fig. 19 alongside the experimental data for the variation of cross-section with wavelength. It will be noted that the position of the minimum and the value of the maximum are well reproduced by the calculation. The value of b for thermal neutrons $0{\cdot}28 \times 10^{-12}$ cm is less than that calculated, probably on account of the influence of other resonances at higher values of energy. At thermal energies the value of b is effectively constant and the calculation shows that there is less than $\frac{1}{3}$ per cent change in b between values of 0 eV and 1000 eV for the neutron energy.

The form of eqn (2.9) means that the scattering amplitude is a real number for neutron energies far removed from the resonance energy, but the amplitude becomes complex near resonance and the net change of phase on scattering is then neither 0° nor 180°. The way in which the phase changes as the neutron energy varies can be represented graphically (following Hughes (1954)) as in Fig. 20, where the circle has a radius of $(1/\kappa).(\Gamma_n/\Gamma)$ and OP = OP′ = R. In diagram (a) the line OA_5 is drawn such that

$$\tan\phi = \frac{\frac{1}{2}\Gamma}{E_r - E}$$

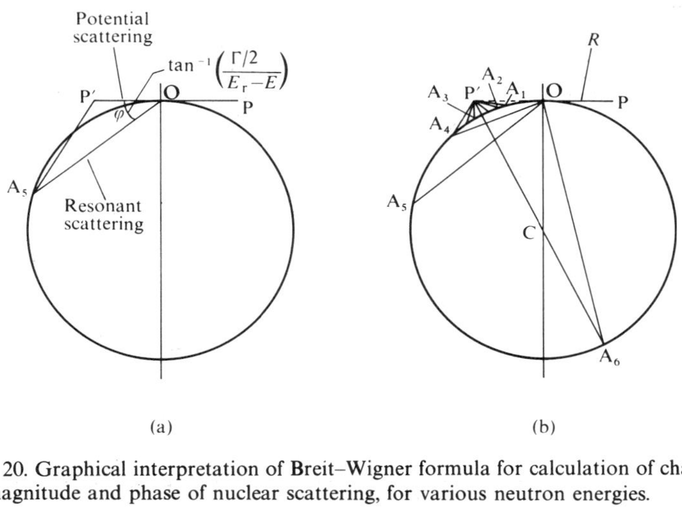

FIG. 20. Graphical interpretation of Breit–Wigner formula for calculation of change of magnitude and phase of nuclear scattering, for various neutron energies.

for some particular value of the neutron energy. It can be shown, from eqn (2.9), that the resonance amplitude will be equal to OA_5 and the net scattering amplitude will be $P'A_5$. In the second diagram (b) we indicate the value of P′A for several different energy values E. As E increases from zero the value of ϕ increases from the small value of $\tan^{-1}(\Gamma/2E_r)$ through $\frac{1}{2}\pi$ to π. Consequently P′A decreases from $P'A_1$, through a minimum value $P'A_3$, to a maximum $P'A_6$, and then decreases again to P′O, which is equal in value to R. The phase angle will be given by the inclination of P′A to the horizontal in the diagram, being very small when $E = 0$, rising to a maximum between A_4 and A_5 and then becoming zero at energies far beyond the resonance. This graphical representation is, of course, fully consistent with Fig. 17.

Bound and free nuclei

So far we have considered the scattering nucleus to be fixed at the origin. In a solid a slow neutron which is scattered elastically transmits no energy to the nucleus, but in the gaseous state where the nucleus is free to move it will recoil under the impact of the neutron. It can easily be shown that the scattering cross-section applicable to the free state is related to the cross-section for the bound state, by the relation

$$\sigma_{\text{free}} = \left(\frac{A}{A+1}\right)^2 \sigma_{\text{bound}}. \tag{2.11}$$

Consequently for hydrogen $\sigma_{\text{free}} = \frac{1}{4}\sigma_{\text{bound}}$, but the difference between the two cross-sections rapidly becomes negligible for the heavier nuclei. The respective scattering amplitudes are to be distinguished as a (free) and b (bound).

The effect of nuclear spin

The discussion so far must be regarded as applying strictly only to a nucleus with zero spin and the simple form of the Breit–Wigner formula given as eqn (2.6) only applies for zero spin. If the scattering nucleus has a spin I, then it may combine with a neutron, of spin $\frac{1}{2}$, for form one of two alternative compound nuclei, having spins of $I+\frac{1}{2}$ and $I-\frac{1}{2}$, respectively. Halpern and Johnson (1939) have shown that different scattering lengths b_+, b_- are associated with these two possible compound systems and that the resultant scattering is of two different types. The total scattering cross-section σ for the nucleus is the sum of two terms, so that

$$\sigma = \mathcal{S} + s, \tag{2.12}$$

where $\mathcal{S}$ is the cross-section for *coherent* scattering, i.e. scattering which is coherent with that by other nuclei and which can therefore produce interference, and s is the cross-section for *incoherent* scattering. We have adopted the symbols $\mathscr{S}$, s here, rather than S, s used by Cassels, in order to avoid any confusion with the use of S as the conventional symbol for the spin quantum number of a magnetic atom, to which we shall have occasion to refer later in this book.

The values of $\mathscr{S}$, s can be expressed in terms of b_+, b_- and w_+, w_-, the latter being the effective weights which have to be ascribed to the two possible compound nuclear states. For a spin state J the number of possible orientations is $2J+1$, so that the number of possible spin orientations for the compound states of spins $I+\frac{1}{2}$, $I-\frac{1}{2}$ will be $2I+2$, $2I$ respectively. Accordingly the weighting fractions of each will be

$$w_+ = (I+1)/(2I+1), \tag{2.13}$$

$$w_- = I/(2I+1) \tag{2.14}$$

It is then found (see Cassels 1950) that

$$\mathscr{S} = 4\pi(w_+b_+ + w_-b_-)^2, \tag{2.15}$$

$$\begin{aligned} s &= 4\pi w_+w_-(b_+ - b_-)^2 \\ &= 4\pi\{(w_+b_+^2 + w_-b_-^2) - (w_+b_+ + w_-b_-)^2\}, \end{aligned} \tag{2.16}$$

$$\sigma = 4\pi(w_+b_+^2 + w_-b_-^2). \tag{2.17}$$

Eqns (2.15) and (2.16) together replace the single expression $\sigma = 4\pi b^2$ which applies to nuclei with zero spin.

Only the scattering represented by $\mathscr{S}$ can produce interference effects. This has been illustrated by Cassels, who considers the resultant effect of two nuclei very close together. The combined scattering cross-section in this case will equal

$$(\sqrt{\mathscr{S}} + \sqrt{\mathscr{S}})^2 + (s+s) = 4\mathscr{S} + 2s,$$

the two terms corresponding to coherent and incoherent scattering respectively.

It is possible for b_+, b_- to have opposite signs with the result, as can be seen from the form of $\mathscr{S}$ given in eqn (2.15), that the coherent scattering cross-section is very much reduced. If w_+b_+ and w_-b_- were equal in value and opposite in sign there would be no coherent scattering at all; this circumstance almost occurs with normal hydrogen, for which b_+ and b_- are equal to $+1{\cdot}04 \times 10^{-12}$ cm and $-4{\cdot}7 \times 10^{-12}$ cm respectively and $I = \frac{1}{2}$. Consequently, although

$\sigma = 81 \times 10^{-24}\,\text{cm}^2$, $\mathscr{S}$ is only equal to $2 \times 10^{-24}\,\text{cm}^2$, and practically the whole of the scattering is incoherent.

The values for w_+, w_- given by eqns (2.13), (2.14) and the subsequent values of $\mathscr{S}$, s are derived on the assumption that the nuclei are unpolarized. If, however, there is a preferred direction along which the nuclear spins are directed, which can be achieved by applying magnetic fields at very low temperatures, then the coherent scattering amplitude will be modified. This has been demonstrated experimentally by Shull (1967), who showed that the scattering amplitude of cobalt could be increased by about 2 per cent at a temperature of 2·2 K. It is found that the change of b is proportional to $1/T$, as illustrated in Fig. 21.

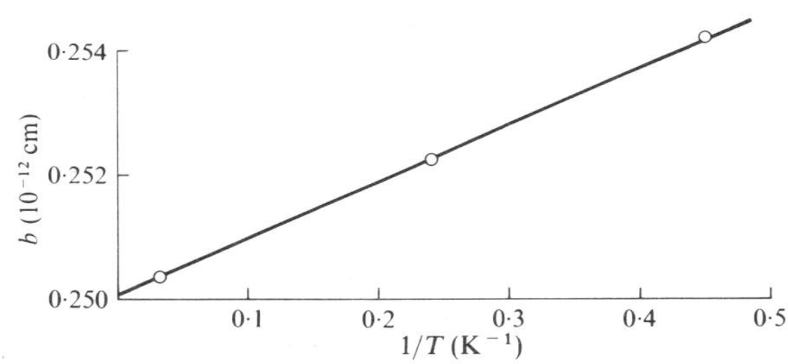

FIG. 21. Modification of the nuclear scattering amplitude of cobalt by nuclear polarization produced by applying a magnetic field at low temperature. (Shull 1967.)

The scattering constants of an individual nucleus

Before proceeding to the study of assemblies of nuclei in solids it is emphasized that the values of the cross-sections σ, $\mathscr{S}$, and s which have been defined above by eqns (2.15), (2.16), and (2.17), refer to a single nucleus, which possesses spin in the general case. They are to be regarded as nuclear constants applicable over the whole range of thermal neutron energies to a nucleus of particular mass and charge. They must be distinguished from the various cross-sections, some varying with neutron energy, which we shall proceed to discuss for the case of the crystalline solid elements which are built up of atoms which themselves may, in turn, be constituted of several nuclear types or isotopes.

2.3. Assemblies of nuclei in solids

Interference effects for nuclei of infinite mass

It was remarked earlier that in the solid state the scattering nuclei were not free to recoil under the neutron impact and that the scattering cross-sections were those applicable to fixed bound nuclei. However, when the nuclei are bound together in the form of a crystal it is possible for them, if not infinitely heavy, to receive energy from an incident neutron and transfer it to the crystal vibrations or, alternatively, receive energy from the latter and pass it on to the neutron, thus increasing the neutron velocity. Thus, if the nuclei are of finite mass the neutron scattering process may be inelastic. Before proceeding to this more general case we shall consider the nuclei to be of infinite mass, thus enabling the interference effects associated with the three-dimensional assembly of nuclei to be considered alone, without the additional complication of the effect of inelastic collisions.

If the incident plane wave of neutrons is, as before,

$$\Psi = e^{i\kappa z},$$

then the resultant wave-function after scattering by the assembly of atoms will be

$$\Psi = e^{i\kappa z} - \sum_{\rho} (b_\rho/r)\, e^{i\kappa r}\, e^{i\boldsymbol{\rho}.\boldsymbol{\kappa}-\boldsymbol{\kappa}'}, \tag{2.18}$$

where $\boldsymbol{\rho}$ is the vector from the origin to the nucleus and $\boldsymbol{\kappa}$, $\boldsymbol{\kappa}'$ are the wave vectors of the neutron before and after scattering. The expression $\exp(i\boldsymbol{\rho}\,.\,\boldsymbol{\kappa}-\boldsymbol{\kappa}')$ is introduced by considering the phase difference between the contributions from the various nuclei and is equivalent to $\exp\{2\pi i(hx/a_0+ky/b_0+lz/c_0)\}$, where x, y, z are the Cartesian coordinates of the nucleus, a_0, b_0, c_0 are the dimensions of the crystallographic unit cell, and (h, k, l) are the Miller indices appropriate to the particular direction $(\boldsymbol{\kappa}-\boldsymbol{\kappa}')$, which is the normal to what is usually regarded in the crystallographic treatment of diffraction as the 'reflecting plane'.

Thus at unit distance from the nuclei the amplitude of the scattered neutron wave will be equal to

$$-\sum_{\rho} b_\rho \exp\{2\pi i(hx/a_0+ky/b_0+lz/c_0)\}, \tag{2.19}$$

and the cross-section per nucleus per unit solid angle, or 'differential cross-section' for scattering in this particular direction, will be

$$G_{hkl} = \frac{1}{N_0}\left|\sum_{\rho} b_\rho \exp\{2\pi i(hx/a_0+ky/b_0+lz/c_0\}\right|^2, \tag{2.20}$$

where N_0 is the total number of nuclei in the crystallite under examination, assuming that the latter is sufficiently small for effects due to extinction and absorption to be ignored.

The effect of isotopes

For many elements in their normal form the nucleus is not a single unique type but may consist of one of several different isotopes each with a defined abundance. These isotopes will each have their own characteristic values of the scattering length b or, in the more general case for isotopes possessing spin, there will be characteristic values of b_+, b_- for each isotope, with resulting values of $\mathscr{S}$ and s as defined by eqns (2.15) and (2.16). In certain cases the values of these constants may be quite different for the different isotopes of an element. For example, the values of the scattering lengths of the two iron isotopes ^{54}Fe and ^{56}Fe, both of which have zero spin, are equal to $0{\cdot}42\times10^{-12}$ cm and $1{\cdot}01\times10^{-12}$ cm respectively.

The different isotopes of an element will be distributed at random among the atomic positions in the crystal, their relative numbers being determined, of course, by the isotopic abundance ratios of the element. Consequently, when the summations of eqns (2.19) and (2.20) are made, the value of b_ρ corresponding to a particular nucleus must be that appropriate to the particular isotope at that point in the crystal. The measured value of the differential scattering cross-section will be the average value of G_{hkl} over all possible distributions of the isotopes among the atomic positions in the crystallite.

We can deduce this average value as follows. At any atomic site the scattering from the particular isotope found there can be considered by separating its scattering length b_ρ into two portions, by writing

$$b_\rho = \overline{b_r}+(b_\rho-\overline{b_r}), \tag{2.21}$$

where $\overline{b_r}$ is the average value of scattering length and equals $\sum w_r b_r$ if w_r is an isotopic abundance. The first term in this expression will be the same for all the atomic sites and will contribute by coherent superposition an *amplitude* given by

$$\overline{b_r} \sum_{\rho} \exp\{2\pi i(hx/a+ky/b+lz/c)\},$$

where the summation is taken over all the sites in a crystallite. On the other hand, the second term will vary randomly over the sites, and there will be no phase correlation between their contributions from it. It is the *intensities* which will be additive. It follows therefore that the total differential scattering cross-section in any direction will be

$$(\overline{b_r})^2 \left| \sum_{\rho} \exp\{2\pi i(hx/a+ky/b+lz/c)\} \right|^2 + \sum_{\rho} (b_\rho - \overline{b_r})^2.$$

But

$$\begin{aligned} \sum_{\rho} (b_\rho - \overline{b_r})^2 &= \sum_{\rho} (b_\rho^2 - 2b_\rho \overline{b_r} + \overline{b_r}^2) \\ &= \sum_{\rho} (b_\rho^2 - \overline{b_r}^2), \end{aligned}$$

since

$$\sum_{\rho} b_\rho = \sum_{\rho} (\overline{b_r}).$$

Hence G_{hkl} per atom will be given by

$$G_{hkl} = \{\overline{b_r^2} - (\overline{b_r})^2\} + \frac{1}{N_0} (\overline{b_r})^2 \left| \sum_{\rho} \exp\{2\pi i(hx/a+ky/b+lz/c)\} \right|^2. \tag{2.22}$$

In the case of isotopes possessing spin there will be two abundance factors for each isotope, formed by multiplying the isotopic abundance w_r by $(I+1)/(2I+1)$ and $I/2I+1$ respectively, and these will be associated with their own appropriate scattering lengths b_+ and b_-.

The two terms in eqn (2.22) correspond to disordered and ordered scattering respectively. The first term $\{\overline{b_r^2} - (\overline{b_r})^2\}$ means a differential disordered scattering cross-section of this magnitude at all angles and hence a total disordered scattering $E(s)$ which will be given by

$$E(s) = 4\pi\{\overline{b_r^2} - (\overline{b_r})^2\}. \tag{2.23}$$

The quantity $4\pi\{\overline{b_r^2}-(\overline{b_r})^2\}$ is defined as the 'disordered scattering cross-section' s of the *element*. In the more general case to be considered later, where inelastic scattering is also possible, it will be seen that two quantities $E_{\text{el}}(s)$ and $E_{\text{inel}}(s)$ can be specified in terms of s.

The second term in eqn (2.22) means that there is a coherent scattering amplitude for the *element* equal to $\overline{b_r}$. It is this quantity, the mean value of the scattering length of the element averaged over the various possible isotopes and where necessary over their positive and negative spin combinations, which will determine the intensities of the Bragg reflections of crystals for neutrons. It is $\overline{b_r}$ which is the equivalent of f_X, the atomic scattering factor for X-rays. Similarly, we can define a coherent scattering amplitude for the *free* atom, to be denoted by $\overline{a_r}$, which will be related to $\overline{b_r}$ according to the expression $\overline{a_r}/\overline{b_r} = A/(A+1)$, where A is the effective mass number. In our subsequent discussions we shall be concerned almost entirely with $\overline{b_r}$, the coherent scattering amplitude for the element in the bound state. Except when we wish to draw particular attention to the fact that this is a value averaged over various isotopes and spin states we shall often write it simply as b.

The quantity $\mathscr{S} = 4\pi(\overline{b_r})^2$ is called the 'coherent scattering cross-section' for the element. The *total* scattering cross-section σ for the element, which is the sum of $\mathscr{S}$ and s, will be equal to $4\pi(\overline{b_r^2})$.

2.4. Experimental values of the scattering constants of elements and isotopes

The values of b

Table 2 shows in particular the values of b which are known at the present time. Data are given for all the naturally occurring

TABLE 2

Neutron and X-ray scattering data for elements and isotopes

					Neutrons			X-rays	
								f_X (10^{-12} cm)	
Element	Atomic number	Atomic weight of natural element	Specific nucleus	Nuclear spin	b (10^{-12} cm)	$\mathscr{S} = 4\pi b^2$ (barns)	σ (barns)	$\sin\theta = 0$	$(\sin\theta)/\lambda = 0{\cdot}5$ Å^{-1}
H	1		^{1}H	$\frac{1}{2}$	−0·374	1·76	81·5	0·28	0·02
			^{2}H	1	0·667	5·59	7·6	0·28	0·02
			^{3}H	$\frac{1}{2}$	0·47	2·77		0·28	0·02
			^{4}H	0	0·30	1·13	1·1	0·56	0·15

TABLE 2 (*continued*)

Element	Atomic number	Atomic weight of natural element	Specific nucleus	Nuclear spin	Neutrons: b (10^{-12} cm)	Neutrons: $\mathscr{S} = 4\pi b^2$ (barns)	Neutrons: σ (barns)	X-rays: f_x (10^{-12} cm), $\sin\theta = 0$	X-rays: f_x (10^{-12} cm), $(\sin\theta)/\lambda = 0{\cdot}5\,Å^{-1}$
He	2		^{4}H	0	0·30	1·13	1·1	0·56	0·15
Li	3	6·94			−0·214	0·57	1·2	0·84	0·28
			^{6}Li	1	0·18+0·025i	0·41		0·84	0·28
			^{7}Li	$\frac{3}{2}$	−0·233	0·68	1·4	0·84	0·28
Be	4		^{9}Be	$\frac{3}{2}$	0·774	7·53	7·54	1·13	0·39
B	5	10·81			0·54+0·021i	3·66	4·4	1·41	0·42
			^{11}B	$\frac{3}{2}$	0·60	4·52		1·41	0·42
C	6		^{12}C	0	0·665	5·56	5·51	1·69	0·48
			^{13}C	$\frac{1}{2}$	0·60	4·52	5·5	1·69	0·48
N	7		^{14}N	1	0·94	11·1	11·4	1·97	0·53
			^{15}N	$\frac{1}{2}$	0·65[1]	5·31		1·97	0·53
O	8		^{16}O	0	0·580	4·23	4·24	2·25	0·62
			^{17}O	$\frac{5}{2}$	0·578	4·20		2·25	0·62
			^{18}O	0	0·600	4·52		2·25	0·62
F	9		^{19}F	$\frac{1}{2}$	0·56	3·94	4·0	2·53	0·75
Ne	10	20·18			0·46	2·66	2·9	2·82	0·96
Na	11		^{23}Na	$\frac{3}{2}$	0·36	1·63	3·4	3·09	1·14
Mg	12	24·3			0·52	3·40	3·7	3·38	1·35
Al	13		^{27}Al	$\frac{5}{2}$	0·35	1·54	1·5	3·65	1·55
Si	14	28·06			0·42	2·22	2·2	3·95	1·72
P	15		^{31}P	$\frac{1}{2}$	0·51	3·27	3·6	4·23	1·83
S	16		^{32}S	0	0·28	0·99	1·2	4·5	1·9
Cl	17	35·5			0·96	11·58	15	4·8	2·0
			^{35}Cl	$\frac{3}{2}$	1·18	17·50		4·8	2·0
			^{37}Cl	$\frac{3}{2}$	0·26	0·85		4·8	2·0
A	18	39·94			0·20	0·50	0·9	5·07	2·2
			^{36}A	0	2·43	74·1		5·07	2·2
K	19	39·1			0·37	1·72	2·2	5·3	2·2
			^{39}K	$\frac{3}{2}$	0·37	1·72		5·3	2·2
Ca	20	40·1			0·47	2·78	3·2	5·6	2·4
			^{40}Ca	0	0·49	3·02	3·1	5·6	2·4
			^{44}Ca	0	0·18	0·41		5·6	2·4
Sc	21		^{45}Sc	$\frac{7}{2}$	1·18	17·50	2·4	5·9	2·5
Ti	22	47·9			−0·34	1·45	4·4	6·2	2·7
			^{46}Ti	0	0·48	2·90		6·2	2·7
			^{47}Ti	$\frac{5}{2}$	0·33	1·37		6·2	2·7
			^{48}Ti	0	−0·58	4·23		6·2	2·7
			^{49}Ti	$\frac{7}{2}$	0·08	0·08		6·2	2·7
			^{50}Ti	0	0·55	3·80		6·2	2·7
V	23		^{51}V	$\frac{7}{2}$	−0·05	0·03	5·1	6·5	2·8
Cr	24	52·0			0·352	1·56	4·1	6·8	3·0
			^{52}Cr	0	0·490	3·02		6·8	3·0
Mn	25		^{55}Mn	$\frac{5}{2}$	−0·39	1·91	2·0	7·0	3·1
Fe	26	55·8			0·95	11·34	11·8	7·3	3·3
			^{54}Fe	0	0·42	2·22	2·5	7·3	3·3
			^{56}Fe	0	1·01	12·82	12·8	7·3	3·3
			^{57}Fe		0·23	0·66	2	7·3	3·3
Co	27		^{59}Co	$\frac{7}{2}$	0·25	0·79	6	7·6	3·4
Ni	28	58·7			1·03	13·33	18·0	7·9	3·6
			^{58}Ni	0	1·44	26·06		7·9	3·6
			^{60}Ni	0	0·28	0·99		7·9	3·6
			^{61}Ni		0·76	7·26		7·9	3·6
			^{62}Ni	0	−0·87	9·51		7·9	3·6
			^{64}Ni		−0·037	0·02		7·9	3·6
Cu	29	63·6			0·76	7·26	8·5	8·2	3·8
			^{63}Cu	$\frac{3}{2}$	0·67	5·64		8·2	3·8
			^{65}Cu	$\frac{3}{2}$	1·11	15·48		8·2	3·8

TABLE 2 (*continued*)

Element	Atomic number	Atomic weight of natural element	Specific nucleus	Nuclear spin	Neutrons: b (10^{-12} cm)	Neutrons: $\mathscr{S} = 4\pi b^2$ (barns)	Neutrons: σ (barns)	X-rays: f_X (10^{-12} cm) $\sin\theta = 0$	X-rays: f_X (10^{-12} cm) $(\sin\theta)/\lambda = 0{\cdot}5$ Å^{-1}
Zn	30	65·4			0·57	4·08	4·2	8·5	3·9
			^{64}Zn	0	0·55	3·80		8·5	3·9
			^{66}Zn	0	0·63	4·99		8·5	3·9
			^{68}Zn	0	0·67	5·64		8·5	3·9
Ga	31	69·7			0·72	6·51	7·5	8·8	4·1
Ge	32	72·6			0·82	8·45	9·0	9·0	4·2
As	33		^{75}As	$\frac{3}{2}$	0·64	5·15	8	9·3	4·4
Se	34	79·0			0·80	8·04		9·6	4·5
Br	35	79·9			0·68	5·80	6·1	9·8	4·7
Kr	36	82·9			0·74	6·88		10·2	4·9
Rb	37	85·5			0·71	6·33	5·5	10·4	5·0
			^{85}Rb	$\frac{5}{2}$	0·83	8·66		10·4	5·0
Sr	38	87·6			0·69	5·98	10	10·7	5·2
Y	39		^{89}Y	$\frac{1}{2}$	0·79	7·84		11·0	5·4
Zr	40	91·2			0·71	6·33	6·3	11·3	5·5
Nb	41		^{93}Nb	$\frac{9}{2}$	0·71	6·33	6·6	11·5	5·7
Mo	42	95·9			0·69	5·98	6·1	11·8	5·9
Tc	43				0·68	5·81		12·0	6·1
Ru	44	101·7			0·73	6·70	6·81	12·5	6·2
Rh	45		^{103}Rh	$\frac{1}{2}$	0·58	4·23	5·6	12·8	6·4
Pd	46	106·7			0·60	4·52	4·8	12·9	6·5
Ag	47	107·9			0·60	4·52	5·5	13·3	6·7
			^{107}Ag	$\frac{1}{2}$	0·83	8·66	10	13·3	6·7
			^{109}Ag	$\frac{1}{2}$	0·43	2·32	6	13·3	6·7
Cd	48	112·4			0·37 + 0·16i	2·04		13·6	6·9
			^{113}Cd	$\frac{1}{2}$	− 1·5 + 1·2i	46·36		13·6	6·9
In	49	114·8			0·39	1·91		13·9	7·1
Sn	50	118·7			0·61	4·67	4·9	13·9	7·1
			^{116}Sn	0	0·58	4·23		13·9	7·1
			^{117}Sn	$\frac{1}{2}$	0·64	5·15		13·9	7·1
			^{118}Sn	0	0·58	4·23		13·9	7·1
			^{119}Sn	$\frac{1}{2}$	0·60	4·52		13·9	7·1
			^{120}Sn	0	0·64	5·15		13·9	7·1
			^{122}Sn	0	0·55	3·80		13·9	7·1
			^{124}Sn	0	0·59	4·37		13·9	7·1
Sb	51	121·8			0·56	3·94	4·2	14·2	7·3
Te	52	127·5			0·58[(2)]	4·23	4·5	14·7	7·6
			^{120}Te	0	0·52	3·40		14·7	7·6
			^{123}Te	$\frac{1}{2}$	0·57	4·08		14·7	7·6
			^{124}Te	0	0·55	3·80		14·7	7·6
			^{125}Te	$\frac{1}{2}$	0·56	3·94		14·7	7·6
I	53		^{127}I	$\frac{5}{2}$	0·53	3·53	3·8	15·0	7·7
Xe	54	130·2			0·48	2·89		15·3	8·0
Cs	55		^{133}Cs	$\frac{7}{2}$	0·55	3·80	7	15·5	8·1
Ba	56	137·4			0·52	3·40	6	15·8	8·3
La	57		^{139}La	$\frac{7}{2}$	0·83	8·66	9·3	16·1	8·4
Ce	58	140·25			0·48	2·89	2·7	16·3	8·6
			^{140}Ce	0	0·47	2·78	2·6	16·3	8·6
			^{142}Ce	0	0·45	2·54	2·6	16·3	8·6
Pr	59		^{141}Pr	$\frac{5}{2}$	0·44	2·43	4·0	16·6	8·8
Nd	60	144·3			0·75[(3)]	7·07	16	16·9	9·0
			^{142}Nd	0	0·77	7·45	7·5	16·9	9·0
			^{144}Nd	0	0·28	0·99	1·0	16·9	9·0
			^{146}Nd	0	0·87	9·51	9·5	16·9	9·0
Pm	61							17·3	9·2
Sm	62	150·4						17·5	9·3
			^{152}Sm	0	− 0·5	3·14		17·5	9·3
			^{154}Sm	0	0·96[(4)]	11·58		17·5	9·3
Eu	63	152·0			0·55	3·80		17·8	9·5

TABLE 2 (*continued*)

Element	Atomic number	Atomic weight of natural element	Specific nucleus	Nuclear spin	Neutrons: b (10^{-12} cm)	Neutrons: $\mathscr{S} = 4\pi b^2$ (barns)	Neutrons: σ (barns)	X-rays: f_X (10^{-12} cm), $\sin\theta = 0$	X-rays: f_X (10^{-12} cm), $(\sin\theta)/\lambda = 0{\cdot}5$ Å^{-1}
Gd	64	157·3			1·5	28·3		18·2	9·7
			^{160}Gd	0	0·91[5]	10·4		18·2	9·7
Tb	65		^{159}Tb	$\frac{3}{2}$	0·76	7·26		18·1	9·8
Dy	66	162·5			1·69	35·89		18·6	10·0
			^{160}Dy	0	0·67	5·64		18·6	10·0
			^{161}Dy	$\frac{5}{2}$	1·03	13·3		18·6	10·0
			^{162}Dy	0	−0·14	0·25		18·6	10·0
			^{163}Dy	$\frac{5}{2}$	0·50	3·14		18·6	10·0
			^{164}Dy	0	4·94	306		18·6	10·0
Ho	67		^{165}Ho	$\frac{7}{2}$	0·85	9·08	13	18·9	10·2
Er	68	167·6			0·79	7·84	15	19·2	10·3
Tm	69		^{169}Tm	$\frac{1}{2}$	0·72	6·51		19·5	10·5
Yb	70	173·0			1·26	19·95		19·8	10·7
Lu	71	175·0			0·73	6·70		20·0	10·9
Hf	72	178·6			0·78	7·64		20·3	11·1
Ta	73		^{181}Ta	$\frac{7}{2}$	0·70	6·16	6	20·5	11·3
W	74	183·9			0·48	2·89	5·7	20·8	11·4
			^{182}W	0	0·83	8·66		20·8	11·4
			^{183}W	$\frac{1}{2}$	0·43	2·32		20·8	11·4
			^{184}W	0	0·76	7·26		20·8	11·4
			^{186}W	0	−0·12	0·18		20·8	11·4
Re	75	186·2			0·92	10·6		21·1	11·6
Os	76	192·2			1·07	14·39	14·9	21·4	11·8
			^{188}Os	0	0·78	7·65		21·4	11·8
			^{189}Os	$\frac{3}{2}$	1·10	15·20		21·4	11·8
			^{190}Os	0	1·14	16·33		21·4	11·8
			^{192}Os	0	1·19	17·79		21·4	11·8
Ir	77	192·2			1·06	14·12		21·7	12·0
Pt	78	195·2			0·95	11·34	12	22·0	12·1
Au	79		^{197}Au	$\frac{3}{2}$	0·76	7·26	9	22·2	12·3
Hg	80	200·6			1·27	20·26	26·5	22·5	12·5
Tl	81	204·4			0·89	9·95	10·1	22·8	12·7
Pb	82	207·2			0·94	11·10	11·4	23·1	12·9
Bi	83		^{209}Bi	$\frac{9}{2}$	0·86	9·29	9·37	23·3	13·1
Po	84	210						23·7	13·3
At	85							24·0	13·5
Rn	86	222						24·3	13·7
Fr	87							24·5	13·9
Ra	88	226						24·8	14·1
Ac	89	227						25·1	14·2
Th	90		^{232}Th	0	1·03	13·33	12·6	25·3	14·4
Pa	91	231			1·30[6]	21·23		25·7	14·6
U	92	238·03			0·85	9·08		25·9	14·8
			^{235}U	$\frac{7}{2}$	0·98	12·07		25·9	14·8
			^{238}U	0	0·85	9·08		25·9	14·8
Np	93				1·055	13·99		26·0	15·5
Pu	94				0·75	7·07		26·0	15·5
			^{240}Pu	0	0·35	1·54		26·3	15·7
			^{242}Pu	0	0·81	8·24		26·3	15·7
Am	95		^{243}Am	$\frac{5}{2}$	0·76[7]	7·26		26·6	15·9
Cm	96		^{244}Cm	0	0·7[7]	6·16		26·9	16·1

b is the coherent scattering amplitude of neutrons, $\mathscr{S}$ is the coherent scattering cross-section and σ is the total scattering cross-section. Complex neutron scattering amplitudes relate to $\lambda = 1$ Å.

1. Kuznietz and Wedgwood (1972).
2. Lindqvist and Lehmann (1973).
3. Schobinger-Papamentellos, Fischer, Vogt, and Kaldis (1973).
4. Koehler and Moon (1972).
5. Koehler, Child, and Cable (1971).
6. Wedgwood and Burlet (1973, unpublished).
7. Mueller, Lander, and Reddy (1974).

elements (which are isotopic mixtures) and in many cases for individual isotopes of the elements. The first list of values was compiled by Shull and Wollan (1951), and some of them remain unaltered, and often inevitably unchecked, since that time. On the other hand, many of the values have been successively checked and refined by more accurate experiment. The compilation in Table 2 is based on the table published by the Neutron Diffraction Commission (1972), with a few recent additions and modifications for which references to original papers are given. For comparison, the table also shows in the final two columns the values of f_X, the atomic scattering factors for X-rays at values of $(\sin\theta)/\lambda$ of 0 Å^{-1} and 0·5 Å^{-1}. The values of f_X, as already stated, vary with Bragg angle θ whereas the neutron coherent scattering amplitude b is the same at all angles.

In addition to the value of b, Table 2 also shows the value of $\mathscr{S} = 4\pi b^2$ which, as defined above, is the coherent scattering cross-section. Alongside this is the value of σ, the total scattering cross-section; the difference between the two $(\sigma - \mathscr{S})$ is equal to s, the disordered scattering cross-section. The actual determination of these quantities will be considered in Chapter 5. For the present we shall examine some of the values given in Table 2 in the light of the discussion which has been given earlier in the present chapter on the effects of nuclear resonance, nuclear spin, and the presence of isotopes.

Variation with atomic weight

In considering the variation of scattering amplitude with atomic weight it will be seen that there is in general a slight increase with A but that there are considerable irregular variations superimposed upon this. This is to be accounted for by the superposition of the random effects of nuclear resonance on the regular variation of potential scattering. This is emphasized in Fig. 22 where the value of b is plotted against atomic weight for some of the elements and their isotopes. The curve appropriate to 'potential' scattering alone is drawn, and it will be clear how resonance effects cause the resultant amplitude to depart, haphazardly, from this curve. For contrast the linear increase of the scattering amplitude for X-rays is also shown.

The same information is presented in a more pictorial form for neutrons in Fig. 23, where the areas of the circles represent the target areas of the atoms for neutron scattering. Thus, the radius

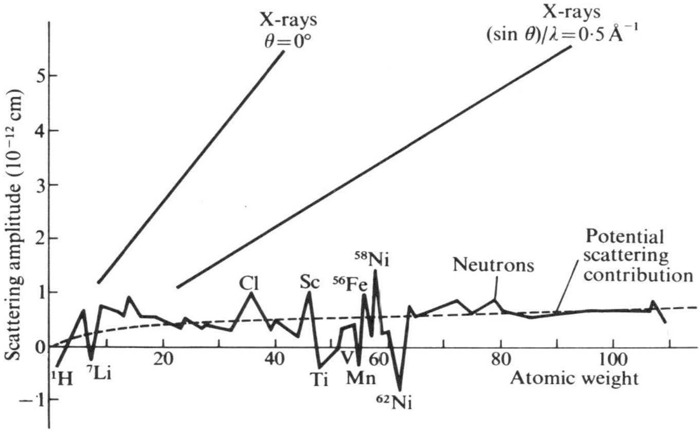

FIG. 22. Irregular variation of neutron scattering amplitude with atomic weight due to superposition of 'resonance scattering' on the slowly increasing 'potential scattering'; for comparison the regular increase for X-rays is shown. (From *Research* (*London*) **7**, 257 (1954).)

of each circle is proportional to b and the area of each circle is proportional to $\mathscr{S}$.

The sign of the scattered amplitude

Most of the values of b are listed in Table 2 as being of positive sign, with a few exceptions such as hydrogen, titanium, and manganese, which are given as negative. In Fig. 23 the negative elements

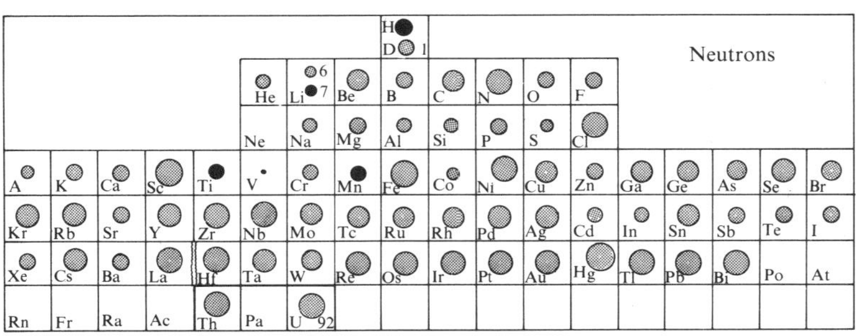

FIG 23. Diagrammatic representation of the neutron scattering cross-section of the elements. The radii of the circles are proportional to b and their areas are proportional to $\mathscr{S}$. The black areas shown for H, Li, Ti, and Mn indicate a negative value for b.

are identified by their distinctive shading. Reference back to eqn (2.3) will show that a positive value of b means that there is a phase change of 180° between the incident and scattered neutron waves; the choice of a minus sign in eqn (2.3) was, in fact, made to ensure that most nuclei would have positive values of b. The general occurrence of a phase shift of 180° is a consequence of the general predominance of potential scattering, for in the case of scattering by an impenetrable sphere the outgoing wave must be exactly out of phase with the incident wave, just as for the reflection of a sound wave at the closed end of a tube. There is a similar phase change of 180° when X-rays are scattered by the extranuclear electrons in an atom, so that if the same convention with regard to sign is used the X-ray scattering amplitudes given in the table are also positive.

As mentioned earlier a negative value of b is to be associated with the presence of a resonance level for which the value of $\frac{1}{2}\Gamma_n^{(r)}/\kappa E_r$ exceeds the value of R. For manganese, scattering resonances were shown by Hibdon, Selove, and Woolf (1950) to exist for neutron energies of both 300 eV and 2400 eV, and it is believed that both of these contribute to the observed negative value of b. Similarly it is thought that the negative value of b for ^{7}Li can be accounted for in terms of a broad scattering resonance at about 1 MeV which has been found by Adair (1950*a*).

Interpretation by the Breit–Wigner equation

The value of b given for ^{62}Ni is of particular interest, not only because it is negative but also because of its relatively large numerical value $-0{\cdot}87 \times 10^{-12}$ cm. Consideration of eqn (2.5) suggests that in this case the resonance energy must be considerably nearer to thermal energy, relative to the neutron width $\Gamma_n^{(r)}$, than is the case for Mn, Li, and Ti. The behaviour of chlorine is of interest in a different connection: here b has a particularly large numerical value, $0{\cdot}99 \times 10^{-12}$ cm, but this time of positive sign. This is what would be expected if there were a resonance neutron energy a little below thermal energies, or indeed, of negative energy. It has, in fact, been shown by Hibdon and Muehlhause (1950) that such a level does exist for ^{35}Cl, corresponding to an E_r value of -75 eV, and that this level can account satisfactorily on the basis of the Breit–Wigner equation for the observed scattering amplitude.

Nuclear spin incoherence

The effect of nuclear spin can be seen most easily by considering the values in Table 2 which refer to single isotopes. For nuclei having zero spin, which may be represented by even charge and even mass number examples such as ^{12}C, ^{54}Fe, ^{56}Fe, ^{58}Ni, there will be no incoherent scattering due to spin and consequently, as seen in the table, the values of σ and $\mathscr{S}$ will be equal within experimental error. On the other hand, the considerable differences between σ and $\mathscr{S}$ for ^{1}H, ^{23}Na, ^{51}V, and ^{59}Co are to be interpreted as due to large differences in the values of b_+ and b_- appropriate, respectively, to parallel and antiparallel spins of the neutron and scattering nucleus. From eqns (2.13) and (2.14) it can be shown that

$$b_+ = \overline{b_r} \pm \left(\frac{\varsigma}{4\pi}\frac{I}{I+1}\right)^{\frac{1}{2}} \tag{2.24}$$

and

$$b_- = \overline{b_r} \mp \left(\frac{\varsigma}{4\pi}\frac{I+1}{I}\right)^{\frac{1}{2}}, \tag{2.25}$$

so that it is not possible to determine b_+, b_- unambiguously from the observed value of $\overline{b_r}$. The presence of the ambiguity of sign means that the experimental values of σ and $\mathscr{S}$ can be satisfied by either of two pairs of values of b_+, b_-. In the case of ^{23}Na, for example, the measured value of $\overline{b_r}$ is $0{\cdot}35 \times 10^{-12}$ cm with σ and $\mathscr{S}$ values of 3·5 barns and 1·5 barns respectively. By substituting in eqns (2.24) and (2.25) and knowing that the spin I is equal to $\frac{3}{2}$ it is found that there are two possible pairs of values, namely,

$$b_+ = +0{\cdot}88 \times 10^{-12}, \qquad b_- = +0{\cdot}02 \times 10^{-12}\ \text{cm}$$

or

$$b_+ = -0{\cdot}20 \times 10^{-12}, \qquad b_- = +0{\cdot}67 \times 10^{-12}\ \text{cm}.$$

Shull and Wollan (1951) have given a graphical method of determining the two values of b_+, b_- from the experimental values of $\bar{b}$, σ, and $\mathscr{S}$. Rewriting eqn (2.17) which expresses the value of σ, the total scattering cross-section, we have

$$\frac{b_+^2}{\sigma/4\pi w_+} + \frac{b_-^2}{\sigma/4\pi w_-} = 1. \tag{2.26}$$

This is the equation of an ellipse with semi-axes equal to $(\sigma/4\pi w_+)^{\frac{1}{2}}$ and $(\sigma/4\pi w_-)^{\frac{1}{2}}$, which can be plotted as in Fig. 24 if the spin of the nucleus is known in order to give the values of w_+ and w_- from eqns (2.10) and (2.11). Further, we have that

$$\overline{b_r} = w_+ b_+ + w_- b_-, \tag{2.27}$$

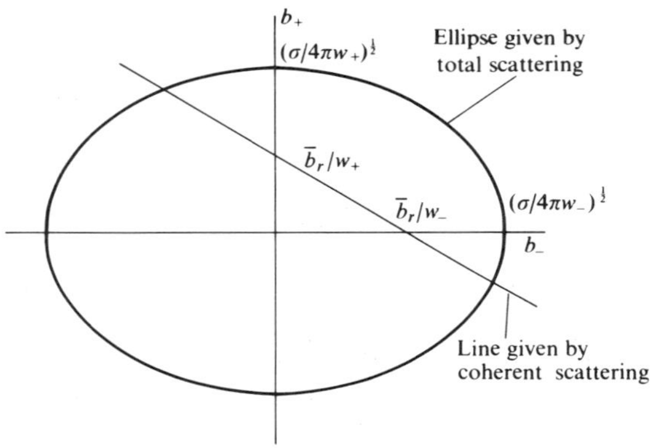

FIG. 24. Shull and Wollan's graphical method for deducing b_+, b_- from the values of $\overline{b_r}$, σ. (After Shull and Wollan 1951.)

which is the equation of a straight line making intercepts of $\overline{b_r}/w_+$ and $\overline{b_r}/w_-$ on the axes of b_+ and b_- respectively. The two points of intersection of the line and ellipse in the figure will give the two possible combinations for b_+, b_-.

The most noteworthy example of spin incoherence is provided by hydrogen, for which the bound coherent scattering cross-section is only 2 barns, corresponding to $b = -0{\cdot}38 \times 10^{-12}$ cm, in comparison with a total bound scattering cross-section of about 81 barns. These cross-sections lead to the following possible values of b for the bound nuclei:

$$b_+ = +1{\cdot}04 \times 10^{-12}, \qquad b_- = -4{\cdot}7 \times 10^{-12} \text{ cm}$$

or

$$b_+ = -1{\cdot}82 \times 10^{-12}, \qquad b_- = +4{\cdot}0 \times 10^{-12} \text{ cm}.$$

In this particular case of hydrogen, however, other considerations enable a distinction to be drawn between the two possible sets of values of b_+, b_-. The compound nucleus formed when the hydrogen nucleus and the neutron have parallel spins is indeed the ground state of the deuteron. The binding energy of the deuteron can be found experimentally; it is positive, and from these data and the results of proton–proton scattering it can be deduced that the scattering length a_+ for hydrogen is positive and equal to $+0{\cdot}55\times\times 10^{-12}$ cm for the free atom. Such a value would account for a free total scattering cross-section of only 3·9 barns, in contrast to the observed value of 20·4 barns. The scattering must therefore be strongly spin dependent, and it is deduced that the *numerical* value of $(a_-)_{\text{free}}$ is $2{\cdot}35\times 10^{-12}$ but no information is obtainable, from these considerations, regarding the sign of a_-. However, Schwinger and Teller (1937) showed that this could be inferred from the results of scattering experiments with *ortho*- and *para*-hydrogen in which the protons spins are parallel and antiparallel respectively. For very slow neutrons and gases at very low temperature it is predicted that the *ortho* scattering cross-section would be about 40 times greater than the *para* cross-section if a_- were negative. On the other hand, if a_- were positive the two cross-sections would be about equal. The experimental investigations of Sutton *et al.* (1947) gave a cross-section ratio of about 30, so that a_- was indeed negative and it was deduced that

$$a_+ = +0{\cdot}522\times 10^{-12}\text{ cm},\qquad a_- = -2{\cdot}34\times 10^{-12}\text{ cm}.$$

It will be seen that these conclusions were entirely confirmed by the neutron diffraction data, it being clear that the first set of values given above for b_+, b_- is the correct one, these (of b for the bound nuclei) being exactly twice those of a just deduced for the free nuclei.

A discussion of the measurements of scattering by *ortho*- and *para*-hydrogen and their relation to theoretical calculations can be found in a book by Halliday (1950). In examining the early papers on this subject it is to be noted that the sign convention used is the opposite to that employed in the later work, which is the one used above.

In general the ambiguity between the two pairs of values of b for the two spin states can only be resolved by measurements in which the spins of the nuclei are aligned, using beams of polarized neutrons, as discussed by Hamermesh and Schwinger (1946) and by Rose (1949).

A direct measurement of the spin-incoherent scattering has been made for vanadium by Shull and Wollan (1951). In the case of vanadium the spin incoherence is very large and contributes a much larger background to the powder diffraction pattern than do thermal diffuse scattering and multiple scattering. Consequently, by correcting the background intensity for the two latter effects quite an accurate measure of the spin incoherent scattering can be obtained. Using vanadium carbide Shull and Wollan showed that the scattering was isotropic and equal to 0·40 barns per unit solid angle, which corresponds to a total incoherent scattering cross-section σ of 5·0 barns. This is consistent with the known σ value of 5 barns and the negligibly small $\mathscr{S}$ value of only 0·029 barns which was later measured by Peterson and Levy (1952*b*).

Since spin incoherence arises from the randomness in the alignment of the spins of nuclei and neutrons it should be possible to eliminate it, with great advantage in the study of hydrogenous compounds, by polarizing both the incident neutrons and the nuclear targets. This has been discussed by Schermer and Blume (1968) and Jauho and Pirila (1970).

Isotopic incoherence

The effect of isotopic incoherence can be seen most readily for elements of even atomic number and with predominant isotopes of even mass number, for in such cases the nuclear spin will be zero and there will accordingly be no spin incoherence to cause confusion.

Some striking examples of the variation of scattering length among the isotopes of individual elements are found in the iron groups of transition elements, and some of these are illustrated pictorially in Fig. 25 for the elements titanium, iron, and nickel. The rare earths, too, show some very large effects, and these are particularly prominent in neodymium and dysprosium, as can be seen from the data in Table 2 (see p. 40). Perhaps the best-known example of all is that afforded by nickel for which the difference between the σ and $\mathscr{S}$ values of 18·0 b and 13·4 b is to be accounted for by isotopic incoherence. The coherent scattering amplitudes have been measured for the various isotopes and are listed together with the isotopic abundances in Table 3.

For ordinary nickel it would be expected that the coherent scattering amplitude would be equal to $\sum w_r b_r$, which according to

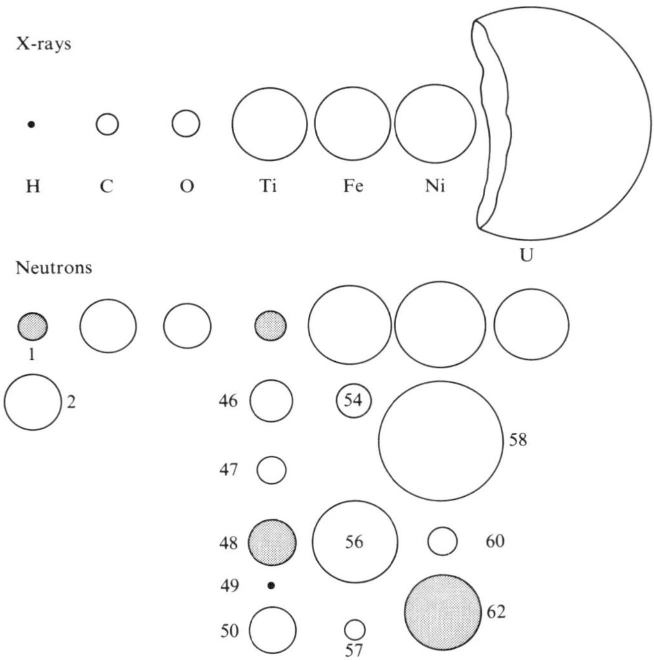

FIG. 25. The visibilities of some atoms and isotopes for X-rays and neutrons. The radii of the circles are proportional to the scattering amplitude b. Negative values of b are indicated by the cross-hatched shading.

TABLE 3

The value of $\overline{b_r}$ for nickel

Isotope	Coherent scattering amplitude b_r (cm)	Abundance ω_r	$\omega_r b_r$ (cm)
^{58}Ni	$1{\cdot}44 \times 10^{-12}$	0·679	$0{\cdot}978 \times 10^{-12}$
^{60}Ni	0·30	0·262	0·079
^{61}Ni	0·76	0·012	0·009
^{62}Ni	−0·87	0·037	−0·032
^{64}Ni	−0·037	0·011	—
			1·034

Table 3 would be equal to $1{\cdot}034 \times 10^{-12}$ cm. This result agrees with the measured experimental value.

In the case of silver the experimental values of b for the separated isotopes ^{107}Ag and ^{109}Ag are equal to $0{\cdot}83 \times 10^{-12}$ cm and $0{\cdot}43 \times 10^{-12}$ cm. The abundances of the two isotopes are 51·3 per cent and 48·7 per cent, from which we calculate that for ordinary silver

$$b = (0{\cdot}83 \times 0{\cdot}513) + (0{\cdot}43 \times 0{\cdot}487)$$

$$= 0{\cdot}63 \times 10^{-12} \text{ cm},$$

in comparison with the experimental value of $0{\cdot}61 \times 10^{-12}$ cm.

Other elements which show marked isotopic incoherence are zirconium, for which the σ and $\mathscr{S}$ values are 7 barns and 4·9 barns, and titanium, for which the values are 6 and 1·8. On the other hand, certain elements such as lead which consist of several isotopes have values of σ and $\mathscr{S}$ which are practically equal. In these cases the predominant isotopes must possess very similar b values; the isotopes with small abundances may, indeed, have quite different values for they would exert little determining effect on σ and $\mathscr{S}$. Tin is a good example of an element for which the values of b for many isotopes have been measured but there are only small differences between the individual values. Elements such as carbon and oxygen which are very nearly mono-isotopic, and whose single nuclear type, being of even atomic number and even mass number, has zero spin will, of course, have equal values of σ and $\mathscr{S}$, and these elements are of particular importance in the actual experimental determinations of σ and $\mathscr{S}$ for other elements, as will be described in Chapter 5.

2.5. The integrated scattering by polycrystalline substances

$E(\mathscr{S})$, $E(\varsigma)$ for a polycrystalline element or compound

The foregoing discussion of the scattering by an element in the form of a three-dimensional array of nuclei has concentrated on the determination of the effective 'coherent scattering amplitude' $\overline{b_r}$, this being the average value of the scattering length b when the average is taken over the various isotopes and the parallel and antiparallel spin states for those isotopes which have a finite nuclear spin. We have *defined* a 'coherent scattering cross-section' for the element as $\mathscr{S} = 4\pi\overline{b_r^2}$ and a 'disordered cross-section' as $s = 4\pi\{\overline{b_r^2} - (\overline{b_r})^2\}$. From eqn (2.22) we saw that the disordered scattering was isotropic and that the total number of neutrons scattered incoherently per atom in the crystal, which was denoted by $E(s)$, was, in fact, equal to the quantity which we defined as s. However, we did not at this point investigate the sum total of neutrons scattered coherently but merely remarked that it was the 'coherent scattering amplitude' $\overline{b_r}$ which would determine the intensities of Bragg reflections from crystals. It will be convenient at the present stage to revert to this matter and consider the total number of neutrons scattered coherently in all directions in the case of a polycrystalline sample of an element. We shall see in Chapter 5 that an experimental determination of this quantity provides a means of arriving at some of the data given in Table 2.

We shall call this total coherent scattering $E(\mathscr{S})$, and it is obtained by integrating the second term of eqn (2.22) over all angles in space, bearing in mind that the microcrystals will be oriented at random to give collectively all possible orientations. The result of this calculation, which has been given by a number of authors such as Weinstock (1944) and Cassels (1950), is that the total coherent scattering per unit cell of any crystal, whether composed of a single element or of several different elements, is given by

$$E(\mathscr{S}) = \frac{\pi N_c}{2\kappa^2} \sum_{hkl} 4\pi F_{hkl}^2 d_{hkl}, \tag{2.28}$$

where the summation is taken over all planes (hkl) which are capable of giving Bragg reflection for the neutron wavelength used—i.e. all planes which have spacings $d \geqslant \lambda/2$. N_c is the number of unit cells per unit volume and F_{hkl}^2 is the square of the structure factor of the unit cell for the (hkl) reflection. Thus

$$F_{hkl}^2 = |\sum b \exp\{2\pi i(hx/a_0 + ky/b_0 + lz/c_0)\}|^2, \tag{2.29}$$

the summation being taken over all the atoms in the unit cell. In the case of compounds the value of b taken for each atomic position is, of course, that appropriate to the particular atom situated there.

Replacing the wave-number κ by $2\pi/\lambda$ we can rewrite eqn (2.28) in the alternative form

$$E(\mathscr{S}) = \frac{N_c\lambda^2}{2} \sum_{\substack{hkl \\ d \geq \lambda/2}} F_{hkl}^2 d_{hkl}, \tag{2.30}$$

the expression specifying, as in (2.28), the total coherent scattering cross-section per *unit cell.* Fig. 26 shows the form of $E(\mathscr{S})$ as calculated by eqn (2.30) for the case of graphite, the ordinate having been

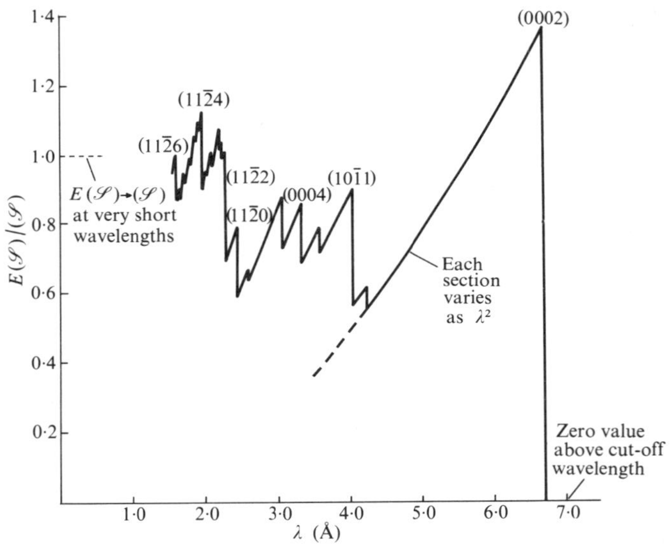

FIG. 26. Variation of $E(\mathscr{S})$ with λ for graphite, neglecting inelastic scattering.

referred to a single nucleus by dividing by the number of nuclei per unit cell. At very long wavelengths, namely those for which λ is greater than twice the largest interplanar spacing in the crystal, $E(\mathscr{S})$ will be zero. In the case of graphite the largest interplanar spacing is that for the (0002) planes and is equal to 3·35 Å. Thus when the neutron wavelength is greater than 6·7 Å the value of $E(\mathscr{S})$ is zero. The Figure shows how the greatest discontinuities occur at wavelengths below which reflection can first take place for some plane of large structure factor in the graphite structure. In carrying out the summation in eqn (2.30) due regard must be paid to the multiplicity factors of the various planes, for example, the $(11\bar{2}2)$ plane must be counted 12 times to allow for the various equivalent planes possessing this same spacing. As the wavelength decreases new families of planes are included, and more and more frequently, in the summation of eqn (2.30) but their contribution to the intensity of the reflected beam gets less and less owing to the presence of the factor λ^2.

At very short wavelengths, which are small in comparison with the interatomic separations, interference phenomena no longer occur and the coherent scattering becomes isotropic. $E(\mathscr{S})$ is then equal to $(\mathscr{S})$, i.e. $4\pi b^2$, for each atom.

It may be noted here that a corresponding peaked variation with wavelength of the integrated scattering, and hence of the transmission cross-section, occurs for X-rays as well as for neutrons. However, in the X-ray case the contribution of the scattering to the total absorption is overwhelmed by other effects, such as Compton scattering and the emission of fluorescent radiation.

The total scattering

From eqns (2.23) and (2.28) we can determine $\{E(\mathscr{S})+E(s)\}$, the total scattering of a polycrystalline element or compound. It will be equal to

$$\frac{\pi N_c}{2\kappa^2} \sum_{\substack{hkl \\ d \geqslant \lambda/2}} (4\pi F_{hkl}^2 d_{hkl}) + \sum (s). \tag{2.31}$$

The summation involved in the second term is made over all the atoms in the unit cell; simple addition of the s values for the individual atoms in the unit cell is, of course, possible since the disordered cross-sections are not subject to interference effects.

At long wavelengths, beyond the cut-off wavelength for coherent scattering, the total scattering will reduce to $\sum s$; at short wavelengths, where both the coherent and incoherent contributions are isotropic, it will be equal to $\sum(\mathscr{S}+s)$, i.e. each atom will contribute its σ value.

The general behaviour is indicated in Fig. 27 which is a generalized version of the type of curve which we showed in Fig. 26, and we are indicating the variation of cross-section as a function of neutron energy, rather than wavelength. Moreover, we are plotting the *total* scattering, having increased the ordered coherent component $E(\mathscr{S})$ by a disordered incoherent component $E(s)$, which, as we have already indicated, is equal to s.

As the neutron energy increases the discontinuities in the curve indicate the points at which Bragg reflection from new families of

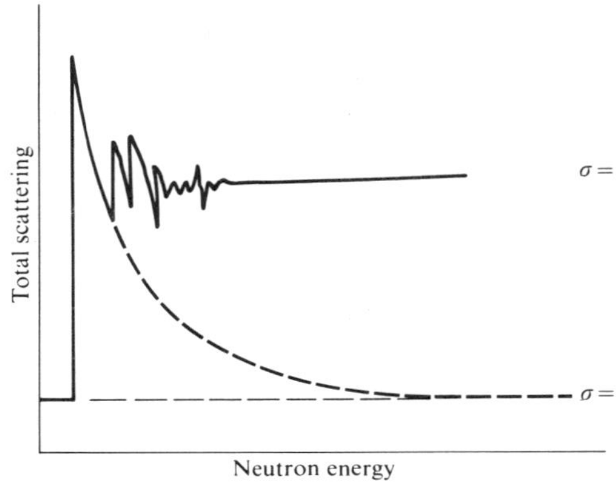

FIG. 27. The variation with neutron energy of the total scattering $E(\mathscr{S}) + E(s)$ for an assembly of static atoms.

planes becomes possible. However, as indicated by the broken line in the curve, the contribution of each family falls off inversely as the neutron energy E, because of the presence of the factor $1/\kappa^2$ in eqn (2.28). Moreover, the contribution of successive new families is less and less because of the influence of the quantity d_{hkl} in eqn (2.28). Accordingly the curve in Fig. 27 settles down at a constant level equal to $\mathscr{S} + s$.

2.6. The effect of finite nuclear mass: inelastic scattering

Elastic and inelastic collisions

The foregoing discussion of the scattering of neutrons by assemblies of nuclei in crystals have been on the underlying assumption that the nuclei were of infinite mass. Under these circumstances the scattering process resembled a classical elastic collision of a particle with a wall, and there was no possibility of any transfer of energy between the neutron and the target nucleus. We made this initial simplifying assumption in order to concentrate our attention on the effects introduced by the three-dimensional periodicities of the crystal structure.

We must now extend our considerations to the case of a real crystal in which the nuclei are not of infinite mass and, instead of being frozen rigidly in position at the atomic sites, they indulge in limited motion because of their thermal energy. It is not true to say that the atoms are in random motion, since there are forces between atoms in close proximity and the motion of any atom will be influenced to some extent by the motion of its neighbours. Indeed the nuclei in a solid take part in collective motions and the pattern of movement is the resultant build-up from a whole spectrum of waves of atomic displacement in the crystal. The neutron, or indeed other colliding particles, can transfer some of its energy to the crystal vibrations or, alternatively, take up energy from these vibrations. In these two cases the scattered neutrons will be of longer or shorter wavelength, respectively, compared with those in the incident beam. The significant feature, to which we shall return later, is that for *neutrons* the changes in wavelength are large enough to be measured easily and, as we shall see, are extremely informative.

The inelastic scattering of neutrons was considered theoretically by Weinstock (1944), well before any experimental work was done on this topic. Weinstock's treatment was restricted to the simplest case of an element which consisted of a single isotope with zero nuclear spin, and therefore giving rise to no incoherent disorder scattering, but it was extended by Cassels (1950) to elements possessing more than one isotope and also possessing nuclear spin. In Weinstock's treatment elastic scattering is treated as a special, zero-order case of inelastic scattering and, because of this, the theory arrives explicitly at the dependence of the *elastic* scattering on the crystal temperature. This dependence is according to the same exponential factor which occurs in the theory of X-ray scattering, in the classical work of Debye (1913, 1914) and Waller (1923, 1925).

Using the Debye model for the crystal vibrations, familiar from discussions of the specific heat of solids, Weinstock showed that the coherent scattering amplitude is reduced by the factor e^{-W}, where

$$W = \frac{6h^2}{m_A k\Theta}\frac{\sin^2\theta}{\lambda^2}\left(\frac{\phi(x)}{x}+\frac{1}{4}\right). \tag{2.32}$$

Here h is Planck's constant, m_A is the nuclear mass, k is Boltzmann's constant, Θ is the Debye temperature of the crystal, x is equal to Θ/T, where T is the absolute temperature of measurement, and $\phi(x)$ is a function of x defined by

$$\phi(x) = \frac{1}{x}\int_0^x \frac{\xi \, d\xi}{e^{\xi} - 1}. \tag{2.33}$$

This reduction factor for the coherent scattering amplitude is exactly the same as that derived for X-ray scattering, for example, by James (1948), whose nomenclature is the same. In the X-ray treatment it is intensity rather than amplitude which is being considered and the reduction factor is e^{-2W}.

The effective coherent neutron scattering amplitude which contributes to the Bragg diffraction peaks is therefore $b\,e^{-W}$, where b is defined as before to take into account the various isotopes, their abundances and nuclear spins. Fig. 28 shows how the value

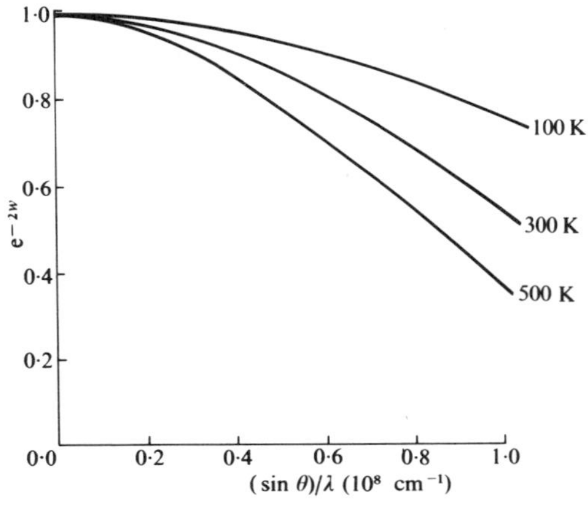

FIG. 28. Dependence of elastic ordered nuclear scattering on crystal temperature for iron.

of e^{-2W}, which will be the relevant factor affecting the intensities, falls off with $(\sin\theta)/\lambda$ for various temperatures in the case of iron whose Debye temperature is 453 K.

The above treatment of the effect of temperature on the elastic scattering can only be strictly applied to a monatomic cubic crystal.

In other cases it is not possible to describe by means of a single Debye temperature the vibrations of the different atoms in a compound or the vibrations of particular atoms in different directions in an anisotropic crystal.

In the general case it is necessary to describe the thermal motion of each atom which is crystallographically different in the unit cell by six parameters. These can be determined from the experimental diffraction intensities if sufficient independent data are available, using modern computational methods. The manner and importance of doing this will be described in Chapter 11 when we examine the application of neutron diffraction methods to the determination of molecular structure.

The disordered part of the elastic scattering decreases with increase of temperature in a similar way to the ordered component, and its cross-section is proportional to e^{-2W}. Accordingly, we have to modify the curves which we showed in Fig. 27, for the total scattering cross-section. So far as the *elastic* scattering is concerned, the behaviour is shown in Fig. 29, where it will be noted that both

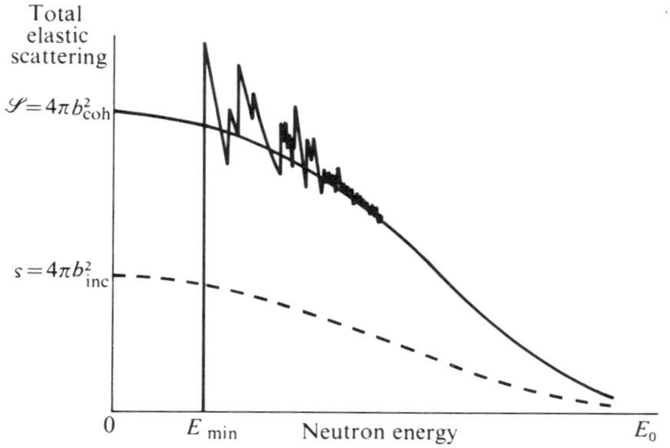

FIG. 29. Variation of the total *elastic* scattering with neutron energy.

the ordered (coherent) and the disordered (incoherent) contributions approach zero as the neutron energy increases.

Before we discuss the quantitative calculation of the *inelastic* scattering it is important to emphasize the significant difference between the inelastic scattering of slow neutrons and of monochromatic X-rays—the latter phenomenon usually being described as diffuse X-ray scattering. If we consider X-rays and neutrons of the same wavelength the energy of the X-ray quantum will be greater than that of the neutron by a factor of the order of 10^5. This follows since the X-ray energy $h\nu$ will equal hc/λ, where c is the velocity of light, giving a value of about $3 \times 10^{18}\,h$ for a wavelength of 1 Å, whereas the neutron energy is $\frac{1}{2}mv^2$, which can be written as $hv/2\lambda$, where v is the neutron velocity, and equals about $1{\cdot}5 \times \times 10^{13}\,h$. The velocity of a neutron of wavelength 1 Å is only 4×10^5 cm s^{-1} in comparison with 3×10^{10} cm s^{-1} for c the velocity of light. The energy in a single quantum of the crystal vibrations is of the order of $10^{13}\,h$, as can be seen most easily by considering the formula $h\nu_m = k\Theta$, which expresses the maximum frequency ν_m of the crystal vibrations in terms of the Debye temperature of the crystal, which will be generally of the order of a few hundred degrees. Consequently when an X-ray quantum is scattered inelastically, the energy lost to the crystal vibrations is negligible and the scattered phonon may be considered to have the same energy as the incident one. On the other hand, for slow neutrons the energy change is of the same order as the incident energy. Thus the inelastically scattered neutrons, but not the X-rays, will have wavelengths which are appreciably different from their initial value. It can be shown that the wavelength change $\delta\lambda$ when a neutron absorbs a quantum of energy δE from the crystal vibrations is

$$\delta\lambda = \frac{m\lambda^3}{h^2}\,\delta E, \tag{2.34}$$

and the same change will be observed when energy is given up to the crystal. Thus, for a given wavelength, the large wavelength change in the case of neutrons may be considered to be a consequence of the large neutron mass.

Experimentally, the first demonstration of the gain of energy when neutrons were scattered inelastically was provided by Egelstaff (1951) using a beam of long-wavelength neutrons from a lead filter. The importance of this behaviour to the study of solids arises because an examination of the detailed energy spectrum of the neutrons after scattering will make it possible to assess the details of the energy spectrum of the crystal, i.e. the acoustic vibrations in the solid. In order to carry out such a study it is necessary to perform an energy analysis of the scattered neutrons, and this means that it is necessary to use more intense neutron beams than those which prove adequate for the study of the Bragg elastic scattering. Consequently the study of inelastic scattering developed several years later than the more straightforward work in neutron crystallography, but this study has become increasingly widespread and important during the last ten years. Perhaps even greater possibilities have been opened up by applying inelastic scattering techniques to the study of liquids. In a liquid the neutron can exchange energy not only with the collective motions of the atoms but also with the motions of individual atoms which constitute liquid diffusion.

The process whereby the neutrons exchange energy with the collective motions in a solid (or liquid) is usually thought of as the creation or annihilation of *phonons*, the phonon being the embodiment, according to the wave–particle theory, of a quantum of energy. At low temperatures and low neutron energies the so-called 'one phonon' processes, in which a single phonon is created or annihilated, are predominant (apart from the zero-phonon process, which is *elastic* scattering), but with increase of temperature and energy multi-phonon processes become increasingly important. For a calculation of the probabilities of occurrence for the various processes the reader is referred to more advanced treatises (e.g. Lomer and Low 1965; Gurevich and Tarasov 1968), and we shall be concerned here only with a general physical description. Fig. 30 shows how, at a given temperature, the importance of second-order and third-order processes increases as the neutron energy increases. If we compute the resultant effect of processes of all types then we arrive at the dotted line in the Figure which, at the highest energies, attains a constant value equal to $4\pi\sigma_{\text{free}}$. It should be noted that it is the cross-section for the free atom (see eqn (2.11)) which is important at these high energies, because the nuclei are effectively free and their chemical bonding is unimportant.

If we combine the information in Fig. 30 for the inelastic scattering cross-section with that in Fig. 29 for the elastic scattering we arrive at the over-all picture for the complete scattering given in Fig. 31. With increase of neutron energy the elastic scattering falls to zero and the inelastic scattering is of overriding importance. Only at low neutron energies (but necessarily above E_{min}, which corresponds to the cut-off wavelength for Bragg scattering) do the interference

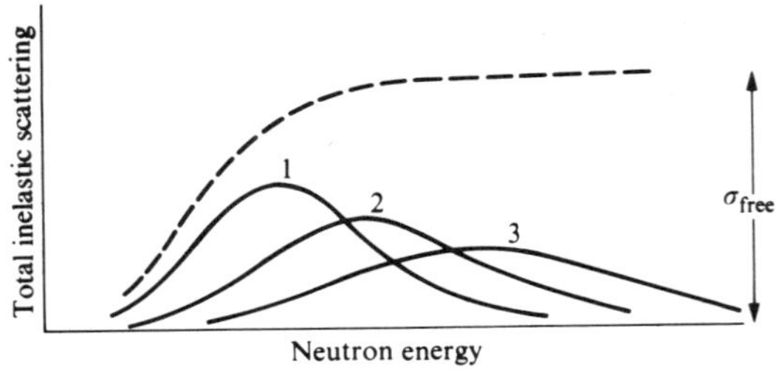

FIG. 30. The variation with neutron energy of the contributions to the neutron inelastic scattering from one (1)-, two (2)-, and three (3)-phonon processes. The total inelastic scattering, indicated by the broken curve, tends at high energies to the value of the scattering cross-section of the *free* atom.

effects predominate. Finally, Fig. 32 indicates the effect of increasing the temperature of the scattering sample, by presenting curves for both elastic and inelastic scattering at two different temperatures. As the temperature increases the inelastic scattering becomes more important and, also, the contribution to the latter of multi-phonon processes becomes more significant. This means that at higher temperatures it becomes more difficult to deduce the details of the lattice dynamics of the crystal from a study of the inelastic scattering. On the other hand, at ordinary temperatures, where one-phonon processes predominate, the interpretation of the neutron spectra

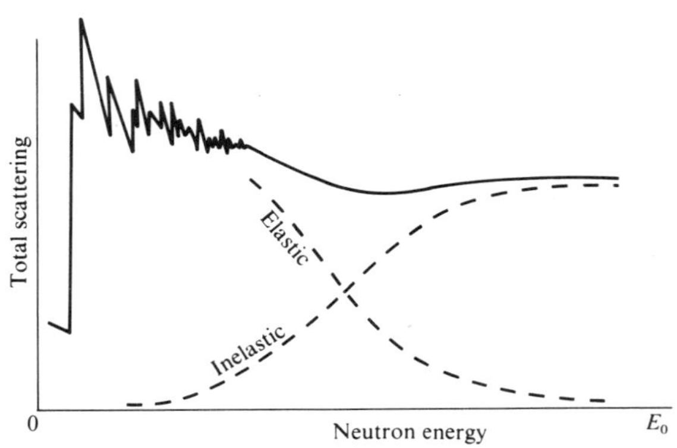

FIG. 31. The variation with neutron energy of the total scattering cross-section, derived by combining the data of Figs 29 and 30 for elastic and inelastic scattering respectively.

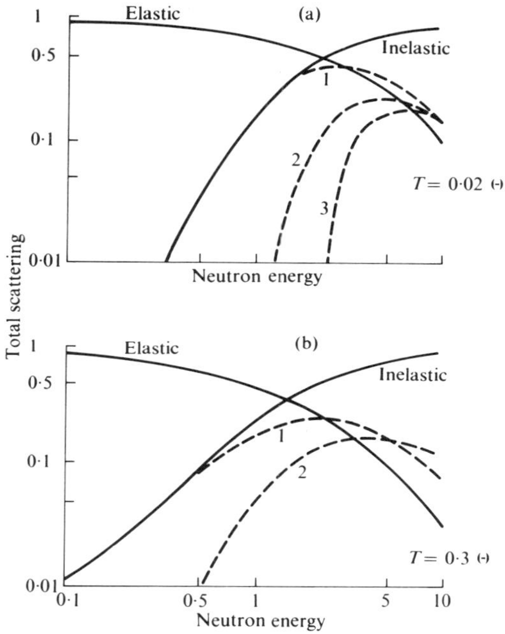

FIG. 32. The elastic and inelastic scattering of neutrons from an assembly of nuclei of mass $10m$, showing the variation with neutron energy at two different temperatures. For the upper curves $T = 0{\cdot}02\Theta$ and for the lower curves $T = 0{\cdot}3\Theta$. The energy scale is logarithmic and the energy is expressed in units of $k_B\Theta$. The broken curves indicate the contribution to the inelastic scattering from one (1)-, two (2)-, and three (3)-phonon processes. (After Gurevich and Tarasov 1968.)

becomes direct and their study is very rewarding, as we shall see in a later chapter.

Experimental measurements are in good agreement with the theoretical calculations. Magnesium provides a very good example for an experimental test because its absorption of neutrons, which would complicate the interpretation of the scattering of the neutrons, is very low. Fig. 33 shows that for neutrons of wavelength 6·25 Å there is agreement to an accuracy of 4 per cent between the measured and calculated cross-sections, if due account is taken of the collisions in which more than one quantum of energy is exchanged. If the calculation is restricted purely to processes in which just one quantum of energy is exchanged, then the result comes out about 30 per cent lower than what is observed in the experiment.

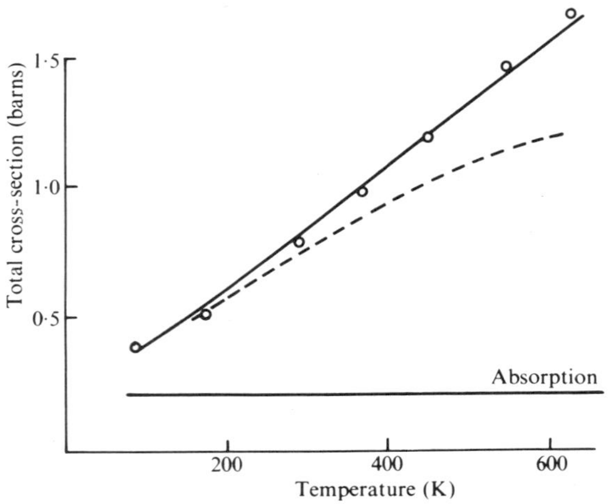

FIG. 33. The variation with temperature of the total neutron cross-section of magnesium. The experimental points are marked ○. The final theoretical curve is shown by ———, while - - - - - - - is the theoretical curve if multi-oscillator processes are neglected. (From Squires 1952.)

We emphasize that the discussion in the foregoing pages has been for polycrystalline samples in which the orientation in space of any particular plane in the crystal structure is completely random. For single-crystal samples the detailed orientation of the crystal planes relative to the incident beam and the direction of scattering will have to be taken into account. An interesting feature of Fig. 31 in this respect, which is of some practical application, is worthy of comment. For neutrons of low energy the Figure indicates how the inelastic scattering is small; for a single crystal the elastic scattering *also* will be quite small since only for very narrow bands of wavelength will the crystal setting be exactly that which gives a Bragg reflection from some particular crystal plane. Thus the high scattering, with the succession of peaks, which is found for the polycrystalline sample will not be observed for a single crystal. Thus a single crystal can act as a filter, permitting the passage of thermal neutrons but removing from the beam the neutrons of high energy. Crystals of bismuth and quartz have been extensively used for reducing in this way the components of unwanted high-energy neutrons in beams from reactors (Brockhouse 1961). Bismuth is particularly valuable because the high atomic number ensures that there is also a substantial removal of γ-rays from the incident beam.

2.7. X-ray and neutron scattering amplitudes

We have already emphasized the relatively small variation with atomic number of the coherent neutron scattering amplitude, in contrast with the proportionality to atomic number of the X-ray scattering amplitude. The variations in the two cases can be readily appreciated by examining columns 6 and 10 of Table 2 (p. 38), the latter column giving the X-ray amplitude for a value of $(\sin\theta)/\lambda$ equal to $0{\cdot}5\times10^8\ \mathrm{cm}^{-1}$. Attention is now drawn to a comparison between the absolute values of the two scattering amplitudes. For hydrogen and deuterium the X-ray amplitudes are smaller than those for neutrons. From lithium onwards the X-ray value shows a steady increase relative to that for neutrons, but it is not until an atomic number of about 50 has been reached that the X-ray amplitude consistently exceeds the neutron scattering amplitude by a factor of 10. It follows therefore that for a great many substances the ratios of the amplitudes of the reflected and incident beams when X-rays and neutrons are scattered by an atom, or by the group of atoms comprising the unit cell of a crystal structure, will be the same within a factor of 10. Consequently the distinction between 'perfect' and 'mosaic' crystals, which is very essential to an understanding of the quantitative intensities of reflection of X-ray by crystals, will be equally significant in the case of neutrons. Indeed for X-rays and neutrons of the same wavelength, the classifications of particular crystals as 'perfect' or 'mosaic' will be very similar.

In order to complete our picture more fully we may comment, in passing, on the absolute values of the scattering amplitude of atoms for *electron* beams. These values are of the order of 10^{-8} cm and are thus of the order of 10^3 times larger than those for neutrons and X-rays. Consequently electron diffraction techniques are applicable to gases, thin films, and the study of surfaces, whereas X-rays and neutrons are concerned with specimens which are of the order of a millimetre in thickness. Indeed neutrons may often be used to examine samples which have a thickness of several centimetres.

Bearing in mind particularly the remarks earlier in this section we shall proceed to a detailed examination of the diffraction of neutrons by single crystals, with the aim of determining how the intensities of the diffracted beams from these are related to the coherent scattering amplitudes of the atoms with which we have been chiefly concerned in the foregoing chapter. It is recalled that our discussion so far of the interference effects produced by the crystallites of a polycrystalline sample (as represented, for example, by the conclusions embodied in eqns (2.20) and (2.30) with the addition of Debye correction factors for thermal vibrations) assumed that the crystallites were sufficiently small for primary and secondary extinction and absorption to be negligible. These factors must now be taken into account in passing to a consideration of diffraction by single crystals which may be of considerable size.

3

DIFFRACTION BY CRYSTALS

3.1. Primary and secondary extinction

WE shall proceed in this chapter to a detailed treatment of the diffraction of neutrons by single crystals, aiming to determine the intensities of the reflected beams in terms of the coherent scattering amplitudes b of the atomic nuclei and their arrangements in the crystal structure.

In section 2.5 we define F_{hkl}, the structure factor of a unit cell of the crystal, by the expression

$$F_{hkl}^2 = |\sum b_r \exp\{2\pi i(hx/a_0 + ky/b_0 + lz/c_0)\}|^2 e^{-2W}, \qquad (3.1)$$

where F_{hkl} is the amplitude of the diffracted neutron beam for the (hkl) reflection, assuming an incident beam of unit amplitude. The exponential term takes account of the effect of thermal vibrations. As a first stage in the calculation of the diffracted beam from an extended crystal, we shall consider the amplitude reflected by the material lying between successive (hkl) planes in a crystal. Fig. 34

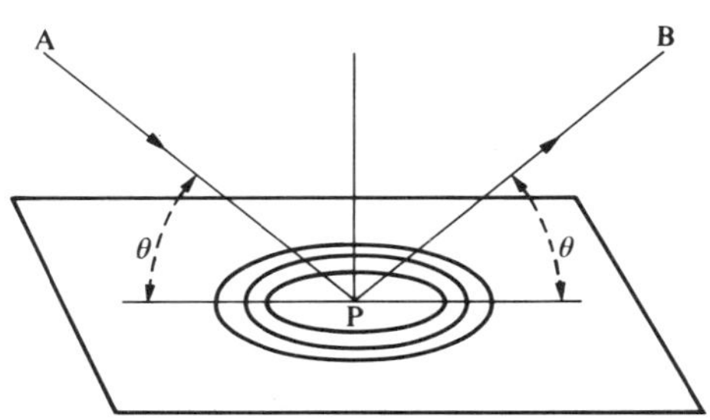

FIG. 34. Reflection of neutron or X-ray beam from A to B by net plane of crystal.

indicates the reflection of neutrons (or X-rays) from a source A to a point B, the point P on the crystal surface being such that the plane APB is normal to the reflecting plane (hkl) and B being chosen so that θ is the Bragg angle for the (hkl) reflection. The amplitude of the beam reflected to B can be deduced by employing the device of the Fresnel spiral which is commonly used in optical problems. It follows that the amplitude is half of that which would be given by the first Fresnel half-period zone, which, in its turn, is $2/\pi$ times the sum of the amplitudes coming from all the unit cells within this first zone. In order to be assured of the latter fact it is necessary to show that each half-period zone covers a large number of unit cells. It can indeed be shown that at least 10^7 atoms are involved, and the condition is therefore well satisfied. With these assumptions it can be shown (James 1948, p. 36) that the amplitude reflected to B is equal to q_{hkl}, given by

$$q_{hkl} = N_c d \frac{\lambda}{\sin\theta} F_{hkl} = 2N_c d^2 F_{hkl}, \tag{3.2}$$

where N_c is the number of unit cells per unit volume and d is the distance between successive (hkl) planes.

Table 4 gives the values of q_{hkl} for a number of different reflections from a few common crystals for the cases of both neutrons and X-rays. From the form of eqn (3.2) it will be seen that these q_{hkl} values are independent of wavelength.

TABLE 4

Values of q_{hkl}, the amplitude reflected per layer plane for neutrons and X-rays

Crystal	Diamond		NaCl		Copper		CaF_2	
Reflection	111	220	111	200	111	200	111	200
q_{hkl} for neutrons (10^{-5})	6·8	3·6	3·0	4·8	5·6	4·2	2·4	2·2
q_{hkl} for X-rays (10^{-5})	7·6	2·9	5·4	19	39	28	21	1·8

From the values given in the table it follows that each successive reflecting plane which the incident neutron beam meets in its passage through the crystal will contribute a reflected amplitude which is less than 1 part in 10^4 of the incident amplitude. The contribution is in general rather less than for X-rays, but it is indeed of the same order. If then the incident neutron beam passes through a crystal of some 500 planes thick, say 1000 Å, then the reflected amplitude will be only 5 per cent and the crystallite may be considered to be uniformly bathed in radiation, assuming that there is no appreciable absorption of the neutrons by nuclear capture processes. Under these circumstances we say that there is negligible 'primary extinction'. It can then be shown that the so-called 'integrated reflection' for the rotating crystal method, in which the crystal is considered to rotate through the Bragg reflection position in a beam of monochromatic radiation, is equal to $Q\,\delta V$, where δV is the volume of the crystallite and Q is the well-known crystallographic quantity defined by

$$Q = \frac{\lambda^3 N_c^2}{\sin 2\theta} F^2. \tag{3.3}$$

Table 5 gives values of Q for neutrons and X-rays for the same crystals and reflections for which the q_{hkl} values were given in

TABLE 5

(*a*) *Values of Q for neutrons* (λ = 1·08 Å) *and X-rays* (λ = 1·54 Å)

Crystal	Diamond		NaCl		Copper		CaF_2	
Plane	111	220	111	200	111	200	111	200
Q_{neutrons} (cm^{-1} 10^{-2})	1·64	2·14	0·08	0·3	1·06	0·98	0·05	0·07
$Q_{\text{X-rays}}$ (cm^{-1} 10^{-2})	4·5	3·3	0·5	10·4	105	85	8·5	0·1

(*b*) *Values of s for neutrons* (λ = 1·08 Å) *and X-rays* (λ = 1·54 Å)

Crystal	Diamond		NaCl		Copper		CaF_2	
Plane	111	220	111	200	111	200	111	200
s_{neutrons} (seconds of arc)	1·2	1·1	0·33	0·60	1·0	0·90	0·27	0·29
$s_{\text{X-rays}}$ (seconds of arc)	2·1	1·5	0·9	3·7	10	9	3·5	0·35

Table 4. The values of Q are, of course, dependent on wavelength and are quoted for a typical neutron wavelength of 1·08 Å and for Cu Kα X-radiation (λ = 1·54 Å).

If, however, we consider reflection by a crystal of thickness considerably greater than 1000 Å, then the formula $Q\,\delta V$ for the integrated reflection will no longer apply, since the incident beam will become appreciably attenuated as a result of the process of reflection by successive planes. The penetration distance in a perfect crystal will be of the order of 10^{-4} cm at the angle of Bragg reflection and further increase of thickness beyond this will contribute little to the reflection. We shall see later in this chapter that the linear absorption coefficient μ for neutrons is of the order of $0{\cdot}2\ \mathrm{cm}^{-1}$, and for X-rays it is of the order of $2000\ \mathrm{cm}^{-1}$ (in each case these values refer to 'true' absorption, not including scattering), so that even for X-rays the true absorption in a distance of 10^{-4} cm is only a few per cent. Consequently in the classical derivation of the expressions for the intensity of reflection for a perfect crystal by Darwin (1914) and by Ewald (1916, 1917) the true absorption could be neglected. The theories show that total reflection, as described above, only occurs over a very small angular range near the Bragg angle, equal to $2s$, where

$$s = \left(\frac{\lambda^2 N_c}{\pi \sin 2\theta}\right) F. \tag{3.4}$$

This amounts to less than a second of arc for neutrons and to a few seconds for X-rays, as can be seen from the values shown in the second portion of Table 5. The form of the reflection curve is illustrated in Fig. 35.

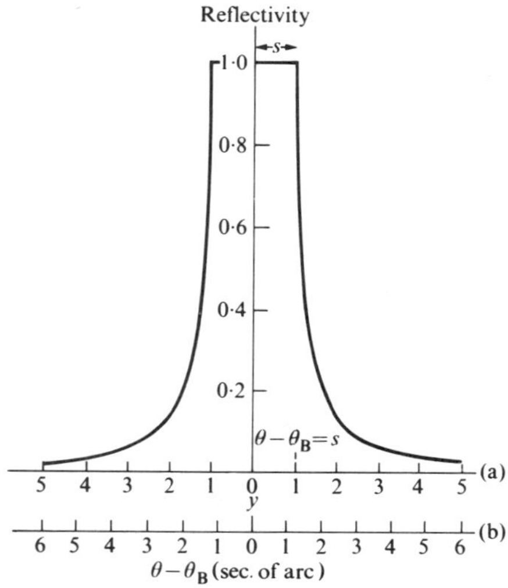

FIG. 35. Reflection curve for thick perfect crystal in terms of (a) $y = (\theta-\theta_B)/s$ and (b) $\theta-\theta_B$, for the particular case of the (111) reflection of diamond at a neutron wavelength of 1·08 Å.

The calculation of the integrated reflection from a crystal depends basically on an assessment of the way in which radiation is transferred between beams in the incident and diffracted directions as the incident beam proceeds through the crystal. If the intensities of the two beams are I_0, I respectively then they must satisfy the fundamental equations

$$\partial I_0/\partial t_1 = -\check{r}I_0 + \check{r}I, \tag{3.5}$$

$$\partial I/\partial t_2 = -\check{r}I + \check{r}I_0, \tag{3.6}$$

where the location of an element of volume in the crystal is specified by t_1, t_2, its depths below the crystal surface as measured along the incident and diffracted beams respectively. The quantity $\check{r}$ is the diffracting power per unit volume and unit intensity and may be regarded as a 'coupling constant' between the incident and diffracted beams. It will depend on the direction of the incident beam and will only differ from zero when the incident direction is close to the Bragg angle. The calculation has only been carried out exactly for a plane parallel crystal plate of infinite extent. The variation with crystal thickness t_0 in this case is shown in Fig. 36. The form of the curve (see Zachariasen (1945), eqns 3.167 and 3.143) is

$$\mathscr{R}_0^\theta = \frac{N_c\lambda^2 F}{\sin 2\theta}\tanh\left(\frac{N_c\lambda t_0 F}{\sin\theta}\right). \tag{3.7}$$

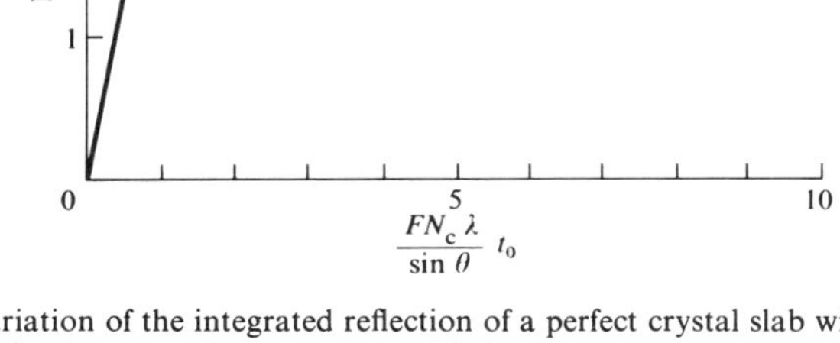

FIG. 36. Variation of the integrated reflection of a perfect crystal slab with thickness t_0 for the reflection case. (Bacon and Lowde 1948.)

In the absence of any primary extinction the integrated reflection would be

$$\mathscr{R}^\theta = Q\delta V = \frac{N_c^2\lambda^3F^2}{\sin 2\theta}\,\frac{t_0}{\sin\theta},$$

so that the 'primary extinction factor' E_p, defined as the ratio $\mathscr{R}^\theta/\mathscr{R}_0^\theta$ will be given by

$$E_p = \frac{\tanh(N_c\lambda t_0 F/\sin\theta)}{N_c\lambda t_0 F/\sin\theta}. \tag{3.8}$$

This expression enables a correction to be applied for the effect of primary extinction.

In general, so-called 'single crystals' do not have a perfectly regular sequence of atomic planes and unit cells throughout the whole volume of the crystal. They are permeated by dislocations which have the effect of dividing the crystal into much smaller regions, of the order of 5000 Å in linear dimensions, which are themselves perfect but which are misoriented from one to another. The extent of the misorientation is usually a few minutes of arc but in more exceptional cases may be half a degree. Such a crystal is said to be 'imperfect', and if the microscopic regions of perfection are sufficiently small for only a negligible amount of 'primary extinction', as defined above, to take place within them then the crystal is said to be 'ideally imperfect'. The components of radiation reflected from the individual small regions, or mosaic blocks as they are called, are not coherent, and it is necessary to sum intensities, rather than amplitudes, in calculating the reflection intensity. The passage of an X-ray or neutron beam through a crystal of this sort will be rather different from the case of a 'perfect crystal'. Individual mosaic blocks will reflect (over the angular range $2s$) at slightly different angles when the whole crystal is considered to be rotated through the mean reflecting position. Consequently the incident beam will be able to penetrate much more deeply into the crystal. There are two factors which will limit its penetration. First, at any particular angular setting of the crystal the beam will eventually reach mosaic blocks identical in orientation with the ones through which the beam travelled at the surface; thus the beam is attenuated by 'secondary extinction' (see Lonsdale 1947) between mosaic blocks. Secondly, the beam will be reduced by true absorption processes, such as photoelectric absorption for X-rays and nuclear capture for neutrons. The relative importance of these two factors in reducing the incident beam will determine the form of the quantitative expression for the intensity of the reflections from mosaic crystals in terms of the crystal thickness and structure. We shall see that, in general, the cases of X-ray diffraction and neutron diffraction represent the two extremes. Broadly we may say that for X-rays the true absorption (i.e. reduction of the incident beam by causes other than scattering) is sufficiently large to be the predominant factor. For neutrons the absorption is usually negligible and the scattering alone very largely determines the reduction of the incident beam in its passage through the crystal.

3.2. The coefficients of true absorption for X-rays and neutrons

The values of the nuclear absorption cross-sections of the elements for neutrons are given in Table 6; these values are those appropriate to the mixture of isotopes which form the naturally occurring variety of the element, except for the few elements such as hydrogen, lithium, and boron, where values are given for individual isotopes. The Table also includes the mass absorption coefficients μ/ρ, where ρ is the density, and the linear absorption coefficients μ of the elements in their normal solid forms, together with corresponding figures for X-rays. It will be seen that, apart from a few exceptions, the absorption coefficients are vastly smaller for neutrons than for X-rays. A typical value of μ for neutrons is $0{\cdot}3\ \text{cm}^{-1}$, in comparison with values of 1000 or so for X-rays. Notable exceptions are boron, cadmium, gadolinium, and other rare earths. The different orders of magnitude of the absorption coefficients for the two radiations may also be appreciated from Table 7, which gives values of the percentage transmission of thicknesses of $\frac{1}{4}$ in (6·3 mm) and 0·1 mm of a number of typical substances. However, the true absorption or 'capture' coefficients for neutrons are even lower relative to X-rays than the experimental figures in Table 7 suggest, since in the case of the polycrystalline samples the apparent absorption is the sum of both true absorption and loss by scattering. For aluminium, lead, and cadmium the scattering cross-sections are much greater than those for true absorption (as can be seen by comparing Tables 2 and 6), whereas in the case of X-ray transmission the energy loss due to scattering is negligible.

TABLE 6

Absorption of the elements for neutrons and X-rays

			Cross-section for true absorption σ (10^{-24} cm^2)		Mass absorption coefficient μ/ρ (cm^2/g^{-1})		Linear absorption coefficient of solid element μ (cm^{-1})	
Element	Symbol	Atomic no.	Neutrons $\lambda = 1{\cdot}08$ Å	X-rays $\lambda = 1{\cdot}54$ Å	Neutrons	X-rays	Neutrons	X-rays
Hydrogen	^{1}H	1	0·19	0·7	0·11	0·43	—	
	^{2}H	1	0·0005	0·7	0·0001	0·43	—	
Helium	He	2	—	2·5	—	0·38	—	
	^{3}He	2	—	2·5	—	0·38		
Lithium	Li	3	40	8·2	3·5	0·72	1·87	0·38
	^{6}Li	3	570	8·2	49	0·72	26	0·38
	^{7}Li	3	—	8·2	—	0·72	—	0·38
Beryllium	Be	4	0·005	22	0·0003	1·5	0·0005	2·7
Boron	B	5	430	43	24	2·4	56	5·6
	^{10}B	5	2300	43	128	2·4	300	5·6
	^{11}B	5		43		2·4		5·6
Carbon	C	6	0·003	92	0·00015	4·6	0·0005	15·2
Nitrogen	N	7	1·1	175	0·048	7·5	—	
Oxygen	O	8	0·0001	305	0·00001	11·5	—	
Fluorine	F	9	0·006	518	0·0002	16·4	—	
Neon	Ne	10	0·2	767	0·006	22·9	—	
Sodium	Na	11	0·28	1150	0·007	30·1	0·007	29
Magnesium	Mg	12	0·04	1560	0·001	38·6	0·0017	67
Aluminium	Al	13	0·13	2180	0·003	48·6	0·008	131
Silicon	Si	14	0·06	2830	0·002	60·6	0·004	141
Phosphorus	P	15	0·09	3810	0·002	74·1	0·004	163
Sulphur	S	16	0·28	4740	0·0055	89·1	0·011	178
Chlorine	Cl	17	19·5	6210	0·33	106	—	
Argon	A	18	0·4	8170	0·006	123	—	
Potassium	K	19	1·2	9250	0·018	143	0·016	123
Calcium	Ca	20	0·25	10 800	0·0037	162	0·0057	251
Scandium	Sc	21	19	13 700	0·25	184	0·75	550
Titanium	Ti	22	3·5	16 500	0·044	208	0·20	994
Vanadium	V	23	2·8	19 700	0·033	233	0·20	1423
Chromium	Cr	24	1·8	22 400	0·021	260	0·15	1869
Manganese	Mn	25	7·6	26 000	0·083	285	0·60	2090
Iron	Fe	26	1·4	28 600	0·015	308	0·12	2420
Cobalt	Co	27	21	32 400	0·21	313	1·87	2790
Nickel	Ni	28	2·7	4450	0·028	45·7	0·25	407
Copper	Cu	29	2·2	5580	0·021	52·9	0·19	474
Zinc	Zn	30	0·6	6550	0·0055	60·3	0·039	430
Gallium	Ga	31	1·8	7850	0·015	67·9	0·089	401
Germanium	Ge	32	1·3	9110	0·011	75·6	0·058	402
Arsenic	As	33	2·5	10 400	0·020	83·4	0·12	478
Selenium	Se	34	7·4	12 000	0·056	91·4	0·27	438
Bromine	Br	35	3·8	13 200	0·029	99·6	—	
Krypton	Kr	36	18	15 000	0·13	108	—	
Rubidium	Rb	37	0·42	16 500	0·003	117	0·0044	179
Strontium	Sr	38	0·7	18 200	0·005	125	0·012	317
Yttrium	Y	39	0·83	19,800	0·006	134	0·031	596
Zirconium	Zr	40	0·10	21 700	0·0006	143	0·0041	933
Niobium	Nb	41	0·63	23 600	0·004	153	0·034	1311
Molybdenum	Mo	42	1·4	25 900	0·009	162	0·08	1650
Technetium	Tc	43	—	28 300	—	172	—	
Ruthenium	Ru	44	1·5	30 700	0·009	183	0·10	2270
Rhodium	Rh	45	90	33 200	0·53	194	6·6	2410
Palladium	Pd	46	4·0	36 400	0·023	206	0·28	2480
Silver	Ag	47	36	39 000	0·20	218	2·0	2290
Cadmium	Cd	48	2650	43 000	14	231	121	2000
	^{113}Cd	48	23 000	43 000	122	231	1050	2000
Indium	In	49	115	46 400	0·6	243	4·4	1780

TABLE 6 (*continued*)

Element	Symbol	Atomic no.	Cross-section for true absorption σ (10^{-24} cm^3) Neutrons $\lambda = 1·08$ Å	X-rays $\lambda = 1·54$ Å	Mass absorption coefficient μ/ρ (cm^2/g^{-1}) Neutrons	X-rays	Linear absorption coefficient of solid element μ (cm^{-1}) Neutrons	X-rays
Tin	Sn	50	0·35	50 500	0·002	256	0·011	1870
Antimony	Sb	51	3·2	54 500	0·016	270	0·10	1810
Tellurium	Te	52	2·7	59 800	0·013	282	0·081	1760
Iodine	I	53	3·7	62 000	0·018	294	0·09	1450
Xenon	Xe	54	18	66 800	0·082	306		
Cesium	Cs	55	17	70 200	0·077	318	0·14	573
Barium	Ba	56	0·6	75 100	0·0026	330	0·01	1155
Lanthanum	La	57	5·3	78 600	0·023	341	0·14	2080
Cerium	Ce	58	0·48	81 900	0·0021	352	0·016	2620
Praseodymium	Pr	59	6·7	85 000	0·029	363	0·19	2440
Neodymium	Nd	60	26	89 700	0·11	374	0·76	2580
Promethium	Pm	61		92 800		386		
Samarium	Sm	62	3500	99 000	14	397	104	2970
	^{149}Sm	62	114 000	99 000	460	397	3420	2970
	^{154}Sm	62	3	99 000	0·012	397	0·09	2970
Europium	Eu	63	1600	107 000	6	425	31	2240
Gadolinium	Gd	64	20 000	115 000	76	439	600	3410
	^{160}Gd	64	1	115 000	0·004	439	0·03	3410
Terbium	Tb	65	26	72 000	0·1	273	0·8	2260
Dysprosium	Dy	66	580	77 000	2·1	286	18	1090
Holmium	Ho	67	50	35 100	0·18	128	1·6	1130
Erbium	Er	68	120	37 200	0·43	134	3·9	1210
Thulium	Tm	69	71	39 300	0·25	140	2·3	1310
Ytterbium	Yb	70	30	42 100	0·10	146	0·7	1020
Lutecium	Lu	71	85	44 400	0·29	153	2·9	1510
Hafnium	Hf	72	61	47 200	0·20	159	2·6	2110
Tantalum	Ta	73	13	49 800	0·043	166	0·7	2750
Tungsten	W	74	11	52 600	0·035	172	0·7	3320
Rhenium	Re	75	50	55 400	0·16	179	3·4	3760
Osmium	Os	76	9	58 700	0·028	186	0·6	4200
Iridium	Ir	77	260	61 600	0·81	193	18	4330
Platinum	Pt	78	5	64 900	0·015	200	0·3	4290
Gold	Au	79	57	68 000	0·17	208	3·3	4170
Mercury	Hg	80	210	71 800	0·63	216	8·5	3040
Thallium	Ti	81	2	75 900	0·006	224	0·07	2650
Lead	Pb	82	0·1	79 700	0·0003	232	0·003	2630
Bismuth	Bi	83	0·02	83 300	0·00006	240	0·0006	2340
Polonium	Po	84						
Astatine	At	85						
Radon	Rn	86		102 000		278		
Francium	Fr	87						
Radium	Ra	88		113 400		304		
Actinium	Ac	89	300		0·79			
Thorium	Th	90	4·1	118 000	0·01	307	0·1	3580
Protactinium	Pa	91						
Uranium	U	92	2·1	121 000	0·005	306	0·1	5800
Neptunium	^{237}Np	93	102		0·26		4·9	
Plutonium	^{237}Pu	94	620	140 000	1·56	353	31	7000

The value of the linear absorption coefficient μ for a compound will be equal to the product of its density and $\sum(\mu/\rho)w$, where the summation is taken over the constituent elements whose proportions by weight are represented by w.

TABLE 7

Transmission factors for neutrons and X-rays

Material	Percentage transmission		
	Neutrons $\lambda = 1{\cdot}08$ Å	X-rays $\lambda = 1{\cdot}54$ Å	
	Thickness $\frac{1}{4}$ in (6·3 mm)	$\frac{1}{4}$ in (6·3 mm)	0·1 mm
Aluminium	94	zero	27
Brass	67	zero	1
Cadmium	zero	zero	zero
Lead	84	zero	zero
Graphite	80	$\frac{1}{4}$	90
CaF_2 (single crystal)	97	zero	5

The values of the absorption cross-section listed in Table 6 are derived from *Nuclear data* (1950), where they are quoted for a wavelength of 1·8 Å (0·025 eV). Except for the few elements which have resonance capture peaks below 0·5 eV all elements absorb neutrons according to $1/v$, where v is the neutron velocity. In Table 6 the cross-sections have been modified according to the $1/v$ law to become appropriate to a neutron wavelength of 1·08 Å, since a shorter wavelength of about this value is generally used for neutron diffraction. In certain cases the value given in the table is the mean of two experimental determinations.

The consequence of the very small value of μ for neutrons is that when a neutron beam is passing through an ideally imperfect crystal at the Bragg reflection position it will be attenuated mainly by scattering; the reduction of intensity by true absorption will be very small and for many purposes may be neglected entirely.

Zachariasen (1945) gave a general treatment of the determination of the integrated intensities for infinite plane slabs of mosaic crystals, for both the symmetrical reflection (Bragg) case and the symmetrical transmission (Laue) case. These two arrangements are contrasted in Fig. 37. Zachariasen then gave a detailed treatment for the case of absorbing materials, i.e. the situation which applies for X-ray diffraction. The application of a similar analysis to materials of very low absorption, i.e. the neutron case, was described by Bacon and Lowde (1948). We shall summarize their conclusions, before proceeding to later theoretical work by Zachariasen and

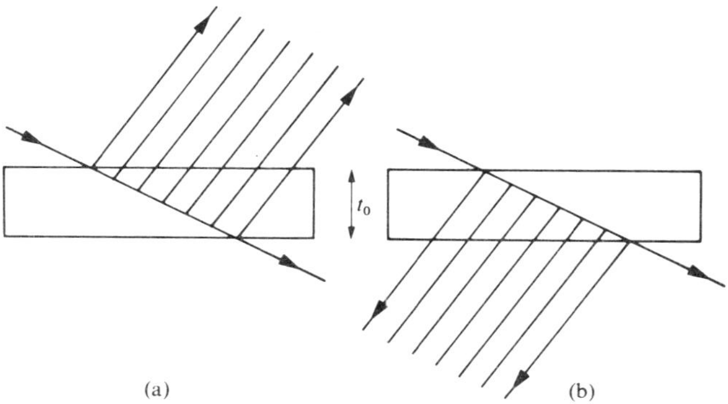

FIG. 37. Distinction between the 'reflection' and 'transmission' cases of diffraction by a parallel crystal slab. (a) Reflection method; (b) transmission method. (Bacon and Lowde 1948.)

others, in order to make clearer the distinction between the diffraction of X-rays and neutrons.

3.3. Ideally imperfect crystals: plane parallel slabs

The mosaic structure of a crystal can be represented by an angular distribution function $W(\Delta)$, defined so that $W(\Delta)\,\mathrm{d}\Delta$ is the fraction of mosaic blocks which have their normals between the angles Δ and $\Delta+\mathrm{d}\Delta$ from the crystal surface. There is good evidence that $W(\Delta)$ has a Gaussian form and can be written

$$W(\Delta) = \eta^{-1}(2\pi)^{-\frac{1}{2}}\exp(-\Delta^2/2\eta^2), \tag{3.9}$$

where η is the standard deviation of the mosaic blocks. It can be assumed that $\eta \gg s$ for (as Table 5 shows) the value of s is only about a second of arc, whereas η is of the order of a few minutes of arc. Bacon and Lowde's (1948) conclusions can be summarized by Fig. 38 in which curves are plotted to show the way in which $(1/\eta)\mathscr{R}^\theta$ varies with the quantity $(1/\eta)(Qt_0/\sin\theta)$. The different curves are appropriate for different values of $\mu t_0/\sin\theta$, where μ is the coefficient of absorption.

The principal features of these curves are as follows.

1. As $(1/\eta)(Qt_0/\sin\theta)$ increases there is an initial region near the origin where $(1/\eta)\mathscr{R}^\theta$ is *equal* to $(1/\eta)(Qt_0/\sin\theta)$, which means that the integrated reflection is equal to QV, where V is the

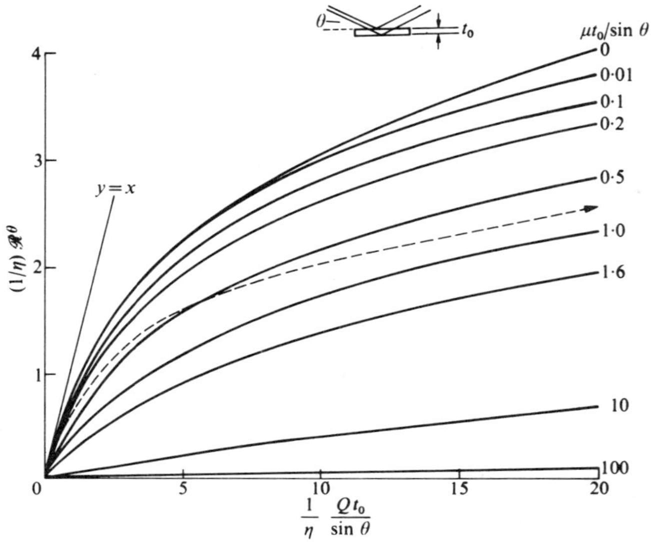

FIG. 38. Theoretical curves for the integrated reflection from ideally imperfect crystal slices (reflection case). Broken curve shows mean line through area of diagram explored in the experimental verification described later. (Bacon 1951.)

volume of crystal being irradiated. In the lowest curves this initial portion, which lies along the line $y = x$, is too small to be distinguished. These lower curves correspond to the X-ray case for which, owing to the high absorption coefficient, only a very thin crystal will give a reflection proportional to its volume.

2. With further increase of $(1/\eta)(Qt_0/\sin\theta)$ it is seen that $(1/\eta)\mathscr{R}^\theta$ becomes *proportional* to this quantity in the case of the lower curves. For a given value of t_0, therefore, $\mathscr{R}^\theta$ will be proportional to Q. Here, again, we expect a result typical of X-ray diffraction and in fact we find that $\mathscr{R}^\theta$ is equal to $Q/2\mu$, as is well known. For example, on the lowest curve, for which $\mu t_0/\sin\theta = 100$, the value of $(1/\eta)\mathscr{R}^\theta$ when $(1/\eta)(Qt_0/\sin\theta) = 20$ is 0·1, which is also found to be the value of $(1/\eta)(Q/2\mu)$.

3. On the other hand, for the curves having small values of $\mu t_0/\sin\theta$, which are the ones in general applicable to *neutron* diffraction, the fall off from the line $y = x$ is followed by a

region where $\mathscr{R}^\theta$ is no longer proportional to Q, for a given value of t_0, but is determined largely by the mosaic spread of the crystal. Here secondary extinction is predominant and the value of $\mathscr{R}^\theta$ is of the order of 3η. With further increase of $(1/\eta)(Qt_0/\sin\theta)$ it is found by calculation that $(1/\eta)\mathscr{R}^\theta$ rises more and more slowly: for the case of $\mu = 0$, $\mathscr{R}^\theta$ increases by a factor of only 2 when the abscissa is increased from 10 to 200.

Certain aspects of the behaviour of the integrated reflection are made clearer if $(1/\eta)\mathscr{R}^\theta$ is plotted for constant values of $(1/\eta)(Q/\mu)$ rather than of $\mu t_0/\sin\theta$. These curves are shown in Fig. 39, and from them we can draw the following main conclusions.

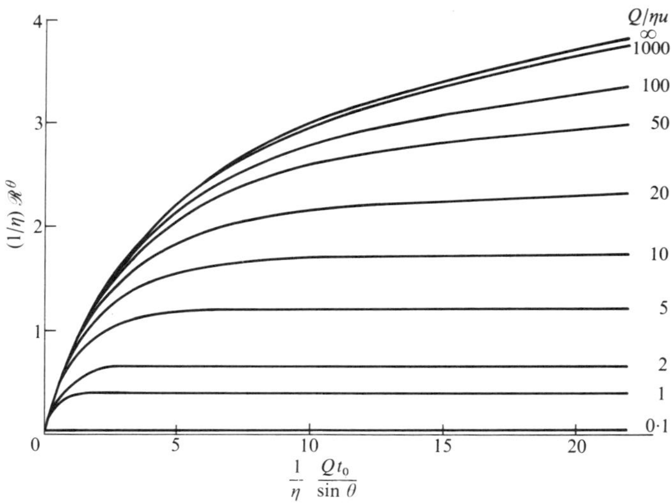

FIG. 39. Theoretical curves for the integrated reflection from ideally imperfect crystal slices (reflection case).

4. With increasing thickness $(1/\eta)\mathscr{R}^\theta$ has fallen away from the value $(1/\eta)(Qt_0/\sin\theta)$ by 5 per cent at an abscissa value which can be shown by calculation to be about 0·25 in the case of a non-absorbing crystal. Thus we have a criterion for a 'thin crystal', i.e. a crystal slab for which extinction will be negligible and whose integrated reflection will be within 5 per cent of

proportionality to Q. The criterion will be

$$\frac{1}{\eta}\frac{Qt_0}{\sin\theta} < \frac{1}{4}. \tag{3.10}$$

The maximum thickness permissible will become less as the linear absorption coefficient μ is increased, but it is clear from the curves that it falls off relatively slowly with reduction of $Q/\eta\mu$.

3.4. Experimental measurements with single crystals

The first experimental measurements of the intensities of reflection by single crystals were reported in an early paper by Fermi and Marshall (1947), who found that their reflectivities were more nearly proportional to the structure factor F rather than to its square F^2; on the basis of analogy with X-ray results proportionality to F^2 was expected for mosaic crystals. However, this near proportionality to F is quite compatible with our earlier discussion, for Bacon and Lowde pointed out that for small values of μ the curves shown in Fig. 38 approximate to parabolae over a wide range of values of $(1/\eta)(Qt_0/\sin\theta)$. In fact the curve for $\mu = 0$ lies within 5 per cent of the parabola

$$\mathscr{R}_0^\theta = 0{\cdot}96(\eta Q t_0/\sin\theta)^{\frac{1}{2}} \tag{3.11}$$

over the range of values 1·9 to 13 for $(1/\eta)(Qt_0/\sin\theta)$. Thus Fermi and Marshall's result is what would be expected for large mosaic crystals.

A quantitative experimental verification of the secondary extinction curves has been made by Bacon (1951) using plane-parallel plates of potassium bromide 5 cm square and ranging from 0·17 cm to 1·38 cm in thickness, cut with their surfaces parallel to the (100) planes. For these crystals the measured value of the linear absorption coefficient was 0·2 cm^{-1}, and the ranges of values of μ, t_0, θ, and η used in the measurements of the (200), (400), and (600) reflections are such as to give values of $\mu t_0/\sin\theta$ ranging from 0·06 to 1·45 and values of $(1/\eta)(Qt_0/\sin\theta)$ varying from 0·04 to 50. The combinations of values are such as to concentrate attention in the portion of Fig. 38 which is close to the dotted line shown there, the line being continued beyond the right-hand edge of the diagram to a value of about 3 for $(1/\eta)\mathscr{R}^\theta$ at a value of $(1/\eta)(Qt_0/\sin\theta)$

equal to 50. Thus the measurements range from the region of the thin crystal, for which $\mathscr{R}^{\theta}$ equals QV, through the region of the approximate parabolic law to the region where secondary extinction is predominant and the integrated reflection is determined mainly by η. The relation between the calculated intensities and those measured, on an absolute scale, by experiment is indicated in Fig. 40, which shows that there is agreement within about 10 per cent.

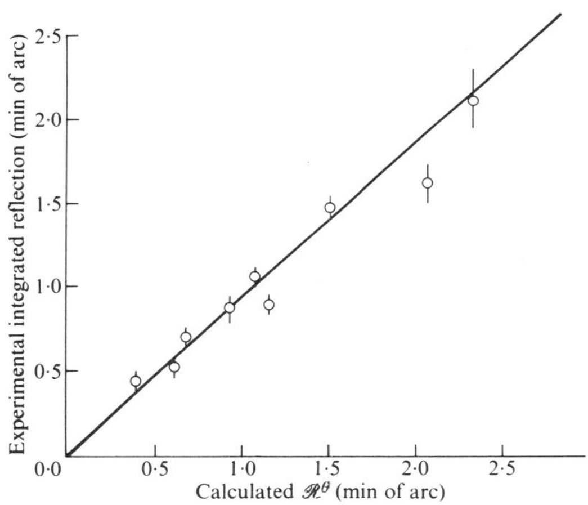

FIG. 40. Integrated reflections of crystal slices; comparison of experimental and calculated values.

The experimental curves showing the variation of $\mathscr{R}^{\theta}$ with $Q/\sin\theta$ obtained in the above measurements with crystals of various thicknesses provide an interesting example of the influence of the mosaic spread. The curves are given in Fig. 41, from which it is seen that the intensity of the strongest reflection (200) is greatest for the thinnest crystal, although this rather suprising behaviour is not found for the weaker reflections. This result is in fact what would indeed be expected if the mosaic spread of the crystals increased as the thickness was reduced. This was established to be the case experimentally by comparing the shapes of the curves shown in Fig. 41 with those calculated theoretically for various values of η. In this way it was possible to assess values of η for the different thicknesses of crystal used as 0·6′, 1′, 1·5′, and 5′ respectively. These values of η were then used to calculate the theoretical intensities

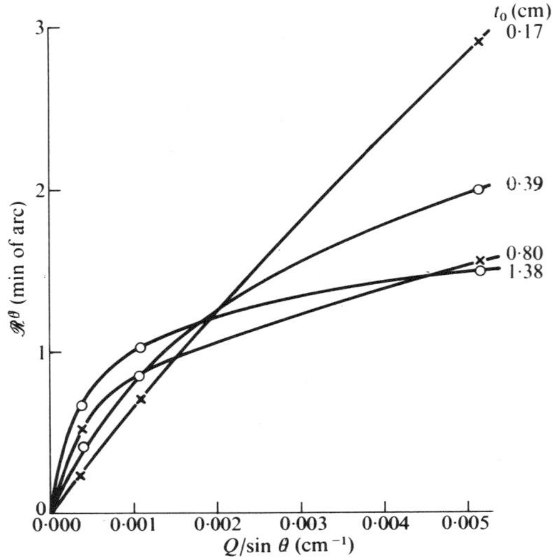

FIG. 41. Experimental measurements of (h00) reflections for crystal slices of various thicknesses showing, from left to right with increasing values of $Q/\sin\theta$, the reflections (600), (400), and (200).

incorporated in Fig. 40. The appreciable increase in η for the thinner plates does not seem unreasonable in view of the strains which must be imposed during the cleaving and working operations. The values of η can only be deduced accurately from these curves when secondary extinction is considerable. As η increases the curves become more nearly linear and in the case of the thinnest plate used above the probable error on the positive side becomes appreciable.

The second conclusion of the above experiments, drawn from the shapes of the curve near the origin, was that the criterion $(1/\eta)(Qt_0/\sin\theta) < \frac{1}{4}$ was satisfactory for ensuring proportionality between $\mathscr{R}^\theta$ and Q. With the crystals of potassium bromide used this was attained with the thinnest slab of thickness $t_0 = 0{\cdot}17$ cm.

3.5. Measurement of Q values for a zone of reflections

A large parallel plate of crystal, such as we have discussed theoretically, is not suitable for structure analysis. However a two-dimensional projection of a structure can readily be obtained by collecting a zone of reflections from a prismatic crystal of square or circular section, and this was the method employed with the early low-flux reactors. In practice, a limitation on size is set by the need to obtain an adequate counting rate with a crystal of side or diameter small enough to give negligible secondary extinction. In eqn (3.10) for the 'thin-crystal' criterion the factor $t_0/\sin\theta$ represents the distance of travel through the parallel slab so that it is reasonable to assume that for a square or cylindrical crystal an acceptable criterion might be

$$(1/\eta)QD < \tfrac{1}{4}, \tag{3.12}$$

where D is the diameter or square edge of the crystal. Support for this is provided by the results in Fig. 42 for square pillars of potassium

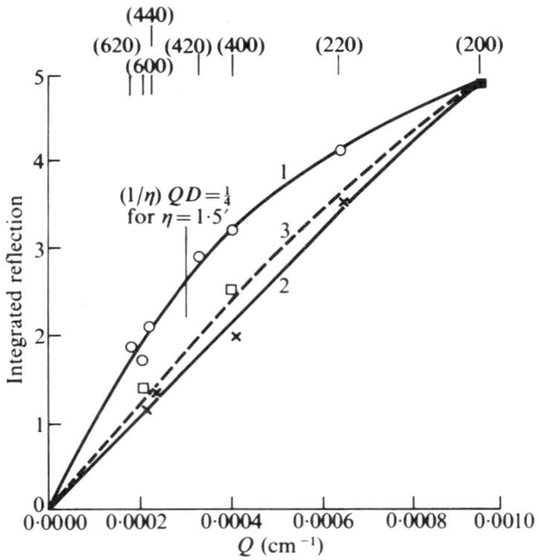

FIG. 42. Intensities of ($hk0$) reflections from crystal pillars. Curve 1. Pillar of KBr 4 mm square. Curve 2. Pillar of KBr 2 mm square. Curve 3. New form of Curve 1 after immersion of crystal in liquid air, increasing mosaic spread.

bromide cut with the length of the pillar parallel to a [100] axis. The experimental intensities have been normalized to a value of unity for the (200) reflection, and curves 1 and 2 relate to pillars 4 mm and 2 mm square respectively, having mosaic spreads of 1·5′ and 5′, which were cut from the large crystal slabs used in the previous measurements. It is seen that the latter pillar is sufficiently small, in conjunction with its large value of η, to give proportionality between $\mathscr{R}^\theta$ and Q. Curve 3 shows the results obtained when the 4 mm pillar has been immersed in liquid air. Clearly the value of η has been increased sufficiently to give an improved range of proportionality between $\mathscr{R}^\theta$ and Q. For $D = 4$ mm and $\eta = 1{\cdot}5'$ the criterion (3.12) is satisfied when $Q = 0{\cdot}0003\ \text{cm}^{-1}$, which is indicated on curve 1. For $D = 2$ mm and $\eta = 5'$ the corresponding value of Q is $0{\cdot}0017\ \text{cm}^{-1}$, which is beyond the range of the diagram, in agreement with the linearity of curve 2 over the whole of the range shown. The criterion (3.12) may therefore be regarded as satisfactory.

Initial progress in neutron crystallography depended on the collection of intensity data for zones of reflections from approximately cylindrical crystals, sufficiently small in dimensions for most of the reflections to be free from substantial errors due to extinction. It is to be noted that from eqn (3.12) the permissible dimension D is inversely proportional to Q. There is a very wide range of Q values among different substances, and Table 8 illustrates how this will

TABLE 8

Values of D satisfying the thin crystal criterion

Substance	Reflection	Q ($\text{cm}^{-1}\ 10^{-2}$)	Value of D (cm) $\eta = 1'$	$\eta = 5'$	$\eta = 10'$
Diamond	111	1·64	0·004	0·022	0·045
	220	2·14	0·003	0·017	0·034
NaCl	111	0·08	0·091	0·45	0·91
	200	0·3	0·024	0·12	0·24
Copper	111	1·06	0·007	0·035	0·07
	200	0·98	0·007	0·035	0·07
CaF_2	111	0·05	0·146	0·73	1·46
	200	0·07	0·105	0·52	1·05

affect the permissible value of D for various values of the mosaic spread parameter η. A further indication of the validity of some of the early measurements of structure factors for single crystals is given in Table 9, which shows a comparison of single-crystal and powder data by Peterson and Levy (1951). For some of the materials the agreement is good, but for LiF, CaF_2, and $CaCO_3$ the single-crystal value for the structure factor is well below that of the powder, indicating the effect of extinction. The last two columns of the Table give the thickness of the crystal in millimetres and the value

TABLE 9

Comparison of structure factors from single-crystal and powder diffraction measurements (Peterson and Levy)

		Structure factor ($\times 10^{12}$ cm)			Thickness D (mm)	
Substance	Reflection	Observed (powder)	Observed (single crystal)	Calculated	Thin crystal limit for $\eta = 5'$	Value employed in experiment
KBr	111	1·29	1·27	1·28	—	—
	200	4·04	4·09	4·08	4·4	1·9
KCl	111	2·40	2·37	2·56	—	—
	200	5·04	4·71	5·36	2·0	1·8
KHF_2	121	3·09	3·08	3·06	2·3	2·8
NH_4Cl	110	1·34	1·27	—	0·8	2·0
LiF	111	2·80	1·63	2·92	0·6	1·8
	200	1·51	1·04	1·48	—	—
	400	1·49	1·19	1·48	—	—
CaF_2	111	1·74	1·57	1·95	—	—
	200	2·45	2·13	2·44	4·5	1·7
	220	6·17	3·17	6·36	—	—
$CaCO_2$	211	3·29	2·39	3·46	1·2	1·6

of D deduced from the criterion (3.12) assuming, in the absence of any measurement, a rather arbitrary value of $5'$ for η.

Since the data in Table 9 were recorded the intensity of neutron beams has increased by about a thousand times, reducing by a corresponding factor the volume of crystal necessary for study and making possible the collection of three-dimensional data from spherical crystals which are sufficiently small in size to be relatively free from extinction. For a flux of 10^{13} neutrons cm^{-2} s^{-1} a spherical crystal with a radius of 1–2 mm is usually adequate, and if the flux is 2×10^{14} neutrons cm^{-2} s^{-1} then the radius may be between 0·5 mm and 1 mm. Consequently, the difficulties associated with extinction corrections have lessened considerably since the days when very large crystals were needed. In many straightforward structural investigations only a few of the reflections are substantially in error, and usually there are sufficient data available for the intense low-angle reflections to be ignored in the analysis. On the other hand, there are other investigations in which much more precise intensity measurements, and much more precise correlation with calculations based on some assumed structural model, are required. For example, investigations of the anisotropy of thermal motion or the close correlation of X-ray and neutron data to give conclusions about electron anisotropies and bonding effects depend on very accurate intensity measurements. Thus interest in the precise study of extinction has increased, rather than decreased, and considerable progress has been made in treating the case of a crystal of particular size and shape rather than the classical, but impractical, case of the infinite plane parallel slab. We shall now describe this later work.

3.6. Zachariasen's calculation

Zachariasen (1967) succeeded, after making a number of approximations, in deriving a general formula for the intensity of reflection from a perfect crystal of finite size and specified shape.

Following Zachariasen's nomenclature we denote the 'primary extinction factor' E_p of eqn (3.8) by the symbol y. Thus y is the factor by which the quantity QV for the crystal must be multiplied in order to give the observed value of the integrated reflection. For a perfect spherical crystal of radius r, Zachariasen shows that

$$y = (1+2x)^{-\frac{1}{2}}, \tag{3.13}$$

where

$$x = \tfrac{3}{2}Qr^2/\lambda. \tag{3.14}$$

The value of y will fall below 0·98 if $x < 0{\cdot}02$ and, as an example, this means that if the primary extinction for the strongest reflection from α-quartz is to be less than 2 per cent with Cu Kα X-rays then the radius of the perfect crystal sphere must be less than $0{\cdot}51\times$ $\times 10^{-4}$ cm.

Zachariasen then proceeds to the calculation of the value of y for a mosaic crystal, making the assumption that the domains of the mosaic are perfect crystals of radius r whose orientations are distributed in angle according to a function $W(\Delta)$ such as we defined in eqn (3.9). Zachariasen's parameter for the angular spread is g, where $4\pi g^2 = 1/\eta^2$, so that

$$W(\Delta) = \sqrt{2}g \exp(-2\pi g^2\Delta^2). \tag{3.15}$$

The analysis then shows that

$$y = (1+2x)^{-\frac{1}{2}}, \tag{3.16}$$

where

$$x = \frac{Qr}{\lambda}\left[\frac{3}{2}r + \frac{\bar{T} - \frac{3}{2}r}{\{1+(r/\lambda g)^2\}^{\frac{1}{2}}}\right], \tag{3.17}$$

where $\bar{T}$ is the mean path length through the mosaic crystal. If this mosaic crystal is spherical then $\bar{T} = \frac{3}{2}R$, where R is the radius of the crystal. Comparison of eqns (3.14) and (3.17) will indicate that in the latter equation the first term within the square brackets represents the primary extinction and the second term is the secondary extinction. Experimental measurements have shown that the primary extinction can practically always be neglected, so that eqn (3.17) becomes

$$x = \frac{Qr\bar{T}\lambda^{-1}}{\{1+(r/\lambda g)^2\}^{\frac{1}{2}}}, \tag{3.18}$$

and for the special case when the mosaic crystal is a sphere we write $\bar{T} = \frac{3}{2}R$ and

$$x = \frac{\frac{3}{2}QrR\lambda^{-1}}{\{1+(r/\lambda g)^2\}^{\frac{1}{2}}}. \tag{3.19}$$

3.7. Further development of Zachariasen's calculation

Zachariasen (1968*a*, *b*, *c*, 1969) has carried out experimental tests of his conclusions with X-rays for several different crystals, and very good agreement with theory has been obtained. Some of these measurements have shown values of y as small as 0·2, thus indicating an 80 per cent reduction of intensity because of extinction. Using neutrons very accurate measurements have been made with spherical and cylindrical crystals of barium fluoride (Cooper, Rouse, and Willis 1968), strontium fluoride, and calcium fluoride (Cooper and Rouse 1971). The general conclusion from these measurements, which are discussed further by Cooper and Rouse (1970), is that for neutrons the Zachariasen theory is adequate for values of y down to 0·6, corresponding to 40 per cent extinction, but that divergencies appear when the extinction is greater than this.

Cooper and Rouse (1970) have shown that for neutrons some of the approximations made in Zachariasen's treatment are not valid, leading to marked deviations between theory and experiment at large angles of scattering and inadequacy for very large degrees of extinction. From eqns (3.16) and (3.18) it follows that

$$1/y^2 = CF^2 \operatorname{cosec} 2\theta, \tag{3.20}$$

where C is a constant, so that $1/y$ can be considered as a function of $(F^2 \operatorname{cosec} 2\theta)^{\frac{1}{2}}$. Thus in Fig. 43, which plots experimental values of $1/y$ deduced from measurements with a cylindrical crystal of CaF_2 3 mm in diameter we should expect all the points to lie on a single

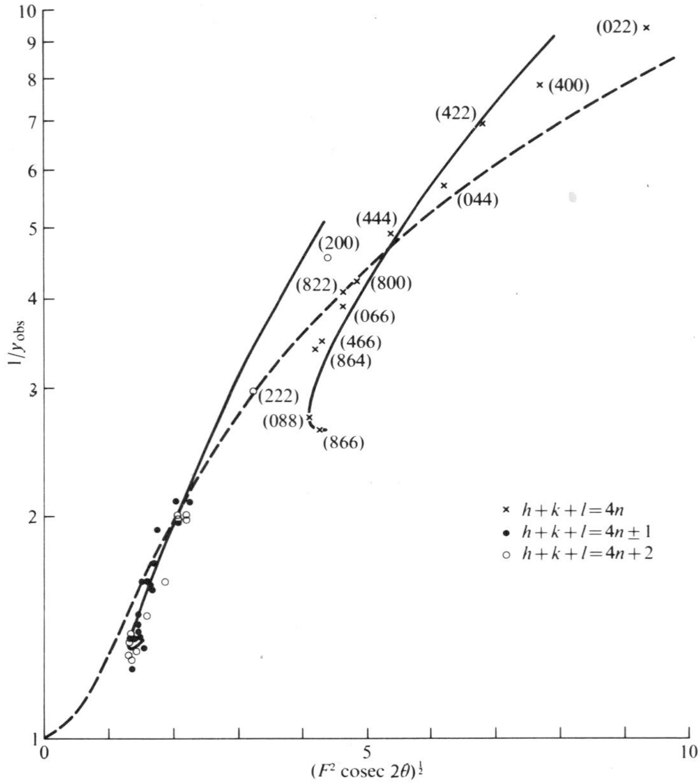

FIG. 43. Variation of the observed value of $1/y$ with $(F^2 \operatorname{cosec} 2\theta)^{\frac{1}{2}}$ for a cylindrical crystal of CaF_2 of diameter 3 mm at $\lambda = 0{\cdot}877$ Å. The broken line represents the value predicted by the Zachariasen theory such that agreement with experiment is obtained for the $4n$ data near $\theta = 45°$, i.e. at the (822) reflection. The full lines are calculated by the more precise theory of Cooper and Rouse (1970).

smooth curve. There are some evident departures from the curve and it can be shown that these occur in a quite systematic way. For the CaF_2 structure the values of F are different for the three classes of reflection given by $h+k+l = 4n$, $4n \pm 1$, and $4n \pm 2$

respectively, being substantially greater for the first class of reflection. Cooper and Rouse interpreted the results in Fig. 43 by showing that departures from Zachariasen's theory occur for neutrons for both large values of F and large values of θ, so that $1/y$ is not solely dependent on $(F^2 \operatorname{cosec} 2\theta)^{\frac{1}{2}}$.

The difficulty arises because of the approximations which have to be made in order that the closed form of expression can be used for y in eqns (3.13) and (3.16), rather than more exact power series. Assumptions have to be made concerning the coupling constant $\check{r}$ which we used in eqn (3.5). It is assumed that $\check{r}\overline{T} < 1$ and, in particular, that the value of $\check{r}\overline{T}$ is small at large angles of scattering. The latter is a satisfactory assumption for X-rays because the atomic form factor ensures that $\check{r}$ becomes small when θ is large, but it is not a good assumption for neutrons. Cooper and Rouse (1970) correct for the angle-dependence of $\check{r}\overline{T}$ by replacing x in eqn (3.16) by $xf(\theta)$ and then replacing y by

$$y' = \left(1 - \frac{1}{f(\theta)}\right) + \frac{1}{f(\theta)} y,$$

so that the extinction factor is

$$\left(1 - \frac{1}{f(\theta)}\right) + \frac{1}{f(\theta)} (1 + 2xf(\theta))^{-\frac{1}{2}}, \tag{3.21}$$

where $f(\theta)$ is a function which can be evaluated numerically and approximates, within $\frac{1}{2}$ per cent, to

$$f(\theta) = 1 + \tfrac{1}{3} \sin^{2 \cdot 5} \theta. \tag{3.22}$$

A number of more intricate functions have also been examined by Cooper and Rouse and a different choice made for cylindrical and spherical crystals in order to give optimum agreement between experiment and calculation for extinctions as high as 80 per cent. In Fig. 43 the two full lines are those predicted in this way by the theory for the different groups of reflections. For the (444) reflection, which lies close to the broken line in the figure, the value of θ is about 34°, whereas for (866), whose intensity is most inadequately predicted by the simpler Zachariasen calculation, θ is about 70°. In a further appraisal of Zachariasen's treatment Becker and Coppens (1974*a*,*b*) have provided a more general solution for both X-rays and neutrons which specifically includes the dependence of the extinction on the angle 2θ. They show that when applied to the neutron data of SrF_2, collected by Cooper and Rouse (1970, 1971) at three different wavelengths, their theory gives better agreement than the empirical formulae previously used. The Zachariasen calculation can be used for neutrons up to an extinction of about 40 per cent, i.e. down to $y = 0{\cdot}6$. A convenient way of making the correction has been described by Bacon and Jude (1973), following the procedure of Cooper, Rouse, and Willis (1968). As we have seen, extinction effects are largest for reflections with large values of Q. Accordingly a number of independent refinements of the crystal structure data are made which ignore all reflections whose Q values are greater than some arbitrary level. If we make this level steadily lower then a point will be reached at which extinction effects are negligible, so that the refined parameters then deduced will give good agreement between observed and calculated intensities. In Bacon and Jude's measurements of α-resorcinol agreement to within 1 per cent was obtained when 116 reflections were set aside out of the 655 independent reflections which were measured. Using the structural parameters obtained in this way F_{calc} is computed for every reflection and compared with the experimental value F_{obs}. If we define the ratio $F^2_{\text{obs}}/F^2_{\text{calc}}$ as y^*, then y^* will be a close approximation to y of eqn (3.16), since our F_{calc} will be very close to F_c and therefore

$$y^* = (1+2c\overline{T}Q)^{-\frac{1}{2}},$$

where

$$c = \frac{r\lambda^{-1}}{\{1+(r/\lambda g)^2\}^{\frac{1}{2}}}.$$

Hence,

$$\begin{aligned} 1/y^{*2} &= 1+2cTQ \\ &= 1+2c^*TF^2_{\text{calc}}/\sin 2\theta, \end{aligned}$$

where

$$c^* = cN^2\lambda^3,$$

since

$$Q = N^2F^2_{\text{calc}}\lambda^3/\sin 2\theta.$$

If we multiply throughout this equation by $y^*(\equiv F^2_{obs}/F^2_{calc})$ then

$$1/y^* = y^* + (2c^*TF^2_{obs}/\sin 2\theta). \tag{3.23}$$

Thus a plot of $\{(1/y^*)-y^*\}$ against $F^2_{obs}/\sin 2\theta$, which is proportional to the observed intensity, will give the value of c^* and accordingly of c. A plot of this kind is shown in Fig. 44 for the results of Titterton (1972, private communication) for a sphere of $CuSO_4 . 5H_2O$ having a diameter of 2·5 mm. To avoid confusion

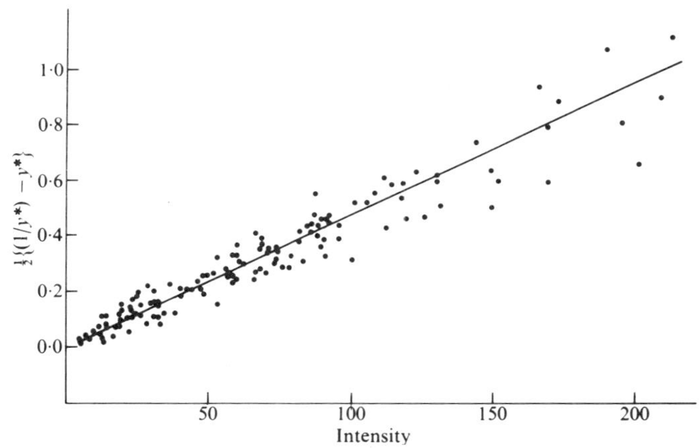

FIG. 44. Correction of intensity data for extinction by plotting $\{(1/y^*)-y^*\}$ against the observed intensity for $CuSO_4 . 5H_2O$. The slope of the line gives a value for c^*. (From the data of D. H. Titterton.)

only a small fraction of the available points are shown on this plot, particularly for the points of low intensity. It may be noted that the maximum value of the ordinate corresponds to a value of y^* of about 0·4. Thus the most intense reflections are reduced by extinction to 40 per cent of their potential values. From the value of c obtained from the plot all the experimental intensities can be corrected for extinction. Subsequently, the value of c can be used to calculate the value of r, the radius of the perfect-crystal domains.

In our discussion of extinction in this chapter we have been considering only non-magnetic materials, for which the nucleus is the only source of neutron scattering. When magnetic crystals are considered there are some additional factors which have to be taken into account in assessing the primary and secondary extinction. For a discussion of this the reader is referred to a paper by Brown (1970).

3.8. Correction for thermal diffuse scattering

We have seen in Chapter 2 that the thermal displacement of atoms from their equilibrium positions leads to a reduction in the intensity of the coherent Bragg reflections. There is a corresponding increase in the diffuse reflection of the neutrons or X-rays and the increase rises to a maximum at the same angular positions as the Bragg reflections. Thus in Fig. 45 the resulting peak at the position

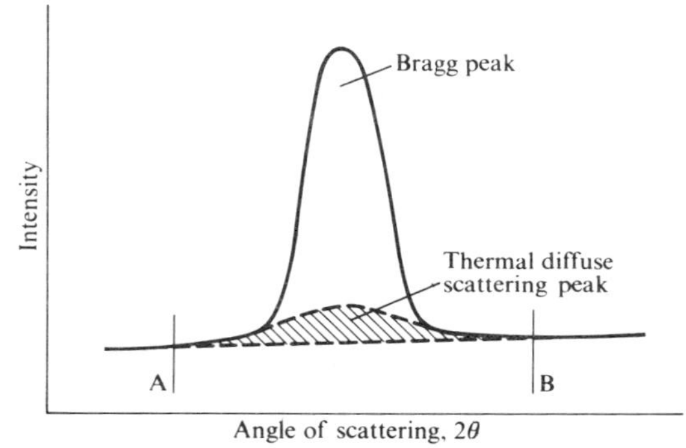

FIG. 45. The resulting intensity at the position of an *hkl* reflection by superposition of the Bragg and diffuse peaks. If the Bragg peak is assessed in the ordinary way by drawing a background line between the θ values A, B then it will be overestimated by the area shown cross-hatched. (Cooper 1970.)

of an *hkl* reflection is the sum of the sharp Bragg peak and the diffuse thermal peak. Any attempt to draw in the background level at, say, points A, B will result in an over-assessment of the size of the Bragg peak because of the inevitable inclusion of at least part of the diffuse contribution. Thus, like absorption and extinction which we have examined earlier, the thermal diffuse scattering is a factor which limits the accuracy of the determination of reflection intensities and, in turn, sets a limit on the accuracy of the structural parameters which we can deduce. In fact, neglect of the diffuse scattering produces errors in the thermal parameters but has very little influence on the atomic coordinates.

A method of correcting the observed intensities for diffuse scattering was given by Nilsson (1957) for cubic crystals. Nilsson showed that the true and observed intensities I_0, I were related by a factor α, where

$$I = I_0(1+\alpha) \tag{3.24}$$

and

$$\alpha = \frac{4\pi kT}{3\lambda^3} \sin 2\theta \sin^2 \theta \, \check{\kappa}\check{\sigma}, \tag{3.25}$$

where $\check{\kappa}$ is a function of the elastic constants of the crystal and $\check{\sigma}$ is obtained by integrating the inverse square of the wave vectors of those phonons associated with the volume in reciprocal space which is comprised in the scan of the peak between points A, B in Fig. 45. In these terms $\check{\sigma}$ is an integral given by

$$\check{\sigma} = \frac{1}{\pi\kappa \sin 2\theta} \int \frac{\mathrm{d}x \, \mathrm{d}y \, \mathrm{d}z}{|q|^2}, \tag{3.26}$$

where $\kappa = 2\pi/\lambda$ for the incident neutrons and $\mathbf{q}$ is the wave vector of the relevant phonon.

Nilsson's treatment derived an analytical expression for $\check{\sigma}$ for a rotating crystal and stationary detector when the detector aperture is taken to be of infinite extent in the vertical direction. Cooper and Rouse (1968) have provided a numerical method which provides an accurate calculation over the precise volume of space covered in the experimental scan, and for the normal $(\theta, 2\theta)$ scan, in which both crystal and counter rotate, which is used in neutron crystallography. The magnitude and importance of the correction for α can be seen in Fig. 46 for some measurements with a crystal of BaF_2 at 673 K. For a Bragg angle of 45° the correction amounts to about 20 per cent. A correction of similar magnitude was found by Duckworth, Willis, and Pawley (1968) in their measurements of hexamethylene tetramine, which we shall describe in Chapter 12.

In this numerical method the integration is carried out over the parallelopiped in reciprocal space which is covered in the experiment. If this volume is approximated by a sphere it follows from eqn (3.26) that $\check{\sigma}$ varies as $1/\sin 2\theta$, and accordingly α in eqn (3.25) will be proportional to $\sin^2 \theta$. Thus when α is small we can write

$$\begin{aligned} I &= I_0 \exp \alpha \\ &= I_0 \exp (2\Delta B \sin^2 \theta/\lambda^2), \end{aligned} \tag{3.27}$$

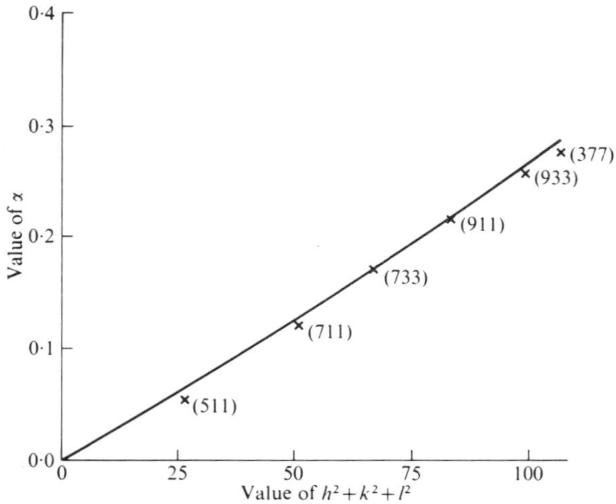

FIG. 46. Determination of the thermal diffuse scattering factor α of $BaCl_2$ at 400 °C by a numerical method. The crosses × are for the *hkl* reflections specified and the full line is an assessment of α as equivalent to $\Delta B = 0{\cdot}16$ Å^2. (After Cooper and Rouse 1968.)

where ΔB is some constant. We have written eqn (3.27) in this form in order that we may compare it with eqn (2.32) which is used to express the reduction of the intensity of coherent Bragg reflections because of thermal motion. It is evident that neglect of the effect of thermal diffuse scattering will result in an underestimation of the Debye factor B by an amount ΔB, with little effect on the values deduced for the atomic coordinates. We emphasize this point in Fig. 46, where the full line denotes an assessment of the thermal diffuse scattering as equivalent to a change in the Debye factor of $\Delta B = 0{\cdot}16$ Å^2.

For further discussion of this topic the reader is referred to articles by Cooper (1970) and Willis (1970*b*).

3.9. The use of powdered crystals

During the early years of neutron-diffraction studies most of the measurements were made with powdered or polycrystalline samples, which produce the diffraction haloes which are familiar from the Debye–Scherrer method of X-ray analysis. Nevertheless, important information was obtained because most of the materials examined were of high crystallographic symmetry and gave simple diffraction patterns. However, the limitations of the powder method of analysis for neutrons were even more marked than for X-rays because of the relative weakness of neutron sources. Only with very high-flux reactors is it usually possible to employ diffractometers of high angular resolution. We shall consider in detail in Chapter 4 the factors which determine this resolution, but we may note for the present that the diffraction lines are generally of the order of a degree wide at low angles and they increase in width fairly rapidly with increase of θ, so that only crystals of high symmetry give the well-resolved non-overlapping lines which are essential for accurate intensity measurement. Single crystals are therefore used whenever possible, but it very often happens, particularly with magnetic materials, that crystals of adequate size are not available, and recourse then has to be made to the powder method. Under these circumstances we shall see, in Chapter 4, the advantages of the so-called 'profile-refinement' technique, first described by Rietveld (1969), which can be used when the main features of a structure are already known.

An outstanding advantage of powdered samples is that extinction effects are negligible so that very precise measurements of structure factors can therefore be made for simple substances. This is of the utmost advantage in the fundamental determination of the scattering lengths of individual elements and isotopes, which are the empirical data already presented in Table 2 (p. 38). On the other hand, single crystals suffer much less disturbance from the incoherent background for materials containing many hydrogen atoms. This scattering is particularly troublesome with powdered samples, since only a small fraction of the material (i.e. that fraction which is correctly oriented) contributes to the diffraction *peaks*, but the whole of the irradiated sample contributes to the background. When a single crystal rotates through its Bragg reflection position the whole volume of the crystal contributes to the diffraction peak, so that the incoherent background is less significant.

4

EXPERIMENTAL TECHNIQUES FOR DIFFRACTION MEASUREMENTS

IN principle, measurements of the intensity and position of the spectra produced by the diffraction of neutron beams by crystal specimens involve similar problems to the corresponding measurements with X-rays. However, for various reasons, some of which have already been mentioned briefly in Chapter 1, there are wide differences in the techniques employed in the two cases and particularly in the actual magnitudes of the apparatus used. With the reactor fluxes which are available and methods of neutron detection, the experimenter is called upon to make a compromise between a large neutron counting rate, which promises rapid measurement, and high resolution, which may be essential in order to obtain discrete spectra free from overlapping. With most materials this problem does not arise to any extent for the worker using X-ray diffraction since the output of an X-ray tube is usually sufficient to give adequate intensity and resolution at the same time. The number of quanta per unit area in a normal X-ray beam will generally be greater than that in neutron beams by a factor of about 10^4.

4.1. Production of the monochromatic beam

The main considerations that determine the experimental arrangement used for obtaining a beam of monochromatic neutrons from the reactor can be followed with the aid of Fig. 47. Let P be a point at the inner end of the collimator which is of width a_c or which, as described earlier on p. 10, is built up of a number of sections each of width a_c. If l_0 is the length of the collimator then the angle β will be given by a_c/l_0. Neutrons of all velocities, or wavelengths, will radiate in all directions from P and those lying within the cone defined by PB, PB′ will emerge from the collimator. For any position XY of the monochromating crystal there will be some particular wavelength which is at the right glancing angle θ to the crystal to undergo Bragg reflection when it is incident along PB. If the crystal is perfect then this particular wavelength when coming from P will only be reflected if it lies within the very small angle s

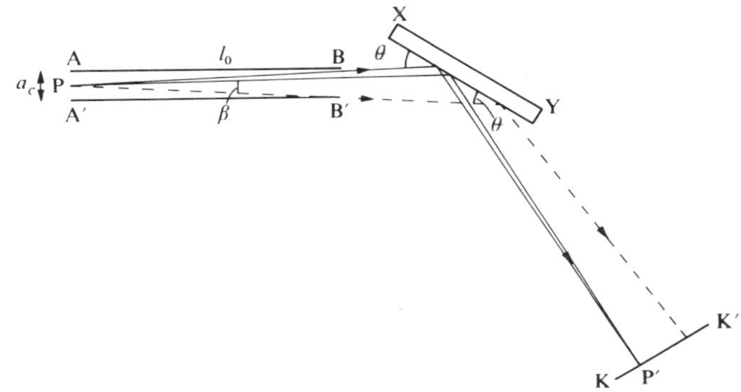

FIG. 47. Production of monochromatic beam.

of the direction PB, this angle having been shown in the previous chapter to be only about 1″ of arc. On the other hand, when the crystal is mosaic, the wavelength λ can be reflected over an angular range about PB of the order of the mosaic spread of the crystal. Over this angular range of a few minutes neutrons of wavelength λ coming from P will all pass through P′, such that the planes AA′ and KK′ are equidistant from the crystal. The corner B′ of the collimator will define an extreme ray PB′, from the point P, which will be incident on the crystal at a glancing angle θ' which is less than θ by an amount $\delta\theta$ equal to β, the collimator angle. For this new glancing angle a different wavelength $\lambda - \delta\lambda$ will be reflected by the crystal. By differentiation of the Bragg equation $\lambda = 2d \sin \theta$ it follows that

$$\delta\lambda = 2d \cos \theta \, \delta\theta = 2\beta d \cos \theta, \tag{4.1}$$

thus giving $\delta\lambda$ in terms of the collimator angle β. In a practical case β may be $\frac{3}{8}^\circ$ which for a typical interplanar spacing of 3 Å and θ equal to 10° gives a value of $\delta\lambda$ equal to 0·04 Å.

If P is imagined to move across the collimator aperture at AA′, it will be seen that the total range of angle of incidence among the neutrons incident on the crystal will be 2β and the range of wavelength in the reflected beam will be $2\delta\lambda$, or about 0·08 Å.

It was mentioned above that any particular wavelength from a single point P would be reflected over an angular range of the order of the mosaic spread of the crystal. A quantitative calculation of $\mathscr{R}^{\lambda}$, the integrated reflection for a fixed crystal in a beam of white radiation, can be made in a similar manner to that of $\mathscr{R}^{\theta}$, the integrated reflection for the rotating crystal method in which a crystal is rotated through a beam of monochromatic radiation, which was given in the previous chapter. It can be shown that

$$\mathscr{R}^{\lambda} = \mathscr{R}^{\theta} 2d \cos \theta. \tag{4.2}$$

$\mathscr{R}^{\lambda}$ thus has the dimensions of a length, being in fact equal to the wavelength range in the incident beam over which reflection may be regarded as complete, just as $\mathscr{R}^{\theta}$ was the angular range over which reflection of the monochromatic beam by a rotating crystal could be considered complete.

Beam intensity

In eqn (1.4) of Chapter 1 we gave the value of $v_{\lambda}\, \mathrm{d}\lambda$, the total number of neutrons having wavelengths between λ and $\lambda + \mathrm{d}\lambda$ for neutrons in thermal equilibrium at the temperature inside the reactor. It follows from eqn (4.2) that the total number of neutrons $\mathscr{N}_2$ reflected by the monochromator will be equal to $\mathscr{R}^{\lambda} v_{\lambda}$. Thus

$$\mathscr{N}_2 = \frac{2\mathscr{N}_1}{\lambda}\left(\frac{E}{kT}\right)^2 \mathrm{e}^{-E/kT} 2d \cos \theta\, \mathscr{R}^{\theta}, \tag{4.3}$$

where $\mathscr{N}_1$ = total number of neutrons of all velocities striking the monochromator per second,
E = energy of neutron of wavelength λ,
θ = Bragg angle.

If the crystal is oriented so that neutrons at the peak of the distribution curve are reflected, then $E = \frac{3}{2}kT$ and

$$\mathscr{N}_2 = \frac{2\mathscr{N}_1}{\lambda}\left(\frac{3}{2}\right)^2 \mathrm{e}^{-\frac{3}{2}} 2d \cos \theta\, \mathscr{R}^{\theta},$$

whence

$$\mathscr{N}_2 \doteqdot \mathscr{N}_1 \cot \theta\, \mathscr{R}^{\theta}. \tag{4.4}$$

It is possible therefore, by means of eqn (4.4), to determine the number of neutrons reflected from the monochromator in a practical case. As an example we shall take a collimator of fairly small aperture, 2 cm × 1 cm, used in the early Harwell reactor BEPO.

At the effective source, which is about 300 cm within the front face of the collimator, the total neutron flux is known to be 6×10^{11} neutrons $cm^{-2}\,s^{-1}$. By evaluating the solid angle subtended at the source by the aperture it is found that each square centimetre of radiating area should produce 7×10^5 neutrons s^{-1} at the aperture. For a radiating area of 3 cm^2 this means that the value of $\mathcal{N}_1$ is of the order of $2 \times 10^6\,s^{-1}$. If we assume typical values of 10°, 15′ for θ, $\mathscr{R}^\theta$ respectively, for a mosaic spread of about 5′ and a large single crystal, we find from eqn (4.4) that $\mathcal{N}_2$ will be equal to about $\mathcal{N}_1/40$, which amounts to approximately 3×10^6 neutrons min^{-1} at the peak of the distribution curve. In fact, when a lead crystal was set to reflect neutrons of wavelength $\lambda = 1{\cdot}08$ Å which, as seen in Fig. 2, is somewhat below the intensity peak, it was found experimentally that there were $1{\cdot}5 \times 10^6$ neutrons min^{-1} in the reflected monochromatic beam.

4.2. The choice of monochromator

As explained in Chapter 3, the large single crystals needed for use as monochromators will have values of $\mathscr{R}^\theta$ equal to a few times the mosaic spread. Consequently the intensity of the reflected beam will roughly be proportional to the mosaic spread of the monochromator. In the earliest experiments by Sturm (1947) and Wollan and Shull (1948) metallic halide crystals were used, particularly LiF, NaCl, and CaF_2, with mosaic spreads measuring a minute or two of arc. Sturm has described comparative measurements with a number of crystals and has shown that with LiF increased reflectivity was obtained by roughening the surface of the crystal. Lithium fluoride has a relatively high absorption, with μ equal to about 4 cm^{-1}, so that the penetration by the neutron beam is small. Consequently it is sufficient to increase the mosaic spread of the surface layers in order to attain increased intensity. On the other hand, it is found that surface roughening of low-absorbing crystals such as CaF_2 and KBr produces only a small intensity increase, and in these cases it is necessary to increase the mosaic spread over an appreciable volume of crystal. It was later shown by Shull and Wollan (1951) that metallic crystals such as lead and copper gave intensities which were greater than those of the halides by 2 or 3 times, and these crystals are now generally employed. Such crystals can be used either in 'reflection' or 'transmission', the latter method having the advantage that a much smaller width of crystal is necessary for intercepting the full width of beam from the collimator, particularly when small values of θ are used at the monochromator.

A great deal of effort has been devoted to the task of producing monochromators of high reflectivity and symmetrical rocking curves for a variety of materials and interplanar spacings. This task is a very important one, from the experimenter's point of view, since an increase in reflectivity from the monochromator will achieve the same purpose as an increase in the reactor flux. Unfortunately we are still far from being able to control the growth of large crystals to produce a monochromator of specified properties. The influence of mechanical treatments, such as bending, hammering, and quenching, on the reflectivities of large crystals of lead, zinc, aluminium, and copper has been described by Modrzejewski and Kobla (1969). Recent work in other laboratories has shown that good metal crystals can be improved by applying small plastic deformations and increases in reflectivity by 2–3 times have been achieved. It is evident that when any new apparatus is being installed it is essential to search carefully for a high-quality monochromator crystal. For some purposes, particularly for experiments with neutrons of higher energy, crystals of beryllium have proved very satisfactory, but continued attempts to reproduce the virtues of certain legendary crystals have not proved successful. Another interesting recent development has been the use of slabs of pyrolytic graphite, which are effectively artificial crystals produced by depositing two-dimensional layers of carbon atoms in a highly parallel orientation, thus producing the effect of a mosaic spread of about 0·4° which provides a good compromise between resolution and high intensity.

The monochromator is normally set at the angle which gives Bragg reflections of neutrons at a wavelength a little over 1 Å. This wavelength, which is well below that giving the peak intensity, is chosen in order to be free from second-order contamination, as discussed in Chapter 1, and also to give diffraction patterns over a wider range of d spacings with the limited angular range of the spectrometer generally available. With the lead crystal used at U.K.A.E.A., Harwell for $\lambda = 1{\cdot}08$ Å the peak intensity due to the second-order component $\lambda = 0{\cdot}54$ Å is only about $\frac{1}{3}$ per cent of that due to the main wavelength. The precise values of λ used by different experimenters are not usually of particular significance, generally arising fortuitously by fixing θ at some particular integral number of degrees for a given crystal, or by subsequently using some chosen angle for a different crystal in order to avoid rebuilding the shielding which surrounds the monochromator. For example, with an experimental arrangement designed to give a wavelength of 1·08 Å by reflection at $\theta = 11°$ from the (111) plane of lead, an appreciably smaller wavelength of 0·81 Å can be attained, and used to advantage to extend the range of interplanar spacings which can be examined, by substituting a copper crystal as monochromator. However, there is a certain advantage to be gained by using a wavelength of 1·066 Å, for it happens that plutonium has an absorption resonance at a wavelength of one-half this value. It is possible, therefore, to use a plutonium foil as a filter to remove the second-order contamination. A suitable foil in practice reduces this $\frac{1}{2}\lambda$ component by about 100 times at the expense of a twofold reduction in the intensity of the fundamental wavelength. If this over-all reduction can be tolerated, the plutonium provides a very useful means of ensuring that weak reflections are not confused by the presence of any second-order component.

An alternative way of avoiding the second-order contamination is to choose as monochromator some material whose internal structure is such that the structure amplitude factor, from a suitable set of layer planes, is zero, or very small, for the second-order reflection. Germanium, which has the same structure as diamond, is a favourite choice and it is easy to calculate that the value of F_c resulting from the eight atoms in the unit cell is zero for the (222) reflection, whereas the first-order reflection from the (111) planes is intense. Unfortunately germanium crystals as normally produced for use as semiconductor material are rather perfect, with mosaic spreads of no more than about 6″ of arc. Barrett, Mueller, and Heaton (1963) and others have shown that it is possible to increase substantially the mosaic spread, and hence the reflectivity, by mechanical deformation. By pressing the crystals at temperatures just below the melting point the mosaic spread can be increased to about 6′ of arc. This is still below the values for the most satisfactory copper and lead crystals, but work is proceeding with the aim of producing further increases. Another crystal which can be utilized in this way, to enable satisfactory use to be made of longer neutron wavelengths, is magnetite Fe_3O_4; the structure of this is such that the (222) reflection is very much weaker than (111).

Experimental adjustment of the monochromator is made with the counter positioned along the line joining the monochromator and counter axes. With the counter in this position, but with no sample in place, a 'rocking curve' is plotted to show the counter reading as a function of the angular position of the monochromator. With a high-flux reactor the incident intensity may be so high that the counter is saturated, and it will then be necessary during these measurements to interpose an absorbing plate in front of the counter. A typical 'rocking curve' is shown in Fig. 48, and the peak

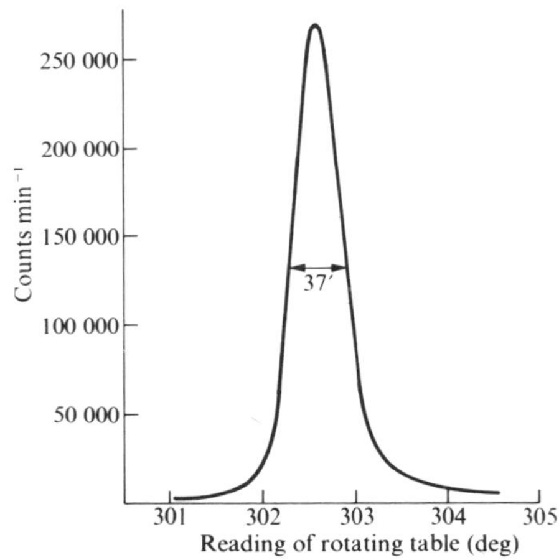

FIG. 48. Typical experimental rocking curve for lead crystal monochromator.

of the curve gives the optimum position of the monochromator. With the crystal in this azimuthal position small adjustments are made to the tilt of the crystal relative to the horizontal plane and the effects of small movements across the neutron beam are examined in order to obtain the maximum possible counting rate. If the monochromator is truly a single crystal the rocking curve will be symmetrical about the peak; crystals which consist of more than one individual may give markedly unsymmetrical and mis-shapen curves. These should be discarded in order not to complicate the interpretation of the subsequent diffraction patterns from samples under investigation.

4.3. Analysis of the monochromatic beam: parallel and antiparallel positions

The wavelength composition of the reflected beam from the monochromator can be examined by means of a second single crystal set at the centre of the second rotating table. This measurement serves to determine the amount of second-order contamination, that is neutrons with a wavelength equal to a half that of the fundamental, which is present in the 'monochromatic' beam. A curve illustrating this has already been given in Fig. 4 and should always be borne in mind in interpreting diffraction patterns, particularly, for example, when weak lines due to superlattices are being sought. If a complete diffraction pattern is plotted for this analysing crystal, by rotating the crystal with the counter following at twice the angular velocity, it will be found that the angular widths and intensities of corresponding lines on the two sides of the zero position are quite different. This is a well-known fact from the theory of double-crystal spectrometry (see Zachariasen 1945; Compton and Allison 1935) and is interpreted most simply when identical crystals are used on the two tables. It can then be shown that in the case of the so-called 'parallel position', for which the incident and twice-reflected beams are parallel, the widths of the spectrum and of the rocking curves of the crystals are proportional to their mosaic spread. On the other hand, in the antiparallel position the widths are much greater, being roughly proportional to the sum of the mosaic spread and the angle of collimation. The practical result of this is shown in Fig. 49, which shows some experimental curves obtained with calcium fluoride crystals for two different degrees of collimation. In the parallel position the width of the rocking curve at half the peak intensity is about 7′ for both the fine and coarse collimation. The collimator angles were $\pm 8'$ and $\pm 22'$ respectively in the two cases. For the antiparallel position the width is equal to 20′ for the fine collimation and has increased to 46′ for the coarse collimation. With a given collimator the integrated reflections in the parallel and antiparallel positions will be equal. It follows therefore that the narrower rocking curve and spectrum on the 'parallel' side of the zero position are associated with proportionally higher peak intensities as can be seen in Fig. 49. Use of the parallel arrangement is therefore very advantageous.

Calculation shows that for identical crystals the rocking curve width on the parallel side is equal to about 3η, where η is the standard

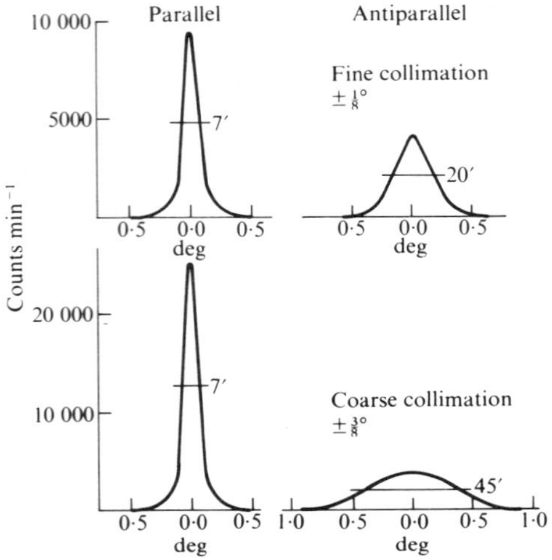

FIG. 49. Parallel and antiparallel rocking curves with fine and coarse collimation.

deviation of their common mosaic spread, as defined in Chapter 3. With different crystals, for which the two Bragg angles θ_1 and θ_2 are not equal, and which may also have different mosaic spreads, the rocking curve width is increased but is still, nevertheless, much smaller than for the corresponding spectrum on the antiparallel side. Equally, it is found that a good deal of focusing action is obtained even when the diffracting sample on the second table is a polycrystalline powder. Fig. 50 illustrates this, showing the parallel

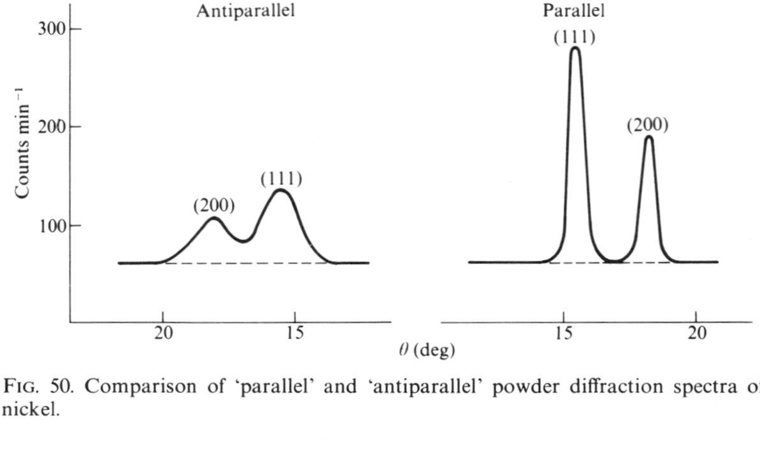

FIG. 50. Comparison of 'parallel' and 'antiparallel' powder diffraction spectra of nickel.

and antiparallel (111) and (200) spectra of a polycrystalline sample of nickel using a lead monochromator. The improved resolution and increased peak-height which is obtained for the parallel position is of great value in practice and it is normally arranged, as in Fig. 9, to record the diffraction patterns on this side.

The characteristics of a crystal C_1 for use as a monochromator can be studied by mounting it on the spectrometer in the normal sample position so that it reflects a monochromatic beam produced by some highly perfect crystal C_2, such as silicon or germanium. Ideally the interplanar spacings of C_1, C_2 should be equal, giving a true parallel position for the reflection which is being examined. The homogeneity of C_1 can then be studied by observing the variation of both its reflectivity and rocking curve as the monochromatic beam is allowed to fall on different portions of the surface.

4.4. The optics of collimating systems

The geometrical optics of collimating systems for neutron spectrometers have been studied in considerable detail because of the importance of achieving the best compromise between intensity and resolution in the diffraction pattern. As a result of the work of Sailor, Foote, London, and Wood (1956), Caglioti, Paoletti, and Ricci (1958, 1960, 1962, 1970), Willis (1960), Shull (1960), Sakamoto, Kunitomi, Motohashi, and Minakawa (1965), Popovici and Gelberg (1966), and others, the principles are now well understood, but there remains a good deal of flexibility in the designs and recipes for collimating systems, depending on the particular kind of measurements for which any particular apparatus is intended.

We shall give an outline of the treatment which has been developed by Caglioti and his colleagues in the papers mentioned above. Fig. 51 illustrates the situation where a monochromatic beam is produced by reflection from a suitable crystal and then falls on a *powdered* sample. The chain-dotted line indicates the path of a neutron which travels centrally down the collimator within the reactor and falls on the monochromator. After reflection by the chain-dotted crystal plane *aa* in the monochromator it continues (being of the correct wavelength λ_0) along the central path from monochromator to sample. This path may be defined merely by two simple slits or by a second collimator. When it reaches the sample a crystal plane with orientation *bb* will reflect it along the remainder of the chain-

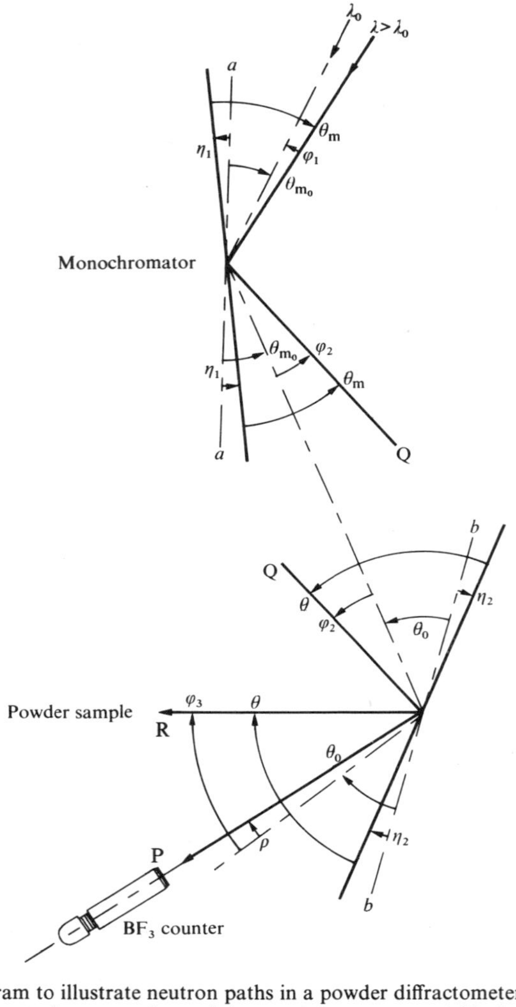

FIG. 51. Diagram to illustrate neutron paths in a powder diffractometer. The broken line *aa* represents the mean plane of the monochromator and *bb* the plane in the powder sample which reflects the central ray from the collimator, indicated by the chain-dotted line. (Caglioti 1970.)

dotted path to P. We emphasize that this particular path is followed only by a ray which is central in the reactor collimator and which has the wavelength λ_0. In practice, neutrons within a range of angle α_1 will be travelling down the reactor collimator, and they will have a wide range of wavelengths, as we indicated earlier in Fig. 2 (p. 4). Different wavelengths will be reflected at different angles by the monochromator and there will be a small range of λ which can effect passage through the slits (or down the collimator of angle α_2) between monochromator and sample, and this range of wavelength will be increased if the monochromator itself has a significant mosaic spread β. The figure indicates by a full line a non-central neutron of wavelength $\lambda > \lambda_0$, which is reflected by a mosaic block in the monochromator different from *aa*. This ray reaches the point Q and then hits the sample where a suitably oriented plane in the powder reflects it to R. The angular distribution of neutrons beyond the sample will depend on the initial collimation α_1, the slit-restriction or collimation α_2, and the mosaic spread of the monochromator. In practice, we examine this angular distribution by traversing a detecting counter which is centred on the sample. This counter will receive neutrons within a certain acceptance angle α_3, determined by the physical width of the counter, by a slit mounted in front of it, or by a further collimator or 'Soller slits' between sample and counter. In a diffraction experiment we require to know, in particular, the total number of neutrons which enter the counter when it traverses a particular reflection and the angular width of the pattern which it records. Caglioti has shown that the former of these quantities depends on the neutron source, the natures of the monochromating crystal and the sample, and the structure amplitude factors of the reflections under consideration, and is proportional to an instrumental factor L called the luminosity and given by

$$L = \frac{\alpha_1\alpha_2\alpha_3\beta}{(\alpha_1^2+\alpha_2^2+4\beta^2)^{\frac{1}{2}}}. \tag{4.5}$$

The angular width of the pattern, defined by its full width measured between the two points of half-maximum intensity, is

$$A_{\frac{1}{2}} = \begin{bmatrix} \{\alpha_1^2\alpha_2^2+\alpha_1^2\alpha_3^2+\alpha_2^2\alpha_3^2+4\beta^2(\alpha_2^2+\alpha_3^2)-4a\alpha_2^2(\alpha_1^2+2\beta^2)+ \\ +4a^2(\alpha_1^2\alpha_2^2+\alpha_1^2\beta^2+\alpha_2^2\beta^2)\}/(\alpha_1^2+\alpha_2^2+4\beta^2) \end{bmatrix}^{\frac{1}{2}}. \tag{4.6}$$

In these expressions α_1, α_2, and α_3 are the collimator angles which we have defined and β is the mosaic spread of the monochromator. The quantity a is the 'dispersion parameter' which compares the dispersion produced by the sample and the monochromator for a given spread of wavelength in the incident beam.

Thus

$$a = \frac{(\mathrm{d}\lambda/\mathrm{d}\theta)_{\mathrm{monochr}}}{(\mathrm{d}\lambda/\mathrm{d}\theta)_{\mathrm{sample}}},$$

and it can be shown, by differentiating the Bragg equation, that

$$a = \frac{(\tan\theta)_{\mathrm{sample}}}{(\tan\theta)_{\mathrm{monochr}}}. \tag{4.7}$$

In arriving at these expressions for L, $A_{\frac{1}{2}}$ it is necessary to assume that the transmission function for each of the collimations α_1, α_2, and α_3, is approximately Gaussian, and it has been shown experimentally by Shull (1960) that this is reasonable.

It is evident that the angular width, in particular, depends on the parameters in a very complicated way, and careful calculations will be needed to assess the precise influence of changes in the parameters, but we can note some more general conclusions. If we take a very simple case for which

$$\alpha_1 = \alpha_2 = \alpha_3 = \beta \equiv \alpha$$

then

$$A_{\frac{1}{2}} = \alpha\left(\frac{11-12a+12a^2}{6}\right)^{\frac{1}{2}} \tag{4.8}$$

and

$$L = \alpha^3/6^{\frac{1}{2}}. \tag{4.9}$$

Thus the line-width is proportional to α, whereas the luminosity is proportional to α^3. Hence the resolution could only be improved by a factor of 2 at the expense of an eightfold reduction of luminosity.

The expression for $A_{\frac{1}{2}}$ can easily be shown to have a minimum value when $a = \frac{1}{2}$. It remains small near $a = 1$ and then at larger values of a it increases roughly proportional to a.

It is also noteworthy that, in a similar way to the above analysis, which relates to the 'parallel' position, we can derive corresponding expressions for the 'antiparallel' position. In this particular simple case the value of L, the luminosity, is unchanged but the sign of the coefficient of a within the expression for $A_{\frac{1}{2}}$ is reversed, giving

$$A_{\frac{1}{2}} = \alpha\left(\frac{11+12a+12a^2}{6}\right)^{\frac{1}{2}}. \tag{4.10}$$

This gives much inferior resolution, as we have already noted in Fig. 50.

Caglioti and Ricci (1962) have confirmed the above analysis with some experimental measurements in a practical case for which the collimator angles were $\alpha_1 = \pm 19{\cdot}1'$, $\alpha_2 = \pm 25{\cdot}8'$, and $\alpha_3 = \pm 34{\cdot}4'$. Fig. 52 shows the diffraction patterns observed from a sample of nickel powder with two different monochromating

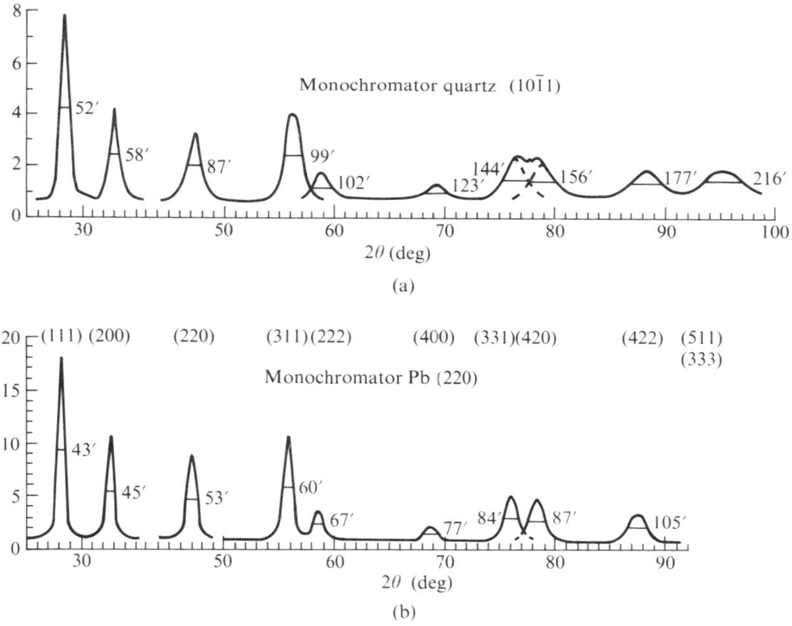

FIG. 52. Powder diffraction patterns at $\lambda = 1$ Å for nickel with two different monochromating systems. Curve (a) is for a quartz crystal and an angle $\theta_B = 8°36'$, whereas (b), for a lead crystal, uses a larger value of $\theta_B = 16°31'$. (After Caglioti and Ricci 1962.)

systems, each yielding the same wavelength of about 1·0 Å. Curve (a) is for a monochromator of quartz, a highly perfect crystal with a mosaic spread of only a few seconds of arc, reflecting the neutrons from a (10$\bar{1}$1) plane at a Bragg angle of 8°36′. Curve (b) is for a lead crystal of mosaic spread $\beta = 12{\cdot}5'$, using the (220) plane at a Bragg angle of 16°31′. The widths of the diffraction peaks are marked in the Figure, and it is noteworthy that the peaks which occur at large values of 2θ are much sharper for the lead monochromator, whereas for the low-angle lines the difference between the two cases is much less marked. The improvement in resolution with lead is a consequence of the lower values of the parameter a for corresponding reflections, brought about by the larger value of $(\tan\theta)_{\text{monochr}}$ for the lead crystal. This is illustrated more directly in Fig. 53, where the width $A_{\frac{1}{2}}$ is plotted against the value of a for

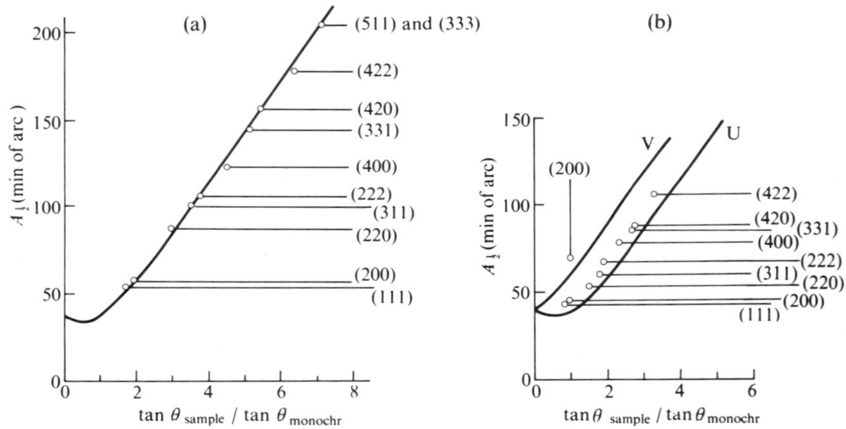

FIG. 53. The variation of the full width at half maximum of the reflections from nickel powder for the quartz (a) and lead (b) monochromators specified in Fig. 52. The abscissa is the ratio of the values of $\tan\theta$ at the sample and at the monochromator. Curves U, V in (b) apply to the parallel and antiparallel positions, respectively. (After Caglioti and Ricci 1962.)

the quartz and lead systems. As an example we may note the (331) and (420) reflections, for which the value of a is greater than 5 for quartz and less than 3 for lead, giving the reduced width in the latter case, which is observed in Fig. 52. This is an example of the improvement in resolution which can be achieved (see Willis 1960) by employing a large take-off angle for the monochromator. This result arises because of the reduction in width of the wavelength band reflected by the monochromator, though the improvement is achieved at the expense of some reduction in integrated intensity because of the occurrence of the terms $\cot\theta_0$ in eqn (4.4). The two curves drawn in Fig. 53(b) for the practical collimator emphasize the point that the best resolution is obtained in the region of $a = 1$ for the parallel focusing position, so that it is advisable, wherever possible, to design apparatus so that the most interesting regions of the diffraction pattern occur near the focusing position. As a practical example of this, we show in Fig. 54 some results by Stoll and Halg (1965) for a powdered sample of spinel $MgAl_2O_4$ using a wavelength of 1·02 Å. The over-all improvement in resolution by using the (311) plane of the aluminium monochromator, instead of the (111) plane, is very marked.

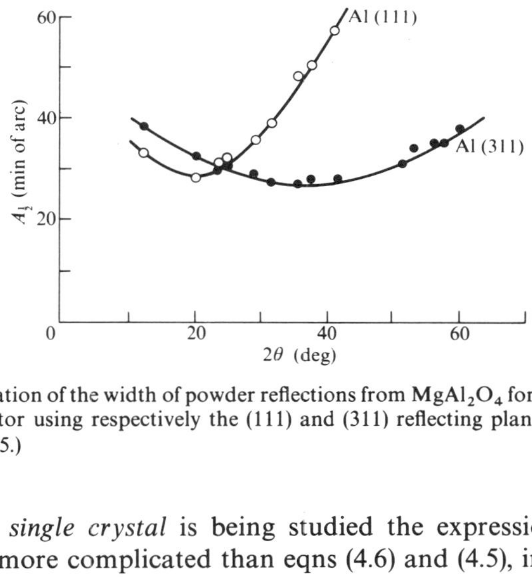

FIG. 54. Variation of the width of powder reflections from $MgAl_2O_4$ for an aluminium monochromator using respectively the (111) and (311) reflecting planes. (After Stoll and Halg 1965.)

When a *single crystal* is being studied the expressions for $A_{\frac{1}{2}}$ and L are more complicated than eqns (4.6) and (4.5), including in addition the mosaic spread of the crystal under examination. It is difficult to give general recommendations to cover a wide variety of situations, but these have been discussed by Caglioti and Ricci (1960, 1962) and others. As with powdered samples it is of advantage to ensure that the important part of the diffraction pattern is in the neighbourhood of the focusing position, and it is generally of value to use a large value of θ, the Bragg angle at the monochromator. When a wide range of values of a is to be covered in the pattern it is usually preferable to arrange that $\alpha_1 < \alpha_2 < \alpha_3$, though a smaller value of α_3 may be preferable at small values of a. It must also be remembered that the background of the diffraction patterns, as well as the luminosity of its lines, is important, and its intensity may be roughly proportional to the product of the three collimation angles $\alpha_1, \alpha_2, \alpha_3$.

Finally, it is worthwhile to consider the effect which a change of wavelength will have on the resolution and intensity of the powder diffraction pattern. Loopstra (1966) has shown that an improvement in over-all resolution occurs if the wavelength is increased at a constant value of the interplanar spacing of the monochromatic crystal. Over-all resolution is defined in terms of the relative change in interplanar spacing $\Delta d/d$ to which a given angular line-width corresponds, since this quantity will determine whether two nearby reflections in a powder pattern can be separately distinguished. Loopstra shows that at a wavelength of 2·6 Å a twofold improvement in resolution is secured, in comparison with his conventional wavelength of 1·15 Å. The choice of the value of 2·6 Å depends on the use of a filter of pyrolytic graphite to give substantial absorption of the second-order component of wavelength 1·3 Å. Fig. 55 indicates how the absorption per atom of such a filter varies with neutron energy; a thickness of 8 cm achieves an improvement by a factor

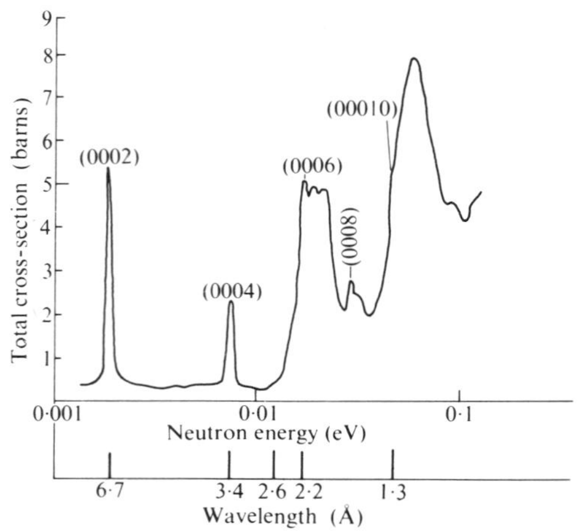

FIG. 55. The change in the transmission cross-section of pyrolytic graphite with neutron energy, when the neutrons are incident normal to the well-aligned graphite planes. The main peaks in the curve correspond to the onset of the various (000*l*) reflections from graphite. The preferential transmission of $\lambda = 2{\cdot}6$ Å, compared with $\lambda = 1{\cdot}3$ Å, is to be noted. (Loopstra 1966.)

of 100 times in the ratio of the two orders, thus reducing the second-order contamination to a fairly acceptable value of $1\frac{1}{2}$ per cent. It is deduced that if the same high resolution were to be achieved at $\lambda = 1{\cdot}15$ Å, by improvements in collimation, then the intensity would be inferior by a factor of 1·7, thus demonstrating a substantial advantage in favour of this longer wavelength technique.

4.5. The powder diffraction method

With a polycrystalline sample there are two geometrical arrangements which permit a ready comparison of the experimentally measured intensities with the results of calculation. In the first of these, illustrated in (a) of Fig. 56, a parallel-sided slab of material, enclosed if necessary in a thin metal box, usually of aluminium, giving negligible absorption and scattering, is placed in the symmetrical transmission position to intercept adequately the whole of the monochromatic beam. The number of neutrons $\mathscr{P}$ diffracted per minute into the detecting counter is then given by

$$\frac{\mathscr{P}}{\mathscr{P}_0} = \frac{\lambda^3 l_s}{4\pi r}\frac{t\rho'}{\rho}\frac{e^{-\mu t \sec\theta}}{\sin^2 2\theta} j N_c^2 F^2 e^{-2W}, \tag{4.11}$$

where

$\mathscr{P}_0$ = number of neutrons per minute in the incident monochromatic beam,
l_s = height of counter slit,
r = distance from specimen to counter,
t = thickness of specimen,
j = number of co-operating planes for the particular reflection being measured,
ρ' = measured density of the specimen,
ρ = theoretical density,
e^{-2W} = Debye temperature correction factor,
N_c = number of unit cells per cm^3,
F = structure amplitude factor per unit cell,
μ = linear absorption coefficient of the sample.

This arrangement permits the correction for absorption in the sample to be calculated easily, being the term $\exp(-\mu t \sec\theta)$ in the above expression. In practice the value of μt is determined by direct measurement of the reduction of the monochromatic beam in the 'straight-through' or zero position which is brought about by insertion of the sample, the ratio of the two neutron counting rates being equal to exp $(-\mu t)$.

In the alternative experimental arrangement, shown at (b) in Fig. 56, the powdered sample is in the form of a vertical circular

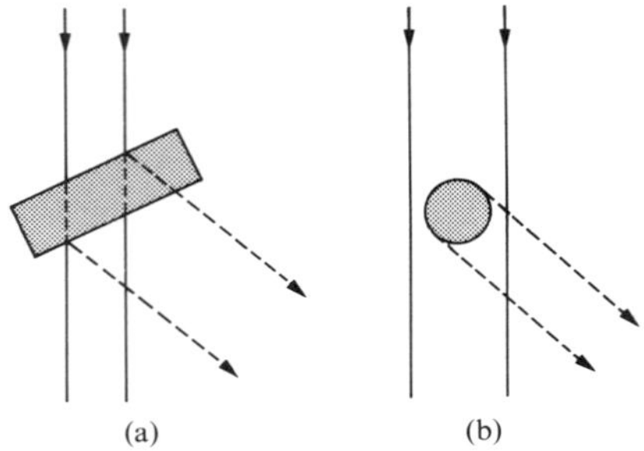

FIG. 56. Alternative forms of specimen for powder diffraction.

cylinder, sufficiently small in diameter to be bathed fully in the neutron beam. Under these conditions the number of neutrons $\mathscr{P}$ diffracted per minute into the counter is given by

$$\frac{\mathscr{P}}{\mathscr{I}_0} = \frac{\lambda^3 l_s}{8\pi r} \frac{V\rho'}{\rho} \frac{jN_c^2 F^2}{\sin\theta \sin 2\theta} e^{-2W} A_{hkl}, \tag{4.12}$$

where

$\mathscr{I}_0$ = number of neutrons *per unit area* hitting the specimen per minute, i.e. $\mathscr{P}_0$/beam area,
V = volume of sample in beam,
A_{hkl} = absorption factor,

and the other quantities have the same significance as in eqn (4.11) above. The expression for the absorption factor A_{hkl} is less simple than for the symmetrical transmission case considered previously. It depends in a rather complicated manner on μR_s, where R_s is the radius of the cylindrical specimen, and on θ, the Bragg angle, in a manner which was originally calculated for use in X-ray diffraction by Claassen (1930) and by Bradley (1935). Suitable tabulations of A_{hkl} can be found in the literature for a wide range of values of μR_s; however, in the neutron case the magnitude of μR_s found in practice is quite small, generally being less than 0·5. Under these circumstances there is little variation of A_{hkl} with angle θ, particularly for angles less than 45°, and it is usually possible to use the same value of A_{hkl} over the whole of the measured portion of the diffraction pattern. The magnitude of A_{hkl} can be appreciated from Table 10,

TABLE 10
Values of absorption factor A_{hkl} for cylindrical specimens

	θ				
μR_s	0°	22½°	45°	67½°	90°
0·1	0·845	0·845	0·847	0·853	0·857
0·2	0·718	0·718	0·719	0·724	0·729
0·3	0·610	0·611	0·612	0·621	0·628
0·4	0·518	0·519	0·528	0·545	0·554
0·5	0·440	0·442	0·458	0·478	0·491
0·6	0·374	0·377	0·397	0·420	0·437
0·7	0·318	0·323	0·345	0·372	0·390
0·8	0·272	0·278	0·304	0·332	0·352

which lists the values over the range of μR_s usually involved. A more extensive tabulation, accurate to four decimal places, has been given by Rouse and Cooper (1970), who also provide an approximate analytical expression which will yield the absorption correction to an accuracy of 0·4 per cent. In order to calculate A_{hkl} the value of μR_s can be measured quite easily experimentally by observing the transmission loss when a very narrow beam passes through the centre of the specimen. This can be done by inserting finite slits, say 0·15 cm in width, at the two ends of the channel between the monochromator and the sample. Under these circumstances the transmission through the specimen will be equal to $\exp(-2\mu R_s)$, thus giving μR_s.

This second form of specimen is often more convenient in practice than the parallel-sided slab. It is more economical in use of material, which is often of importance considering the relatively large samples needed for neutron diffraction, thicknesses and diameters of 1 cm often being used. With a cylindrical sample it may also be possible sometimes to reduce effects due to preferred orientation by rotating the cylinder about its vertical axis.

It is to be noted that precise calculation of the absorption correction factor for either of these forms of specimen can be done much more accurately with the relatively large specimens which are used than is the case for the small specimens characteristic of X-ray powder diffraction. In the latter case the specimen may be of non-uniform diameter and density.

A further factor which increases the relative accuracy of neutron diffraction when experimental intensities are being compared with the result of calculation for a given structure is the absence of any variation of scattering factor with angle θ. On the other hand, with X-rays the shape of the atomic scattering curve has to be assumed, on the basis of theoretical prediction. Uncertainty in this may often limit the accuracy of the conclusions which can be drawn from intensity data.

Two main considerations determine the size of the samples used. First, the sample should be large enough to give an adequate counting rate, and secondly it should not be so large as to increase unduly the width of the diffracted beam, since this would further impair the limited resolution inherent in the relatively poorly collimated beam. Fig. 56 shows how the specimen (a) increases the diffracted beam width.

With the earlier reactors and a flux of about 2×10^{12} neutrons $cm^{-2}\,s^{-1}$ a diameter or thickness of one or even two centimetres was used. With a flux of 2×10^{13} neutrons $cm^{-2}\,s^{-1}$ a diameter of about 0·6 cm is adequate and the powder is often contained in an aluminium can. If a length of 3 cm is filled with powder then the over-all volume of the sample will be about 1 cm^3. A wall thickness of 0·25 mm is commonly used for the aluminium can, but a thinner wall, down to 0·1 mm, is of great advantage in order to avoid confusion of weak lines in the diffraction pattern with those due to the aluminium. In either case it may be necessary to record a pattern with the empty can alone and subtract it from the pattern obtained with the material under examination. This latter correction procedure is almost always found to be necessary in measurements at low temperature where the neutron beam not only passes through the two walls of the specimen tube but also through four thicknesses of aluminium which constitute the walls of a vessel, like a Dewar flask, holding liquid air. Such apparatus is described in more detail later in the chapter. An alternative possibility is to manufacture the specimen tubes from metallic vanadium, which has negligible coherent scattering power and produces no diffraction lines to confuse the pattern under investigation.

The problems and restrictions set by the materials used as sample holders, and as constructional materials through which the neutron beam passes, are much more troublesome when high-flux reactors are being used. If advantage is taken of the higher neutron intensity to use smaller samples, then the thickness of supporting material in the beam must be reduced proportionally if the diffraction pattern is not to be impaired. The use of vanadium has advantages but the difficulties of working and machining it, in comparison with the ease of using aluminium, are a big drawback. Other materials with attractive possibilities are the nul-matrix alloys which give zero coherent scattering and to which attention was drawn by Sidhu, Heaton, Zauberis, and Campos (1956). An example is an alloy of 60 atomic per cent of titanium with 40 atomic per cent of zirconium, in which the negative scattering amplitude of the titanium atoms is cancelled out by the positive amplitude of the zirconium. This alloy has been used for making sample holders and in the construction of associated apparatus.

The quality and value of the diffraction pattern will depend immensely on the intensity of the neutron beam which is available. Some examples are given in Fig. 57. Curves (a), (b) are both recorded with the same sample of diamond powder, enclosed in an aluminium can of diameter 1·5 cm and utilizing about 8 cm^3 of material. The two curves are for the reactors BEPO and DIDO respectively, and the increased counting rate in (b), which is about 20 times as great as in (a), is brought about by the flux improvement from 2×10^{12} neutron $cm^{-2}\,s^{-1}$ to 4×10^{13} neutrons $cm^{-2}\,s^{-1}$. The third pattern in the Figure is measured at the higher flux, but for a much smaller sample measuring 0·3 cm in diameter.

For any material which is being investigated it is possible to express the diffracted intensities on an absolute scale most simply by comparing them with those for a sample of identical size but constituted of some standard material, such as diamond, of known structure and containing nuclei of well-known scattering amplitudes. From eqn (4.12) it follows that for cylindrical samples of standard dimensions and under uniform experimental conditions the diffracted power

$$\mathscr{P} = \text{constant} \times \frac{\rho'}{\rho} \frac{jN_c^2 F^2}{\sin\theta \sin 2\theta} e^{-2W} A_{hkl}, \qquad (4.13)$$

so that for comparison purposes only the factors in eqn (4.13) need be evaluated. Eqn (4.11) also may be simplified similarly when samples of a standard size are always used.

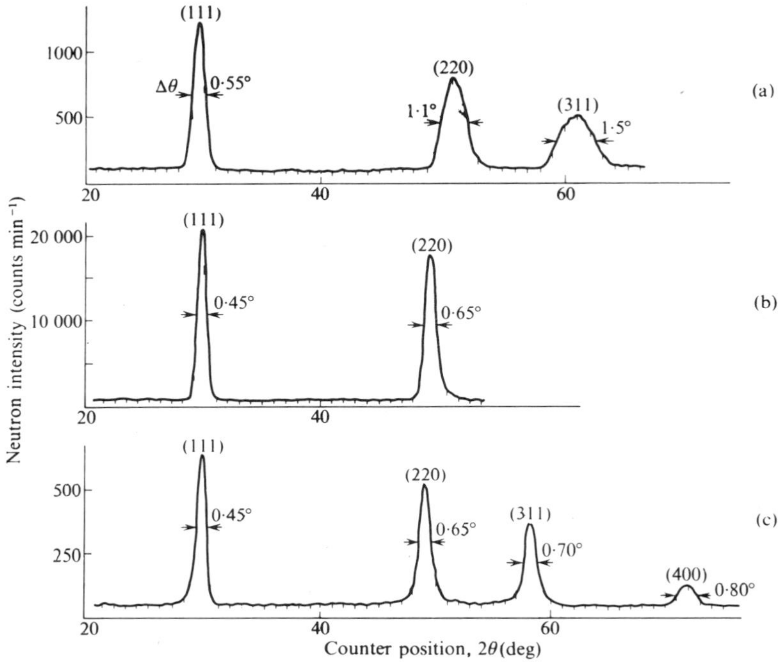

FIG. 57. A comparison of diffraction patterns for powdered diamond using cylindrical samples (a) for diameter 1·5 cm at the reactor BEPO, (b) of diameter 1·5 cm at the reactor DIDO, and (c) of diameter 0·3 cm at the reactor DIDO.

The examples of diffraction patterns which we showed in Fig. 57 were recorded by the simple device of connecting the output of a ratemeter to a pen recorder, so that a continuous record of the rate of arrival of neutrons was achieved as the counter moved steadily in azimuth around the sample under examination. Nowadays, impelled by the need for automatic control and programming, it is more usual for the counter to be moved in discrete steps, sufficiently small for the diffraction peak to require ten or more steps for complete coverage. This process has the advantage that variations in reactor power can be allowed for by arranging that the counting time at each step is controlled by a small fission chamber which monitors the main beam. The fission chamber normally consists of a rectangular box, about 1 cm in thickness, which contains gaseous UF_6 and is placed across the monochromatic beam.

Alternatively, a coating of metallic uranium is deposited on the counter wall. The fission fragments which are produced by the neutrons produce intense ionization, which is detected electronically. The attenuation of the neutron beam by its passage through the counter is only 1 or 2 per cent. It is then arranged that the main counter remains at each stepping point for the interval of time necessary to accumulate some stated number of monitor counts. An example of such a record appears in Fig. 58. This is a pattern for nickel powder obtained with a diffractometer of high resolution at the Harwell reactor PLUTO, using a wavelength of 1·14 Å. The instrument used a germanium crystal as monochromator, mechanically worked to give an increased mosaic spread as described earlier in this chapter, and employing a relatively large take-off angle of $2\theta = 47{\cdot}5°$ to give Bragg reflection from the (400) planes, which have an interplanar spacing of 1·405 Å. Counting was carried out at intervals of 2θ of 0·1° and the period of counting at each position was about 30 s. The improved angular resolution and narrow line-width out to large values of 2θ is noticeable in this pattern.

We have already mentioned in Chapter 1 the increased rate of data collected that can be achieved by using multiple counters, as in Fig. 15, where each counter records, say, 10° or 20° of 2θ. The range of each counter has sufficient overlap with the coverage of its neighbours to simplify correlation of the information. Experience has shown that, although speed of operation is achieved in this way, great precautions must be taken if a loss in accuracy of measurement of the intensities is to be avoided. It is evident that there are substantial differences in the efficiencies of individual, nominally similar, counters. Differences of up to 20 per cent have been found, with some evidence that the ratios between pairs may change with time by a few per cent. A typical set of sensitivities for seven counters was 0·84 : 1 : 1·08 : 1·03 : 0·93 : 1·07 : 1·00. Frequent calibration may be needed therefore if the highest accuracy of intensity measurement is to be achieved.

When high-flux reactors are being used it becomes particularly important that the results of an experiment can be assessed quickly and, preferably, continuously. Fig. 59 shows a section of a powder diffraction pattern as displayed on a cathode-ray oscilloscope at the Institut Laue–Langevin at Grenoble, France. This portion of the powder pattern includes the (200) reflection from iron. By

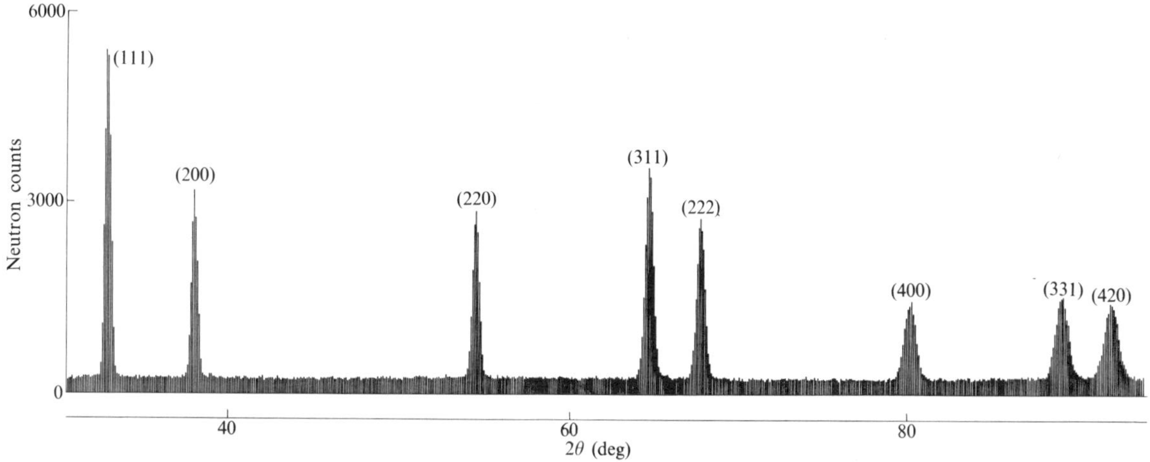

FIG. 58. A section of a powder diffraction pattern for nickel at a wavelength of 1·14 Å recorded on the PANDA diffractometer at A.E.R.E.-Harwell using a germanium monochromator. Counts are made at intervals of 0·1° of 2θ. (Courtesy of R. F. Dyer.)

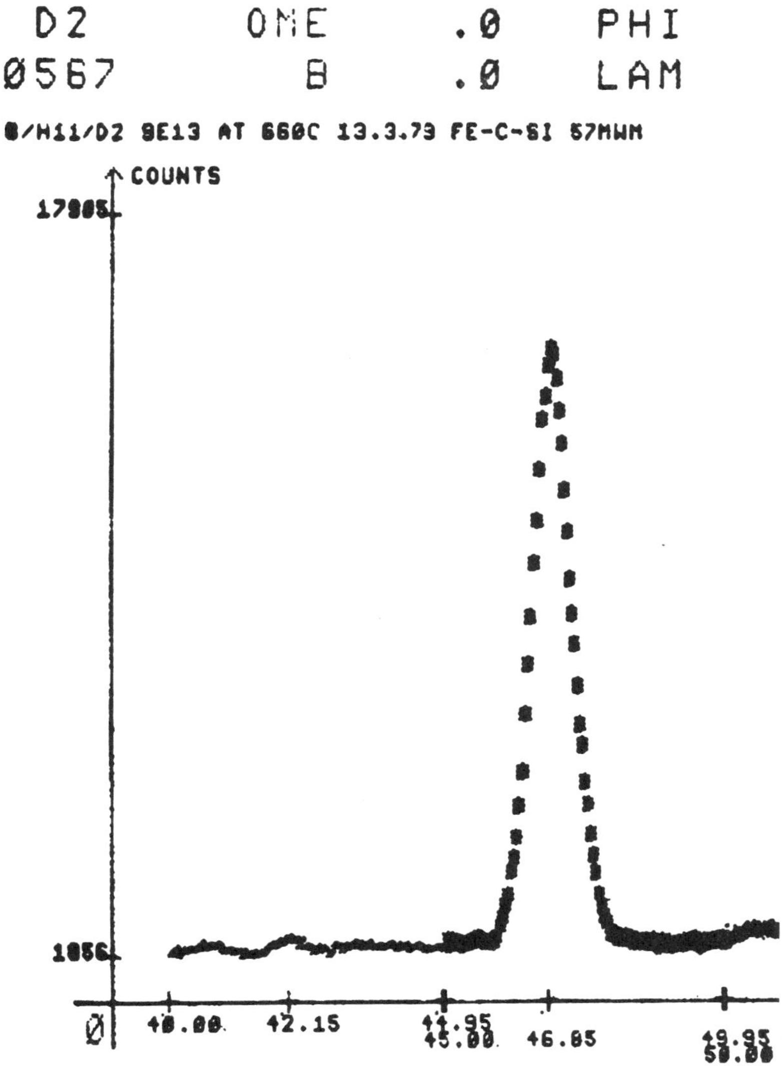

FIG. 59. A recording of a portion of the powder diffraction pattern of iron at the Institut Laue–Langevin, Grenoble. (N. Cowlam.)

moving an electronic strobe across the pattern (via a teleprinter) it is possible to print out the angular positions of particular features; for example, the peak of the (200) reflection is recorded as $2\theta = 46{\cdot}85°$. It is also possible to show a smaller section of the pattern on an enlarged scale, as for the (200) reflection in Fig. 60, and to record the peak count, the background count, and the integrated area under the peak. Permanent copies, on paper, of the oscilloscope display can be called for by suitable instructions from the tele-typewriter, and Figs 59 and 60 are taken directly from these copies.

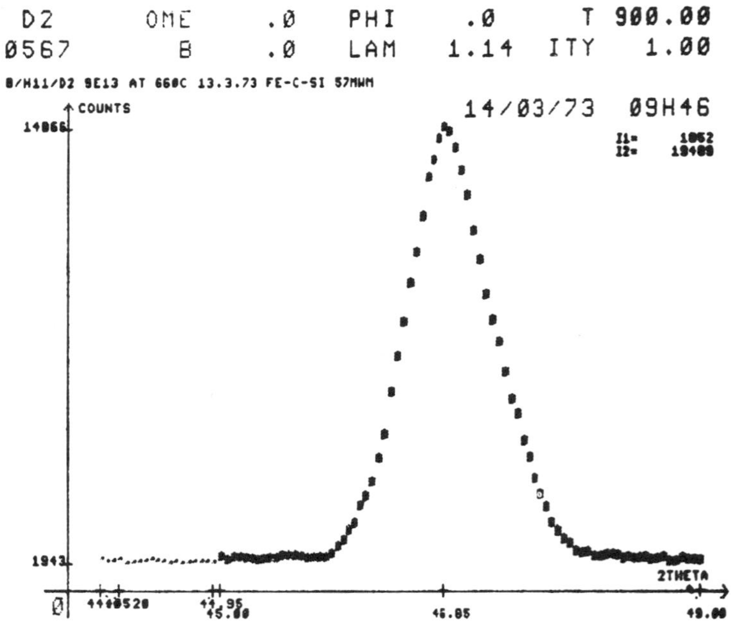

FIG. 60. An enlargement of the (200) reflection of iron shown in Fig. 59. (N. Cowlam.)

We have emphasized already that it is the peak counting rate and the angular width of a diffraction line which will determine the latter's visibility. More exactly we must acknowledge that it is the counting rate relative to the background which is important, particularly as the background of the pattern will be subject to statistical fluctuations. This means that it is essential to reduce the background neutron intensity as far as possible. A proportion of the background is inherent in the diffraction process, with an intensity proportional to the volume of the sample and the strength of the incident beam, and cannot be avoided; there are contributions from inelastic scattering, from the disordered scattering caused by the nuclear spin and isotope effects discussed in Chapter 2, and from multiple scattering. The effect of disordered scattering will be particularly marked for hydrogenous material. Both the disordered and the multiple scattering will be isotropic. Much of the total background, however, may be regarded as instrumental. Some neutrons will enter the counter from among those forming the general background in the neighbourhood of the reactor, but the largest number will arise from diffuse scattering of the initial beam at the monochromator and by air scattering of that part of the beam (about 99 per cent) which is not reflected by the monochromator but proceeds unhindered along the line of the reactor collimator. In order to prevent scattered neutrons of this sort from entering the counter the whole of the chosen path from reactor to counter is channelled in material which will absorb any neutrons incident upon it, first slowing them down with hydrogenous material if necessary. The scattering which, in spite of these precautions, still enters the counter will not, of course, be isotropic but will vary with counter position. The origin of a troublesome contribution at low angles is indicated in Fig. 61. Because of the large size of the counter aperture, its shielding, and any collimator, part of the undeviated monochromatic beam may enter the counter when the latter is quite remote from its zero position. Thus there will be a large contribution, which extends to several degrees of 2θ as indicated in the curves in the Figure. The effect will be worse if the front of the collimator is close to the sample, as in position and curve (a), but can be reduced by cadmium plating the collimator tube and counter shield. When low-angle lines are being studied there is a substantial advantage in increasing the distance from sample to counter. A relatively flat background can then be achieved down to angles of scattering 2θ as small as 1°.

The background due to inelastic thermal diffuse scattering may be calculated from the formula $\sum \{1-\exp(-2W)\}b^2$, where W is the Debye factor defined in eqn (2.32) and b is the coherent scattering amplitude. The summation is taken over the various atoms present.

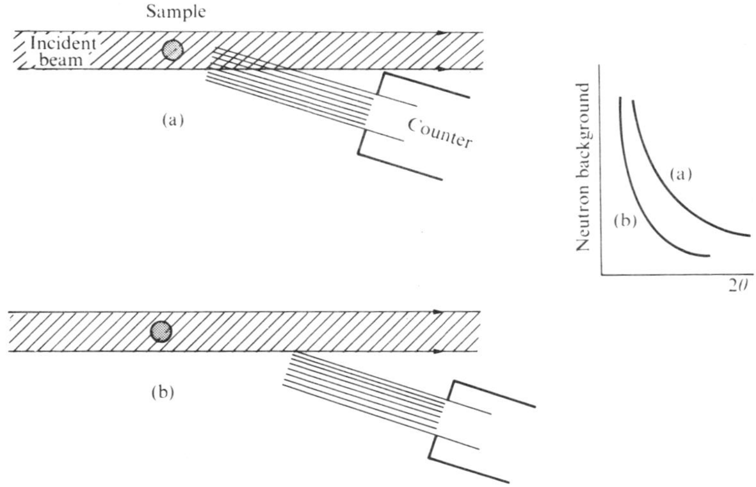

FIG. 61. Reduction of neutron background near the angular position of the main beam can be achieved by placing the counter further from the diffracting sample.

This part of the background scattering will increase with Bragg angle θ owing to the dependence of W on $(\sin\theta)/\lambda$.

Multiple scattering

Multiple scattering is a good deal more important with neutrons than for X-rays because of the large samples which are used in the former case and the low coefficients of absorption. Precise calculation of the effect on the background is difficult, but an approximate calculation is easily made, and the result is consistent with what is observed in practice. Consider a neutron beam giving 5×10^5 counts $\min^{-1}$ to be incident on a sample of diamond powder 1 cm thick. A fraction $NE(\mathscr{S})$ of the neutrons will be scattered by the sample, and in the absence of any subsequent scattering or absorption would emerge as the diffraction haloes from the powder. Here N is the number of nuclei cm^{-3} and $E(\mathscr{S})$ is defined by eqn (2.30). However, these neutrons will have a chance of suffering a second scattering process before leaving the sample and, indeed, a fraction equal to about $\frac{1}{2}\{NE(\mathscr{S})\}^2$ will be re-scattered in this way and give rise to an isotropic background. The value of $NE(\mathscr{S})$ will be about 0·5, so that a fraction 0·12 of the neutrons will enter the background.

A counter with an aperture of 5 cm^2 at 60 cm from the sample would receive about 7 neutrons min^{-1}.

Vineyard (1953) has calculated the multiple scattering for the case of a plane parallel slab of material and shown how it depends on the ratio of the cross-sections for scattering and absorption-plus-scattering. This has been extended to the more useful case of a cylindrical sample of radius R and height h by Blech and Averbach (1965). The differential cross-section for multiple scattering is given by

$$\frac{\mathrm{d}\sigma^{\mathrm{M}}}{\mathrm{d}\Omega} = \frac{1}{4\pi}\frac{\sigma_{\mathrm{s}}(\sigma_{\mathrm{s}}/\sigma_{\mathrm{t}}) \,.\, \delta}{1-(\sigma_{\mathrm{s}}/\sigma_{\mathrm{t}})\delta} \tag{4.14}$$

thus depending on the ratio of the scattering cross-section σ_{s} to the *total* cross-section σ_{t} and on the parameter δ . δ depends on μR and R/h in the manner indicated in Fig. 62. The influence of the ratio $\sigma_{\mathrm{s}}/\sigma_{\mathrm{t}}$ is seen by comparing both experimental and calculated results for vanadium and copper. Fig. 63 shows the variation of the scattering per unit solid angle for a cylinder of radius 1 cm, as a function of the height h of the cylinder. For copper and a height of 2 cm the multiple scattering amounts to about 85 per cent of the total diffuse scattering. For vanadium, which has a large amount of

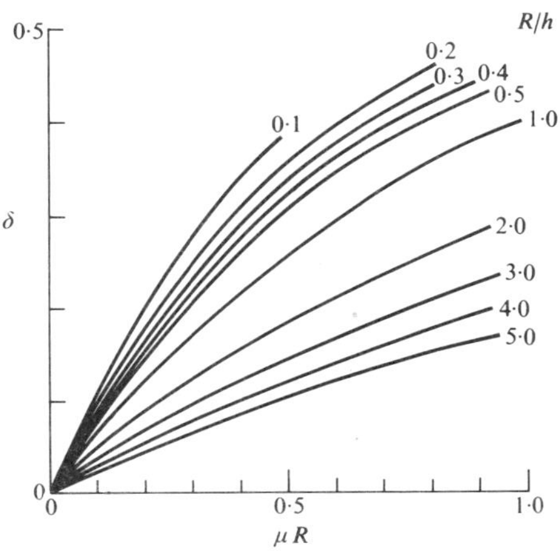

FIG. 62. Calculation of the multiple-scattering parameter δ for a cylindrical sample of radius R, height h, and linear absorption coefficient μ. (Blech and Averbach 1965.)

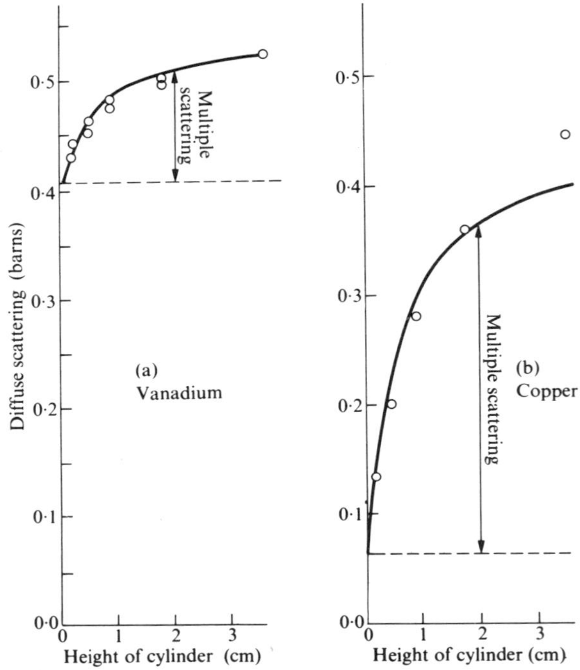

FIG. 63. Multiple scattering from (a) vanadium and (b) copper. The broken horizontal lines show the calculated values of the sum of the incoherent and temperature-diffuse scattering. Addition of the calculated multiple scattering gives the full-line curves which are in good agreement with the experimental points shown. (After Blech and Averbach 1965.)

spin-incoherent scattering, the multiple scattering accounts for only 25 per cent of the total.

Wells (1971) has considered the case of a spherical specimen of radius R and derived an expression for δ in terms of μR. When μR is small, say less than 0·4, δ may be approximated as $\frac{2}{3}\mu R$.

4.6. Single-crystal measurements

In the early days of neutron diffraction studies it was common to use the same diffractometers for both powdered samples and single crystals. In due course, as interest in single-crystal work developed, special designs were produced for this technique which were smaller in size but little different in their main principles. The two factors which mainly determined the large size of the powder instrument, namely large samples to give adequate counting rates and a large sample–counter distance to give good resolution are no longer significant. In particular, large crystals cannot be used because of the difficulties caused by secondary extinction. From a practical point of view the interest in the size of crystal to be used centres on the fact that it is the available neutron flux which determines whether an adequate counting rate will be forthcoming from a crystal which is sufficiently small to be satisfactory theoretically. A single crystal, of course, will give a counting rate which is enormously greater than that from an equal volume of powder, since when in its reflecting position the whole of the reflected neutrons will enter the counter, whereas a powder sends its diffracted radiation into conical haloes and only about 1 per cent of it is detected. Quite as important perhaps as the reduced size of the apparatus is the improved resolution and freedom from the trouble of overlapping spectra which can be achieved. With the rotating crystal method of operation the counter can be placed much closer to the single crystal than for the powdered sample. As the crystal rotates, a diffracted beam for any particular set of planes only appears over a small range of angular position of the crystal, and the fact that the aperture of the counter (rotating at twice the angular velocity) has a finite size does not impair the resolution. On the other hand, with the powdered specimen, each diffracted beam is always present and will be recorded throughout the passage of the counter aperture across it, thus making it advisable for this aperture to subtend at the sample an angle somewhat less than the spread of the diffracted beam.

The early designs of single-crystal apparatus usually employed shorter counters of about 15 cm in length and positioned roughly 20 cm from the crystal about which they rotated. The fact that the counter had to be surrounded by several centimetres of shielding meant that the apparatus was still very bulky. Usually a crystal of roughly cylindrical shape was used, with its axis vertical, and movement of the counter was restricted to the horizontal plane, thus restricting the observed reflections to those from planes parallel to a single axis, e.g. those with indices *hk*0. Adjustments of the crystal and counter were made by hand and the collection of intensity data was slow and laborious. The task of measuring two hundred reflections would occupy about 20 days, and from this information, after conversion of the numbers of neutrons into values of structure amplitude factors, a projection of the neutron scattering density on a plane could be obtained. The early analyses of KHF_2, KH_2PO_4, and $Na_2CO_3 . NaHCO_3 . 2H_2O$, for example, to which we shall refer again later, were carried out in this manner. The use of a cylindrical crystal in this way stemmed from the need to utilize an adequate volume of crystal while retaining short paths for the neutrons travelling through the crystal.

The first stage of real advancement occurred between 1953 and 1960, when reactors giving higher neutron fluxes appeared. Then, for the first time, it became possible to use a *spherical* crystal whose volume gave an adequate counting rate but whose diameter was sufficiently small to avoid large secondary extinction errors on all but a few of the most intense reflections. For many substances a crystal measuring 2 mm in diameter would satisfy these requirements, and this permitted the development of three-dimensional analysis. The motion of the counter was again restricted to a single plane, usually horizontal, but the crystal itself was mounted on a suitable holder to permit orientation in three dimensions. Such an apparatus, described as a four-circle diffractometer, is illustrated in Fig. 64 (see between pp. 10–11) and a sketch of the vertical circle which carries the crystal appears as Fig. 65. The main axis of the instrument is formed by two concentric shafts, the outer of which carries the detecting counter and whose position measures 2θ and the inner one connects to the assembly shown in Fig. 65, whose angular position about the vertical axis is recorded as 2ω. The crystal can then be oriented in any direction in space, yet remain centred in the neutron beam and on the instrument axis, by employing two further motions. These are the χ motion, around the vertical circle, and the ϕ motion, about the axis of the crystal support itself. In typical designs the four rotations can be achieved both manually and automatically, and the mechanical rotations can be programmed on paper or magnetic tape to provide the necessary sequence of crystal motions to permit the desired reflections to be scanned in turn. Prior to each scan the controls will place the crystal in the correct orientation for Bragg reflection to take place, satisfying the usual reciprocal lattice conditions, and will then ensure that the angular position 2θ of the counter advances twice as fast as the crystal position ω. Usually the counter and crystal rotate in small steps of, say, 4′ of arc, and the count at each step is continued long enough to accumu-

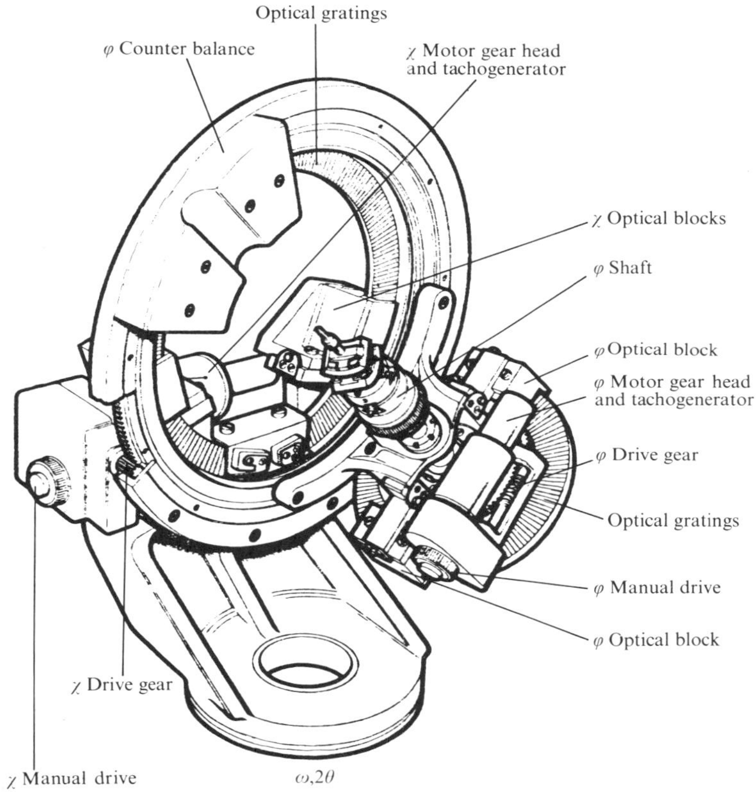

FIG. 65. A detailed sketch of the Hilger and Watts vertical circle, which incorporates the χ and ϕ motions and carries the crystal in the diffractometer of Fig. 64. (R. F. Dyer.)

late a stated number of monitor counts, as in the procedure which we have already described for the powder diffractometer. The angular positions can be set, and measured, to about 0·01°, and various devices, including stepping motors and Moiré fringe systems, have been employed in different designs in order to achieve this accuracy. A variety of different methods have been used to record the results of the diffractometer's operation. Essentially the information is desired to be available in three ways: first, a printed record of intensities, alongside angular positions and crystallographic indices; secondly, a record on tape or punched cards which can be used for subsequent analysis, including the conversion of intensities to structure factors; and, thirdly, a visual record of the diffraction curves on a chart or oscilloscope which will enable the operator to be sure that the equipment is performing satisfactorily. Further ancillary apparatus may permit the operator to amend the programme of the diffractometer as a result of his assessment of the information which it is currently presenting. Fig. 66 shows a portion of the printed record of the diffraction pattern from a single crystal of salicyclic acid $C_6H_4OH\,.\,COOH$ in the neighbourhood of the

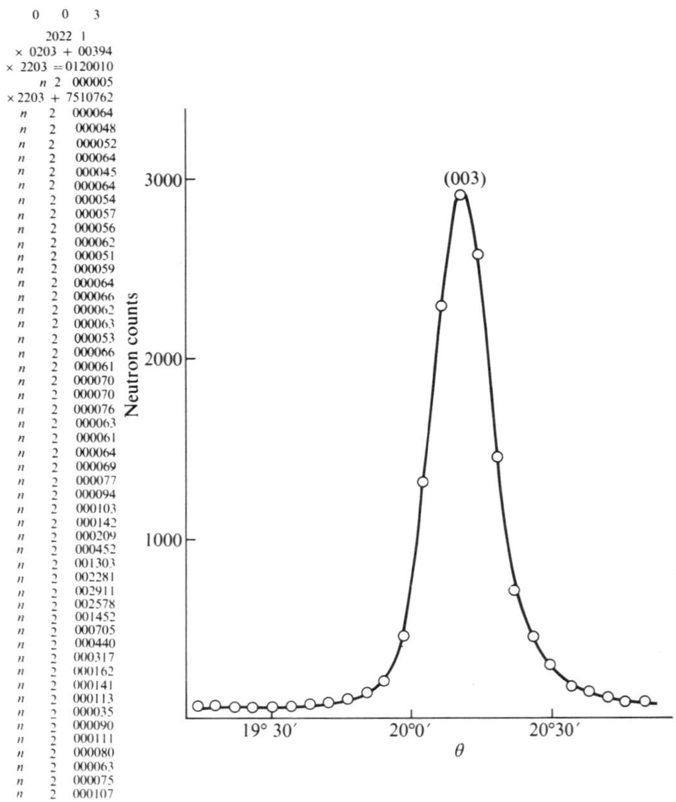

FIG. 66. The (003) reflection from a single crystal of salicylic acid with, on the left, a portion of the printed record. Successive lines of the record correspond to increments of 4 min of arc in the Bragg angle θ and the final column gives the number of neutrons counted at the corresponding position. (R. J. Jude.)

(003) spectrum. Successive lines in the record correspond to increments of 4′ of arc in the position of the crystal, and the figures in the final columns give the numbers of neutrons which were counted during the time interval determined by the monitor. These numbers have been used to plot out the diffraction pattern shown on the right of the Figure.

One of the essential aims behind the development of fully automatic instruments is that of ensuring the utmost use of the neutron beams from the reactor, bearing in mind the very high cost of installing and operating the reactor. Another contribution to full use of the beam is a scheme for selecting two distinct bands of wavelength from a single collimated beam. As we emphasized in Chapter 1, only about 1 per cent of the incident neutrons are reflected by the monochromator. Two monochromators may therefore be placed in series in Fig. 67, or one above the other (Willis 1962), to reflect beams of slightly different wavelength to left and right.

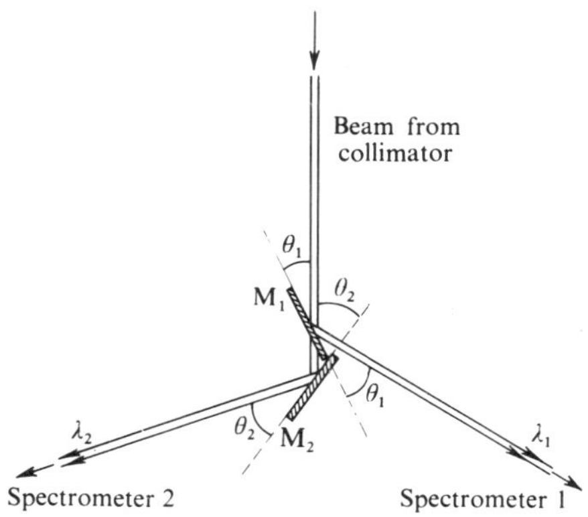

FIG. 67. An arrangement of two monochromating crystals M_1 and M_2 in series to permit the employment of two spectrometers on a single collimated beam from a reactor.

Some difficulties are found when seeking to utilize fully those beams of neutrons which emerge from high-flux reactors in channels which are inclined to the horizontal. It has been shown by Haywood (1974) that it is satisfactory for a vertical monochromating plane to be followed by a horizontal scattering plane in the case of a single-crystal diffractometer, but not normally for a powder instrument.

4.7. Profile refinement in powder diffraction

It will have become evident from our discussion of single-crystal and powder methods that the former is the more powerful and direct, because it examines quite separately neutrons which have traversed the crystal in specific directions. In the powder method all the directional information is lost and the data provided relate purely to interplanar separations. For all but the crystal systems of high symmetry, contributions from several different crystal reflections will arrive at any given position 2θ in the diffraction pattern, bearing in mind that each reflection will have an angular width of the order of a degree of arc. This width depends in detail upon the collimating system, the nature of the monochromator, the neutron spectral distribution, and the shape and crystalline perfection of the sample, but it is found in practice that each reflected spectrum is almost exactly Gaussian in shape, except at very low angles. This is emphasized in Fig. 68, where a measured peak is compared with a Gaussian shape (Rietveld 1969). Although it may not be practicable to separate out from the pattern the discrete peaks, and hence the structure factors for individual reflections, it becomes

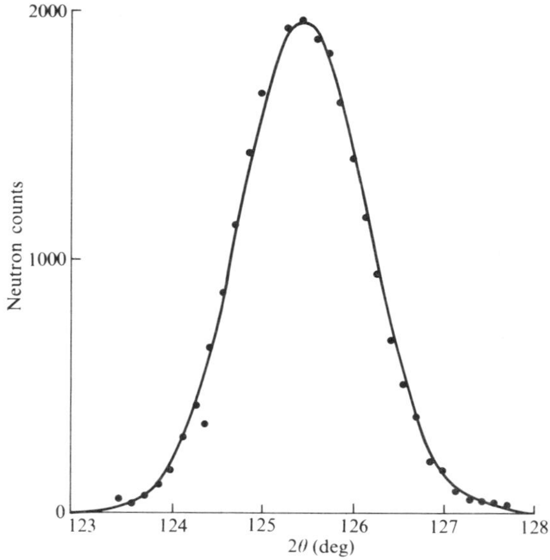

FIG. 68. The curve indicates how a calculated Gaussian shape is in very good agreement with the measured points of a diffraction peak. (Rietveld 1969.)

possible to calculate at any angular position the resultant of the overlapping contributions in terms of a series of parameters, assuming that the main details of the crystal structure are already known. A least-squares refinement of the parameters can then be carried out to give optimum agreement between calculation and measurement at, say, 1000 different values of 2θ made up by moving in steps of 0·1° from 10° to 110°. A suitable program for computer refinement has been developed by Rietveld, in which the count at each position is weighted by the factor $1/\sigma_i$, where σ_i is the standard deviation of the count. The parameters which are refined fall into two groups, those describing the characteristics of the diffractometer and those which depend on the crystal structure. The first group consists of the neutron wavelength λ, the zero position of the counter, and three parameters U, V, W which describe the variation with 2θ of the angular width H of the Gaussian curve, according to the equation

$$H^2 = U \tan^2 \theta + V \tan \theta + W. \tag{4.15}$$

There is also a correction for asymmetry of the reflection curves which is necessary below $2\theta = 30°$. This asymmetry is caused by the fact that the sample has a finite height and cannot be regarded simply as a point on the axis of the diffractometer. The structural parameters are the usual ones, namely, a scale factor, the dimensions and angles of the unit cell, and the coordinates and thermal parameters of the individual atoms. There is also a correction to take account of preferred orientation of the crystallites in the powder.

A typical example of the use of this technique to refine structures is provided by Hewat (1973*a*) in the determination of the displacements of the potassium, niobium, and oxygen atoms in $KNbO_3$ from the ideal cubic perovskite positions, when this material undergoes phase changes. A second example (Hewat 1973*b*) determines the hydrogen atom positions in $NH_4H_2PO_4$, a ferroelectric material which is difficult to obtain in single-crystal form below the transformation temperature. We show in Fig. 69 a comparison of the computed profile and the experimental points for $NH_4H_2PO_4$. This substance can be contrasted with KH_2PO_4, which can be examined as a single crystal in a very straightforward way, as we shall discuss later in Chapter 10.

We emphasize that the Rietveld method is a process of refinement in terms of a postulated model. Clearly only those parameters and

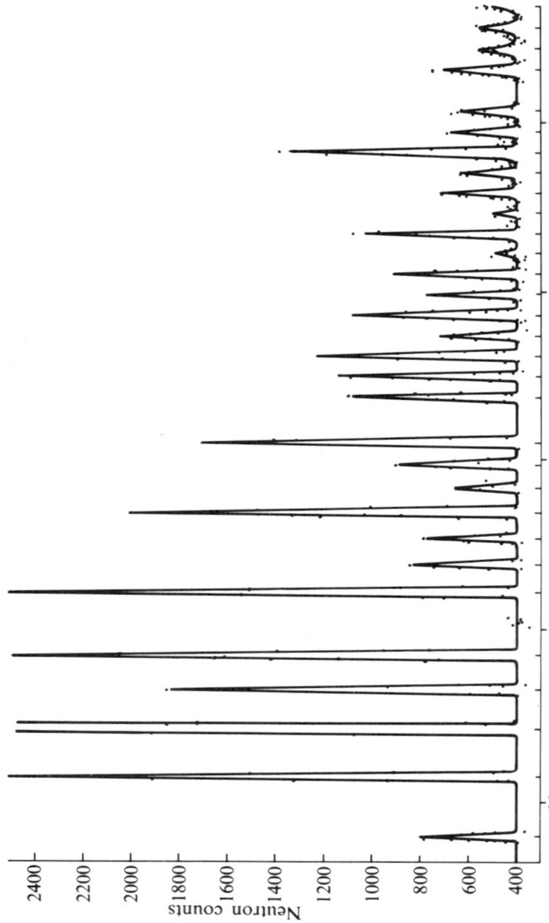

FIG. 69. A comparison of experimental points in the powder diffraction pattern of $NH_4H_2PO_4$ with the computed pattern from profile analysis. (Hewat 1973*b*.)

features which are postulated in the model can appear in the result. Certainly when the structural investigation has proceeded to this stage then the necessary experimental measurement of the powder diffraction pattern is very much simpler than the conventional collection of single-crystal intensity data in three dimensions. This reduction of experimental time is very important when a high-flux reactor is being used. Finally, it may be noted that the profile refinement technique can be applied with neutrons, because in general the reflections have a Gaussian shape which can be described over a wide range of 2θ by the simple function in eqn (4.15). In contrast, much more complex expressions are needed to describe the profiles of X-ray reflections (Wilson 1963).

4.8. Measurements at low and high temperatures

In many applications, particularly the study of magnetic substances, it is desirable to make measurements at reduced temperatures, quite commonly down to 1 K and, in certain cases, even lower. In general, the problems in the construction of apparatus for low-temperature work are much less serious than is the case for X-ray diffraction, because of the very low absorption coefficients for neutrons of materials employed for construction. As a result, the neutron beam can be passed through thin sheets or foils of metals such as aluminium or copper without significant loss of intensity. It must be remembered though that these materials will give a diffraction pattern of their own and this will be particularly troublesome, as mentioned on p. 115, when small polycrystalline samples are being used with high-intensity neutron beams. With single-crystal samples the problems are much less because of the more intense diffraction peaks which are normally being observed.

The general type of construction which has usually been followed for low-temperature equipment can be seen in Fig. 70. Fig. 70(a) shows a Dewar vessel for operation at temperatures down to that of liquid nitrogen, based on an original design of Shull and Wollan. The inner container is filled with liquid nitrogen, automatically replenished at intervals by a pump, and the sample is supported beneath it. Conduction between the outer and inner cases is reduced by evacuating the intervening space with a diffusion pump and by making the inner container of thin stainless steel. Radiation losses are reduced by surrounding the sample with a radiation shield of copper or aluminium foil. Fig. 70(b) shows a more elaborate

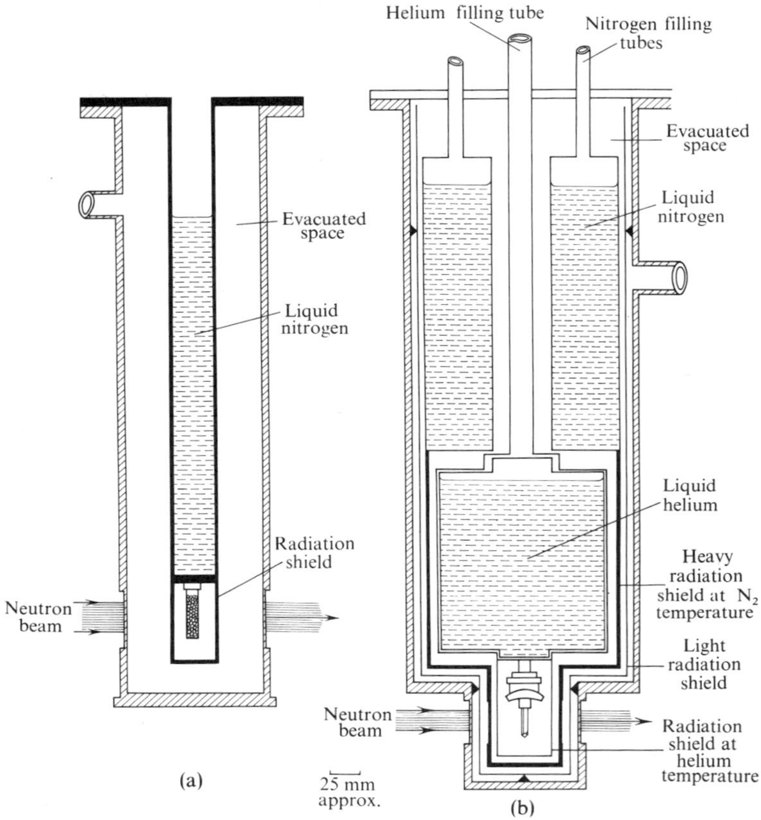

FIG. 70. The general construction of simple cryostats for measurements at (a) liquid-nitrogen temperature, and (b) liquid-helium temperature, following original designs by Shull and Wollan.

arrangement which has an inner reservoir containing liquid helium surrounded by an outer reservoir for liquid nitrogen. Such an apparatus can be used for temperatures down to about 1·9 K and has been described in more detail by Erickson (1953) and Hastings and Corliss (1956*a*). The first low-temperature 'dewars' were designed for using powdered samples, usually without rotation of the sample. They have been developed for using single crystals, incorporating angular movements for aligning the crystals, and for the application of magnetic fields during the measurements

(Wollan and Koehler 1955; Wollan, Koehler, and Wilkinson 1958*a*).

Fig. 71(a) illustrates a design of the Societé T.B.T. of Grenoble in which the sample is actually immersed in the liquid helium. With

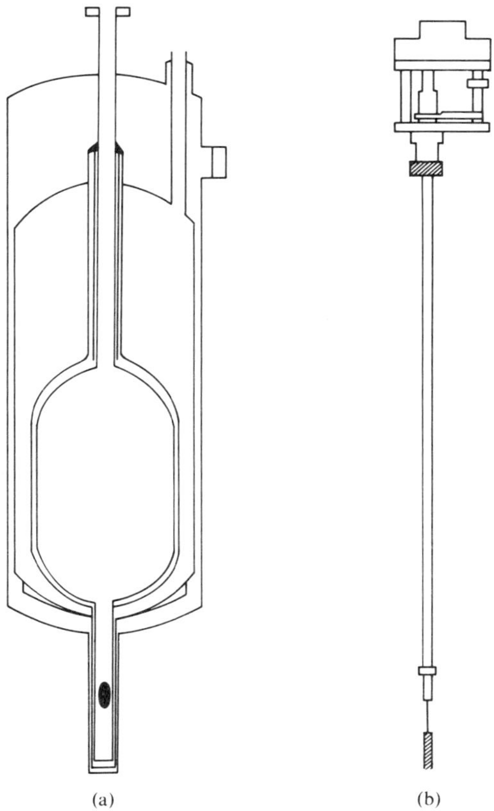

FIG. 71. Sketch (a) shows a liquid helium cryostat manufactured by the Societé T.B.T. in which the sample is immersed in the liquid. Sketch (b) shows a device for rotating the sample at 4 revolutions min^{-1}. (N. Cowlam.)

this arrangement there is some scattering of the incident beam by the helium, but this is not large enough to be troublesome and the design has the distinct advantage that the sample can readily be changed without any dismantling of the cryostat and, therefore, without much loss of the liquid helium. This is conveniently done by mounting the sample at the end of a thin-walled tube as shown in Fig. 71(b). The tube carries a small electric motor which enables the sample to be rotated in the liquid helium at about 4 rev min^{-1} in order to reduce effects due to any preferred orientation of the crystallites. The sample can then be removed from the cryostat when the measurements are complete.

An obvious advantage of designs of cryostat which employ a very narrow tail at the lower end is that the tail can be accommodated between relatively closely-spaced poles of a magnet, thus simplifying the production of a high magnetic field. When this arrangement is adopted it is highly desirable to construct the tail of the cryostat with vanadium, which gives very little coherent scattering and, accordingly, no observable spurious diffraction peaks. On the other hand, when it is not desired to insert the cryostat between the poles of a magnet it will generally pay to make the vacuum jackets and radiation shields of large diameter. The reason for this is made clear in Fig. 72. The incident neutron beam will illuminate the cylindrical surfaces only in the neighbourhood of points A and B, and the diffracted beam from these points will not enter the counter, which is directed radially to the central axis at C. Under these circumstances it is better to construct the cylinder of aluminium rather than vanadium, since the large incoherent scattering from the latter would enter the counter and be troublesome at low angles.

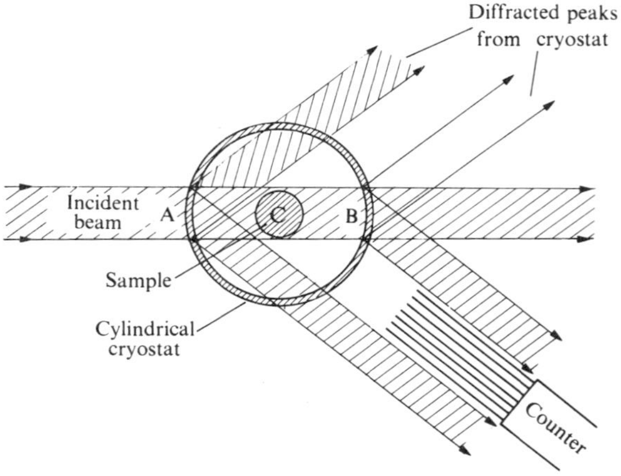

FIG. 72. Spurious diffraction peaks from a cryostat are not detected by the counter if the cylindrical wall is much larger in diameter than the sample.

The apparatus shown in Fig. 70(a) can also be used, without any disturbance of the sample, at elevated temperatures, by replacing the liquid nitrogen with an oil and immersing a heating element in it. This arrangement is very simple for temperatures of up to about 200 °C, but for higher temperatures it is more convenient to mount the sample in a vertical furnace tube with heating elements above and below it. A suitable arrangement is indicated in Fig. 73, where the sample is oscillated from below by a shaft driven by an electric motor and a thermocouple is attached to the sample from above,

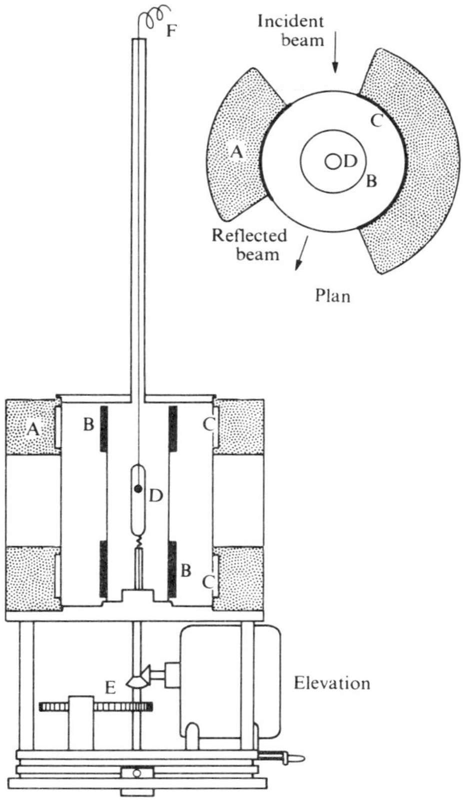

FIG. 73. Sketch of simple furnace for neutron diffraction measurements at elevated temperatures. A indicates calcium silicate insulation; B, C heating elements; D sample on rotating shaft; E reversing drive; F thermocouple. (N. Cowlam.)

to give a direct measurement of the temperature. As for the low-temperature cryostat, it will usually be preferable to make the containing cylinders of large diameter.

Heaton, Mueller, Adam, and Hitterman (1970) have described a cryo-orienter for single crystals which combines a double Dewar cryostat with a three-circle goniometer and which has a temperature stability of 0·01° between liquid helium and room temperatures. A full angular rotation of 360° is permitted about the vertical axis and a tilt of ±50° is possible before trouble is experienced through spilling of the coolant. This allows three-dimensional data to be collected over three-quarters of all reciprocal space. The high accuracy of temperature control means that this apparatus is of great value in studying phase changes. Fig. 74 shows, as an example, the variation with temperature of the (110) reflection of uranium phosphide UP which undergoes a magnetic transformation near 22 K.

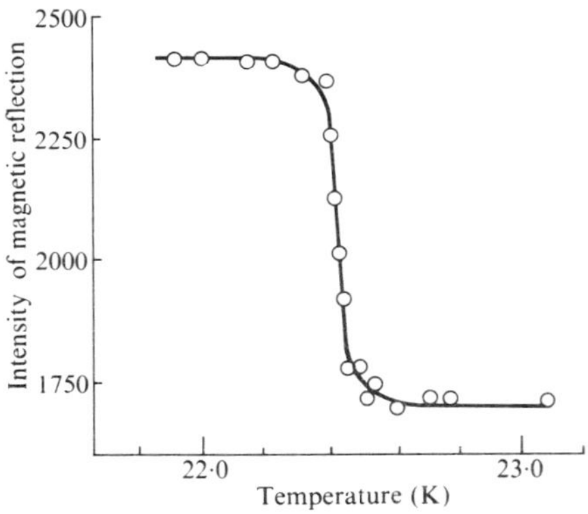

FIG. 74. Intensity variation of (110) reflection of uranium monophosphide with temperature. (Heaton *et al.* 1970.)

4.9. Use of 'cold' and 'hot' sources

We have discussed earlier how the spectrum of thermal neutrons depends on the temperature of the moderator in the reactor. Heavy- or light-water research reactors have a moderator temperature of about 313 K, and graphite moderators are usually run at a higher temperature of about 423 K to avoid difficulties due to Wigner energy storage brought about by radiation. In either case the peak of the wavelength spectrum occurs in the region 1·0–1·5 Å, and the distribution falls off rapidly outside this wavelength range. Fig. 75 is a typical spectrum for a heavy-water reactor, and it will be seen that the number of neutrons has fallen below 10 per cent of the peak value for wavelengths outside the range of about 0·7–

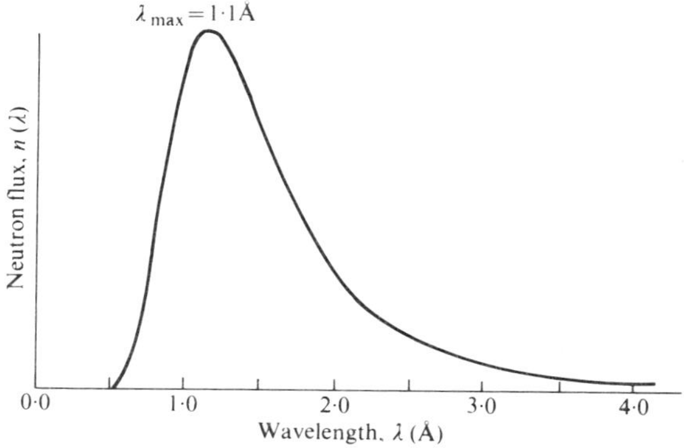

FIG. 75. Typical neutron spectrum from a heavy-water moderated reactor. (Stirling 1973.)

2·8 Å. For some applications it is of great advantage to use wavelengths outside this range, and it becomes essential to increase the number of neutrons of either short or long wavelengths if worthwhile experiments are to be performed. In particular, long-wavelength neutrons are of immense value in the study of both inelastic scattering and the nature of defects, which we shall discuss in Chapters 9 and 17 respectively. The number of long-wavelength, or 'cold', neutrons can be substantially increased by inserting in the reactor a 'cold source' maintained at low temperature. Thus a volume of liquid hydrogen at 20 K, typically of about 300 cm^3 in size and contained in a hemispherical vessel of diameter 10 cm, will increase the number of neutrons in the region of 5–10 Å by an order of magnitude. Experimental assemblies have been discussed by many authors (e.g. Webb and Pearce 1963; Cocking and Webb 1965; Davies *et al.* 1968), and much investigation has been made of the use of different materials for the cold source. Some results by Jacrot (1962) (Fig. 76) show how the enhancement for various selected wavelengths depends on the relative proportions of hydrogen and deuterium in a liquid

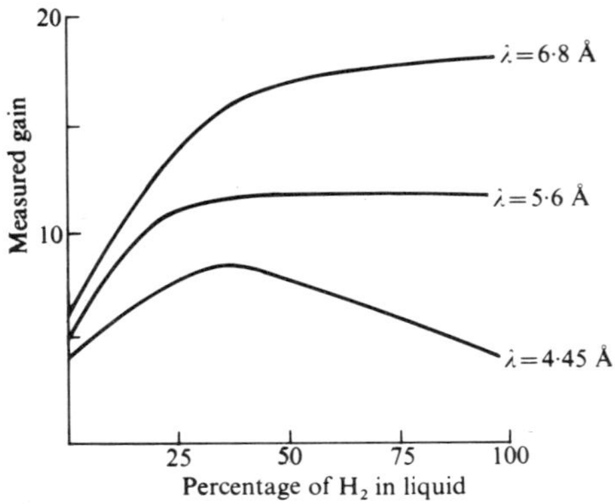

FIG. 76. Enhancement of selected wavelengths for neutrons emitted by 'cold sources' of liquid hydrogen–deuterium. (Jacrot 1962.)

cold source. At the high-flux reactor at Grenoble a volume of 25 l of deuterium constitutes the cold source.

At the other end of the wavelength scale 'hot' sources can be used to increase the number of short-wavelength neutrons. This is of great advantage when excitations and vibrations of relatively high energy are being studied by inelastic scattering or when the double-differential scattering cross-section of liquids is being examined at high values of momentum transfer, as we shall note later on in Chapter 16. Abeln, Drexel, Glaser, Gompf, Reichardt, and Ripfel (1968) have described a hot source consisting of a cylinder of graphite 18 cm in diameter and 18 cm in height. At a temperature of 1700 K the neutron intensities are expected to be improved below a wavelength of about 0·9 Å, corresponding to an energy of 0·1 eV, with a maximum increase by a factor of seven times at a wavelength of 0·5 Å (0·3 eV). At Grenoble the hot source runs at about 2000 K, being maintained at this temperature by the γ-ray heating of the reactor core.

4.10. Techniques using 'white' radiation: time-of-flight crystallography

We have already mentioned the inefficiency of the standard method of producing diffraction patterns by using a monochromating crystal to select a very narrow band of wavelength. This method utilizes only about 1 per cent of the available neutrons from the reactor.

From the very early days of neutron studies attempts have been made to remedy this situation. Wollan, Shull, and Marney (1948) were the first to attempt the use of photographic recording to produce the neutron-equivalent of an X-ray Laue photograph. A collimated beam, about 0·5 cm square, passes through a crystal plate placed at the collimator aperture and the diffracted beams are detected by an X-ray film which is covered by an indium foil. The blackening of the photographic film is caused by the β-particles which are produced by neutron capture in the indium; the X-ray film is quite insensitive to neutrons in the absence of the activating foil. Laue patterns have been published for a number of substances such as quartz, LiF, calcite, and $NaNO_3$, and exposures of about 10 hours were required for crystals a few millimetres thick.

A much-improved photographic method was developed both by Smith and Peterson, of Oak Ridge National Laboratory, and by Wang and Shull (1962). In each case a scintillating screen made of ^{6}LiF and ZnS is used, backed by high-sensitivity Polaroid film. This combination is several thousand times faster than the original indium-foil method and Laue photographs can be taken with exposure times of a few minutes. The sensitivity is also good enough to permit photographs of powder patterns from polycrystalline samples in a monochromatic beam. Apart from the potential value of the photographic method for recording diffraction patterns, particularly in the early stages of an investigation, it has proved extremely useful in the preliminary lining-up of spectrometers and samples. For example, it is only a matter of a few minutes' work to examine the uniformity of the intensity distribution in a monochromatic beam or to adjust the height and centring of a sample in a metal cryostat. The Polaroid film is of particular advantage in this type of work because of its on-the-spot method of development.

Lowde (1951) has described a method of using white radiation with very small single crystals. In this case a small counter (Lowde 1950) of about 20 per cent efficiency, containing boron-coated foils as the detecting element, is employed. The experimental layout is shown in Fig. 77. A conical collimator of apex angle almost 3° is used and, in the absence of any monochromator, neutrons in the whole of this solid angle reach the counter. For each reflection to be measured the crystal and counter are set, in turn, at the

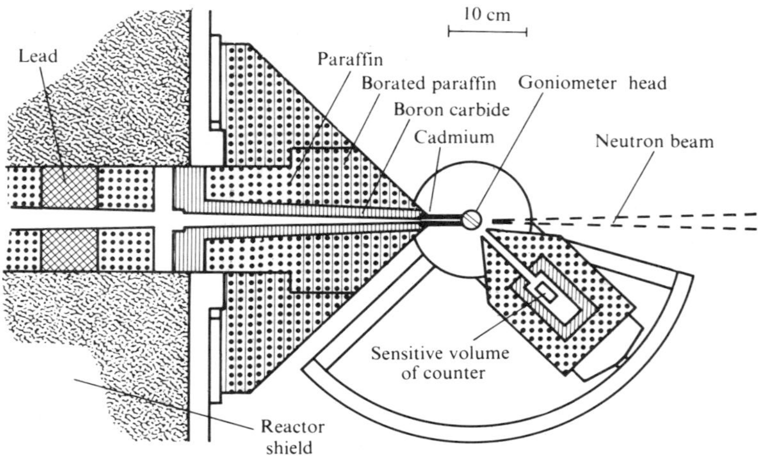

FIG. 77. Single-crystal spectrometer using white radiation. (Lowde 1951.)

appropriate angles for a particular wavelength. The wavelength chosen is that which, determined by the reactor spectrum, gives the maximum counting rate. Over a wide range of interplanar spacings *d* this is independent of *d* and in the experimental work described was 1·45 Å. The actual measurement of integrated reflection is performed by plotting a rocking curve for the crystal after having placed the counter in its nominally correct position by calculation. It can be shown that the peak counting rate of the rocking curve gives the integrated reflection. The fundamental weakness of this method is that the successive orders of reflection from a given set of planes are all reflected, for different wavelengths, to the same counter position. The shape of the reactor spectrum tends to discriminate against higher and lower orders, in comparison with the particular order for which the counter has been set, but large corrections may be necessary in certain cases, depending, of course, on the relative structure factors of the different orders. The method is limited in scope but may be useful for substances with simple structures, since it offers a means of obtaining quite large intensities with very small single crystals. In comparison with the use of an ordinary crystal monochromator in the incident beam there is an intensity increase by a factor of about 10 as a result of the geometrical advantage of the relaxed collimation.

With the development of computers the technique just described offers possibilities as a method for *refining* structures. A crystal is set up in turn to reflect a neutron beam through an angle 2θ from a specific plane such as (111), as in Fig. 78. In general, the counter will, at any one moment, receive several reflected beams which satisfy the Bragg equation, such as (111) for λ_0, (222) for $\frac{1}{2}\lambda_0$, (333) for $\frac{1}{3}\lambda_0$, and so on. For a structure which is known to a reasonable accuracy it is possible to calculate the resulting total number of reflected neutrons which are received by the counter if the spectrum of the incident 'white beam' is accurately known. In Fig. 78 the intensity is measured at intervals of $\frac{1}{2}°$ as θ is increased from 5° to

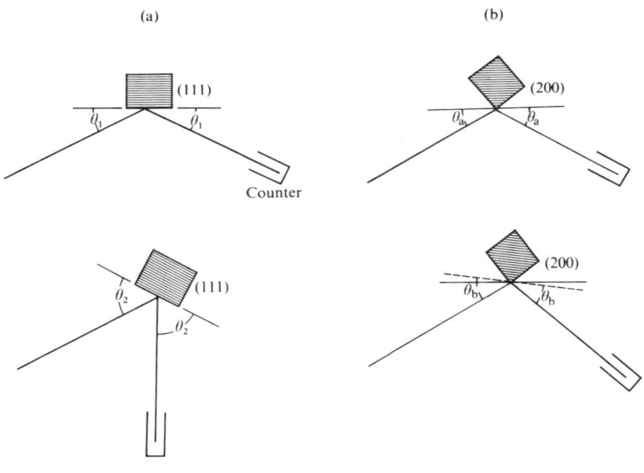

FIG. 78. Composite diffraction patterns of (a) the reflections (111), (222), (333), etc. and (b) (200), (400), (600), etc., using white radiation and rotation of crystal and counter in θ, 2θ synchronism.

60° by rotating the crystal, the counter following to give a θ, 2θ scan. If similar measurements are made with planes such as (111), (200), (220), (311), (331), and (420) for a face-centred structure, there will be more than 600 intensities from which the parameters of the structure can be refined. This procedure has been followed by Hubbard, Quicksall, and Jacobson (1972) and by Wilson and Cooper (1973), the latter applying the method to KCl and hexamethylenetetramine. The conclusions are that, although the method is attractive because of the speed with which the measurements at the reactor can be made it is likely to be much less accurate than the conventional technique with a monochromatic beam. The difficulties arise from uncertain knowledge of the incident spectrum and of the precise way in which absorption, extinction, and thermal diffuse scattering depend on the neutron wavelength.

Time-of-flight crystallography

Apart from providing an extremely useful photographic method of aligning diffraction apparatus, as we have already noted, none of the methods just described made any significant contribution to actual structure investigations before the work of Buras in the early 1960s. This work effectively combined the principles of the Laue diffraction method and the time-of-flight method of measuring the wavelength of neutrons and it was encouraged by a number of new developments. First the growth of studies of *inelastic* scattering, which we shall discuss later, had yielded great advances in time-of-flight technique, and secondly, the appearance of pulsed reactors (as distinct from steady continuous-powered reactors) offered a possibility of applying these new techniques to the study of transient phenomena. Consideration of the possibility of using accelerator sources of neutrons has still further encouraged the development of these methods.

We shall examine first the use of the time-of-flight method of diffractometry for studying powdered materials. The principle is

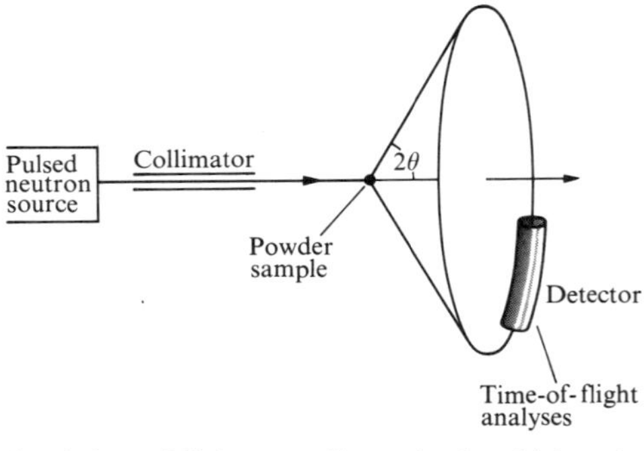

FIG. 79. Principle of time-of-flight crystallography in which a detector collects a substantial proportional of the neutrons in a Debye–Scherrer halo at some constant value of scattering angle 2θ.

proceeds direct from the reactor, without any monochromatization, and falls on a powdered sample. The detector is fixed in position, at some constant angle of scattering 2θ, and may be constructed in such a way that it collected the whole, or a substantial part of, a Debye–Scherrer cone. The detector will collect a diffracted halo from each set of interlayer planes (hkl), but the operative wavelength will be different in each case, since the variables in the Bragg equation are now λ, d instead of θ, d. We shall be able to determine the value of λ for any particular reflection by carrying out a time-of-flight analysis, assuming that the incident beam of neutrons is pulsed and not continuous. While travelling the distance from the chopper, via the sample, to the counter the neutrons are spread out in time and arrive at the counter after the elapse of an interval of time which is proportional to their wavelength. The wavelength λ, of course, is related to the neutron velocity by the equation $\lambda = h/mv$ and v is measured by L/t, where L is the length of the flight path from the chopper to the detector and t is the time taken by the neutrons to traverse this path. The wavelength resolution $\Delta\lambda$, and hence the resolution for the diffraction peaks, will depend on the time-width Δt of the initial pulse, so that

$$\frac{\Delta\lambda}{\lambda} = \frac{\Delta t}{t} = \frac{\Delta t}{L}\frac{h}{m\lambda} = \frac{\Delta t}{L}\frac{h}{2md\sin\theta}. \tag{4.16}$$

Thus high resolution is achieved for large values of the interplanar spacing d. Fig. 80 shows a typical result achieved with a sample of powdered silicon. It will be evident that the angular resolution is best for the larger interplanar spacings, which utilize neutrons of longer wavelength.

We have already shown that the integrated reflection $\mathscr{R}^{\lambda}$ for a stationary crystal in a beam of white radiation is related to the value of $\mathscr{R}^{\theta}$ for a crystal rotating through the reflecting position in a monochromatic beam by

$$\mathscr{R}^{\lambda} = \mathscr{R}^{\theta} 2d\cos\theta = \mathscr{R}^{\theta}\lambda\cot\theta.$$

It then follows from eqn (4.12) that the number of neutrons which enter the counter per minute from a cylindrical powder sample will be

$$\mathscr{P} = \mathrm{i}(\lambda)\frac{\lambda^3 l_{\mathrm{s}}}{8\pi r}\frac{V\rho'}{\rho}\frac{jN_{\mathrm{c}}^2 F^2}{\sin\theta\sin 2\theta}\lambda\cot\theta\,\mathrm{e}^{-2W}A_{hkl},$$

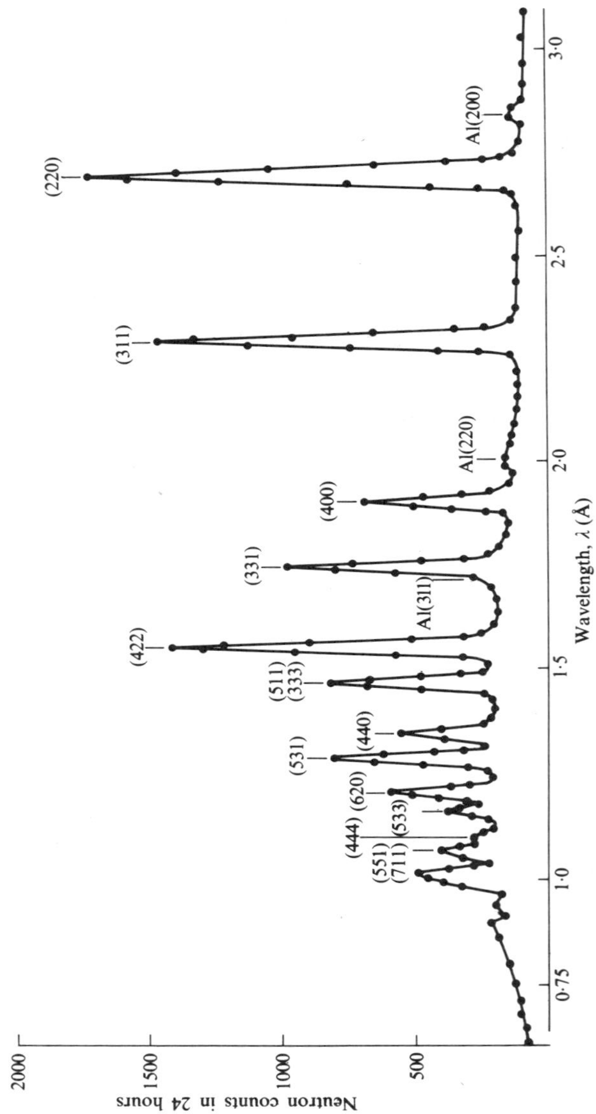

FIG. 80. Diffraction pattern of silicon powder, measured at a scattering angle of 90° with a time-of-flight spectrometer. (Lebech and Mikke 1967.)

where $i(\lambda)\,d\lambda$ is the number of neutrons with wavelengths between λ and $\lambda+d\lambda$ which hit unit area of the sample per minute. This expression reduces to

$$\mathscr{P} = i(\lambda)\frac{\lambda^4 l_s}{8\pi r}\frac{V\rho'}{\rho}\frac{jN_c^2F^2}{2\sin^3\theta}e^{-2W}A_{hkl} \tag{4.17}$$

if the neutrons are collected by a slit of length l_s. If, as is possible, arrangements are made to collect all the neutrons within a Debye–Scherrer ring, then we must multiply $\mathscr{P}$ by $2\pi r\sin 2\theta/l_s$, thus giving

$$i(\lambda)\lambda^4\frac{V\rho'}{\rho}\frac{jN_c^2F^2\cot\theta}{4\sin\theta}e^{-2W}A_{hkl}. \tag{4.18}$$

It is important to notice that not only does a factor λ^4 appear in these expressions, thus giving favourable response for the large interplanar spacings which we have already commented on as receiving good resolution, but also that the neutron spectrum $i(\lambda)$ is also important. It is indeed necessary to know the form of this spectrum in order to compute the structure amplitude factor from the observed intensities. It will also be necessary to make corrections for absorption, and possibly extinction, which are dependent on wavelength, and these features constitute the main disadvantage of this technique.

In Fig. 81 the full curve represents the measured spectral distribution of the incident beam, which can be converted into an effective spectrum, shown by the broken curve, by multiplying by the factor

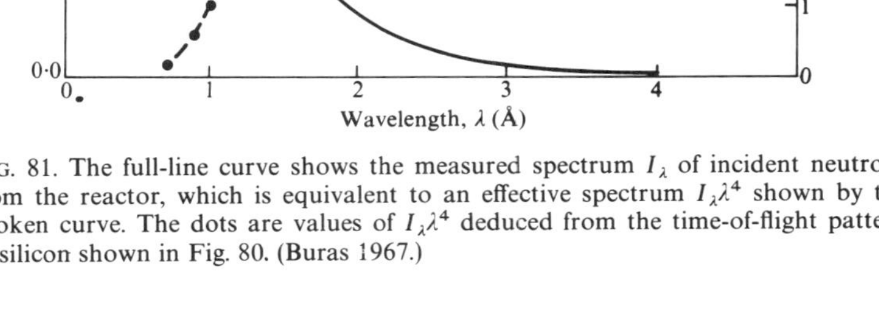

FIG. 81. The full-line curve shows the measured spectrum I_λ of incident neutrons from the reactor, which is equivalent to an effective spectrum $I_\lambda\lambda^4$ shown by the broken curve. The dots are values of $I_\lambda\lambda^4$ deduced from the time-of-flight pattern of silicon shown in Fig. 80. (Buras 1967.)

λ^4 which appears in the above equation. The points on this latter curve are the experimentally determined intensities for the peaks in the diffraction pattern of powdered silicon, giving good agreement with the foregoing interpretation.

In a convenient experimental arrangement at Risö the value of 2θ is chosen to be 90°, so that the scattering position is at right-angles to the incident beam. The layout is indicated in Fig. 82,

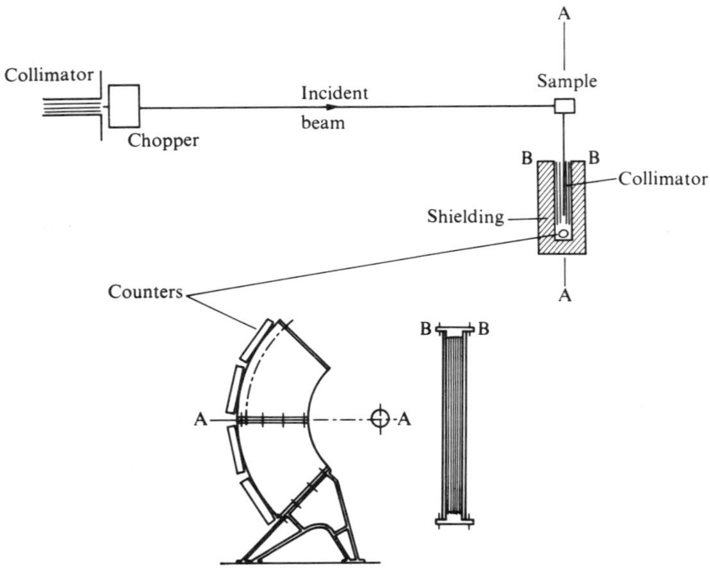

FIG. 82. Diagram of the time-of-flight spectrometer at Risö, in which the diffracted neutrons are observed at $2\theta = 90°$ using an assembly of four counters mounted around the quadrant of a circle and collimated by a set of Soller plates.

and it will be seen from the vertical section at the foot of the diagram that four pairs of counters are used, arranged circumferentially around the Debye–Scherrer halo to intercept a quarter of the latter. For the case of $2\theta = 90°$ the cone is transformed into a plane perpendicular to the incident neutron beam, and it is then an easy matter to construct a secondary collimator which covers the 90° in the vertical plane.

Success of the time-of-flight method of carrying out powder diffraction measurements has led to a study of the adaption of the method for single-crystal investigations. The layout is shown in Fig. 83, where a pulsed and collimated beam falls on a stationary single-crystal sample. Arrangements are made so that all the neutrons which are scattered within a range of angle from $2\theta_{min}$ to $2\theta_{max}$ will enter the detecting counter. We may also say that a wavelength range from λ_{min} to λ_{max} is likewise available. It follows therefore that all the reciprocal lattice points which lie within the shaded area of Fig. 84 will give rise to a diffracted beam which enters the counter. If the acceptance angle of the counter is restricted to the

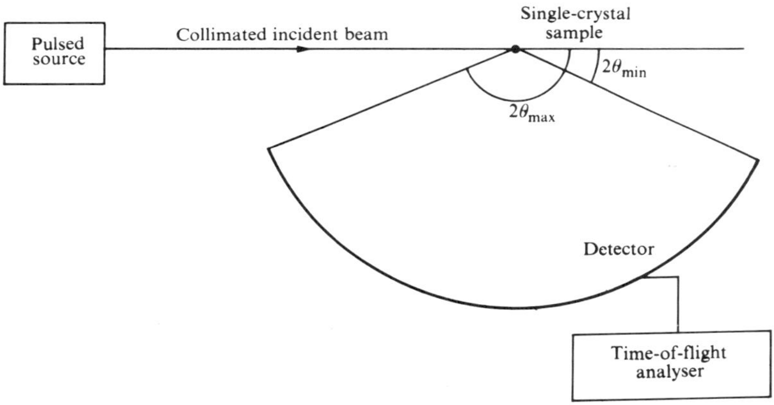

FIG. 83. Principle of a time-of-flight diffractometer for single crystals. (Turberfield 1970.)

neighbourhood of the plane of the paper then it will simply be the planar layer of reciprocal lattice points which are effective, but in principle a three-dimensional counter would collect diffracted spectra corresponding to a three-dimensional volume of reciprocal space. The neutrons which are detected will be within a wide range of wavelength and travelling anywhere within the acceptance angle of the counter, and in order to identify the individual spectra and to measure their intensities it is necessary to determine both the wavelength and the Bragg angle θ.

The wavelength can be determined by the time-of-flight technique, but the determination of θ is more troublesome. In principle it can be determined with a 'position-sensitive' detector and a number of such instruments are under development for use with this and

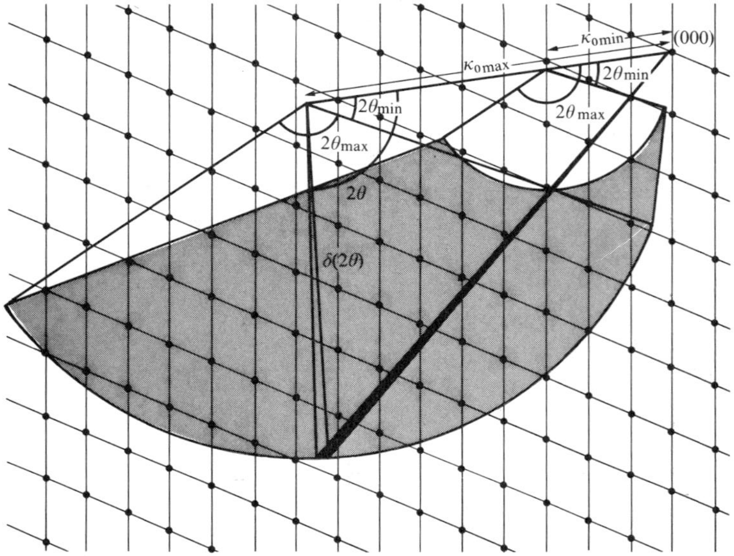

FIG. 84. The shaded area indicates the portion of the reciprocal lattice from which diffracted neutrons can enter the counter in Fig. 83. $2\theta_{min}$ and $2\theta_{max}$ are the angles specified in Fig. 83, and κ_{0min} and κ_{0max} are the extreme wave-numbers $2\pi/\lambda$ of the neutrons available in the incident beam. (Turberfield 1970.)

other techniques. The principle of one such design described by Haywood and Leake (1972) is illustrated in Fig. 85. It may be regarded as a multi-cell BF_3 counter. The over-all length of the counter is about 4 m, consisting of 400 cells of width 1 cm, and it is

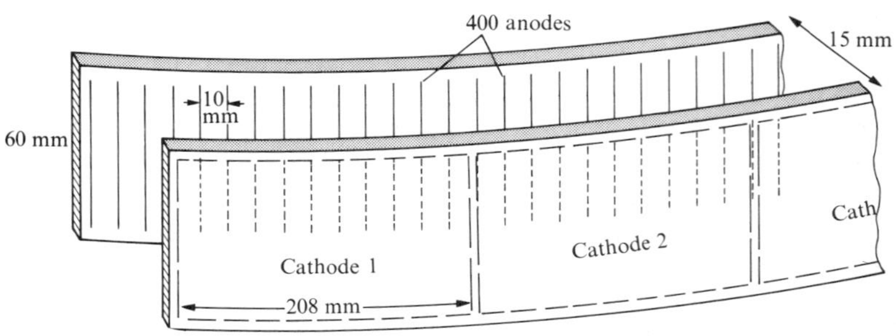

FIG. 85. Sketch of a multi-cell BF_3 counter which locates the angular position 2θ of an arriving neutron. 400 narrow anodes are deposited on the rear plate, facing 20 broad cathodes on the front plate. (After Haywood and Leake 1972.)

arranged circumferentially about the diffracting sample so that each cell subtends about 0·2°, and thus the counter measures the diffraction pattern at intervals of 0·2°. There are 20 cathodes, each with its own amplifier system, and each cathode spans a group of 20 anodes. The latter are connected together in groups of 20 so that numbers 1, 21, 41, 61 ... etc. and 2, 22, 42, 62 ... etc. form groups, each with an amplifier system. If then a coincidence is observed between an anode pulse and a cathode pulse it will be possible to say in which of the 400 cells the neutron arrived. A similar design in operation on one of the powder spectrometers at the Institut Laue–Langevin, Grenoble, utilizes 300 cells in a ^{3}He counter. These are arranged around the circumference of a circle of radius 1·5 m, covering an arc of 60° so that readings of the diffracted intensity are secured at intervals of 0·2° in 2θ. Another design under development utilizes a scintillator screen, made of a lithium-loaded glass and coupled by a lens to a channel-plate electron-multiplier. The resulting electrons are collected by an array of horizontal and vertical crossed wires and an observation of which wires receive pulses at identical times will determine the x- and y-coordinates of an arriving neutron. A simpler arrangement which locates the neutron position in only one dimension can be made by stacking together thin strips of scintillator, optically coupled to a photocathode and followed by a channel plate and an array of metal collection strips. This arrangement promises a resolution of 1 mm, with 100 channels covering a sensitive area of 10 cm × 10 cm.

Meanwhile, other more cumbersome methods have been employed to determine θ and thus to demonstrate the practicability of the time-of-flight method for single crystals. One method is to rotate the crystal through a small angle $\Delta\theta$ about the zone axis and measure the change in wavelength $\Delta\lambda$ for each peak. Since

$$\lambda = 2d \sin \theta,$$

$$\Delta\lambda = \lambda \cot \theta \,.\, \Delta\theta$$

thus giving the value of θ. Accordingly the value of d can be determined and the indices of the reflection identified.

As before, for the powder case, the integrated intensity in a Laue spot from a single crystal will be

$$\mathscr{R}^{\lambda} = \mathscr{R}^{\theta} \lambda \cot \theta,$$

so that in the absence of extinction we have, from eqn (3.3),

$$\mathscr{R}^{\lambda} = i(\lambda)\lambda^4 \frac{N_c^2 F^2}{\sin 2\theta} \cot\theta$$

$$= i(\lambda)\lambda^4 \frac{N_c^2 F^2}{2\sin^2\theta}. \tag{4.19}$$

Thus it is again necessary to know the form of the neutron spectrum, and the calculation of the corrections for secondary extinction, which will vary with wavelength and may be very substantial for the more intense reflections, will be very important.

The advantages of time-of-flight methods from the point of view of fully utilizing the available neutrons are only realized with pulsed reactors, and if the future development of very high flux reactors leads to the pulsed type, then advances in time-of-flight technique may be expected. For a reactor working at constant power the efficiency of the conventional method is reasonably matched to the time of flight. In the former, about 1 per cent of the reactor spectrum appears in the monochromatic beam and can be substantially reflected into the detector by a single crystal plane at any one time. If a continuous beam is chopped then the resulting white beam will again contain about 1 per cent of the neutrons, but these are spread over a wide range of wavelength. Only about 1 per cent of these will appear in any individual diffraction peak, but they will be present throughout the duration of an experiment, as distinct from just for the small fraction of time for which the mechanical rotation of the crystal enables them to satisfy the diffraction condition. If 10 or 20 beams can be collected simultaneously in an extended detector then the time-of-flight method will be at no disadvantage.

Apart from the consideration of efficiency the time-of-flight technique offers two significant advantages. First, when used with a pulsed reactor, it lends itself to the study of relaxation phenomena if the neutron pulses are synchronized with any applied electric or magnetic field which is producing phase changes in a material. Secondly, the advantages of working at a constant angle of scattering are very considerable when measurements are being made at very high or very low temperature or under high pressure. The design of the vessels which hold the sample is greatly simplified since only small windows need be provided to permit the entrance and exit of the neutrons at the single chosen angle of scattering. Some applications have been described by Brugger, Bennion, and Worlton (1967) and Worlton, Brugger, and Bennion (1968).

Correlation techniques

As we have already made clear, any chopping technique is wasteful with a reactor which operates continuously because the chopper is only open for 1 per cent or less of the operating time. The duty cycle can be increased considerably, to approach 50 per cent, by a correlation technique. The principle is to modulate the reactor beam by a chosen pseudo-random sequence of pulses. The observed diffraction spectrum will be a function both of the modulating function and of the normal time-of-flight spectrum, and the latter can be disentangled by suitable analysis from the observed neutron records. A number of different ways are available for applying the modulating function to the initial beam. In one method the circumference of a rotor made of weakly absorbing material, such as an aluminium–magnesium alloy, is impregnated by a random sequence of sections of intense absorber, such as a resin containing gadolinium oxide. A typical arrangement described by Gompf, Reichardt, Glaser, and Beckurts (1968) is shown in Fig. 86, which illustrates a rotor of 51 cm in diameter and bearing two sets of slits which provide 127 basic steps. A second method utilizes the

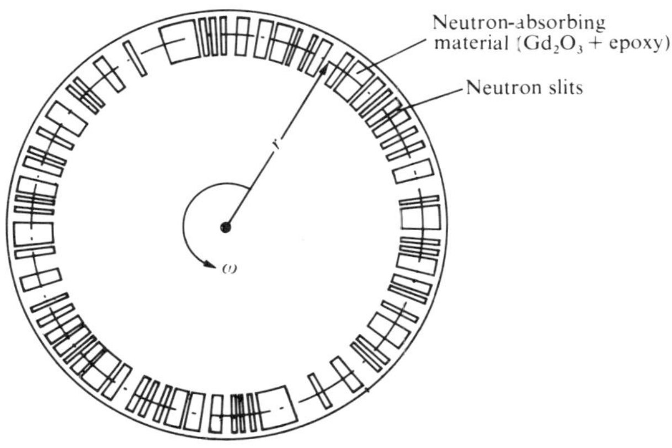

FIG. 86. Plan of the slit system on the rotor of a correlation chopper. The areas outlined are coated with a highly absorbing mixture of Gd_2O_3, the dimensions being chosen to produce a random sequence of pulses. (Gompf *et al.* 1968.)

spin-flip method of varying the polarization of a beam which we shall describe later in Chapter 6. If the electrical oscillator which controls the ratio-frequency field is excited in a suitable random way then the output of polarized neutrons will vary in a corresponding manner.

Other pulsed sources

The attraction of the time-of-flight technique for certain applications has encouraged interest in pulsed neutron sources, and this has again drawn attention to the possibilities of using a linear accelerator instead of a reactor as the primary source of neutrons. For example, pulses of electrons may be accelerated by a voltage of 100 MeV to fall on a target of a heavy element such as tungsten or mercury. Through the intermediary of γ-rays pulses of fast neutrons are produced and can be suitably moderated. An even more efficient way is to utilize beams of protons, which yield neutrons by spallation reactions which they undergo with heavy elements. The more elaborate, and potentially efficient, of these proposals would require immense technological development to bring to fruition, and they have not progressed beyond the design stage. However, a number of actual diffraction measurements have been made with existing accelerators in order to assess the possibility of these techniques (Moore *et al.* 1968; Kimura *et al.* 1969; Sinclair *et al.* 1974). The latter have examined particularly the use of a linear accelerator in studies of amorphous materials.

It is generally considered that, at the present time, the nuclear reactor still remains the more versatile source of neutrons for diffraction and scattering experiments. Nevertheless there are certain experiments for which accelerators offer advantages, giving, for example, improved resolution in powder diffraction patterns for interplanar spacings less than 0·5 Å.

5

FUNDAMENTAL MEASUREMENTS OF SCATTERING AMPLITUDES

WE shall describe in this chapter the fundamental measurements of the scattering amplitudes of the elements. These provide the basis for the interpretation of the neutron diffraction patterns in terms of atomic arrangement, as will be seen in the later chapters devoted to the applications of neutron diffraction techniques. Three main methods of investigation have provided the data on the signs and magnitudes of the scattering amplitudes which have already been listed in Table 2 (see p. 38). These are determinations of refractive index, measurements of cross-section by transmission experiments, and the measurement of the intensities of the coherent Bragg peaks in diffraction patterns. We shall see that it is often easiest to determine the sign of the scattering amplitude by one method and the numerical magnitude by another. It will be convenient to consider first the methods of general application, following these by discussion of particular techniques which have been developed for certain restricted or individual cases.

5.1. General determination of the sign of $\bar{b}$

Measurements of refractive index

The earliest determinations of the sign of the scattering amplitudes were based on measurements of the refractive indices of materials for slow neutrons.

It can be shown that the refractive index n of a material for neutrons is given by

$$n = 1 - \frac{\lambda^2 N\bar{b}}{2\pi}, \tag{5.1}$$

where N is the number of nuclei cm^{-3} and $\bar{b}$ is the average value of their bound coherent scattering amplitude. More explicitly $N\bar{b}$ will be equal to $\sum N_r \bar{b}_r$, where the summation is taken over the various types of nucleus, and N_r is the number of nuclei cm^{-3} for the rth nucleus. Eqn (5.1) expresses the fact that, so far as the refractive index is concerned, the neutron interacts with the whole medium rather than with individual nuclei, and even in the cases of gases it can be shown (Kleinman and Snow 1951) that it is the bound coherent scattering amplitude which is important. The equation can be derived as follows, using the simple method given by Fermi (1950).

Consider a slab of material of infinite width and thickness t, which is much greater than λ_0 the neutron wavelength *in vacuo*, as illustrated in Fig. 87. Let there be N identical nuclei cm^{-3} of scattering length b.

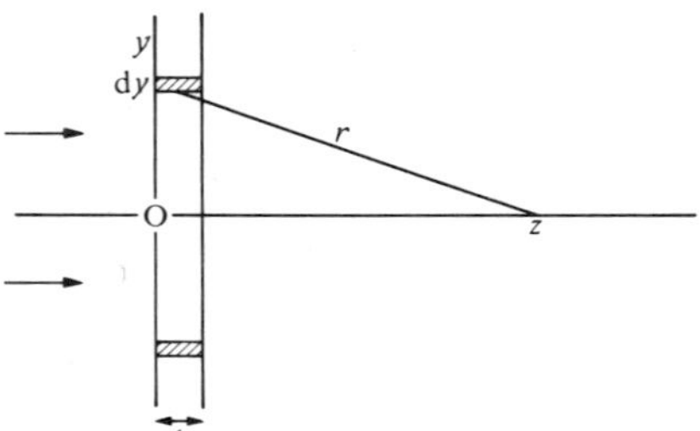

FIG. 87. Calculation of refractive index.

Let $e^{i\kappa z}$ be the incident wave for which the wave-number *in vacuo* is $\kappa = 2\pi/\lambda_0$ and which will be $n\kappa$ in the slab of refractive index n. After passing through the slab, which is assumed to be sufficiently thin for the attenuation to be negligible, the wave will become $\exp[i\{n\kappa t+\kappa(z-t)\}]$. The wave can also be defined by the sum of the initial and scattered amplitudes,

i.e. as $$\exp(i\kappa z)-\int_0^\infty (b/r)\exp(i\kappa r)Nt2\pi y\,dy.$$

Equating the above two expressions and writing $y\,dy = r\,dr$ we have

$$\exp(i\kappa z)\exp\{i\kappa t(n-1)\} = \exp(i\kappa z)-2\pi bNt\int_z^\infty \exp(i\kappa r)\,dr. \quad (5.2)$$

The integral is indeterminate at $r = \infty$, but it is permissible to include a factor $\exp(-\beta^2 r)$ in the integrand, subsequently taking the limit as $\beta \to 0$. In this way the integral is evaluated as

$$(\mathrm{i}/\kappa)\exp(\mathrm{i}\kappa z).$$

Hence (5.2) becomes

$$\exp\{\mathrm{i}\kappa t(n-1)\} = 1-\mathrm{i}2\pi bNt/\kappa,$$

and, equating the imaginary parts in this equation,

$$\sin \kappa t(n-1) = -2\pi bNt/\kappa,$$

which, when t is very small, reduces to

$$\kappa t(n-1) = -2\pi bNt/\kappa,$$

whence

$$n = 1-\frac{\lambda^2 N}{2\pi}\bar{b},$$

as given in eqn (5.1), replacing b by $\bar{b}$ for the general case.

It follows from eqn (5.1) that when the sign of $\bar{b}$ is positive, as indeed it is for all but a few elements as shown in Table 2, then the refractive index will be less than unity. Consequently, when neutrons are incident at a solid surface in air or, more strictly, in a vacuum, they will be passing into a less dense medium and will be capable of undergoing total internal reflection if the glancing angle θ is less than a certain critical value θ_c. In fact the refractive indices of materials are only slightly different from unity for normal wavelengths, $(1-n)$ being of the order of 10^{-6}, so that the critical glancing angle θ_c is only of the order of 10′ of arc for thermal neutrons. For such a small angle it follows from eqn (5.1) that

$$\theta_c = \lambda(N\bar{b}/\pi)^{\frac{1}{2}}. \tag{5.3}$$

Thus a measurement of the critical glancing angle which just gives total reflection of a monochromatic neutron beam at a surface will determine the mean value of b for the nuclei which constitute the reflecting medium. As the glancing angle is increased beyond the critical value the reflectivity falls off extremely rapidly, according to the formula

$$R = \left\{\frac{(n^2-\cos^2\theta)^{\frac{1}{2}}-\sin\theta}{(n^2-\cos^2\theta)^{\frac{1}{2}}+\sin\theta}\right\}^2$$

$$\simeq \left\{\frac{1-(1-\theta_c^2/\theta^2)^{\frac{1}{2}}}{1+(1-\theta_c^2/\theta^2)^{\frac{1}{2}}}\right\}^2. \tag{5.4}$$

The form of this relationship is indicated in Fig. 88 for a value of 10′ for the critical glancing angle θ_c.

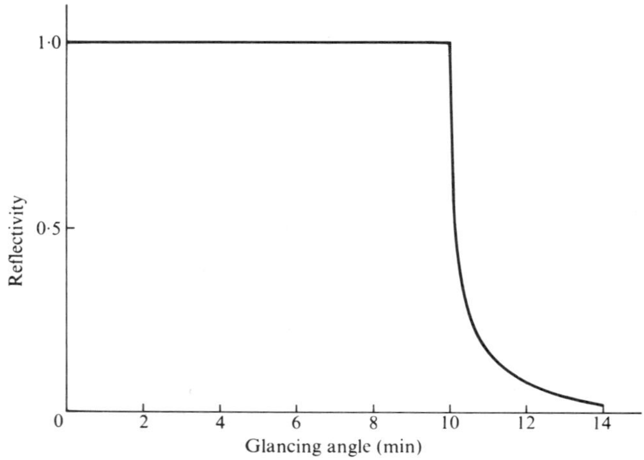

FIG. 88. The variation of reflectivity in the neighbourhood of critical reflection, calculated from eqn (5.4) for a critical glancing angle of 10′.

The first experimental demonstration of these conclusions was provided by Fermi and Zinn (1946). These early experiments were made in 1946 with a beam of thermal neutrons having a Maxwell distribution of velocities, before neutron sources of sufficient intensity to permit monochromatization were available. In this case no limiting angle can be observed, for the successive wavelengths of the Maxwell spectrum drop out of the reflected beam in turn as the glancing angle is increased. However, from the large intensity found in the reflected beam at very small angles it was clear that total reflection was taking place; the small reflection which would have been observed if the refractive index had been *greater* than unity would have been very much smaller. It was therefore inferred from these measurements that the elements Be, Cu, Zn, Ni, Fe, C, of which mirrors were made and examined, had positive values of b corresponding, as has already been emphasized in Chapter 2, to a phase change of 180° on scattering.

The experiment was repeated later by Fermi and Marshall (1947) using a monochromatic beam of wavelength 1·873 Å produced by reflection from a crystal of CaF_2. Before being allowed to fall on the mirror being examined the monochromatic beam was collimated to a width of a few minutes by two cadmium slits, as indicated in Fig. 89, which illustrates the experimental arrangement. Both the direct and reflected beam could be measured by tracking the counter from A to B for any particular setting θ of the mirror surface relative to the incident beam. For small values of θ an intense reflected beam is observed. With increase of θ an angle is reached at which the reflected beam falls rapidly to zero, as we have already shown in Fig. 88. It was thus established conclusively that the six elements listed above has positive values of $\bar{b}$, and from the observed values of θ_c the magnitude of $\bar{b}$ was deduced.

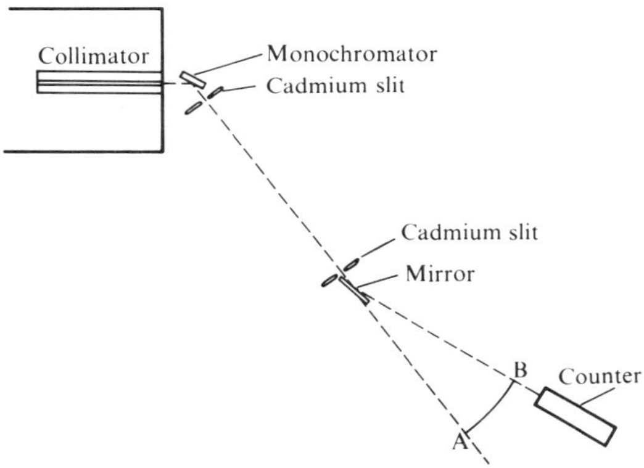

FIG. 89. Fermi and Marshall's demonstration of the total reflection of a monochromatic neutron beam by a mirror. (Fermi and Marshall 1947*a*.)

Neutron guide tubes

The total reflection of neutrons within the critical glancing angle, in accordance with eqn (5.4), is also of importance in the design of collimators and the planning of the layout of neutron instruments at the reactor. In Fig. 90 neutrons emerging from a point A to travel down the collimator AB in diagram (a) might at first be thought to be restricted in angular divergence to those within the shaded cone of $\pm\alpha$. However, if the angular divergence of this cone is less than the critical angle θ_c, then it is possible for other neutrons to be guided out of the tube by total internal reflection. The collimation which is achieved will not be better than $\pm\theta_c$. Fig. 90(b) indicates how the number of emerging neutrons is increased by

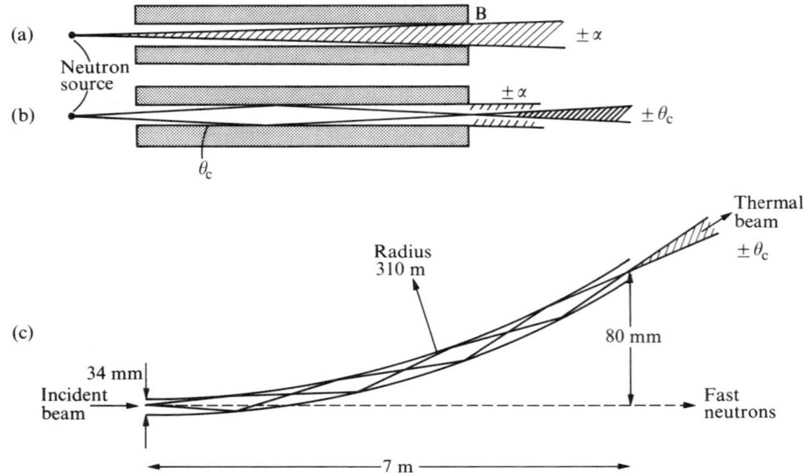

FIG. 90. Diagrams (a), (b) indicate how total internal reflection of neutrons in a collimating tube may increase the angular divergence of the emerging neutrons from $\pm\alpha$ up to $\pm\theta_c$, where θ_c is the critical angle. Diagram (c) shows how in a bent tube the neutron beam, proceeding by reflection at the tube walls, emerges free from contamination by fast neutrons.

this effect but only at the expense of increasing their angular divergence so that they emerge over a solid angle of $4\theta_c^2$.

On the other hand, advantage may be taken of the total reflection to provide a guide tube which will transfer neutrons, without much loss, to locations quite distant from the reactor face. By successive reflections a long tube will permit transmission of neutrons within a solid angle $4\theta_c^2$, instead of within the normal angle hw/L^2 for a tube of height h, width w, and length L; thus a tube with well-polished reflecting surfaces has an effective gain of $4\theta_c^2L^2/hw$ compared with a non-reflecting tube. We note from eqn (5.3) that θ_c is proportional to λ, so that the solid angle of the emerging neutrons, and hence their number, is proportional to λ^2. For example, calculation shows that for copper $\theta_c = 4{\cdot}8'$ for $\lambda = 1$ Å and $\theta_c = 48'$ for $\lambda = 10$ Å.

Of more importance from a practical point of view is an advantage offered by a *curved* tube, since this permits the production of a thermal beam free from contamination by fast neutrons. The neutrons in Fig. 90(c) proceed along the curved tube making successive reflections, at constant glancing angle to the tube wall, and emerge within a solid angle of $4\theta_c^2$ but well-displaced from the direction of the incident beam. Only a negligible proportion of the fast neutrons are reflected and the beam which emerges from the tube is not contaminated. For the tube shown in the figure, with a length of 7 m and a radius of curvature of 310 m, the guided beam is displaced by 80 mm from the centre line of the incident beam. Since the solid angle is proportional to λ^2 these curved guide tubes are very efficient for long-wavelength 'cold' neutrons and in these circumstances they permit scattering experiments to be done under ideal conditions of very low background. At the same time the use of guide tubes permits greater flexibility in the disposition of diffractometers and other scattering apparatus around a reactor. These techniques have been described by Maier-Leibnitz and Springer (1963) and have been much exploited in the design of the layout of the neutron instruments at the Institut Laue–Langevin at Grenoble. Fig. 91 (see between pp. 10–11) shows a group of ten guide tubes which leaves the thermal column of the high-flux reactor and serve a group of more than 20 instruments which are spaciously accommodated in a neutron-guide hall which is quite separate from the main hall containing the reactor. The guide tubes provide a variety of wavelengths. Those for $\lambda = 0{\cdot}88$ Å have a radius of curvature of 27 000 m and serve instruments which are almost 90 m from the face of the reactor. Several tubes for $\lambda = 2{\cdot}8$ Å have a curvature of 2700 m and lengths of about 30 m, and, at the extreme end of the spectrum of neutrons from the 'cold source' in the reactor, there is a tube of length 10 m and radius 25 m which delivers neutrons of wavelength 29 Å. The flux of neutrons received at any individual instrument is naturally less than it would be if the instrument were located close to the reactor face, but, bearing in mind the number of experiments which can be carried out simultaneously and with low neutron backgrounds, the arrangement ensures very full use of the available neutrons.

It will be instructive to conclude this discussion of the refractive index of neutrons with a mention of the production of so-called *ultra-cold* neutrons. Although the term $\lambda^2 N\bar{b}/2\pi$ in eqn (5.1) is normally very small, nevertheless it would become greater than unity if λ were greater than about 700 Å, for typical values of N, $\bar{b}$. Under these circumstances the refractive index actually becomes complex and the situation is formally analogous to the reflection of light from a metal surface. Discussion of the complex refractive index is given by Goldhaber and Seitz (1947) and Lax (1951). The practical conclusion is that for these very long wavelengths the neutron wave penetrates the surface by no more than about one wavelength and is, in effect, totally reflected at all angles. Such neutrons would therefore be unable to escape from a closed vessel and would remain until their disappearance by β-decay. In experiments at Dubna in the U.S.S.R. following earlier observations by Lushikov, Pokotilovsky, Strelkov, and Shapiro (1969), ultra-cold neutrons, in densities of 10 neutrons per litre, have been retained for at least 100 s. From the point of view of the nuclear physicist these experiments are of immense interest as they provide a means of greatly increasing the accuracy of measurement of the possible electric-dipole moment of the neutron.

The relative intensities of Bragg reflections

In simple compounds of two elements and of known structure it is usually possible to infer the relative phases of the two elements by observing the relative intensities of various spectra in the neutron diffraction pattern. For example, if a compound AB has the NaCl type of structure then it would be expected that the odd orders of reflection from the (111) planes would be weak if A and B had scattering lengths of the same sign, with the even orders strong. On the other hand, if A and B have opposite signs then the odd orders would be enhanced relative to the even orders. By examining suitable simple compounds Fermi and Marshall (1947*a*) were able to show that C, O, Fe, Mg, Ba, Ca, S, F, Pb, N, Na, K, Cl, Br, and I all had the same sign, whereas Li and Mn had the opposite sign. Recalling the fact that the mirror reflection experiments had established the signs of Fe and C to be positive it follows that the above group of 15 elements all have a positive sign for the scattering amplitude, whereas Li and Mg have a negative sign.

5.2. General methods of determining the magnitude of $\bar{b}$

It has been mentioned that by measuring the magnitude of the critical angle in a mirror reflection experiment the value of the scattering amplitude can be deduced. For most materials, however, other methods of measuring the numerical value of $\bar{b}$ are more convenient.

Determination of total scattering cross-sections

We have referred in Chapter 2 to the integrated scattering, summed over all directions in space, for a polycrystalline sample, and have described how the total scattering is composed of elastic and inelastic, ordered and disordered terms. We have seen that these various components depend on the σ, $\mathscr{S}$ values for the elements and that they will vary with temperature and wavelength. For a given set of temperature and wavelength conditions the total scattering of all types can be measured by a simple transmission experiment in which is determined the reduction in intensity of a monochromatic beam of neutrons when a known sample of an element is placed in its path. If the beam intensities with and without the sample in position are $\mathscr{I}$, $\mathscr{I}_0$ respectively, then

$$\mathscr{I}/\mathscr{I}_0 = \exp(-\sigma_{\text{eff}} Nt), \tag{5.5}$$

where t is the sample thickness, N is the number of nuclei cm^{-3}, and σ_{eff} is the effective cross-section per nucleus which will be the sum of the scattering terms $E_{\text{el}}(\mathscr{S})$, $E_{\text{in}}(\mathscr{S})$, $E_{\text{el}}(s)$, and $E_{\text{in}}(s)$ together with σ_{a}, the cross-section for true absorption appropriate to the wavelength and material. The inclusion of σ_{a} follows, of course, since true absorption, as well as scattering of all types, will be effective in removing neutrons from the initial incident beam.

Most of the measurements of this type have been made with neutron beams from velocity selectors which permit data to be obtained very readily over a very wide range of neutron energies, from the thermal region upwards. Detailed investigations for many elements have been described in papers such as those of Havens, Rainwater, Wu, and Dunning (1948) and of Bendt and Ruderman (1950), and they have been reviewed by Goldsmith, Ibser, and Feld (1947) and by Adair (1950*b*). Detailed data on cross-sections of all kinds, over a wide range of neutron energies are given in the various volumes and supplements of the Brookhaven National Laboratory publication BNL325, supplied by the U.S. Government Printing Office.

A typical experimental curve is illustrated in Fig. 92, which shows the results of some measurements by Bendt and Ruderman for nickel. For our present purpose the results of this type of measurement can be interpreted most simply at short wavelengths where crystal diffraction effects average out and the coherent scattering is

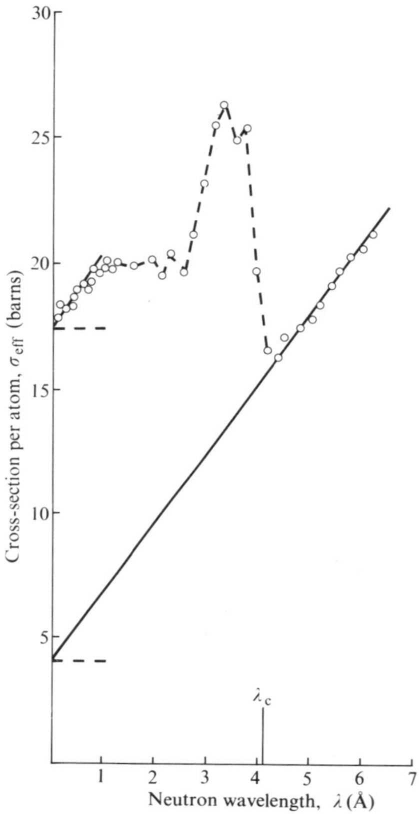

FIG. 92. Variation of cross-section of nickel with neutron wavelength, as measured by transmission experiments. (After Bendt and Ruderman 1950.)

practically isotropic. Moreover, at these short wavelengths, such as $\lambda = 0{\cdot}25$ Å corresponding to a neutron energy a little greater than 1 eV, true absorption, as distinct from scattering, has become negligible, since it falls off according to a $1/v$ law. It follows then that the effective cross-section σ_{eff} which is determined from eqn (5.4) will indeed be equal to the total cross section σ for the element which constitutes the sample. First, however, one correction must be made, for at these and higher neutron energies the nuclei in solids must be regarded as being *free*. Thus the value deduced for σ in this way is that applicable to a free atom and must be increased by the factor $(A+1)^2/A^2$ to give the cross-section for the *bound* atom, which is the appropriate quantity so far as neutrons of ordinary diffraction wavelengths of about 1·1 Å are concerned.

For elements which can show no isotope or spin incoherence we can proceed further immediately, for in these cases

$$\sigma = \mathscr{S} = 4\pi b^2,$$

giving the scattering amplitude b of their single nuclear type. Thus we can determine the value of b for ^{12}C from cross-section measurements with ordinary carbon, it being possible to confirm experimentally (Koehler and Wollan 1952) that the effect of the small percentage of ^{13}C is negligible. In this way Shull and Wollan (1951) have determined absolute values of $\mathscr{S}$, b for ^{12}C, ^{58}Ni, ^{40}Ca, and Th from measurements with diamond, ^{58}Ni, ^{58}NiO, ^{40}CaO, and ThO_2.

For the general case of elements which show spin and isotope incoherence it is possible to determine $\mathscr{S}$ from transmission measurements of σ only if the incoherent scattering cross-section ς can be found. The method can be illustrated by the experimental results of Bendt and Ruderman (1950) for nickel, which have already been shown in Fig. 92. In the short-wavelength region below about 0·7 Å the effect of crystal diffraction on the total cross-section is small, and the measured cross-section σ_{eff} is a linear function of the wavelength. It can be expressed empirically by the equation

$$\sigma_{\text{eff}} = 17{\cdot}4+2{\cdot}77\lambda, \tag{5.6}$$

in which the cross-section is in barns and the wavelength in ångströms. σ_{eff} is to be interpreted as the sum of two quantities, first 17·4 barns, which is the ordinate on the curve for $\lambda = 0$ and will therefore be the *free* scattering cross-section for nickel, and secondly, an absorption term $2{\cdot}77\lambda$ barns, which shows the expected proportionality to λ or $1/v$. This absorption term corresponds to an absorption cross-section of about 2·8 barns for a neutron wavelength of 1 Å, in agreement with the value recorded earlier in Table 6. From the σ value of 17·4 barns for a free nickel atom we infer that σ for the *bound* atom will be 18·0 barns. On the other hand, it is found in the long-wavelength region, beyond the cut-off wavelength for ordered elastic scattering which is seen from Fig. 92 to be about 4 Å for nickel, the experimental cross-section can be expressed as

$$\sigma_{\text{eff}} = 4{\cdot}1+2{\cdot}77\lambda. \tag{5.7}$$

If the effect of inelastic scattering is neglected the cross-section in this region will be the sum of the disordered scattering s and the absorption term, $2{\cdot}77\lambda$, as before. We infer therefore from the form of eqn (5.6) that $s = 4{\cdot}1$ barns. Subtracting s from the determined value of 18·0 barns for σ it follows that the coherent scattering cross-section $\mathscr{S}$ for nickel is equal to 13·9 barns.

When the absorption cross-section of an element is as high as for the case of nickel the accuracy of the determination of s is limited by the long extrapolation back to zero wavelength from beyond λ_c, the cut-off wavelength. The effect of inelastic scattering can be avoided most easily by making the cross-section measurements at reduced temperature, such as at liquid air temperature, since both $E_{in}(\mathscr{S})$ and $E_{in}(s)$ fall off with reduction of temperature. Cassels (1951) by some very accurate measurements of this kind at 90 K has deduced that the value of s for aluminium is zero, so that the value of $\mathscr{S}$ in this case is equal to the σ value of 1·45 barns. Squires (1952) has deduced a value of $s = 0{\cdot}1$ barns for magnesium by the same method.

The determination of the value of $\mathscr{S}$ by measuring both the σ value at short wavelengths and the s value at long wavelengths is particularly valuable for elements which form few compounds of simple structure. In these cases the more generally used method of finding $\mathscr{S}$ from the neutron diffraction intensities may not be very accurate. Before passing to a discussion of this latter method we note for completeness that at the very shortest wavelengths the experimental value of σ_{eff} may increase again owing to the effect of resonance at very high neutron energies. This behaviour cannot be appreciated from curves such as those in Fig. 92 where cross-section is plotted against wavelength but can be readily seen in a plot of cross-section against energy. Numerous curves of this type have been given by Rainwater, Havens, Dunning, and Wu (1948).

Determination of b from crystal diffraction intensities

The great majority of the original data for the coherent scattering amplitudes of the elements were obtained by Shull and Wollan (1951) from their measurements of the powder diffraction intensities of the elements themselves and, by making intercomparisons of various elements, from suitable compounds. Many of the later and more accurate measurements have used the same methods.

In eqns (4.5) and (4.6) of the previous chapter we gave expressions for the diffracted intensities of any reflection plane (*hkl*) for the two commonly used forms of powder diffraction specimen. Since all the quantities other than F^2 in these two expressions can be measured it follows that the value of F^2 can be deduced on an absolute scale. In practice, however, it is easier not to attempt to make an absolute measurement of F^2 in this way but to use the simplified form of equation such as (4.13), determining the instrumental constant by making a similar measurement with a substance of known scattering cross-section. Suitable substances are those containing elements with no isotope or spin incoherence for which the coherent cross-section $\mathscr{S}$ is identical with σ, which can be found directly from a transmission experiment at short wavelengths, as described above. Shull and Wollan (1951) have found that ThO_2, diamond, ^{58}NiO, ^{58}Ni, and ^{40}CaO serve as convenient calibrating substances, giving few and well-spaced reflections and mutual agreement with each other for the value of the instrumental constant. By comparison with these substances ordinary nickel powder has been found to give a coherent cross-section of 13·9 barns and provides a convenient sub-standard for determining the constant, giving particularly intense lines on account of the high value of the cross-section.

Using a calibrating substance of known cross-section in this way it is unnecessary to make dimensional measurements of the counter and spectrometer or to have knowledge of the intensity distribution across the incident beam. In using eqn (4.13), or the corresponding form for a parallel-sided transmission specimen, a diffracting sample of standard dimensions is, of course, always used. From the values deduced for F^2 for the various reflections it is then possible to obtain $\bar{b}$ directly in the case of an element whose structure is known. For a compound a value of $\bar{b}$ can be obtained for one of the elements if the $\bar{b}$ values of all the other constituents are known. In structure determinations where there is uncertainty about the accuracy of $\bar{b}$ for one of the atoms it is usual to include $\bar{b}$ as one of the parameters refined in the final least-squares analysis.

5.3. Subsidiary methods of determining $\bar{b}$

The methods described so far served to provide the first data on the sign and magnitude of $\bar{b}$ for the great majority of the elements and of b for such separated isotopes as were available. In certain cases particular techniques have been developed where the orthodox methods were inaccurate or inapplicable.

McReynolds (1951) described a development of the mirror reflection technique to determine the refractive index of gases and thus measure the scattering amplitudes of the inert gas elements helium and argon. A value for nitrogen was also determined in this way; the ordinary crystal diffraction method is rather inaccurate for nitrogen because of the absence of well-defined nitrogen compounds of simple structure and purity. McReynolds' method depended on measuring the change of intensity in a neutron beam, reflected at a glancing angle from a liquid surface, when a gas is introduced above the surface. Liquid hydrocarbons, such as ethylene glycol and triethylene glycol, were chosen because of the very small difference of their refractive index from unity, brought about because the positive scattering amplitude contributed by the carbon atoms is almost cancelled by the negative contribution from hydrogen. McReynolds and Weiss (1951) applied a similar method to find the scattering amplitude of vanadium, which was known from crystal diffraction measurements to have a value of b smaller by an order of magnitude than that of any nucleus studied. The experiment consisted of reflecting neutrons from a vanadium mirror in nitrogen at various pressures and extrapolating to determine the pressure at which the refractive index of nitrogen was the same as that of vanadium. They deduced a value of $(0{\cdot}028 \pm 0{\cdot}005)$ barns for $\mathscr{S}$ and concluded that the scattering amplitude of vanadium was positive.

Particular interest attaches to the experimental determination of the coherent scattering amplitude of hydrogen, not only because of its importance in determining the intensity of neutron scattering in organic compounds but also because of its significance in making deductions concerning the range of forces in the neutron–proton interaction (Blatt 1948; Blatt and Jackson 1949; Bethe 1949). A value was obtained by Shull, Wollan, Morton, and Davidson (1948) from crystal diffraction measurements with sodium hydride. The result obtained, $\bar{b} = (-0{\cdot}396 \pm 0{\cdot}020)10^{-12}$ cm was limited in accuracy by the difficulty of making precise correction for thermal vibrations. Earlier, Sutton *et al.* (1947) had deduced $\bar{b} = (-0{\cdot}395 \pm$ $\pm 0{\cdot}012)10^{-12}$ cm from measurements of the scattering of very slow neutrons by *ortho*- and *para*-hydrogen at low temperature, a method of investigation which had been discussed by Schwinger and Teller (1937) as mentioned in Chapter 2. Although these two measurements were in agreement the result was too high to be satisfactory theoretically. In view of the possible error in the second measurement on account of *ortho*-contamination of the *para*-hydrogen a further investigation was made by Burgy, Ringo, and Hughes (1951) using the mirror reflection technique. Compared with the crystal diffraction measurement the mirror method has the particular advantage in this case that the effect of thermal vibrations is negligible, since the glancing angle is so small that scattering takes place practically in the forward direction. Thus no knowledge of the Debye temperature is required. Moreover, the measurement of critical angle depends only on the coherent scattering, and there is no interference from the much larger contribution from incoherent scattering. The latter amounts to 79 barns, in contrast to about 2 barns from coherent scattering.

Use of a mirror of hydrogen itself was not practicable since not only would internal reflection, with serious absorption, have been necessary on account of the negative value of $\bar{b}$ but also it would have been necessary to determine the critical angle of about 10′ to an accuracy of 0·05′ in order to obtain an accuracy of 1 per cent in the value of the scattering amplitude. The principle of the method of Burgy, Ringo, and Hughes was to use a liquid hydrocarbon mirror having a refractive index slightly less than unity. Two variations of the method were employed. In the first experiment, described in greater detail in an earlier paper (Hughes, Burgy, and Ringo 1950), a beam of 'white' neutrons was reflected at a critical angle of 5·3′ from a surface of triethylbenzene $C_{12}H_{18}$, this compound being chosen because $\bar{b}$ for carbon is slightly greater than 1·5 times the numerical value of the negative amplitude for hydrogen. The cut-off wavelength λ_c was about 8 Å and was measured accurately by analysing the reflected beam by further reflection at a beryllium mirror. As the latter is rotated, with increase of glancing angle, the reflected intensity remains constant until the critical angle for reflection of the wavelength λ_c by beryllium is reached. The value of λ_c can accordingly be deduced by an application of eqn (5.3). Having found λ_c and knowing θ_c for the liquid mirror reflection, the value of $\bar{b}$ for hydrogen is then obtained from the equation

$$\theta_c = \lambda_c\left\{\frac{N}{\pi}(\bar{b}_C + 1{\cdot}5\bar{b}_H)\right\}^{\frac{1}{2}}, \tag{5.8}$$

where N is the number of carbon atoms cm^{-3}. To obtain 1 per cent accuracy in the value of $\bar{b}_H$ requires only about 3 per cent accuracy in measuring θ_c.

In the second method, the necessity for making an absolute measurement of λ_c was avoided, and the results were independent of any assumption concerning the wavelength distribution of the incident neutrons. The procedure adopted was to determine the values of the incident glancing angle θ which gave a constant reflected intensity, corresponding to a constant critical wavelength, for several liquids having different ratios of hydrogen and carbon atoms. As for eqn (5.8) it follows that

$$\theta^2 = \lambda_c^2\left\{\frac{N}{\pi}\left(\bar{b}_C+\frac{H}{C}\bar{b}_H\right)\right\}, \qquad (5.9)$$

where H/C is written for the ratio of the number of hydrogen and carbon atoms in the liquid. Thus, for any given value of the reflected intensity, i.e. for a constant critical wavelength, θ^2/N will be a linear function of the ratio H/C. Fig. 93 shows the experimental results obtained, using three different values for the constant reflected intensity. The liquids used were triethylbenzene, cyclohexene, and a mixture of benzene and cyclohexane having a H/C ratio of 1·70. The ordinate is the value of θ^2 corrected for the variation of N from liquid to liquid. The straight lines drawn in the Figure all intercept the axis of abscissae at the H/C value of $1{\cdot}748 \pm 0{\cdot}005$. Since, from eqn (5.9), θ will equal zero when $(\bar{b}_C+(H/C)\bar{b}_H)$ is zero, it follows that

$$\bar{b}_C/\bar{b}_H = -1{\cdot}748 \pm 0{\cdot}005.$$

The most recent value of $\bar{b}_C$ is $(0{\cdot}665 \pm 0{\cdot}003)10^{-12}$ cm, which gives

$$\bar{b}_H = (-0{\cdot}379 \pm 0{\cdot}002)10^{-12} \text{ cm}.$$

Within the last few years greatly improved accuracy has been obtained in the determination of scattering amplitudes by a further development of the mirror reflection method, employing a technique first suggested by Maier-Leibnitz in 1962 and usually known as the 'gravity mirror' method. We shall describe this method in some detail in the following section.

5.4. The gravity mirror refractometer: Christiansen filters

In its most precise form this method measures the mean scattering amplitude of the atoms in a liquid spread in a layer about 2 mm

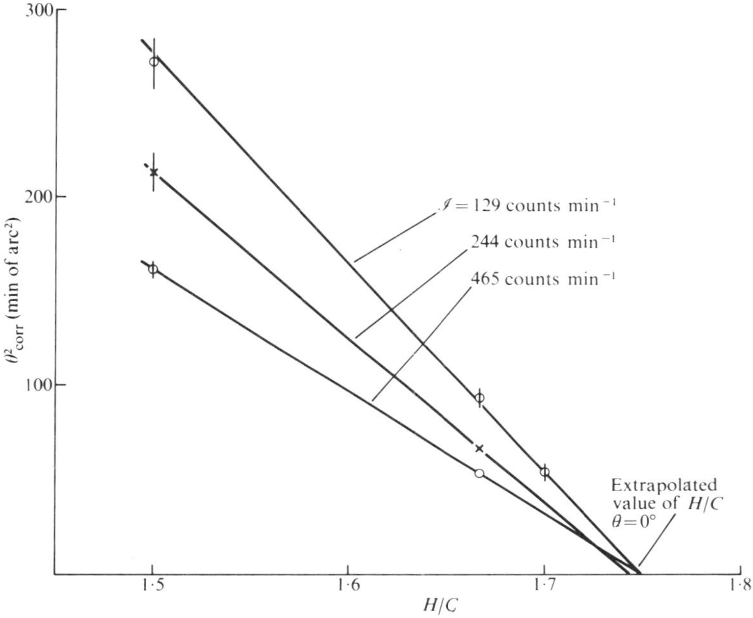

FIG. 93. Variation of θ^2, corrected for changes in N compared with triethylbenzene, with the ratio H/C. The lines are drawn for three different values of the intensity of the neutron beam reflected from the liquid surface. (Burgy, Ringo, and Hughes 1951.)

thick on a glass plate, to cover an area of about 80 cm × 80 cm. In Fig. 94 this mirror is mounted on a substantial support whose height can be adjusted with great precision, within 0·01 mm, and whose surface can be aligned to be very accurately horizontal. The mirror is placed to receive a neutron beam which emerges in a horizontal direction from a collimator in a reactor and then falls under the action of gravity, to describe a parabolic path. The neutrons travel in an evacuated tube over a distance of about 100 m. If the height of the mirror is such that the neutron beam falls upon the mirror at the small glancing angle which just achieves total internal reflection then the refractive index, and hence the scattering amplitude, can be determined. The success of the technique depends on the precision with which the total reflection condition can be defined.

For a flight of length l in Fig. 94 the time of transit to the mirror will be l/v, where v is the neutron velocity. The vertical height fallen

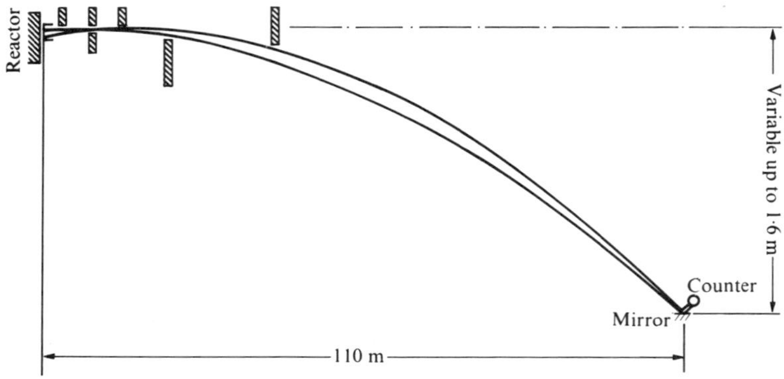

FIG. 94. Diagram of the gravity mirror refractometer. The neutrons fall under the action of gravity as they follow a parabolic path in an evacuated tube of length about 110 m. (Maier-Leibnitz 1962.)

under gravity by the neutron during this period of time will be

$$h_0 = \tfrac{1}{2}gl^2/v^2,$$

and the vertical velocity acquired will be gl/v. Consequently the glancing angle β at which the neutron strikes the horizontal mirror surface is given by

$$\beta = \frac{gl}{v}\bigg/ v = 2h_0/l.$$

Therefore†

$$\beta^2 = 2gh_0/v^2 = 2g(\lambda^2 m^2/h^2)h_0. \tag{5.10}$$

If h_γ is the height which corresponds to critical reflection then from eqn (5.3)

$$Nb/\pi = (2gm^2/h^2)/h_\gamma. \tag{5.11}$$

We note that this value of the fall-height h_γ for total reflection, is independent of the neutron wavelength λ. However, the value of the flight path l at which this fall is achieved certainly does depend on λ and is indeed inversely proportional to it, since l is proportional to v. In the experimental geometry which is usually chosen, with a view to increasing the accuracy of determination, the wavelength of the neutrons which are actually being observed is about 15 Å.

† Note that h is Planck's constant.

For heights less than h_γ the reflectivity will be total, but as the height is increased by lowering the mirror surface the reflected intensity will decrease rapidly, in accordance with eqn (5.4). The latter equation can, by virtue of the above expression for β^2, be rewritten as

$$R_{h_0,h_\gamma} = \left\{\frac{1-(1-h_\gamma/h_0)^{\frac{1}{2}}}{1+(1-h_\gamma/h_0)^{\frac{1}{2}}}\right\}^2 \tag{5.12}$$

for an actual height h_0 under circumstances for which the critical height is h_γ.

Fig. 95 shows some experimental results by Koester (1965) and a calculated curve, from eqn (5.12), for mercury. Although the

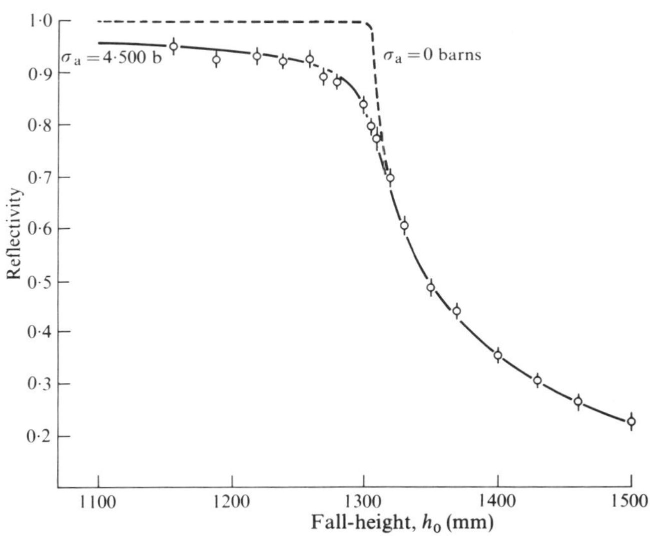

FIG. 95. Gravity mirror data for mercury showing the variation of reflected intensity with the height of fall h_0. The broken line, with a very sharp drop of intensity, is calculated from eqn (5.12) for $h_0 = 1311$ mm; the full line takes account of absorption $\sigma_a = 4500$ barns, as given by eqn (5.13). The circles indicate the experimental measurements. (Koester 1965.)

intensity curve falls away very sharply, and agrees with theory below a reflectivity of 70 per cent, the cut-off is markedly rounded and the reflectivity at fall-heights smaller than critical is significantly less than unity. These effects are due mainly to absorption and incoherent scattering in the material of the mirror: mercury has indeed, as Table 6 has shown, a large absorption coefficient. If these factors are taken into account then eqn (5.12) is modified to

$$R_{h_0,h_\gamma,A} = \left\{\frac{1-(1-h_\gamma/h_0+\mathrm{i}A/h_0)^{\frac{1}{2}}}{1+(1-h_\gamma/h_0+\mathrm{i}A/h_0)^{\frac{1}{2}}}\right\}^2, \tag{5.13}$$

and both refractive index and scattering length have effectively real and imaginary components. The parameter A is equal to $N(\sigma_a+\sigma_{\mathrm{incoh}})(hl/4m)\surd(2/gh_0)$, where σ_a, σ_{incoh} are the cross-sections for absorption and incoherent scattering respectively. Utilization of this equation modifies the calculated curve in Fig. 95 and gives very good agreement with the observed experimental points. In this way an absolute value of the scattering length of mercury was deduced

$$b = (1{\cdot}269 \pm 0{\cdot}002 \times 10^{-12})\ \mathrm{cm}.$$

This technique has been applied by Koester (1967) to measure the mean scattering lengths of the atoms in a variety of organic liquids. In particular by making measurements with *p*-xylene (C_8H_{10}), diphenylmethane ($C_{13}H_{12}$), carbon tetrachloride (CCl_4), tetrachlorethylene (C_2Cl_4), and chlorobenzene (C_6H_5Cl) it has been possible to deduce very accurate scattering lengths for carbon, hydrogen, and chlorine, as follows

$$\text{hydrogen } b = -0{\cdot}372 \times 10^{-12}\ \mathrm{cm},$$

$$\text{carbon } b = 0{\cdot}663 \times 10^{-12}\ \mathrm{cm},$$

$$\text{chlorine } b = 0{\cdot}958 \times 10^{-12}\ \mathrm{cm},$$

all believed to be accurate to three significant figures. Extension to other liquids, including heavy and light water, has yielded values for oxygen and deuterium. Fig. 96 shows some typical experimental results, for tetrachlorethylene. The measurements were done under two different conditions, first in air and secondly in helium, and experimental results and a theoretical curve are shown for both conditions. The theoretical curves, which show good agreement with experiment, are computed for critical heights h_γ of 775·58 mm for air and 770·77 mm for helium. The exactness of agreement leads to belief in an accuracy of $\pm 0{\cdot}2$ mm in these heights, amounting to 0·03 per cent, and an approximately equal accuracy in the value of the scattering lengths. In this particular case the wavelength of

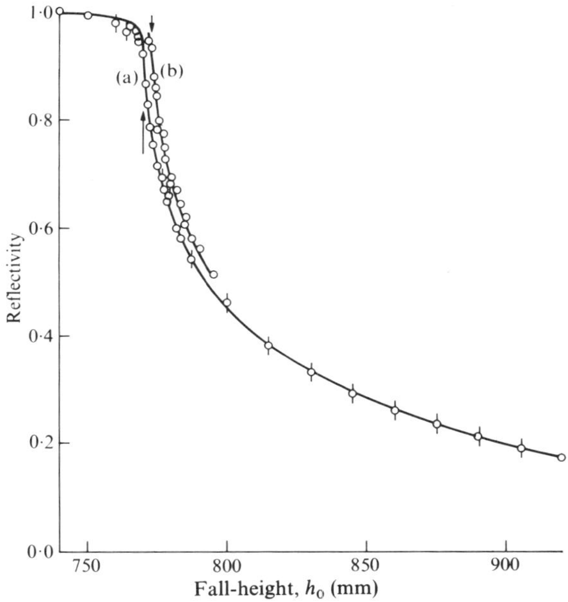

FIG. 96. The variation of neutron reflectivity with fall-height h_0 for tetrachlorethylene. Curve (a) shows measurements in air at 20 °C, and (b) is for helium at 26 °C. The rounding of the cut-off curves (which are calculated) is caused by the absorption of the chlorine atoms, for which $\sigma_a = 300$ barns at $\lambda = 16$ Å. The experimental points are in very good agreement with the calculated curves. (Koester 1967.)

the neutrons which are critically reflected is about 16 Å, and the rounding of the cut-off curve, which is well interpreted by the theory, is due mainly to the absorption cross-section of 300 barns for each chlorine atom for neutrons of this wavelength.

The precision achieved is only possible with liquid mirrors for which the surface is kept precisely horizontal by gravity and compensation is applied to avoid disturbance of the surface by waves due to oscillatory movements of the earth.

Christiansen filters

Koester and Ungerer (1969) have adapted the gravity mirror method to measure the scattering length of the atoms in a *solid* by utilizing Christiansen filters (1884), which consist of a mixture of a powder, with a grain diameter of the order of 10 μm, and a liquid.

When a well-collimated beam of X-rays or neutrons is passed through such a mixture the width of the beam is increased by refraction and diffraction by the individual particles. This broadening of the beam by particles whose linear dimensions are several orders of magnitude greater than the incident wavelength is to be contrasted with the coherent diffracted spectra associated with the various interplanar distances, of the order of a wavelength, and whose intensities are determined by the crystalline structure within the particles.

For neutrons, the details of this small-angle scattering process were first discussed fully by Weiss (1951), who developed a method of using it to measure the phase and amplitude of nuclear scattering. A later discussion was given by Vineyard (1952). The determining factor, as we shall see in more detail in Chapter 17, is ϕ, the difference in radians between the phase change for a neutron wave traversing a particle and for one traversing, in the case of a Christiansen filter, an equal distance in the liquid.

As an example we can consider a mixture of tungsten powder in one of the organic liquids such as C_2Cl_4, which we have considered in the previous section for study in the gravity-fall refractometer. If such a mixture, of thickness about 1·5 mm, is inserted near the entrance slit of the refractometer, as in Fig. 97, it will be found that the reflection curve becomes substantially altered near the critical height h_γ. The effect is illustrated in Fig. 98, from which it will be seen that the reduction of intensity Δr is a maximum at a height h_0, which is about 2 mm less than the critical height h_γ. It can be shown that Δr depends on the phase angle ϕ defined above and given by

$$\phi = \frac{2\pi}{\lambda} . 2R(n_F - n_P),$$

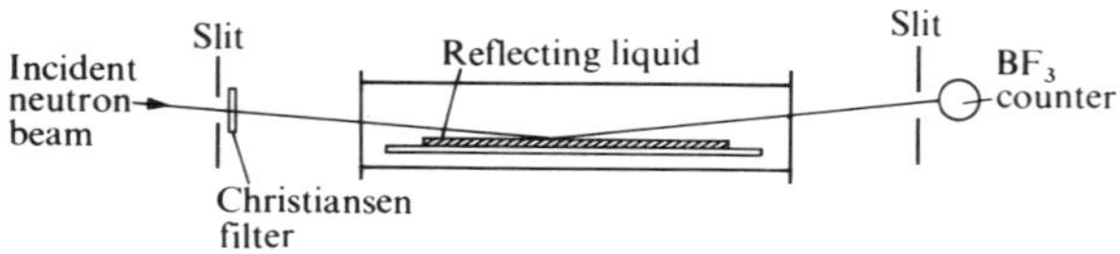

FIG. 97. Measurement of scattering amplitude of a powder by the Christiansen filter method. The filter, consisting of the powder in an organic liquid, is inserted in the neutron beam before it falls on the liquid mirror of the gravity refractometer. The variation of reflectivity with fall-height h_0 is then measured in the conventional way.

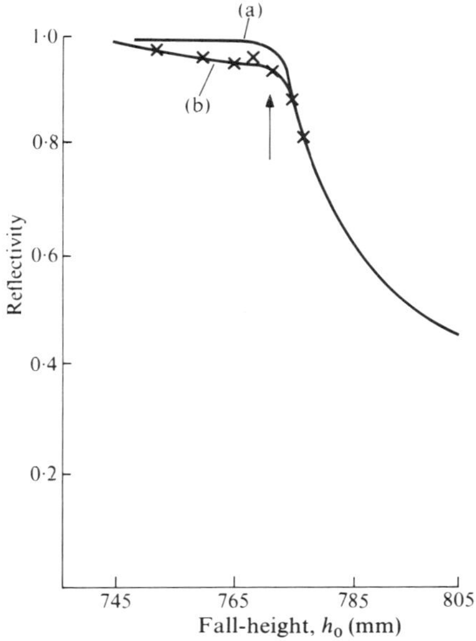

FIG. 98. Gravity-mirror reflection curves for perchlorethylene. Curve (a) is a normal measurement, whereas curve (b) is distorted by the small-angle scattering brought about by first passing the incident beam through a Christiansen filter. The reduction of reflectivity Δr is a maximum for the height h_0 indicated by the arrow, which is a few millimetres less than the normal critical height. (Koester and Ungerer 1969.)

where R is the particle radius and n_F, n_P are the refractive indices of liquid and solid, respectively. It follows from eqn (5.1) that

$$\phi = 2R\lambda(N_F b_F - N_P b_P), \tag{5.14}$$

where N_F, N_P are the numbers of liquid and solid nuclei cm^{-3} and b_F, b_P are their mean scattering lengths. It follows therefore that the reduction of intensity Δr will have a minimum value when the values of $N_F b_F$ and $N_P b_P$ are equal. If therefore we carry out measurements at a fixed height h_0 for a series of liquids, covering a suitable range of values of $N_F b_F$, we shall be able to identify the minimum value of Δr and conclude that for the solid

$$N_P b_P = (N_F b_F)_{min}. \tag{5.15}$$

For tungsten powder, which we took as a suitable example, it is found that mixtures of light and heavy water provide a suitable range of liquids, and Fig. 99 indicates the results of some measurements carried out in the way which we have described. It can be shown that the curve for Δr is symmetrical on the two sides of the minimum, so that these can be extrapolated to give a very accurate value of the H_2O–D_2O ratio at the minimum. The value of $N_F b_F$ for such a mixture will be known from the gravity-fall measurements made as described earlier in this section, and the value of $N_P b_P$ for the powder can then be equated to it.

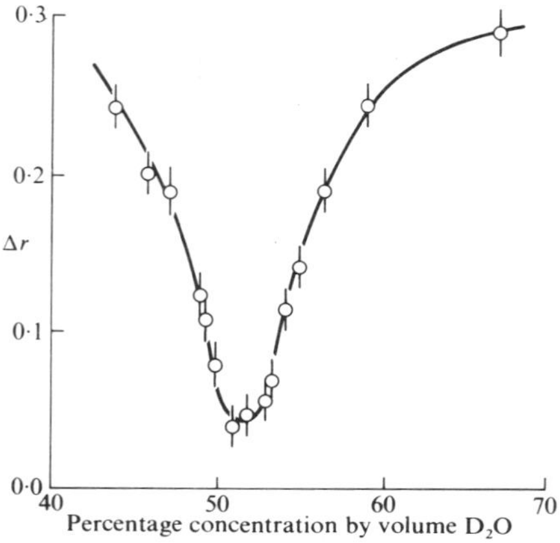

FIG. 99. Variation of the reduction of reflectivity Δr brought about by small-angle scattering with the percentage composition of a light–heavy water mixture, to determine the scattering length of tungsten. Powder of size 6 μm is suspended in the H_2O–D_2O mixture to form the Christiansen filter. (Koester and Ungerer 1969.)

Koester and Knopf (1971) have made a detailed study of the small-angle scattering process and, using powders of the elements, have determined scattering lengths for Al, Cr, Sb, and Bi to an accuracy estimated at $\pm 0{\cdot}3$ per cent. Later work (Koester and Knopf 1972) with powders of halides and sulphates has given good values for the alkali metals and the halogens.

5.5. *Pendellösung* fringes

Another very precise method of determining b for crystals of very small absorption and incoherent scattering has been developed by Shull (1968) from an application of Ewald's treatment of the dynamical theory of scattering to the case of neutrons. This can be appreciated in the following outline; the reader is referred for further details to an article by James (1963).

In Fig. 100 an incident neutron beam falls at an angle θ, close to the Bragg angle θ_B, on the reflecting planes of a crystal slice which is in the symmetrical transmission position and gives rise,

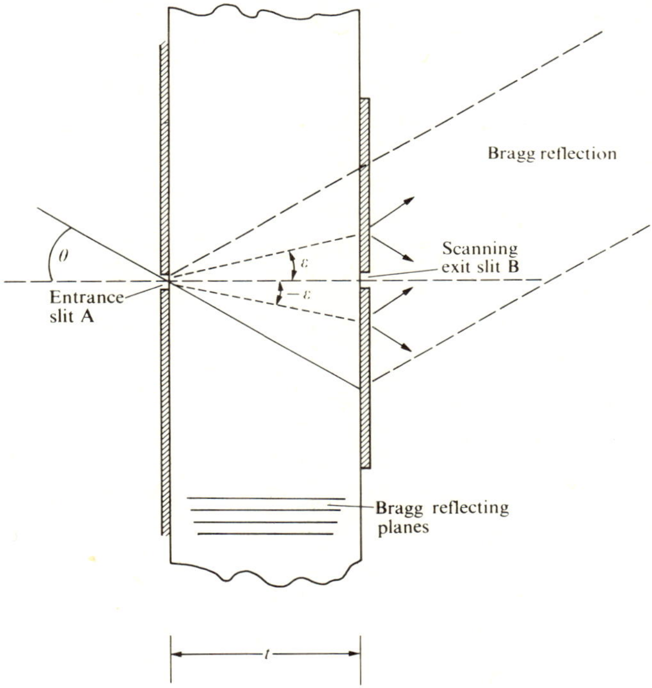

FIG. 100. Observation of *Pendellösung* fringes with scanning exit slit at the surface of a crystal slice. The Bragg reflection planes from which reflection is taking place are perpendicular to the crystal surface. (Shull 1968.)

on the far side of the crystal, to both a reflected and transmitted beam. The reflecting planes which give rise to the Bragg reflection are perpendicular to the faces of the crystal. Within the crystal the direction of energy flow depends on the value of ε, which equals $\theta - \theta_B$, and on the depth below the crystal surface. At the Bragg reflection position the energy flow is along the crystal planes, and the emerging energy oscillates between the transmitted and reflected beams, depending on the thickness of the crystal, thus leading to the name of 'pendulum solution'—*Pendellösung*—for the phenomenon. If the incident beam is more than a few seconds in angular divergence, to cover the whole of the Bragg reflection, and is allowed to fall on the crystal via a narrow slit at A, then the intensity of the reflected beam emerging from the right-hand face of the crystal will vary sinusoidally over the face. This variation can be studied experimentally by scanning the reflected beam with a narrow slit **B** which traverses the surface. It can be shown that the intensity I varies according to the expression

$$I = C(1-\gamma^2)^{-\frac{1}{2}} \sin^2 \{(\pi t/\Delta_0)(1-\gamma^2)^{-\frac{1}{2}}\}, \tag{5.16}$$

where C is a normalization constant, $\gamma = \tan \varepsilon/\tan \theta_B$, and $\Delta_0 = \pi \cos \theta/N\lambda F_{hkl}$. Fig. 101 shows Shull's measured intensity

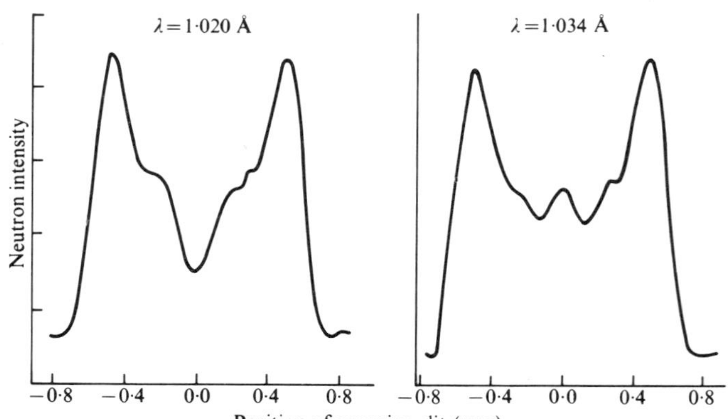

FIG. 101. The intensity distribution within the (111) reflection from silicon plotted against the displacement of the exit slit from the symmetrical position for two different values of the neutron wavelength. (Shull 1968.)

distribution for the (111) reflection across a crystal face of silicon for two different wavelengths of 1·020 Å and 1·034 Å. Much more striking, however, is the variation of intensity at the central zero position of the slit as a function of the neutron wavelength. This is indicated in Fig. 102, which shows results for three different thicknesses of crystal. As the wavelength is varied the tilt of the crystal relative to the incident beam has to be continuously adjusted to maintain the Bragg reflection condition when the slits A, B are in fixed positions on the crystal. For this geometry $\varepsilon = 0$, and from the equations above the intensity will vary according to $\sin^2(\lambda NtF_{hkl}/\cos\theta)$. Thus the change of wavelength from fringe to fringe will be inversely proportional to the thickness of the crystal, as is evident from the Figure. By measuring the θ values of the

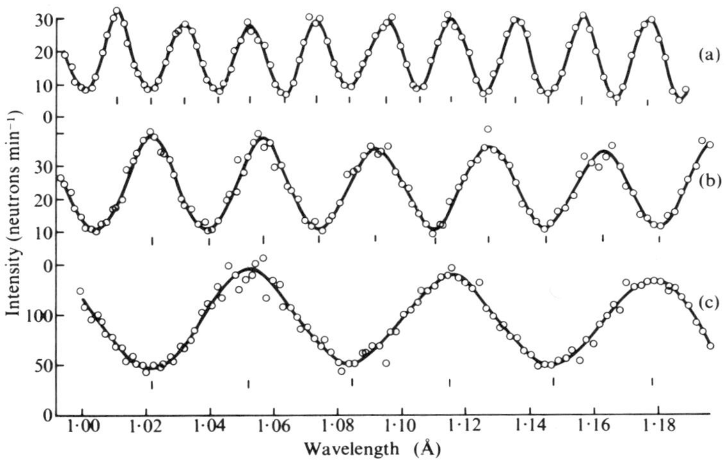

FIG. 102. Change of fringe intensity at the centre of the Bragg reflection with variation of neutron wavelength, the crystal being steadily tilted to maintain the reflection conditions. The curves are for different thicknesses of crystal (a) 1·0000 cm, (b) 0·5939 cm, and (c) 0·3315 cm. (Shull 1968.)

individual fringes in the experiment and knowing accurately the interplanar spacing for silicon it is possible to derive a very accurate value for the structure factor F_{hkl} and, in turn, the scattering length b for silicon. It is necessary to correct for the effect of thermal motion, and uncertainty in this is the primary source of error in the measurement. The corrected value for silicon is $b = (0{\cdot}41646 \pm 0{\cdot}00022) \times \times 10^{-12}$ cm.

In considering a determination of this accuracy it is necessary to consider exactly what is being measured. This value of b is an *atomic* scattering amplitude and will include small terms arising from the Foldy interactions for both the nuclear charge and the electron charge. The electron correction depends on the form factor f and for silicon, which has 14 electrons, will equal $(14 \times 1{\cdot}4 \times \times 10^{-16} f)$ cm, from the discussion on p. 23. This amounts to $0{\cdot}00140 \times 10^{-12}$ cm for the (111) reflection. It follows therefore that the *nuclear* scattering amplitude, including the Foldy nuclear term, is $b_{\text{nucl}} = 0{\cdot}41786 \times 10^{-12}$ cm.

5.6. Complex scattering amplitudes

In section 2.2 we showed that, in the neighbourhood of a scattering resonance, the scattering amplitude becomes a complex quantity, with the consequence that the change of phase on scattering is not 0° or 180° but takes some intermediate value. For most elements the resonances are sufficiently far away from thermal energy for this effect to be unimportant to us, but in a few exceptional cases resonances do occur close to thermal energy, and in these cases there are very significant consequences. The best-known example is the cadmium isotope ^{113}Cd, which has an abundance of about 12 per cent in ordinary cadmium and has a neutron resonance at an energy of 0.178 eV, which corresponds to neutrons of wavelength of 0·68 Å.

It follows from eqn (2.8) that the scattering length is given by

$$b = R + \frac{\frac{1}{2}\Gamma_{\text{n}}^{(\text{r})}/\kappa}{(E-E_{\text{r}}) + \frac{1}{2}\text{i}\Gamma}, \tag{5.17}$$

where R is the nuclear radius, $\Gamma_{\text{n}}^{(\text{r})}$ is the width of the resonance for re-emission, and Γ is the total width of the resonance. E is the energy of the incident neutron and E_{r} is the energy which would give resonance. This expression for b can be expressed as the sum of a real and an imaginary part, by

$$b = \left\{R + \frac{\frac{1}{2}(E-E_{\text{r}})\Gamma_{\text{n}}^{(\text{r})}/\kappa}{(E-E_{\text{r}})^2 + \frac{1}{4}\Gamma^2}\right\} - \text{i}\left\{\frac{\frac{1}{4}\Gamma\Gamma_{\text{n}}^{(\text{r})}/\kappa}{(E-E_{\text{r}})^2 + \frac{1}{4}\Gamma^2}\right\}. \tag{5.18}$$

In the case of a nucleus which possesses nuclear spin I the resonance will occur for one of the two possible spin combinations in the compound nucleus, either $J = I + \frac{1}{2}$ or $J = I - \frac{1}{2}$, and allowance must be made for the appropriate weighting of the two spin states discussed in Chapter 2.

The values of the resonance parameters E_{r}, $\Gamma_{\text{n}}^{(\text{r})}$, and Γ have been found experimentally by making measurements of the variation of σ_{a}, σ_{s} with neutron energy, and for cadmium this has been done by

Brockhouse (1953b). For ^{113}Cd he found that $E_r = 0{\cdot}178$ eV, $\Gamma_a = 0{\cdot}113$ eV, $\Gamma_n^{(r)} = 0{\cdot}00068$ eV, and $R = 0{\cdot}70 \times 10^{-12}$ cm. The nucleus has a spin of $\frac{1}{2}$, and it was found that the resonance corresponded to a compound nucleus of spin $\frac{1}{2}+\frac{1}{2}$, i.e. a spin of unity. At resonance $\lambda = 0{\cdot}68$ Å, leading to a value of $\Gamma_n^{(r)}/\kappa$ equal to $0{\cdot}73 \times 10^{-12}$ eV cm. Therefore we can calculate the real and imaginary components of the scattering amplitude from the above equation, bearing in mind that only a fraction $(I+1)/(2I+1)$ of the nuclei take part in the resonance; this fraction amounts to 0·75. The result of this calculation is given in Fig. 103, which plots the real and

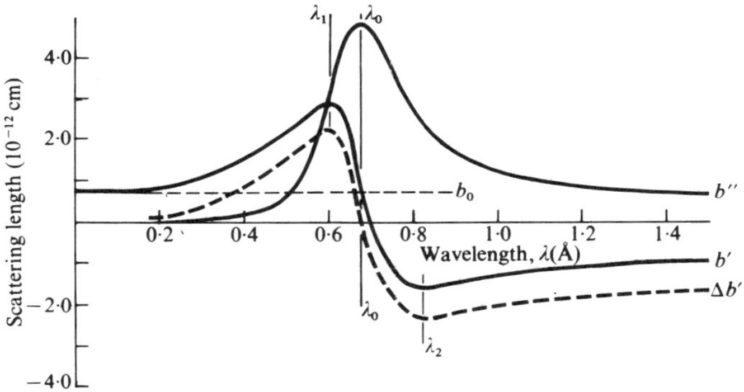

FIG. 103. Calculated values of the real and imaginary components b', b'' of the scattering amplitude of ^{113}Cd in terms of neutron wavelength, for the parameters measured by Brockhouse (1953*b*). The broken-line curve gives $\Delta b'$, the resonance contribution to b', by subtraction of b_0.

imaginary parts b', b'' of the scattering amplitude as a function of the neutron wavelength. b', which is represented by the first group of terms in eqn (5.18), may be regarded as compounded from b_0, equal to the nuclear radius R, and a resonance contribution $\Delta b'$. The latter is shown separately by the broken curve in Fig. 103. At short wavelengths, when E becomes much greater than E_r, $\Delta b'$ becomes zero and $b' = b_0$. It will also be noticed that $\Delta b'$ is equal to zero at the resonance wavelength and is positive and negative, respectively, at lower and higher wavelengths. On the other hand, the imaginary component b'' does not change sign and has its maximum value at the resonance wavelength.

The Breit–Wigner expression for the *absorption* cross-section, complementary to eqn (2.6) which gives the scattering cross-section, is

$$\sigma_{\mathrm{a}} = \frac{\pi}{\kappa}\cdot\frac{\Gamma_{\mathrm{a}}^{(\mathrm{r})}\Gamma_{\mathrm{n}}^{(\mathrm{r})}/\kappa}{(E-E_{\mathrm{r}})^2+\frac{1}{4}(\Gamma_{\mathrm{a}}^{(\mathrm{r})}+\Gamma_{\mathrm{n}}^{(\mathrm{r})})^2}, \tag{5.19}$$

from which it follows, by comparing eqn (5.18), that

$$\begin{aligned} b'' &= \sigma_{\mathrm{a}}\kappa/4\pi \\ &= \sigma_{\mathrm{a}}/2\lambda, \end{aligned} \tag{5.20}$$

so that the imaginary term in the scattering amplitude can be calculated very readily from the absorption cross-section. Moreover, this imaginary term will be negligible if the terms in the first set of brackets in eqn (5.18) are much greater than the term in the second set of brackets,

$$\text{i.e. } b'' \text{ is negligible if } b \gg \frac{\frac{1}{4}\Gamma\Gamma_{\mathrm{n}}^{(\mathrm{r})}/\kappa}{(E-E_{\mathrm{r}})^2+\frac{1}{4}\Gamma^2}; \tag{5.21}$$

thus

$$\frac{\sigma}{4\pi} \gg \frac{1}{16}\left\{\frac{\Gamma\Gamma_{\mathrm{n}}^{(\mathrm{r})}/\kappa}{(E-E_{\mathrm{r}})^2+\frac{1}{4}\Gamma^2}\right\}^2.$$

Therefore from (5.19)

$$\sigma/4\pi \gg \frac{1}{16}\left(\frac{\sigma_{\mathrm{a}}\kappa}{\pi}\right)^2$$

or

$$4\pi\sigma/\kappa^2\sigma_{\mathrm{a}}^2 \gg 1. \tag{5.22}$$

For most nuclei $\sigma, \sigma_{\mathrm{a}} \sim 10^{-24}\ \mathrm{cm}^2$, so that when $\lambda = 1$ Å, for which $\kappa \sim 10^9\ \mathrm{cm}^{-1}$, the expression on the left-hand side of the above inequality is approximately equal to 10^7. Thus the inequality certainly holds and the imaginary contribution to the scattering amplitude is negligible. On the other hand, at the resonance wavelength, the value of σ_{a} for ^{113}Cd is equal to $7750\times 10^{-24}\ \mathrm{cm}^2$ and the left-hand side of expression (5.22) is equal to about two units. Consequently the imaginary contribution to the scattering is very important, as we have already seen. Under these circumstances we can

TABLE 11

Element or Isotope	Abundance	Resonant λ (Å)	b'' (10^{-12} cm)	
			At resonance	At $\lambda = 1$ Å
Cd		0·68	0·58	0·15
^{113}Cd	0·12	0·68	4·70	1·2
Sm		0·92	0·88	0·7
^{149}Sm	0·14	0·92	6·30	5·1
Eu		0·6	1·31	0·07
^{151}Eu	0·48	0·6	2·74	0·15
Gd		1·8	1·2	0·8
^{157}Gd	0·16	1·8	6·60	

readily calculate the value of b'' from eqn (5.20). Table 11 gives some values both for the individual isotopes in which resonances occur and also for the natural elements in which these isotopes are present.

It may be recalled that similar anomalous scattering effects are observed with X-rays in the neighbourhood of the absorption edges of the electron shells, but these effects are very much smaller than for neutrons. In fact the magnitudes of $\Delta b'$ and b'' for X-rays are about equal and amount to about 20 per cent of the value of the scattering amplitude which is effective at wavelengths remote from resonance. This is to be contrasted with the case of ^{113}Cd for neutrons, which we have just examined and where the value of b'' is 7 times as great as b_0, the scattering amplitude at short wavelengths, which has a real value of $0{\cdot}70 \times 10^{-12}$ cm. From a practical point of view the virtue of the large neutron value of b'' is that it provides a method of phase-determination in non-centro-symmetric structures. Some readers may wish to be reminded that in the technique of crystal-structure analysis by diffraction methods a basic difficulty is that the experimental methods give only diffracted *intensities*; it is necessary to convert these into values of amplitude and phase angle by indirect methods in order to pursue the usual process of Fourier synthesis, whereby we build up a picture of the atomic content of the unit cell.

The method of phase determination which the study of the anomalous scattering provides rests on the breakdown of Friedel's law, which states that even in a non-centro-symmetric structure the reflections (hkl) and ($\bar{h}\bar{k}\bar{l}$) from opposite sides of a crystallographic plane will be of equal intensity. The equality no longer holds

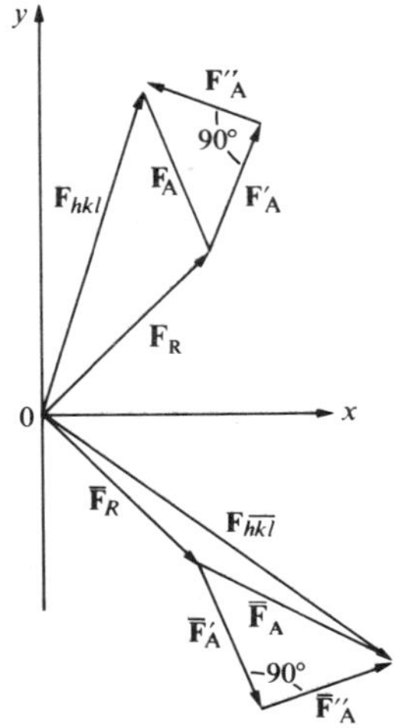

FIG. 104. Addition of the structure amplitude vectors from atoms in a non-centrosymmetric structure will produce different resultants for the reflections (hkl) and ($\overline{hkl}$) if one of the amplitudes is complex. $\mathbf{F}_R$ is the contribution from those atoms whose scattering is completely real.

when the structure includes an atom which has a complex scattering amplitude. Fig. 104 shows how the addition of two structure amplitude vectors will produce different resultants for F_{hkl}, $F_{\overline{hkl}}$ if one vector is complex, with real and imaginary components. Very often the coordinates of an anomalous scatterer, such as cadmium or samarium, in a structure will be known, so that F_A in Fig. 104 will be known in both magnitude and phase. In order, therefore, to determine the phase of F_{hkl}, $F_{\overline{hkl}}$ we have to solve the vector equation

$$\mathbf{F}_{hkl} - \mathbf{F}_{\overline{hkl}} = 2\mathbf{F}''_A. \tag{5.23}$$

Fig. 105 shows how this equation may be solved graphically, but only to the extent of leaving an ambiguous choice between two possible solutions with magnitudes and phase angles F_1, α_1 and F_2, α_2 respectively. The ambiguity can be resolved by making measurements at two different wavelengths, preferably one either side of the resonance position in order to take advantage of the change of sign of $\Delta b'$ which occurs there. Fig. 106 shows a vector diagram, of the type given in Fig. 105, for three different wavelengths. λ_0 is the resonant wavelength, and λ_1, λ_2 are the values which give the maximum and minimum values in the curve of Fig. 103 for $\Delta b'$. For λ_0, $\Delta b'$ is zero, and the two values OA, OB for F_{hkl}, $F_{\bar{h}\bar{k}\bar{l}}$ are compounded simply by adding F_R to either GA or GB. For λ_1 however the two values OC, OD for F_{hkl}, $F_{\bar{h}\bar{k}\bar{l}}$ are derived by compounding $F_R+(b')_1$ with either HC or HD, and similarly for λ_2 we compound $F_R+(b')_2$ with either KE or KF. Thus we have to draw our pairs of circles about A, B for λ_0, about C, D for λ_1, and about E, F for λ_2, and the right choice of the two possibilities is the one for which all three circles intersect at a single point O, such that OG gives the vector F_R for the real-scattering atoms.

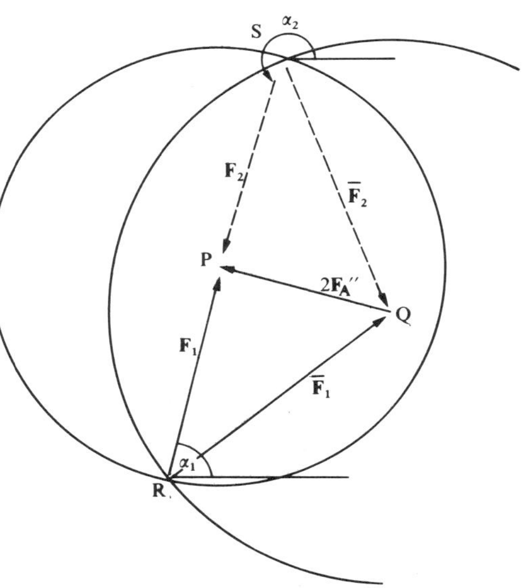

FIG. 105. If the position of the anomolously scattering atom is known then the vector $2\mathbf{F}''_A$ is known in magnitude and phase and can be marked off as PQ in the diagram, zero phase being taken as horizontal. If circles are drawn about P and Q with radii of F_{hkl} and $F_{\bar{k}\bar{h}\bar{l}}$ respectively they will intersect at two points R and S. At each of these points the equation $\mathbf{F}_{hkl}-\mathbf{F}_{\bar{h}\bar{k}\bar{l}} = 2\mathbf{F}''_A$ will be satisfied. For either solution the phase angles can then be read from the diagram. (Dale and Willis 1966.)

Dale and Willis (1966) have discussed how this method could be applied to the further refinement of the structure of insulin, whose structure was first announced by Hodgkin and her colleagues (Adams,

Blundell, Dodson, Dodson, Vijayan, Baker, Harding, Hodgkin, Rimmer, and Sheat, 1969). The intention is to incorporate the ^{149}Sm isotope, listed in Table 11, in the insulin molecule.

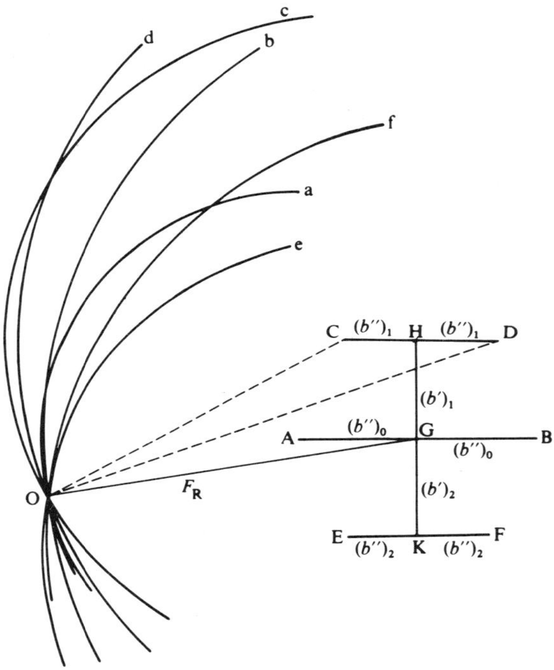

FIG. 106. Composite vector diagram for the F_{hkl}, $F_{\overline{hkl}}$ data from the three wavelengths λ_0, λ_1, and λ_2 marked in Fig. 103. Circles of the kind shown in Fig. 105 have to be drawn from points A, B, C, D, E, F, which are positioned by the quantities F_R, $(b'')_0$, $(b')_1$, $(b'')_1$, $(b')_2$, and $(b'')_2$. The circles are indicated respectively by a, b, c, d, e, and f and all intersect at O. F_R is the contribution from the atoms whose scattering is completely real.

6
THE PRINCIPLES OF MAGNETIC SCATTERING

6.1. Introduction

AT the beginning of Chapter 2 we explained that, whereas in general the scattering of neutrons by atoms was a nuclear process, nevertheless an exception occurred with magnetic atoms, in the case of which there is additional scattering on account of an interaction between the neutron magnetic moment and the magnetic moment of the atom. We return in this chapter to a discussion of the principles of magnetic scattering and will proceed in the following two chapters to a detailed account of magnetic form-factors and of the way particular types of magnetic structure are revealed.

The determination of the magnetic structures of materials is a task which can achieved only by making measurements of the scattering of neutrons. Indeed it is true to say that our whole conception of the existence of a 'magnetic architecture' within the atomic structure of solids has arisen from observations with neutrons.

The elements of the first transition series which includes iron, cobalt, and nickel, have incomplete 3d shells, as indicated in Table 12 which lists the electronic structures of the free atoms with atomic numbers 19–30. The arrangements of the 3d and 4s shells of some free atoms and ions are shown in Table 13 which also gives the number of unpaired electrons and the spectroscopic ground terms of the ions. These unpaired electrons give rise to a resultant magnetic moment. Interaction of this with the magnetic moment of the neutron, which has a spin quantum number of $\frac{1}{2}$ and a magnetic moment of 1·9 nuclear magnetons, produces neutron scattering which is additional to that produced by the nucleus.

It is worth noting in passing that the fact the neutron, which is an uncharged particle, possesses a magnetic moment is anomalous. A likely explanation is that a neutron spends part of its time dissociated into a proton and a negative π-meson, and although the centres of their respective positive and negative charges do coincide yet the negative charge is more diffuse. This would cause the neutron

TABLE 12

Electronic configuration of atoms in their free states

Element	Atomic number	K	L		M			N
		1s	2s	2p	3s	3p	3d	4s
K	19	2	2	6	2	6	—	1
Ca	20	2	2	6	2	6	—	2
Sc	21	2	2	6	2	6	1	2
Ti	22	2	2	6	2	6	2	2
V	23	2	2	6	2	6	3	2
Cr	24	2	2	6	2	6	5	1
Mn	25	2	2	6	2	6	5	2
Fe	26	2	2	6	2	6	6	2
Co	27	2	2	6	2	6	7	2
Ni	28	2	2	6	2	6	8	2
Cu	29	2	2	6	2	6	10	1
Zn	30	2	2	6	2	6	10	2

TABLE 13

3d and 4s shell structures of free atoms and ions

		3d					4s	Number of unpaired electrons	Spectroscopic ground terms	S	L	J
Vanadium	V	↓	↓	↓	—	—	⇅	3	—	—	—	—
	V^{2+}	↓	↓	↓	—	—	—	3	$^4F_{\frac{3}{2}}$	$\frac{3}{2}$	3	$\frac{3}{2}$
Chromium	Cr	↓	↓	↓	↓	↓	↓	6	—	—	—	—
	Cr^{2+}	↓	↓	↓	↓	—	—	4	$^5D_0$	2	2	0
	Cr^{3+}	↓	↓	↓	—	—	—	3	$^4F_{3/2}$	$\frac{3}{2}$	3	$\frac{3}{2}$
Manganese	Mn	↓	↓	↓	↓	↓	⇅	5	—	—	—	—
	Mn^{2+}	↓	↓	↓	↓	↓	—	5	$^6S_{5/2}$	$\frac{5}{2}$	0	$\frac{5}{2}$
	Mn^{3+}	↓	↓	↓	↓	—	—	4	$^5D_0$	2	2	0
	Mn^{4+}	↓	↓	↓	—	—	—	3	$^4F_{3/2}$	$\frac{3}{2}$	3	$\frac{3}{2}$
Iron	Fe	⇅	↓	↓	↓	↓	⇅	4	—	—	—	—
	Fe^{2+}	⇅	↓	↓	↓	↓	—	4	$^5D_4$	2	2	4
	Fe^{3+}	↓	↓	↓	↓	↓	—	5	$^6S_{5/2}$	$\frac{5}{2}$	0	$\frac{5}{2}$
Cobalt	Co	⇅	⇅	↓	↓	↓	⇅	3	—	—	—	—
	Co^{2+}	⇅	⇅	↓	↓	↓	—	3	$^4F_{9/2}$	$\frac{3}{2}$	3	$\frac{9}{2}$
Nickel	Ni	⇅	⇅	⇅	↓	↓	⇅	2	—	—	—	—
	Ni^{2+}	⇅	⇅	⇅	↓	↓	—	2	$^3F_4$	1	3	4
Copper	Cu	⇅	⇅	⇅	⇅	⇅	↓	1	—	—	—	—
	Cu^{2+}	⇅	⇅	⇅	⇅	↓	—	1	$^2D_{5/2}$	$\frac{1}{2}$	2	$\frac{5}{2}$
Zinc	Zn	⇅	⇅	⇅	⇅	⇅	⇅	0	—	—	—	—
	Zn^{2+}	⇅	⇅	⇅	⇅	⇅	—	0	$^1S_0$	—	—	—

to behave as though it is a *negative* charge which is rotating in the direction of the spinning neutron top, i.e. the magnetic moment is negative.

There is also magnetic scattering by atoms and ions of the rare-earth group, which have magnetic moments by virtue of an incomplete 4f shell. These N-shell electrons are screened from neighbouring atoms by an intervening O-shell, and we shall see later that this causes an important distinction between the magnetic scattering from the iron group and the rare-earth group.

This *magnetic* scattering was first discussed by Bloch (1936) who was particularly concerned with the behaviour of ferromagnetic materials. Later Halpern and Johnson (1939) in a comprehensive treatment of magnetic scattering showed that it should be possible to demonstrate it most clearly with paramagnetic materials.

6.2. Paramagnetic scattering

In a true paramagnetic substance the magnetic moments of the atoms, which arise from their unpaired electronic moments, are completely uncoupled to each other and are randomly oriented in direction. Under the influence of an external magnetic field they will tend to align themselves in the field direction, although this tendency will be opposed by the effect of thermal vibrations, with the result that the magnetic susceptibility will be proportional to $1/T$, where T is the absolute temperature, according to the well-known Curie law. In general it is found experimentally that the susceptibility is proportional to $1/(T+\Delta_C)$, where Δ_C is the Curie–Weiss constant. This form of variation is to be regarded as evidence of a degree of interaction between the magnetic fields of neighbouring atoms and the interaction will be greater the larger the observed value of Δ_C. The interpretation of the paramagnetic scattering is simplest when the interaction is negligible, as, for example, in the case of manganous fluoride for which Δ_C is only 113 K, in contrast to the case of manganous sulphide for which Δ_C is 827 K.

Halpern and Johnson (1939) have shown that for completely randomly oriented paramagnetic ions the differential magnetic scattering cross-section, i.e. cross-section per unit solid angle, is

$$\mathrm{d}\sigma_{\mathrm{pm}} = \tfrac{2}{3}S(S+1)\left(\frac{e^2\gamma}{mc^2}\right)^2 f^2 \tag{6.1}$$

per atom. Here m is the *electron* mass, e is the charge, and c is the velocity of light. S is the spin quantum number of the scattering atom, γ is the magnetic moment of the neutron expressed in nuclear magnetons, and f is an amplitude form factor. Halpern and Johnson's treatment was originally carried out for ions having angular momentum due to electron spin only. We shall see later how it is extended to cases where there is orbital momentum also. The occurrence of the form factor f determining the *magnetic* scattering, in contrast to the absence of any such factor for the *nuclear* scattering, arises since the electrons which determine the magnetic moment will be distributed over a volume of space having linear dimensions comparable with the neutron wavelength. Thus this form factor is somewhat similar to the electronic form factor of an atom for X-ray scattering, although the two are by no means identical since it is only a few electron orbits in an outer shell of the atom which contribute to the magnetic moment and, hence, to the form factor f for the magnetic scattering of neutrons. It would be expected therefore that the form factor for magnetic neutron scattering would fall off more rapidly with angle than does the factor for X-ray scattering. This is indeed found to be the case, as illustrated in Fig. 110, to which we shall refer later in this chapter. From eqn (6.1) we can also derive a total cross-section for paramagnetic scattering. This will be given by

$$\sigma_{\mathrm{pm}} = \frac{8\pi}{3} S(S+1)\left(\frac{e^2\gamma}{mc^2}\right)^2 \overline{f^2}, \tag{6.2}$$

where $\overline{f^2}$ is the integral form factor for intensity, which will be the average value of f^2 when this is integrated over all directions in space.

Quantitative calculation from the above two equations shows that the cross-sections for magnetic scattering may exceed those for nuclear scattering in certain cases. For example, in the case of the Mn^{2+} ion, which occurs in paramagnetic salts such as manganous fluoride, the value of S is $\frac{5}{2}$, arising from five electrons in the 3d shell with their spins all parallel as indicated in Table 13, and in this case $d\sigma_{\mathrm{pm}}$ is equal to 1·69 barns per unit solid angle in the forward direction, where f will be unity. On the other hand, the value of $\bar{b}^2$ for manganese, which will be the differential cross-section for nuclear scattering, is only equal to 0·14 barns, so that the magnetic scattering in the forward direction at least is very much greater in this case. This is by no means always so, and as a contrast we may consider the case of Ni^{2+} for which S is equal to 1. Here $d\sigma_{pm}$ is only equal to 0·39 barns per unit solid angle in the forward direction in comparison with the much larger value of 1·06 barns for $\bar{b}^2$.

Measurement of total cross-section

Equally, if we consider the *total* cross-section σ_{pm}, for paramagnetic scattering by manganese we find that it would have the large value of 21 barns if $\overline{f^2}$ were unity, in comparison with a nuclear scattering cross-section σ of 1·7 barns. This value of the cross-section σ_{pm} should be approached as the neutron wavelength becomes large compared with the size of the ion. Earlier experiments by Whitaker and co-workers (1937, 1938, 1940, 1941) sought to establish this paramagnetic scattering by demonstrating the existence of excess scattering by a compound, such as MnS, compared with the sum of the cross-sections of its two constituents. These experiments were inconclusive for a number of reasons, in particular the weakness of the radium–beryllium neutron sources which were the only sources then available, and the fact that the neutron beams were not monochromatic. Later experiments by Ruderman (1949), consisting of transmission measurements with anhydrous crystals of MnF_2 and MnO using monochromatic neutrons from a cyclotron, did confirm, however, the theoretical predictions. Fig. 107 shows the observed variation with wavelength of the cross-section of manganous fluoride, in comparison with a theoretical curve which allows for absorption and scattering (though without making detailed allowance for crystal diffraction effects such as were illustrated in Fig. 26) but not for paramagnetic scattering. The observed excess scattering is therefore to be interpreted as giving σ_{pm}. In Fig. 108 are shown experimental values of σ_{pm} obtained in this way at various wavelengths, excluding the range from 4 Å to the cut-off wavelength of 5·5 Å over which diffraction effects will be appreciable. It will be seen that these experimental values are in good agreement with the curve shown in the Figure, which is calculated from eqn (6.2), incorporating a calculation of $\overline{f^2}$ from Halpern and Johnson's theoretical work. The latter authors show that when spin coupling between ions is negligible the scattering is entirely elastic and $\overline{f^2}$ is a function of $1/R_0\lambda$, where R_0 is the most probable radius of the paramagnetic ion and λ is the neutron wavelength. In deriving the curve shown in Fig. 108, Ruderman (1949) takes a value of R_0

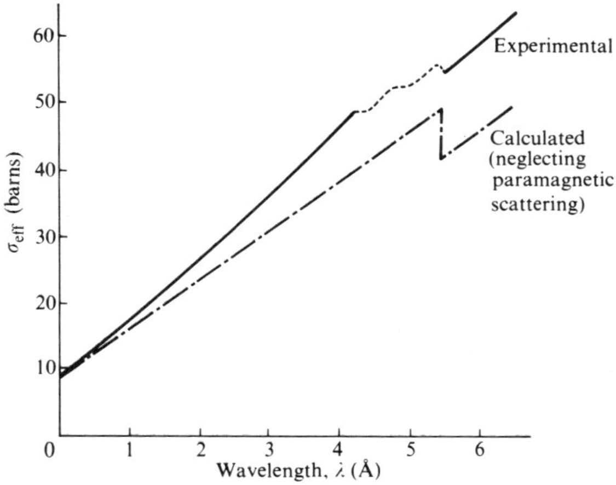

FIG. 107. Observed variation with wavelength of the total cross-section of MnF_2, compared with a theoretical curve which takes into account nuclear absorption and scattering but not paramagnetic scattering. The latter curve does not make detailed allowance for crystal diffraction effects except at the cut-off wavelength. (After Ruderman 1949.)

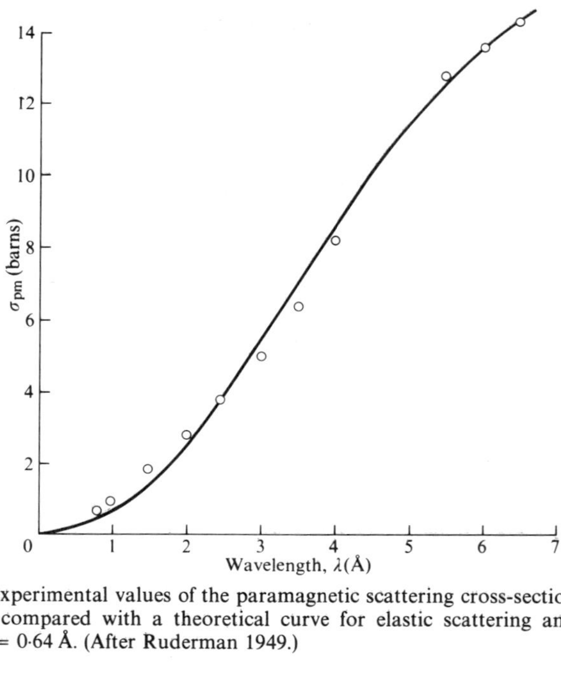

FIG. 108. Experimental values of the paramagnetic scattering cross-section of Mn^{2+} in MnF_2, compared with a theoretical curve for elastic scattering and an ionic radius $R_0 = 0{\cdot}64$ Å. (After Ruderman 1949.)

equal to 0·64 Å in order to obtain the best fit with the experimental points. This is to be compared with 0·80 Å which is the empirical ionic radius of Mn^{2+} in crystals as given by Pauling (1940).

Paramagnetic diffuse scattering in diffraction patterns

The transmission measurements of Ruderman (1949) that we have just described established unambiguously the truth of Halpern and Johnson's (1939) predictions concerning the paramagnetic scattering. However, we shall find that the complementary evidence first presented by Shull, Strauser, and Wollan (1951) from a study of the background scattering in their neutron-diffraction powder patterns will be of more direct significance in our subsequent discussion of magnetic effects in other materials.

When the ions have their magnetic moments oriented in an entirely random manner, as in a paramagnetic salt, the magnetic scattering will be entirely incoherent and will contribute to the background scattering of a powder pattern. Shull, Strauser, and Wollan, by calculating how much of the background was due to other effects, namely spin incoherence, thermal diffuse scattering and multiple scattering, were able to ascribe the residue to paramagnetic scattering. This work was done with very pure and anhydrous samples of MnO, $MnSO_4$, and MnF_2, all of which contain the Mn^{2+} ion, and special steps were taken to exclude all traces of water from the samples, in order that the very large spin incoherent scattering by hydrogen nuclei would not be mistaken for paramagnetic scattering. In practice it was found that for small values of the angle θ the paramagnetic scattering accounted for about three-quarters of the total background intensity and could therefore be evaluated quite accurately. Fig. 109 shows the experimental results for the three salts at room temperature, expressed as the variation of the differential scattering cross-section $d\sigma_{pm}$ in barns as a function of the angle θ. The curves for MnF_2 and MnO were considered to be the most reliable and, of these, that of MnF_2 shows the expected variation of $d\sigma_{pm}$ with the angular form factor, whereas the peaked curve found for MnO suggests that at room temperature the magnetic moments are not completely randomly aligned and there exists a short-range magnetic ordering. The angular position of this peak is consistent with the magnetic structure of MnO at low temperatures which will be deduced in Chapter 14.

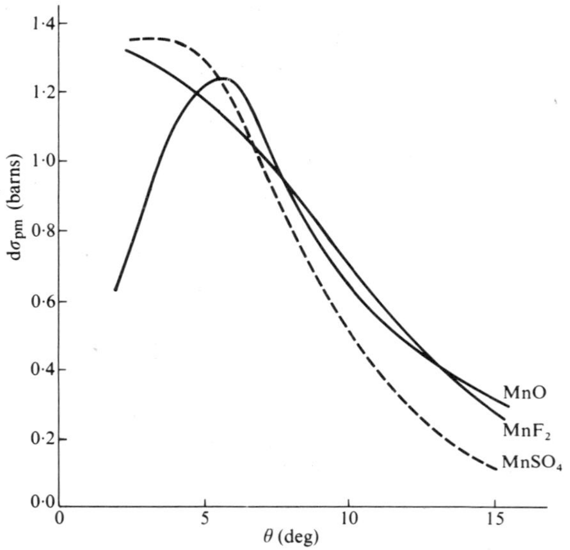

FIG. 109. The variation with angle θ of the differential cross-section of Mn^{2+} for paramagnetic scattering for a number of manganous salts, measured at room temperature. (After Shull, Strauser, and Wollan 1951.)

It was concluded that the curve for MnF_2 represented the true variation of the magnetic form factor, particularly as the curve underwent no change when the sample temperature was raised to 400 °C, and it is of significance that the extrapolated value of $d\sigma_{pm}$ when θ is zero was within about 10 per cent of the calculated value of 1·69 barns per unit solid angle referred to above, the calculation assuming that $S = \frac{5}{2}$. The experimental accuracy is scarcely sufficient to decide whether this value of 10 per cent, by which the experimental intensity is low, represents a real discrepancy with the theory. Fig. 110, which is derived from the cross-section curve for MnF_2 shown in Fig. 109, gives the variation of the amplitude form factor f with $(\sin\theta)/\lambda$. The vertical lines on the figure indicate the estimated accuracy of the experimental results. The figure also shows the form factor for X-ray scattering by a manganese atom as calculated by James and Brindley (1932). As already mentioned, the neutron magnetic scattering by the Mn^{2+} ion falls off much more rapidly with increase of the angle θ.

The form factor and the electronic distribution are related as Fourier transforms, and if $U(r)\,dr$ is the number of electrons within a spherical shell of radius r and thickness dr then it can be shown (Compton and Allison 1935) that

$$U(r) = (2r/\pi)\int_0^\infty xf(x)\sin rx\,dx, \tag{6.3}$$

where $x = 4\pi \sin\theta/\lambda$ and $f(x)$ is the form factor. In Chapter 7 we shall discuss the experimental determination of the function $U(r)$ and give some results for Fe, Co, and Ni in Fig. 130 (see p. 230).

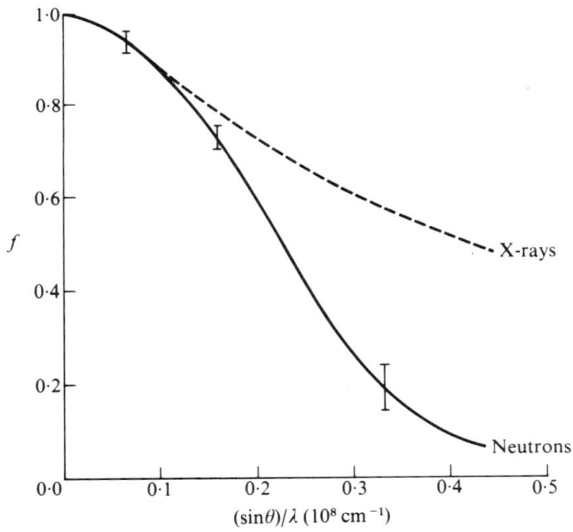

FIG. 110. Variation of the magnetic amplitude form factor f for the Mn^{2+} ion with $(\sin\theta)/\lambda$, as deduced from experimental measurements of paramagnetic scattering by MnF_2. For comparison the more slowly varying curve for X-ray scattering by a manganese atom is also shown. (After Shull, Strauser, and Wollan 1951.)

As more intense neutron beams have become available it has become possible to measure the paramagnetic scattering with increasing accuracy. Fig. 111 shows some results by Atoji (1961) for the cerium ion in CeC_2. By subtraction of the estimated contributions shown, for incoherent nuclear scattering, multiple scattering, and thermal diffuse scattering, Atoji obtained a curve for paramagnetic scattering which extrapolated to 0·311 barns per unit solid

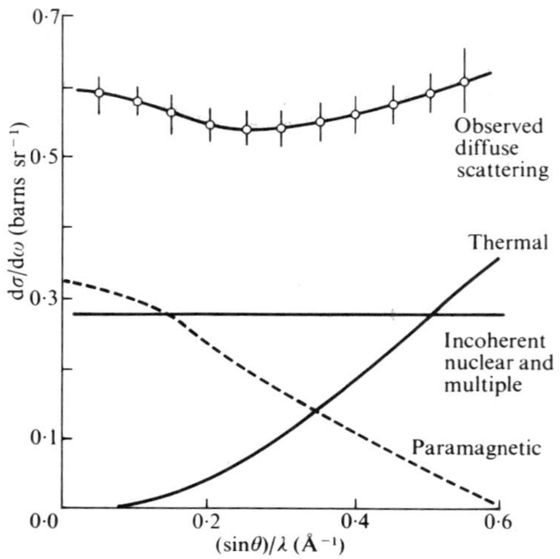

FIG. 111. Diffuse scattering by CeC_2. The paramagnetic scattering is obtained by subtracting the calculated values of the thermal diffuse scattering, incoherent nuclear, and multiple scattering from the observed diffuse scattering, giving the broken-line curve shown. (After Atoji 1961.)

angle at $\theta = 0°$. This corresponds to expectation for a cerium ion in the Ce^{3+} state. Many of the rare-earth elements are attractive for demonstrating the paramagnetic scattering because of the large values of their magnetic moments. As a more extreme example we may examine the measurements of Wilkinson, Wollan, Koehler, and Cable (1962), who have sought for evidence that localized paramagnetic moments exist on atoms of metallic chromium above 310 K. A sample composed solely of the isotope ^{52}Cr was used in order to reduce the amount of incoherent diffuse scattering. A small residue of diffuse scattering was found, but the precision of the observation was sufficient for drawing the conclusion that this showed no angular variation of form-factor type and was probably due to multiple scattering. It was therefore concluded that the small localized moment of 0·4 Bohr magnetons (μ_B) which exists on chromium atoms in the ordered antiferromagnetic state does not persist into the paramagnetic region.

6.3. Spin and orbital coupling

We have already mentioned that the fact that the susceptibilities of paramagnetic salts obey the Curie–Weiss law, rather than the simple Curie law, is to be interpreted as showing the effect of interaction between neighbouring ions. We meet further evidence of interaction when we investigate the magnetic susceptibilities of ions which possess orbital angular momentum in addition to the angular momentum associated with the electron spins. Just as for the study of magnetic *neutron scattering* (for which Halpern and Johnson's theory was originally limited to the case of ions with spin momentum only) so also is the interpretation of magnetic *susceptibilities* simplest when there is no orbital momentum to be considered.

When an atom or ion possesses both spin and orbital angular momentum these are normally combined according to the Russell–Saunders (1925) coupling. The several l vectors of the electrons combine to form a resultant **L** and the **s** vectors give a separate resultant **S**. **L** and **S** then combine to form a resultant **J** which represents the total angular momentum of the whole atom. For an atom with a resultant spin quantum number S, the spin magnetic moment will be

$$\mu_S = 2\sqrt{\{S(S+1)\}}\,\mu_B. \tag{6.4}$$

If the resultant orbital quantum number is L then the orbital magnetic moment is

$$\mu_L = \sqrt{\{L(L+1)\}}\,\mu_B. \tag{6.5}$$

The resultant spin and orbital angular momenta are then coupled to give a resultant total angular momentum characterized by the quantum number J which may take any of the values $(L+S)$, $(L+S-1)$, $(L+S-2), \ldots, (L-S+2)$, $(L-S+1)$, $(L-S)$ and the resultant paramagnetic moment is

$$\mu_J = g\sqrt{\{J(J+1)\}}\,\mu_B, \tag{6.6}$$

where g is the Landé splitting factor given by

$$g = 1+\frac{J(J+1)+S(S+1)-L(L+1)}{2J(J+1)}. \tag{6.7}$$

In the case of the salts of the rare earths there is good agreement between the experimentally determined paramagnetic susceptibilities and those calculated from eqns (6.6) and (6.7) using the values of L, S, and J known from spectroscopic data. This is shown in Table 14, which gives some data quoted by Syrkin and Dyatkina (1950).

On the other hand, in the case of the transition elements the experimentally determined magnetic moments of paramagnetic ions are by no means in agreement with the calculated values of $g\sqrt{\{J(J+1)\}}$. In fact agreement is rather better if it is assumed that only the magnetic moment from the electron *spin* is effective. This is illustrated in Table 15, also derived from Syrkin and Dyatkina, which compares the experimental values of the magnetic moments

TABLE 14

Paramagnetic moments of ions of the rare-earth elements, determined for the sulphates

Ion	Electronic configuration of 4f shell	S	L	J	Paramagnetic moment (μ_B) Calculated	Paramagnetic moment (μ_B) Experimental
Ce^{3+}	f^1	$\frac{1}{2}$	3	$\frac{5}{2}$	2·54	2·50
Pr^{3+}	f^2	1	5	4	3·58	3·55
Nd^{3+}	f^3	$\frac{3}{2}$	6	$\frac{9}{2}$	3·62	3·59
Gd^{3+}	f^7	$\frac{7}{2}$	0	$\frac{7}{2}$	7·94	8·03
Tb^{3+}	f^8	3	3	6	9·7	9·3
Dy^{3+}	f^9	$\frac{5}{2}$	5	$\frac{15}{2}$	10·6	10·55
Er^{3+}	f^{11}	$\frac{3}{2}$	6	$\frac{15}{2}$	—	—

TABLE 15

Paramagnetic moments of salts of the transition elements

Ion	Electronic configuration of 3d shell	Number of unpaired electrons	S	L	J	μ_J	μ_{exp}	μ_s 'spin only'
Cr^{3+}	d^3	3	$\frac{3}{2}$	3	$\frac{3}{2}$	0·78	$CrCl_3$, 3·81 $Cr_2O_3 . 7H_2O$, 3·85	3·88
Mn^{3+}	d^4	4	2	2	0	0	$Mn_2(SO_4)_3$, 5·19 $MnCl_3$, 5·08	4·90
Mn^{2+}	d^5	5	$\frac{5}{2}$	0	$\frac{5}{2}$	5·91	$MnSO_4$, 5·90	5·91
Fe^{3+}	d^5	5	$\frac{5}{2}$	0	$\frac{5}{2}$	5·91	$Fe_2(SO_4)_3$, 5·86	5·91
Fe^{2+}	d^6	4	2	2	4	6·76	$FeCl_2$, 5·23 $FeSO_4$, 5·26	4·90
Co^{2+}	d^7	3	$\frac{3}{2}$	3	$\frac{9}{2}$	6·68	$CoCl_2$, 5·04 $CoSO_4$, 5·04–5·25	3·88
Ni^{2+}	d^8	2	1	3	4	5·61	$NiCl_2$, 3·24–3·42 $NiSO_4$, 3·42	2·83
Cu^{2+}	d^9	1	$\frac{1}{2}$	2	$\frac{5}{2}$	3·56	$CuCl_2$, 2·02 $CuSO_4$, 2·01	1·73

with those calculated from both eqns (6.6) and (6.4), corresponding in the two cases to the resultant angular momentum and the 'spin-only' momentum respectively being effective.

The difference in behaviour between the salts of the transition elements and the rare earths is explained as due to a 'quenching' of the orbital momentum in the former by the 'crystalline field', which is the term used to describe the electric field produced by the particular arrangement of atoms and ions around the ion under consideration. This quenching was first postulated by Stoner (1929). For the transition elements it is the 3d electrons forming the outermost shell of the ion which give rise to the magnetic moment and it is reasonable to suppose that these are affected by the surrounding field, in contrast with the immunity of the 4f electrons in the rare earths which are shielded by the intervening O shell. Van Vleck (1932) has shown theoretically that the non-spherical asymmetric electric fields in a crystal would be expected to exert a quenching action of this type, with the result that when an external field is applied the orbital momenta remain fixed in position under the influence of the *internal* field, in preference to attempting to align themselves with the *external* field. The spin momenta are not affected by internal electrostatic fields of this type, although they are indeed affected by the exchange interaction field between paramagnetic ions, which may produce ferromagnetism.

Just as it is the *effective* value of the total angular momentum, reduced wholly or partially to the spin-only value by quenching in the case of the transition elements, which determines the susceptibility and gyromagnetic ratio, so it is this effective quantum number which has to be inserted in Halpern and Johnson's equation, given as eqn (6.1) for the magnetic scattering of neutrons. It should be possible to correlate the neutron scattering cross-section curve for various ions, such as was shown in Fig. 109 for Mn^{2+} which has spin momentum only, with their effective magnetic moments as determined by susceptibility measurements and accounted for by quenching of the orbital momentum. Measurements of the diffuse background scattering in the neutron diffraction patterns of a number of paramagnetic materials have already been mentioned. However, when ions occur in antiferromagnetic and ferromagnetic compounds it is much simpler and more accurate to measure the coherent magnetic diffraction peaks which arise. Therefore we now pass to a general consideration of scattering by antiferromagnetic and ferromagnetic materials and will refer again subsequently to the evidence for the quenching of the orbital contribution to the magnetic moment.

6.4. Scattering by antiferromagnetic and ferromagnetic materials

In the case of antiferromagnetic and ferromagnetic materials the magnetic moments of the individual ions are oriented in a defined manner, in contrast with their random orientation in space for a paramagnetic material. Thus in a ferromagnet all the moments in a single domain are aligned parallel. In an antiferromagnet the atoms may be considered to lie on two sublattices whose spins are oppositely directed. All the atoms lying on one sublattice have their magnetic moments parallel to a given direction, whereas those on the other sublattice are antiparallel. The intense electrostatic field which maintains these two kinds of alignment is accounted for by quantum mechanics as being an 'exchange force' which simulates a coupling between the spins of neighbouring atoms equivalent to a potential energy V_{ij} of the form

$$V_{ij} = -2J_{ij}S_iS_j,$$

where S_i, S_j are the spin quantum numbers of the two atoms and J_{ij} is the 'exchange integral'. In ferromagnetic materials this integral is positive. This can be correlated with the fact that the ratio of the separation distance of neighbouring atoms to the radius of the electron shell containing the uncompensated spins must exceed a certain minimum value if ferromagnetism is to appear (Slater 1930). On the other hand, in antiferromagnetic materials the exchange integral is negative and the molecular field tends to direct the spins of the atoms on the two sublattices in antiparallel directions. Above a certain temperature, the Néel temperature, this tendency is overcome just as the wholly parallel alignment in a ferromagnetic substance disappears above the Curie temperature. In the antiferromagnetic case reduction of temperature below the Néel point makes it progressively harder for an external field to line up the spins in a single direction, it being borne in mind that they are endeavouring to arrange themselves alternately in opposite directions. Thus the susceptibility, which will be small, *falls* as the temperature is lowered, giving a characteristic peaked susceptibility–temperature curve for an antiferromagnet. A peak will be found at the Néel temperature, as first shown for MnO by Bizette, Squire, and Tsai (1938), in contrast with the steady decrease of susceptibility with increase of temperature shown by an ordinary feeble paramagnetic material.

Coherent magnetic diffraction spectra

Compared with the manner in which neutrons are scattered by paramagnetic materials there are two immediate differences when we consider scattering by atoms whose magnetic moments are in fixed orientations—e.g. in the simplest case, when the moments are parallel or antiparallel. In the first place, eqn (6.1) for the differential magnetic scattering cross-section per unit solid angle is replaced by

$$\mathrm{d}\sigma_{\mathrm{m}} = q^2 S^2 \left(\frac{e^2\gamma}{mc^2}\right)^2 f^2, \qquad (6.8)$$

where $\mathbf{q}$ is the magnetic interaction vector defined by

$$\mathbf{q} = \boldsymbol{\varepsilon}(\boldsymbol{\varepsilon}\,.\,\mathbf{K}) - \mathbf{K}, \qquad (6.9)$$

where $\mathbf{K}$ is a unit vector in the direction of the atomic magnetic spin and $\boldsymbol{\varepsilon}$ is a unit vector in the direction perpendicular to the effective 'reflecting' planes, i.e. the so-called 'scattering vector', as indicated in Fig. 112. The other quantities in eqn (6.8) are the same as those defined for eqn (6.1). The appearance of the factor q^2 in (6.8) and the replacement of $S(S+1)$ by S^2 are a consequence of the rigid alignment of the magnetic spins in an antiferromagnetic or ferromagnetic substance. It follows from the definition of $\mathbf{q}$ provided by eqn (6.9) that $\mathbf{q}$ lies in the plane of $\boldsymbol{\varepsilon}$ and $\mathbf{K}$ and is perpendicular to

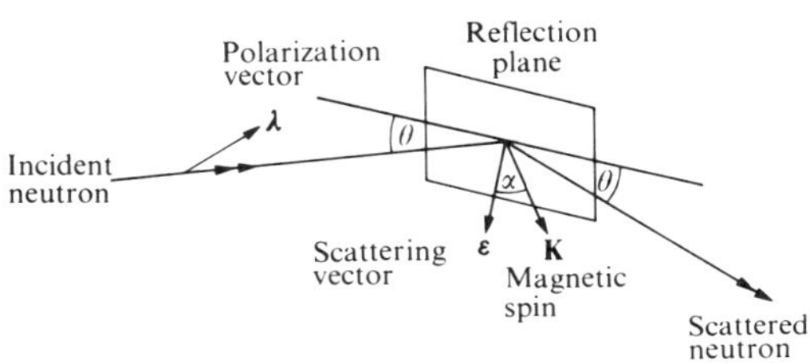

FIG. 112. Identification of the unit vectors $\boldsymbol{\varepsilon}$, $\mathbf{K}$, and $\boldsymbol{\lambda}$, used in the discussion of magnetic scattering, together with the angles α and θ. The magnetic interaction vector $\mathbf{q}$ is defined as $\boldsymbol{\varepsilon}(\boldsymbol{\varepsilon}\,.\,\mathbf{K}) - \mathbf{K}$, so that $\mathbf{q}$ lies in the plane of $\boldsymbol{\varepsilon}$ and $\mathbf{K}$ and is perpendicular to $\boldsymbol{\varepsilon}$ and of magnitude $\sin\alpha$.

$\boldsymbol{\varepsilon}$ and of magnitude sin α. Thus

$$|\mathbf{q}| = \sin \alpha. \tag{6.10}$$

We emphasize that eqn (6.9), and the definition of $\mathbf{q}$ which it provides, uses the original notation of Halpern and Johnson, and we point out that $\mathbf{K}$ is the direction of the *spins* in the magnetic materials. This vector is in the opposite direction to that of the applied field or the magnetization in the material, simply because the spin moment arises from negative electrons. Elsewhere in the literature alternative definitions are often found. For example, Lomer and Low (1965) define a vector $\mathbf{N}$ by

$$\mathbf{N} = \boldsymbol{\mu} - (\boldsymbol{\varepsilon} \cdot \boldsymbol{\mu})\boldsymbol{\varepsilon}, \tag{6.11}$$

where $\boldsymbol{\mu}$ is the magnetization. It will be seen that this vector $\mathbf{N}$ is identical with our $\mathbf{q}$, allowing for the interchange of sign between the two terms, in comparison with eqn (6.9), and for the fact that the magnetization direction is opposite to that of the atomic spin.

Secondly, and of great importance, the existence of defined orientations for the magnetic moments of the ions means that there is coherence between the neutrons scattered by the various individual atoms. The atoms may be regarded as having a magnetic scattering amplitude, usually denoted by the symbol p rather than the original symbol D used by Halpern and Johnson, such that

$$p = (e^2\gamma/mc^2)Sf. \tag{6.12}$$

It will be noted that the factor e^2/mc^2, which occurs in this expression, is the classical radius of the electron. Thus p, the magnetic scattering length, is of the order of the electron radius, and if we substitute the numerical values of the constants into eqn (6.12) we obtain

$$p = 0{\cdot}54Sf \times 10^{-12}\ \text{cm},$$

so that the magnitude of p is comparable with that of the nuclear scattering length b. This scattering amplitude is the magnetic counterpart of the nuclear scattering amplitude b and gives rise to the build-up of coherent diffraction peaks dependent on the three-dimensional magnetic structure of a domain, which will be the counterpart of the crystallite for ordinary nuclear scattering. It must be emphasized that p can be calculated from magnetic and spectroscopic data, in contrast with b which, in the present state of knowledge at least, can only be found empirically by experiment. The value of p will be different for different valence states of an atom, such as Fe^{2+}, Fe^{3+} for which the S values are 2, $\frac{5}{2}$ respectively. Eqn (6.12) can be rewritten as

$$p = (e^2\gamma/2mc^2)2Sf, \tag{6.13}$$

containing the factor $2S$ which, for an atom in which the orbital moment is completely quenched, is equal to the magnetic moment expressed in Bohr magnetons. This interpretation of the equation is quite general, and for an atom in which both spin and orbital moments are operative we merely replace $2S$ by the corresponding value of the magnetic moment, namely, gJ, where g is the Landé splitting factor. Thus, in the general case we have

$$p = (e^2\gamma/2mc^2)gJf. \tag{6.14}$$

Later we shall discuss further the significance of the form factor f in the general case where the magnetic moment is accounted for by both spin and orbital contributions.

It is of crucial importance to consider the interplay of the magnetic scattering, which we have just discussed, and the nuclear scattering which we examined in Chapter 2 and which occurs for all atoms. In examining this feature the polarization state of the neutrons is of great significance. At the outset the general effect of the polarization can be seen from Halpern and Johnson's (1939) expression for the differential scattering cross-section in the case of a magnetic system in which all the magnetic moments are aligned parallel or antiparallel to a single direction, i.e. in a ferromagnetic or a simple antiferromagnetic material. This cross-section

$$\mathrm{d}\sigma = b^2 + 2bp\boldsymbol{\lambda}\cdot\mathbf{q} + p^2q^2, \tag{6.15}$$

where $\boldsymbol{\lambda}$ is a unit vector in the direction of the spin of the incident neutron. We see that when the neutron beam is unpolarized, i.e. when the unit vector $\boldsymbol{\lambda}$ can take all possible directions, then the middle term in this expression will average to zero and we have, for an unpolarized beam,

$$\mathrm{d}\sigma = b^2 + p^2q^2. \tag{6.16}$$

However, the detailed treatment of the neutron polarization is much more complicated than eqn (6.15) suggests, particularly for non-collinear arrangements of spins such as are found in helimagnetic materials. It proves to be necessary and profitable to consider not only the polarization of the incident beam but also the changes of polarization which may occur during the scattering process and the information concerning the system of scatterers which may be deduced from such changes. Essentially, we may think of the cross-section as consisting of five separate terms,

$$\mathrm{d}\sigma = \begin{array}{l}\text{Coherent}\\ \text{nuclear}\\ \text{scattering}\\ \\ \\ \end{array} + \begin{array}{l}\text{incoherent}\\ \text{isotope and}\\ \text{nuclear spin}\\ \text{scattering}\\ \\ \end{array} + \begin{array}{l}\text{nuclear–}\\ \text{magnetic}\\ \text{interference}\\ \text{term}\\ \\ \end{array} + \begin{array}{l}\text{purely}\\ \text{magnetic}\\ \text{scattering}\\ \\ \\ \end{array} + \begin{array}{l}\text{polarization}\\ \text{dependent}\\ \text{term in}\\ \text{complicated}\\ \text{structures.}\end{array} \qquad (6.17)$$

For a full discussion the reader is referred to an article by Blume (1963) and a full and later account in the book by Marshall and Lovesey (1971). Another useful article, related more specifically to the interpretation of experimental data, is by Brown (1970). We shall return again to this topic in Chapter 15 when we consider the experimental technique of polarization analysis.

At this stage we shall consider separately the cases of (a) unpolarized and (b) polarized beams of neutrons.

6.5. Use of unpolarized neutrons

In the more usual case where the neutron beam contains all directions of spin the nuclear–magnetic interference term averages to zero and the square of the structure amplitude factor F_{hkl} for the (hkl) reflection is given by the sum of two terms which represent respectively the nuclear and magnetic intensity

Thus

$$|F_{hkl}|^2 = \left|\sum_n b_n \exp\{2\pi \mathrm{i}(hx_n/a + ky_n/b + lz_n/c)\}\right|^2 + \\ + \left|\sum_n \mathbf{q}_n p_n \exp\{2\pi \mathrm{i}(hx_n/a + ky_n/b + lz_n/c)\}\right|^2. \qquad (6.18)$$

We consider the first term as F_{nucl} and the quantity

$$\mathbf{F}_{\text{magn}} = \sum_n \mathbf{q}_n p_n \exp\{2\pi \mathrm{i}(hx_n/a + ky_n/b + lz_n/c)\} \qquad (6.19)$$

can be regarded as the magnetic structure amplitude factor. Particularly in relation to co-operative magnetic materials in which the magnetic moments point in more than one direction, we note that $\mathbf{F}_{\mathrm{magn}}$ is a vector quantity. More correctly, the terms in eqns (6.18) and (6.19) should include a Debye–Waller factor. The essential feature of eqn (6.18) is that there is no coherence between the nuclear and magnetic scattering with unpolarized neutrons and the two intensity components are additive.

From a practical point of view our essential task is to identify $\mathbf{F}_{\mathrm{magn}}$ from the diffraction patterns of magnetic materials and then to use knowledge of it to determine the details of the magnetic structure. The simplest case is that of a ferromagnetic material, in which, within a domain, all the magnetic moments point in the same direction, as in Fig. 113 (a). In this case the factor q_n in the second term of eqn (6.18) is the same for all atoms in the unit cell. Consequently, peaks in the second term of eqn (6.18) will occur at exactly the same angular positions as for the first term.

Thus the magnetic scattering will, as in Fig. 114(a), merely reinforce the nuclear peaks; it is shown in the Figure as a shaded

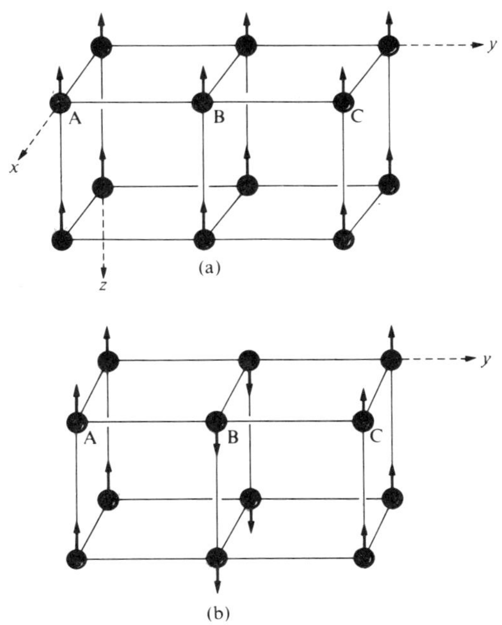

FIG. 113. Diagrams of (a) ferromagnetic and (b) antiferromagnetic structures. The directions of the arrows indicate the directions of the magnetic moments.

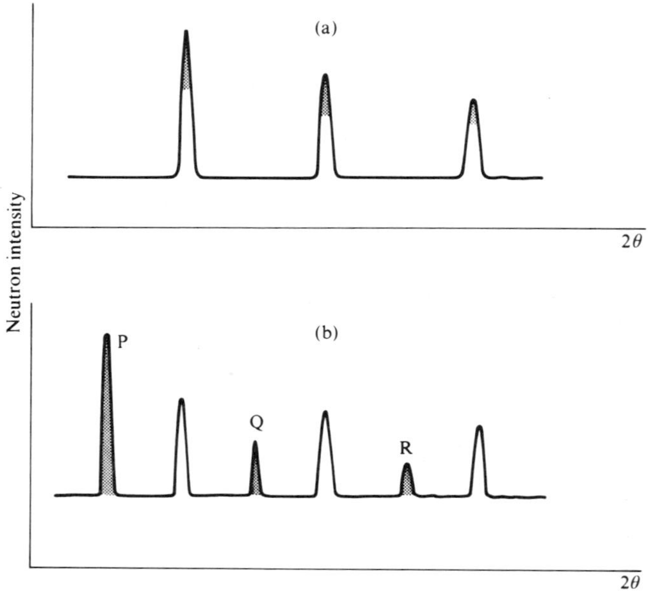

FIG. 114. Contribution of magnetic scattering to the diffraction pattern. For a ferromagnetic material (a) additional scattering, shaded, contributes to the nuclear peaks; for an antiferromagnetic material (b) additional scattering appears at the new positions P, Q, R.

addition. An important feature is that it falls off relatively quickly with increasing angle of scattering because the form factor f occurs in the expression for p, the effective magnetic scattering amplitude.

The situation is quite different for a simple antiferromagnetic material of the kind illustrated in Fig. 113(b). In this case the magnetic moments point alternately up and down, so that their values of p are alternatively $+$ve and $-$ve and the magnetic unit cell is twice as long as the chemical cell in the y-direction. Accordingly, at any angle, set by the indices hkl, for which the nuclear b_n contributions build up a reflection, the p_n contributions will be alternately $+$ve and $-$ve and will add up to zero. More important, it is seen that at angles whose sines are a half of those for these nuclear angles the magnetic contributions from atoms A, B, and C will reinforce one another and give an intense reflection. Thus an antiferromagnetic material of this kind will give a pattern of the type shown in Fig. 114(b), displaying extra reflections of purely magnetic origin, which appear at $\sin\theta$ values which are a half of those for the expected 'chemical' angles. In particular, instead of the reflection of lowest angle being indexed as 100 it will index as $\frac{1}{2}00$. The half-integral indices occur simply because we are indexing the pattern in terms of a unit cell with edge AB, whereas from a magnetic point of view the length of the edge of the unit cell in the y-direction is AC, which is twice as great. Thus the neutrons give direct evidence of the doubling of the unit cell, which is the fundamental feature of the antiferromagnetic structure which we have illustrated. In other structures the unit cell may be doubled in two or three dimensions, leading to more general groupings of half-integral indices amongst the reflections.

We may note here that a small amount of magnetic scattering occurs when low-energy *electrons* are scattered. In experiments on surfaces of NiO using electrons of energy about 50 eV Palmberg, De Wames, and Vredevoe (1968) observed two-dimensional reflections which indexed as $\frac{1}{2}0$, corresponding to a doubling of the magnetic unit cell. However, these reflections are very weak and amount to only about 2 per cent of the non-magnetic scattering from these surface layers.

The unique antiferromagnetic lines which we have just described for neutrons will decrease in intensity as the temperature is increased and will become zero at the Néel temperature where the antiferromagnetic order disappears. For a ferromagnetic material the magnetic intensity will fall to zero in the same way as the Curie temperature is approached, but this will be much less striking because the nuclear intensity will remain unchanged (apart from a small Debye thermal effect) at these positions. In either case the magnetic material will become paramagnetic at the higher temperature and, as we have already seen, the magnetic-scattered neutrons will contribute rather inconspicuously to the background, in accordance with eqn (6.1).

The technique of magnetic structure determination rests largely in correlating the experimental measurements of intensity with calculations of structure amplitude factors for various model structures, very often searching for agreement by trial-and-error methods. A number of points which often arise in the calculations are worthy of further mention here.

Most of the samples which are examined, whether of polycrystalline or single-crystal form, will consist of many magnetic domains, and it is only within a single domain that the prescribed regular arrangement of moments will be maintained. In an adjacent domain the preferred direction may be a different one, though often crystallographically equivalent, and there will be similar differences amongst the fragments of a powder which are suitably oriented for reflection. The resultant intensity will be determined by the average value of q^2 which results from these possible arrangements. For unmagnetized iron the direction of common alignment within a domain may be any one of the six cube axes and the average value of q^2 which results is $\frac{2}{3}$. The same value would be found if *any* crystallographic direction was permissible for the unique directions within a domain. On the other hand, if the sample of iron is magnetized, and in such a way that the magnetic moments are parallel to the reflecting plane which we are considering, then α will be 90° and q^2 is unity; if the moments are perpendicular to the reflecting plane then α is zero and q^2 also is zero. In the case of a compound such as CrSb, which has a hexagonal layer structure in which the magnetic moments are parallel or antiparallel to the hexagonal c-axis, q^2 will be zero for the $(000l)$ reflections and equal to unity for the $(hki0)$ reflections.

When the contributions of nuclear and magnetic intensity are being calculated it is important that the calculation is done for the same number of 'molecules'. This point is important for substances such as MnO, in which the chemical and magnetic cells are of different size and for which, rather misleadingly, the nuclear and magnetic reflections are occasionally indexed in terms of their own different unit cells. It seems preferable to index all reflections in terms of the chemical cell but to evaluate the structure factors for the number of molecules which constitute the larger, magnetic cell.

It often happens that the symmetry of the magnetic unit cell is lower than that of the chemical cell, with the result that the magnetic contributions for the individual reflections which constitute a *form* $\{hkl\}$ of reflections may no longer be identical. An example of a material for which this occurs is MnO, which we shall describe in detail in Chapter 14. Of the eight faces which together constitute the $\{111\}$ form, only the reflections (111) and $(\bar{1}\bar{1}\bar{1})$ give finite magnetic intensities. In this case therefore the effective multiplicity of the planes which contribute to the (111) powder reflection is reduced from 8 to 2.

If accurate determinations of magnetic structures are to be made it will be necessary to have an accurate knowledge of the variation of the amplitude form factor f with $(\sin\theta)/\lambda$. We have already illustrated in Fig. 110 the form factor for the manganese ion Mn^{2+} which was deduced by Shull, Strausser, and Wollan (1951) from measurements of paramagnetic scattering, and this curve has been used as a first approximation for many other atoms and ions. Later refinements have enabled some discrimination to be made between the form factors applicable to individual atoms and ions, and we shall examine these in the following chapter. We note that the form factor depends on the spatial distribution of the unpaired magnetic electrons, so that careful measurements may serve as a check and guide on theoretical measurements of these distributions.

We have already commented that the numerical value of p is of the same order as the nuclear amplitude b. Table 16 lists some

TABLE 16

Comparison of nuclear and magnetic scattering amplitudes

Atom or ion	Nuclear scattering amplitude b (10^{-12} cm)	Effective spin quantum number, S	Magnetic scattering amplitude p (10^{-12} cm)	
			$\theta = 0$	$(\sin\theta)/\lambda = 0{\cdot}25$ Å^{-1}
Cr^{2+}	0·35	2	1·08	0·45
Mn^{2+}	−0·37	$\frac{5}{2}$	1·35	0·57
Fe (metal)	0·96	1·11	0·60	0·35
Fe^{2+}	0·96	2	1·08	0·45
Fe^{3+}	0·96	$\frac{5}{2}$	1·35	0·57
Co (metal)	0·28	0·87	0·47	0·27
Co^{2+}	0·28	2·2	1·21	0·51
Ni (metal)	1·03	0·3	0·16	0·10
Ni^{2+}	1·03	1·0	0·54	0·23

values for a number of atoms and ions. The values of p are given both for the forward direction, where $\theta = 0°$, and also for the direction at which $(\sin\theta)/\lambda = 0{\cdot}25 \times 10^8\ \text{cm}^{-1}$. It is important to recognize the rapid decrease of the form factor f with increase of the Bragg angle θ, leading as it does to a severe limitation on the range of θ values over which significant magnetic intensities can be detected. From the typical form factor in Fig. 110 it can be deduced that the magnetic *intensity* will have fallen to a half the value for forward scattering at $(\sin\theta)/\lambda = 0{\cdot}15 \times 10^8\ \text{cm}^{-1}$. For a wavelength of 1 Å this corresponds to the quite small Bragg angle of 8·5°, at an interplanar spacing of 3·3 Å. Finally, in this general discussion of magnetic measurements with unpolarized neutrons we mention that from studies of polycrystalline samples, as distinct from single-crystal or single-domain samples, it is often not possible to determine the orientation relative to the crystallographic axes of the common direction of alignment of the magnetic spins in an antiferromagnetic or ferromagnetic material. This point has been discussed by Shirane (1959), who shows that no information about orientation can be obtained if the symmetry is cubic. For uniaxial crystals it is possible to define the angle between the spin direction and the unique axis, but nothing more. For example, Erickson (1953) in his study of NiF_2 was able to say that the moments were inclined at 10° to the tetragonal axis but was unable to specify the direction of the projected component on the (0001) plane.

Diffraction by magnetized samples

From the form of eqn (6.16) it is clear that the intensity of scattering in any direction will depend on the magnetic interaction vector **q**, and from (6.10) in turn is seen the dependence of q on the angle between the scattering and magnetization vectors. In the case of unmagnetized ferromagnetic samples the magnetic moments of the various domains will be randomly oriented in space but on application of a magnetic field the moments will be forced round into the field direction, as expressed by the well-known magnetization curve for a ferromagnetic material. With increase of field there will be an approach to saturation, this being attained when all the magnetic moments are aligned in the field direction. If the field direction is parallel to the scattering vector, as indicated in Fig. 115(a), then

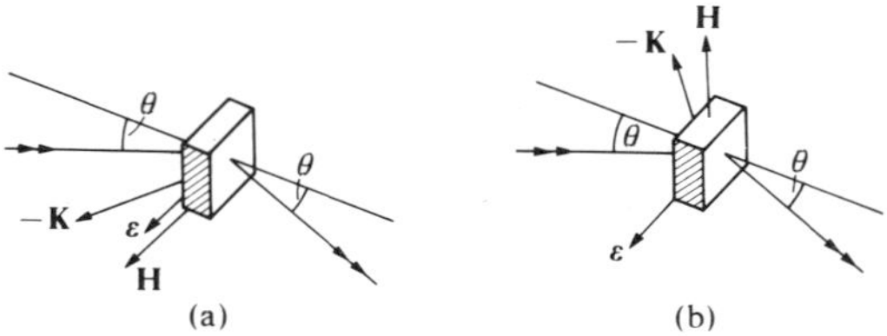

FIG. 115. Arrangements for investigating the variation of the magnetic diffraction intensity with field direction for a magnetized sample. In (a) the field **H** is parallel to **ε**, i.e. perpendicular to the reflection plane. As saturation is approached the direction of $-\mathbf{K}$ approaches **ε** and q^2 becomes zero. In (b) the field **H** is perpendicular to **ε**. As saturation is approached **K** becomes perpendicular to **ε** for all domains, this making q^2 equal to unity.

q^2 will be zero for all the domains when saturation is achieved, with the result that the intensity of a reflection will be given, from eqn (6.18), by F^2_{nucl} alone. On the other hand, when the field direction is perpendicular to the scattering vector, q^2 will be unity for all the domains at saturation and the reflection intensity will be $F^2_{\text{nucl}}+F^2_{\text{magn}}$. These two extreme values are to be contrasted with $F^2_{\text{nucl}}+\frac{2}{3}F^2_{\text{magn}}$, which will be the intensity in the unmagnetized condition of a material like magnetite or iron in which the magnetic moments of a domain can be directed along any one of the cube axes, or for a material in which they are directed randomly with respect to the crystallographic axes.

The above variation in the diffracted neutron intensity for a ferromagnetic when the direction of the magnetizing field is varied is shown in Fig. 116, which illustrates the results obtained by Shull,

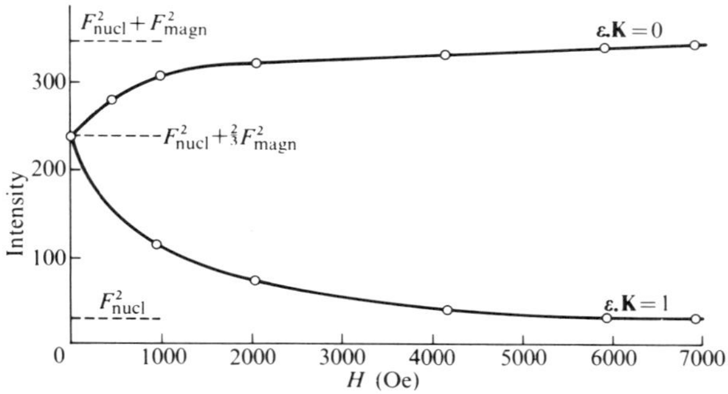

FIG. 116. The variation with field strength of the diffracted intensity of the (111) reflection from magnetite. The magnetization directions of the individual domains become aligned in the field direction. Experimental points are shown for the two cases in which the field is perpendicular and parallel, respectively, to the scattering vector. (Shull, Wollan, and Koehler 1951.)

Wollan, and Koehler (1951) for the (111) reflection of magnetite. This example shows the effect of variation of the field direction particularly well, since the value of F^2_{nucl} for this particular reflection is very small in comparison with the value of F^2_{magn}.

6.6. Polarized-neutron beams

Let us now return to a discussion of the relation between the nuclear and magnetic scattering for the case of a beam of polarized neutrons. This proves to be of very great importance because it leads to a very effective technique for actually producing a polarized beam.

In Fig. 117 let a vertical magnetic field be applied to a ferromagnetic material which is oriented suitably in space to give a Bragg reflection from the (*hkl*) plane which is shown shaded in the Figure and let the field be sufficiently intense to produce saturation

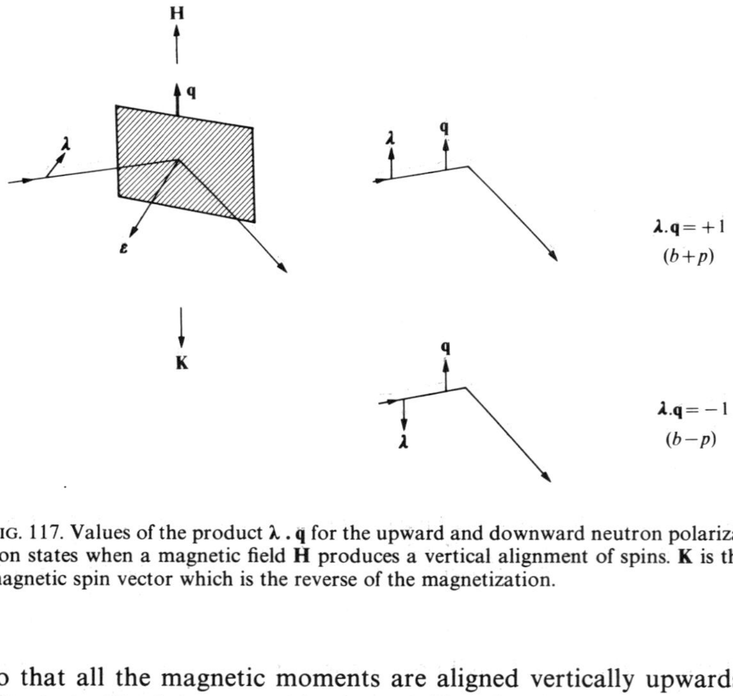

FIG. 117. Values of the product $\boldsymbol{\lambda}\cdot\mathbf{q}$ for the upward and downward neutron polarization states when a magnetic field **H** produces a vertical alignment of spins. **K** is the magnetic spin vector which is the reverse of the magnetization.

so that all the magnetic moments are aligned vertically upwards. The *spins* **K** of the magnetic atoms will point downwards, as we explained in our discussion of eqn (6.9). In these circumstances an unpolarized beam may be regarded as the sum of two completely polarized beams, one with the neutron spin directed upwards and the other with downward spin. The vectors $\boldsymbol{\lambda}$ and $\mathbf{q}$ for these two beams will therefore be as shown in the centre of Fig. 117, leading to values of $+1$ and -1 for the product $\boldsymbol{\lambda}\cdot\mathbf{q}$ for the upward and downward spin vectors respectively. It follows from eqn (6.15) that the scattering amplitudes for the upward- and downward-pointing spins will be $b+p$ and $b-p$ respectively, corresponding to intensities of $(b+p)^2$ and $(b-p)^2$. Two conclusions of particular interest arise from this. First, the over-all intensity obtained by adding together both spin states in the unpolarized beam, is b^2+p^2, indicating incoherence between the nuclear and magnetic scattering. Secondly, if we could observe the two states *separately* then we should find, in each case, *coherence* between the nuclear and magnetic scattering, indicated by the amplitudes $b+p$ and $b-p$. In fact this circumstance provides us with a means of separating the two spin states, thus enabling us to produce a polarized beam which we can use for further experiments. We shall find that a polarized beam has some very important practical advantages to the experimenter.

The key to producing the polarized beam is to find some material for which $b = p$, since under these circumstances the scattering amplitude for neutrons of downward-pointing spin will be zero and the reflected beam will consist only of neutrons of upward pointing spin. We must remember that the expression for the magnetic scattering amplitude p includes the form factor f, so that the equality of b and p will only hold for some particular reflection, corresponding to a particular value of $(\sin\theta)/\lambda$. Suitable crystals are magnetite Fe_3O_4, for which the (220) reflection is used, and a face-centred cubic Co–Fe alloy containing 92 per cent of Co; for the latter the (111) or (200) reflection can be used. For magnetite the structure factor for the (220) reflection is determined solely by the trivalent Fe^{3+} ions on the tetrahedral A sites, as discussed in more detail on p. 387. When $\theta = 0°$ the magnetic scattering amplitude of this ion is $1{\cdot}35\times10^{-12}$ cm, as listed in Table 16. For the (220) reflection the value of $(\sin\theta)/\lambda$ is equal to approximately $0{\cdot}17\times10^{8}$ cm^{-1} and the form factor f has fallen to about 0·7, so that the magnetic scattering amplitude p at this angle will closely match the value of 0·95, which is the nuclear scattering amplitude b for iron. In practice, values of 95–8 per cent are found for the polarizing efficiency of crystals of magnetite. This efficiency P is defined by

$$P = (I_1-I_2)/(I_1+I_2), \tag{6.20}$$

where I_1, I_2 are the numbers of neutrons in the two spin states. Since I_1 and I_2 are proportional to $(b-p)^2$ and $(b+p)^2$, respectively, it follows that

$$P = \frac{(b+p)^2-(b-p)^2}{(b+p)^2+(b-p)^2} = \frac{4bp}{2(b^2+p^2)}$$

$$= \frac{2p/b}{1+(p/b)^2}. \tag{6.21}$$

The disadvantages of magnetite crystals are that the absolute reflectivity is rather low and the second-order contamination, from the component of the neutron spectrum of wavelength $\frac{1}{2}\lambda$, is rather high. This arises because the structure factor for the (440) reflection of magnetite is very much larger than for (220), on account of the fact that all the iron ions, on both the A and B sites, contribute to the higher-order reflection.

For the alloy $Co_{0\cdot92}Fe_{0\cdot8}$ it can be shown, using the values for the pure metals given in Table 16, that $b = 0{\cdot}307 \times 10^{-12}$ cm and that p for $\theta = 0°$ is $0{\cdot}480 \times 10^{-12}$ cm. The values of the atomic form factor f for the (111) and (200) reflections have been shown by Nathans and Paoletti (1959) to be 0·781 and 0·652 respectively, so that $p_{111} = 0{\cdot}375 \times 10^{-12}$ cm and $p_{200} = 0{\cdot}312 \times 10^{-12}$ cm. It follows from eqn (6.21) that the expected polarization efficiencies will be 98·1 per cent for the (111) reflection and almost 100 per cent for the (200). It will be evident from this calculation that a high efficiency is achieved even when, as for the (111) reflection, b and p differ quite substantially.

However, even if b and p were exactly equal the efficiency of polarization in a practical case is unlikely to be 100 per cent. The crystal will not be completely saturated by the applied magnetic field, and there will be magnetic inhomogeneities dependent on the domain structure and the mosaic spread. These factors lead to continual depolarization of the beam as it passes through the crystal and have been discussed in some detail by Nathans *et al.* (1959). Thus the value of P given above is reduced by a depolarizing factor which can be written, in terms of a linear depolarization coefficient μ_P and the path length t in the crystal, as

$$P/P_0 = \exp(-\mu_P t). \tag{6.22}$$

On balance, $Co_{0\cdot92}Fe_{0\cdot8}$ turns out to be a much more satisfactory polarizing crystal than magnetite. In particular a much smaller applied field will achieve saturation, as illustrated by the two curves in Fig. 118. The main disadvantage of the Co–Fe crystal is that although its reflectivity is higher than for magnetite it is limited by the relatively large absorption coefficient of Co, for which σ_a, listed in Table 6 (p. 72) is about 21 barns. This means that the reflectivity cannot be improved by increasing the thickness of the polarizing

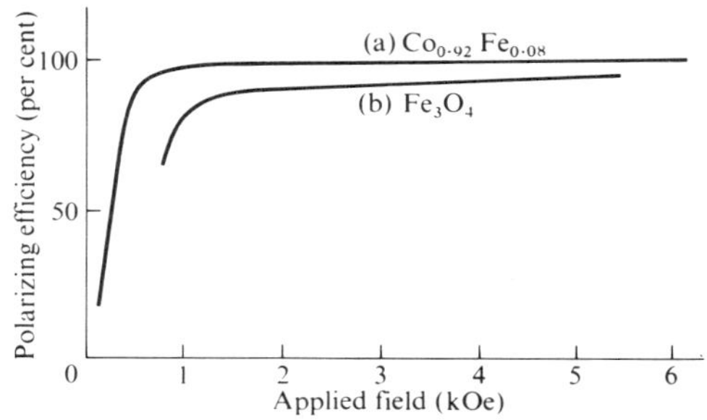

FIG. 118. Variation of polarizing efficiency of (a) $Co_{0\cdot92}Fe_{0\cdot8}$ and (b) magnetite with the value of the applied magnetic field. (Nathans *et al.* 1959.)

crystal, as it can for the ordinary metallic monochromators, such as Cu, Pb, Al, and Ge, which we discussed in Chapter 4. The practical result is that polarized beams have much less than a half of the intensity of a corresponding unpolarized beam; we should of course expect a reduction to a half because we have rejected one of the two spin states. The polarized beams which are normally available have intensities which are lower by 5 or 10 times than ordinary unpolarized beams. Efforts are being made to find better materials which do not suffer from the drawback of the high absorption of cobalt, and the Heusler alloy Cu_2MnAl, Fe_3Si, and ^{57}Fe, together with other materials, are being studied for possible use as polarizing crystals.

We emphasize that the Fe_3O_4 or $Co_{0\cdot92}Fe_{0\cdot8}$ crystal will act as a monochromator for the incident beam of 'white' neutrons as well as performing the function of polarizer. The need to equalize the parameters b and p specifies the value of $(\sin\theta)/\lambda$. There remains flexibility in specifying the wavelength of the polarized beam according to the choice of θ, the inclination of the (220) or (111) planes to the incident beams in the respective cases.

It is of interest to note that the direction of neutron polarization obtained with an Fe_3O_4 polarizer is the opposite to that for the

Co–Fe crystal. This happens because with the former it is the Fe^{3+} ions which are responsible for the (220) reflection and these, being in a minority for this ferrimagnetic substance, have their moments directed opposite to the applied field; their *spin* direction $\mathbf{K}$ would be parallel to the field. Thus in Fig. 117 the direction of $\mathbf{q}$ would be vertically downwards and (for $b = p$) it would be neutrons for which $\boldsymbol{\lambda}$ is directed downwards which would be reflected. On the other hand, for the Co–Fe crystal the spins $\boldsymbol{\lambda}$ of the reflected neutrons will be directed upwards.

When such a beam of almost completely polarized radiation has been produced it can be used for accurate measurements of small magnetic scattering amplitudes. As a typical example we may consider the measurement of the (111) reflection for magnetized nickel with the incoming neutrons polarized first parallel and then antiparallel to the magnetic field on the crystal. It follows from eqn (6.15) that the intensities in the two cases will be $(b+p)^2$ and $(b-p)^2$ respectively, where $b = 1{\cdot}03 \times 10^{-12}$ cm and p for the (111) reflection is equal to $0{\cdot}10 \times 10^{-12}$ cm. As a result, the ratio of intensities is 150:100 for the two directions of polarization. On the other hand, with unpolarized neutrons we could merely observe a change from $(b^2 + \frac{2}{3}p^2)$ to b^2, which is a change of only 0·6 per cent when a magnetic field is switched on along the scattering vector. Fig. 119 summarizes these conclusions regarding the intensities

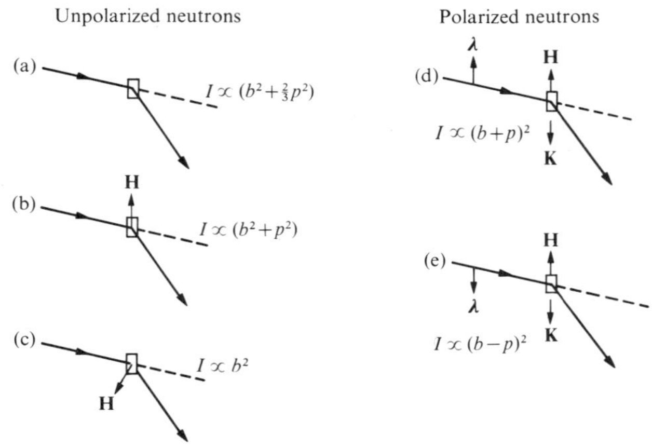

FIG. 119. Comparison of the intensities of reflection given by unpolarized and polarized neutrons for different arrangements of magnetic field on the diffracting sample.

which will be obtained under the various conditions. Polarized neutrons thus offer a very great increase in accuracy for magnetic measurements. As an application which takes advantage of this we mention the use of a polarized beam for determining form-factor curves. The ability to measure weak magnetic intensities means that the form-factor curves can be extended to much larger values of $(\sin\theta)/\lambda$, as, for example, in Nathans and Paoletti's (1959) measurements with cobalt. At the extreme range of $(\sin\theta)/\lambda = 0{\cdot}8$, for the (440) reflection, they were able to measure a form factor as low as 0·022 with an accuracy of 0·002.

Polarized-neutron spectrometer

In principle the definitive measurements which have to be made with a polarized neutron beam are indicated in Fig. 120. The crystal B under examination is subjected first to a magnetic field which is vertically upwards, and secondly to a downward pointing field, thus enabling the quantities $(b+p)$ and $(b-p)$ to be deduced in turn. In achieving the second of these arrangements there will inevitably be a discontinuity in the magnetic field along the neutron beam. The field will change from being upwards at the polarizing crystal A

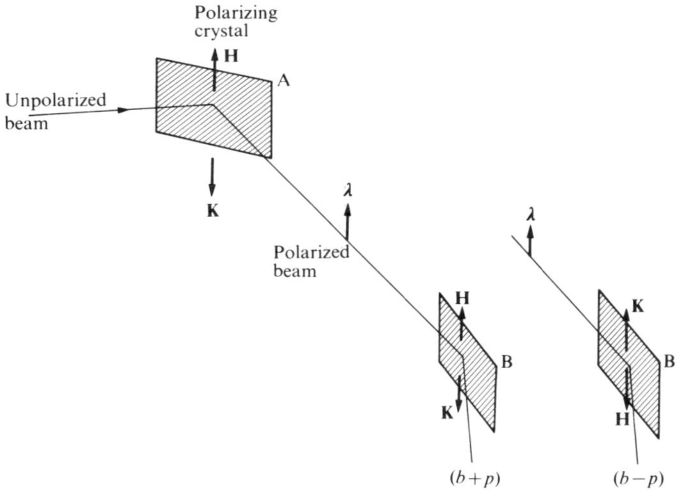

FIG. 120. Individual measurement of $(b+p)$ and $(b-p)$ by reflection of polarized beam by sample with applied magnetic field **H**, respectively, vertically upward and vertically downward. The directions of the magnetic spins **K** are, conversely, downward and upward in the two cases.

to downwards at the sample crystal B. In fact the neutron would respond to such a discontinuity by reversing its direction of polarization and would therefore arrive at B with its spin vector still pointing in the field direction. This difficulty can be avoided by following a practice which is well known in atomic-beam work. The field direction at B is always maintained to agree with that at A but the spin direction is turned around by just 180° between A and B, by applying a suitable *axial* magnetic field of radio-frequency. This technique was first utilized by Nathans, Shull, Shirane, and Andresen (1959), and the components of their spectrometer which achieved the desired result are indicated in Fig. 121.

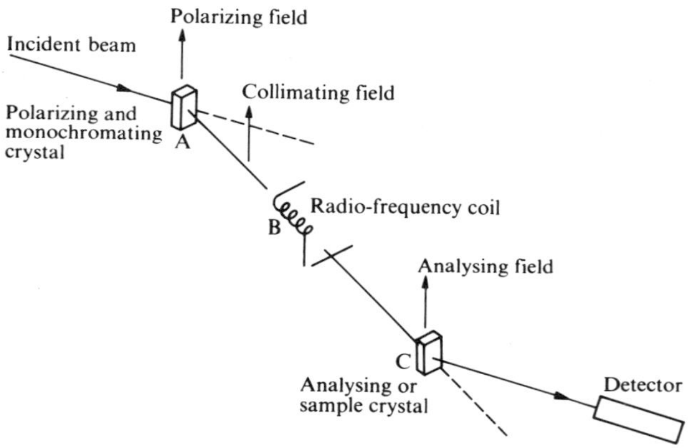

FIG. 121. A schematic diagram of a polarized-neutron spectrometer. (From Nathans *et al.* 1958.)

The initial polarizing crystal is magnetized in a field of 2200 Oe and the reflected polarized neutrons pass along a magnetic collimator which maintains a uniform field H_0 of 150 Oe in the same direction. A field of 7500 Oe, again in the same direction, is applied to the sample crystal which is under investigation. Between the two crystals is inserted the 'spin-flip' coil B which applies an axial radio frequency field $H_1 \cos \omega t$; a typical frequency will be 300 kHz. It can be shown (Alvarez and Bloch 1940; Rabi, Zacharias, Millman, and Kusch 1936) that the probability $\check{p}$ that the spin will be exactly reversed when a neutron travels for a time $\check{t}$ in this radio-frequency field, superimposed on the normal field H_0, is given by

$$\check{p} = \frac{\sin^2 [\gamma H_1 \check{t}/2h\{1+(2\Delta H/H_1)^2\}^{\frac{1}{2}}]}{1+(2\Delta H/H_1)^2}, \tag{6.23}$$

where γ is the magnetic moment of the neutron. In this expression ΔH has been written for $(H_0 - H_0^*)$, where H_0^* is the value of the field H_0 for which the neutron would have a Larmor precession frequency equal to the frequency ω of the applied radio-frequency field. Thus H_0^* is equal to $h\omega/4\pi\gamma$. It follows from eqn (6.23) that two conditions must be satisfied if complete reversal of the spin direction is to be achieved. First, ΔH must be zero so that the denominator in the expression can have its minimum value of unity, and secondly, the value of the factor $\gamma H_1 \check{t}/2h$ must equal an odd multiple of $\frac{1}{2}\pi$. Thus the experimental conditions to be satisfied are that the radio-frequency ω shall resonate with the Larmor precession frequency in the steady field H_0 and that the amplitude of H_1 shall be properly adjusted to suit the neutron transit time through the coil. The value of $\check{p}$ will then be unity.

These conditions mean that

$$\omega = (4\pi\gamma/h)H_0 \tag{6.24}$$

and

$$H_1 = \pi h/\gamma\check{t}. \tag{6.25}$$

Fig. 122 shows a view of the various magnets in the spectrometer at the Trombay reactor in India, described by Satya Murthy, Samanathan, Begum, Srinivasan, and Murthy (1969). In this arrangement the value of the collimating field H_0 was 156 Oe, leading to a value of about 452 kHz for the frequency of the radio-frequency field, by applying eqn (6.24). The achievement of the resonance condition by varying the frequency of the radio-frequency field is shown in Fig. 123, which illustrates how the neutron counting rate after reflection from the analysing crystal varies with the frequency. In practice the method proves very satisfactory and polarization reversal efficiencies of 99 per cent can be achieved. The reversal of polarization direction is produced as required, simply by switching on the radio-frequency field.

6.7. Scattering by rare-earth atoms

In the case of ions of the rare-earth series the orbital momenta are not quenched, and both the spin and orbital momenta contribute to the magnetic moment and, in turn, to the magnetic scattering. Fig. 124 is a vector diagram which indicates how the angular momentum and the magnetic moment are built up from the orbital and spin contributions.

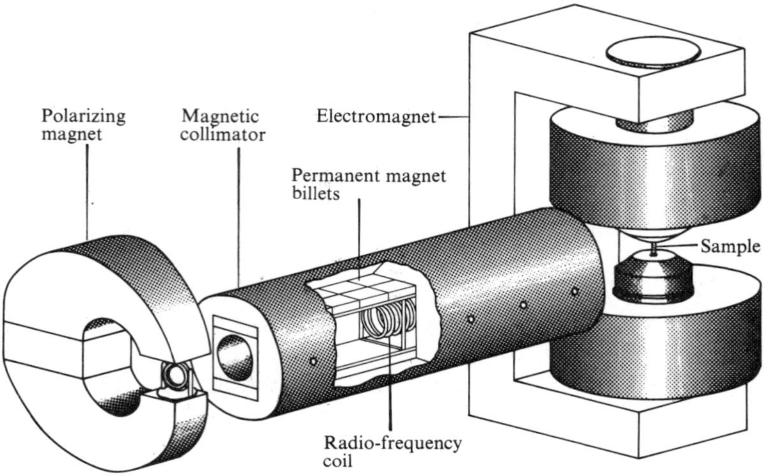

FIG. 122. A perspective view of the magnets in the polarized-neutron spectrometer at Trombay. The cut-away section reveals the radio-frequency coil used for reversing the neutron polarization. (Satya Murthy *et al.* 1969.)

Triangle OAD at the right-hand side of the diagram illustrates how the orbital momentum vector **L** and the spin vector **S** produce the resultant momentum vector **J**. The magnitudes of these momenta are respectively

$$\frac{h}{2\pi}\sqrt{\{L(L+1)\}}, \qquad \frac{h}{2\pi}\sqrt{\{S(S+1)\}}, \quad \text{and} \quad \frac{h}{2\pi}\sqrt{\{J(J+1)\}},$$

where L, S, and J are the quantum numbers and they are represented by the three sides of the triangle ODA. The magnetic moments to which these momenta give rise are indicated by triangle OEB on the left of the diagram and they are, respectively,

$$\mu_L = \frac{h}{2\pi}\sqrt{\{L(L+1)\}}\frac{e}{2mc}, \tag{6.26}$$

$$\mu_S = \frac{h}{2\pi}\sqrt{\{S(S+1)\}}\frac{e}{mc}, \tag{6.27}$$

and their resultant μ_J.

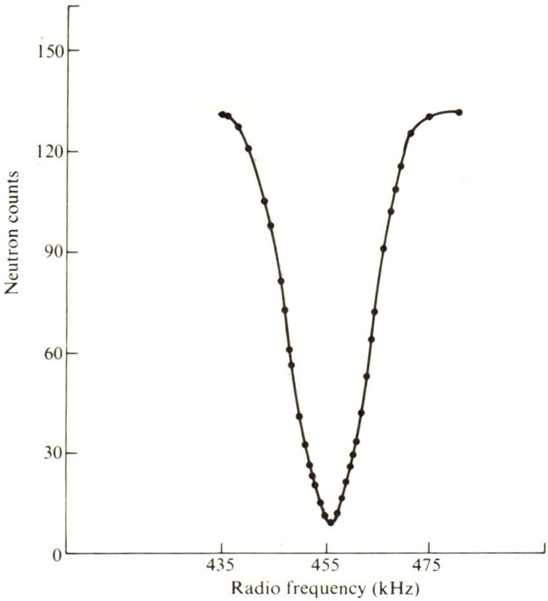

FIG. 123. Achievement of resonance condition for reversal of direction of neutron spin. A Co–Fe analysing crystal similar to the polarizing crystal is placed in the sample position and the reflected neutron intensity is plotted as a function of the radio-frequency field. The polarizing and analysing fields are in the same vertical direction. (Satya Murthy *et al.* 1969.)

It is important to note that the multiplying factor for the spin moment is twice as large as for the orbital moment, thus producing a resultant magnetic moment μ_J along OB which is *not* in line with the resultant angular momentum along OA. The result is that μ_J

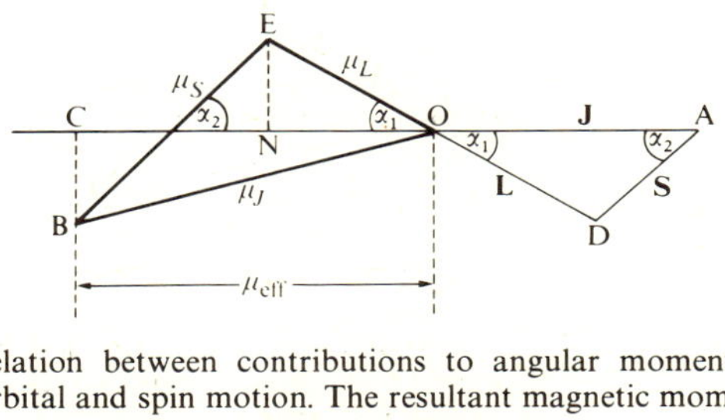

FIG. 124. Correlation between contributions to angular moment and magnetic moment from orbital and spin motion. The resultant magnetic moment μ_J precesses about AC to produce an *effective* magnetic moment μ_{eff}.

precesses around the direction OA, and the effective value of magnetic moment which can be measured experimentally is the projection of μ_J on to the direction OA, measured by OC in the diagram and identified as μ_{eff}.

We can calculate the value of μ_{eff} as the sum of the two projections NO, CN, thus

$$\mu_{\text{eff}} = \text{OE} \cos \alpha_1 + \text{BE} \cos \alpha_2, \tag{6.28}$$

and the values of $\cos \alpha_1$ and $\cos \alpha_2$ can be found from our knowledge of the lengths of the sides of triangle ODA. It is easy to show that

$$\cos \alpha_1 = \frac{-S(S+1)+J(J+1)+L(L+1)}{2\sqrt{\{L(L+1)\}}\sqrt{\{J(J+1)\}}}$$

and

$$\cos \alpha_2 = \frac{-L(L+1)+J(J+1)+S(S+1)}{2\sqrt{\{S(S+1)\}}\sqrt{\{J(J+1)\}}}.$$

Combining these expressions with eqns (6.26) and (6.27), eqn (6.28) gives

$$\mu_{\text{eff}} = \frac{eh}{4\pi mc}\sqrt{\{J(J+1)\}}\left\{\frac{S(S+1)+3J(J+1)-L(L+1)}{2J(J+1)}\right\}. \tag{6.29}$$

The factor $eh/4\pi mc$ is the Bohr magneton (μ_{B}), and the factor in the large braces is the Landé splitting factor g (eqn (6.7)), i.e.

$$\mu_{\text{eff}} = g\sqrt{\{J(J+1)\}}\ \mu_{\text{B}}, \tag{6.30}$$

in contrast to

$$\mu_{\text{eff}} = 2\sqrt{\{S(S+1)\}}\ \mu_{\text{B}} \tag{6.31}$$

for a 'spin-only' moment.

In line with this deduction, Trammell (1953) has shown that for the differential scattering cross-section our earlier spin-only equation

$$\mu_{\text{eff}} = 2\sqrt{\{S(S+1)\}}\mu_{\text{B}} \tag{6.31}$$

becomes

$$\text{d}\sigma_{\text{pm}} = \tfrac{2}{3}g^2\frac{J(J+1)}{4}\left(\frac{e^2\gamma}{mc^2}\right)^2 f^2. \tag{6.32}$$

At the same time the equation for the coherent scattering length

$$p = S\left(\frac{e^2\gamma}{mc^2}\right)f = \left(\frac{e^2\gamma}{2mc^2}\right)2Sf, \tag{6.12}$$

becomes

$$p = \frac{gJ}{2}\left(\frac{e^2\gamma}{mc^2}\right)f = \left(\frac{e^2\gamma}{2mc^2}\right)gJf,$$

which we have already quoted as eqn (6.14).

It remains for us to consider the form factor f in this situation where the magnetic moment is accounted for by both spin and orbital contributions. The effective form factor will be a composite value which depends on two individual form factors f_S and f_L computed for the pure spin moment and pure orbital moment respectively. Such calculations have been performed by Blume, Freeman, and Watson (1962), and we shall refer to them again later. Here we merely wish to note the statistical weights which we have to give to f_S and f_L in order to arrive at a value of f for insertion in eqns (6.13) and (6.14). The correct proportions can be shown to be equal to the proportional contributions which the spin and orbital magnetic moments make to the over-all magnetic moment, i.e. the fractions CN/CO and NO/CO in Fig. 124, which provide the two terms in eqn (6.28).

It is easily found from our previous analysis that

$$\frac{\mathrm{CN}}{\mathrm{CO}} = \frac{J(J+1)+L(L+1)-S(S+1)}{3J(J+1)+S(S+1)-L(L+1)} \quad \text{or} \quad \frac{J(J+1)+L(L+1)-S(S+1)}{2gJ(J+1)}$$

and

$$\frac{\mathrm{NO}}{\mathrm{CO}} = \frac{2\{J(J+1)+S(S+1)-L(L+1)\}}{3J(J+1)+S(S+1)-L(L+1)} \quad \text{or} \quad \frac{2\{J(J+1)+S(S+1)-L(L+1)\}}{2gJ(J+1)}.$$

Hence

$$f = \frac{\{J(J+1)+L(L+1)-S(S+1)\}f_L + 2\{J(J+1)+S(S+1)-L(L+1)\}f_S}{3J(J+1)+S(S+1)-L(L+1)}. \tag{6.33}$$

7

MAGNETIC FORM FACTORS

As we have already indicated, the study of the atomic form factor for magnetic scattering is of importance from two points of view. First, a knowledge of the angular variation of the form factor is essential if magnetic structures are to be deduced from intensity measurements and, secondly, it is possible to make detailed deductions about electron distributions in space if the form factor can be measured by experiment.

7.1. The iron group of transition metals

Mn, Fe, Co, Ni

We showed in Fig. 110 an early experimental curve for the form factor of the ion Mn^{2+}, deduced from measurements of the paramagnetic scattering by MnF_2. Fig. 125 shows a collection of fairly early data by Shull and Wollan (1956) from the Bragg diffraction peaks of a number of manganese compounds, including a number of different ionic states. Within the accuracy of the measurements the use of a single form-factor curve for these different ionic states can be justified.

More recently the introduction of polarized-beam measurements, giving high accuracy of determination for the scattering factors of ferromagnetic substances, has led to much more precise knowledge of form factors and has justified careful comparison of these with theoretical calculations. Fig. 126 presents results for iron. The individual points are for metallic iron from Shull and Yamada's (1962) work with polarized neutrons. The two full-line curves are calculated by Wood and Pratt (1957) for a free iron atom for, respectively, the five 3d electrons of +ve spin and the single 3d electron of −ve spin. The two curves are different because the exchange interaction between the 3d electrons leads to a slightly different distribution of the +ve, −ve electrons in the presence of the resultant +ve spin. The +ve electrons are less diffuse, giving a slightly broader *f* curve. As a result of the different distribution of +ve, −ve spins there is also a difference between the radial

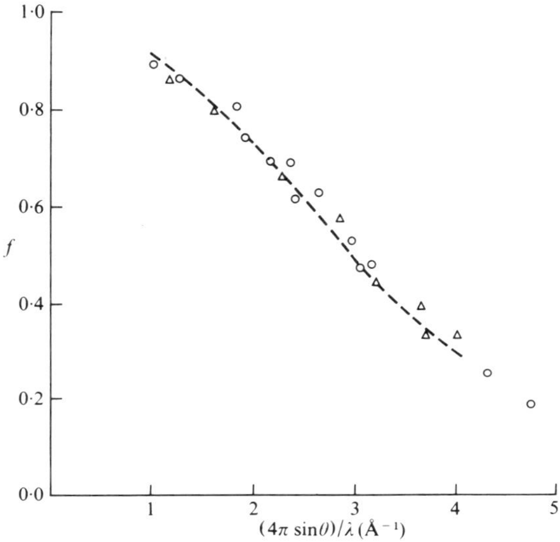

FIG. 125. The magnetic form factor for manganese showing experimental data from MnO ▲, MnF_2 ○, Mn_2Sb ●, and $Mn(La_{0\cdot 65}Ca_{0\cdot 35})O_3$ △. The dotted curve is calculated for Mn^{2+}, using Pauling and Sherman's method. (After Shull and Wollan 1956.)

distribution of charge density and spin density. For comparison, the broken line curve in the Figure shows the experimental results for the Fe^{3+} ion in magnetite, studied by Nathans, Pickart, and Alperin (1960). The relative position of this curve suggests that the unpaired electrons in the ionic oxide are more spread out than in the metal itself. Equally it may be deduced from the Figure that the radial distribution of the 3d electrons in the metal is not very different from that in the free atom.

Comprehensive results have also been obtained for nickel, and these are displayed in Fig. 127 which contrasts the data of Mook (1966) for the metal with those of Alperin (1961) for Ni^{2+} in NiO. It is again found by calculation that there is a difference between the radial distribution curves for the +ve, −ve electrons, but this is substantially less marked than for iron. This is illustrated very clearly in Fig. 128 from the calculations of Weiss and Freeman (1958). Similarly Fig. 129 gives results for cobalt and compares measurements for the cubic form of the metal (Nathans and Paoletti 1959), the hexagonal form of the metal (Moon 1964), and

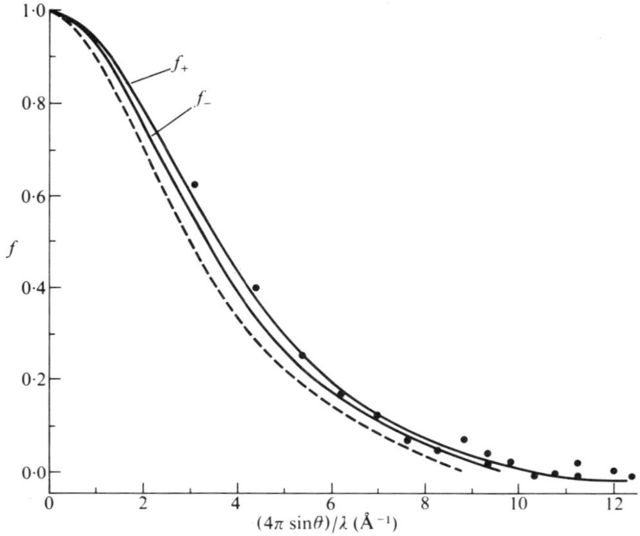

FIG. 126. Magnetic form factor curves for iron. The full-line curves show the calculation by Wood and Pratt for, respectively, the five 3d electrons of +ve spin and the single 3d electron of −ve spin for a free iron atom. The points are the experimental measurements of Shull and Yamada for metallic iron. The broken-line curve contrasts the experimental data for the Fe^{3+} ion in magnetite.

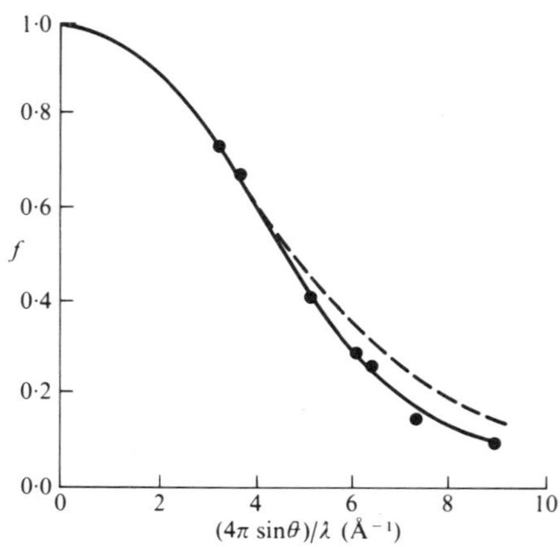

FIG. 127. Experimental form-factor data for nickel, contrasting measurements of Mook for the metal (points and full-line) with those for Ni^{2+} in NiO (Alperin) indicated by the broken-line curve.

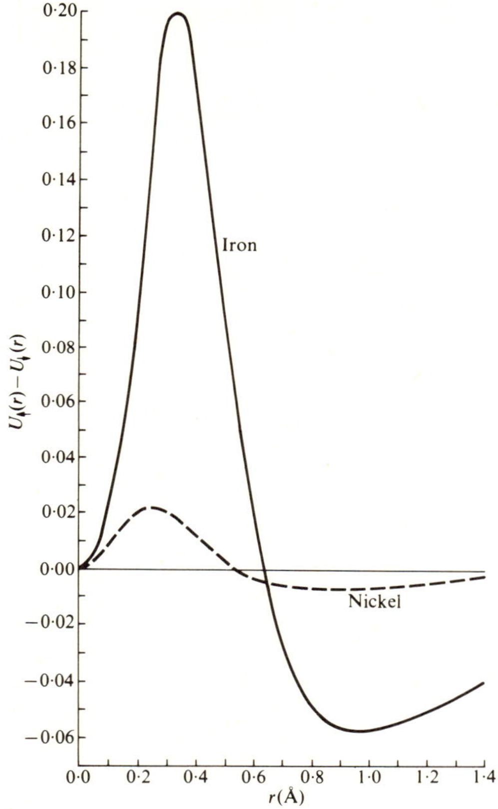

FIG. 128. Difference between the radial distribution curves for the electrons of positive and negative spin. The full-line curve is for a free iron atom; the broken-line curve is for nickel. (Weiss and Freeman 1959.)

the Co^{2+} ion in CoO (Khan and Erickson 1970); there are some clear differences for the ion. It will be noticed both in Fig. 127 for nickel and in Fig. 129 for cobalt that the curve for the divalent ion lies above the curve for the metal, suggesting that the unpaired electrons in the oxide are less spread out than in the metal—a contrary conclusion to that which was reached for iron.

A useful comparison of the experimental data for the three metals Fe, Co, and Ni can be made by calculating radial distributions for the unpaired 3d electrons in each case, utilizing the Fourier transform relation of eqn (6.3). The results are illustrated in Fig. 130 and show a clear compression of the electron distribution as the atomic number increases.

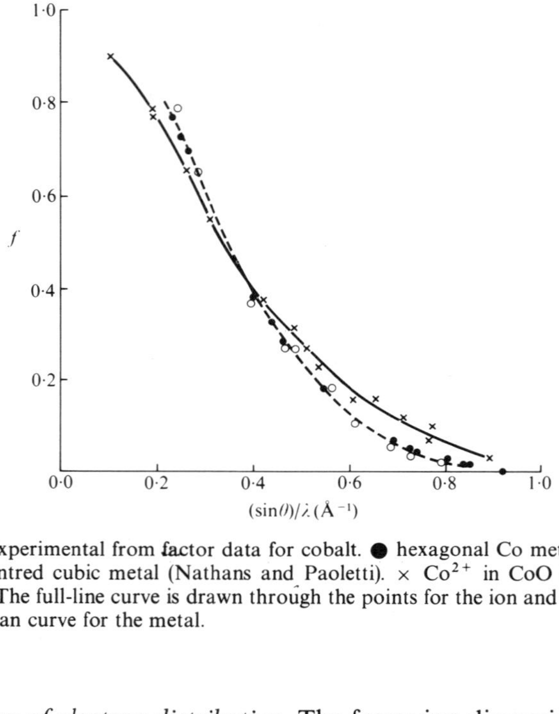

FIG. 129. Experimental from factor data for cobalt. ● hexagonal Co metal (Moon). ○ Face-centred cubic metal (Nathans and Paoletti). × Co^{2+} in CoO (Khan and Erickson). The full-line curve is drawn through the points for the ion and the broken line is a mean curve for the metal.

Anisotropy of electron distribution. The foregoing discussion of the electron distributions and form factors of the 3d metals must be regarded only as a first approximation to the truth. It assumes that the electron distribution has spherical symmetry, with the inference that the form-factor curve is not dependent on the orientation of the scattering vector with respect to the crystallographic axes of the material. Refined experiments with polarized neutrons have shown that these assumptions are not true. Moreover, these experiments also give strong support to the conclusion that there is negative polarization of the conduction electrons. The first evidence for this came from the form-factor curve for iron of Shull and Yamada (1961). The experimental points, some of which were

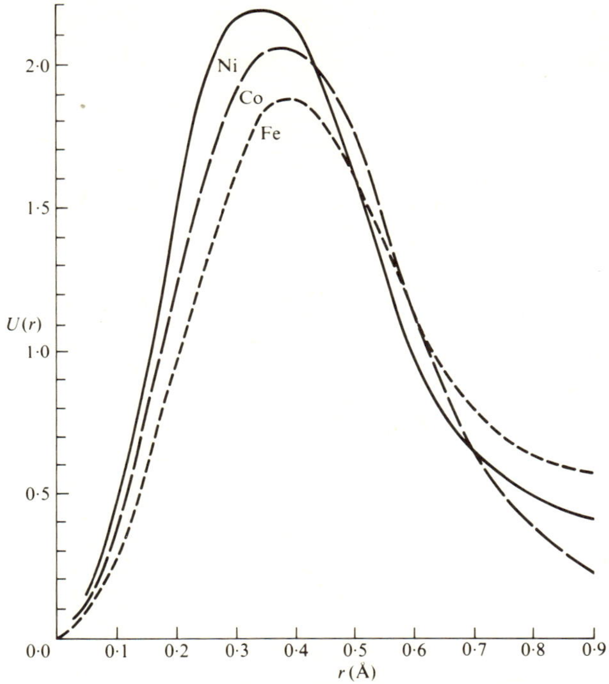

FIG. 130. Normalized radial distribution functions for the unpaired 3d electrons in metallic iron, cobalt, and nickel, produced by Fourier transform of the neutron form-factor data. (Nathans and Pickart.)

shown in Fig. 126, were for the following 26 reflections:

110	321	510, 431	620
200	400	521	541
211	411, 330	440	622
220	420	433, 530	
310	332	442, 600	
222	422	611, 532	

It will be noted that there are five pairs of reflections for which the scattering angle is the same within the pair. Inspection of Fig. 131, which emphasizes the high-angle section of the experimental data, will show that there are substantial differences between the values of f deduced for the two members of each pair; the differences are far greater than can be ascribed to experimental inaccuracies.

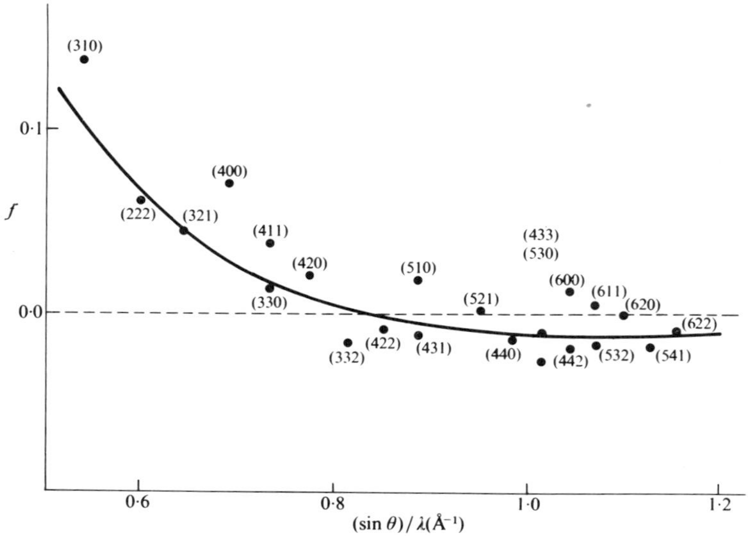

FIG. 131. The high-angle portion of the Shull and Yamada form-factor data for metallic iron, emphasizing the differences between five pairs of reflections with the same value of scattering angle.

The reason for these differences is that the form factor is sensitive to direction because the unpaired-electron density is not symmetrical about the nucleus. At the same time the measurements lead to the conclusion that the net spin in the metallic case is not entirely due to 3d electrons; the 4s conduction electrons are partially polarized, in a direction opposite to that of the 3d polarization.

For a ferromagnetic crystal under an applied field the three-dimensional distribution function of the unpaired electrons can be synthesized, by applying the technique which is conventionally employed in X-ray structure analysis to the magnetic scattering intensities measured with polarized neutrons.

Thus, the unpaired-electron density at the point x, y, z is

$$\rho(x, y, z) = \frac{1}{V} \sum \sum \sum (F_{magn})_{hkl} \cos 2\pi(hx/a + ky/b + lz/c), \qquad (7.1)$$

where $\rho(x, y, z)$ is the electron density in Bohr magnetons per cubic ångström and the summation is taken over all combinations of indices hkl; V is the volume of the unit cell.

From the results of this computation it is possible to plot the electron distribution for any plane in the crystal. Fig. 132 shows such a plot on the (100) basal plane of the crystal, and it will be seen that the contours are not circular but indicate that there is an excess of density along the [100] axes and a deficiency along [110]. A similar plot on a (110) plane shows that there is an even greater deficiency along the [111] direction. The conclusion is

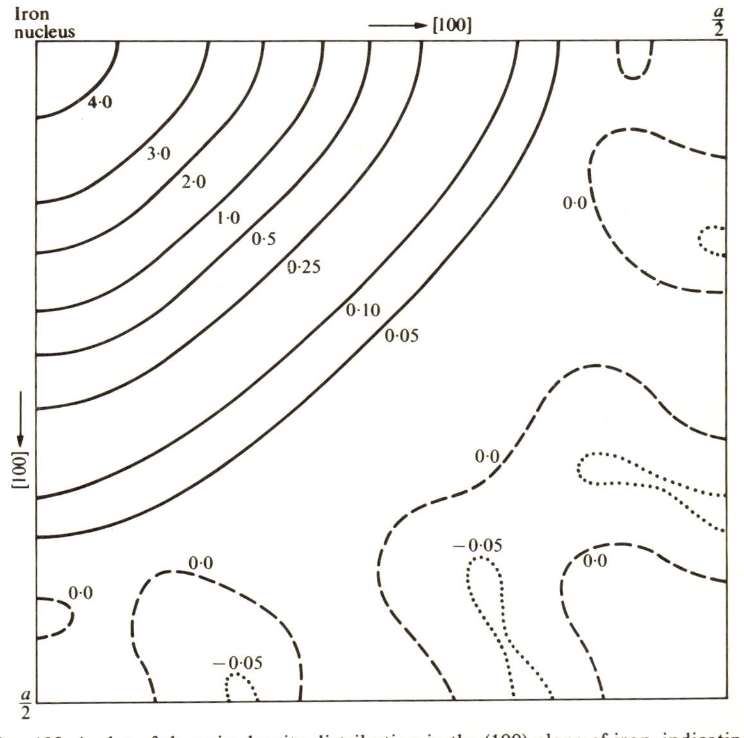

FIG. 132. A plot of the spin-density distribution in the (100) plane of iron, indicating an excess along the [100] axis relative to that along [110]. Contours are in units of μ_B Å^{-3}. (Shull and Yamada 1962.)

evident in a more striking way in Fig. 133 which is a plot of the *excess* magnetic electron density observed experimentally over that calculated for a spherically symmetrical free atom. There is an evident excess at points on the [100] axes and a deficiency in the

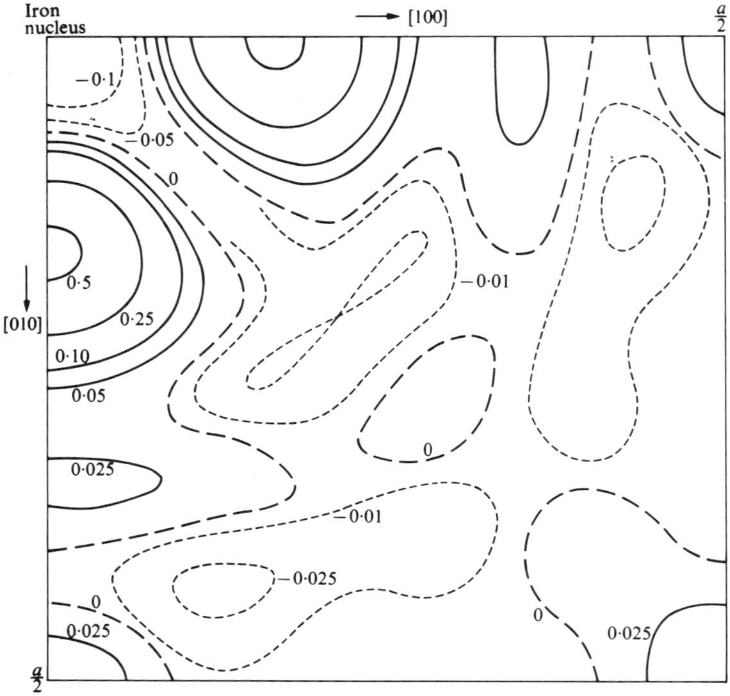

FIG. 133. A plot of the excess spin density in iron, measured experimentally, compared with that expected for a spherically symmetrical free atom. Contours are in units of μ_B Å^{-3}. (Shull and Yamada 1962.)

face- and space-diagonal directions. These results are interpreted in terms of the relative occupancy of E_g and T_{2g} orbitals. The former give spin density along the [100] cube edges and the latter along the [111] diagonals. Spherical symmetry would be attained for 60 per cent occupancy of T_{2g} and 40 per cent of E_g. For iron it is concluded from the neutron results that the E_g orbitals are favoured, with an occupancy of 53 per cent. In further refinement of these data Shull and Mook (1966) have developed a method of assessing the residual magnetization at points in the unit cell which are remote from the atomic sites, such as the mid-points of the cube edges. There is, in general, a small negative magnetization, of opposite sign to that from the 3d electrons, and it is distributed in the form of interlocking rings, centred on points such as $\frac{1}{2}00$

and forming a three-dimensional chain structure. This is shown in Fig. 134 with the addition of the asymmetrical 3d magnetization which is concentrated along the cube edges. Shull and Yamada suggested that the negative magnetization was contributed by the 4s electrons, amounting in total to $-0{\cdot}21\ \mu_B$, which together with $+2{\cdot}39\ \mu_B$ from the 3d electrons accounted for the observed total moment of iron of $+2{\cdot}18\ \mu_B$.

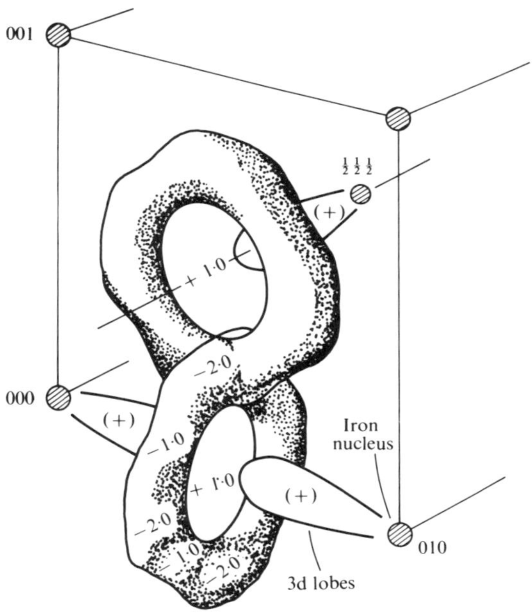

FIG. 134. The residual magnetization (kG) in iron as deduced experimentally by Shull and Mook (1966). The positive excess of 3d magnetization along the cube edges is over-emphasized in comparison with the interlocking rings of negative magnetization. The rings shown are centred on $0\frac{1}{2}0$, $0\frac{1}{2}\frac{1}{2}$.

More recently, metallic nickel has been studied in a similar way by Mook and Shull (1966) and Mook (1966), and some very accurate results have been obtained, in spite of the fact that the magnetic scattering from nickel is, as Table 16 shows, substantially smaller than that for iron. Fig. 135 indicates the magnetic moment distribution in the (100) and (110) planes. In contrast with the case for iron, it is the T_{2g} orbitals which are favoured beyond the proportions

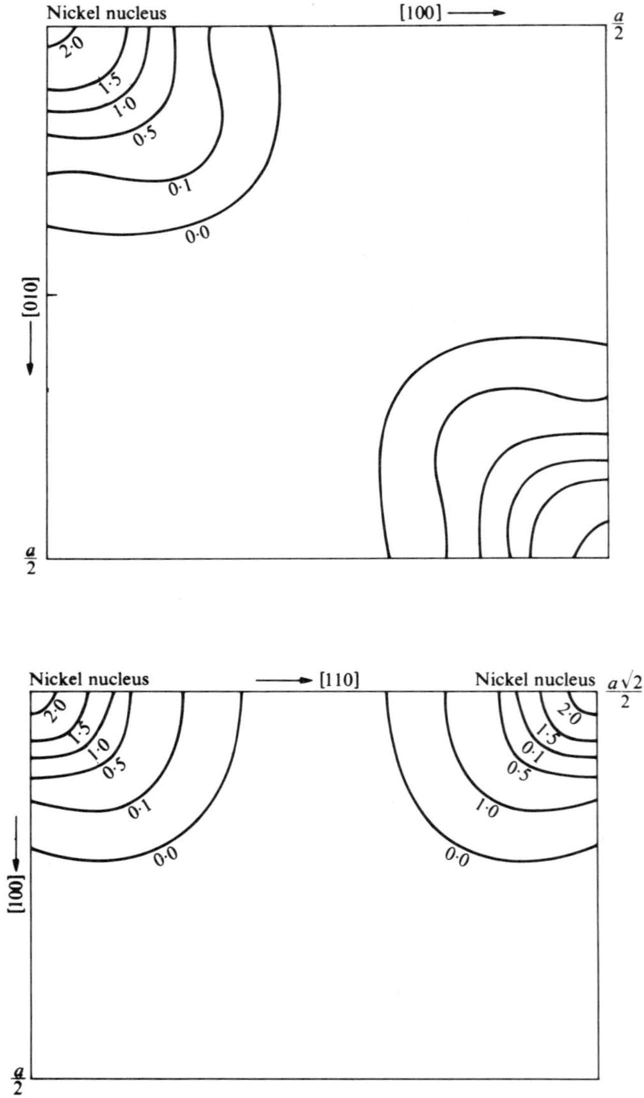

FIG. 135. Distribution of magnetic moment in the (100) and (110) planes of nickel. The contour values indicate units of $\mu_B\,Å^{-3}$ and the general background measures $-0{\cdot}0085\,\mu_B\,Å^{-3}$. (Mook 1966.)

required for spherical symmetry. In fact 81 per cent of the 3d electrons occupy T_{2g} orbitals, giving a marked excess of electron density along the cube diagonals in Fig. 135. The second important conclusion from the Fourier plot of the unpaired-electron density is that there is a region of almost constant negative spin density between neighbouring nickel atoms, as illustrated in Fig. 136. The density of this background is estimated at $-0{\cdot}0085\ \mu_B\ \text{Å}^{-3}$, which,

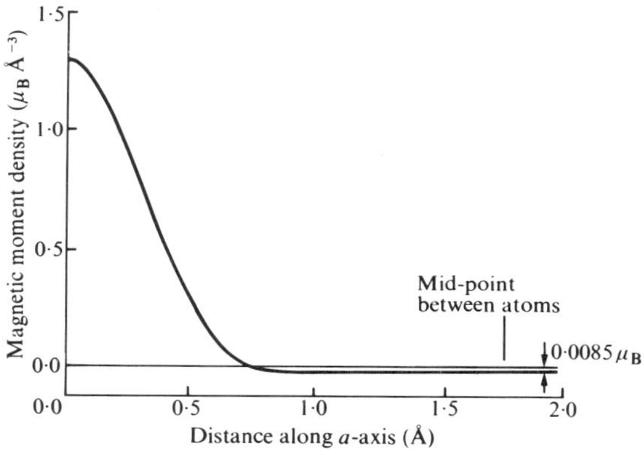

FIG. 136. The variation of magnetic moment density along the [100] direction between next-nearest neighbour atoms in nickel. (Mook 1966.)

in a unit cell of edge 3·52 Å containing four atoms, amounts to an integrated value of about 0·09 electrons per atom. This conclusion is in good agreement with the implications of the calculations of Watson and Freeman (1960, 1961) based on the measured form-factor curves. An extremely good fit between theory and experiment is given by a model which distributes the measured magnetic moment of nickel, amounting to $0{\cdot}606\ \mu_B$, in the following way: 3d spin $0{\cdot}656\ \mu_B$, 3d orbit $0{\cdot}055\ \mu_B$ (appropriate to a g value of 2·20), and 4s $-0{\cdot}105\ \mu_B$. This agreement is very evident in Fig. 137 in which the asymmetry of the electron distribution is emphasized by the substantial difference in form factor for the pairs of reflections (511, 333), (600, 442), (711, 551), and (731, 533).

Cobalt also has been examined, in the hexagonal form by Moon (1964) and as face-centred cubic cobalt by Nathans and Paoletti (1959). Moon's experiments show that, in contrast to the results for iron and nickel, there is almost complete spherical symmetry

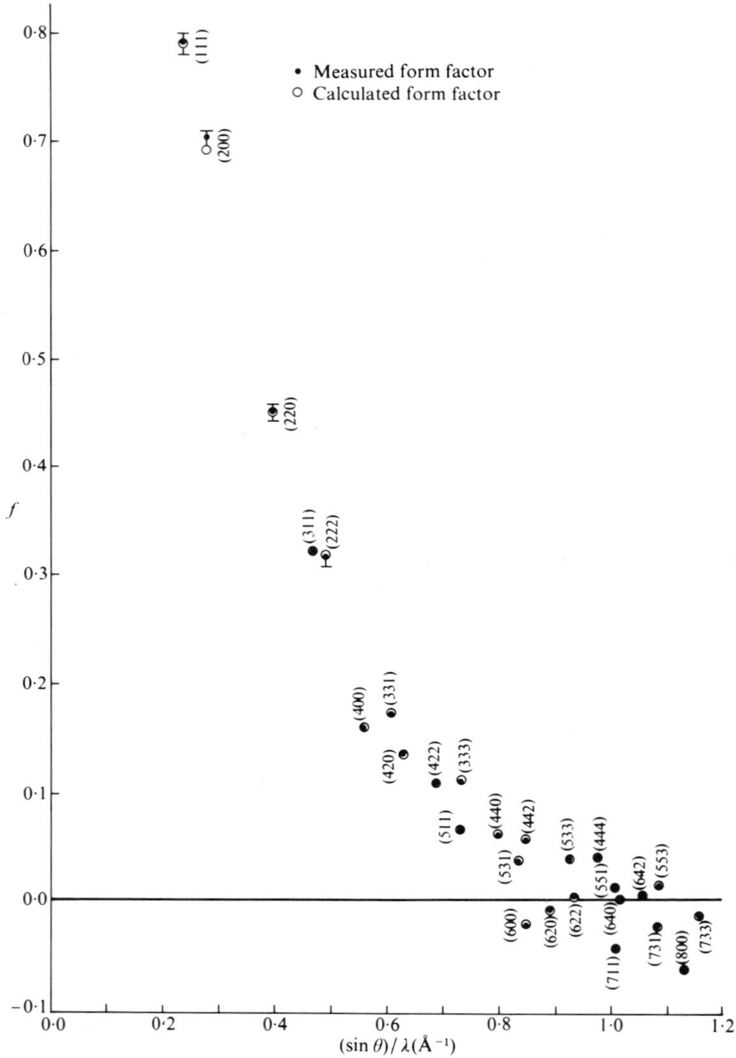

FIG. 137. Comparison of the measured and calculated values of the form factors for nickel, emphasizing the dependence of f on the orientation of the reflecting planes. (Mook 1966.)

in the distribution of the spin electrons for hexagonal cobalt, though again with an almost constant negative spin density at distances remote from the atoms. The allocation of moments is 3d spin $1{\cdot}86\,\mu_B$, 3d orbit $0{\cdot}13\,\mu_B$, and 4s $-0{\cdot}28\,\mu_B$. The symmetry of the distribution is evident in Fig. 129, where all the points for hexagonal cobalt lie on a single smooth curve. This is not the case for the face-centred cubic metal, whose points lie less satisfactorily on a curve and also indicate some asymmetry.

Ti, V, Cr

Information on the form factors of the other 3d metals—titanium, vanadium and chromium—is much more limited, mainly because they do not exist in the ferromagnetic form which can be so accurately investigated with polarized neutrons.

Chromium exists below 313 K in a complex antiferromagnetic structure to which we shall refer later. However, by addition of traces of manganese it can be caused to take a simple antiferromagnetic structure in which corner and centre atoms of the body-centred cube have their magnetic moments oppositely directed, and in this form single crystals have been studied by Moon, Koehler, and Trego (1966). Considerable departures from spherical symmetry were found, as indicated, for example, by the different values of form factor for the (221) and (300) reflections, and it was concluded that the T_{2g} levels were occupied by 79 per cent of the 3d electrons, compared with the 60 per cent required for spherical symmetry.

There is no known magnetically ordered state of vanadium, but the polarized neutron technique has been used to assess the extent to which the paramagnetic atoms are aligned in a strong magnetic field. For a molar susceptibility χ_m and an applied field H the effective magnetic moment per individual atom will be $\chi_m H/N_c$, where N_c is Avogadro's number. This moment plays the same part as the quantity $2Sf\,\mu_B$ in eqn (6.12), so that the latter equation becomes

$$p = 0{\cdot}484\chi_m Hf \times 10^{-16}\ \text{cm} \qquad (7.2)$$

when the numerical values of the constants are inserted. Shull and Ferrier (1963) determined the value of p experimentally by measuring the polarization ratio for the reflections (110) and (220) for a series of values of the applied magnetic field. This ratio R is the ratio of the observed intensities for the two neutron spin states, up and down respectively, relative to the applied field. Thus

$$R = \frac{(b+p)^2}{(b-p)^2} = 1 + \frac{4p}{b} \tag{7.3}$$

when p is small. By plotting the value of R against the magnetic field H we obtain a measure of the form factor f. The variation of p/b with $(\sin\theta)/\lambda$ is indicated in Fig. 138 together with theoretical

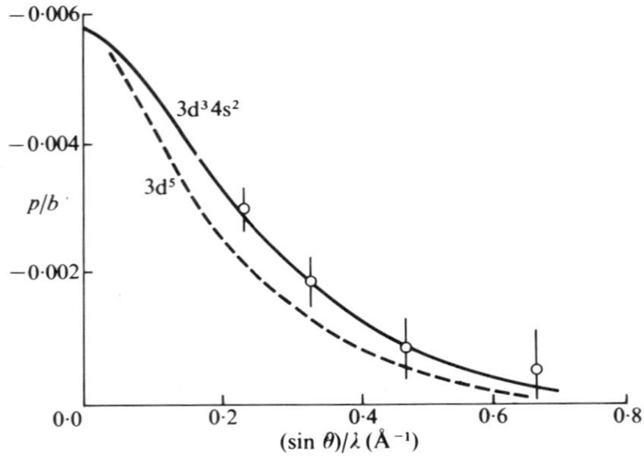

FIG. 138. Variation with scattering angle of the magnetic scattering amplitude of vanadium for the paramagnetic atoms aligned in magnetic fields up to 24 kOe. Theoretical curves are shown in relation to the experimental points. (Shull and Ferrier 1963.)

form factors, from Watson and Freeman, for both $3d^5$ and $3d^54s^2$ groupings of electrons. Good agreement is obtained with the latter curve when the susceptibility is distributed among the following components: 3d orbit $+1{\cdot}76$, 3d spin $-0{\cdot}68$, 4s spin $+0{\cdot}30$, diamagnetic $-0{\cdot}15$, totalling the observed value of $2{\cdot}60 \times 10^{-4}$ e.m.u. mol^{-1}.

A redetermination of the form factor of Mn^{2+} has recently been made by Jacobson, Tofield, and Fender (1973) from powder diffraction measurements of MnO. The results are presented in Fig. 139, where the crosses indicate the experimental results, which are in very good agreement with the full-line curve which represents the calculation of Watson and Freeman (1961) for the free Mn^{2+} ion. The broken line in the Figure corresponds to the earlier curve which we gave in Fig. 125. Bearing in mind that intensities are

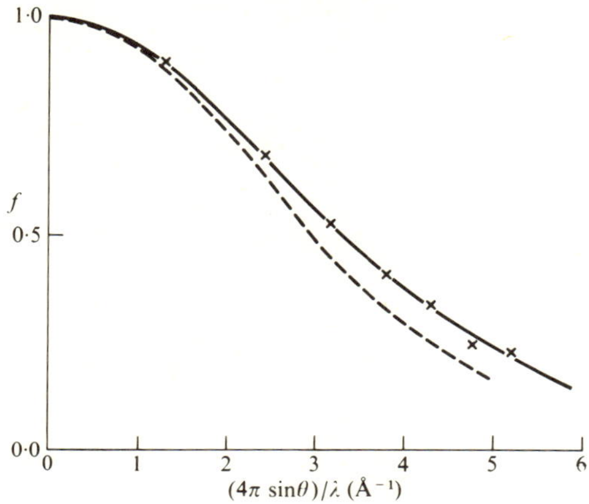

FIG. 139. Determination of the form-factor curve of Mn^{2+}. The experimental points × of Jacobsen *et al.* (1973) for MnO are in good agreement with the calculated curve ——— of Watson and Freeman (1961) for the free Mn^{2+} ion. An early experimental curve ---- of Shull and Wollan (1956) is substantially different.

proportional to f^2 it will be clear that the new curve implies substantial changes in intensity at large angles of scattering.

No measurements have been reported on metallic titanium, but some information on the Ti^{3+} ion in Ti_2O_3 has been given by Abrahams (1963).

7.2. The 4d metals

The 4d metals rhodium, palladium, and zirconium do not occur as ordered magnetic materials, but certain of their alloys are ferromagnetic and single crystals have been studied with polarized neutrons, in order to determine the magnetic form factors of the elements.

Shirane *et al.* (1964) have examined Fe-Rh at the composition 48 per cent rhodium, which is ferromagnetic; with additional rhodium the alloy becomes antiferromagnetic. The shape of the form factor for rhodium can be seen in Fig. 140. It is much narrower than the curve for Fe, which is also shown, corresponding to the fact that the 4d electrons are much more spread out in space than are the 3d electrons. A Fourier plot of the magnetic moment density on the (110) plane for the alloy, in which the iron and rhodium atoms are almost completely ordered chemically, is shown in

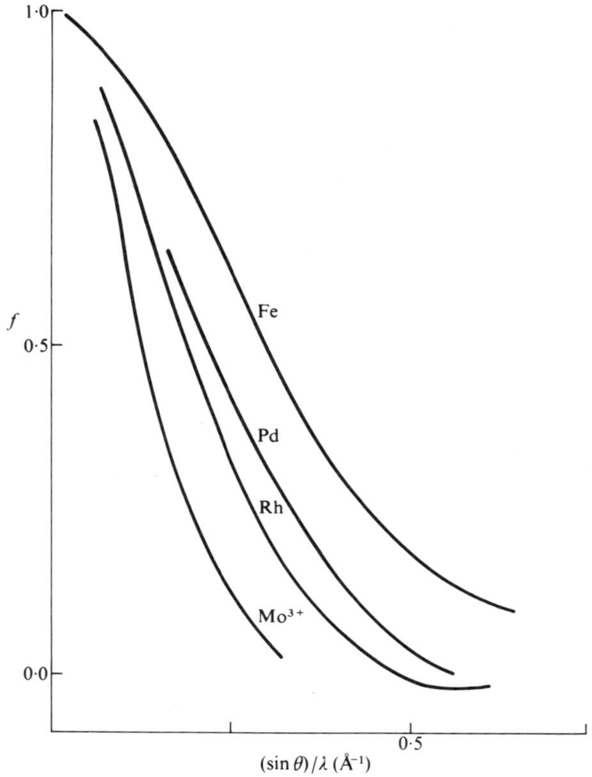

FIG. 140. Experimental form-factor curves for Fe, Rh, Pd, and Mo. (Shirane *et al.* 1964.)

Fig. 141. The iron atom has almost spherical symmetry but the rhodium atom has an excess of electron density along the [001] axis. This corresponds to a preponderance of E_g modes among the 4d electrons. Calculation shows the presence of 62 per cent of E_g symmetry, compared with the 40 per cent required for a spherical distribution.

Palladium has been studied in the alloy Pd_3Fe, in which the iron and palladium atoms are very largely confined to the corners and face centres, respectively, of the face centred cubic unit cell. The magnetic structure factors lie on two separate curves in Fig. 142, depending whether the indices of the reflection are either (1) all even or all odd or (2) mixed, since these two types depend on $p_{Fe}+3p_{Pd}$ and $p_{Fe}-p_{Pd}$ respectively. The distinction between the shapes of the two curves indicates the difference between the individual form-factor curves for iron and palladium. As Fig. 140 indicates, the curve for palladium is very much sharper and is quite similar to that for rhodium, which we discussed above. Comparison of the structure factors of pairs of reflections which occur at identical values of $(\sin\theta)/\lambda$, such as (500, 340) and (511, 333),

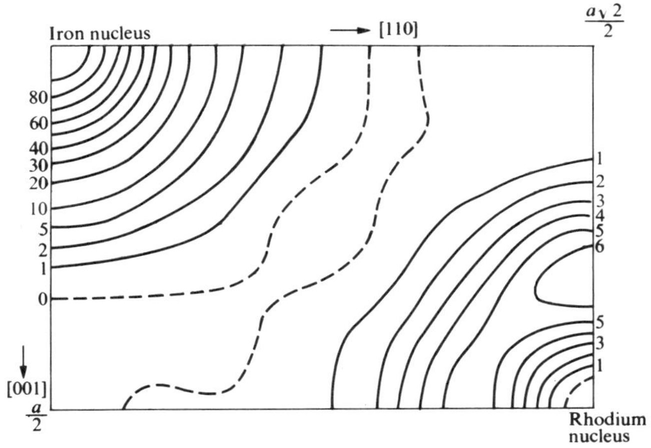

FIG. 141. A plot of the density of magnetic moment in the (110) plane of RhFe indicating asymmetry for the rhodium atom. (Shirane *et al.* 1964.)

indicates that there is little departure of the electron distribution from spherical symmetry. It is estimated that the E_g character of both the iron and palladium atoms is within ± 5 per cent of the 40 per cent required for spherical symmetry. Later measurements on the equiatomic composition PdFe by Antonini, Felcher, Mazzone, Menzinger, and Paoletti (1964) suggest a slight excess of E_g, but an alloy containing only 1–2 per cent of iron is reported to show definite T_{2g} symmetry.

Zirconium has been studied in the alloy $ZrZn_2$, which is ferromagnetic with a magnetic moment of 0·18 μ_B per molecule. Neither zirconium nor zinc is known to have a localized moment in any other compound, so that the detailed location of the moment in $ZrZn_2$ is of particular interest. The measurements are less accurate

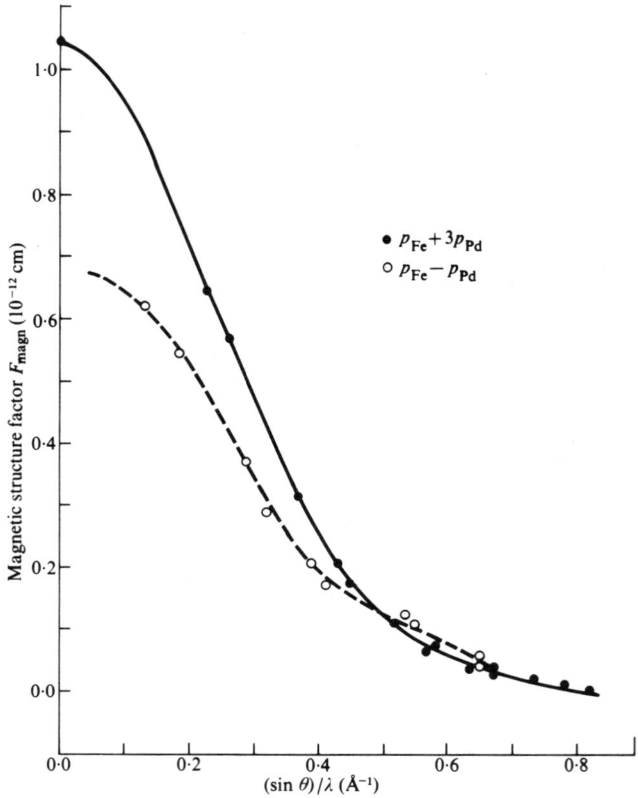

FIG. 142. Angular variation of magnetic structure factor for two different groups of reflections in Pd_3Fe. For the full-line curve the structure factor is $p_{Fe}+3p_{Pd}$, whereas the broken-line curve depends on $p_{Fe}-p_{Pd}$. (Shirane *et al.* 1964.)

than those which we have described for other elements because of the small value of the moment. They show that the moment is almost entirely associated with the zirconium atoms but is distributed very anisotropically, with a considerable concentration along the line which joins near-neighbour atoms of zirconium. This is indicated in Fig. 143, which plots the spin density along this line and contrasts the observed result with a calculated curve for spherical Zr^{2+} ions, which predicts a deep minimum at the centre point. There is possibly a very slight magnetic moment, less than 0·002 μ_B, associated with the zinc atoms.

7.3. The rare-earth elements

We have shown in the previous chapter how the form factors for the rare-earth elements include two components, representing the contributions to the magnetic moment from the spin and the orbital momenta of the electrons. Eqn (6.33) has expressed the resulting form factor in terms of the two separate factors f_L and f_S. Expressions for these have been calculated by Blume, Freeman, and Watson (1962), and the curves are of substantially different shape for the orbital and spin components. Fig. 144 shows the two curves for

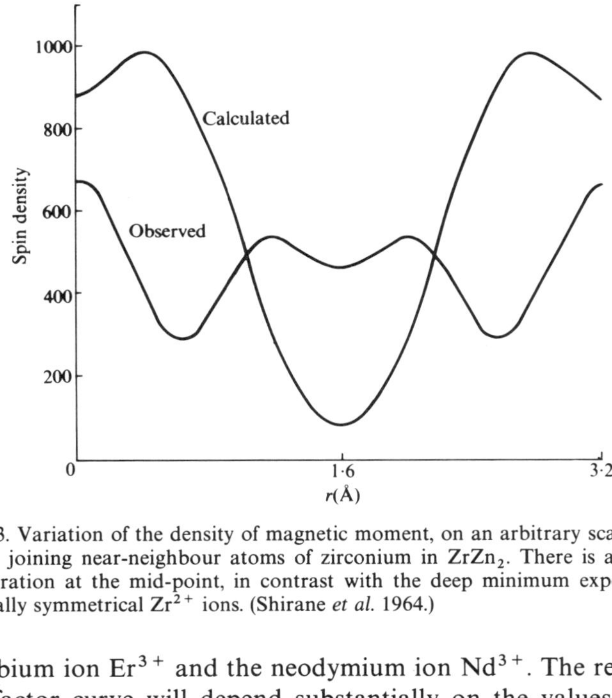

FIG. 143. Variation of the density of magnetic moment, on an arbitrary scale, along the line joining near-neighbour atoms of zirconium in $ZrZn_2$. There is a marked concentration at the mid-point, in contrast with the deep minimum expected for spherically symmetrical Zr^{2+} ions. (Shirane *et al.* 1964.)

the erbium ion Er^{3+} and the neodymium ion Nd^{3+}. The resultant form-factor curve will depend substantially on the values of the weighting factors which are the coefficients of f_L and f_S in eqn (6.33), particularly when these coefficients are of opposite sign. Table 17 gives the values of the coefficients for a number of rare-earth ions. Fig. 145 shows the calculated values† of the *resultant* form factor

† There are errors in the calculated form factors for Nd^{3+} and Er^{3+} in Blume, Freeman and Watson's (1962) paper due to the use of incorrect weighting factors. This was pointed out in Blume, Freeman and Watson (1964).

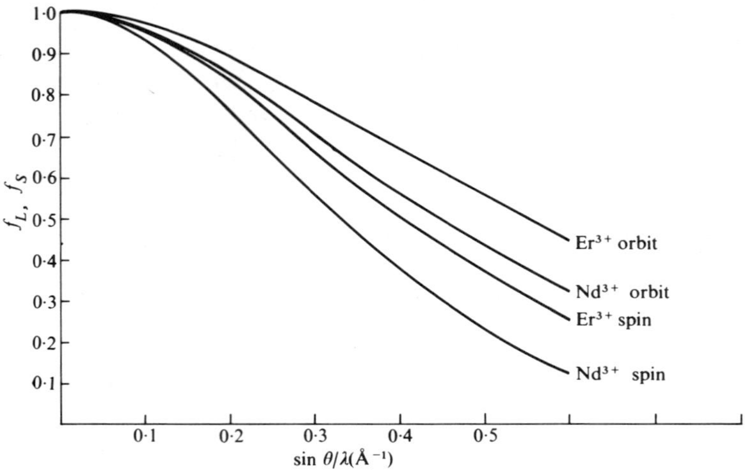

FIG. 144. Calculated values of the separate orbital and spin form factors f_L, f_S for Nd^{3+} and Er^{3+}, from the data of Blume, Freeman, and Watson (1962).

for neodymium and erbium, together with the experimental values obtained from Koehler and Wollan's (1953) original measurements of the paramagnetic scattering of the rare-earth oxides Nd_2O_3 and Er_2O_3. For erbium in particular the experimental form factor is significantly sharper than the calculated curve.

An interesting case is provided by samarium for which, as Table 17 shows, the two coefficients are almost equal in magnitude and of opposite sign. When the two terms in eqn (6.33) are compounded the resulting form-factor curve shows a substantial initial rise as $(\sin\theta)/\lambda$ increases, passing through a maximum at a value of $(\sin\theta)/\lambda \simeq 0{\cdot}4$, as illustrated in Fig. 146. Experimental support for

TABLE 17

Ion	S	L	J	Orbital coefficient	Spin coefficient
Pr^{3+}	1	5	4	1·50	−0·50
Nd^{3+}	$\frac{3}{2}$	6	$\frac{9}{2}$	1·75	−0·75
Sm^{3+}	$\frac{5}{2}$	5	$\frac{5}{2}$	6·00	−5·00
Tb^{3+}	3	3	6	0·33	0·67
Ho^{3+}	2	6	8	0·60	0·40
Er^{3+}	$\frac{3}{2}$	6	$\frac{15}{2}$	0·67	0·33

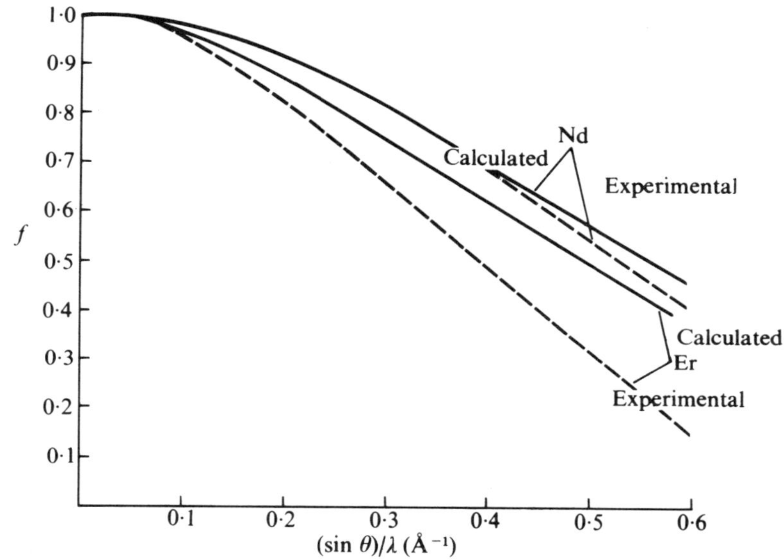

FIG. 145. The *resultant* calculated form factors (full line) for Nd^{3+}, Er^{3+} together with experimental curves (broken line) from Koehler and Wollan's measurements of the paramagnetic scattering of Nd_2O_3 and Er_2O_3.

this unusual form-factor curve is provided by Koehler and Moon's (1972) measurements with a single crystal of ^{154}Sm.

Since Koehler and Wollan's measurements of the paramagnetic scattering of Nd_2O_3 and Er_2O_3 several studies have been made of the rare-earth elements themselves using polarized neutrons. However, a complication arises which limits the advantage which might be expected from the polarized-neutron experiments. For the 3d metals the magnetic moment is contributed by the spins S and the anisotropy of this spin density, which we have shown to be significant, is not affected by applying a magnetic field. The spin direction follows the applied field, but its distribution in space, which is determined by the crystal field, does not change. On the other hand, for the rare earths there is a large spin–orbit coupling, and when the resultant moment directions are turned around by a field there is a change in the moment distribution in space. It follows therefore that the polarized-neutron experiments measure a

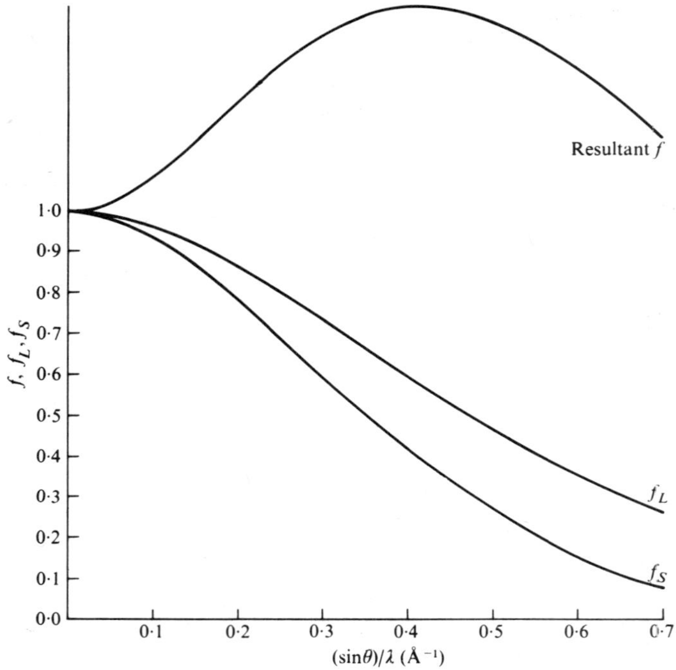

FIG. 146. The curves f_L, f_S show the computed values of the orbital and spin form factors for samarium. The resultant curve passes through a maximum value when $(\sin\theta)/\lambda \simeq 0{\cdot}4$ Å^{-1}.

different distribution from that to be observed in the absence of an applied field. Steinsvoll *et al.* (1967) have considered this problem from both the experimental and theoretical points of view for the metal terbium, which is ferromagnetic at 4·2 K, where the measurements were carried out. They show that the form factors applicable to polarized and unpolarized measurements can be written as

$$f_{\text{polarized}} = f_{\text{spherical}} - \Delta_{\text{p}} \tag{7.4}$$

and

$$f_{\text{unpolarized}} = f_{\text{spherical}} - \Delta_{\text{u}}. \tag{7.5}$$

The aspherical contributions Δ_{p} and Δ_{u} are both small and are listed together with $f_{\text{spherical}}$, the main contribution to the form factor, in Table 18. The value of Δ_{p} depends solely on $(\sin\theta)/\lambda$,

TABLE 18
Theoretical calculated form factor for terbium

hkl	$f_{\text{spherical}}$	Δ_p	Δ_u
100	0·890 }		−0·002
002	0·873 }	0·003	0·003
100	0·865 }		0·000
102	0·785	0·006	0·002
103	0·673	0·009	0·005
200	0·660	0·010	−0·006
201	0·640	0·011	−0·004
004	0·612	0·012	0·012
203	0·517	0·015	0·003
105	0·443 }	0·018	0·015
300	0·428 }		−0·010
302	0·387 }	0·021	−0·006
006	0·368 }		0·021
110	0·722	0·008	−0·004
112	0·643	0·009	−0·001
114	0·470	0·016	0·007
220	0·340 }	0·022	−0·011
222	0·310 }		−0·008
116	0·293	0·023	0·016
224	0·243	0·025	−0·001

whereas Δ_u, which is effective in the normal state with no applied field, depends also on the orientation of the scattering vector to the crystallographic axes and is, for example, quite different in value for the reflections (302) and (006). The extent of the agreement between theory and experiment is indicated in Fig. 147. Up to a value of $(\sin\theta)/\lambda = 0{\cdot}5$ there is quite good agreement between the polarized measurements and the calculated value of $f_{\text{polarized}}$. At larger values of $(\sin\theta)/\lambda$, where data can be obtained only with polarized neutrons, the experimental points are substantially lower than theory indicates. This can be explained if the process of normalizing the data has been invalidated by the fact that, of the observed magnetization of 9·34 μ_B, a moment of 0·34 μ_B is contributed by the conduction electrons. This inference is supported by Fourier plots of the magnetic moment density, such as we discussed earlier for Fe, Co, and Ni, which suggest a uniform distribution of density over the unit cell which amounts on integration to about this same value. It seems reasonable to conclude therefore that, bearing in mind the limitations of the experimental data, there is good support for the accepted view of the influence of the spin and orbital contributions.

Among recent measurements of paramagnetic scattering is the study of Gd_2O_3 by Child, Moon, Raubenheimer, and Koehler (1967). Gadolinium is not typical of the rare-earth elements because the ion Gd^{3+} has no orbital contribution to the magnetic moment,

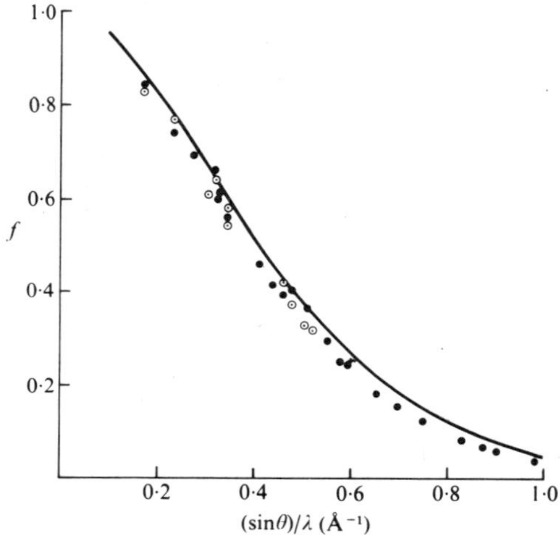

FIG. 147. Form-factor measurements for terbium for polarized ● and unpolarized ○ neutrons. The curve shows a calculation for the polarized case. (Steinsvoll *et al.* 1967.)

since $L = 0$ and $J = S = \frac{7}{2}$. It is, however, satisfying to note that good agreement is found over the limited experimental range of $(\sin\theta)/\lambda$, with the form factor calculated by Blume, Freeman, and Watson (1962) for the free atom. This work is also of interest, because it employed the low-absorbing isotope ^{160}Gd, in order to avoid the prohibitively high absorption $\sigma_a = 20\,000$ barns of natural gadolinium at a wavelength of about 1 Å.

7.4. Uranium

Several studies have been made of the form factor of uranium and they are of particular interest for the light which they throw on the 5f and 6d contributions to the electron distribution of uranium ions. Some early measurements were made by Shull and Wilkinson

(1956) for the hydride UH_3, which is ferromagnetic. The data were of limited accuracy but were able to show that the form factor fell off far too slowly with $(\sin\theta)/\lambda$ to be accounted for by 6d electrons. Fig. 148 indicates the extreme narrowness of an expected 6d form factor compared with that for 5f electrons. The latter curve is one computed by Johnston (1966) for the U^{3+} ion $5f^3$. Curry (1965, 1966) has studied UN and UP. The former was examined as single crystals, and the experimental form-factor curve was in quite good agreement with the $5f^3$ curve in Fig. 148. Results for the polycrystalline UP were much less precise but compatible with the same curve.

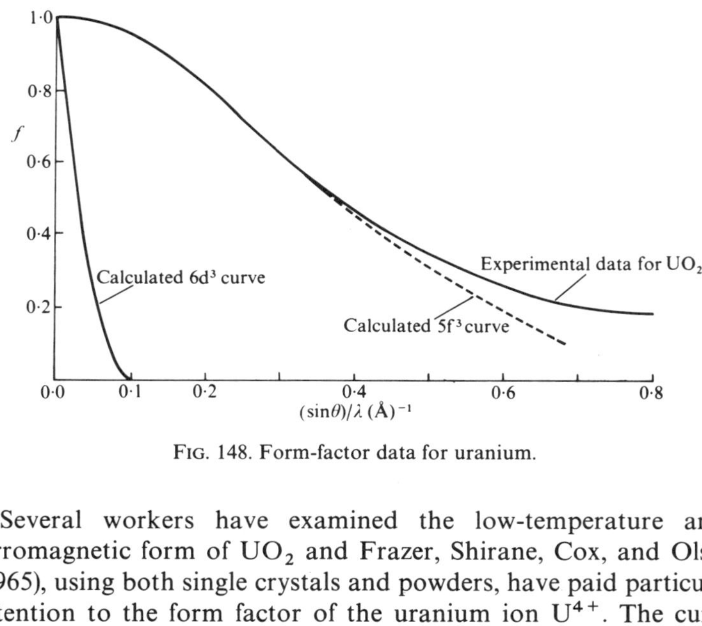

FIG. 148. Form-factor data for uranium.

Several workers have examined the low-temperature antiferromagnetic form of UO_2 and Frazer, Shirane, Cox, and Olsen (1965), using both single crystals and powders, have paid particular attention to the form factor of the uranium ion U^{4+}. The curve found experimentally is also shown in Fig. 148 and is in good agreement with what is expected for a $5f^2$ ion, with $S = 1$, $L = 4$, and $J = 5$ and having an effective magnetic moment of about $1{\cdot}8\ \mu_B$.

7.5. Covalency effects in salts

We have seen earlier that there is a broad similarity between the form factors of elements in their metallic state and in ionic compounds, but it became evident, as more precise experimental data became available, that significant differences exist. Early measurements related to the ions Fe^{3+}, Mn^{2+}, and Ni^{2+}, and it appeared that for the ions Fe^{3+} and Mn^{2+} in a crystal the spin density was more diffuse than that calculated for a free atom, but that for the ion Ni^{2+} the reverse was the case; for the latter ion the observed form-factor curve was broader than expected.

In our subsequent study of the form factors of the metals themselves we have shown that precise measurements with polarized neutrons have revealed angular dependencies in the form-factor curve which could be correlated with the details of the electronic distribution. Ionic compounds which show co-operative magnetism are generally antiferromagnetic, and the quantitative study of their atomic form factors is much more difficult. However, it became evident in the 1960s that a number of discrepancies existed. The most outstanding discovery was that of Nathans *et al.* (1963), who observed magnetic scattering in forbidden directions in MnF_2. This and other observations can be accounted for by the assumption that these salts are not completely ionic but show a certain amount of covalency. Thus some part of the electronic charge, which in an ideal ionic bond is completely transferred from cation to anion, is returned to the cation. In the case of a magnetic material there is a transfer of electron spin. We shall see that this can result in modification both of the observed form factors and the observed magnetic intensities.

The principle involved in this discussion is illustrated in Fig. 149, which relates to a magnetic salt constituted of a magnetic cation and an anion such as O^{2-} or F^-. For an ideal ionic bond there

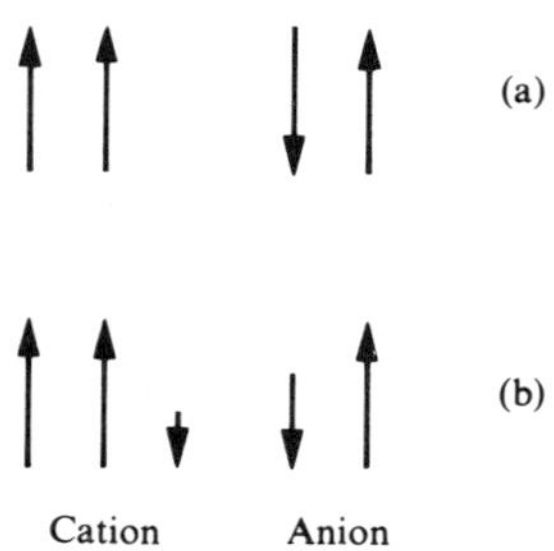

FIG. 149. Transfer of electron charge from anion to cation in covalent bond (b). For an ideal ionic bond (a) there will be no unpaired spin density on the anion.

will be, as in Fig. 149 (a), no unpaired spin density on the anion. If, however, some transfer to a covalent bond occurs, then it will be an electronic charge whose spin is opposite to that of the cation; thus (as in Fig. 149(b)) the moment of the cation is reduced and a spin of opposite sign to that of the transferred charge appears on the anion. In a crystal, as distinct from an isolated pair of ions, the situation may be different. In most antiferromagnetic salts, such as NiO, neighbouring cations have oppositely directed magnetic moments, and each anion is surrounded by equal numbers of +ve, −ve moments, so that the *net* unpaired spin produced on an anion by this exchange is zero. In other cases, and MnF_2 is an example of this, the distribution of magnetic cations around an anion is asymmetrical; in such cases cancellation does not occur and there remains a net amount of magnetic moment on the anion. This will be evident from Fig. 150, which shows the environment of a fluorine

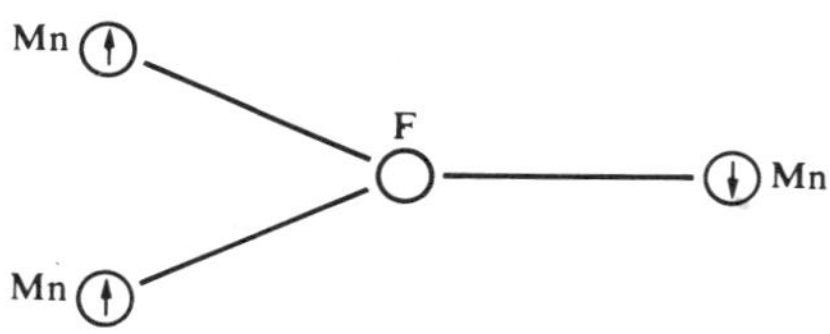

FIG. 150. Asymmetrical distribution of magnetic cations (Mn) around the anion (F) in MnF_2.

atom in MnF_2; although the three manganese neighbours are at almost equal distances two of them have upward pointing spins and the third one points downward.

The effect on the neutron scattering will be different for the two cases between which we have just distinguished. In the first case, where the symmetry ensures cancellation, the intensity of the anti-ferromagnetic reflections will be less than what would be calculated if there were no covalency. It can be shown (e.g. Marshall and Lovesey 1971) that the cation moment is reduced to

$$1-3A_\sigma^2-3A_S^2 \quad \text{for} \quad E_g \text{ orbitals} \tag{7.6}$$

$$1-4A_\pi^2 \quad \text{for} \quad T_{2g} \text{ orbitals} \tag{7.7}$$

and $1-1{\cdot}2(A_\sigma^2+A_S^2+2A_\pi^2)$ for the spherically symmetrical mixture of 40 per cent E_g and 60 per cent T_{2g}, where A_σ, A_S, and A_π are three covalency parameters. For NiO, Alperin (1962) estimated a magnetic moment of $(1{\cdot}81 \pm 0{\cdot}20)\ \mu_B$, by extrapolation of the form-factor curve to the forward direction at $\theta = 0°$, in comparison with an expected moment for Ni^{2+} of $2{\cdot}2\ \mu_B$. Thus there is an apparent reduction factor of $(0{\cdot}82 \pm 0{\cdot}10)$. More recent observations by Fender, Jacobson, and Wedgwood (1968) gave a reduction factor of $(0{\cdot}886 \pm 0{\cdot}005)$, which from eqn (7·6) gives

$$A_\sigma^2 + A_S^2 = (3{\cdot}8 \pm 0{\cdot}2) \text{ per cent.}$$

This compares well with estimates of $A_\sigma^2 = 3{\cdot}7$ per cent and $A_S^2 = 0{\cdot}5$ per cent from nuclear magnetic resonance data.

Furthermore, it can be shown that there is a redistribution of spin density due to an overlap-charge contribution which affects the shape of the form-factor curve. The detailed shape depends on the signs of A_σ, A_S, and A_π. For ligands such as F^-, O^{2-}, and Cl^- the parameters are all positive, and this results in a spin density which falls to zero at some point approaching the anion, then takes a negative value with further approach to the anion and then becomes zero again at the anion itself. Under these circumstances the form factor is flatter than would be the case in the absence of covalency. NiO is an example of this situation, and Fig. 151 shows that the experimental data are in much better agreement with theory when allowance is made for covalency. In fact there is considerable additional broadening of the form factor through the influence of a contribution from orbital moment, which is incompletely quenched by the crystal field. On the other hand, A_π would be expected to have a negative sign for cyanide complexes, leading to a more rapid fall-off of the form factor.

In the general case of an antiferromagnetic structure such as MnF_2 whose structure is illustrated later in Fig. 153 (p. 256), there is a resultant spin density on the fluoride ion. In Fig. 150 it can be shown that the fluorine ion bears a moment equal to $A_\sigma^2 + 2A_\pi^2 + A_S^2$. This additional spin density is at a distance of about 2 Å from the manganese ion and will produce a very sharp forward peak to the form factor. For substances with a very large unit cell there could be a Bragg reflection within this enhanced forward peak, but this is not usually the case and with MnF_2, for example, there is no Bragg peak within this region and the form-factor curve will appear to extrapolate to what is a low value.

The two contrasting cases of high symmetry and low symmetry are analysed in Fig. 152, which is due to Rimmer. There are three

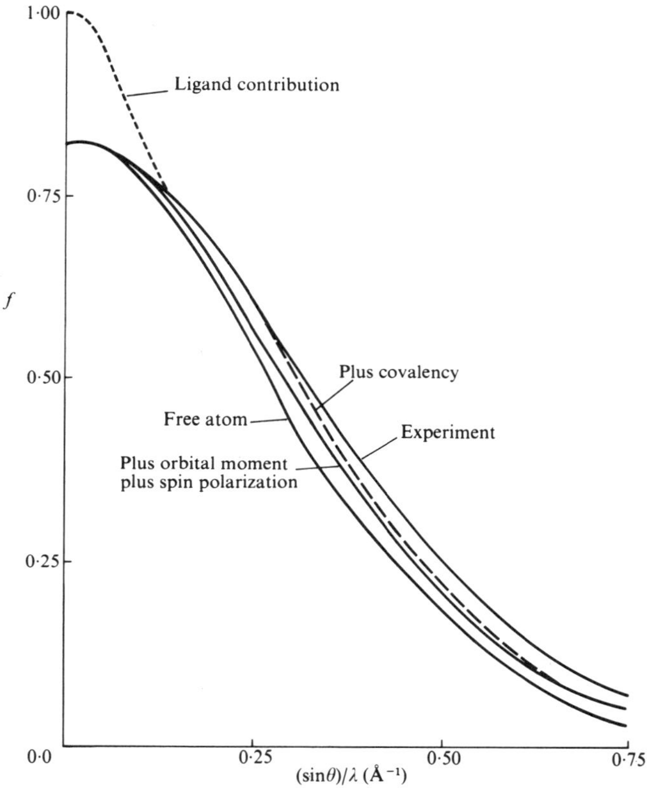

FIG. 151. Experimental and theoretical form factors for NiO.

types of term in the form-factor curve of an antiferromagnet. First, a 3d form factor of reduced amplitude for the metal ion; secondly, an overlap term of negative sign, and thirdly—and only for the asymmetrical case—a contribution from the ligand. For high-symmetry structures the f curve is both reduced and broadened. For low symmetry there is the sharp forward peak which usually escapes detection because there is not usually a Bragg reflection at such a small value of $(\sin\theta)/\lambda$. However, the presence of ligand spin, which accounts for this forward peak, can be deduced from other effects to which it gives rise. These are the occurrence of magnetic scattering in forbidden directions, first observed for MnF_2 by Nathans *et al.* (1963). MnF_2 has the rutile structure which is illustrated in Fig. 153,

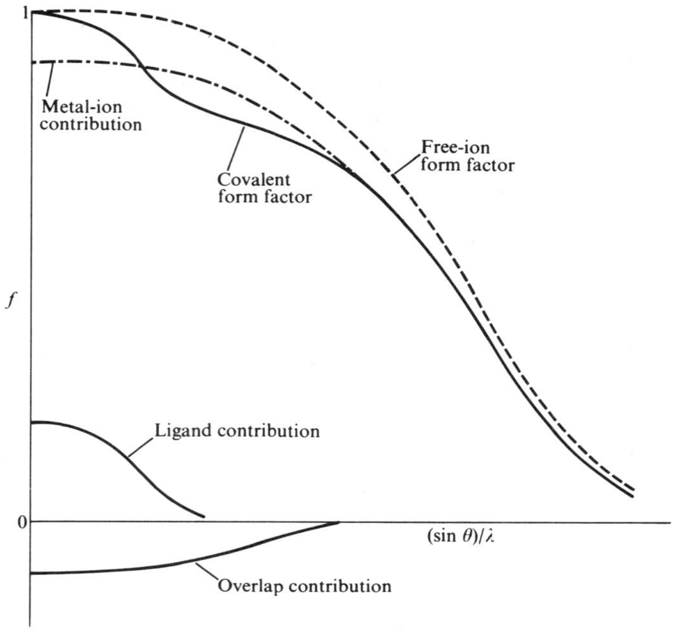

FIG. 152. The three contributions to the form-factor curve of an antiferromagnet: the reduced metal ion contribution, a negative overlap term, and (for asymmetrical structures) a ligand contribution. (Rimmer 1970.)

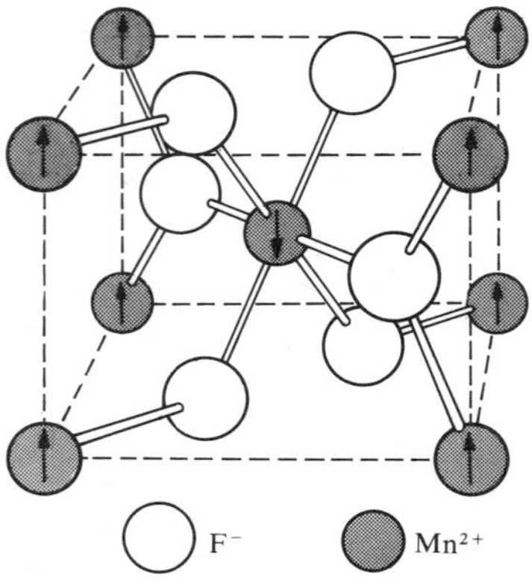

FIG. 153. The antiferromagnetic structure of MnF_2 in which the magnetic moments of the manganese atoms at the corners and body centres of the unit cell are oppositely directed.

and it was shown first by Erickson (1953) to be antiferromagnetic with the magnetic moments of the corner and body-centring atoms oppositely directed, as indicated in the Figure. Under these circumstances it can be shown that the diffraction spectra are of three different types:

(1) ($0kl$ reflections for which $k+l$ is odd will be purely magnetic;

(2) all other reflections for which $h+k+l$ is odd will be mixtures of nuclear and magnetic intensity and the nuclear component will come entirely from the fluorine atoms;

(3) reflections for which $h+k+l$ is even will be purely nuclear.

Alperin *et al.* (1961) have shown that there is a polarization dependence of intensity for antiferromagnetic structures in which there is no translational symmetry in the magnetic environment. MnF_2 is a structure of this type, and it is accordingly possible to make accurate measurements of the intensities of antiferromagnetic reflections with polarized neutrons. Such measurements of MnF_2 by Nathans *et al.* (1963) revealed that reflections with indices of type (3) above *did* display a magnetic component. A plot of the observed structure factors is given in Fig. 154, and it is possible to

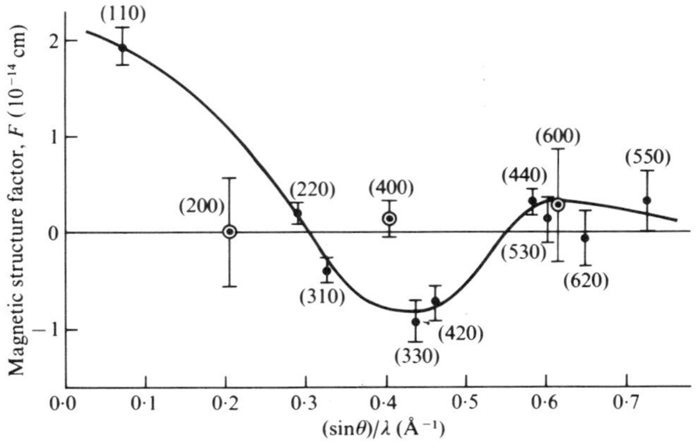

FIG. 154. The magnetic structure factors observed experimentally for MnF_2. The points (200), (400), and (600) which are enclosed in circles correspond to reflections for which the magnetic scattering would be expected to be exactly zero. (Nathans *et al.* 1963.)

show that this behaviour would be expected if part of the magnetic moment was *not* distributed spherically about the manganese atoms. The contribution observed is only small, amounting to a scattering length of about 2×10^{-14} cm, but it was possible to arrive at a fairly detailed picture of its distribution in the unit cell. A plot of the distribution of this covalent spin density showed it to be localized in maxima and minima on the line joining the Mn and F ions and near to the latter. The suggested distribution is shown in Fig. 155, and it will be seen that there is a *net* spin in the

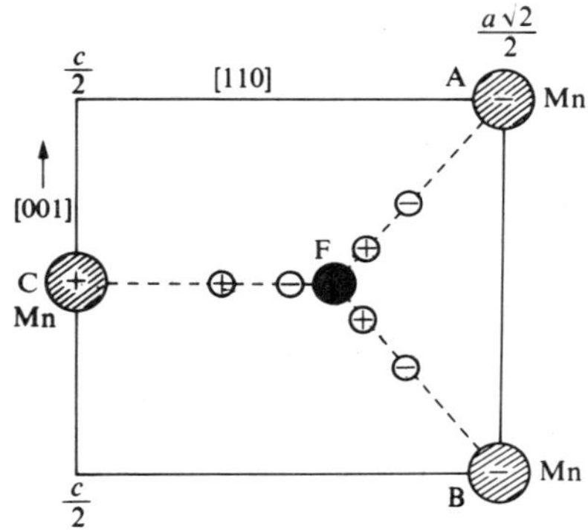

FIG. 155. Localized maxima and minima of covalent spin on the lines joining the Mn and F ions in MnF_2 shown in the (110) plane. There is a *net* spin near the F ion of opposite sign to the spin of its two manganese neighbours A and B.

neighbourhood of the fluorine ion with a direction which is opposite to that of the spin on two of the three manganese neighbours A, B, i.e. the fluorine spin is in the same direction as that of the single manganese neighbour C. This net spin was estimated to be about $0{\cdot}1\ \mu_B$.

8

OBSERVATION OF MAGNETIC SCATTERING

IN this chapter we shall discuss the main principles which are important in the determination of the magnetic structures of materials by indicating how the different types of magnetic arrangement reveal themselves in the scattering pattern. We shall see that some arrangements give rise to very striking and distinctive features in the pattern whereas others can be distinguished only by lengthy and careful measurement.

8.1. Paramagnetic materials

In a paramagnetic material the directions of the magnetic moments vary randomly from atom to atom and there is no magnetic contribution to the Bragg diffraction peaks. The magnetic scattering is distributed in the background of the diffraction pattern, falling off with increase of the angle θ because of the influence of the form factor f, in accordance with eqn (6.32),

$$d\sigma_{\text{pm}} = \tfrac{2}{3}g^2J(J+1)\left(\frac{e^2\gamma}{2mc^2}\right)^2 f^2. \tag{8.1}$$

It will be possible to measure this contribution to the background if the other contributions can be assessed. If the magnetic scattering is fairly large then this assessment can be done with sufficient accuracy, as we have already shown in Fig. 111 (p. 198), which illustrated the measurements by Atoji (1961) for CeC_2. By subtraction of calculated contributions from thermal diffuse scattering, incoherent nuclear scattering, and multiple scattering, a curve for the magnetic scattering alone was deduced. From this curve the scattering at $\theta = 0°$ was assessed as equal to 0·31 barns per unit solid angle. This is in very good agreement with what would be expected if the Ce^{3+} ion is in the state ${}^2F_{\frac{5}{2}}$; for this ion the values of S, L, and J for the 4f electrons will be $\frac{1}{2}$, 3, and $\frac{5}{2}$ respectively, and the calculated value of the differential scattering cross-section at $\theta = 0°$

is equal to 0·311 barns. In the same way, Atoji found good agreement with the measured scattering for TbC_2 on the assumption that the Tb^{3+} ion is in the state $^7F_8$. On the other hand, no paramagnetic scattering was detected for the carbides CaC_2, YC_2, LaC_2, or LuC_2, and this is what would be expected if the ions present are Ca^{2+}, Y^{3+}, La^{3+}, and Lu^{3+} respectively.

8.2. Ferromagnetism

For materials in which there is a correlation between the magnetic moment directions of neighbouring atoms the magnetic scattering will appear as sharp diffraction peaks, with distinctive features which will identify ferromagnetic, antiferromagnetic, and helimagnetic substances. We shall consider these first of all for unpolarized beams of neutrons for which, as we have already seen in Chapter 6, there is no coherence between the contributions from nuclear and magnetic scattering.

For a ferromagnetic material, whose structure was illustrated earlier in Fig. 113(a) (p. 207), all the magnetic moments within a single domain point in the same direction and the unit cell, from both a magnetic and a chemical point of view, is the cube of side AB. Accordingly, as we saw in Chapter 6, the peaks of magnetic scattering will appear in exactly the same angular positions as the peaks from the nuclear scattering. Thus the latter peaks are 'boosted' by a magnetic component which becomes insignificant at larger angles of scattering because of the influence of the form factor f. If the temperature of the magnetic material is raised then the magnitude of the magnetic component will diminish, becoming zero at the Curie temperature. At the same time there will be a steady increase in the level of the background scattering, and above the Curie temperature the whole of the magnetic scattering will appear in the background, in accordance with the description of a paramagnetic material which we have given above.

8.3. Antiferromagnetism

In the typical antiferromagnetic structure of Fig. 113(b) the important feature is that when magnetism is taken into account, i.e. when the directions of the magnetic moments are considered, the length of the unit cell in the Oy direction is no longer AB, but AC. The result of this is that additional peaks appear in the diffraction pattern, such as $(0\frac{1}{2}0)$, $(0\frac{3}{2}0)$, $(0\frac{5}{2}0)$, etc. There will be no magnetic contribution at the positions (010), (020), etc. because at these angular positions the contributions from successive planes in the structure will cancel out, on account of the opposite phases of the neutrons scattered from the upward- and downward-pointing moments. A typical example of an antiferromagnetic structure of this type is provided by the alloy AuMn, whose structure is illustrated in Fig. 156. By contrast with the appearance of diffraction peaks at the positions ($0\frac{1}{2}0$), ($0\frac{3}{2}0$), ($0\frac{5}{2}0$), there will *not* be any peaks at positions

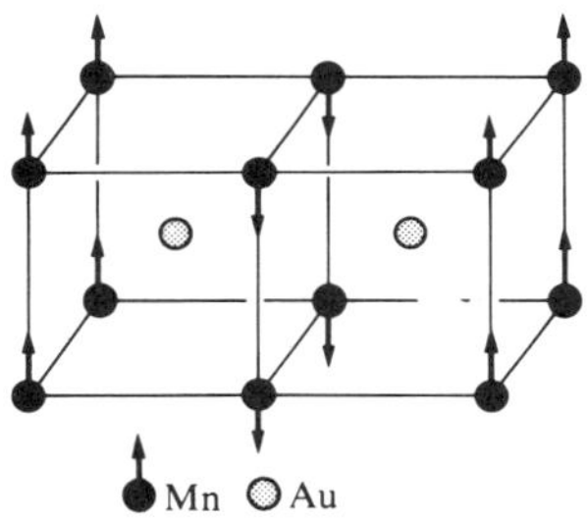

FIG. 156. The antiferromagnetic structure of AuMn.

such as ($\frac{1}{2}00$) or ($00\frac{1}{2}$), because there is no doubling of the unit cell along the Ox and Oz axes for this structure. However, this limitation to a single axis does not always occur, and for manganese oxide MnO, for example (to which we refer later on p. 444), the unit cell is doubled in all three directions when magnetism is considered. In this case we observe reflections such as ($\frac{1}{2}\frac{1}{2}\frac{1}{2}$).

8.4. Spiral-spin arrangements: helimagnetism

A distinctive class of antiferromagnetic structures whose existence has been revealed by measurements with neutrons is the so-called 'helimagnetic', in which the direction of the magnetic moments traces a helical spiral. An example of such a structure is provided by the Au–Mn alloy of composition Au_2Mn. The essential feature of the structure is indicated in Fig. 157, which shows the ferromagnetic sheets of atoms, perpendicular to the direction Oy. As we advance through the structure, from left to right in the Figure, the direction of the magnetic moments within a sheet rotates progressively about the axis Oy. In the particular case of Au_2Mn there is a rotation of about 50° between successive sheets. In the diffraction pattern of such a material the basic nuclear peaks are accompanied

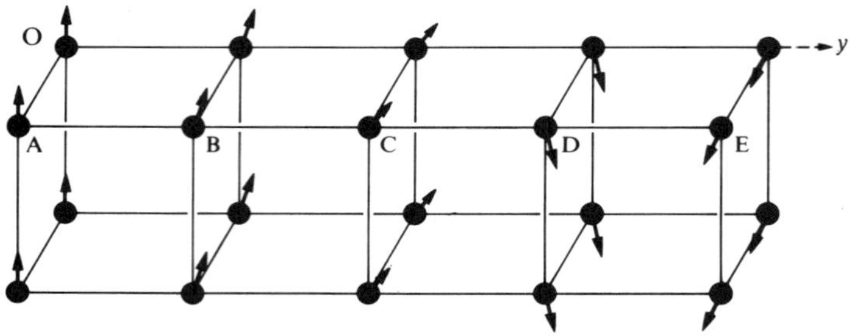

FIG. 157. The structure of a helimagnet. There are ferromagnetic sheets of atoms, perpendicular to Oy, and the moment direction within a sheet rotates progressively by about 60°, for the example shown in the Figure, from sheet to sheet.

by pairs of satellites of magnetic origin, as shown in Fig. 158, which also illustrates the detailed structure of Au_2Mn, in particular the way in which the gold atoms are sandwiched between sheets of manganese atoms. We can determine both the direction of the spiral axis and the magnitude of the rotation by studying the displacement of the satellite peaks from the fundamental reflections, in the following manner.

From eqn (6.19) we can express the magnetic structure factor for the unit cell as the vector quantity,

$$\mathbf{F}_{\text{magn}} = \frac{e^2\gamma}{mc^2}\sum_n \mathbf{q}_n S_n f_n \exp\{2\pi i(hx_n/a + ky_n/b + lz_n/c)\}, \qquad (8.2)$$

where we again have omitted the Debye temperature factor for simplicity and where the summation is made over all the magnetic atoms in the unit cell. In the general case, where we permit the n magnetic spins to take individual directions, we can write a corresponding vector equation by replacing the vector $\mathbf{q}$ by its fuller expression of $\boldsymbol{\varepsilon}(\boldsymbol{\varepsilon}\,.\,\mathbf{K}) - \mathbf{K}$ given in eqn (6.9). Thus

$$\mathbf{F}_{\text{magn}} = -\frac{e^2\gamma}{mc^2}\sum_n \{\mathbf{K} - \boldsymbol{\varepsilon}(\boldsymbol{\varepsilon}\,.\,\mathbf{K})\} S_n f_n \exp\{2\pi i(hx_n/a + ky_n/b + lz_n/c)\}.$$

We can then simplify this expression by replacing $\mathbf{K}S_n$ by its equivalent $\mathbf{S}_n$, which is the magnetic spin of the nth atom expressed as a vector, and therefore describes the spin in both magnitude and

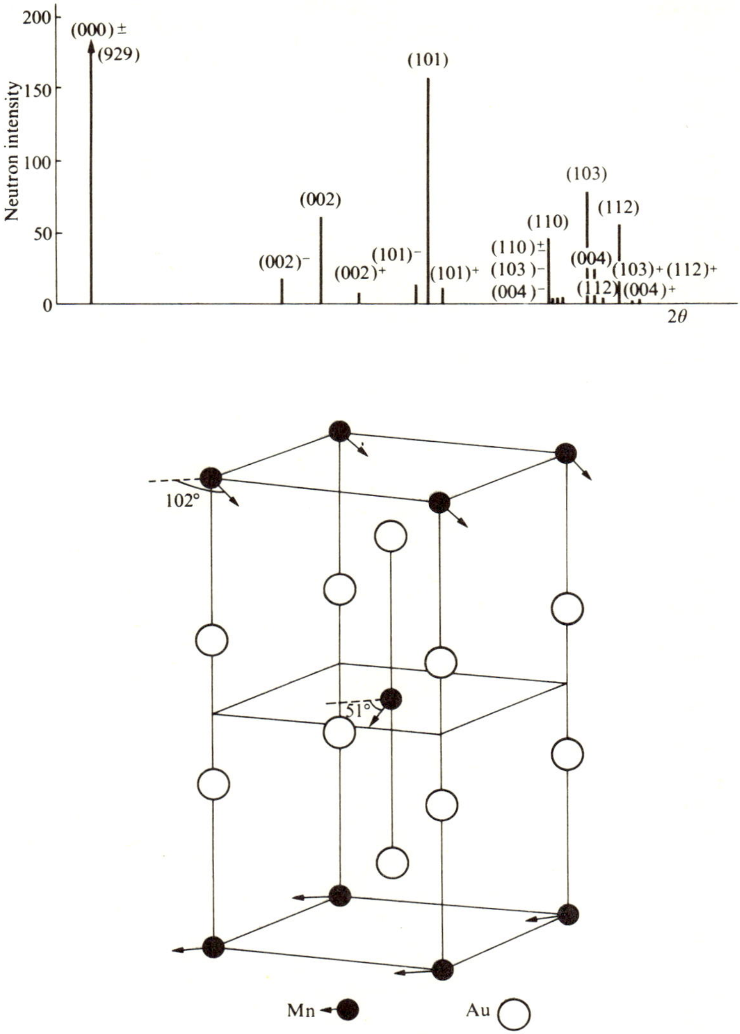

FIG. 158. The lower diagram illustrates the structure of Au_2Mn in which double layers of gold atoms are sandwiched between layers of manganese atoms. The direction of the moment in the manganese layers spirals around the *c*-axis. The upper Figure shows how the diffraction pattern of Au_2Mn includes pairs of magnetic satellite reflections which accompany the nuclear reflections. Thus, (002) is accompanied by the satellites marked as (002^-) and (002^+), which are found to index as $(00\frac{12}{7})$ and $(00\frac{16}{7})$. Similarly, (101) is accompanied by $(10\frac{5}{7})$ and $(10\frac{9}{7})$.

direction, i.e.

$$\mathbf{F}_{\text{magn}} = -\frac{e^2\gamma}{mc^2}\sum_n \{\mathbf{S}_n - \boldsymbol{\varepsilon}(\boldsymbol{\varepsilon}\,.\,\mathbf{S}_n)\} f_n \exp\{2\pi i(hx_n/a + ky_n/b + lz_n/c)\}. \tag{8.3}$$

We emphasize that this structure amplitude factor is a vector quantity, the square of whose modulus will represent the magnetic intensity.

The quantity in the braces can be simplified if we resolve the spin vector $\mathbf{S}_n$ into two components $\mathbf{S}_\tau$, $\mathbf{S}_\mathrm{p}$ along and perpendicular respectively to the unit scattering vector $\boldsymbol{\varepsilon}$; the latter lies in the same direction as the reciprocal lattice vector $\boldsymbol{\tau}$ in Fig. 159. Thus

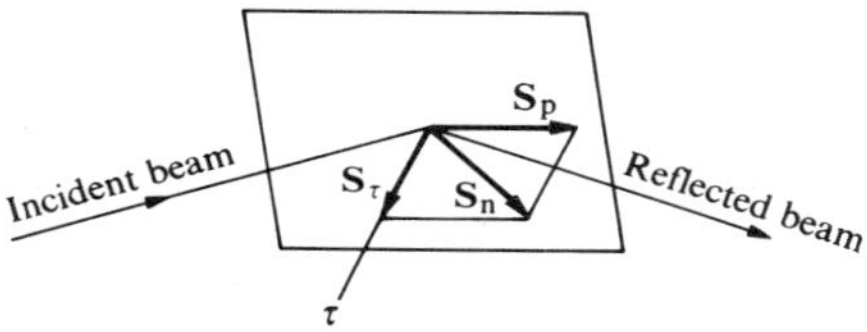

FIG. 159. Resolution of spin vector $\mathbf{S}_n$ into two components $\mathbf{S}_\tau$, $\mathbf{S}_\mathrm{p}$ along the perpendicular to the reciprocal lattice vector $\boldsymbol{\tau}$, which is in the same direction as the unit vector $\boldsymbol{\varepsilon}$.

$\mathbf{S}_\tau$ is normal to the reflecting planes which we are considering and $\mathbf{S}_\mathrm{p}$ lies in those planes. For the component $\mathbf{S}_\tau$ the expression $\{\mathbf{S}_\tau - \boldsymbol{\varepsilon}(\boldsymbol{\varepsilon}\,.\,\mathbf{S}_\tau)\}$ becomes simply $(\mathbf{S}_\tau - \mathbf{S}_\tau)$ which is zero. For the component $\mathbf{S}_\mathrm{p}$, $(\boldsymbol{\varepsilon}\,.\,\mathbf{S}_\mathrm{p})$ is zero, so that the equivalent expression in the braces is simple $\mathbf{S}_\mathrm{p}$.

Accordingly, eqn (8.3) becomes

$$\mathbf{F}_{\text{magn}} = -\frac{e^2\gamma}{mc^2}\sum_n \mathbf{S}_\mathrm{p} f_n \exp\{2\pi i(hx_n/a + ky_n/b + lz_n/c)\}, \tag{8.4}$$

where $\mathbf{S}_\mathrm{p}$ is, as we have said, the resolved component of the spin vector on to the reflection plane which we are considering. We can apply this equation to the general case of a helimagnetic structure illustrated in Fig. 160, where the reflection plane (hkl) which is under consideration is at an angle ϕ, to the mágnetic sheets. There is thus an angle ϕ between the spiral axis and the normal to the reflection plane.

In order to proceed with the summation we shall consider the components of $\mathbf{S}_\mathrm{p}$ in two orthogonal directions OX, OY within the reflection plane as indicated in the spherical diagram of Fig. 161.

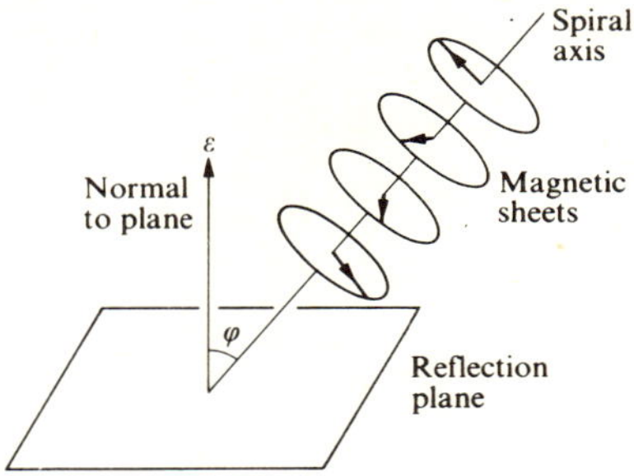

FIG. 160. The general case of a helimagnetic structure in which the spiral axis, perpendicular to the magnetic sheets, is inclined at an angle ϕ to ε the normal to the reflecting planes under examination.

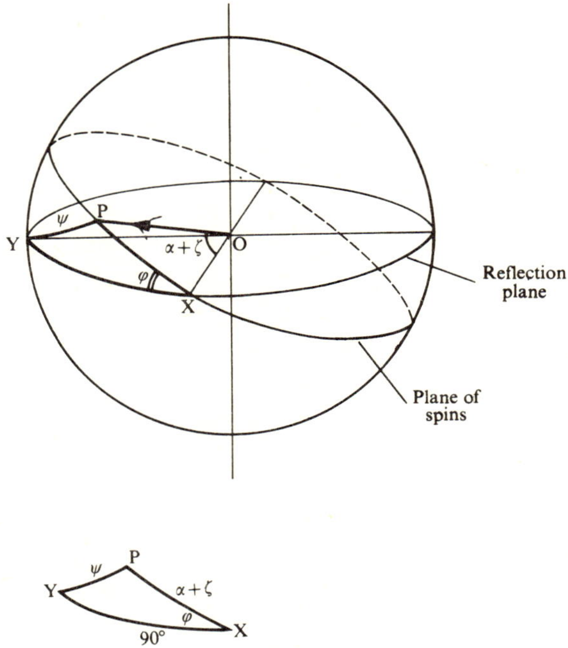

FIG. 161. Calculation of intensities of reflection for a helimagnetic structure. A magnetic spin S along OP is resolved into two components within the reflection plane, parallel to the two orthogonal axes OX, OY respectively. α is an arbitrary constant and the direction of OP rotates successively by an angle ζ from sheet to sheet.

Within the ferromagnetic sheets the magnetic-moment direction OP will be at an angle $\alpha+\zeta$ to OX, where α is an arbitrary constant and ζ is the rotation from sheet to sheet of the spiralling moments.

For a particular spin OP its projection on the axis OX will be

$$S_x = S_p \cos(\alpha+\zeta).$$

The projection on the axis OY will be $\mathbf{S}_p \cos\psi$, where ψ is indicated in Fig. 161 and $\cos\psi$ can be deduced by applying Napier's rule to the spherical triangle PXY in the Figure, as

$$\cos\psi = \sin(\alpha+\zeta)\,.\,\cos\phi,$$

thus

$$S_y = S_p \cos\phi \sin(\alpha+\zeta).$$

Hence the expression for $\mathbf{F}_{\text{magn}}$ becomes

$$\mathbf{F}_{\text{magn}} = -\frac{e^2\gamma}{mc^2}S_p\Big[\mathbf{x}\sum_n \cos(\alpha+\zeta)\,.\,\exp\{2\pi i(hx_n/a+ky_n/b+lz_n/c)\}+$$
$$+\cos\phi\,\mathbf{y}\sum_n \sin(\alpha+\zeta)\exp\{2\pi i(hx_n/a+ky_n/b+lz_n/c)\}\Big]$$
$$= -\frac{e^2\gamma}{2mc^2}S_p\Big(\mathbf{x}\sum_n[\exp\{i(\overline{\alpha+\zeta})\}+\exp\{-i(\alpha+\zeta)\}]\times$$
$$\times\exp\{2\pi i(hx_n/a+ky_n/b+lz_n/c)\}-$$
$$-\mathbf{y}\cos\phi\sum_n i[\exp\{i(\overline{\alpha+\zeta})\}-\exp\{-i(\overline{\alpha+\zeta})\}]\times$$
$$\times\exp\{2\pi i(hx_n/a+ky_n/b+lz_n/c)\}\Big). \tag{8.5}$$

By summing the squares of the **x**, **y** components we have

$$\left|\mathbf{F}_{\text{magn}}\right|^2 = \left(\frac{e^2\gamma}{mc^2}\right)^2\frac{S_p^2}{4}\Big(\Big[\sum\exp\{i(\alpha+\zeta)+2\pi i(hx_n/a+ky_n/b+lz_n/c)\}+$$
$$+\exp\{-i(\alpha+\zeta)+2\pi i(hx_n/a+ky_n/b+lz_n/c)\}\Big]^2+$$
$$+\cos^2\phi\Big[\sum\exp\{i(\alpha+\zeta)+2\pi i(hx_n/a+ky_n/b+lz_n/c)\}-$$
$$-\exp\{-i(\alpha+\zeta)+2\pi i(hx_n/a+ky_n/b+lz_n/c)\}\Big]^2\Big). \tag{8.6}$$

This expression is the counterpart of

$$\left[\sum b_n \exp\{2\pi i(hx_n/a+ky_n/b+lz_n/c)\}\right]^2$$

for nuclear scattering, and we require to know whether we can choose a direction $[hkl]$ such that the magnetic contributions from all the atoms will reinforce, in the way that the nuclear contributions do at the ordinary Bragg peak. The indices hkl for such a direction will not usually all be integers. Inspection of the expression in eqn (8.6) above shows that the spin angle, which increases by ζ from sheet to sheet, alters the phase of each contribution to the two parts of the expression.

We shall get reinforcement if the value of
either

$$(\alpha+\zeta)+2\pi(hx_n/a+ky_n/b+lz_n/c)$$

or

$$(\alpha+\zeta)-2\pi(hx_n/a+ky_n/b+lz_n/c)$$

increases by an integer from sheet to sheet (1)

and is constant for all atoms in a magnetic sheet (2). (8.7)

The second of these two conditions will be satisfied if the reciprocal lattice point, corresponding to the direction $[hkl]$, is displaced from a nuclear reciprocal lattice point by a component parallel to the spin spiral axis, since such a displacement will not produce any differential phase changes for atoms within a plane which is perpendicular to this direction. Condition (1) will then be satisfied if the size of the displacement is such that the increment in $2\pi(hx_n/a+ky_n/b+lz_n/c)$ from sheet to sheet is equal to $\pm\zeta$.

As an example we may consider the case of the alloy Au_2Mn, for which the spin axis is the c-axis and the spins lie in the x,y planes, successive planes being separated by a distance $\frac{1}{2}c$. The magnetic satellites will therefore be displaced from the nuclear positions by a reciprocal lattice vector $(00\Delta l)$, where

$$\zeta = 2\pi\,.\,\Delta l\,.\,\tfrac{1}{2},$$

i.e.

$$\Delta l = 2\zeta/2\pi.$$

It is found experimentally that Δl is approximately $\frac{2}{7}$, so that the value of ζ is deduced to be about 51°. The (002) nuclear reflection is accompanied, as Fig. 158 shows, by magnetic satellites which index as $(00\frac{12}{7})$ and $(00\frac{16}{7})$; the (101) nuclear reflection is accompanied by $(10\frac{5}{7})$ and $(10\frac{9}{7})$. As a further example we mention the helimagnetic form of Au_2MnAl (Bacon and Mason 1967), which is illustrated in Fig. 162. Here the screw axis is a [100] direction, and the direction of the magnetic moment of the manganese atoms rotates by about 45° for successive magnetic sheets. These sheets are spaced at a distance of $\frac{1}{2}a_0$, so that Δl is the reciprocal lattice vector $(\frac{1}{4}00)$. Thus the (200) reflection would have the magnetic satellites $(\frac{7}{4}00)$ and $(\frac{9}{4}00)$ and (020) would be accompanied by $(\frac{1}{4}20)$ and $(\bar{\frac{1}{4}}20)$.

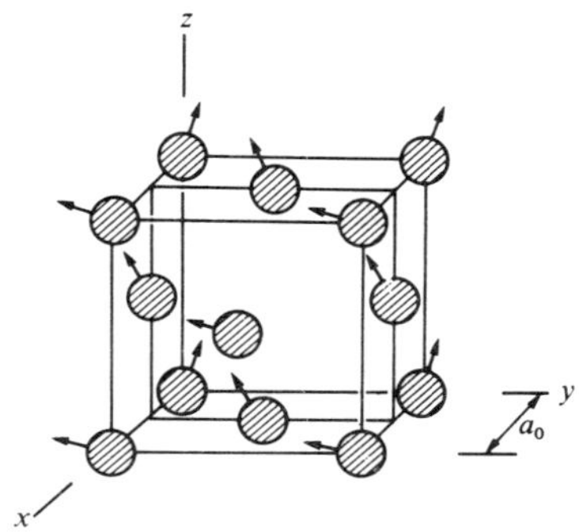

FIG. 162. The helimagnetic structure of Au_2MnAl, consisting of ferromagnetic sheets perpendicular to Ox and separated by $\frac{1}{2}a_0$. The direction of the magnetic moments rotates by 45° between successive sheets.

The intensities of the satellites will be determined by eqn (8.6) and will therefore be proportional to

$$\tfrac{1}{4}(1+\cos^2\phi),$$

where ϕ is the angle between the reflection plane and the plane of the spins. If, for example, $\phi = 0°$ then each satellite will have an intensity that is 50 per cent of that which a ferromagnetic arrangement of spins would contribute. On the other hand, if $\phi = 90°$ then each satellite has an intensity which is only 25 per cent of that for a ferromagnetic arrangement. These relationships are illustrated in Fig. 163, together with the corresponding patterns for other types

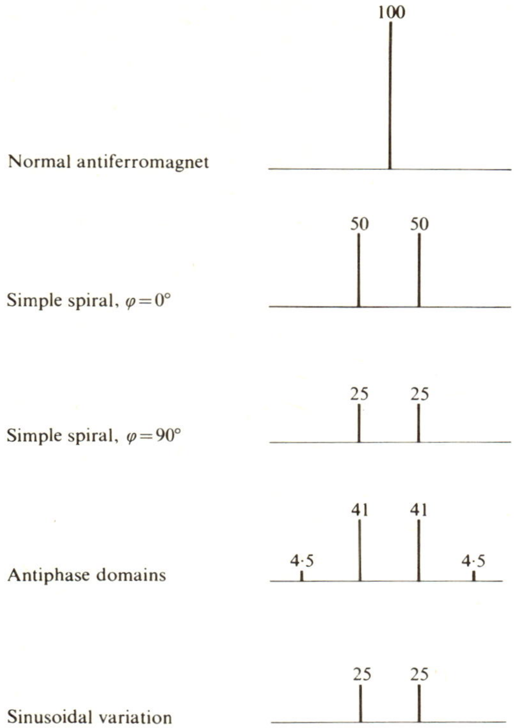

FIG. 163. The expected intensities of magnetic satellites for various postulated structures.

of structure which are characterized by satellites, which we shall consider below.

The expression for the displacement vector between the nuclear peaks and the magnetic satellites applies to the zero-order reflection, as well as to reciprocal lattice points with finite values of the integers h, k, l. There is, accordingly, a reflection which indexes as $(00\frac{2}{7})$ for Au_2Mn and as $(00\frac{1}{4})$ for Au_2MnAl. Such a reflection appears at a very small value of 2θ, the angle of scattering, which will be very close to the undeviated main beam. Indeed, it is normally necessary to fit very fine collimating slits in front of the detecting counter in order to resolve the very low-angle reflection. Fig. 164 shows this reflection, occurring at $2\theta = 2°$, for the alloy Au_2MnAl, which was

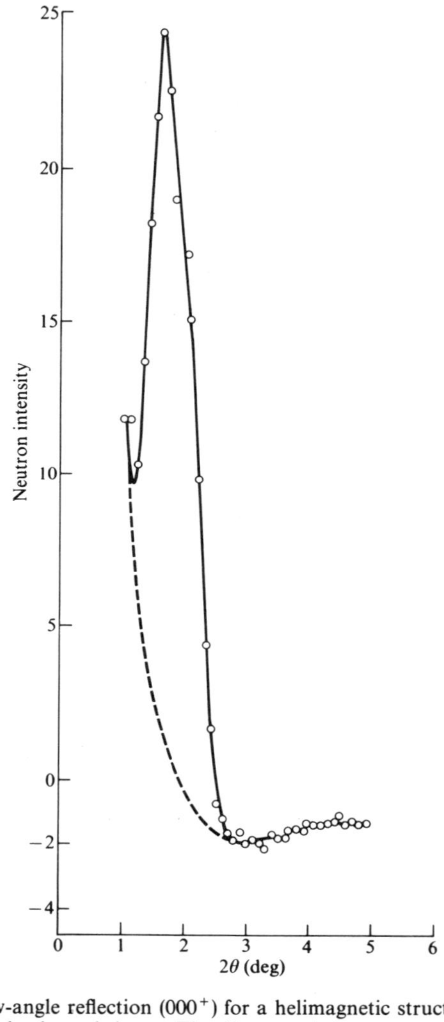

FIG. 164. The low-angle reflection (000^+) for a helimagnetic structure which occurs close to the straight-through beam. The diagram shows an experimental curve for Au_2MnAl for which the reflection at $2\theta = 2°$ indexes as $(00\frac{1}{4})$.

observed by subtraction of the diffraction pattern at room temperature, where no magnetic order remains, from that at 4·2 K where the spiral-spin structure exists. Both of the patterns were measured with collimating slits of 16′ angular divergence in front of the counter.

For further detailed discussion of intensity formulae relevant to spiral structures and conical spirals the reader is referred to the Appendix in a paper by Cox *et al.* (1963), to which we shall refer again in Chapter 14.

8.5. Antiphase domains and sinusoidal variations of moments

A main feature of a helimagnetic arrangement such as we have just described is that it involves a large superstructure which is superimposed on the ordinary unit cell of the material. In the case of Au_2Mn, for which the phase angle ζ is approximately 51°, the direction of the magnetic moments will repeat at intervals of about 7 planes, or $3\frac{1}{2}$ unit cells. We note however that this is only an approximate number and what we can regard as the magnetic unit is, strictly, not commensurate with the ordinary unit cell.

There are other types of structure in which a superstructure of this kind exists. One of these is the arrangement of antiphase domains which was once postulated for metallic chromium. It is no longer believed that such an arrangement exists for chromium, and there is possibly no indisputable evidence that it occurs in any other material, but it is instructive to examine it in detail because of the way it assists in the understanding of the diffraction patterns from other arrangements and variations of magnetic moments, such as that in which a sinusoidal variation is present.

The magnetic structure of chromium is undoubtedly based on the antiferromagnetic arrangement shown in Fig. 165, where the magnetic moments of the atoms at the corners and body centres of the cubes are oppositely directed. The observation of satellite reflections in the diffraction pattern indicates that the precise structure is more complicated than Fig. 165 suggests and there has been considerable speculation concerning its actual form. Corliss, Hastings, and Weiss (1959) proposed an antiphase-domain structure in which the $+-$ sequence of magnetic moments continued regularly for about 13 unit cells and then reversed for a further 13 unit cells; this succession of reversals was continued, giving a superstructure with a period of 26 unit cells. The arrangement is illustrated in Fig. 166, but with a repeating unit of only 10 cells in order to improve the visibility of the Figure. If we consider the reflection of neutrons from the group of cells shown at C there will be zero intensity at the position θ normally expected for the (100) magnetic reflection because of the mutual cancellation of the sections PQ, QR. There will however be partial reinforcement at slightly larger and slightly smaller values of the Bragg angle $\theta \pm \delta\theta$, where $\delta\theta$ is such as to give an additional change of path length of $\pm\lambda$ between the contributions from P and R. This condition is expressed by writing

$$N\,\delta(2d\sin\theta) = \pm\lambda,$$

where N is the number of unit cells within the larger repeating unit and $\delta(2d\sin\theta)$ represents the increment in $2d\sin\theta$, therefore

$$2dN\cos\theta\,.\,\delta\theta = \pm\lambda$$

i.e.

$$\mathrm{PR} \equiv Nd = \pm\lambda/2\cos\theta\,.\,\delta\theta; \tag{8.7}$$

thus expressing the distance PR in terms of the angular displacement of the satellites. So far as the domain PR is concerned, the

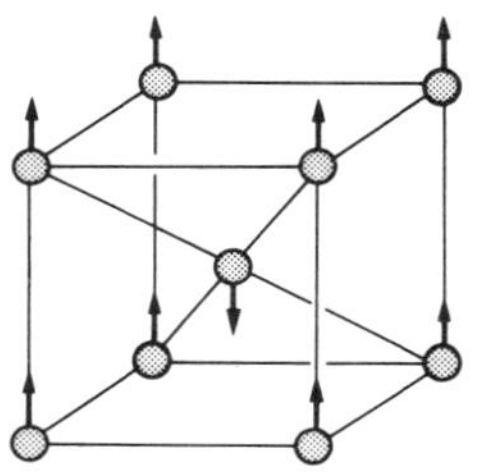

FIG. 165. The basic antiferromagnetic structure of chromium.

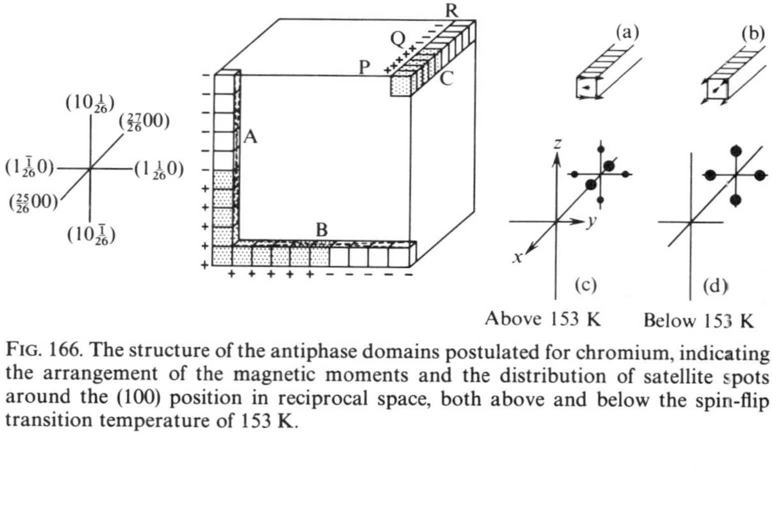

FIG. 166. The structure of the antiphase domains postulated for chromium, indicating the arrangement of the magnetic moments and the distribution of satellite spots around the (100) position in reciprocal space, both above and below the spin-flip transition temperature of 153 K.

(100) reflection is replaced by two satellites which index as $(\frac{25}{26}00)$ and $(\frac{27}{26}00)$.

In order to maintain over-all cubic symmetry domains of types A and B, in the Figure, will also exist and these will contribute satellites such as $(1\frac{1}{26}0)$, $(1\overline{\tfrac{1}{26}}0)$. The net result is that the (100) spot in reciprocal space is replaced by the grouping of six closely-spaced spots which is shown in (c) of the Figure. This distribution is found experimentally at room temperature, and it is significant that two of the spots are twice as intense as the other four; from this fact it is deduced that the magnetic moments are parallel to the domain wall, as drawn in (a). Below 153 K only four spots, all of equal intensity as illustrated in (d), constitute the satellite pattern, and it is deduced that the magnetic moments are perpendicular to the domain walls, as in (b).

More detailed investigation reveals that this model of antiphase domains is not completely in accord with the experimental data for chromium. In Fig. 167 the full-line curve indicates the contribution from a half-domain PQ in the neighbourhood of the (100) reflection, calculated in the manner which is typically used for optical diffraction gratings; $\delta\theta_0$ is equal to $\lambda/2\mathrm{PQ}\cos\theta$. The

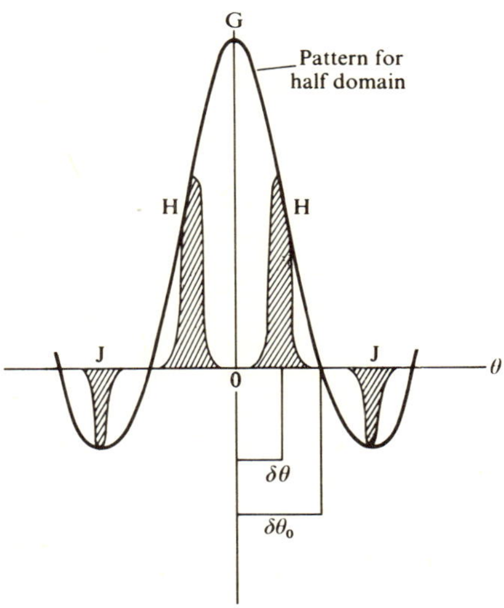

FIG. 167. Diagram to illustrate the calculation of satellite intensities from antiphase domains.

contribution from a *series* of half-domains PQ, QR, ... is indicated by the shaded peaks in the Figure, such as the predominant peaks H which are found on each side of 0, the normal Bragg angle, at a position $\delta\theta$; we have already shown in eqn (8.7) that $\delta\theta$ is equal to $\lambda/2\text{PR}\cos\theta$, so that it is equal to $\frac{1}{2}\delta\theta_0$. However, as well as the shaded peak at H we should also expect to observe a third-order peak at J, for which the displacement from 0 is 3 times as great as for H. The relative amplitudes at G, H, and J can be shown to be 1·00, $2/\pi$, and $\frac{2}{3}\pi$, which amount to 1·00, 0·64, and 0·21, corresponding to intensities 100 per cent, 41 per cent, and 4 per cent. The third-order satellites J, which are expected therefore to have intensities a tenth of the fundamental satellites H, have been looked for very carefully experimentally for chromium but have not been found. Accordingly, Shirane and Takei (1962) suggested a model in which the value of μ_B does not vary discontinuously along the modulation axis, but changes in a regular and sinusoidal way as illustrated in Fig. 168. The arrangement which the Figure shows,

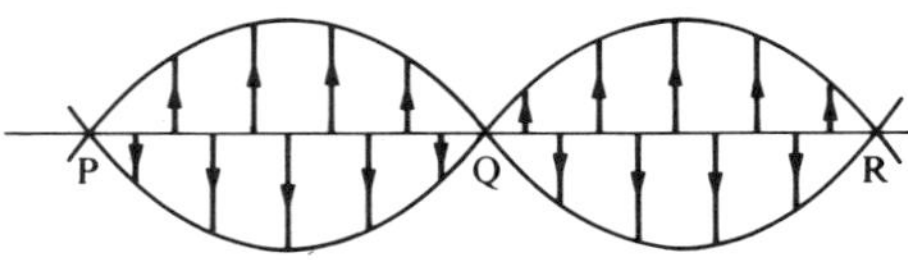

FIG. 168. Sinusoidal variation of magnetic moments of corner and body-centring atoms in chromium with distance along PQR.

with the moments directed perpendicular to the modulation axis, is believed to exist at room temperature; below the transition temperature of 153 K the moments lie along the axis. A sinusoidal variation of moments of this kind does *not* give any third- or higher-order satellites. The reader may recall that this fact is a statement of a general principle in the Fourier treatment of diffraction theory, namely, that a sinusoidal distribution yields *only* a first-order spectrum. It is important to note, however, that the magnitude of the intensity which would be expected for the *first*-order satellite H in this sinusoidal model is different from that for the antiphase-domain arrangement. Calculation shows that the amplitude in the sinusoidal case is a half of what would be observed if the maximum moment existed on all the atoms. Thus the *intensities* of the satellites are only 25 per cent of what would be given by a simple conventional alignment of moments, compared with the 41 per cent that we deduced earlier for the antiphase-domain model. Conversely, when we interpret the experimental data for chromium we deduce a maximum moment of 0·59 μ_B when we use the sinusoidal model, compared with 0·45 μ_B if we assume antiphase domains.

In order to summarize these intensity relationships the reader is referred back to Fig. 163, which indicates the relative intensities of the satellites which would be expected for various models. It should be noted that the simple spiral model would not be very acceptable for chromium because of the difficulty in accounting for the phase change at 153 K.

We have considered the case of chromium at length because of the way in which it has contributed to our general understanding of the types of arrangement of moments. The reader will have noticed the degree of sophistication which is required in interpreting the magnetic structure of what is, at first sight, a potentially straightforward body-centred cubic metal.

8.6. Magnetic symmetry

From the foregoing examples it will have been realized that the determination of a magnetic structure involves the calculation of the magnetic scattering, and its distribution in space, for some postulated arrangement of magnetic moments and then a correlation of this scattering with the experimental data. This correlation is essentially an indirect process, in contrast with the straightforward calculation of structure factors for an assumed structure. It is indeed, so far as magnetic structures are concerned, a more indirect process than the corresponding problem in X-ray structure analysis, for two reasons. First, it is necessary to study single-domain samples, and not merely single-crystal samples, if all the individual reflections are to be identified separately and unambiguously. This means that the regular magnetic arrangements, such as were shown in Figs. 113 and 158 have to be maintained through a single crystal which measures a few millimetres in size. In certain circumstances, by the application of magnetic fields or mechanical strains, this can be achieved, but more usually a single crystal will consist of many domains. Very often, of course, only polycrystalline material is available, particularly for alloys and materials in which several phase changes occur as the temperature is varied; here the interpretation has to be made largely by trial and error.

A further disadvantage is that it is much more difficult to apply space group theory to the determination of magnetic symmetry and structure than to the conventional determination of atomic positions by either X-rays or neutrons. Thus the structure of AuMn in Fig. 156 appears, experimentally, to be tetragonal but must in reality be orthorhombic because of the distribution and orientation of the magnetic moments. No doubt there is a very small difference in the magnitudes of the x- and y-axes, but it is too small to be observed experimentally and the pseudo-symmetry is misleading. In a similar way in apparently cubic materials there may be very small changes to tetragonal and rhombohedral symmetry which are not revealed.

Moreover, the total number of magnetic reflections that can be studied is limited severely by the fact that the magnetic form factor falls off rapidly with increase of the angle of scattering. This means that it is not usually possible to isolate sufficient magnetic reflections to utilize the formal knowledge of the magnetic space groups which has been built up in recent years. This formal development has been brought about, largely by Russian crystallographers, by adding to the normal symmetry operations a new operation, designated R, which reverses the direction of a magnetic moment. In this way the normal 230 space groups are increased by a further 1421 groups which are applicable to the structures of ferromagnetic and antiferromagnetic materials, giving a total of 1651 so-called 'Shubnikov groups'. However, this treatment of symmetry cannot be applied to the helimagnetic structure, whose successive spins are rotated through an arbitrary angle ζ, which does not have to be any simple fraction of 360°, and where the spiral superstructure is not commensurate with the chemical unit cell.

We have already mentioned that the magnetic unit cell may be doubled in one or more directions relative to the chemical unit cell. This idea can be viewed more formally in terms of the concept of anti-translation, whereby a directional translation from a point in space arrives not at an identical point but at a point where the magnetic moment, and all neighbouring moments, is reversed. In these terms Belov, Neronova, and Smirnova (1957) have shown

FIG. 169. The 36 Bravais lattices which are possible in magnetic materials where the concept of symmetry is extended from single colour to black and white by inclusion of the operation of anti-translation. Black and white circles in the diagrams correspond to antiparallel orientation of the magnetic moments.

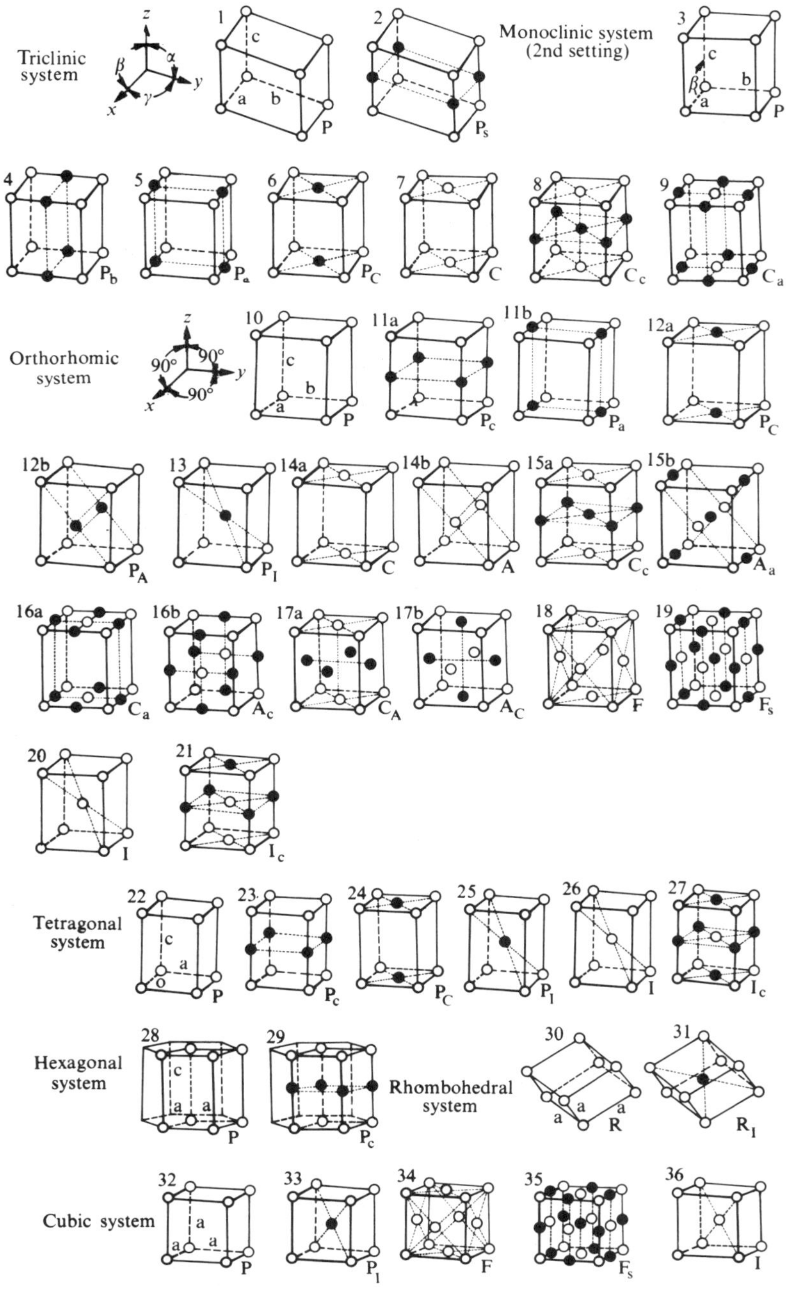
Triclinic system
z
α
β
y
x
γ
1
c
a
b
P
2
P_s
Monoclinic system
(2nd setting)
3
c
β
b
a
P
4
P_b
5
P_a
6
P_C
7
C
8
C_c
9
C_a
Orthorhomic system
z
90°
90°
y
x
90°
10
c
b
a
P
11a
P_c
11b
P_a
12a
P_C
12b
P_A
13
P_I
14a
C
14b
A
15a
C_c
15b
A_a
16a
C_a
16b
A_c
17a
C_A
17b
A_C
18
F
19
F_s
20
I
21
I_c
Tetragonal system
22
c
a
o
P
23
P_c
24
P_C
25
P_I
26
I
27
I_c
Hexagonal system
28
c
a
a
P
29
P_c
Rhombohedral system
30
a
a
a
R
31
R_I
Cubic system
32
a
a
a
P
33
P_I
34
F
35
F_s
36
I

that the 14 conventional Bravais lattices of ordinary crystallography are extended to 36 magnetic Bravais lattices, each of which will be associated with a particular pattern of extinctions for magnetic reflections. These are illustrated in Fig. 169, where the black and open circles correspond to antiparallel orientation of the magnetic moments. It will be noticed that the 36 lattices include 14 in which the circles are all of the same kind, namely, open; these are the ordinary Bravais lattices. In the development of the Shubnikov groups from these magnetic lattices we introduce the concepts of anti-rotation axes, screw anti-rotation axes, and anti-glide planes to describe the symmetry operations. Examples of these, showing the way in which the magnetic moments are related, are given in Fig. 170. It is found that two kinds of symmetry operation can be distinguished, and these are described as of the 'first and second kind'. In considering the operation of the symmetry elements it is essential to regard the magnetic moment as a circular current; after operation of the symmetry element on this current the direction of the new moment can then be defined. It will then be seen that for elements of the 'first kind', comprising translations and rotations, the moment transforms as a true polar vector; the anti-rotation axis $2'$ provides an example of this. On the other hand, after an operation of reflection (or inversion), such as the reflection plane m, the moment vector—as distinct from the circular current—will not always appear to have been transformed.

We give in Fig. 171 typical diagrams for a magnetic space group $Pn'm'a$ showing both the general positions of the atoms and the conventional representation of the symmetry elements. In an analogous way to the operation of the symmetry elements in the 230 ordinary crystallographic space groups, each of the magnetic symmetry elements will lead to individual extinctions among particular types of magnetic reflection. The details have been tabulated by Izyumov and Ozerov (1970).

Because of the practical difficulties which we have mentioned earlier, there are very few examples of the practical use of symmetry representations for determining actual magnetic structures. The first such determination was that of chalcopyrite, $CuFeS_2$, by Donnay, Corliss, Donnay, Elliott, and Hastings (1958), and a more recent example is the study of $LiCuCl_3 . 2H_2O$ by Abrahams and Williams (1963). X-ray measurements show that the space group of this substance is $P\ 2_1/c$, and the neutron diffraction pattern at

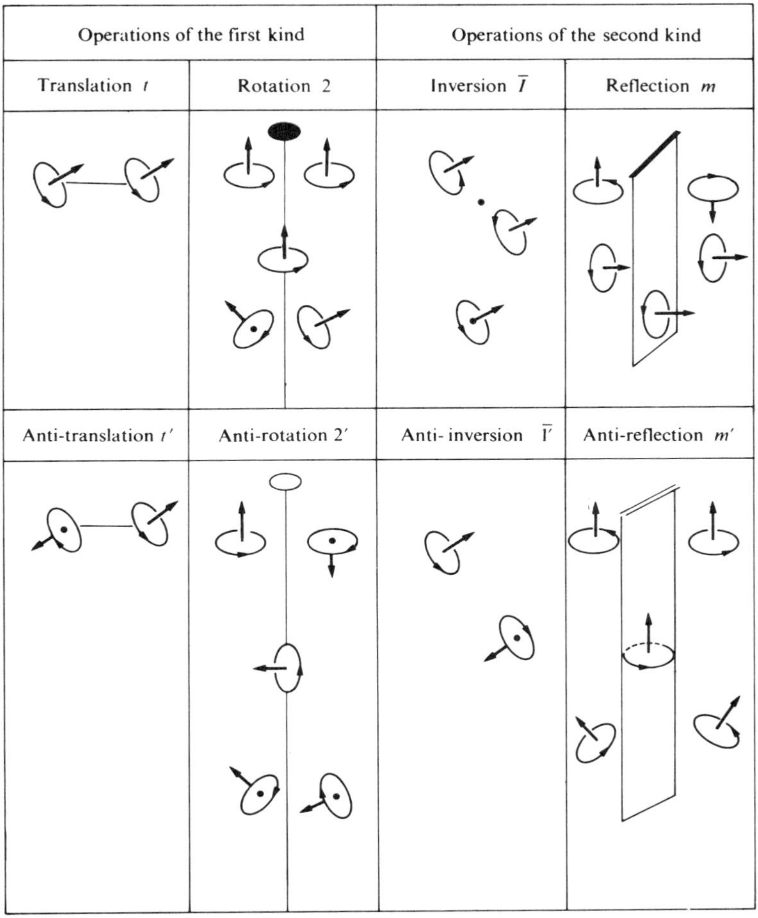

FIG. 170. Examples of the correlation between magnetic moments related by various permissible types of symmetry element.

room temperature is consistent with this. There is a transition to antiferromagnetism at a temperature of 6·7 K, and neutron measurements of $(0kl)$ reflections were made down to 1·5 K. The new magnetic reflections which appeared at the lower temperature were solely those $(00l)$ reflections for which $l = 2n+1$ and which are absent from the nuclear pattern; in particular no new $(0k0)$ reflections appear. These facts were considered in terms of what would be

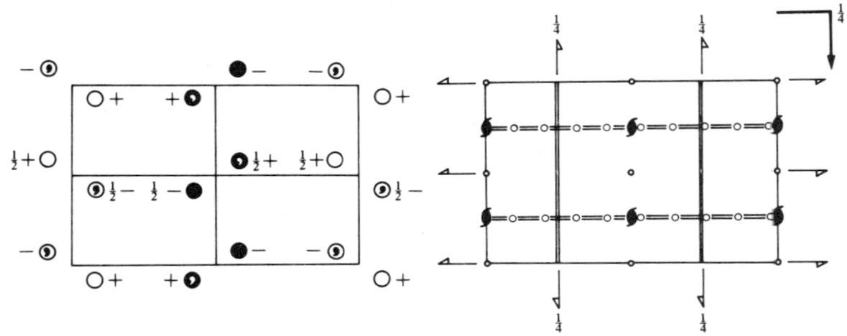

FIG. 171. General atomic positions and symmetry elements in $Pn'm'a$.

expected for each of the ten Shubnikov groups which are derived from the space group $P2_1/c$. All but three of these may be eliminated, and a choice in favour of group $P2_1'/c$ may then be made by comparing the actual values of the magnetic intensities found at the lowest temperature. The model chosen for the magnetic structure is illustrated in Fig. 172. The magnetic moments of the copper atoms lie close to the *ac*-plane, i.e. almost perpendicular to the *b*-axis, and inclined at about 50° to the *c*-axis. Within experimental error it is possible that the moments point along the line which joins pairs of copper atoms.

In the case of $LiCuCl_3$ the chemical and magnetic unit cells are

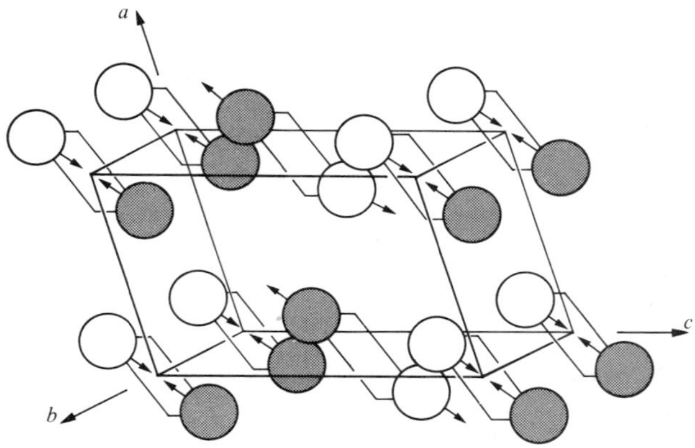

FIG. 172. The magnetic structure of $LiCuCl_3 . 2H_2O$. The spin directions of the copper atoms lie in the ac-plane at 49° to the *c*-axis, i.e. close to the Cu—Cu direction. Parallel and antiparallel moments belong to white and shaded atoms respectively. The linkages shown between pairs of copper atoms represent the $Cu—Cl^-$ bonds. (After Abrahams and Williams 1963.)

the same. For calcium vanadite, CaV_2O_4, which is another antiferromagnetic material whose magnetic structure has been discussed in terms of magnetic space groups (Hastings, Corliss, Kunumaura, and La Place 1967), the unit cell is doubled in the *b*-, *c*-directions. In this case the chemical space group is *Pnam*, whereas the magnetic reflections from a powdered sample are compatible with three possible structural models which belong to the magnetic space groups P_a2_1 and C_ac; these possible alternatives could be distinguished with single crystals.

For further information and additional references concerning magnetic space groups the reader is referred to articles by Belov *et al.* (1957) and Atoji (1965).

Multiplicity

Before concluding this general discussion of the principles of interpreting the neutron scattering in terms of magnetic structure we return to one or two topics which we have mentioned briefly earlier. A point of major importance in calculating the intensities to be expected from any postulated magnetic model can be illustrated by considering Fig. 288, which appears later (p. 447) when we discuss structural models for MnO. At room temperature the symmetry of MnO is cubic, but below 120 K, which is the Néel temperature for the antiferromagnetic transition, there is a slight distortion as the unit cell becomes rhombohedral. To a first approximation nevertheless we can discuss the magnetic reflections in terms of the usual cubic indices. When for model (a) in the Figure we calculate the intensity of the magnetic reflection (111) we must note that the four pairs of planes which make up the $\{111\}$ form are no longer identical. Only the faces (111) and $(\overline{1}\overline{1}\overline{1})$ of this form will contribute any magnetic intensity, for they alone are populated by atoms having magnetic moments all of the same sign. Other planes, such as $(\overline{1}1\overline{1})$, contain equal numbers of positive and negative magnetic moments and will therefore contribute nothing to the neutron intensities. As a result the multiplicity for the (111) reflection in a powder pattern is reduced from 8 to 2. Thus, when Roth (1960) applied stress to a single crystal of NiO and produced effectively a single domain, he found that 6 of the 8 faces of the $\{111\}$ form gave zero intensity. However, single crystals of such pseudo-cubic structures as this will usually consist of many domains, and among them there will be four possible choices for the special [111] direction associated with the antiferromagnetic arrangement. In less-symmetrical systems, such as the rhombohedral, there will be a unique antiferromagnetic axis if the structure requires the magnetic moments to align themselves along the unique crystallographic axis, and this will then have a common direction throughout a single crystal. In such a case it is also possible to confirm this axis directly by making measurements with polarized neutrons and observing the way in which the state of polarization of the beam is altered by the scattering process. It can be shown that the change of polarization depends on the cosine of the angle between the magnetic interaction vector $\mathbf{q}$ and the neutron polarization vector $\boldsymbol{\lambda}$, becoming zero when $\mathbf{q}$ and $\boldsymbol{\lambda}$ are parallel. In this way Nathans *et al.* (1958) were able to confirm the postulated moment directions with single crystals of α-Fe_2O_3 and Cr_2O_3.

A general discussion of the limitations of powder data for determining magnetic structures has been given by Shirane (1959), with particular reference to the non-equivalence, magnetically, of different faces of a crystallographic form, which results in identical intensities being given by different magnetic models. Shirane derives expressions for the powder intensities appropriate to magnetic structures of various symmetries, using the concept of 'configurational symmetry', which is the effective symmetry which is displayed when we assign merely + or − signs rather than vectors to the magnetic atoms. On this basis the symmetry of MnO and the other related oxides are rhombohedral, although lower than this when vector directions are taken into account, and a ferromagnetic material such as iron remains cubic. This concept is not applicable to non-collinear structures such as helimagnetics. Shirane's discussion shows that when the configurational symmetry is cubic no information about the spin directions can be obtained from powder data, whereas in the uniaxial systems it is possible to deduce the angle which the spin direction makes with the unique axis of the magnetic cell, but nothing more. As three typical examples we have: (1) iron, for which we can make no deduction; (2) MnO for which the spins are at 90° to the [111] axis, but in some unknown direction within the (111) planes; (3) FeO for which the spins are along the [111] axis, giving a special case for which the solution is completely defined.

9

INELASTIC SCATTERING

9.1. Introduction

WE have previously mentioned in Chapter 2 the importance of inelastic scattering for neutrons. Inelastic scattering occurs because the atoms in a solid are of finite mass and they can gain or lose energy when neutrons, or any kind of radiation, collide with them. We might at first imagine that the motions which the individual atoms undergo are random, but this cannot be the case in a solid (as distinct from a gas) since the atoms are close enough for strong forces to exist between near neighbours. Thus if one atom is displaced from its mean position, then all the interatomic forces in the neighbourhood will be affected and all the nearby atoms will undergo displacement of some kind as a repercussion of the initial displacement which we postulated. The motions of the atoms within the solid are therefore not individual but collective, and we may analyse them into a spectrum of thermal vibrations which travel as waves through the crystal.

The energy associated with these waves is quantized, and we regard each quantum of energy as a 'phonon' just as in the theory of radiation a quantum of energy is regarded as a 'photon'. When neutrons, or X-rays, are inelastically scattered in their passage through a solid these phonons are generated or annihilated according as the neutrons lose or gain energy. Study of the process is of particular value with neutron beams because the phonon energies are about the same as the kinetic energy of the thermal neutrons (0·01–0·1 eV), so that absorption or loss of the energy of a phonon will produce a big change in the velocity or energy of a neutron and this will be relatively easy to measure. On the other hand, the energy of an X-ray is about 10 000 eV and absorption or loss of a phonon will be quite negligible.

The aim of the measurements of inelastic scattering which we shall first describe is to find the details of the phonon spectrum in the solid, as described by first the shape of the frequency distribution function $g(\omega)$, which is defined by saying that $g(\omega)\,d\omega$

is the number of phonons with frequencies lying within the interval from ω to $\omega + d\omega$, and secondly the dispersion law, which describes the dependence of the angular frequency ω on the wave vector $\mathbf{q}$, which is numerically equal to $2\pi/\lambda$, for any direction in the crystal.

Before examining the way in which neutrons can be used to determine the frequency-distribution function and the dispersion law by experiment it may be helpful to consider the expected form of these in terms of various simple models.

In Einstein's model of vibrations in a crystal it was assumed that they all vibrated with the same frequency which is equivalent to regarding the vibrations of individual atoms as being independent. Debye's model was based essentially on consideration of a crystal as continuous in nature, like an elastic solid, but took its periodic nature partly into account by assuming that the spectrum of vibration was cut off at some characteristic frequency ω_{max}, indicating the fact that waves of length less than about twice the distance between atoms will not propagate. The number of waves within a band of frequency of width $d\omega$ increases as $\omega^2\, d\omega$, and the frequency spectrum is that shown in Fig. 173(a). The dispersion curve is shown at (b) in this Figure. For this model the curve is linear, which implies that the velocity of the vibrations is a constant and not dependent on their wavelength. This is the same conclusion as is reached for waves on a continuous string or rod, for which there are constant velocities of $\sqrt{(T/m)}$ and $\sqrt{(E/\rho)}$ respectively, where T is the string

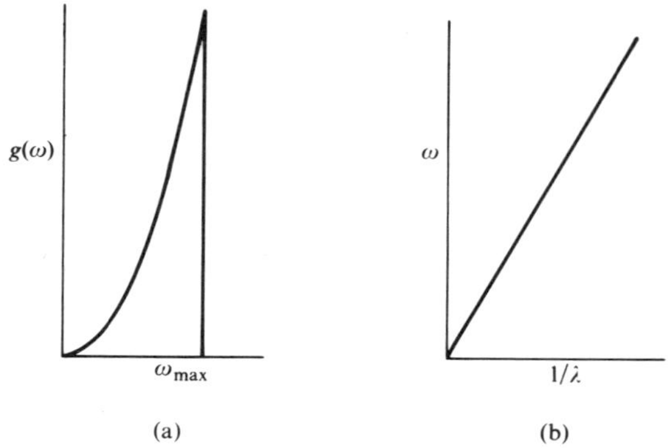

FIG. 173. (a) shows the frequency spectrum $g(\omega)$ according to the Debye model of a solid, with a sharp cut-off at ω_{max}. The proportionality between ω and $1/\lambda$ in (b) indicates a constant wave velocity (the velocity of sound) in the crystal.

tension and m its mass per unit length, E is Young's modulus for the rod, and ρ is its density. The slope of the straight line in Fig. 173 (b) is the velocity of sound in the material. The next stage in the development of a realistic model is to take into account the interactions among the atoms and the calculation becomes progressively more difficult as we deal with solids containing several different kinds of atoms and as we assume that interatomic forces extend beyond the immediate neighbours of an atom to its second, third, fourth, . . . , nth neighbours. A relatively simple case is shown in Fig. 174, which is for a linear chain of atoms of two kinds, alternately

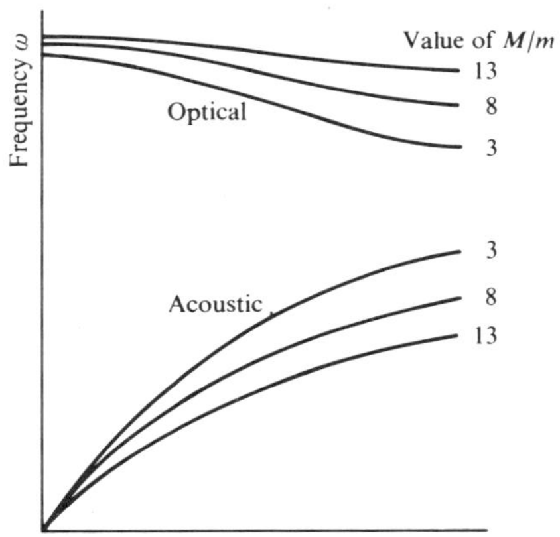

FIG. 174. The dispersion curve showing the relation between frequency ω and wave-number $2\pi/\lambda$ for a linear chain of two kinds of nuclei of masses m, M.

of masses m, M, and for which only nearest-neighbours interactions are assumed. The precise curves will depend on the ratio of M/m, and there will be two branches to the curves, labelled as 'acoustic' and 'optical', for which the two types of atom are vibrating in the same and in opposite directions respectively. Moreover, in a three-dimensional anisotropic crystal the form of the curves will be different for different directions of propagation of the waves.

The importance of neutron measurements is that they provide a means of determining the actual form of the $g(\omega)$–ω curve and the ω–q curve by experiment. Thus Fig. 175 shows the distribution function for the vibrations in metallic sodium, as found by Dixon *et al.* (1963).

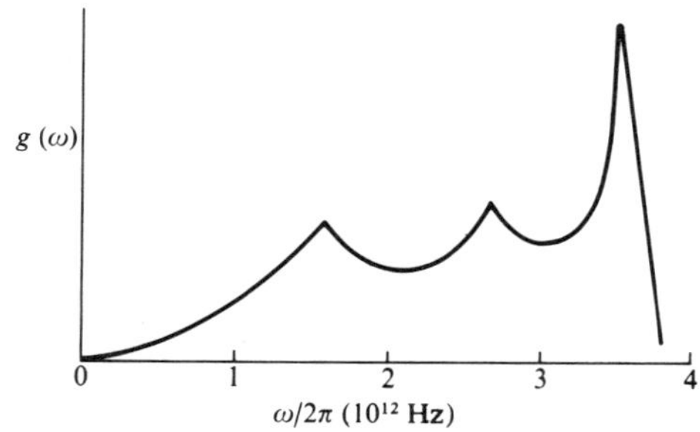

FIG. 175. The distribution function for the vibrations in sodium. (Dixon, Woods, and Brockhouse 1963.)

9.2. Energy and momentum relations

The analysis of the neutron scattering is best done in terms of the reciprocal lattice, extending the treatment from that used in the analysis of elastic scattering. The latter case is illustrated in Fig. 176(a) which indicates Ewald's original construction. From the origin O of the reciprocal lattice a point C is defined by drawing the vector $\boldsymbol{\kappa}_0/2\pi$ parallel to the direction of the incident neutrons and of magnitude $1/\lambda$. From C we then draw a vector of the same length parallel to the direction of the emerging neutrons, to terminate in the point P. Only when P happens to lie at a reciprocal lattice point (hkl) will a coherent Bragg reflection take place, since it then follows from the Figure that

$$\mathrm{OP} = 2\frac{1}{\lambda}\sin\theta,$$

i.e.

$$\frac{1}{d_{hkl}} = 2\frac{1}{\lambda}\sin\theta,$$

so that

$$\lambda = 2d_{hkl}\sin\theta,$$

which means that the Bragg equation is satisfied for the (hkl) planes which correspond to P and a reflection will accordingly be produced.

In the above example of elastic scattering there is no change of neutron energy and we have made $\boldsymbol{\kappa}_0$ and $\boldsymbol{\kappa}$ of equal length, thus

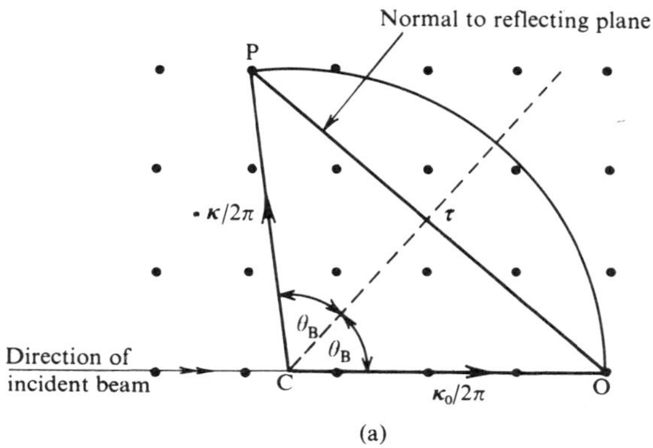

FIG. 176. Reciprocal lattice representations for (a) elastic and (b) inelastic scattering. In each case $\boldsymbol{\kappa}_0$, $\boldsymbol{\kappa}$ are the wave vectors of the incident and scattered neutrons and $\boldsymbol{\tau}$ is a reciprocal lattice vector. In (b) $\mathbf{q}_1$ is the wave vector of the phonon given up to the crystal vibrations or, in the alternative case indicated by the dotted lines, $\mathbf{q}_2$ is the wave vector of an annihilated phonon.

indicating that the kinetic energy E, for which

$$E = \tfrac{1}{2}mv^2 = \frac{\hbar^2\kappa^2}{2m},$$

is the same before and after scattering. However, there will be a change of momentum, which is taken up by the crystal lattice, and which is equal to

$$m\mathbf{v}_1 - m\mathbf{v}_2 = \frac{h}{2\pi}(\boldsymbol{\kappa}_0 - \boldsymbol{\kappa}).$$

From the Figure it follows that we can express the conservation of momentum by the equation

$$(\boldsymbol{\kappa} - \boldsymbol{\kappa}_0) = 2\pi\boldsymbol{\tau}.$$

In the case of inelastic scattering the situation is modified to that which is illustrated in Fig. 176 (b), where the magnitudes of $\boldsymbol{\kappa}$, and $\boldsymbol{\kappa}_0$ are no longer equal. There has been a change in the neutron energy and a phonon of frequency ω and wave vector $\mathbf{q}$ has been generated. The energy equation accordingly becomes

$$\frac{\hbar^2\kappa_0^2}{2m} - \frac{\hbar^2\kappa^2}{2m} = \hbar\omega, \tag{9.1}$$

and the momentum equation will be

$$\mathbf{Q} \equiv \boldsymbol{\kappa} - \boldsymbol{\kappa}_0 = 2\pi\boldsymbol{\tau} + \mathbf{q}. \tag{9.2}$$

Here $\mathbf{Q}$ is called the momentum-transfer vector. $\hbar 2\pi\boldsymbol{\tau}$ will be the momentum imparted to the crystal as a whole (just as for Bragg scattering) but there will also be the phonon with momentum $\hbar\mathbf{q}$. In addition to the energy and momentum equations the phonon will also have to satisfy the dispersion law

$$\omega = \omega_i(\mathbf{q}), \tag{9.3}$$

which specifies ω in terms of polarization i and wave vector $\mathbf{q}$.

For a given orientation of crystal and a chosen direction of incidence and observation there are (in the simplest case of a single acoustic branch) two values of $\mathbf{q}$ which give solutions to the above set of three equations, corresponding to creation and annihilation respectively of a phonon and indicated by $\mathbf{q}_1$ and $\mathbf{q}_2$ in Fig. 176 (b).

9.3. Experimental measurements and apparatus

Early experimental measurements of inelastic scattering were restricted in scope because of the limited intensity of neutron beams available, and were devoted to confirming certain aspects of the theory. For example, Egelstaff (1951) and Brockhouse and Hurst (1952) demonstrated the change in neutron energy after scattering by polycrystalline samples, and Cassels (1951) and Squires (1952) showed that the cross-sections of Mg, Al, and Ni were in substantial agreement with theory. Lowde (1952, 1954) studied the angular dependence and temperature variation of the diffuse streaks in the Laue diffraction pattern of a single crystal of iron and showed that the results were in agreement with the theory, as formulated by Froman (1952*a*,*b*) and Waller and Froman (1952). As Fig. 177 shows, there is a diffuse background streak of inelastically

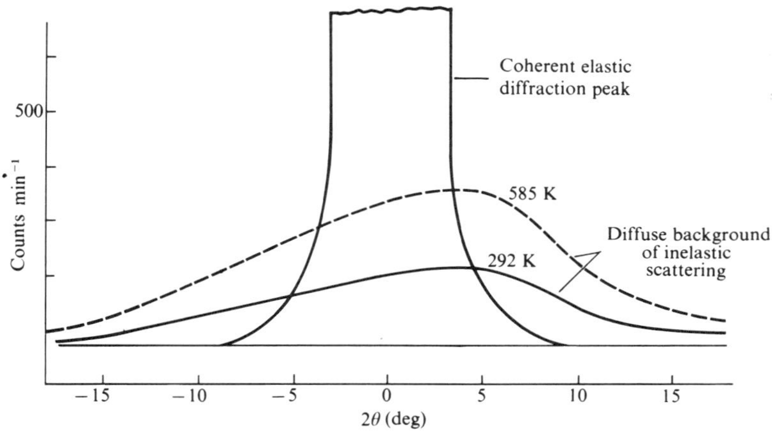

FIG. 177. The scattered intensity found experimentally in the neighbourhood of the (110) Laue reflection from a single crystal of iron, resolved into a coherent diffraction peak and a diffuse background of inelastic scattering. The latter is shown for temperatures of 292 K and 585 K, the peak total intensity in the two cases being 35 400 counts min^{-1} and 34 300 counts min^{-1} respectively. (After Lowde 1954.)

scattered neutrons superimposed upon the Bragg peak for elastic scattering. Over the temperature range of 0–300 °C the inelastic scattering was found to be closely proportional to the absolute temperature.

In later years with the availability of more intense neutron beams and much improved techniques much more refined experiments have become possible. The principle of these experiments is that a beam of monochromatic neutrons of known wavelength is incident on a crystal at a defined angle, and arrangements are made to permit determination of the wavelength of the neutrons after they have been scattered through a known angle. Several different experimental methods have been used to achieve this.

One of the simplest methods is illustrated in Fig. 178, where a beam of thermal neutrons is passed through a block of polycrystalline beryllium. All the neutrons with wavelengths shorter than a cut-off value of 3·96 Å, which is twice the largest interplanar spacing discussed on p. 52, are scattered out of the beam, leaving a residue consisting almost entirely of neutrons between 4 Å and 5 Å, as indicated in Fig. 179. These fall on the sample, and the scattered neutrons are observed at an angle of 90° and energy-analysed. The

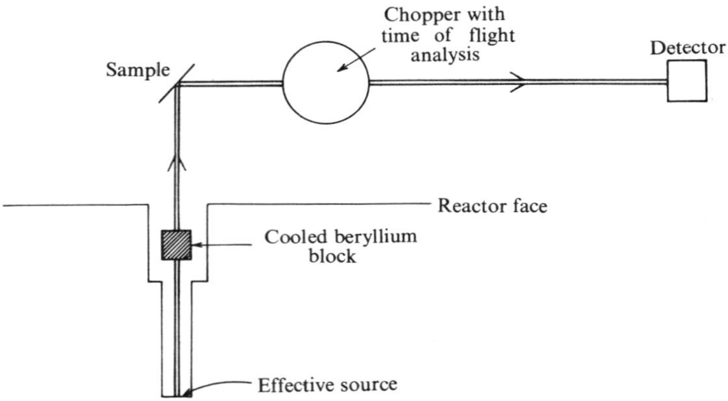

FIG. 178. Diagrammatic sketch of apparatus for measuring the change of neutron energy on scattering.

energy analysis is achieved by chopping the beam and then making a time-of-flight measurement with a multi-channel analyser. Fig. 180 shows how the neutrons which arrive at the detector at any particular time t after the chopper opened will be restricted to those having a velocity $v = s/t$, where s is the length of the path between the chopper and the detector. Thus the slow neutrons arrive late and a time-analysis of the beam scattered by the sample will provide an energy analysis of the beam. In general the chopper consists of a rotating disc, with a diameter of about 20 cm, traversed by a slit which may be along a diameter or curved in a variety of ways.

Rotating discs bearing slits of various sorts may also be used to produce the initial monochromatization of the neutron beam. In

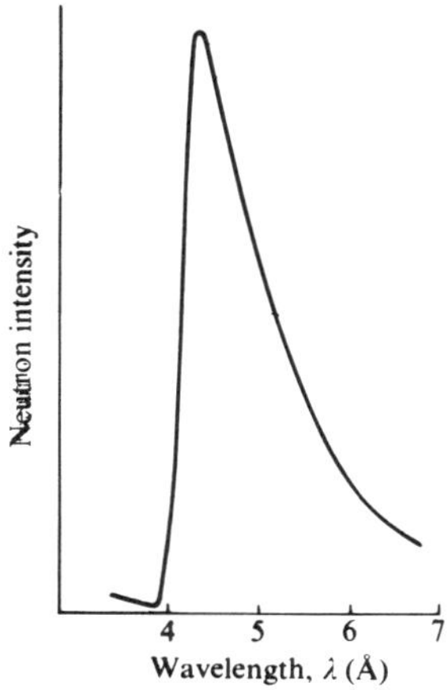

FIG. 179. The neutron spectrum from a reactor, after transmission through a block of beryllium.

order that a suitable range of neutron energies may be covered, high speeds of rotation up to 40 000 rev min^{-1} are employed, and considerable technical developments have been involved in arriving at suitable materials for the rotors. A typical spectrometer, installed at the Harwell reactor PLUTO and described by Low and Dyer (1961), is illustrated diagrammatically in Fig. 181. A collimated beam of neutrons from the reactor falls on two rotating discs arranged in line. Slots are cut in the discs, which rotate at 30 000 rev min^{-1} in a horizontal plane, so that only neutrons of a given velocity can traverse the distance between the leading edge of the slot in the first disc and the trailing edge of the slot in the second disc. The resulting monochromatic beam, which has a wavelength of a few ångströms, then falls on a scattering sample and the neutrons scattered at some chosen angle are energy-analysed in a 500-channel electronic time-of-flight unit. The design principles of a mechanical velocity selector of this kind have been described by Lowde (1960), and for a later account of some of the mechanical and other practical problems involved the reader is referred to an article by Brugger (1965). In order to increase the rate at which data are accumulated a bank of several counters is often employed, for concurrent collection of information at several different angles of scattering. Very substantial shielding has to be employed in order to maintain a

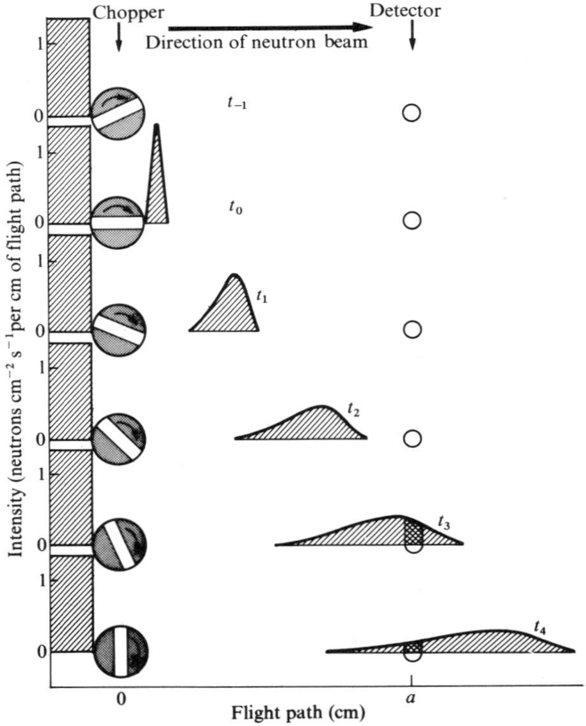

FIG. 180. Principle of time-of-flight detection. The chopper is open at time t_0, and the shaded area indicates the spatial distribution of the neutrons at successive later times. Neutrons of successively smaller velocities, shown in heavy shading, will be detected at succeeding moments t_3, t_4, etc. (Brugger 1965.)

sufficiently low level of background scattering to permit accurate measurement of very low rates of counting.

Another interesting type of spectrometer, developed by Brockhouse (1958), uses a rotating crystal. If, for example, a hexagonal crystal is rotated about its *c*-axis then, by reflection from the various faces of the form $\{10\bar{1}0\}$, six pulses of neutrons could be obtained during each rotation. A beam of filtered neutrons was allowed to fall on the rotating crystal and the reflected pulses were then scattered by a sample and subsequently energy-analysed by a time-of-flight method.

A typical spectrum of the scattered neutrons is shown in Fig. 182 for a single crystal of silicon. The neutrons received by the detector

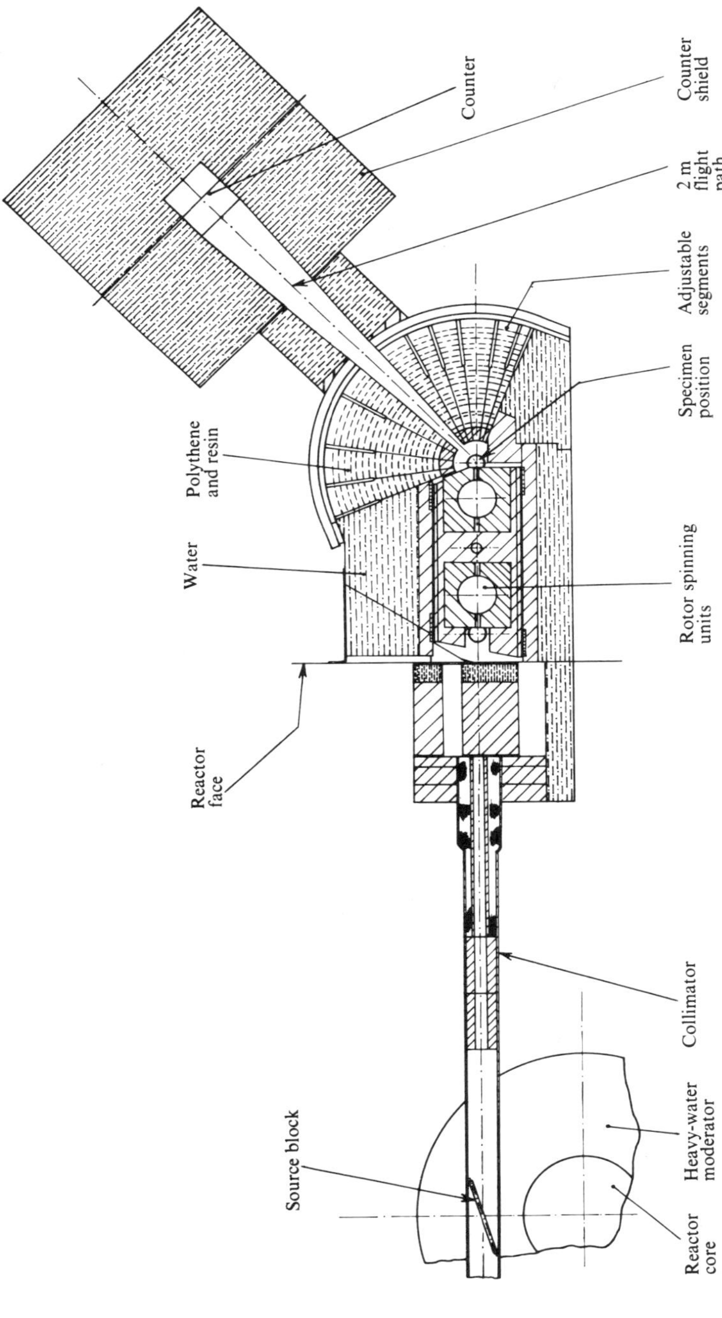

FIG. 181. A plan of the twin-rotor time-of-flight spectrometer installed at the Harwell reactor PLUTO. (Low and Dyer 1961.)

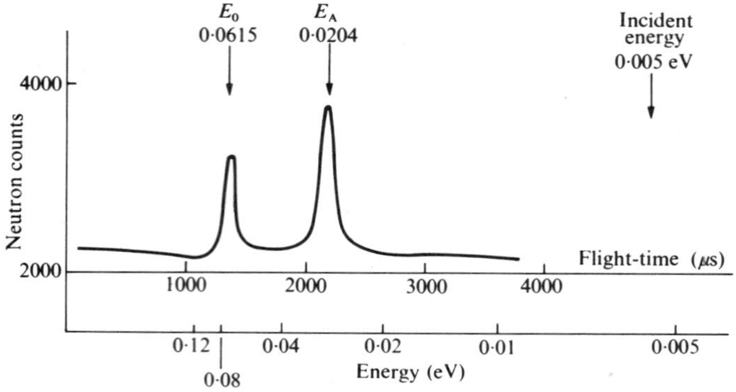

FIG. 182. The energy spectrum of neutrons scattered at 90° from a single crystal of silicon, for an incident energy of 0·005 eV.

are identified by their flight-time and their energy in electron volts. The spectrum shows that for the particular settings of crystal and detector in the experiment the required conditions for conservation of momentum and energy could be satisfied by neutrons which had taken up energies of either 20·4 meV or 61·5 meV from the crystal vibrations. For either of these peaks we are able to deduce the magnitude and direction of the momentum vector $\boldsymbol{\kappa}$ for the scattered neutron and then use the reciprocal lattice construction of Fig. 176(b) to determine the phonon wave vector $\mathbf{q}$. The data can then be plotted to show the shape of the scattering surface, as for aluminium in Fig. 183; this is for the $(hk0)$ plane in the neighbourhood of the reciprocal lattice points (020) and (220). The surface consists of three branches for neutrons which have gained energy; two of these will be transverse but they are too close together to be resolved in this experiment and the third, which is well resolved and is the upper curve in Fig. 183, is the longitudinal branch. From this Figure we can then select the phonons travelling in some particular direction, say [110], and for each one determine its value of ω from eqn (9.1). A series of measurements will then provide a curve which relates the frequency ω and the wave vector magnitude q; such a set of curves, for the longitudinal and transverse branches in aluminium, is given in Fig. 184. This Figure also shows the result of Squires' (1956) calculation. Referring again to Fig. 183 it should be noted that for a given direction of scattering, such as 90° from the incident direction, only those neutrons for which $\boldsymbol{\kappa}$ corresponds to a

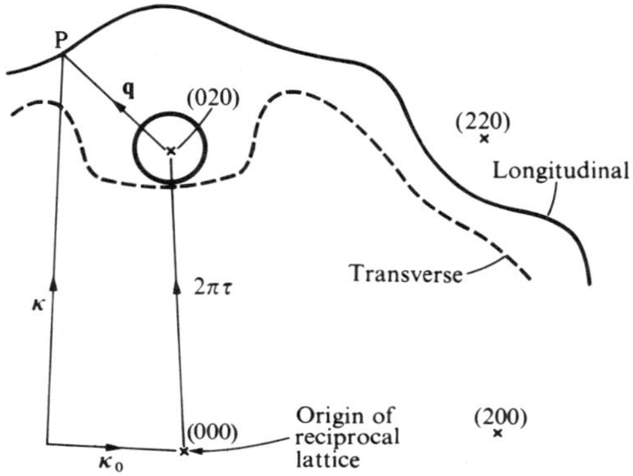

FIG. 183. The scattering surface for aluminium in reciprocal space, derived by energy-analysis of neutrons scattered through 90°. The full-line curve, including the small circle around (020), is the longitudinal branch. The broken-line curve represents two unresolved transverse branches. The derivation of a typical point P from the orthogonal vectors $\boldsymbol{\kappa}$, $\boldsymbol{\kappa}_0$ is shown. The wave-number vector $\mathbf{q}$ is then deduced. (From Carter, Palevsky, and Hughes 1957.)

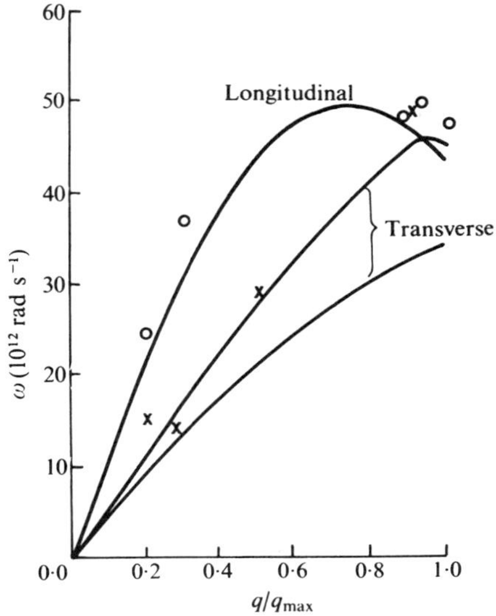

FIG. 184. The frequency versus wave-number relation for the [110] direction in aluminium. The experimental points ○ and × from neutron scattering observations are shown in relation to Squires's calculation for the longitudinal and transverse vibrations. (After Carter, Palevsky, and Hughes 1957.)

point on the scattering surface, such as P, will appear in the scattered beam. Consequently the energy in the scattered beam is sharply peaked. It should be emphasized that in a corresponding kind of experiment with X-rays it is equally possible to find the value of q for a phonon in a diffuse spot, but $\mathbf{\kappa}$ and $\mathbf{\kappa}_0$ are so close together in numerical magnitude that it is impracticable to use the difference eqn (9.1) to give ω with any accuracy. In fact ω can only be determined indirectly, making various assumptions, from the intensity of the thermal diffuse scattering and only for waves travelling in directions of high symmetry in the crystal.

A disadvantage of the experimental arrangement of Fig. 178 is that the use of a fixed angle of 90° between $\mathbf{\kappa}$ and $\mathbf{\kappa}_0$ means that the points obtained for the scattering surface consist of a very random solution of pairs of values of ω and q scattered about the reciprocal lattice. A more flexible method is provided by the use of a triple-axis spectrometer which is shown diagrammatically in Fig. 185 and a

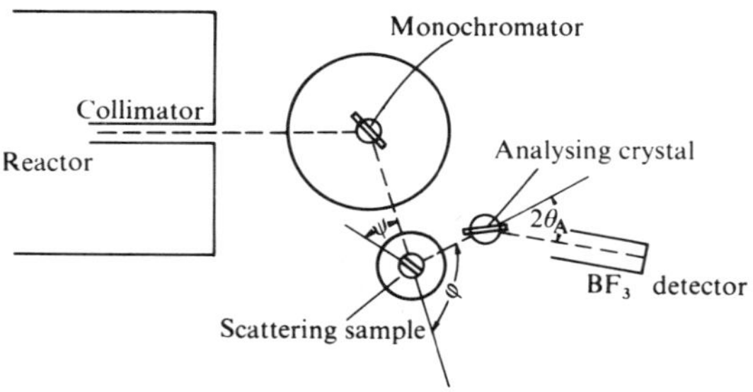

FIG. 185. Diagram of a triple-axis spectrometer. The neutrons are scattered through an angle ϕ and subsequently energy-analysed.

photograph of the instrument at U.K.A.E.A. Harwell appears at Fig. 186 (facing p. 11). The energy analysis is performed by using reflections from an analysing crystal, and there is freedom of choice of both the setting ψ of the scattering crystal with respect to the incident beam and of ϕ, the angle at which the inelastic scattering is observed. In the method of operation, due to Brockhouse, which is usually known as 'constant-Q' a given point in reciprocal space defined by $\mathbf{Q} = \mathbf{\kappa} - \mathbf{\kappa}_0$ is chosen. Then ψ, ϕ are chosen for a succession of values of $\mathbf{\kappa}_0$ for a fixed $\mathbf{\kappa}$ or for a succession of values of $\mathbf{\kappa}$ at fixed $\mathbf{\kappa}_0$. The spectrometer automation is programmed to take up the appropriate positions one by one, and counting is continued for a fixed time interval or a given number of monitor counts. Fig. 187 indicates how the frequency of a phonon in the [0001] direction for magnesium is obtained in this way, incorporating 22 measurements at a single point in reciprocal space. The use of the triple-axis spectrometer in this way ensures very accurate and precise analysis, although the rate at which data can be collected is not so high as in the time-of-flight method.

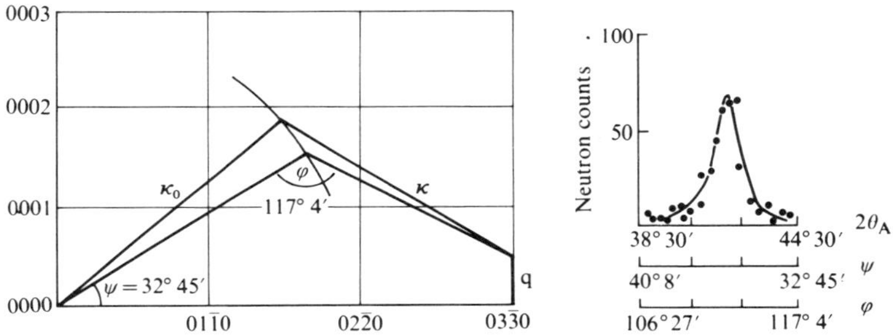

FIG. 187. Reciprocal lattice diagram of magnesium to illustrate observation of phonons by 'constant-Q' method. $\mathbf{Q}$ is the momentum-transfer vector. (Iyengar 1965.)

A third method of experiment described by Stiller and Danner (1961) and Woods, Brockhouse, Sakamoto, and Sinclair (1961) employs what is usually called the 'beryllium filter' method of analysis. The detecting counter is covered by about 10 cm of polycrystalline beryllium which transmits, in the way which we have described on p. 290, only those neutrons whose wavelength is greater than twice the maximum interplanar spacing of beryllium. This means that the only neutrons detected are those with an energy which is less than 0·005 eV. If the sample is irradiated by neutrons of variable energy E and wavelength λ, either by using a crystal reflection or a chopper and time-of-flight device, it will be possible to plot the number of neutrons scattered with an energy change of $(E - 0{\cdot}005)$. Only neutrons which have *lost* energy are detected. The advantage is that a scattered beam of large solid angle is collected, increasing the counting rate for experiments where the resulting loss of momentum resolution is not detrimental.

Yet a further alternative is the so-called MARX spectrometer, which derives its name from the multi-angle reflecting crystal. The principle is indicated in Fig. 188, which makes clear that it is really a development of the triple-axis spectrometer in which the usual analysing crystal and detector have been replaced by a much larger analysing crystal and a linear position-sensitive detector. This analysing crystal accepts neutrons which leave the sample over a wide range of angle of scattering and of energy. The value of the distance x at which the neutron arrives in the detector determines

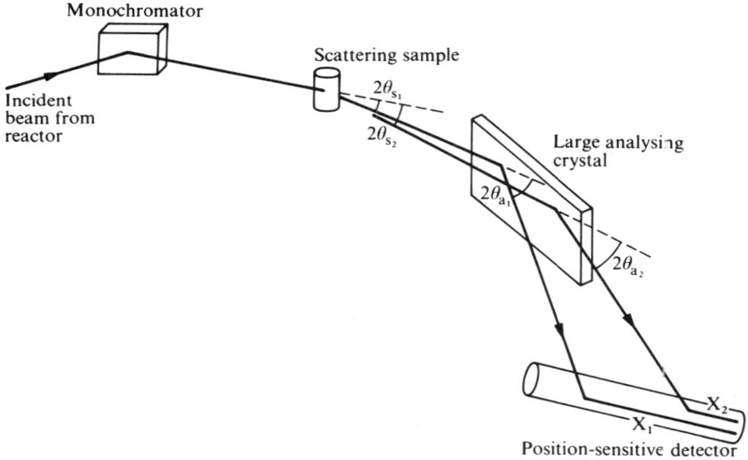

FIG. 188. Principle of the MARX spectrometer in which neutrons are accepted by the analysing crystal over a large angle and subsequently located in a position-sensitive detector.

uniquely both the value of $2\theta_s$, the scattering angle at the sample, and of $2\theta_a$, the scattering angle at the analysing crystal and, hence, of the neutron energy. The rate at which data are accumulated is therefore much increased, although the neutrons will range considerably in both Q and energy.

For some further details of experimental methods and apparatus the reader is referred to an article by Stirling (1973).

9.4. Incoherent scattering

We have shown at the beginning of this chapter that the paths of the inelastically scattered neutrons can be determined by studying two conditions which have to be satisfied, namely the conservation of energy and of momentum as represented by eqns (9.1) and (9.2) respectively, bearing in mind that these are correlated by eqn (9.3), which expresses the dispersion law for the phonons within the crystal.

In general this view of inelastic scattering is adequate, but it requires considerable amendment for substances which contain nuclei that give a substantial amount of incoherent scattering, arising from either isotope or spin coherence. Examples of such substances would be those which contain hydrogen or vanadium. For incoherent scattering there is no phase relation between the contributions from neighbouring nuclei and eqn (9.2), which expresses the conservation of momentum and depends on correlations between nuclei at neighbouring positions, is no longer applicable. Indeed a scattering event which involves a given change E in the neutron energy involves a phonon of equal energy E but any value of wave-number q is acceptable. Therefore the scattering pattern, which expresses the relative occurrence of energy changes of various values E, will be directly proportional to $\mathcal{N}(E)$, where $\mathcal{N}(E)\,\mathrm{d}E$ is the number of phonons which have energies within a narrow range from E to $E+\mathrm{d}E$. This quantity is directly related to $g(\omega)$, which defines the number of oscillators per unit range of angular *frequency*, close to the frequency $\omega = E/\hbar$. It can be shown in fact that the differential scattering cross-section per unit solid angle and per unit range of energy change centred about E is

$$\frac{\mathrm{d}^2\sigma}{\mathrm{d}\Omega\,\mathrm{d}E} = \frac{\kappa}{\kappa_0} b_{\mathrm{inc}}^2 \frac{\hbar(\boldsymbol{\kappa}-\boldsymbol{\kappa}_0)^2}{6ME}\mathrm{e}^{-2W}\frac{1}{\mathrm{e}^{(E/kT)}}\frac{1}{N}g\left(\frac{E}{\hbar}\right)$$

for cases where a phonon is annihilated, and

$$\frac{\mathrm{d}^2\sigma}{\mathrm{d}\Omega\,\mathrm{d}E} = \frac{\kappa}{\kappa_0} b_{\mathrm{inc}}^2 \frac{\hbar(\boldsymbol{\kappa}-\boldsymbol{\kappa}_0)^2}{6ME}\mathrm{e}^{-2W}\frac{\mathrm{e}^{(E/kT)}}{\mathrm{e}^{(E/kT)}-1}\cdot\frac{1}{N}g\left(\frac{E}{\hbar}\right)$$

for processes where a phonon is created. In these expressions M is the mass of the nucleus, $\exp(-2W)$ is the Debye factor, N is the number of atoms in the target, and b_{inc}^2 is $s/4\pi$. The two formulae given above are applicable only to the simple case of cubic symmetry and with a single atom at each Bravais-lattice point. Together they may be replaced by

$$\frac{d^2\sigma}{d\Omega\,dE}=\frac{\kappa}{\kappa_0}b_{\text{inc}}^2\frac{\hbar(\boldsymbol{\kappa}-\boldsymbol{\kappa}_0)^2}{6ME}\,e^{-2W}\left\{\frac{1}{e^{E/kT}-1}+\tfrac{1}{2}(1\pm 1)\right\}\frac{1}{N}g\left(\frac{E}{\hbar}\right), \tag{9.4}$$

where the + and − signs are taken for creation and annihilation of phonons respectively. They are applicable to the case of vanadium, which has a body-centred structure and for which, as Table 2 shows, the scattering is almost completely incoherent. It can be shown, moreover, that for a cubic material the expression applies not only to a single crystal but also to a polycrystalline sample. Vanadium was investigated first by Eisenhauer, Pelah, Hughes, and Palevsky (1958) and, more recently, by Zemlyanov, Kagan, Chernoplakov, and Chicherin (1963). The results of the latter are shown in Fig. 189, where the full-line curve indicates the function

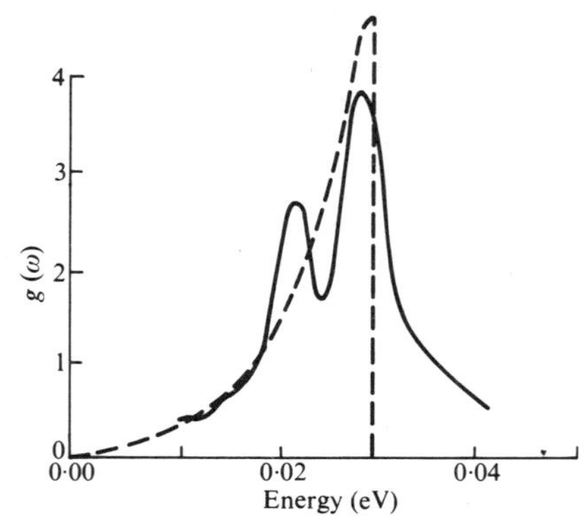

FIG. 189. The phonon spectrum of vanadium deduced from incoherent neutron scattering. (Zemlyanov *et al.* 1963.)

$g(\omega)$, deduced from the neutron observations, and the dotted curve represents a simple Debye spectrum, as calculated for a Debye temperature of 338 K.

As we have stated $g(\omega)$ in the above equations represents the number of oscillators per unit range of angular frequency and hence $\int g(\omega)\,d\omega = 3N$. Alternatively a normalized frequency distribution

$Z(\omega)$, for which $\int Z(\omega)\,d\omega = 1$, is often used and then

$$\frac{d^2\sigma}{d\Omega\, dE} = \frac{\kappa}{\kappa_0} b_{\text{inc}}^2 \frac{Q^2}{2M} e^{-2W} \frac{Z(\omega)}{\omega} N \left\{ \frac{1}{\exp(E/kT)-1} + \tfrac{1}{2}(1 \pm 1) \right\} \tag{9.5}$$

for a sample containing N atoms. Here we have also substituted $\mathbf{Q}$, the momentum transfer vector of eqn (9.2), for $\boldsymbol{\kappa} - \boldsymbol{\kappa}_0$.

Thus the cross-section for this simple case where there is one atom for each Bravais lattice point will be proportional to b_{inc}^2, $1/M$ and $Z(\omega)$. In multi-atom and molecular structures each of these factors will 'weight' the vibrations in which the different atoms take part, and there will be no direct correlation between the neutron spectrum and any single frequency-distribution function. In addition it turns out that each vibration will also be weighted by the square of the atomic displacement u_ν of the atom taking part: this is a parameter which appears in the analysis of the simple case only as a unit vector and later disappears in the process of averaging over all directions. As a result eqn (9.5) becomes

$$\frac{d^2\sigma}{d\Omega\, dE} = \frac{\kappa}{\kappa_0} \sum_\nu (b_\nu)_{\text{inc}}^2 \frac{Q^2 u_\nu^2}{2M_\nu} e^{-2W_\nu} \left\{ \frac{1}{\exp(E/kT)-1} + \tfrac{1}{2}(1 \pm 1) \right\} N \frac{Z(\omega)}{\omega}, \tag{9.6}$$

where the summation is made over the ν atoms in the molecule. For a full justification of this expression and the way the normalization conditions are determined the reader is referred to Chapter 4 of the book by Marshall and Lovesey (1971).

We shall return to eqn (9.6) when we discuss the determination of molecular structure in Chapter 11. We shall see how the weighting factors which we have just discussed lead to unique advantages for neutron spectroscopy in comparison with infrared and Raman techniques. Meanwhile we mention some studies of ammonium salts which provided early examples of the use of neutron spectroscopy. Fig. 190 shows spectra of $(NH_4)_2SO_4$ above and below the ferroelectric change point of 224 K. At both temperatures there is evidence of torsional oscillations of the two ammonium ions close to energies of 40 meV and 25 meV respectively and an acoustic vibration at 11 meV, but no evidence of any substantial change in the spectrum at the transition temperature. It seems likely that the transition is accompanied by shifts in atomic positions, rather than any onset of free rotation. For ammonium perchlorate, on the

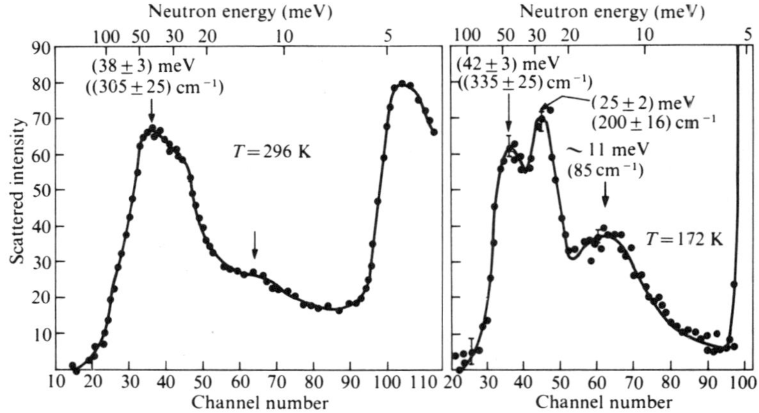

FIG. 190. Inelastic incoherent neutron scattering spectra from $(NH_4)_2SO_4$ at 296 K and 172 K respectively for neutrons scattered at 90°. (Rush and Taylor 1965.)

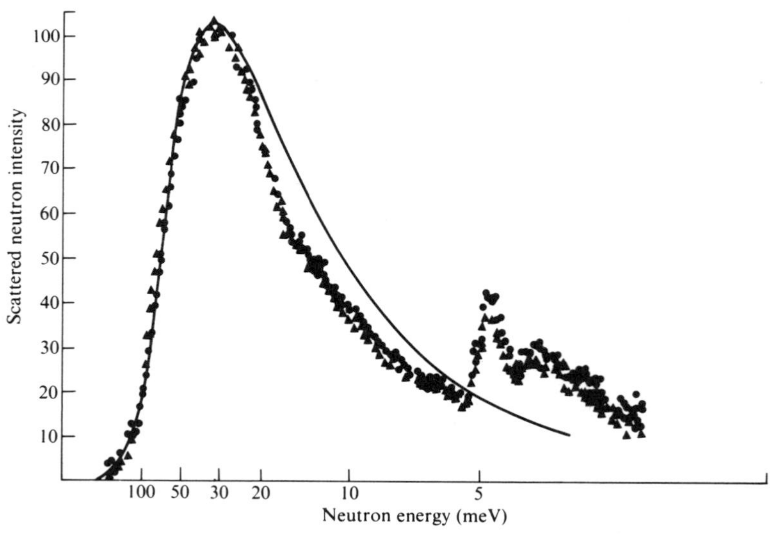

FIG. 191. Inelastic incoherent neutron scattering from ammonium perchlorate, NH_4ClO_4. The solid line is the theoretical curve for a free rotor with an effective mass of 2·1 m, in terms of the theory of Krieger and Nelkin (1957). (Janik *et al.* 1964.)

other hand, the pattern of Fig. 191 shows only a very broad peak. This is interpreted by assuming that the ammonium ion is rotating freely and there is a very wide range of closely spaced energy levels between which transitions can take place. By contrast, the pattern for phosphonium iodide, PH_4I (Fig. 192), shows a very sharp peak, indicating that the PH_4^+ ion is very tightly bound.

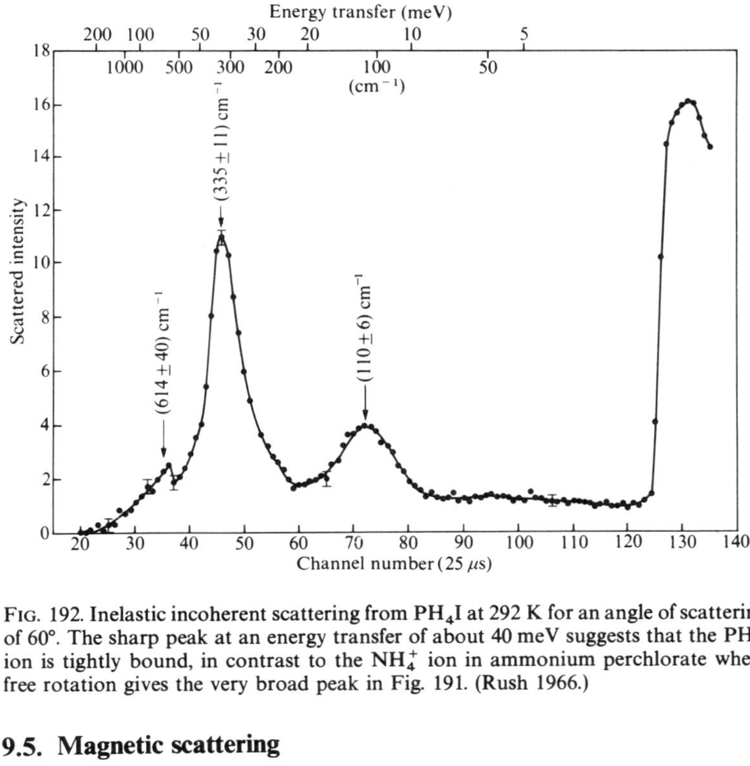

FIG. 192. Inelastic incoherent scattering from PH_4I at 292 K for an angle of scattering of 60°. The sharp peak at an energy transfer of about 40 meV suggests that the PH_4^+ ion is tightly bound, in contrast to the NH_4^+ ion in ammonium perchlorate where free rotation gives the very broad peak in Fig. 191. (Rush 1966.)

9.5. Magnetic scattering

What is possibly an even more valuable field of research opens up when we consider the inelastic contributions to the magnetic scattering. By a suitable choice of sample materials and of experimental conditions and methods we can gain information not merely about the magnetic moments but also about the energy-level structure and we can do this for both isolated, uncoupled, atoms and also for the coupled atoms which form the basis of those materials which display co-operative magnetic effects.

Uncoupled systems

Ideally, an uncoupled system is represented by a few magnetic ions which are present dilutely in a non-magnetic salt, but this can be approximated by the room-temperature state of some material which has a very low Curie temperature, of the order of 1 K. In such a case the energy levels of each magnetic ion are determined by the crystalline electric field due to its neighbours. When a monochromatic neutron beam is scattered by such an assembly, exchange of energy can take place between the neutron and these energy levels. This was first demonstrated by Cribier and Jacrot (1960) for the rare-earth oxides. An example is provided in Fig. 193 for Ho_2O_3

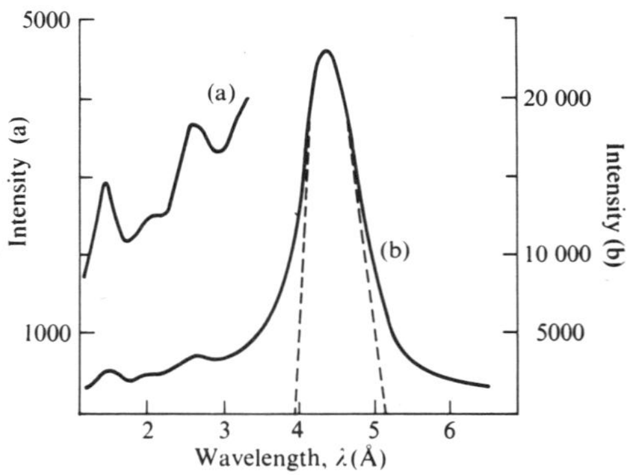

FIG. 193. Wavelength-analysis of a neutron beam of initial wavelength 4·35 Å after scattering by Ho_2O_3 at 27 °C. The magnified version at (a) for the shorter wavelengths in the scattered beam reveals well-defined energy gains of 0·002 eV, 0·010 eV, and 0·037 eV respectively. (Cribier and Jacrot 1960.)

where the peaks in the wavelength spectrum of the scattered neutrons correspond to energy gains of 0·002 eV, 0·010 eV, and 0·037 eV respectively. These energies represent the splitting of the lowest-J state of the holmium ions by the crystalline field.

Coupled systems

Coupled systems are best represented by ferromagnetic materials below their Curie temperature or by antiferromagnetics below the Néel temperature. First, however, we can examine such a material above its transition temperature, where the spin system is completely disordered because the coupling forces are inadequate to compete with the disordering effect of the thermal motion. Nevertheless, these forces are sufficient to result in a variation of susceptibility with temperature which is of the form $\chi = c/(T-\Delta_C)$,† where the Curie–Weiss constant Δ_C may be positive or negative, in contrast with the simple reciprocal relationship $\chi = c/T$ which occurs for an ideal paramagnet. The theory of the inelastic scattering by such an assembly of coupled ions was first given by Van Vleck (1939), who concluded that there would be a Gaussian spread of energy in the scattered neutron beam with a root-mean-square change of energy equal to $\delta E = 2\{\frac{2}{3}zS(S+1)J^2\}^{\frac{1}{2}}$, where S is the spin quantum number of the paramagnetic ions and J is the exchange integral between near neighbours, of which there are a number z for each atom. If we express J in terms of the Curie–Weiss constant Δ_C, using the expression of Van Vleck,

$$\Delta_C = -\tfrac{2}{3}zJS(S+1)/k, \tag{9.7}$$

we have

$$\delta E = k\Delta_C\left\{\frac{6}{zS(S+1)}\right\}^{\frac{1}{2}}. \tag{9.8}$$

The essential truth of this interpretation was demonstrated by Brockhouse (1955) in an experimental comparison of results for $MnSO_4$ and Mn_2O_3. These results are shown in Fig. 194, which contrasts the difference in spectral width of the neutron beam after scattering by the two materials. For Mn_2O_3 the value of Δ_C is -176 K and the Néel temperature T_N is 80 K, and the strong coupling gives substantial broadening of the neutron spectrum at room temperature. For $MnSO_4$, on the other hand, Δ_C is -24 K and T_N is less than 14 K, so that the ions are effectively uncoupled and there is no significant broadening of the neutron spectrum. de Gennes (1958) has shown that eqn (9.8) is valid only for large angles of scattering and has developed a more general theory which relates the spread of the energy spectrum to the value of the scattering angle 2θ. Effectively, the expression given above for δE is multiplied by an additional term $\{1-(\sin \mathbf{Q}\,.\,\boldsymbol{\delta}/\mathbf{Q}\,.\,\boldsymbol{\delta})\}^{\frac{1}{2}}$, where $\boldsymbol{\delta}$ is the interatomic separation. This term approaches unity when $\mathbf{Q}$ is large. This theory has been confirmed in experiments with MnF_2 by Cribier, Ericson, Jacrot, and Gobalakichena (1959) and Cribier and Jacrot (1963).

† Some authors write $T+\Delta_C$ here, leading to reversed sign for Δ_C.

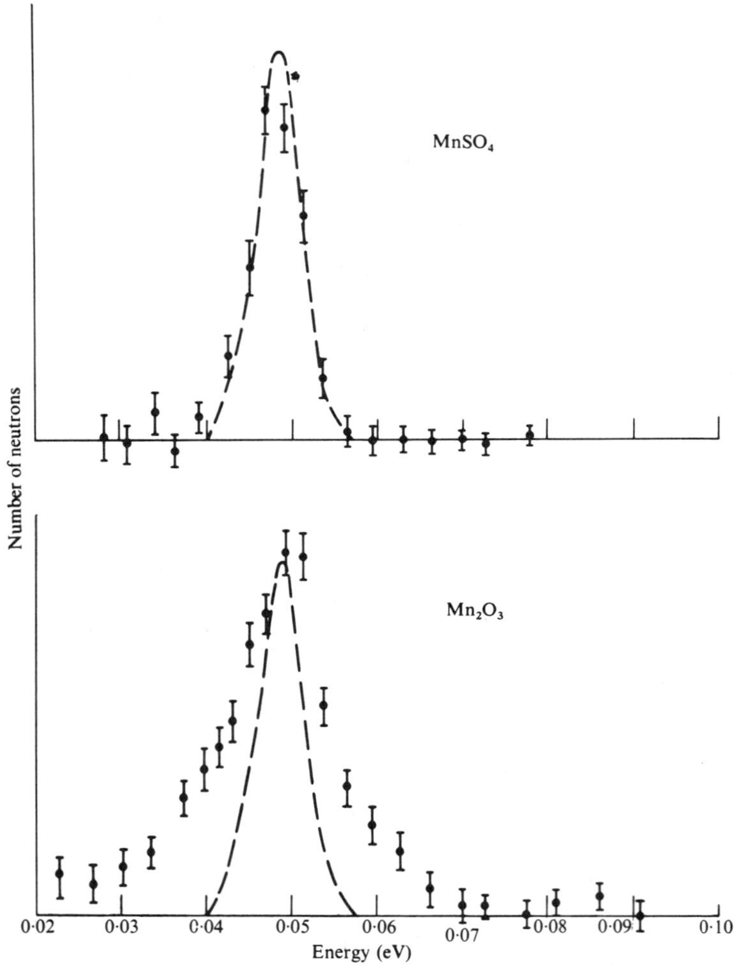

FIG. 194. The energy distribution curves for neutrons of initial wavelength 1·28 Å after scattering through an angle of 11° 20′ by $MnSO_4$ and Mn_2O_3 respectively. The broken curves represent the geometrical beam spread. (After Brockhouse 1955.)

At lower temperatures the ordering forces due to the exchange interactions between the neighbouring magnetic ions overcome the disordering tendencies of the thermal motion, and the spin system then displays a high degree of order. A study of the inelastic scattering of the neutrons enables us to study the departure from perfect order in terms of 'spin waves'. We shall see that in the quantized form of 'magnons' these account for the changes in direction of magnetic spins in an analogous way to that in which phonons account for the coordinate displacements of atomic positions in ordinary materials.

The concept of a spin-wave may be understood with the aid of Fig. 195. Diagram (a) indicates the ground state of a simple ferromagnet in which all the magnetic spins are perfectly aligned.

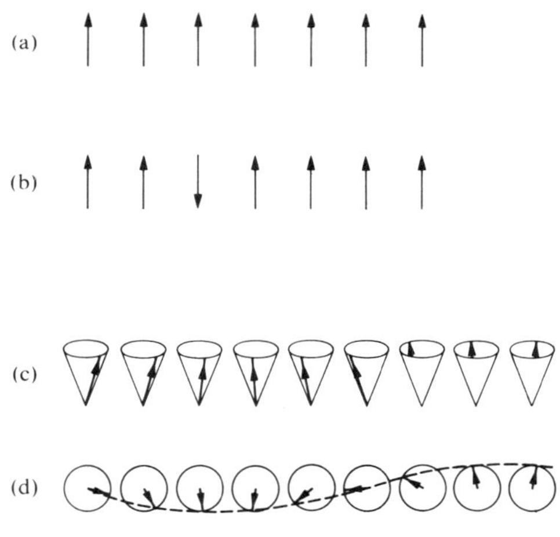

FIG. 195. Diagram (a) for the spin alignment in a simple ferromagnet is modified to (b) by completely reversing an isolated spin. A lower energy is achieved by sharing the disturbance between neighbouring atoms as in (c). The spin-wave which describes their moment directions is made evident in the plan view shown at (d).

Diagram (b) indicates an excited state in which one isolated spin is completely reversed. This is an unlikely situation and an excitation of much lower energy is achieved at (c), where the reversal is shared between a number of neighbouring atoms, so that the variation of spin direction takes a waveform. Any particular form of disturbance can be compounded from a series of such waves, in a similar way to that in which the displacements of atoms can be compounded from a series of waves of displacement. Likewise each spin-wave will possess a wavelength and a quantized energy, and there will be some definite relation between frequency ω and the wave-number q.

Before describing the form of relation between ω and q which is expected for different types of magnetic material, and which can be investigated experimentally, it is necessary to consider the way in which the scattering due to spin-waves can be isolated experimentally from that arising from other causes. In particular we have to distinguish it from the magneto-vibrational scattering which arises because the motion of the atoms under the influence of thermal energy involves translational motion of their spins, in contrast to *rotation* of the spins, which is what the spin-waves describe. Thus spin-wave 'magnon' scattering arises from rotational disorder of the spins, whereas magneto-vibrational scattering occurs through distortion of the framework on which the spins lie. The two can be distinguished in a ferro- or ferrimagnetic material by noting the contrasting effects when magnetic fields are applied in various directions.

It can be shown that the intensity of spin-wave scattering is proportional to $1+(\boldsymbol{\varepsilon}\,.\,\mathbf{K})^2$, where $\boldsymbol{\varepsilon}$ is the unit scattering vector, defined by $(\boldsymbol{\kappa}-\boldsymbol{\kappa}_0)/|\boldsymbol{\kappa}-\boldsymbol{\kappa}_0|$, and $\mathbf{K}$ is the direction of the magnetization within the domain. The spin-waves describe, as in Fig. 195, the coherent motions of the x, y components of spin which are normal to the magnetization direction $\mathbf{K}$. When $\mathbf{K}$ is parallel to $\boldsymbol{\varepsilon}$, the contribution to the coherent magnetic reflections, which we discussed on p. 203, is zero but the contribution from the x, y components which constitute the spin-wave will be a maximum. On the other hand, the contributions to the magneto-vibrational scattering, dependent on $1-(\boldsymbol{\varepsilon}\,.\,\mathbf{K})^2$, will, like the coherent Bragg reflections, be equal to zero.

Ferromagnetic materials

We can see the practical application of the above principle to the case of ferromagnetic materials by considering a particular example and we shall take the determination of the dispersion relation for the spin-waves in a metal. This was first done by Sinclair and Brockhouse (1960), using not a pure metal but a face-centred cubic alloy of cobalt containing 8 per cent of iron. The choice of this material was made to provide a simple structure, a high Curie temperature (1300 K), a large magnetic moment of $1{\cdot}84\,\mu_B$, and a small nuclear scattering cross-section which would improve the accuracy with which the magnetic scattering could be assessed. A typical set of data is illustrated in Fig. 196(a) in which a fixed change of energy between the incident and scattered neutrons, amounting to 0·00921 eV, has been employed in the experiment and for various settings ψ of the crystal the angle of scattering has been chosen to ensure that the wave vector **q** was always in a fixed direction, namely, in the [011] direction through the (200) reciprocal lattice point. The geometry of the scattering process is indicated in Fig. 196(b),

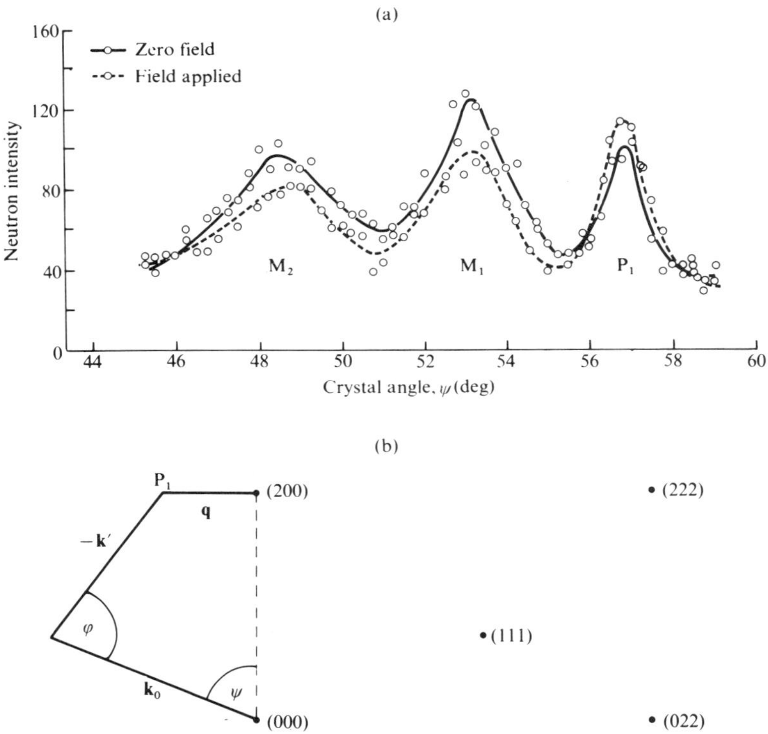

FIG. 196. Survey of magnons and phonons travelling along the [011] direction in $Co_{0\cdot92}Fe_{0\cdot08}$. Neutrons detected are restricted to those which have made an energy gain of 0·00921 eV. As indicated in (b) a suitable choice of scattering angle ϕ for any crystal orientation ψ will ensure the correct direction for **q**. The neutron intensity is plotted against ψ in (a), revealing three neutron groups. The full-line curve in (a) is the normal reading which becomes the dotted curve on application of a magnetic field in the [011] direction. The decrease produced in the neutron groups M_1, M_2 is evidence that they represent spin-wave scattering. (Sinclair and Brockhouse 1960.)

and it is the crystal orientation ψ which appears as the abscissa of the curve in Fig. 196 (a) which gives the variation of neutron intensity. The latter curve shows that there are evidently three neutron groups within the range of q covered, and it can be demonstrated by applying a magnetic field along the [011] direction, i.e. at right-angles to the scattering vector, that the first two groups represent spin-wave peaks, whereas the right-hand group represents magneto-vibrational scattering. This conclusion follows because, as seen in the Figure, application of this magnetic field gives a reduction of the first two peaks but an increase of the third one.

From the measured positions of the peaks in the curve of Fig. 196 (a) we can calculate the frequency ω and the wave-number q of the magnons involved, and in this way we can construct the dispersion curve of Fig. 197 for the Co–Fe crystal. This curve includes points for the [100], [110], and [111] directions in the crystal and, to a first approximation at least, the dispersion curve is isotropic.

It has been shown by Kranendonk and Van Vleck (1958) that the dispersion law for a cubic ferromagnet may be written

$$\hbar\omega = C+2JS\left(Z-\sum_{l}\cos \mathbf{q}\,.\,\mathbf{l}\right), \tag{9.9}$$

where C is a small constant which allows for the external and anisotropy fields, J is the exchange interaction, and S is the magnetic spin of the ions, each of which possesses Z nearest neighbours. The vector $\mathbf{l}$ is the vector which joins an atom to one of its neighbours, and the summation is made over all the nearest neighbours. If we evaluate this expression for a face-centred cubic cell, for which $Z = 12$, then we find that for magnons in the [111] direction, for example,

$$\hbar\omega = C+12JS\left(1-\cos\frac{qa}{\sqrt{3}}\right), \tag{9.10}$$

where a is the side of the unit cell. The results of Fig. 197 lead, assuming that $S = 0{\cdot}92$, to a value for J of $(16\pm 1{\cdot}6)$ meV and a value for C of 1·3 meV.

It also follows from eqn (9.10) that for small values of q

$$\hbar\omega = C+2JSa^2q^2, \tag{9.11}$$

which is of parabolic form, clearly evident in Fig. 197.

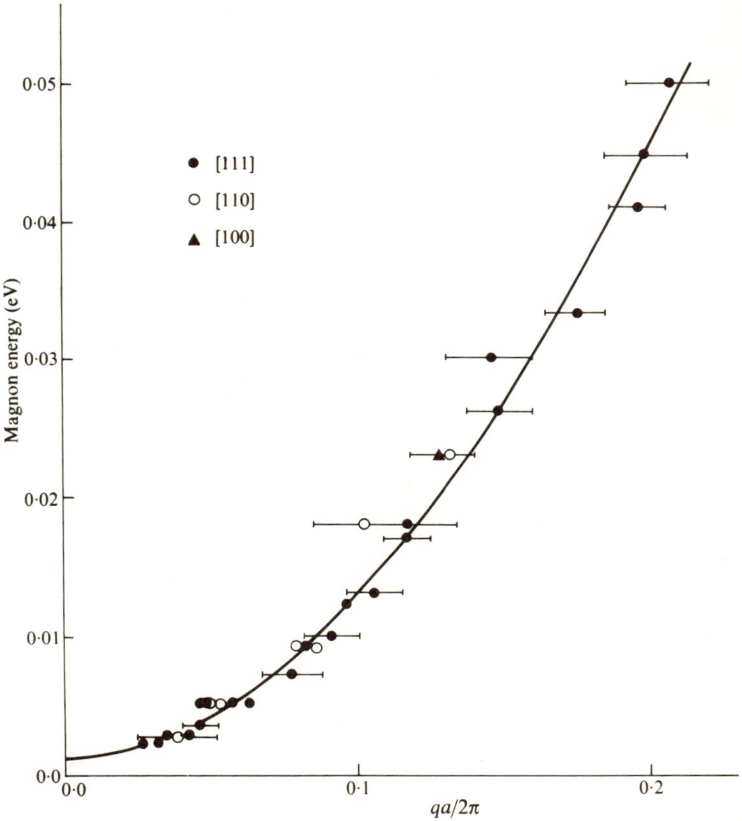

FIG. 197. The magnon spectrum of the alloy $Co_{0\cdot 92}Fe_{0\cdot 08}$ for the directions [100], [110], and [111] constructed from observations of the type indicated in Fig. 169. (Sinclair and Brockhouse 1960.)

Antiferromagnetism

The expected form of dispersion curve for an antiferromagnetic material is different from that which we have just discussed for ferromagnetism. Near the origin, where the wave vector **q** is small, the curve is linear instead of parabolic. The difference can be seen most simply by deriving expressions for the cases of one-dimensional arrays of moments (Kittel 1971) with only nearest-neighbour interactions. For ferromagnetism, as above,

$$\hbar\omega = 4JS(1-\cos qa), \tag{9.12}$$

where a is the spin separation, and $\hbar\omega = 4JS \sin qa$ for antiferromagnetism. For small values of q these expressions become

$$\hbar\omega = 2JSa^2q^2 \qquad \text{for ferromagnetism} \tag{9.13}$$

and

$$\hbar\omega = 4JSaq \qquad \text{for antiferromagnetism.} \tag{9.14}$$

A very convincing experimental measurement of the dispersion curve for an antiferromagnet has been made by Windsor and Stevenson (1966) for $RbMnF_3$. This salt has the cubic perovskite structure which we shall discuss and illustrate later (p. 477) and in which the manganese atoms occupy simple cubic positions, each with six nearest neighbours. The experimental data relating the magnon energy and the vector **q** are illustrated in Fig. 198. The points

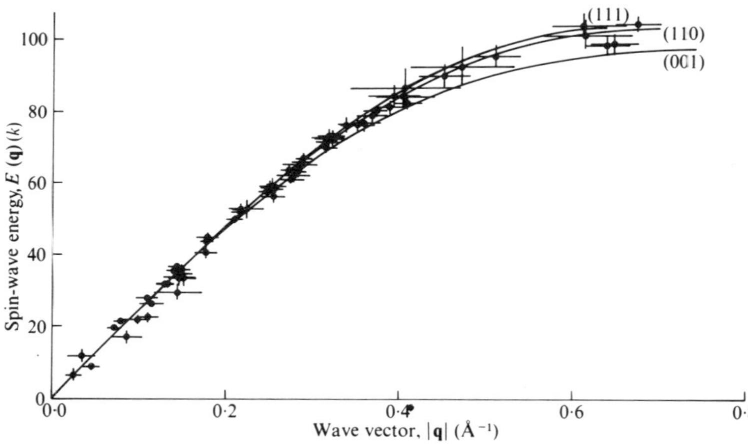

FIG. 198. Dispersion data for spin-waves in $RbMnF_3$, for magnons in various directions within the (110) plane. The three curves drawn have been derived for the directions [111], [110], and [001] from a least-squares analysis of the experimental results. The energy is expressed in units of Boltzmann's constant k. (Windsor and Stevenson 1966.)

which are included correspond to various directions within the (110) plane and are shown in relation to the slightly different calculated curves for the directions [111], [110], and [001]. The curves were derived by a least-square analysis in which the exchange interactions for both first J_1, second J_2, and third J_3 neighbours were allowed to vary. It is concluded that J_2, J_3 are not significantly different from zero and that the value of J_1 is 0·30 meV, which corresponds to a temperature J_1/k, where k is Boltzmann's constant, of (3·4 ± 0·3) K. The very small value of the intercept on the axis of ordinates, when q is zero, indicates that the anisotropy field is very small.

Windsor (1966) has examined the scattering from $RbMnF_3$ at room temperature, where the material is paramagnetic, in terms of the theory of Van Vleck (1939) and de Gennes (1958) which describes the broadening of the energy spectrum found in the scattered beam as discussed on p. 305. It is deduced that J_1/k is equal to (3·3 ± 0·3) K, in very good agreement with the results of the measurements in the antiferromagnetic condition.

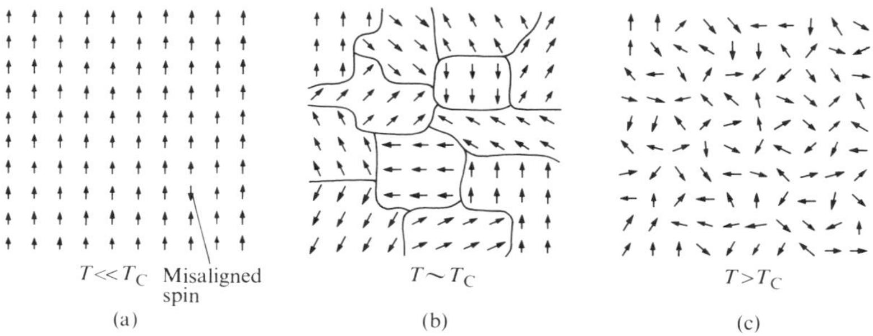

FIG. 199. A schematic diagram of the spin-correlation regions in iron. The ordered regions are dynamic in nature and must be considered to move continuously through the magnetic domain. The three diagrams show conditions (a) far below the Curie temperature T_C, (b) near T_C, and (c) above T_C. (Wilkinson and Shull 1956.)

9.6. Critical scattering

As the temperature of a magnetic material is increased the extent of the disorder among the magnetic spins increases and the spontaneous magnetization within a domain decreases. As a result the coherent elastic magnetic scattering, which gives the Bragg peaks in the diffraction pattern, decreases and the spin-wave scattering increases. At the Curie or Néel temperature there is a balance between the exchange interaction forces, which tend to order the spins, and the disordering forces which arise from the thermal energy of the atoms. The outcome of this balance is that there are localized regions in the material which are more or less ordered and which, dynamically, go through large fluctuations in time and position.

Instantaneously the regions of spin correlation near T_C can be indicated by Fig. 199 (b), but these regions are fluctuating rapidly in both time and position. The relaxation time may be of the order of 10^{-10} s near T_C, and it is important that this is many times larger than the time spent by a neutron in passing through a 'correlated' region. For a neutron of wavelength 4 Å and a distance of 10 Å this time is only 10^{-12} s. Further away from T_C, as Fig. 199 (a) and (c) shows, normal long-range order and disorder will, respectively, develop. Below the Curie temperature the mean magnetization of a ferromagnetic will fluctuate about a finite value, determined by the temperature; above the Curie temperature the magnetization will fluctuate about zero value. As the temperature departs more and more from the critical value of Curie temperature or Néel temperature the preference for order or disorder, as appropriate, will increase and the fluctuations will have a shorter relaxation time. This behaviour gives rise to strongly peaked scattering of neutrons at the critical temperature T_C which has been observed in a variety of experiments. It was first observed by Palevsky and Hughes (1953) and Squires (1954) in experimental measurements of the total cross-section of iron for long-wavelength neutrons, which indicated a very large increase in scattering at the Curie temperature. The same effect is shown more strikingly in Fig. 200, which illustrates more recent measurements by Cribier *et al.* (1962) for nickel. A detailed interpretation was given by Wilkinson and Shull (1956), who measured, in particular, the temperature-dependence of the small-angle scattering for iron at various temperatures in the neighbourhood of T_C. These results are illustrated in Fig. 201, from which it was deduced that just below T_C the size of the fluctuating ordered regions ranged from about 12 Å to 25 Å. Following Van Hove (1954) many workers have contributed to a quantitative theoretical treatment of this critical scattering (see Elliott and Marshall 1958; de Gennes 1959; de Gennes and Villain 1960).

At T_C itself the fluctuations have their maximum range and their maximum time of relaxation. Since the forces of order and disorder are so closely balanced, the interchange of energy between the neutron and the solid, accompanied by transition of the latter between different energy states, will be very small, i.e. the inelasticity

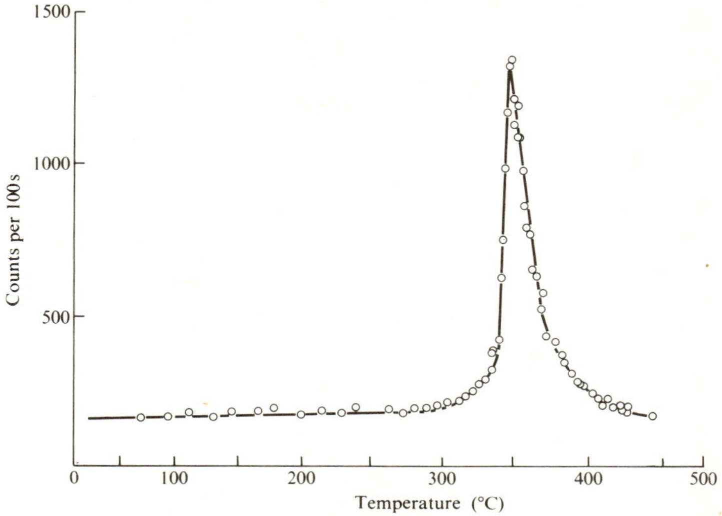

FIG. 200. The variation of neutron scattering with temperature for nickel at an angle of 2°, for neutrons of wavelength 4·75 Å, indicating the sharply peaked critical scattering. (Cribier *et al.* 1962.)

of the scattering is a minimum at T_C. As Fig. 200 indicates, the peak in the scattering is located very sharply in temperature, and the temperature has to be extremely well defined for accurate interpretation of the data. It was at first thought that the relaxation time would be infinite at T_C, giving elastic scattering, but this is not the case. The point is made clear in the work of Jacrot, Konstantinovic, Parette, and Cribier (1963), who measured the increase in the spread of energy for an incident beam of neutrons which had a wavelength centred on 4·75 Å. Fig. 202 shows the broadening of the spectrum for a scattering angle of 9°. From measurements of this type it is possible to determine the variation of both the correlation range and the relaxation time with the temperature. Fig. 203 shows how the correlation range varies. The innermost point for iron corresponds to a temperature which is about 10° above T_C and indicates a range of about 25 Å. At T_C itself the experiments suggest a range of at least 140 Å.

The relaxation time τ_q is related to the spin-fluctuation vector $\mathbf{q}$ by the relation

$$1/\tau_q = \Lambda q^2, \tag{9.15}$$

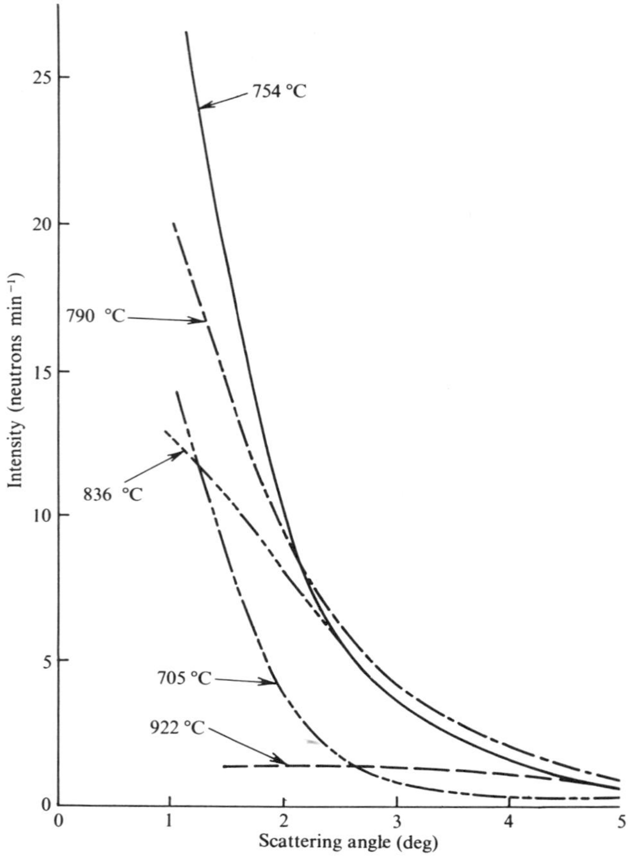

FIG. 201. The small-angle scattering of iron at various temperatures in the neighbourhood of the Curie temperature, as measured by Wilkinson and Shull (1956). The intensity plotted is the excess over that observed at room temperature.

where Λ is a parameter which varies with temperature. By integrating the theoretical expression for the scattering cross-section over the measured incident spectrum it is possible to calculate the scattered spectrum at different angles of scattering for various values of Λ. By choosing the value of Λ which gives the best fit with experiment Λ has been determined for iron by both Jacrot *et al.* (1963) and Passell, Blinowski, Brury, and Nielsen (1964, 1965), finding values of 7 and 11·5 respectively for the dimensionless parameter $2m\Lambda/h$

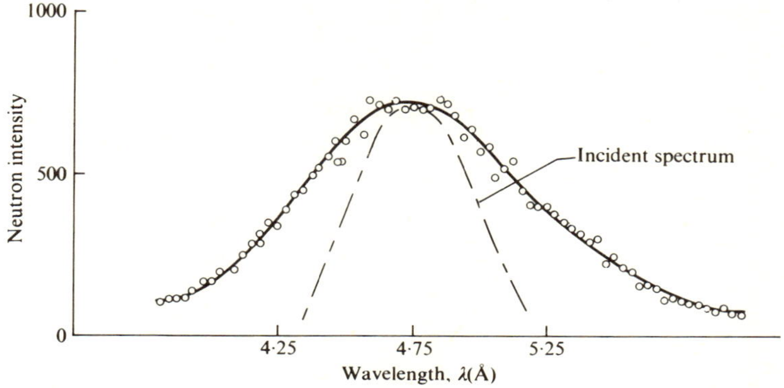

FIG. 202. The broadening of the wavelength spectrum of a neutron beam scattered at an angle of 9° by iron at the Curie temperature. (Jacrot *et al.* 1963.)

at the Curie temperature. The results suggest that the relation above should be replaced by

$$1/\tau_q = D\{a(T - T_C) + q^2\}, \tag{9.16}$$

where a and D are constants, so that the relaxation time is infinite only when both $(T - T_C)$ and q are zero.

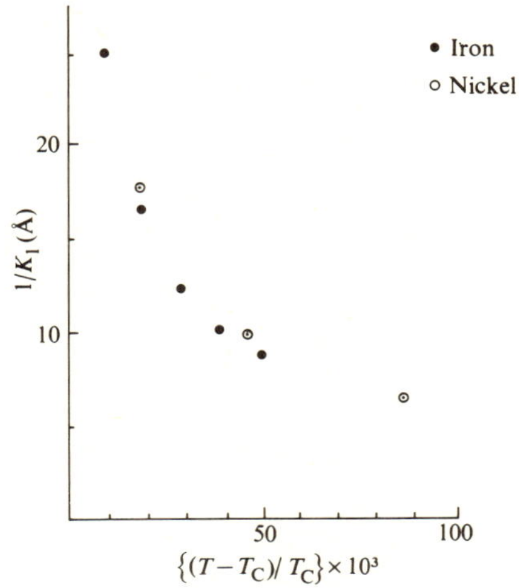

FIG. 203. The correlation range $1/K_1$ in iron and nickel determined from observations of critical scattering near the Curie point. (Jacrot *et al.* 1963.)

SUMMARY OF PRINCIPLES AND FIELDS OF APPLICATION OF NEUTRON DIFFRACTION

In the foregoing chapters we have discussed mainly the techniques and principles of neutron diffraction, making reference to the distinctions which can be drawn when comparison is made with X-ray diffraction. For convenience we now summarize these distinctions in Table 19 overleaf.

In the following chapters we shall give some examples of the various investigations to which neutron diffraction measurements have contributed significant information. We shall find that the main applications of our subject fall into five classes:

(1) structural investigations of solids which aim to discover the positions of light atoms, particularly hydrogen atoms;

(2) problems, such as are often met with in alloy systems, which require distinction to be made between atoms of neighbouring atomic number, and which therefore have very similar scattering amplitudes for X-rays;

(3) investigations of magnetic materials, in which advantage is taken of the additional scattering of neutrons which occurs for atoms which possess magnetic moments.

(4) vibrations, including magnetic vibrations, which can be studied by inelastic scattering;

(5) studies of departures from perfect order—liquids, gases, and defects.

TABLE 19

Comparative properties of X-rays and neutrons for diffraction studies

Property	X-rays	Neutrons
Wavelength	Characteristic line spectra such as Cu $K\alpha$ 1·54 Å.	Wavelength band such as (1·1 ± 0·05) Å separated out from Maxwell spectrum by crystal monochromator. Small second-order component of $\lambda/2$, less than 1 per cent in intensity.
	Absorption prohibitive for long λ.	Long wavelengths often of advantage.
Energy, for $\lambda = 1$ Å	$10^{18}\,h$.	$10^{13}\,h$, i.e. the same order as energy quantum of crystal vibrations.
General nature of scattering by atoms	Electronic.	Nuclear.
	Form-factor dependence on $(\sin\theta)/\lambda$.	Isotropic, no angular-dependent factor.
	Angular-dependent polarization factor.	
	Regular increase of scattering amplitude with atomic number, calculable from known electronic configurations.	Irregular variation with atomic number. Dependent on nuclear structure and only determined empirically by experiment.
	No differences among isotopes.	Amplitude is different for different isotopes and depends also on nuclear spin, giving isotope and spin incoherence.
	Phase change of 180° on scattering.	Phase change is 180° for most nuclei but H, ^{7}Li, Ti, V, Mn, ^{62}Ni give zero phase change.
		Anomalous scattering by e.g. ^{113}Cd.

Magnetic scattering	No additional scattering.	Additional scattering by atoms with magnetic moments (1) diffuse scattering by paramagnetic materials (2) coherent diffraction peaks for ferromagnetic and antiferromagnetic materials. Amplitude of scattering falls off with $(\sin\theta)/\lambda$. Amplitude is calculable from magnetic moment and is different for ions with different spin quantum numbers, e.g. Fe^{2+}, Fe^{3+}.
Absorption coefficient	Very large, true absorption being much greater than scattering $\mu \sim 10^2$–10^3, increasing with atomic number.	Absorption usually very small and less than scattering $\mu \sim 10^{-1}$. There are marked exceptions, e.g. B, Cd, and rare earths which have large absorptions. Varies with isotopes.
Thermal effects	Reduction of coherent scattering by Debye exponential factor.	
Inelastic scattering	Negligible change of wavelength.	Appreciable change of wavelength; frequency-wave-number relation can be found for lattice vibrations and magnetic spin waves.
Single crystal reflection	Perfect crystal reflection limited by primary extinction. Mosaic crystal element gives integrated reflection equal to QV.	
	Secondary extinction in thick crystals is of secondary importance.	Secondary extinction predominant in thick crystals. 'Thin crystal' criterion.
	$\mathscr{R}^\theta$ (thick crystal) $= Q/2\mu$	$\mathscr{R}^\theta \sim 3\eta$.

TABLE 19 (*continued*)

Property	X-rays	Neutrons
Normal methods of detection	Photographic film. Geiger counter.	BF_3, ^3He counters.
Absolute intensity measurement	Difficult Interpretation depends on precise knowledge of atomic scattering factor curves.	Straightforward, particularly by powder methods.

10

DETERMINATION OF THE ATOMIC POSITIONS OF LIGHT ELEMENTS

10.1. Introduction

THE first structural investigations to which neutron diffraction methods were applied involved the determination of the positions of light elements. The fact that the amplitude of X-ray scattering by an element is proportional to its atomic number sets a severe limitation on the amount of information which can be obtained by X-ray investigations of substances containing hydrogen or of oxides and carbides of heavy elements. In these cases the very high precision necessary in intensity measurement in order to distinguish accurately the small contribution made by the light element in face of the overwhelming amplitude from the heavy constituent is not generally attainable in practice. With neutrons, however, the light element is not usually at any marked disadvantage; for example, oxygen, with a scattering amplitude of $0{\cdot}58 \times 10^{-12}$ cm, can readily be detected in the presence of tungsten, gold, or lead which have values of b of $0{\cdot}48 \times 10^{-12}$ cm, $0{\cdot}76 \times 10^{-12}$ cm, and $0{\cdot}94 \times 10^{-12}$ cm respectively. Similarly the scattering amplitude of hydrogen, $-0{\cdot}37 \times 10^{-12}$ cm is not greatly inferior numerically to that of most of the other elements.

In principle, the whole field of organic chemistry is filled with problems which can be solved by neutron diffraction investigation of the position of hydrogen atoms. Equally, there are wide fields of investigation in the crystal chemistry of the compounds of heavy elements.

In the investigations of the compounds to be described the unit cell and atomic positions of the elements other than hydrogen, or light elements, are known from X-ray studies which lead to one or more space groups which describe the symmetry of the structure so far as the heavier elements are concerned. If no additional spectra appear in the neutron diffraction pattern it follows that the unit cell found by X-rays is indeed the true unit cell when the hydrogen atoms also are considered. Modern investigations of crystal structures with X-rays, involving intensity measurements of high accuracy, will usually locate the hydrogen atoms approximately, and it is the aim of the neutron measurements to refine these positions. With the most up-to-date neutron data the accuracy of determination of location and movement of hydrogen atoms approaches that which can be obtained, using X-rays or neutrons, for heavier atoms. In general, the standard routine for this kind of work is the collection of intensity data in three-dimensions for a single crystal, followed by least-squares and Fourier analysis. However the earlier classical studies of light elements with neutrons were of a quite different type, being mainly studies of powdered materials of cubic symmetry, because of the difficulty of interpreting the effects of secondary extinction in the large single crystals which were needed for reactors of low neutron flux. We shall examine some of the early studies in detail, because of the way in which they illustrate some of the general problems which arise in neutron work.

10.2. Studies with polycrystalline samples

Sodium hydride and deuteride

X-ray diffraction studies of alkali–metal hydrides, such as LiH, NaH, and KH, had already shown that the structures were face-centred cubic, and in the case of LiH the contribution from hydrogen relative to lithium, was sufficiently great to establish that LiH had the NaCl-type of structure. From a consideration of the size of the unit cells of NaH, and NaD it was concluded that only structures of type B1 (NaCl) or B3 (ZnS) would be possible. From the neutron diffraction patterns, of which the low-angle portions are shown in Fig. 204, Shull *et al.* (1948) were able to show that the NaCl structure alone gave satisfactory agreement with the observed intensities. A striking feature of the patterns is that for NaH the (111) peak is much more intense than the (200), whereas for NaD the converse is observed. Since the structure factors for a simple AB compound of NaCl type are proportional to $b_A - b_B$ and $b_A + b_B$ for the (111) and (200) reflections respectively, it follows that sodium and hydrogen have scattering amplitudes of opposite sign, whereas the amplitudes for sodium and deuterium are of the same sign.

Apart from the differences in the relative intensities of the (111) and (200) reflections which the two curves in Fig. 204 show, the patterns are noteworthy for the significant difference in the background intensities in the two cases. The large background scattering in the case of the hydrogen-containing sample is due to the large spin-incoherence for hydrogen which varies from 20 barns for a free hydrogen atom to 81 barns for a bound atom. The effective binding depends somewhat on the nature of the material but mainly on the neutron wavelength. Fig. 205 shows the results of measurements by Melkonian (1949) on hydrogen gas and a variety of hydrocarbons. In the neighbourhood of a wavelength of 1 Å there is little variation

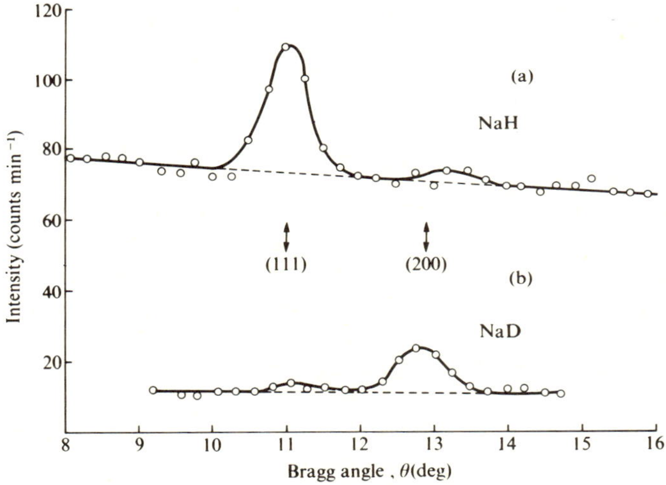

FIG. 204. Diffraction patterns for NaH and NaD illustrating the different relative intensities of the (111) and (200) reflections and of the peak–background ratios. (Shull, Wollan, Morton, and Davidson 1948.)

of σ with choice of material. σ can be measured readily in a transmission experiment using the method of Lampi, Freier, and Williams (1950).

Typical measurements are 36·2 barns for hexamethylenetetramine at $\lambda = 1{\cdot}04$ Å, 35·6 barns for Vitamin B_{12} at $\lambda = 1{\cdot}45$ Å (Willis 1968, private communication), 35 barns at 1·09 Å for hydrated α-oxalic acid (Coppens 1968, private communication), and 36·6 barns at $\lambda = 1{\cdot}04$ Å for polythene (Jude 1971). A mean value of $\sigma = 36$ barns per hydrogen atom seems appropriate for assessing the linear absorption coefficient μ of any hydrogen-containing material, using the relation

$$\mu = N\sigma,$$

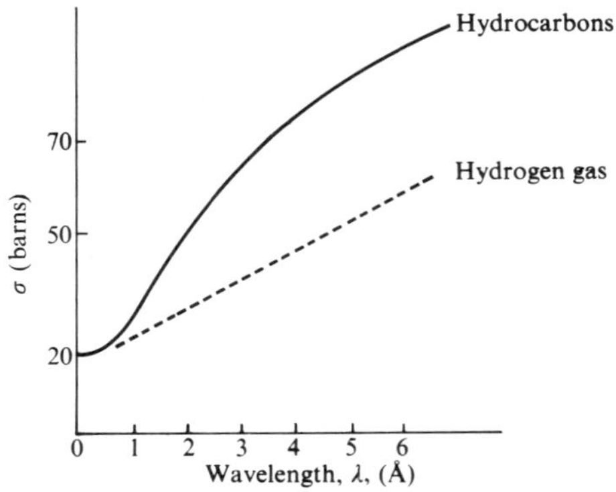

FIG. 205. The variation of the total scattering cross-section of hydrogen with neutron wavelength for hydrogen gas and a variety of hydrocarbons. (Melkonian 1949.)

where N is the number of nuclei cm^{-3}. At the same time the isotropic incoherent scattering can also be calculated from the knowledge of σ.

For deuterium, on the other hand, the total and coherent scattering cross-sections are 7·4 barns and 5·4 barns respectively, ensuring a greatly reduced amount of spin-coherent scattering in the background of the pattern. Thus it will be much easier to interpret powder diffraction patterns if deuterated material is available, and this will be particularly so when weak reflections are being examined. This advisability of deuteration is unnecessary with single crystals.

Structure of ice

Ice was first studied as a powder by Wollan, Davidson, and Shull (1949) and then as a single crystal by Peterson and Levy (1953). More recently progress has been made in elucidating the structures of the polymorphs of ice formed at high pressures. Results have been reported with both powdered and single-crystal samples in which the structures have been 'frozen-in' at low temperatures, in order to permit examination at normal pressure (Kamb *et al.* 1971; Arnold *et al.* 1971). A summary of the information is given by Speakman (1973).

Heavy-element compounds

Preliminary measurements were made with powders of a number of heavy-element compounds of high symmetry. In particular we mention uranium hydride (Rundle 1951), hydrides of thorium and zirconium (Rundle, Shull, and Wollan 1952), and U_3O_8 (Andresen 1958). Some later work by Leciejewicz and Padlo (1962) for PbO_2 and by Atoji and Medrud (1959) with CaC_2 and the rare-earth carbides is interesting because these are simple tetragonal structures in which one structural parameter has to be found. In the carbides, for example, which have the structure shown in Fig. 206,

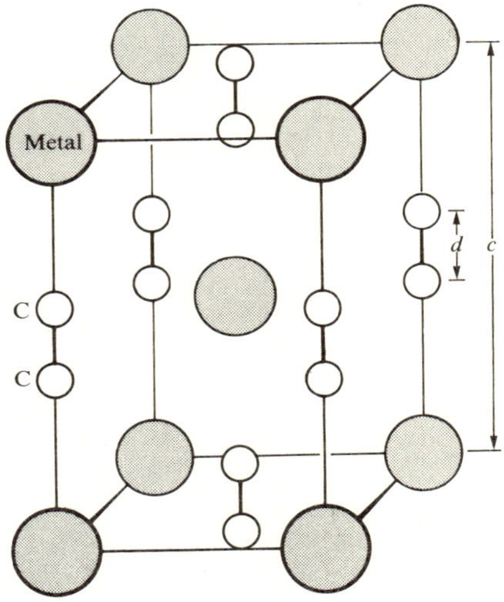

FIG. 206. The CaC_2-type of structure found for the heavy-element carbides. The only structural parameter to be determined is the distance d between the pairs of carbon atoms.

it is necessary simply to determine the distance d which separates the two carbon atoms. With the powdered samples very few of the reflections were completely resolved, but it was possible to calculate the total intensity of several groups of lines as a function of d. A discrepancy factor $R \equiv \sum |I_{obs} - I_{calc}| / \sum I_{obs}$, which is a measure of the extent of disagreement between the actual structure and the postulated model, was then calculated. The parameter d was chosen to give a minimum value of the discrepancy factor. Some results are given in Fig. 207 which shows the deduction of the C—C distance for CaC_2 and UC_2.

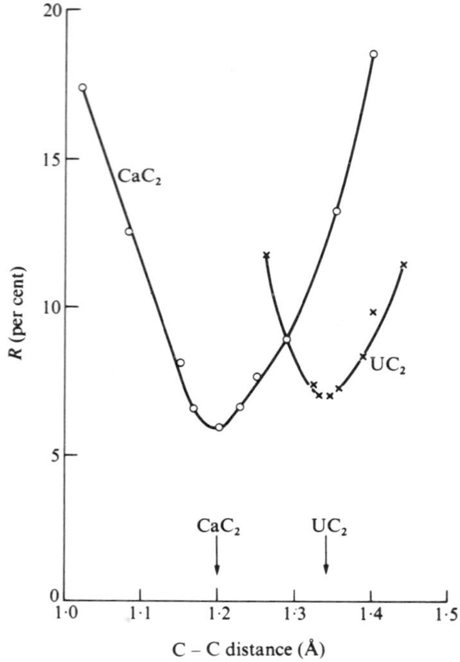

FIG. 207. Determination of the C—C distance in CaC_2 and UC_2 from the position of the minimum in the discrepancy factor $R = \sum |I_{obs} - I_{calc}|/\sum I_{obs}$. R is calculated from the experimentally measured neutron intensities and the calculated values for various values of the C—C distance. (From Atoji and Medrud 1959.)

10.3. The development of single-crystal methods

Potassium bifluoride

An investigation of potassium bifluoride by Peterson and Levy (1952*a*) with the aim of determining the positions of the hydrogen atoms in this compound first showed the advantages of single-crystal methods and was the forerunner of many studies of this kind.

X-ray studies of KHF_2 by Bozorth (1923) and Helmholz and Rogers (1939) had demonstrated that the F—H—F ion was linear, with the two fluorine atoms separated by 2·26 Å. They were not able, however, to decide whether the hydrogen atom was midway between the two fluorines or distributed at random in two possible positions closer to one or other of these atoms. Peterson and Levy's first measurements were made with powdered deuterated material, to avoid the intense incoherent scattering by hydrogen, but experienced trouble owing to preferred orientation of the crystallites in the powdered sample. This caused intensity anomalies, although these measurements were adequate for demonstrating that the unit cell was identical with that deduced by X-rays, so that the hydrogen positions did not call for an enlarged unit cell.

By the use of a single crystal the difficulties caused by orientation were overcome; it was no longer necessary to use deuterated material and much more accurate measurement of intensities was possible. A crystal 1·4 mm × 2·6 mm × 3·6 mm was employed, with the smallest dimension parallel to the c_0-axis. It was shown that correction for secondary extinction was only necessary for intense reflections, namely (310) and (400), measured with the c_0-axis vertical. This conclusion was consistent with predictions from the criterion of Bacon and Lowde (1948) which we have discussed in Chapter 3.

The diffraction pattern was recorded automatically as crystal and counter arm were rotated in synchronism and, in addition, the cumulative count across each reflection was recorded with a scalar as indicated in Fig. 208. A second recording made with the crystal turned out of the reflection position gave the cumulative

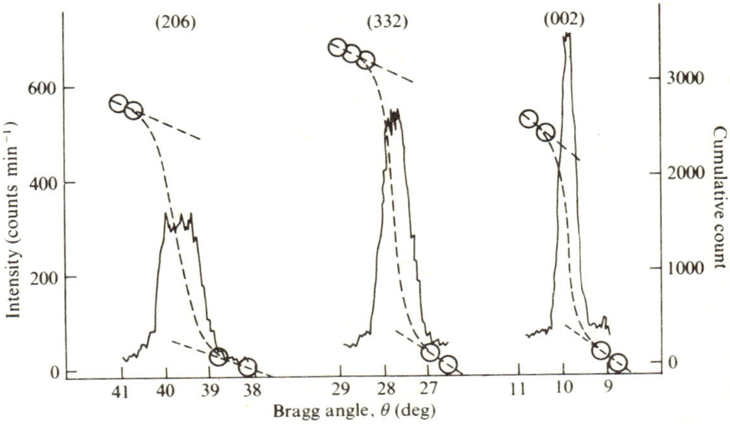

FIG. 208. Some of the first examples of neutron diffraction reflections from a single crystal (of KHF_2). The full-line curves are those traced by a recorder connected to a counting-rate meter. The circled points show the observed cumulative count. (Peterson and Levy 1952*a*.)

background count; subtraction of the latter therefore gave the integrated count $\mathscr{E}$ for the reflection. The integrated reflection $\mathscr{R}^\theta$ is then equal to $\mathscr{E}\bar{\omega}/\mathscr{I}_0$, where $\bar{\omega}$ is the angular velocity of rotation of the crystal and $\mathscr{I}_0$ is the neutron intensity in the incident beam, which is established most easily by comparative measurements with a small crystal of sodium chloride. Knowing the value of $\mathscr{R}^\theta$ we can determine the structure factor F, when extinction is negligible, by the equation

$$\mathscr{R}^\theta = QV = F^2\frac{N_c^2\lambda^3}{\sin 2\theta}V \tag{10.1}$$

as has already been described in Chapter 3. In this equation V is the volume of the crystal which is being radiated. The value of $\mathscr{E}$ used for calculating the integrated reflection $\mathscr{R}^\theta$ is first corrected for absorption.

The structure factors determined in this way were compared with the results of calculation both for a centrally situated hydrogen atom and also for various degrees of asymmetry, represented by the displacement parameter x. Satisfactory agreement could only be obtained with x equal to zero, $\pm 0{\cdot}01$, and it was also necessary to postulate different Debye temperature factors for the K, H, and F atoms. With these provisions very good agreement was obtained between the experimental intensities and those calculated for a symmetrical linear ion. The experimental accuracy is such as to place the hydrogen atom in the central position within 0·1 Å.

The conclusion that the hydrogen atom was effectively centred in this very short hydrogen bond of 2·26 Å led to a tendency to assume that all very short bonds were centred. However, it has since been demonstrated that this is by no means always the case, and a short bond may be markedly asymmetric if the local environment is asymmetric. Thus in the compound *p*-toluidinium bifluoride Williams and Schneemeyer (1973) have found a very short bond of 2·260 Å but with two very unequal H–F sections, measuring 1·025 Å and 1·235 Å respectively.

Of particular interest in connexion with the investigation of KHF_2 is the projection of neutron scattering density on the (001) plane which was constructed by Peterson and Levy from their structure factors for (*hk*0) planes. This projection is illustrated in Fig. 209 and represents the first example of the application of

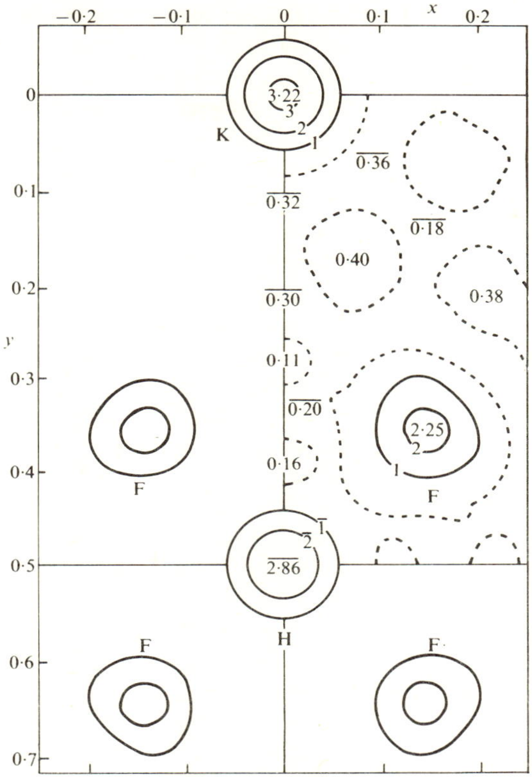

FIG. 209. A projection on the (001) plane of the neutron scattering density in KHF_2. The numbers indicate the values of the density at the contours and the extremes in units of 10^{-12} cm Å^{-2}. The dashed lines represent zero contours. (Peterson and Levy 1952*a*.)

Fourier synthesis methods in neutron crystallography. As expected, the hydrogen atoms appear as areas of negative density on account of their negative scattering amplitude. Diffraction rings are apparent around the atoms, particularly around the potassium atom which has the smallest Debye temperature factor, on account of the non-termination of the Fourier series. This arises not only from the limited number of terms taken but also from the fact that absence of any form factor for nuclear scattering means that the *F* values fall off only slowly with angle θ, by the effect of thermal vibrations.

10.4. Application of Fourier-synthesis methods to single-crystal data

The determination of crystal and molecular structures by neutron diffraction methods has followed the pattern of X-ray diffraction techniques, whereby substantial progress has only been achieved with the development of Fourier-synthesis methods and the use of least-squares analysis for the final refinement of atomic positions and thermal parameters.

In recent years considerable progress has been made in the development of 'direct methods' in order to overcome the basic problem of phase, or sign, determination for structure factors. This arises because experimental measurements with X-rays or neutrons are measurements of *intensity*, whereas the quantity which is needed in order to make a Fourier summation of electron density (using X-rays) or nuclear scattering density (using neutrons) is the scattering *amplitude*. Numerically this quantity is represented by the square root of the intensity, but it is necessary to know also its phase angle. In general this may take any value between 0° and 360°, though in centro-symmetrical structures only the alternatives of 0° and 180° are possible. At the present time indirect methods, ranging from the Patterson synthesis to quite general deductions from physical and chemical properties, are used to suggest a possible starting structure which can be used for a trial-and-error comparison between observed and calculated intensities. Successive trials lead to a structure which becomes increasingly correct and serves for calculation of the signs which have to be attached to the numerical values of the experimental structure factors. At a later stage the positional and thermal parameters are defined by least-squares analysis.

The investigator using neutrons generally starts with a considerable advantage, for it is likely, indeed almost essential, that his material has already been investigated with X-rays. As a result the positions of most of the atoms in the structure will have been determined already and the signs for the structure amplitudes of most of the reflections in a centro-symmetric structure will be correctly known at the outset. The signs of a relatively few structure amplitudes, which are sensitive to the precise positions of the hydrogen or other light atoms, will remain to be determined as the study proceeds. On the other hand, the Fourier synthesis of scattering matter produced from neutron data is much more likely to be confused by false detail arising from diffraction rings which appear around individual atoms when only a finite number of terms is taken in the Fourier series. The introduction of spurious detail in this way is analogous in optics to the observation of an object with a microscope of finite aperture. The image of a point source will appear as a spot of finite width surrounded by a number of diffraction rings of steadily decreasing intensity and associated with vibration amplitudes which are alternately negative and positive in sign. It is to be expected therefore that the nuclei in a structure, which are point sources of neutron scattering, will appear in a Fourier projection as finite areas of scattering density surrounded by positive and negative diffraction rings. This is found in practice as illustrated in Fig. 210 (Bacon and Pease 1953) which shows how the scattering density varies along either the x- or y-axis in a Fourier projection on to the (001) plane for KH_2PO_4. Work with this substance, both in the U.S.A. and in England, established the value and potential possibilities of high-accuracy methods of data analysis in neutron diffraction work. The American work which was carried on independently and concurrently with that of Bacon and Pease, which we shall describe in more detail, is reported in two papers by Peterson, Levy, and Simonsen (1953, 1954). In each case great advantage was taken of a previous study with X-rays by Frazer and Pepinsky (1953). Essentially Fig. 210 shows a cross-section of the apparent scattering density across superimposed potassium and phosphorus nuclei, although with increasing distance from the origin the pattern becomes affected by diffraction ring contributions from the other K and P nuclei. Correction has already been made for the effect of oxygen and hydrogen nuclei. The intensity data from which this pattern was constructed include reflections up to (14, 20) and the expected diffraction pattern when the series of Fourier terms is terminated at this point is indicated by the broken line in the figure, representing a function of the form $J_1(x)/x$, where $J_1(x)$ is the first-order Bessel function. Thus practically the whole of the apparent width of the atoms can be accounted for by diffraction. The further increase of width, apparent when the two curves are compared, is due to the effect of thermal vibrations. These have the result of smearing out some of the false detail in the pattern by partially merging the regions of extreme positive and negative amplitude. The calculated resultant, using a determined value of the

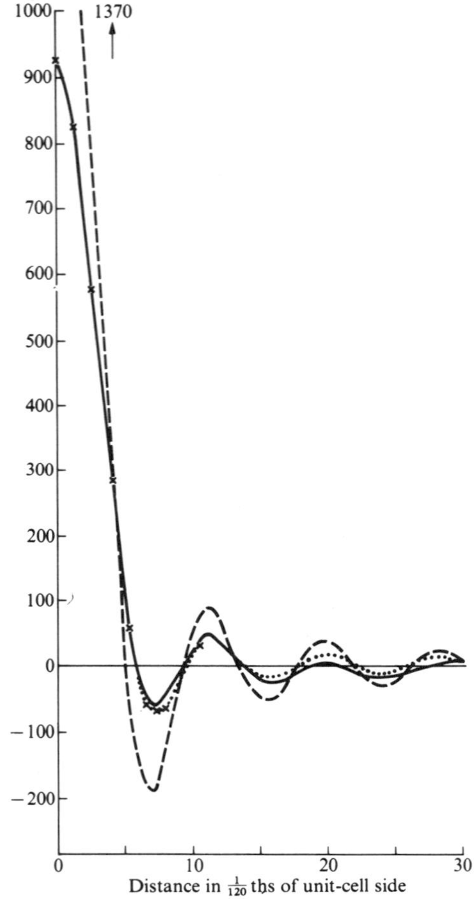

FIG. 210. The apparent scattering density of superimposed potassium and phosphorus nuclei as revealed in the Fourier projection on the (001) plane for KH_2PO_4. The full-line curve shows the distribution of density found experimentally. The broken line is the function $J_1(x)/x$, appropriate to diffraction alone, and the dotted curve, together with the points '×', shows the correction of this function to allow for thermal vibrations.

Debye temperature, is the dotted curve in Fig. 210 which amply accounts for the observed width of the experimental curve.

In Fourier projections of electron density from X-ray diffraction data the situation is rather different. The atoms are no longer to be regarded as point scatterers, since the scattering unit is now an electron cloud of appreciable dimensions, and the effective radius will be of the order of 1 Å, certainly larger than the peak to zero width of the diffraction curves in Fig. 210. Thus the diffraction rings will be smeared out by the overlapping of the contributions from individual parts of the electron cloud and the spurious detail will be practically unobserved and not troublesome. Alternatively, we may say that the extended nature of the scattering atoms means that the atomic form factor falls off with angle θ, as already displayed in curve (a) of Fig. 16 (p. 22), with the result that the structure factors of the reflections fall off rapidly with angle θ, or decreasing interplanar spacing. Consequently the error introduced into the Fourier synthesis when only a finite number of terms is included is much less than for the neutron case, where there is no regular fall-off of the magnitude of the terms with increasing angle θ.

Our present concern is to establish the extent to which precise information can be obtained from a Fourier projection, which means ascertaining how far the spurious detail is misleading in practice and considering the manner in which correction for it may be made. It is interesting to note that the apparent width of a nucleus as seen by neutrons, after being smeared out by the effects of thermal vibrations and diffraction, is very similar to that of the electronic cloud of an atom as revealed by X-rays. This is illustrated in Fig. 211, which compares a potassium atom in KH_2PO_4 as shown by neutrons (Bacon and Pease 1953) and X-rays (West 1930) at room temperature. Even at low temperatures, and particularly for light atoms, the zero-point motion will still give a large apparent size to a nucleus, quite apart from the effect of broadening due to diffraction.

Fig. 212 shows a number of Fourier projections of KH_2PO_4 to which we shall refer. The projection (a) is a direct projection on the (001) plane from intensity data taken at room temperature. This projection is shown on an enlarged scale in Fig. 213 where the diffraction rings around the superimposed potassium and phosphorus atoms are clearly evident. They result in appreciable effects at the position of the hydrogen nucleus, particularly as the latter has a very diffuse distribution and its peak amplitude is only about a quarter of that for oxygen, whereas the value of b for hydrogen is about two-thirds as great as for oxygen. In arriving at this projection from the intensity data it is necessary to correct the most intense

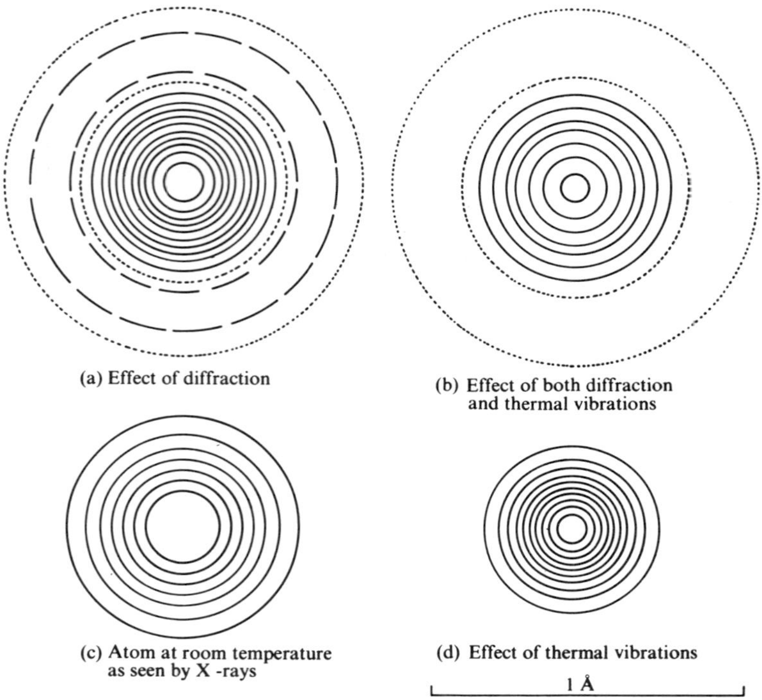

FIG. 211. A point-scattering nucleus is revealed by neutron diffraction as the pattern shown in (a). Under the influence of thermal vibrations the true scattering density is that shown in (d), which because of diffraction appears as in (b). The apparent appearance to X-rays of the electron cloud of an atom is shown for comparison in (c). In each case contours are drawn at equal intervals—full lines are positive, broken lines are negative, and zero contours are dotted. (Bacon and Pease 1953.)

reflections for extinction and to take particular experimental care not to miss the higher-angle reflections. In the latter case, because of the increasing angular width of the reflections at large values of θ, it may happen that a reflection with a large structure factor will give only a small value for the peak count above background.

The projection can be refined in a number of ways in order to give improved values of the atomic parameters. Fig. 214 shows first the result of correcting the contours in the neighbourhood of the hydrogen atom for the diffraction rings produced by neighbouring atoms, utilizing diffraction curves of the type shown in

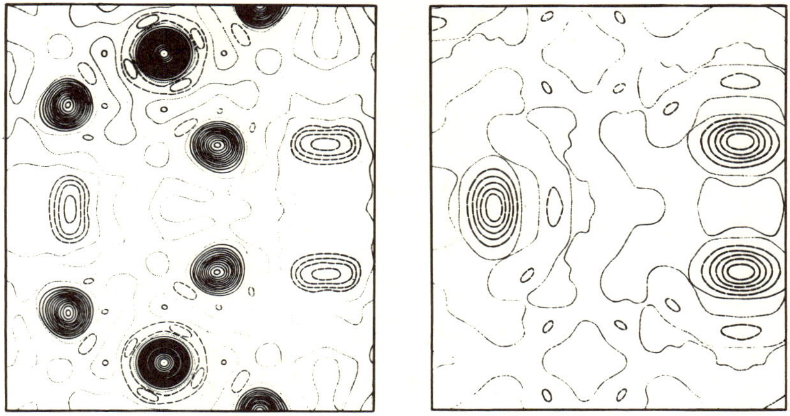

(a) Synthesis of F_{obs} (b) Synthesis of $F_{obs}-F_{KPO}$

(Structure at 20 °C tetragonal)

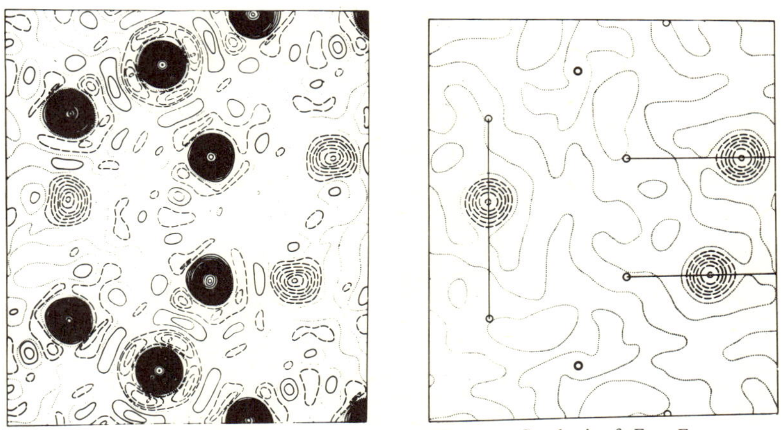

(c) Synthesis of F_{obs} (d) Synthesis of $F_{obs}-F_{KPO}$

Structure at −180 °C (orthorhombic, ferroelectric)

FIG. 212. Potassium dihydrogen phosphate, KH_2PO_4—projections of the neutron scattering density on the (001) plane for both the room-temperature tetragonal form (a), (b), and the low-temperature ferroelectric orthorhombic form (c), (d). (a) and (c) are direct projections which show all the atoms in the structure, the most intense peaks being those of superimposed potassium and phosphorus atoms. The other peaks are oxygen atoms (full lines) and hydrogen atoms (broken lines, indicating their negative scattering amplitude). (b) and (d) are so-called 'difference projections', in which only the hydrogen atoms appear. The latter shows the details of the hydrogen bond free from the distortion produced by the diffraction ripples of neighbouring atoms. (After Bacon and Pease.)

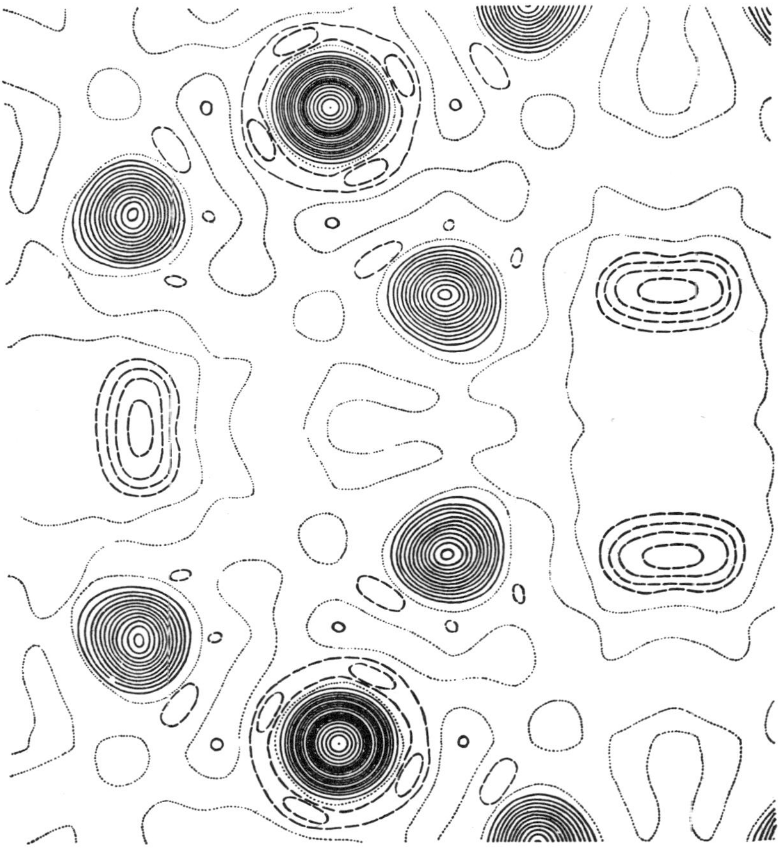

FIG. 213. A Fourier projection of the scattering density on the (001) plane of KH_2PO_4 at room temperature. Contours are at intervals of 50 units, with additional contours at -75 and -125. Full lines are positive, broken lines are negative, and dotted lines are zero contours. The most intense peaks are superimposed K and P. The other positive peaks are O; the negative peaks are H. (Bacon and Pease 1953.)

Fig. 210; the lower portion of Fig. 214 shows a synthesis of the difference function $F_{obs} - F_{KPO}$, where F_{KPO} is the calculated value of the contributions of all the atoms other than hydrogen. In this latter case diffraction effects due to the K, P, and O atoms are eliminated and the contour of the hydrogen is only affected by diffraction effects due to itself and other hydrogen atoms. The close

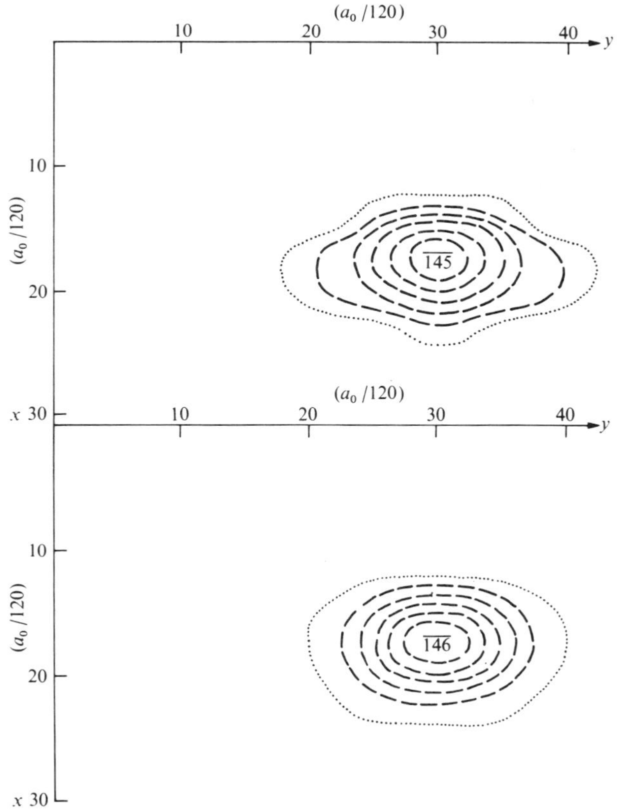

FIG. 214. The hydrogen atom in KH_2PO_4 are seen in the projection on the (001) plane when corrected for series termination errors. In the upper plot correction has been made by assessing the diffraction contributions from neighbouring atoms. The plot below is a synthesis of the function $F_{obs}-F_{KPO}$. Contours are drawn at equal intervals of 25 units. The units of abscissae and ordinates are $\frac{1}{120}$ths of the unit-cell side a_0, which is equal to 7·434 Å.

similarity between the two plots shown in Fig. 214 enables some confidence to be placed in the position of the hydrogen as deduced in this way and in the anisotropy of the observed distribution.

In addition to information on the location of hydrogen atoms the foregoing investigations offer an example of the precision with which oxygen atoms can be located by neutrons on account of their large scattering amplitude, a feature to which we shall refer again in

Chapter 12 when we consider the oxygen parameter in ferrites. For KH_2PO_4 it is considered that the oxygen positions are accurate to about 0·002°. The fractional co-ordinate in the *z*-direction was deduced to be $0{\cdot}126 \pm 0{\cdot}001$ compared with the previously accepted value of 0·139 from West's X-ray work, and results in the PO_4 group being much more closely a regular tetrahedron in shape than was formerly believed.

It may be noticed, in the light of the above remarks, that the use of deuterated compounds would offer advantages even for single-crystal analysis, since the numerically larger scattering amplitude would mean that diffraction effects due to neighbouring atoms would be relatively less disturbing and, further, there would be rather less smearing of the pattern due to thermal displacements. For KH_2PO_4 at room temperature the elongation of the hydrogen atom along the O—H—O bond is one of the most interesting features of the projections. It may be interpreted as due either to very anisotropic thermal motion or to the superposition of two 'half-hydrogens', one each side of the centre of the bond, as a result of a disordered arrangement of hydrogen atoms among pairs of possible positions closer to one or other of the two oxygen atoms. The neutron data cannot distinguish between the two interpretations, but show that if the latter one is correct then the two positions cannot be further from the bond centre than $\pm 0{\cdot}175$ Å. Below 120 K KH_2PO_4 becomes orthorhombic and ferroelectric, and Bacon and Pease (1955) have shown that the hydrogen atoms then appear circular in the projection and lie asymmetrically between the oxygens, in an ordered arrangement which satisfies the reduced symmetry, as supposed by Slater (1941). In Fig. 212 we compare the projections of the structure on the (001) plane both above and below the temperature at which the ferroelectric transition takes place. The Figure shows both the direct projections and also the so-called 'difference projections' which display the hydrogen atoms alone as a synthesis of the function $F_{obs} - F_{KPO}$ which we mentioned above. In order to obtain the projection at low temperature showing an ordered arrangement of hydrogen atoms, it is necessary to make sure that when the ferroelectric change takes place then the whole crystal orders as a single domain. This can be ensured by applying an electric field of about 8000 V cm^{-1} along the *c*-axis. If the electric field is reversed then many of the diffraction intensities change considerably in magnitude as illustrated in Fig. 215 (b). If a projection

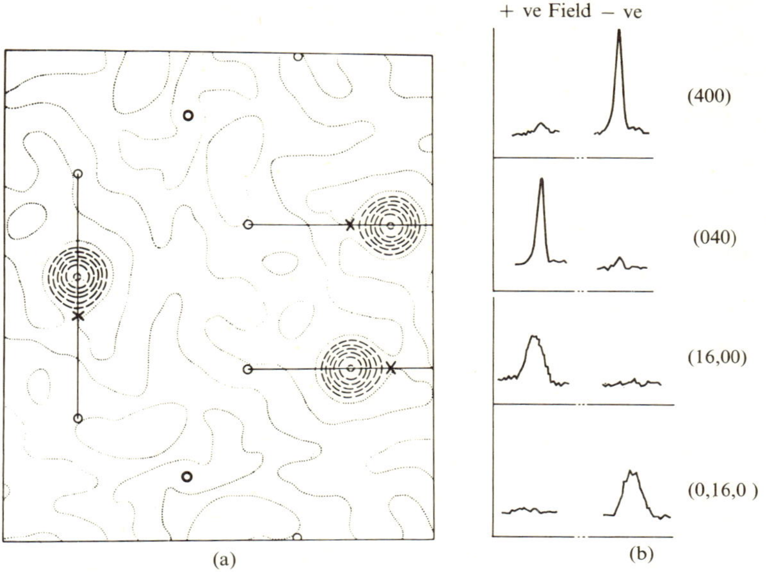

FIG. 215. The movement of the hydrogen atoms in the ferroelectric form of KH_2PO_4 under the influence of an electric field. The left-hand picture (a) is a projection of the hydrogen atoms when a crystal is maintained as a single domain by applying a positive field along the *c*-axis. It has been obtained using intensity data represented by the first column of traces in the right-hand picture (b). On reversing the direction of the electric field the intensities of many of the spectra change considerably, as represented by the second column of traces, and the resulting synthesis then shows that the hydrogen atoms have moved from their initial positions to the positions which are marked in the left-hand picture by ×.

of scattering density is now constructed using the new set of intensities it will be found that the protons have moved from one set of ordered positions to the other, marked by the crosses in Fig. 215 (a). At the same time, as shown by Frazer and Pepinsky (1953), there are small movements of the potassium and phosphorus atoms which account for the observed polarization of the material.

Data analysis and direct methods

In practice the final assessment of the parameters of a structure depends on a process of *refinement* in which the choice of parameters for refinement and their starting values rests on X-ray information and any Fourier projections obtained with neutron data. Indeed as the starting data from X-ray work have, over the years, become more and more accurate the tendency has been to dispense entirely with Fourier projections and to proceed directly to refinement by least-squares analysis. Such a procedure has its dangers, since refinement can only refine those parameters of whose existence the investigator is already aware, and the production of Fourier projections and of Fourier sections, which can be produced very readily from three-dimensional data with the aid of automatic graph-plotters, provides valuable safeguards. Indeed for optimum understanding the processes of Fourier production and least-squares refinement should proceed concurrently.

In the past twenty years very great advances have been made in the use of 'direct methods' in X-ray crystallography as a way of solving the phase problem, utilizing information which is implicit in the $|F_0|$ values themselves. The foundation of these methods is twofold: first, the electrons which scatter X-rays have negative charges and electron density cannot change sign, and, secondly, electron density can only exist in atomic aggregates of linear dimension about 1 Å. In the case of neutrons the condition for scattering density of single sign does not hold for substances containing hydrogen, and it was at first believed that direct methods would not be applicable to neutron data if more than about a quarter of the atoms were hydrogen. In practice the restriction appears to be less severe and the structure of potassium tri-hydrogen succinate, which contains 40 per cent of hydrogen, has been determined by Dunlop and Speakman (1973). Moreover, the structure of glycollic acid, with 45 per cent of hydrogen, has been solved (Ellison, Johnson, and Levy 1971) via a fictitious intermediary substance whose neutron scattering density is the square, which is always positive, of that in the actual material under study. It is, of course, always possible to simplify the analysis by deuteration. The general conclusion, both for centro- and non-centrosymmetric structures, is that the range of applicability of direct methods is very similar for neutrons and X-rays.

11

THE STUDY OF MOLECULAR STRUCTURE

THE realization that methods of Fourier analysis were fully applicable to neutron diffraction data has enabled neutron diffraction to take its place alongside X-ray analysis as a primary method of determining the structure of molecules. The years 1952–62 showed a steady production of two-dimensional projections of the structures of materials which included hydrogen atoms, mainly structures which included hydrogen bonds. Extension to studies in three dimensions had to await the appearance of nuclear reactors giving a higher flux and the development of automatically controlled four-circle diffractometers which are now in regular use. Indeed most of the structural studies which have been published in the last few years have been three-dimensional. However, the earlier two-dimensional work is not only of interest for its contribution to the chemical and physical understanding of particular substances but also because the two-dimensional projections of scattering density, which usually summarized these studies, are particularly easy to appreciate. Therefore we shall present first some typical examples of these.

11.1. Hydrates

Following earlier work on the structure of ice, many studies have been made of hydrated compounds, with a view to detailing the shape and orientation of the molecules of water in these substances and to ascertaining the important part which they play in building up and holding together the crystal structure. The kind of information which can be obtained is illustrated well by Fig. 216 which is a projection of sodium sesquicarbonate, $Na_2CO_3 . NaHCO_3 . 2H_2O$. Here the water molecules are linked at each end by hydrogen bonds which join them to oxygen atoms of the carbonate groups. In turn, pairs of CO_3 groups are linked by a hydrogen atom at the centre of symmetry, with the result that continuous chains of hydrogen-bonded groups extend through the structure. A detailed examination of the positions of the atoms, such as H_1 and H_2, in the water molecules reveals that the hydrogen bonds, for example, in the grouping $O—H_1 \cdots O$, are not quite linear. They are bent in such a way that the angle H_1OH_2 within the water molecule is more closely equal to the tetrahedral value than is the angle O—O—O, which is significantly greater. A more pronounced example of the way in which a water molecule maintains its favoured shape, in spite of the environment, is provided by Bacon and Gardner's (1958) work on chromium potassium alum, for which the positions of the atomic neighbours of the two types of water molecule are indicated in Fig. 217. In the case of the water molecules which

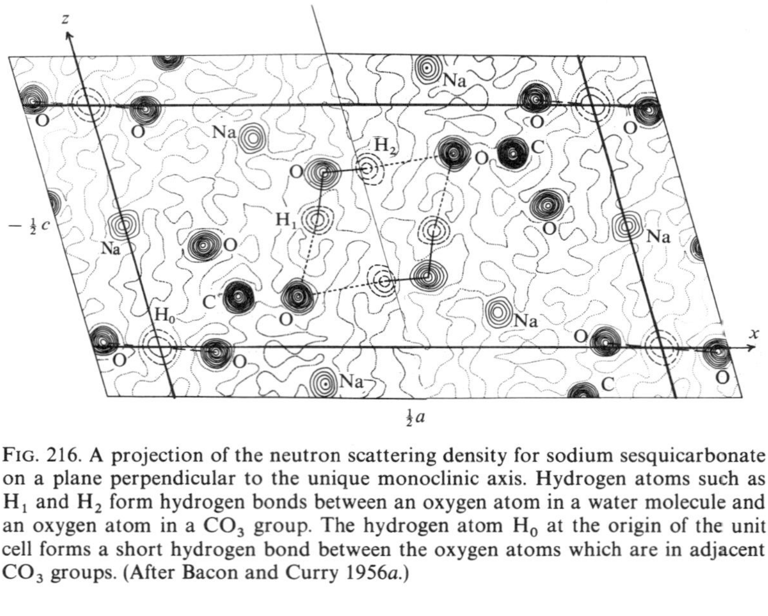

FIG. 216. A projection of the neutron scattering density for sodium sesquicarbonate on a plane perpendicular to the unique monoclinic axis. Hydrogen atoms such as H_1 and H_2 form hydrogen bonds between an oxygen atom in a water molecule and an oxygen atom in a CO_3 group. The hydrogen atom H_0 at the origin of the unit cell forms a short hydrogen bond between the oxygen atoms which are in adjacent CO_3 groups. (After Bacon and Curry 1956*a*.)

surround the potassium atom the angle O—O—O is as small as 94°, but in spite of this the H—O—H angle within the water molecule is not reduced below 103°. Another inorganic hydrate which has been studied is $CuCl_2 . 2H_2O$ which contains a very weak, and long, hydrogen bond. The length, between a chlorine atom and the oxygen atom of a water molecule, is about 3·2 Å. Peterson and Levy's (1957) projection of this substance shows that the O—H distance within the water molecule is only 0·95 Å, very similar to that found in steam. On the other hand, for both ice and sodium sesquicarbonate where the O—O separation is much shorter (about 2·76 Å) the O—H distance is significantly greater than in steam, amounting to about 1·01 Å. A further example is provided by gypsum, $CaSO_4 . 2H_2O$, where the O—O distance is 2·82 Å and the neutron diffraction study by Atoji and Rundle (1958) gives an O—H distance of (0·99 ± 0·03) Å. The increase in the O—H distance as the total

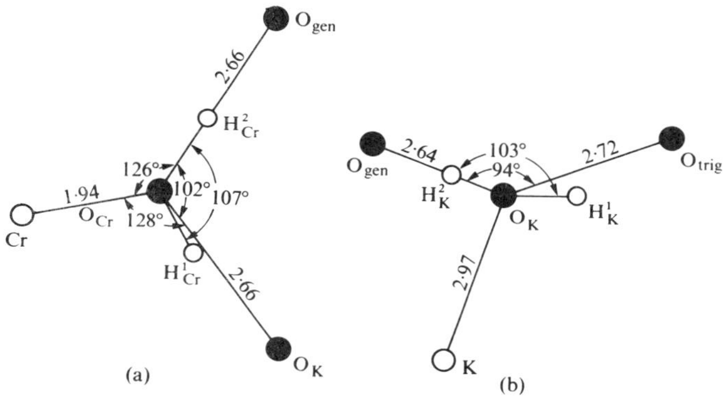

FIG. 217. The water molecules which surround (a) the chromium atoms, and (b) the potassium atoms in chromium–potassium alum, indicating the hydrogen bonds which link them to neighbouring oxygen atoms. (Bacon and Gardner 1958.)

O—H—O distance gets shorter, and the hydrogen bonds get stronger, is observed quite generally. A particularly good example is provided by oxalic acid dihydrate, $(COOH)_2 . 2H_2O$, (Garrett 1954) in which both short and long bonds are present in a single structure. A projection is shown in Fig. 218. There is a short strong bond of length 2·52 Å between a water molecule and a hydroxyl group within the same complex and for this the O—H distance is 1·06 Å. By contrast the total length of the bond between a water molecule and a carbonyl group in a neighbouring complex is much larger 2·85 Å, and the O—H distance is reduced to 0·96 Å which is the value found in isolated water molecules.

Table 20 summarizes some of the information which has been obtained about bond angles in hydrates by comparing the H—O—H angle in the water molecule with the environmental angle O—O—O which the neighbouring oxygen atoms subtend at the centre of the water molecule. The data for $CuSO_4 . 5H_2O$ (Bacon and

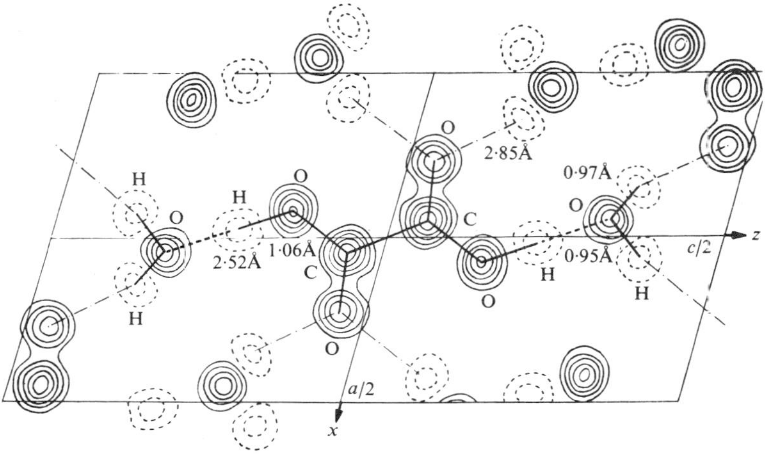

FIG. 218. A projection of the structure of oxalic acid dihydrate, showing the two types of hydrogen bond. The short hydrogen bond (2·52 Å) joins a hydroxyl group to the oxygen atom of a water molecule within the same hydrate complex. The long hydrogen bond (2·85 Å) joins a carbonyl oxygen atom to the oxygen atom of a water molecule in an adjoining complex. (After Garrett 1954.)

TABLE 20
Bond angles in hydrates

Substance	Angle (deg) H—O—H	Angle (deg) O—O—O
D_2O, ice	109·1	109·5
	109·9	109·4
$Na_2CO_3 . NaHCO_3 . 2H_2O$	107	114
$K_2SO_4 . Cr_2(SO_4)_3 . 24H_2O$ (O_K)	103	94
(O_{Cr})	107	102
$(COOH)_2 . 2H_2O$	106	84
$CuCl_2 . 2H_2O$	108	97 (O—Cl)
$CaSO_4 . 2H_2O$	106	108
$CuSO_4 . 5H_2O$	114, 111, 106, 109, 109	119, 121, 122, 130, 105
$BeSO_4 . 4H_2O$	—	120
$NiSO_4 . 6H_2O$	—	99, 125, 132
$MgSO_4 . 4H_2O$	110, 111, 111 109, 108	105, 147, 92 137, 114
$FeSO_4 . 7H_2O$	—	121, 123, 114, 110, 103, 121, 145

Curry 1962) show that the water molecule largely resists attempts to open out the H—O—H angle just as it resists closing up.

The results which we have described so far show that the H—O—H angle in hydrates is always close to the tetrahedral angle of 109·5° and hence there is often substantial bending of the hydrogen bonds. Many of the data are not sufficiently accurate to be much more precise than this. By the middle of 1972 about 50 hydrates had been studied by neutron diffraction and a statistical analysis of about 40 of them, involving 90 water molecules, was made by Ferraris and Franchini-Angela (1972). These authors proposed a classification for water molecules in hydrates of five classes, according to the number of co-ordinated cations and to the position of these cations with respect to the lone-pair orbitals. This classification is a development of proposals by Chidambaram, Sequeira, and Sikka (1964), Hamilton and Ibers (1968), and Baur (1962, 1965).

Progress in correlating the details of the hydrogen bonds with the local atomic environment is likely to come from three-dimensional studies. For example, comparison of $CuSO_4 . 5H_2O$ and $CuSO_4 . 5D_2O$ by Bacon and Titterton (1975) seems to establish the reliability of the variation of H—O—H angle, ranging from about 106° to 112°, among the five water molecules in this salt. We emphasize the value of the higher accuracy obtainable with high-flux reactors by describing results of a re-examination of oxalic acid dihydrate by Coppens and Sabine (1969). Two forms, α and β, of the deuterated compound exist and both were examined with neutrons. The two forms are not isomorphous but the shapes of the molecule and the interatomic distances are quite similar in the two cases, as illustrated in Fig. 219. The most noteworthy difference is that in the β form the deuterium atom D(1) lies almost in the plane of the water molecule D(2)O(3)D(3), whereas this is by no means the case for the α structure. The distinction can be appreciated more clearly in the two insets (a), (c) in the Figure; for the α form the bond D(1)—O(3) makes an angle of 42·3° with the plane of the water molecule, but in the β form this angle is only 1·4°. The inset (b) shows the directions of the lone-pair orbitals of the water molecules, and it follows that one of the orbitals is directed towards the hydrogen-bond donor group D(1) in the α form, but not in the β form, where the bond D(1) lies symmetrically between the two orbitals. In the latter case the angle between the lone pairs closes, to increase the total interaction with the donor group, and would lead

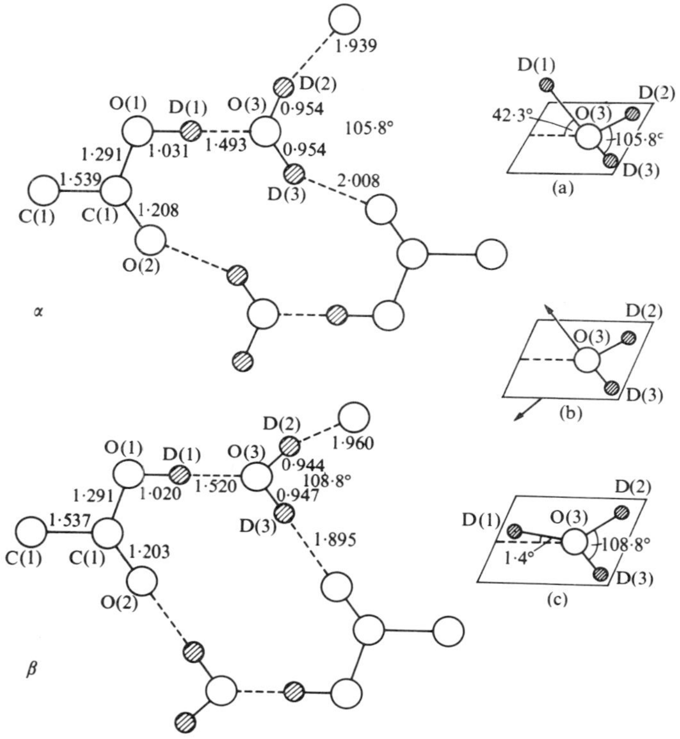

FIG. 219. The structures of the α and β forms of deuterated oxalic acid dihydrate. The inset (a) for the α form shows that the bond D(1)—O(3) is inclined at about 42·3° to the plane of the water molecule, whereas in the β form (c) this angle is only 1·4°. In (b) the arrows indicate the directions of the lone-pair orbitals of the water molecule, indicating that one of these is directed close to D(1) in the α form.

(see Coulson 1961) to a corresponding opening of the angle D—O—D. This is exactly what is observed experimentally where there is a measured increase from 105·8° to 108·8° in the latter angle. At the same time a strengthening of the O—D bonds to D(2) and D(3) is produced, indicated by the reduction of the bond length from 0·954 Å in the α structure to 0·945 Å in the β form accompanied by a corresponding increase in the length of the bond from O(3) to D(1). These results offer great promise of detailed correlation of bond angles, distances, and environment in other substances.

11.2. Hydrogen bonds in aromatic compounds

Some of the best examples of the use of neutrons to study hydrogen atoms and molecular structure are to be found among the fairly simple derivatives of benzene.

Benzene itself has been studied in the solid form at both 218 K and 138 K and the two-dimensional plot (Fig. 220) of the scattering density

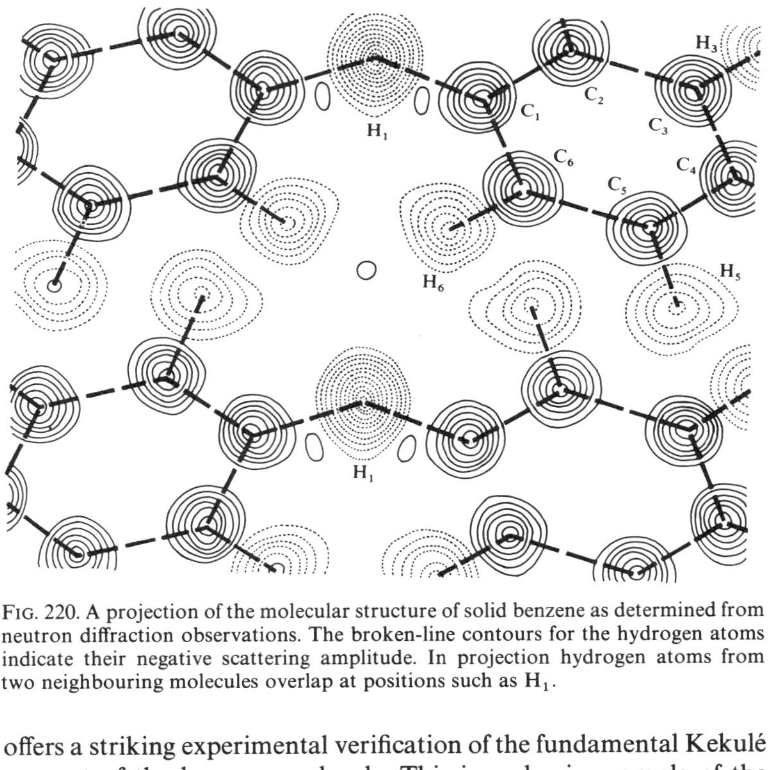

FIG. 220. A projection of the molecular structure of solid benzene as determined from neutron diffraction observations. The broken-line contours for the hydrogen atoms indicate their negative scattering amplitude. In projection hydrogen atoms from two neighbouring molecules overlap at positions such as H_1.

offers a striking experimental verification of the fundamental Kekulé concept of the benzene molecule. This is a classic example of the power of neutron diffraction and the work was carried out in 1958 at a low-flux reactor, with a flux of only about 10^{12} neutrons cm^{-2} s^{-1}. Analysis of the results showed that there was an oscillatory angular motion of the molecule as a whole, in its own plane, of 7·9° and 4·9° at the two temperatures (Bacon, Curry, and Wilson 1964).

A typical example of hydrogen bonds in a small molecule is provided by α-resorcinol, m-$C_6H_4(OH)_2$. This was in fact the first aromatic molecule to be studied with neutrons, and the two-dimensional projection (Fig. 221) which was obtained by Bacon and Curry (1956) should be compared with Robertson's (1936) earlier picture with X-rays, which was obtained at a rather similar stage in the development of techniques. Fig. 221 shows, in particular, the projection of the spiral of hydrogen bonds which links together the hydroxyl groups from neighbouring molecules and plays the most important role in building up the three-dimensional crystal structure. The precision of the measurements was sufficient to give a preliminary idea of the thermal motion of the molecule as a whole. It was clear that the movement of carbon atom C_2 is larger than that of the atoms C_1 and C_3, which in turn is greater than for the atoms C_4, C_5, and C_6, and the same sequence of magnitudes is observed for the motion of the associated hydrogen atoms. It

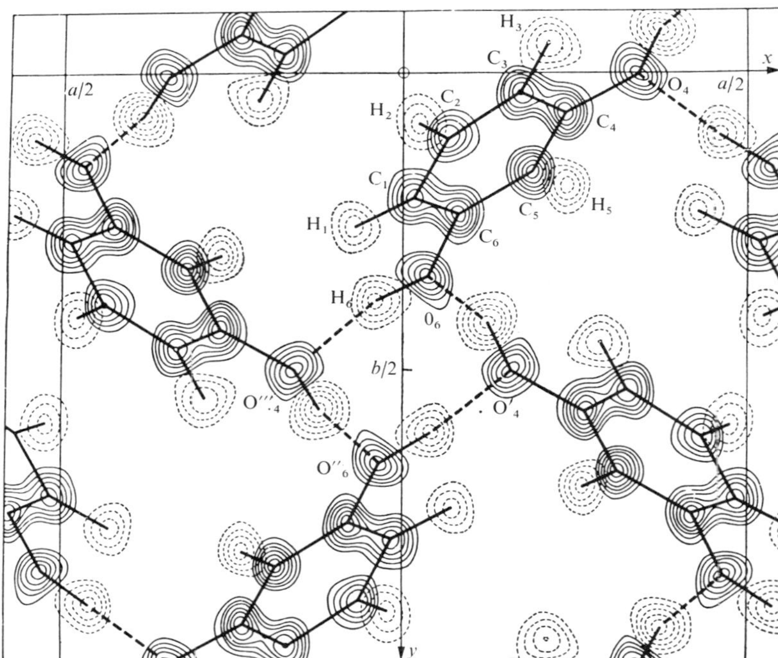

FIG. 221. A projection of the neutron scattering density of α-resorcinol on the (001) plane, constructed from two-dimensional data. In three dimensions the rectangular linkages such as O_6, O'_4, O''_6, and O'''_4 represent spirals of hydrogen bonds.

therefore appeared that the carbon atoms C_4 and C_6 are relatively firmly anchored by the hydrogen bonds and there is motion of the molecule as a whole about the axis which joins these pivots.

More recently α-resorcinol has been examined in three dimensions by Bacon and Jude (1973), with the aid of an automatic single-crystal diffractometer with which 1300 reflections were measured. This work is typical of what can be achieved at the present time with a medium-flux reactor. A crystal measuring roughly 2 mm × × 2 mm × 3 mm was used, and about one-fifth of the reflections required a correction for secondary extinction, the correction ranging in magnitude from 1 per cent to 50 per cent. The data were analysed in two ways, first by a conventional least-squares analysis in which individual anisotropic temperature factors were postulated for each atom and, secondly, in terms of a rigid-body motion of molecule as a whole. The first analysis produced a discrepancy factor R of 3·6 per cent and standard deviations of the atomic coordinates amounting to 0·002 Å for the carbon and oxygen atoms and 0·005 Å for the hydrogen atoms. These standard deviations may be regarded as typical for this kind of work. Fig. 222 illustrates

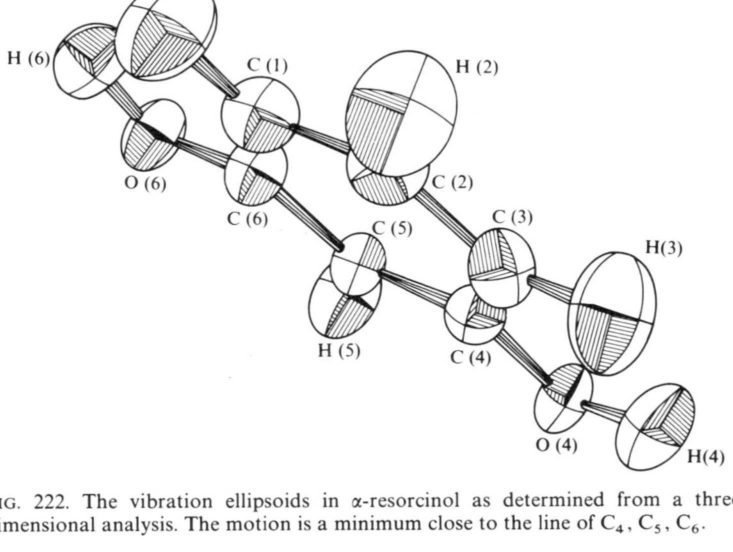

FIG. 222. The vibration ellipsoids in α-resorcinol as determined from a three-dimensional analysis. The motion is a minimum close to the line of C_4, C_5, C_6.

the conclusions of this unconstrained analysis for the thermal vibrations by drawing the vibration ellipsoids. The results confirm earlier conclusions of Bacon and Curry regarding the relative motion of the various atoms, but give much more accurate information about this. The rigid-body analysis resulted in an *R* value of 6·3 per cent, substantially worse than the unconstrained analysis and thus indicating that substantial bending and stretching of bonds takes place within the molecule itself. The root-mean-square amplitudes of translation were deduced to be 0·19 Å, 0·16 Å, and 0·17 Å respectively along the three principle axes, and the amplitudes of libration about these axes were 4·37°, 2·27°, and 2·68°. It is noteworthy that the ability of neutron diffraction to determine accurately these angles of oscillation of the molecule rests substantially on its ability to locate the hydrogen atoms accurately. The hydrogen atoms tend to be on the outside of the molecule, remote from the axes of libration, and their larger radial distances produce larger linear distances of movement for a given angular rotation, which can be assessed quite accurately.

11.3. Correlation of X-ray and neutron data

As a further informative example of the scope of neutron diffraction methods we examine the study of hexamethylenetetramine, $C_6N_4H_{12}$. This substance is unusual among organic compounds in possessing a structure which has cubic symmetry and involves an asymmetric unit consisting of only three atoms. This feature greatly limits the number of independent parameters which have to be determined, and accordingly increases the accuracy with which the atomic coordinates and the thermal motion can be assessed. Like α-resorcinol, hexamethylenetetramine is a much-studied compound, the investigation culminating in an X-ray study by Becka and Cruickshank (1963) and a neutron study by Duckworth, Willis, and Pawley (1969, 1970). In the neutron work both unconstrained and rigid-body models were used for refining the thermal motion and they resulted in values for the discrepancy factors *R* of 2·3 per cent and 3·2 per cent respectively. The standard deviations which were estimated for the atomic coordinates were 0·003 Å, 0·002 Å, and 0·004 Å for carbon, nitrogen, and hydrogen respectively. However, the most interesting feature of this work is that the authors also carried out a joint refinement in which their neutron data and the earlier X-ray data were examined together. This procedure will be seen to be of particular importance when we examine exactly what is revealed by the two sets of measurements. As we have already emphasized, X-rays locate the electrons in an atom, and it is the centre of gravity of the electrons which is identified as the atomic position. When atoms are involved in the formation of covalent bonds or have lone-pair electrons this centre of gravity may not be identical in position with the nucleus of the atom, which is the feature of the atom which is identified specifically by the neutrons. Accordingly the joint refinement was made in terms of two sets of positional parameters—one for the neutron data and the other for the X-ray data—but a single set of thermal parameters. Two separate scale factors, of course, were also required. Because of the high symmetry of the structure there are only four atomic coordinates to determine and the two sets are given in Table 21. The carbon atom is at $(00u)$ and its equivalent positions, the nitrogen is at (vvv), and the hydrogen is at (xxz).

TABLE 21

Position parameters and standard deviations from joint refinement for hexamethylene-tetramine

	Neutron	X-ray
$u(\mathrm{C})$	$0{\cdot}2381 \pm 0{\cdot}0008$	$0{\cdot}2375 \pm 0{\cdot}0003$
$v(\mathrm{N})$	$0{\cdot}1221 \pm 0{\cdot}0005$	$0{\cdot}1237 \pm 0{\cdot}0002$
$x(\mathrm{H})$	$0{\cdot}0899 \pm 0{\cdot}0011$	$0{\cdot}0802 \pm 0{\cdot}0020$
$z(\mathrm{H})$	$-0{\cdot}3258 \pm 0{\cdot}0012$	$-0{\cdot}3292 \pm 0{\cdot}0030$

From the table it will be seen that the values of u for the carbon atom agree within 1 standard deviation, if we take the larger value of this deviation as given by the neutron data. This is what would be expected for carbon because the valence electrons are in four orthogonal hybrid orbitals, thus giving an electron cloud which is symmetrical about the carbon nucleus. For the nitrogen atom, on the other hand, the difference between the two values of v is 3 times as great as the neutron standard deviation and indicates a displacement of the electronic centre of the atom by 0·018 Å from the position of the nucleus. This can be explained readily because, of the five electrons in the valence shell of a nitrogen atom, three electrons participate in bonding and the other two are in lone-pair orbitals which are displaced from the nucleus in a direction away from the centre of the molecule. Calculations by Coulson (1970) have indicated that this displacement should amount to about 0·016 Å, in very good agreement with the experimental observation. The experimental conclusion is indicated rather strikingly in the Fourier plot of Fig. 223. This plot has been obtained by synthesizing the function

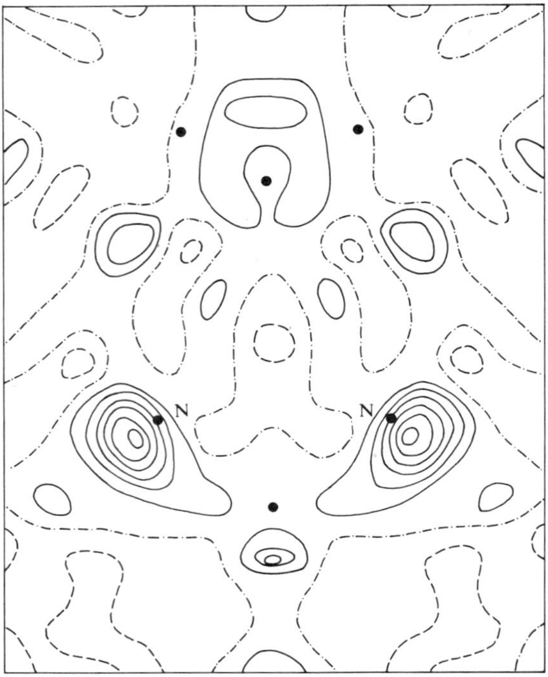

FIG. 223. A Fourier synthesis for hexamethylenetetramine of the function $F_X - F'$, where F_X is the experimentally determined structure factor for X-rays and F' is a calculated value for X-rays using the position coordinates given by *neutrons*. The atomic nuclei are represented by dots and the two large peaks near the nitrogen atoms N indicate the lone-pair electrons of nitrogen. (From Duckworth, Willis, and Pawley 1970.)

$F_X - F'$, where F_X is the experimentally determined value of the structure factor for X-rays and F' is the calculated value expected for X-rays using the position coordinates given by the neutrons. Such a synthesis will indicate the extent to which the centre of gravity of the electron density is displaced from the nucleus. The projection in Fig. 223 is made on the (110) plane of the crystal and shows that, as expected, the lone-pair electrons of the nitrogen atom are displaced from the nucleus in the direction away from the centre of the molecule.

It can be noted also from Table 21 that there are significant differences in the parameters of the hydrogen atoms found with X-rays and neutrons. Although the standard deviations of the X-ray positions for hydrogen are necessarily rather large there is undoubtedly a significantly larger value of the length of the C—H bond as given by neutrons. At present the best available values are (1·11 ± 0·01) Å for neutrons and (1·06 ± 0·01) Å for X-rays. It would be of considerable interest to confirm this difference with other materials but, so far, there are very few substances which have been examined with such high accuracy using both radiations. Some data have been given by Hamilton and La Placa (1968) for the anti-leukaemia drug $C_5N_8H_{12}.2HCl.H_2O$, in which the average values of N—H, O—H, and C—H bond lengths are about 0·1 Å longer for neutrons than X-rays, but the limited accuracy of the X-ray location of the hydrogen atoms restricts the drawing of precise conclusions regarding the significance of the differences. Table 22 compares the mean values of the different types of bond as

TABLE 22

X-ray and neutron bond-lengths (Å) *in* $C_5H_{12}N_82HCl.H_2O$

	X-ray		
	Free atom	Best spherical atom	Neutron
N—$H_{terminal}$	0·91(5)	0·97(6)	1·016(7)
N—H	0·82(17)	0·85(22)	1·003(34)
O—H	0·84(10)	0·88(7)	1·006(27)
C—H_{methyl}	0·94(12)	0·94(15)	1·090(27)
C—H	0·87(14)	0·84(14)	1·145(36)

found with neutrons with the results from two different X-ray refinements. In the first of these the scattering factor for a free atom of hydrogen is used, but the second uses the 'best spherical scattering factor' of Stewart, Davidson, and Simpson (1965). On the whole it seems that the latter model gives a better description of the hydrogen atom, but it seems evident that neither model gives an adequate picture of the electron density distribution.

Coppens (1970) has emphasized the advantages of determining all the parameters, both for position and temperature factor, from the neutron data, followed by an analysis of the X-ray measurements to determine the electron density. It then becomes directly evident whether electron density has migrated into the interatomic bonds or into regions of lone-pair electrons, as in the example of the nitrogen atom. Some very accurate studies have recently been made with both X-rays and neutrons of cyanuric acid (Coppens and Vos 1971) and α-deuterated oxalic acid dihydrate (Coppens, Sabine, Delaplane, and Ibers 1969). These have identified incontrovertibly the shift in the X-ray position of the atom relative to the nucleus and have also established the apparent shortening of O—H bonds for X-rays to an accuracy of 0·02 Å. These conclusions are summarized in Tables 23 and 24, which are taken from the latter paper.

TABLE 23

Shift of X-ray position relative to nuclear position, given by neutrons

Compound	Atom	Shift	Direction	Reference
Sym-triazine	C in ring	0·015(8)	Away from C—H bond	Coppens (1967)
	N in ring	0·009(6)	Towards lone pair	
Cyanuric acid	O(1) in C=O	0·006(1)	Towards lone pairs	Coppens and Vos (1971)
	O(2) in C=O	0·003(1)	Towards lone pairs	
Hexamethylene tetramine	N	0·021(7)	Towards lone pair	Duckworth *et al.* (1969)
α-deuterated oxalic acid dihydrate	O(1) in C—O—D	0·008(2)	Away from O—D	Coppens *et al.* (1969)
α-oxalic acid dihydrate	O(1) in C—O—H	0·008(2)	Away from O—H bond	Coppens *et al.* (1967)

Coppens has demonstrated that in X-ray refinements systematic errors in both the positional parameters and the temperature factors occur because of the non-spherical distributions of the electrons. As a result the very good agreement typically obtained

TABLE 24

Comparison of X-ray and neutron lengths for X—H bonds (Å)

Compound	Bond	Length Neutron	X-ray	Difference
α-oxalic acid dihydrate	O(1)—H(1)	1·026(7)	0·89(2)	0·14(2)
	O(3)—H(2)	0·964(7)	0·84(2)	0·12(2)
	O(3)—H(3)	0·956(9)	0·79(3)	0·17(3)
α-deuterated oxalic acid dihydrate	O(1)—D(1)	1·031(2)	0·86(2)	0·17(2)
	O(3)—D(2)	0·954(2)	0·83(2)	0·12(2)
	O(3)—D(3)	0·954(2)	0·78(2)	0·17(3)
Sym-triazine	C—H	1·045(16)	0·92(4)	0·13(4)
Cyanuric acid	N(1)—H(1)	1·026(2)	0·79(3)	0·24(3)
	N(2)—H(2)	1·037(1)	0·87(1)	0·17(2)

between the observed and calculated values of X-ray intensities is partly illusory. The effects can be illustrated pictorially by plotting Fourier maps of the function $F_X - F$, where F_X is the observed X-ray intensity and F is the calculated value for X-rays when the position and thermal parameters obtained purely from a neutron study are used. The distinction between this procedure and that described earlier for hexamethylenetetramine is that the X-ray data play no part in determining the thermal parameters. Fig. 224 shows such a map in the plane of the molecule of *sym*-triazine (Coppens 1967) and emphasizes that electrons have migrated from the centres of the carbon and nitrogen atoms into the C—N bonds and also into the lone-pair region of the nitrogen atom. The more recent and more accurate work on α-deuterated oxalic acid dihydrate (Coppens *et al.* 1969) shows, in Fig. 225, the concentration of electron density at the centres of the C—C bond and the C—O bonds and both the two lone pairs around the carbonyl oxygen atom and the lone pair behind the oxygen atom in the hydroxyl group. It is important to note that these departures from spherical symmetry are falsely interpreted in the usual X-ray refinement as anisotropic thermal motion. This is indicated in Fig. 226 which shows the difference in magnitude of the ellipsoids of thermal motion for X-rays and neutrons as found by Coppens *et al.* (1969) for α-deuterated oxalic acid dihydrate. Thus, X-rays suggest extra motion of the carbon atoms in the direction of the C—C bond and enhanced movement of the various atoms in the direction of their lone pairs.

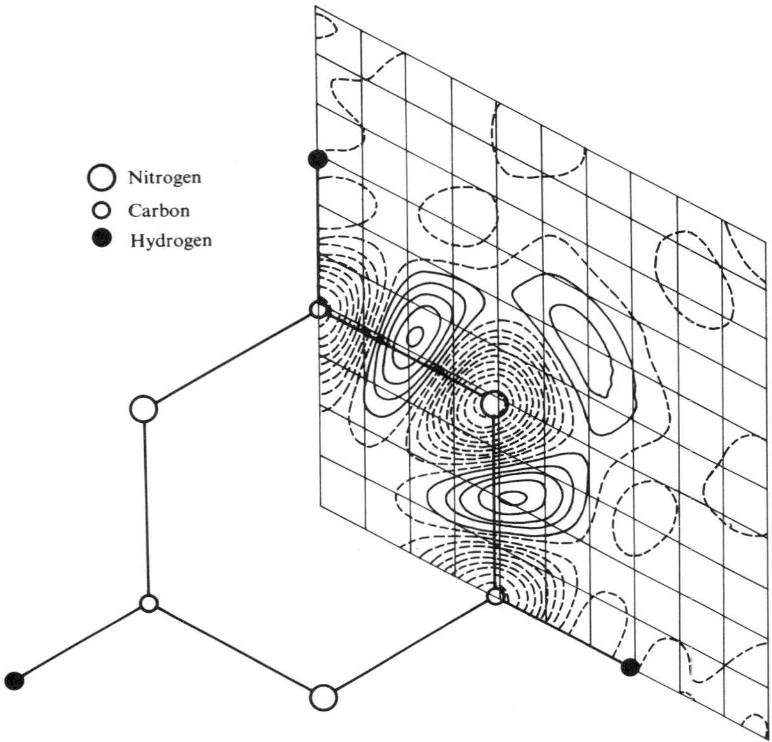

FIG. 224. A difference projection for *sym*-triazine, using the function $F_X - F$, where F_X is the experimentally determined structure factor for X-rays and F is a calculated value for X-rays using both position and thermal parameters determined with neutrons. (Coppens 1970.)

11.4. Biological materials

Up to the present time the structures of very few biological substances have been studied with neutrons, simply because of the difficulty experienced in securing crystals with dimensions as large as 2 mm, which have been needed for reactors of medium flux. It is important to realize that biological materials have very large unit cells, and this means that the diffracted intensities per atom tend to be low. However, the potential advantages of neutron work are very considerable, and it is likely that the operation of the new high-flux reactor at Grenoble will give a big impetus to this kind of work. Apart from the ability of neutrons to locate the

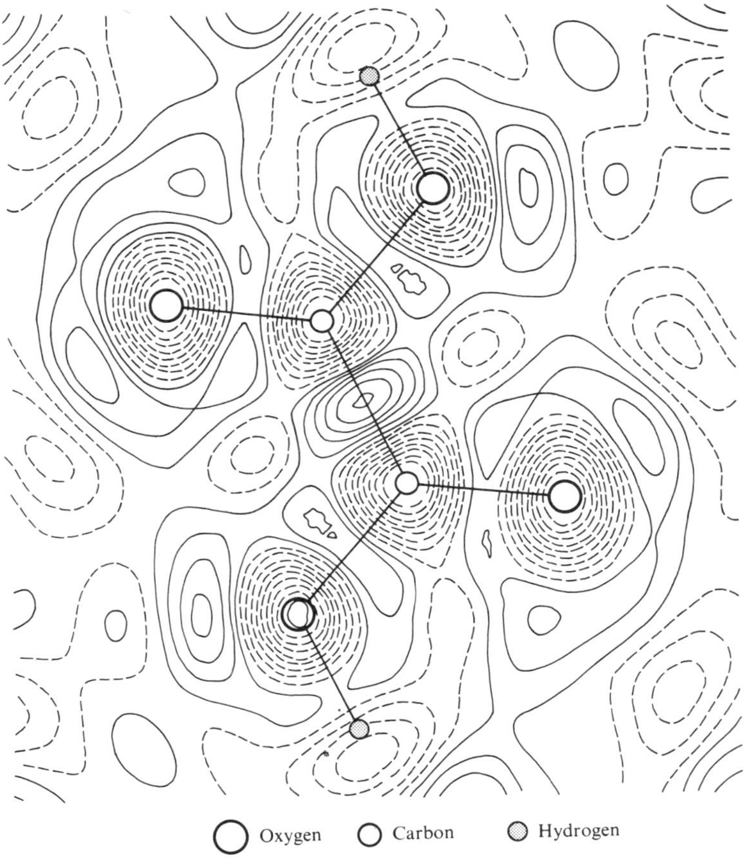

FIG. 225. A difference projection for α-deuterated oxalic acid dihydrate showing concentrations of electron density at the centres of the C—C and C—O bonds and lone pairs near the carbonyl oxygen atoms and the oxygen atoms of the hydroxyl groups. (Coppens *et al.* 1969.)

hydrogen atoms there are other significant advantages which depend on differences of scattering amplitude. For example, it is very much easier to distinguish nitrogen from carbon or oxygen with neutrons than it is with X-rays; the scattering amplitudes for these three atoms are $0{\cdot}94 \times 10^{12}$ cm, $0{\cdot}66 \times 10^{12}$ cm and $0{\cdot}58 \times$ $\times 10^{-12}$ cm respectively. Often, because of the large interplanar distances involved, structural work has to be attempted with limited

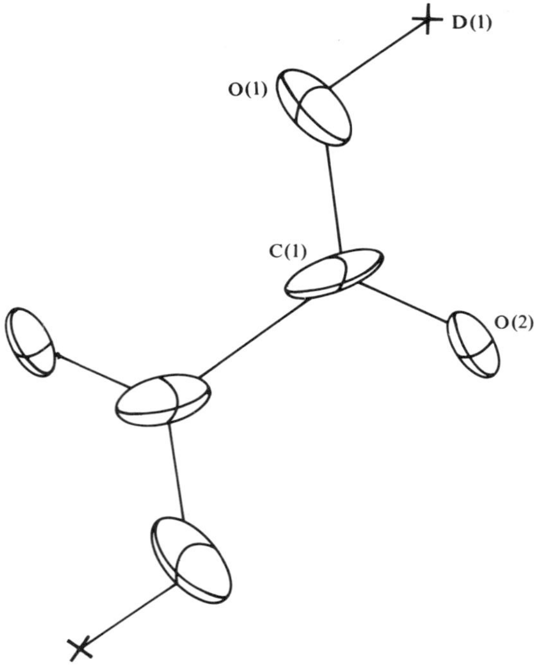

FIG. 226. Ellipsoids showing the difference between the thermal motion of the atoms in α-deuterated oxalic acid dihydrate as assessed by X-rays and neutrons. Comparison with Fig. 225 shows that X-rays falsely suggest extra motion of the carbon atoms along the C—C bond and of the oxygen atoms in the direction of their lone pairs. (Coppens and Sabine 1969.)

resolution to determine the positions of groups of atoms, such as a water molecule, rather than individual atoms. In such cases as water the immense difference in the scattering amplitudes of hydrogen and deuterium is of great value, particularly when individual atoms and groups can be deuterated. Deuteration of a water molecule would produce a very notable change.

The best idea of present progress and future possibilities can be seen in the study of the monocarboxylic acid derivative of vitamin B_{12}, $C_{63}H_{87}O_{15}N_{13}PCo \cdot 16H_2O$, described by Moore, Willis, and Hodgkin (1967) and Hodgkin, Moore, O'Connor, and Willis (1972, unpublished). This has a monoclinic unit cell of dimensions $a = 14{\cdot}9$ Å, $b = 17{\cdot}5$ Å, $c = 16{\cdot}4$ Å, and $\beta = 104°$; the space group is $P2_1$ so that the structure does not contain a centre of symmetry and it is necessary to determine the phase angles for all the spectra. A neutron wavelength of 1·53 Å was used; this is greater than usual and was chosen in order to increase the angular resolution of the diffraction pattern and to make it easier to collect data for the low-angle reflections which occur as a consequence of the large unit cell. For example, an interplanar spacing of 15 Å will give a diffraction peak near $\theta = 3°$, where there will be very heavy background scattering caused by the tail of the incident, undiffracted beam. There are 228 atoms in the asymmetric unit, and the first stage of the investigation was the measurement of about 1500 reflections, going down to an interplanar spacing of 1·3 Å, and analysis of the data to determine 820 parameters. At a later stage the range of interplanar spacing was extended down to 1 Å, requiring the measurement of about 3000 reflections, covering scattering angles up to $2\theta = 100°$. The extended range of reflections gave a substantial improvement in resolution and, using a combination of Fourier and least-squares methods, produced a final discrepancy factor R of 18 per cent. The general accuracy obtained can be appreciated from the Fourier projections in Fig. 227. In Fig. 227 (a) is shown a plot of the neutron scattering density over the corrin nucleus, and it will be noticed immediately that hydrogen atoms are clearly distinguished. It is interesting to note that the central atom of cobalt has only a few contours. For neutrons it is effectively a 'light' atom, whereas for X-rays it would be by far the most prominent atom in the structure. In Fig. 227 (b) is shown the detail of a methyl group CH_3, whose possible rotation about the single bond which links it to the main molecule had been questioned. It is evident from the three sharply defined hydrogen atoms that the group is not rotating. A further detail of the molecule, shown in Fig. 227 (c), represents a group of atoms whose identity remained ambiguous from X-ray work as either $CO.NH_2$ or COOH. The neutron plot, which both shows clearly the separate hydrogen atoms and also distinguishes the nitrogen atom by its enhanced scattering, compared with oxygen, identifies the group conclusively as $CO.NH_2$. A detailed comparison of the Fourier plots obtained in the two sets of measurements shows that limitation of the data to 1·3 Å, rather than 1·0 Å, causes substantially more confusion between the hydrogen atoms and the diffraction ripples from more heavily scattering atoms, which arise as we discussed in Chapter 10.

FIG. 227. Portions of the projection of neutron scattering density of a monocarboxylic acid derivative of Vitamin B_{12}. (a) emphasizes the 'light' cobalt atom at the centre of the corrin nucleus; (b) indicates a methyl group, clearly *not* in rotation; (c) conclusively identifies a group of atoms as $—CH_2CONH_2$ rather than $—CH_2COOH$. (Moore, Willis, and Hodgkin 1967.)

We mentioned in Chapter 5 the application of the study of anomalous neutron scattering to the problem of determining phase angles. Attempts have been made by Hodgkin and Willis to complete the structure of insulin by neutron measurements for a crystal containing ^{149}Sm, which has a resonance at a wavelength of 0·92 Å.

Comparison has been made of (hkl) and ($\bar{h}\bar{k}\bar{l}$) reflections at wavelengths of 1·5 Å, 1·35 Å, and 1·45 Å, but no conclusions have yet been reached because of low intensities and the small proportion of samarium which was introduced. This investigation is expected to benefit enormously from a higher reactor flux now available at Grenoble.

11.5. Thermal motion and bond-length determination

In dealing with the motion of molecules and atomic groupings in substances of the type which we have been discussing in this chapter it is important to distinguish between the equilibrium separation of two atoms and the separation between the centroids of the spaces which, respectively, they occupy under the influence of their thermal energy. When the motions of two atoms are substantially different as, for example, when oscillation of the molecule as a rigid body occurs or when the motion of a particular atom is markedly anisotropic, it is found that the apparent bond length between them has to be corrected in order to obtain the true equilibrium value; there is an apparent shortening of the bond below its equilibrium value.

Consider the atoms A,B in Fig. 228. If their thermal motion is uniform in all directions and quite random then the separation of the centroids of their average positions, i.e. the mean value of the vector S, will be the same as the equilibrium value S_0 of this vector. However, the same result does not apply to the atoms A′ and B′. Here the atom A′ is practically stationary, and atom B′ moves around an arc of a circle centred on A. The centroid G′ of the volume covered by B′ in its motion is closer to A′ than is the equilibrium position G. The displacement is larger than the accuracy with which bond lengths can now be determined by diffraction methods, and it is important to correct the apparent length of the bond for it; for example, in the case of α-resorcinol, which we discussed earlier in this chapter, there is a correction of 0·007 Å to some of the C—C bonds.

The amount of bond-shortening will depend on the type of correlation which exists between the motion of the two atoms concerned—e.g. between the O and H atoms of a hydroxyl group. Unfortunately this correlation cannot be determined directly and will only emerge when a correct model of the motion of the structure as a whole has been devised, and this will be the resultant of rotational

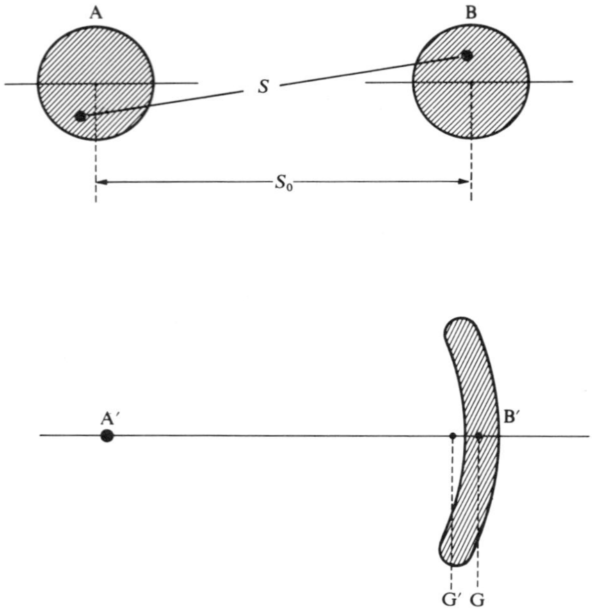

FIG. 228. Reduction of the effective bond length between two atoms when the motion of one of them is markedly anisotropic.

and translational movements of the molecules as rigid bodies together with motion which is due to the bending and stretching of individual bonds. The effect has been discussed at length by Busing and Levy (1964), particularly in relation to their study of calcium hydroxide, $Ca(OH)_2$ (Busing and Levy 1957). If in Fig. 229, S represents the instantaneous separation of two atoms whose equilibrium separation is S_0 then it can be shown that to a first approximation

$$S = S_0 + z + \omega^2/2S_0, \tag{11.1}$$

where z is the instantaneous displacement in a direction parallel to the bond and ω is the radial component, i.e. the relative vector displacement of the two atoms projected on the plane which is normal to the line of their mean positions. When an average is taken over all possible atomic positions we have

$$S = S_0 + \overline{\omega^2}/2S_0, \tag{11.2}$$

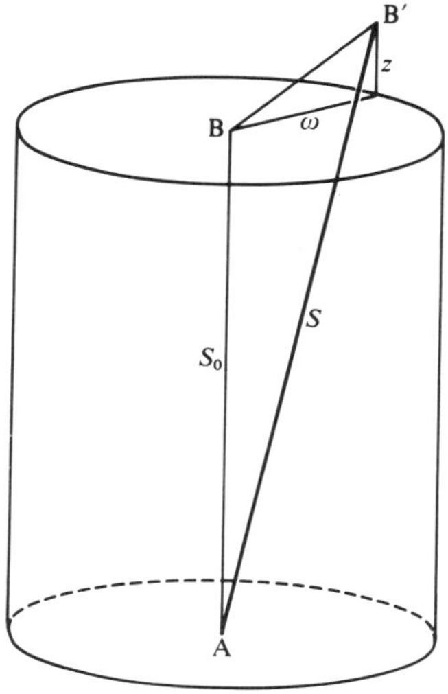

FIG. 229. Calculation of bond-shortening between two atoms for which the instantaneous displacement, relative to the equilibrium value, is a radial component ω and a component z in the direction of the bond.

where $\overline{\omega^2}$ is the average value of ω^2, bearing in mind that the average value of z will be zero.

The smallest value of $\overline{\omega^2}$ occurs when the motions, ω_A and ω_B of the two atoms are correlated parallel in phase with each other. Then

$$\overline{\omega^2} = \{(\overline{\omega_B^2})^{\frac{1}{2}} - (\overline{\omega_A^2})^{\frac{1}{2}}\}^2.$$

At the other extreme, when ω_A and ω_B are correlated antiparallel we have the largest possible value of $\overline{\omega^2}$, then

$$\overline{\omega^2} = \{(\overline{\omega_B^2})^{\frac{1}{2}} + (\overline{\omega_A^2})^{\frac{1}{2}}\}^2.$$

Another possibility is that the two atoms move completely independently of each other, then

$$\overline{\omega^2} = \overline{\omega_B^2} + \overline{\omega_A^2}.$$

Finally, a likely type of motion for such cases as the hydroxyl group is the 'riding motion' in which the light hydrogen atom moves relative to the oxygen atom in a manner which is not dependent on the particular position in which the oxygen atom happens to be. In this case

$$\overline{\omega^2} = \overline{\omega_B^2} - \overline{\omega_A^2},$$

and this situation produces an intermediate degree of bond-shortening.

These various possibilities are summarized in Fig. 230, which also gives the value of the correction in each case for the particular example of $Ca(OH)_2$. The corrections range from 0·016 Å for correlated-parallel motion to 0·125 Å for correlated-antiparallel motion. The corrected values of the bond lengths are also indicated in the Figure.

If it is possible to demonstrate that a molecule moves as a rigid body then the shortening of the bond lengths can be calculated in a straight forward way. In Fig. 231 (a) consider a plane containing the line of length S_0 which is rotating about an axis AB which is perpendicular to the plane. Let rotation take place through a small angle α which equals the root-mean-square angular oscillation, so that at (b) the triangle ACD takes up the new position AC′D′. In (c) the angle between the original and displaced vectors, S_0 and S, will also equal α, and hence the radial component of displacement, which we defined as ω in eqn (11.1), will be equal to αS_0. It follows from eqn (11.2) that the shortening of the bond of length S_0 will be equal to $\frac{1}{2}\alpha^2 S_0$. If S does not lie in the plane of rotation, so that S_0 and AB are inclined at some general angle ϕ instead of at 90°, then it can easily be shown that the shortening of the bond is reduced to

$$\tfrac{1}{2}\alpha^2 S_0 \sin^2 \phi.$$

In a more general case where a body undergoes independent oscillations about three principal axes the three shortening effects are additive and the correction is

$$\tfrac{1}{2}S_0 \sum (\alpha^2 \sin^2 \phi). \tag{11.3}$$

As an example we can calculate the correction for one of the C—C bonds in salicyclic acid which was studied by Bacon and Jude (1973). The analysis of the structure in terms of rigid-body vibrations shows that the molecule executes vibrations about the axes P,Q in Fig. 232 and about a third axis R which is perpendicular to the plane of the diagram, and it was deduced that the root-mean-square angular oscillations were 2·8°, 5·2°, and 4·0° respectively.

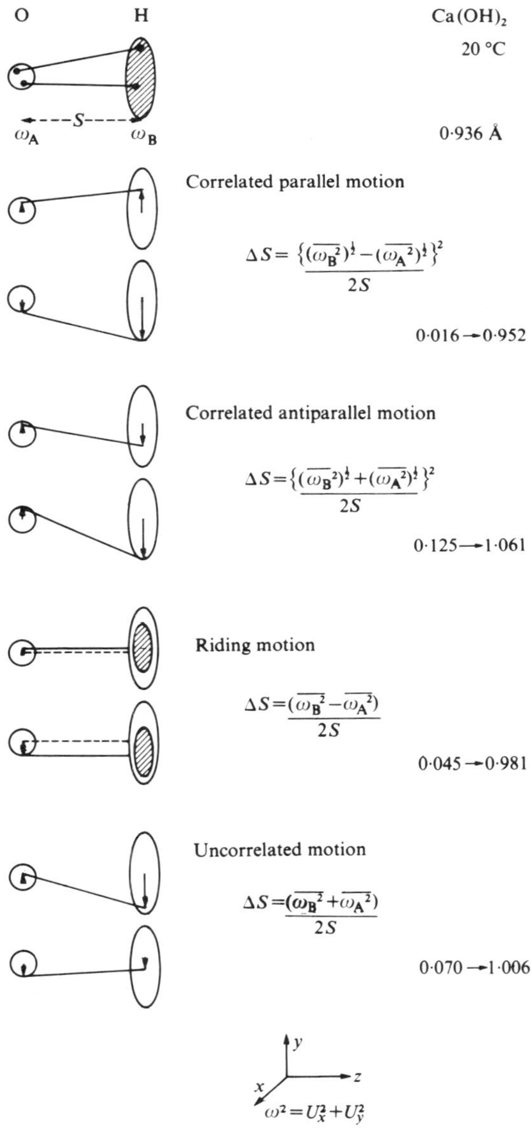

FIG. 230. A summary of possible correlations of the thermal motion for two atoms, such as the O and H atoms in the hydroxyl groups of $Ca(OH)_2$. The numbers on the right of the diagram show the appropriate value of the bond-length correction in each case and the corrected value which would be deduced, using the observed thermal motion data and measured bond length of 0·936 Å for $Ca(OH)_2$.

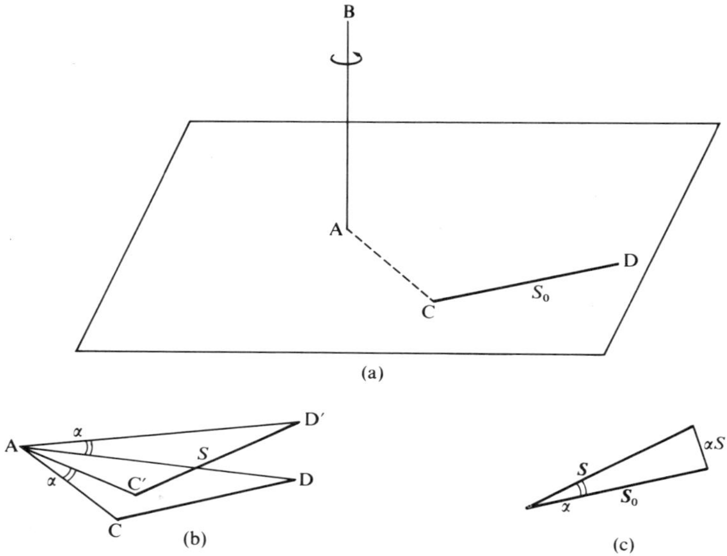

FIG. 231. Calculation of the bond-shortening for a molecule which moves as a rigid body.

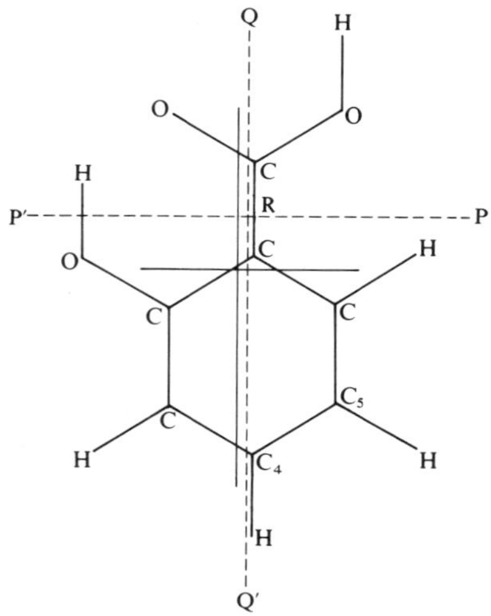

FIG. 232. Diagram of the molecule of salicylic acid $C_6H_4OH.COOH$ showing two of the axes PP′, QQ′ about which the molecule moves approximately as a rigid body. A third axis through R is perpendicular to the plane of the diagram.

The uncorrected length of the bond C(4)—C(5) was 1·408 Å. In Fig. 232 the bond C_4—C_5 makes angles of 30°, 60°, and 90° respectively with the axes P,Q, and R. It follows therefore from eqn (11.3) that the apparent shortening of the bond length will be

$$\tfrac{1}{2}\times 1{\cdot}408((2{\cdot}8^2 \sin^2 30° + 5{\cdot}2^2 \sin^2 60° + 4{\cdot}0^2)/57{\cdot}3^2)\ \text{Å},$$

which equals 0·008 Å. Thus, the corrected length of this bond is 1·416 Å.

11.6. Study of inelastic incoherent scattering

A quite different kind of information about molecular structure comes from studies of inelastic incoherent neutron scattering which we outlined in Chapter 9. This provides a new branch of molecular spectroscopy for which the results are of immense value when viewed alongside the spectra which are conventionally obtained by infrared absorption and Raman methods. Fig. 233 compares the infrared absorption spectrum of $HCo(CO)_4$ with a neutron spectrum (White and Wright 1970). The latter spectrum was obtained by illuminating a sample with a beam of neutrons, consisting of wavelength greater than 4 Å (i.e. energies less than 5 meV), and then making an energy-analysis of the neutrons scattered at a particular angle (90°). The neutrons have gained energy by inelastic scattering in the solid, thus increasing their energy by quanta of molecular vibration energy, so that the experiment is a neutron analogue of a Raman scattering experiment which observes only anti-Stokes lines. The incident neutrons have such low energies that they are not able to *excite* molecular motions. It is evident that the infrared and neutron spectra in Fig. 233 show both similarities and differences, and we shall examine the reasons for these, with the aim of showing the particular contribution which neutron spectroscopy can make to the study of molecular structure.

We shall find that the most important feature of our scattering sample of $HCo(CO)_4$ is the single hydrogen atom which it contains. Incoherent scattering contributes to the pattern and a single hydrogen atom will provide 80 barns, bearing in mind that we are using long-wave incident neutrons for which the hydrogen atom, as we discussed for Fig. 205 on p. 326, is effectively bound. The other atoms in the molecule have a total scattering power of about 45 barns, so that the incoherent scattering of the single hydrogen atom will be dominant. We recall from eqn (9.6) that

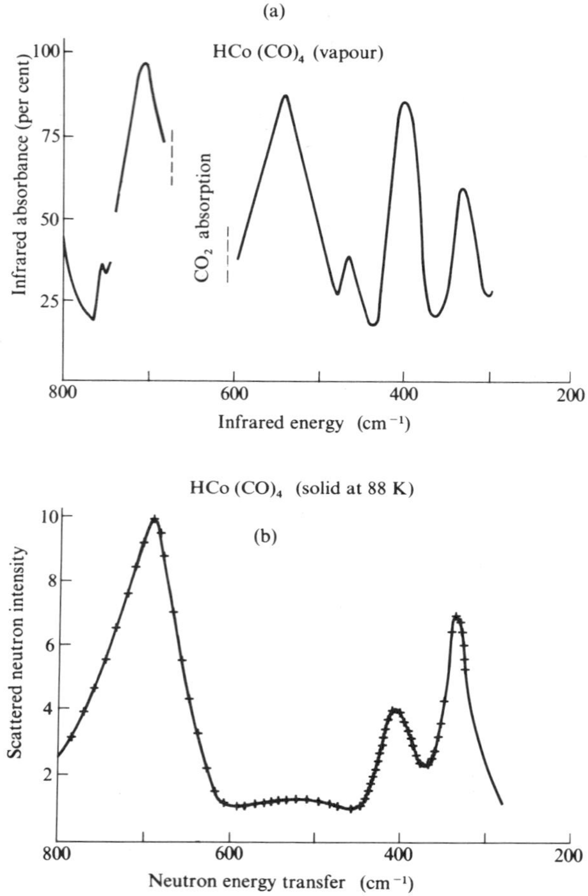

FIG. 233. Comparison of the infrared (a) and incoherent inelastic neutron spectra (b) of $HCo(CO)_4$. Only hydrogen modes contribute to (b). (White and Wright 1970.)

$$\frac{d^2\sigma_{\text{incoh}}}{d\Omega\, dE} =$$

$$\frac{\kappa}{\kappa_0}\sum_\nu (b_\nu)^2_{inc}\frac{Q^2\langle u_\nu^2\rangle}{2M_\nu}\,e^{-2W_\nu}\left\{\frac{1}{\exp(E/kT)-1}+\tfrac{1}{2}(1\pm 1)\right\}N\frac{Z(\omega)}{\omega} \qquad (11.4)$$

where $\hbar\omega$ is the energy transfer which has taken place and $Z(\omega)$ is the density of states in the molecular spectrum for the frequency ω.

The other factors in this expression are κ_0, κ which, as usual, are the wave-numbers of the incident and scattered neutrons, the vector $\mathbf{Q}$, defined by $\mathbf{Q} = \boldsymbol{\kappa} - \boldsymbol{\kappa}_0$ so that the momentum transfer is $\hbar\mathbf{Q}$, and $\langle u_\nu^2 \rangle$ which is the mean-square amplitude of vibration of the atom ν. N is the number of molecules. Inspection of this expression will show that the spectrum of the scattered neutrons, although being dependent on all the modes of vibration in the molecule and on the scattering from all the atoms within the molecule, is preferentially weighted by the following:

(1) the scattering cross-section of each atom, with the result that hydrogen motions are preferentially revealed because of the enormous incoherent scattering of hydrogen,

(2) the mean-square amplitude of vibration $\langle u_\nu^2 \rangle$, which will be particularly large for hydrogen;

(3) the factor $1/M_\nu$, where M_ν is the atomic mass, which again favours hydrogen.

We also note that the occurrence of Q in eqn (11.4), both as the term Q^2 and also within the Debye factor, means that the profile of the neutron spectrum may change markedly with θ.

The neutron spectra are thus heavily weighted to display those molecular vibrations and rotations in which hydrogen atoms take part. This discrimination in favour of the hydrogen atoms is one of the two primary advantages of the neutron technique. The second advantage is to be seen if we recall a predominant feature of the conventional infrared and Raman spectra. Although, as we have pointed out, *all* the molecular motions contribute to the neutron cross-section given by eqn (11.4), this is not the case for the scattering of electromagnetic radiation. Thus, for infrared measurements the only vibrations which can be detected are those which involve a change in the dipole moment of the molecule and in Raman spectra the only motions which can be observed are those which involve a change in the polarizability of the molecule. In this way the so-called 'selection rules' determine which motions are revealed in these optical spectra. In these terms the neutron spectra are quite different because they are free from such selection rules, although it might well be considered that the preference for hydrogen really constitutes a rule of a different kind. We can summarize by saying that the optical and neutron spectra favour different features of the molecule, and it is this factor which is of primary importance and value when the two techniques are used to complement each other.

The example of tetramethylammonium bromide in Fig. 234 (White 1973) illustrates some of these points. The peak at 70 cm^{-1} is very intense in the infrared pattern because it depends on relative motion between the $N(CH_3)_4^+$ and the Br^- ions which produces a very large change in the electric dipole moment; however, the amplitude of this motion is small so there is a relatively small contribution to the neutron pattern. On the other hand, the main neutron peak at 110 cm^{-1} corresponds to torsional vibrations of the $N(CH_3)_4$ ion which cause large displacements of the hydrogen atoms but give little infrared intensity.

An early example of neutron spectroscopy shown in Fig. 235 gives the results of Boutin *et al.* (1963) for KHF_2 and solid HF. The former shows three particular features. The sharp peak at a neutron energy gain of 147 meV corresponds to the bending motion of the F—H—F ion in which the two fluorine atoms move in phase but the hydrogen moves in the opposite direction, as in diagram A of the inset to the Figure. The very small peak at 75 meV corresponds to the stretching motion of the F—H—F ion; the two fluorine atoms move in opposite directions along the axis of the molecule, as shown at B, while the hydrogen atom at the centre remains at rest, so that there is little contribution to the neutron spectrum. The broad peak at 13 meV (100 cm^{-1}) is mainly due to a librational motion (C) of the HF_2 ion, which involves substantial hydrogen motion. The Raman spectra in fact shows a further line at 90 cm^{-1} due to motion of the potassium ion, but the neutron spectrum is not sensitive to this. In comparing the spectra of KHF_2 and HF in Fig. 235 the positions of the peaks of highest energy are noteworthy. For KHF_2 the energy gain of the neutron is 147 meV in contrast with only 67 meV for HF. This is because the hydrogen bond is much shorter (and stronger) in the former, being 2·26 Å in KHF_2 and 2·4 Å in solid HF, so that the frequency of the bending motion is higher.

Fig. 236 compares some results for solid hydrogen at 12 K and liquid hydrogen at 15 K, from the work of Egelstaff, Haywood, and Webb (1967). Apart from the intense peak, corresponding to elastic scattering, the main feature of the spectrum of the solid is a sharp peak at an energy transfer of 15 meV (120 cm^{-1}). The width of this peak is not very dependent on the angle of scattering

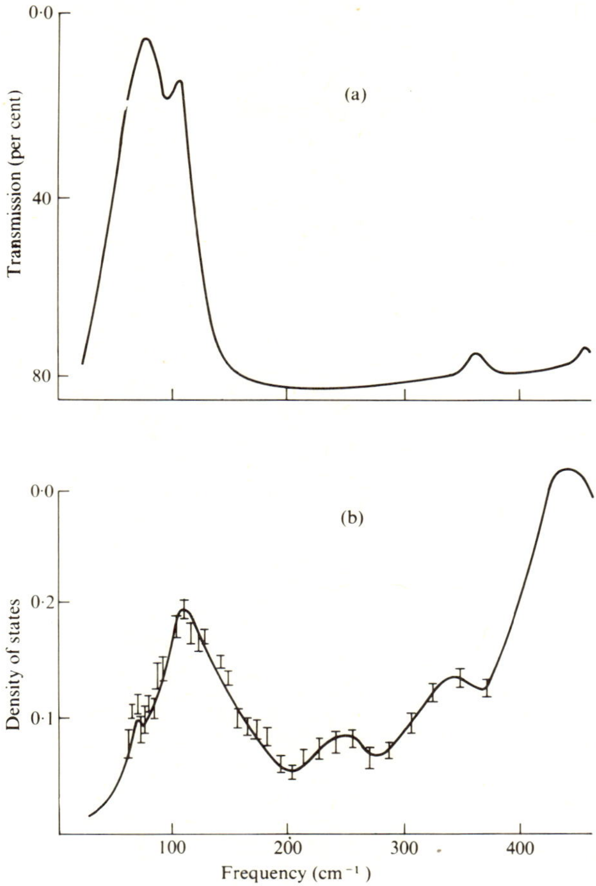

FIG. 234. Comparison of infrared (a) and neutron spectra (b) for $N(CH_3)_4Br$, contrasting the proportions of the peaks at 70 cm^{-1} and 110 cm^{-1}. (White 1973.)

and it corresponds to a loss of one quantum of rotational energy by the hydrogen, giving a change from $J = 1$ to $J = 0$. This transition is not observable by either infrared or Raman spectroscopy, being forbidden by the selection rules in each case. It will be noticed that rotational peak is still visible in the liquid, indicating that the rotations of the molecules are not quenched significantly by the intermolecular collisions which occur in the liquid. On the other hand, the elastic peak is weaker in the liquid, and it becomes appreciably weaker and broader as the angle of scattering increases. It is in fact no more

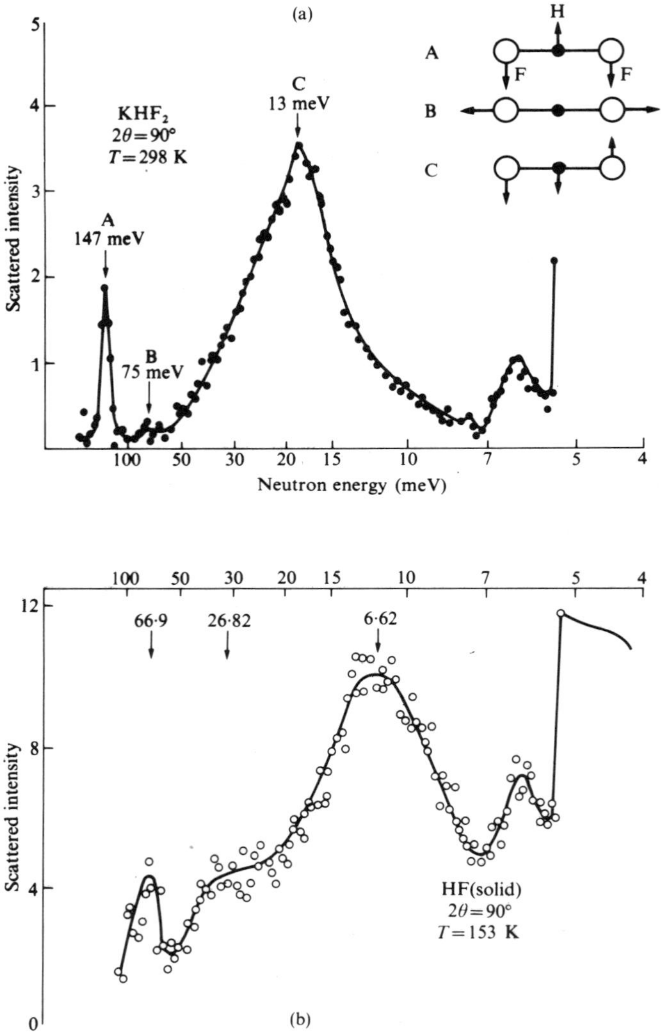

FIG. 235. A comparison of the spectra of inelastically scattered neutrons from (a) KHF_2 and (b) solid HF. The higher energy peak of 147 meV in the former, compared with 67 meV, is due to the shorter and stronger hydrogen bond. The diagrams at the right indicate the nature of the motion of the H and F atoms in KHF_2 which give rise to the peaks correspondingly labelled A, B, and C. (After Boutin *et al.* 1963.)

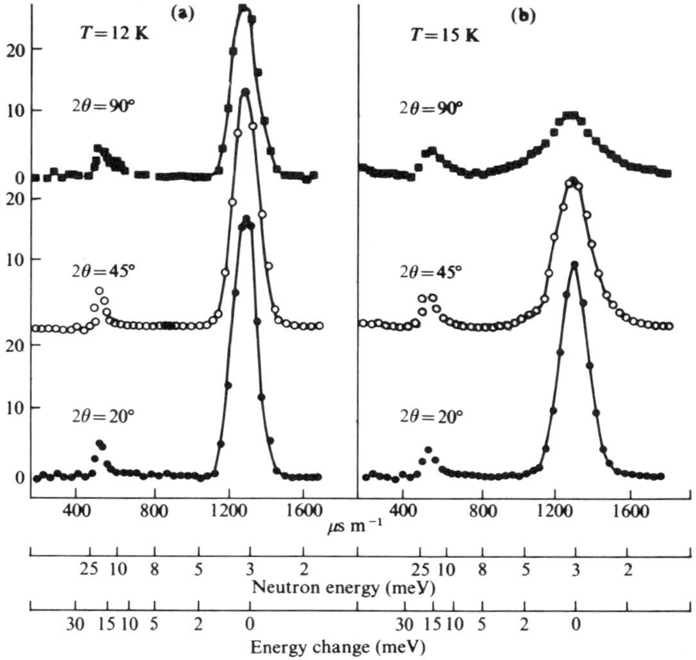

FIG. 236. The spectra of scattered neutrons from (a) solid hydrogen at 12 K and (b) liquid hydrogen at 15 K, measured in turn at scattering angles of 90°, 45°, and 20°. The units on the abscissae indicate the time of flight, the neutron energy and the *change* of neutron energy. (Egelstaff *et al.* 1967.)

than 'quasi-elastic', because the translational movements of the molecules give rise to Doppler shifts of the neutron's velocity, and it is possible to determine the diffusion of the molecules in the liquid by studying the variation of the width and intensity of this peak.

It will be evident that in all spectroscopic work one of the main problems is the identification of individual peaks with particular molecular motions. In neutron spectroscopy this task may be considerably assisted by making isotopic substitutions, particularly of deuterium for hydrogen to take account of the fact that deuterium, unlike hydrogen itself, does not possess a large incoherent scattering cross-section. We can illustrate this by examining the measurements of Aldred *et al.* (1967) on liquid samples of CH_3OH, CD_3OH, and CH_3OD. Three sets of spectra, for scattering angles of 20° and 90°, are shown in Fig. 237 and consist of the quasi-elastic peak and a much broader peak having a maximum near to a neutron energy of 20 meV. The broad peak is noticeably reduced in relative intensity, particularly at higher angles of scattering, when deuterium is substituted for hydrogen in the methyl group, but there is much less change when the deuterium is inserted in the hydroxyl group. It is deduced therefore that the energy transfer of 20 meV is associated with the methyl group.

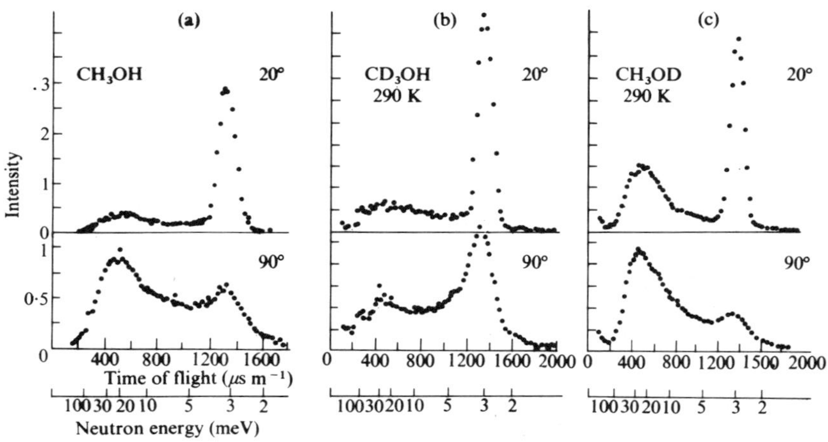

FIG. 237. Neutron inelastic scattering spectra for incident neutrons of wavelength 5·3 Å (0·003 eV). The spectra are measured at scattering angles of 20° and 90° for (a) CH_3OH (b) CD_3OH, and (c) CH_3OD at 290 K. (Aldred *et al.* 1967.)

Another good example of the value of deuterium substitution is provided by the comparison of the spectra of $CsHCl_2$ and $CsDCl_2$ in Fig. 238 which comes from the work of Stirling, Ludman, and Waddington (1970). The neutron data clarified some unexpected features of the infrared spectra. The peak at 675 cm^{-1} for $CsHCl_2$ is much reduced in intensity and shifted in frequency for $CsDCl_2$ and is assigned to a bending mode. The symmetrical stretching vibration at 199 cm^{-1} is almost absent in the deuterated compound,

FIG. 238. Comparison of the inelastic neutron scattering spectra of $CsHCl_2$ and $CsDCl_2$ at 293 K and an angle of scattering of 74°. The abscissae are shown in turn as (1) reciprocal of neutron velocity in μs m^{-1}, (2) neutron energy gain in meV, and (3) energy gain in cm^{-1}. (Stirling *et al.* 1970.) (See opposite.)

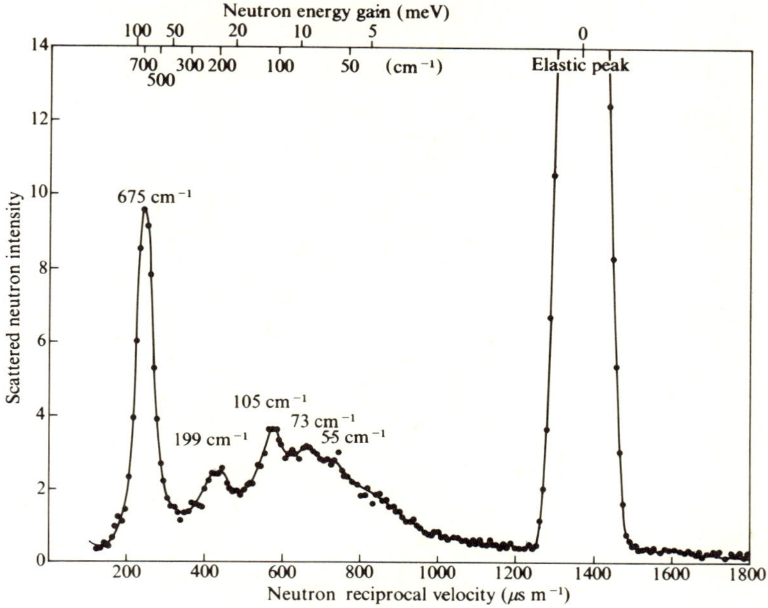

Neutron energy gain (meV)
100 50 20 10 5 0
700 300 200 100 50 (cm⁻¹)
500
Elastic peak
675 cm⁻¹
199 cm⁻¹
105 cm⁻¹
73 cm⁻¹
55 cm⁻¹
Scattered neutron intensity
0 2 4 6 8 10 12 14
200 400 600 800 1000 1200 1400 1600 1800
Neutron reciprocal velocity (μs m⁻¹)

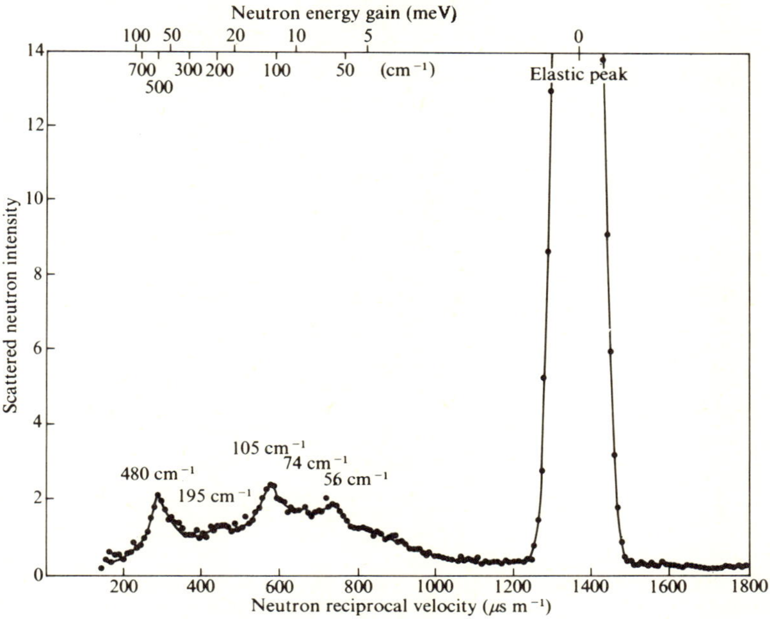

Neutron energy gain (meV)
100 50 20 10 5 0
700 300 200 100 50 (cm⁻¹)
500
Elastic peak
480 cm⁻¹
195 cm⁻¹
105 cm⁻¹
74 cm⁻¹
56 cm⁻¹
Scattered neutron intensity
0 2 4 6 8 10 12 14
200 400 600 800 1000 1200 1400 1600 1800
Neutron reciprocal velocity (μs m⁻¹)

and its occurrence in $CsHCl_2$ must be due to asymmetry of the HCl_2^- ion which permits motion of the hydrogen atom.

In some cases where the substitution of deuterium for hydrogen is difficult it is possible to replace the hydrogen by fluorine, which has a small scattering cross-section. If the two compounds concerned have the same crystal structure then it may be possible to draw conclusions which will identify the peaks in the inelastic spectrum. An example is acetic acid, CH_3COOH, which shows a broad peak

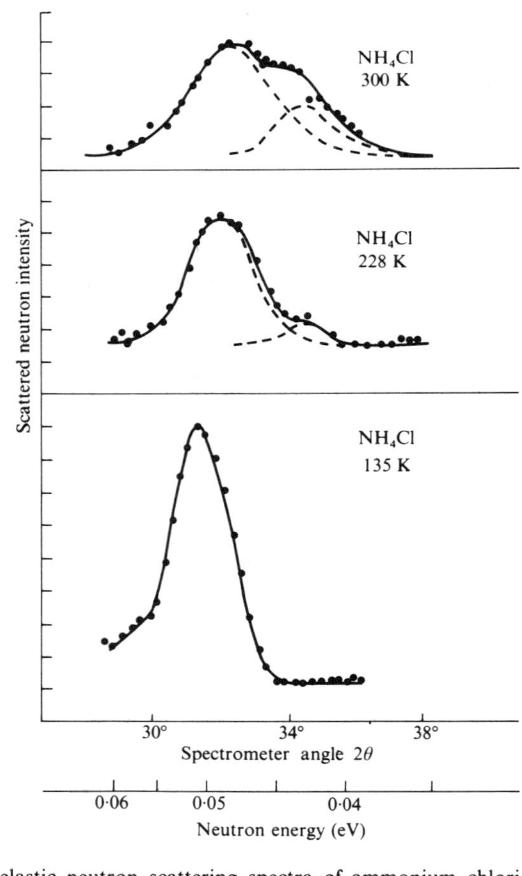

FIG. 239. Inelastic neutron scattering spectra of ammonium chloride, indicating torsional oscillations of the ammonium ion. (Venkataraman *et al.* 1964.)

similar to that which we showed for methanol in Fig. 237. It is found that this peak is completely absent in the compound CF_3COOH, so that it is evident that the peak is again associated with the methyl group.

Several applications of neutron spectroscopy methods have been made in the study of rotations of ions in salts. Brajovic, Boutin, Safford, and Palersky (1963) studied $(NH_4)_2S_2O_8$ at 297 K and 347 K, and from a comparison of the spectra inferred that free rotation of the ammonium ions was taking place at the higher temperature. A series of measurements of ammonium chloride has been made by Venkataraman, Usha Deniz, Iyengar, Vijayaraghavan, and Roy (1964), who have observed hindered rotations of the ammonium ions which are not directly observable by optical spectroscopy. Spectra recorded at 135 K, and 228 K, and 300 K are illustrated in Fig. 239 and are noteworthy for the split peak, with a substantial second component at 300 K which becomes much reduced in magnitude at 228 K and is not observable at the lowest temperature. The two components of the peak are associated respectively with transitions from the ground state to the first excited state and from the first to the second excited states of the torsional oscillator. At the lowest temperature the first excited state is no longer adequately populated, so that no transition upwards from it can be observed. If the torsional motion was harmonic the two peaks would have the same frequency, so that the splitting of the peak provides evidence of anharmonicity.

12

STRUCTURAL INVESTIGATIONS INVOLVING DISTINCTION BETWEEN ATOMS OF NEIGHBOURING ATOMIC NUMBER

THE scattering amplitude of atoms for X-rays increases regularly with atomic number as we have already pointed out. On the basis of X-ray diffraction data alone therefore it is not possible by normal methods to distinguish between neighbouring elements such as iron and cobalt when these occur together in a compound. With neutrons, however, the scattering amplitudes of such elements may be quite different, as consideration of Table 2 (p. 38) shows, and in these cases a distinction can be made between them from the neutron intensity data. By supplementing the structural knowledge already available from X-ray studies interesting information has been obtained in this way for a number of materials.

12.1 Alloys of the transition elements

Many alloy systems contain substitutional solid solutions in which the two constituents are distributed at random among the available atomic sites in the structure. In certain systems suitable thermal treatment may cause the development of a 'superlattice' in which the two types of atom take up an ordered arrangement with respect to each other. Fig. 240 illustrates the behaviour found in many body-centred cubic alloys of the composition AB. In the disordered phase the structure is that shown in Fig. 240 (a) with the two types of atom distributed at random over the corners and body-centres of the unit cells. The structure factor for the (100) reflection for the unit cell will be $b_{\mathrm{I}}-b_{\mathrm{II}}$, where b_{I}, b_{II} are the atomic scattering factors for the corner and centre positions, and in the case of the random arrangement

$$b_{\mathrm{I}} = b_{\mathrm{II}} = \tfrac{1}{2}(b_{\mathrm{A}}+b_{\mathrm{B}})$$

and the intensity of the (100) reflection is zero. For the ordered arrangement shown in Fig. 240 (b) the structure factor will be

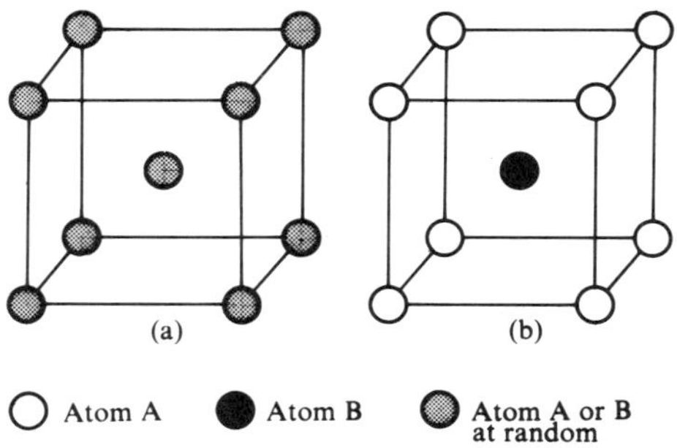

FIG. 240. Disordered (a) and ordered (b) arrangements of body-centred cubic structure of composition AB.

$b_A - b_B$, and this will have a finite value unless $b_A = b_B$. Thus the appearance of order will produce extra lines in the diffraction patterns for X-rays or neutrons and these will be detected the more readily if the values of the scattering amplitudes of A and B are appreciably different. With X-rays the superlattice lines are easily observed in alloys such as FeAl, in which the iron and aluminium atoms have quite different scattering powers. Similarly the appropriate superlattice lines which develop when ordering of a face-centred structure such as Cu_3Au occurs can also be observed. A survey of studies of this sort has been given by Nix and Shockley (1938) and by Barrett (1943).

In the case of alloys of the transition elements such as FeCo and CuZn the scattering amplitudes of the two components for X-rays are so nearly equal that no superlattice lines can be detected, although the likelihood that ordering takes place below a certain temperature may be inferred (see Sykes 1935) from discontinuities in the specific heat curve. In both of these unfavourable cases the presence of ordering has subsequently been established (Jones and Sykes 1937; Ellis and Greiner 1941) with X-rays by special choice of wavelength, taking advantage of the anomalous scattering in the neighbourhood of an absorption edge. Using neutrons, however, the detection of the superstructure in these cases is quite straightforward, for the scattering amplitudes of the A, B atoms are quite different. Comparison of the scattering amplitudes of some metals for X-rays and neutrons is given in Table 25, the X-ray amplitudes being quoted for $(\sin\theta)/\lambda = 0{\cdot}3 \times 10^8\ \mathrm{cm}^{-1}$. Shull and Siegel (1949)

TABLE 25

X-ray and neutron scattering amplitudes for some metals

Atom or Isotope	X-rays $(\sin\theta)/\lambda = 0{\cdot}3$ Å^{-1}	Neutrons
Mn	$4{\cdot}2\times10^{-12}$ cm	$-0{\cdot}39\times10^{-12}$ cm
Fe	4·4	0·95
^{54}Fe	4·4	0·42
^{56}Fe	4·4	1·01
^{57}Fe	4·4	0·23
Co	4·6	0·25
Ni	4·8	1·03
^{58}Ni	—	1·44
^{60}Ni	—	0·28
^{62}Ni	—	−0·87
Cu	5·1	0·76
^{65}Cu	—	1·11
Zn	5·3	0·57
Au	16·1	0·76

have demonstrated the ordering which takes place in FeCo when the material is very slowly cooled from 1023 K, and Fig. 241 is taken from their results. In addition to the appearance of the superlattice lines (100), (111), and (210) there is evidence of a reduction of background scattering, by 20 per cent, for the ordered sample. This is in reasonable agreement with the decrease to be expected.

A further typical example is the alloy system Ni_3Mn, in which the order–disorder transformation has been studied by Marcinkowski and Brown (1961). This is a particularly attractive system to study with neutrons because the amplitude of the superlattice lines is proportional to $(b_{Mn}-b_{Ni})$ and is therefore enhanced by the negative scattering amplitude of manganese. By comparison of the intensities of the fundamental and superlattice reflections it is possible to determine the long-range order parameter S for the alloys. Fig. 242 indicates how S varies with temperature. The measurements were made with samples for which a near-equilibrium state was attained at each temperature, by slow cooling for several days followed by quenching. The direct effect of changes in order on the diffraction pattern is shown in Fig. 243. Here patterns are shown for a sample which was kept at 698 K for, in turn, periods of 0·4 hours, 82 hours, and 303 hours and then quenched. With increase of time the growth of the (110) reflection indicates the

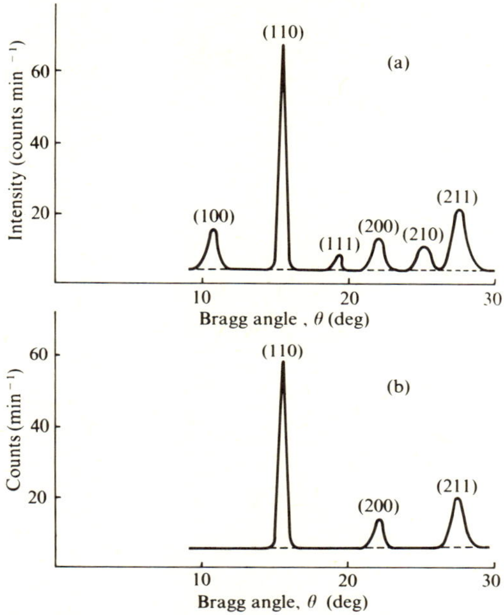

FIG. 241. Neutron diffraction patterns of ordered (a) and disordered (b) samples of FeCo. (Shull and Siegel 1949.)

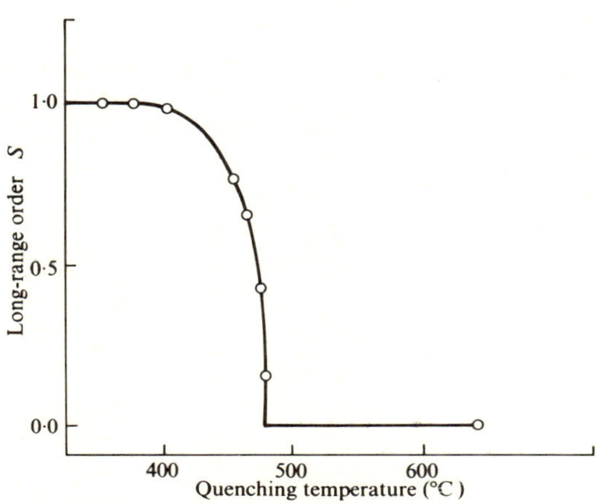

FIG. 242. The change of the long-range order parameter of Ni_3Mn with temperature. The samples were cooled very slowly for several days, to obtain a near-equilibrium state at each temperature, and then quenched. (Marcinkowski and Brown 1961.)

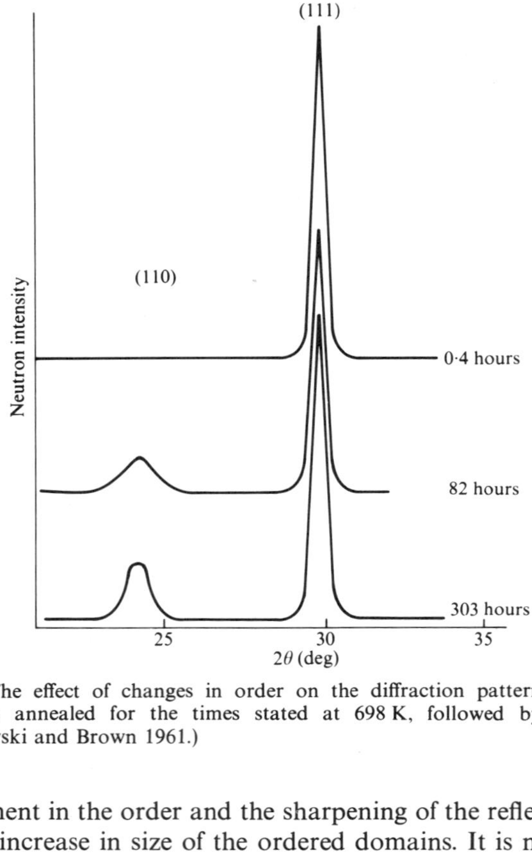

FIG. 243. The effect of changes in order on the diffraction pattern of Ni_3Mn, for samples annealed for the times stated at 698 K, followed by quenching. (Marcinkowski and Brown 1961.)

improvement in the order and the sharpening of the reflection indicates the increase in size of the ordered domains. It is noteworthy that after annealing for 303 hours the long-range order parameter has only reached 0·71, in contrast to the value of 0·96 at the same temperature of 698 K for the much longer anneal in Fig. 242.

In principle it is possible to study superstructures in certain alloy systems particularly accurately by suitable choice of isotopes. Table 25 indicates the range of variation which is possible among the isotopes of iron and nickel. An alloy of $Mn^{60}Ni$, for example, would show normal diffraction lines which were very weak in comparison with the intensities of lines from the superlattice, since Mn and ^{60}Ni have scattering amplitudes for neutrons which are roughly equal in magnitude but of opposite sign. Another good example of the use of suitable isotopes is provided by Walker and

Keating's (1963) study of short-range order in β-CuZn. This is an intractable problem with X-rays, bearing in mind that Cu and Zn are neighbouring elements. For neutrons the prospects are much better because the scattering lengths are $0{\cdot}76 \times 10^{-12}$ cm and $0{\cdot}57 \times 10^{-12}$ cm respectively. However, the experimental problem is the measurement of broad diffuse lines of low intensity, and this is immensely simplified by using the copper isotope ^{65}Cu, which has a scattering length of $1{\cdot}11 \times 10^{-12}$ cm. The intensities of the important lines are proportional to the quantity $(b_{Cu} - b_{Zn})^2$, and this quantity increases by a factor of about 8 times when ^{65}Cu is used in place of ordinary copper. The investigation described shows that the order–disorder transformation in β-CuZn at about 733 K shows the general characteristics of critical scattering of neutrons such as we referred to, in the magnetic case, in Chapter 9. At room temperature this alloy has an ordered structure of the CsCl-type, but as the temperature increases the long-range order decreases, becoming zero at the critical temperature T_C. The diffuse scattering, which is shown in Fig. 244 in plots along the [100] axis near the (100) reciprocal lattice point at several temperatures above T_C, is quite sharp, thus indicating that the short-range order exists over substantial distances. Indeed quantitative analysis shows that even

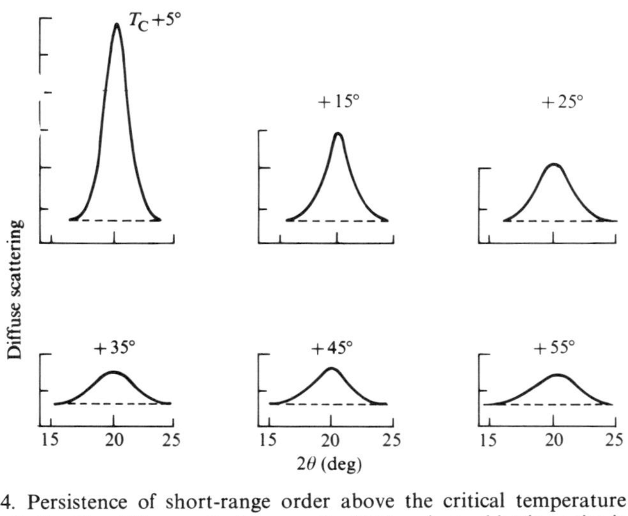

FIG. 244. Persistence of short-range order above the critical temperature T_C for β-brass, CuZn. The diffuse scattering near the (100) reciprocal lattice point is plotted along the y-axis for various temperatures above T_C. (Walker and Keating 1963.)

at 75 K above T_C the short-range order is significant out to at least the 13th neighbours. Accordingly there are too many order parameters for them to be determined uniquely from the observed data, but it is possible to show good agreement with a model based upon theoretical predictions for a second-order transformation. The accuracy with which this can be done depends very substantially on the higher intensities which were achieved by using the ^{65}Cu isotope.

In contrast to the advantage of neutrons over X-rays in studying alloys such as FeCo, Ni_3Mn, and CuZn, they are ineffective in such a classic X-ray case as the alloy Cu_3Au, for it happens that Cu and Au both have the same neutron scattering amplitude of $0{\cdot}76 \times \times 10^{-12}$ cm. Indeed Shull and Siegel (1949) examined a sample of Cu_3Au, which was known by X-ray measurement to be ordered, but were unable to find any evidence of this with neutrons.

In general we can say that for a given ratio of the scattering amplitudes of the A and B atoms it is more difficult to detect superlattice lines with neutrons than with X-rays, because of the inferior angular resolution of the former and the high discrimination which photographic recording provides for X-rays. This is another example where the continuing improvement in angular resolution brought about by increased neutron fluxes is of much importance. In principle the neutron data are more powerful than the X-ray measurements when the *degree* of ordering in alloys is being determined, since it is not necessary to assume knowledge of any angular variation of atomic scattering factor. This is a limitation with X-rays, particularly when it is necessary to incorporate corrections which depend on theoretical assumptions concerning the magnitude of K-electron dispersion.

12.2 Spinel structures

A study of magnesium aluminium oxide, $MgAl_2O_4$, by Bacon (1952), involving a distinction between the atomic positions of magnesium and aluminium atoms, provides a further example of the ability of neutrons to differentiate in certain cases between atoms of closely similar atomic number.

Many metal oxides of the type XY_2O_4 have the spinel structure which was first investigated with X-rays by W. H. Bragg (1915) and by Nishikawa (1915). This structure consists of a close-packed arrangement of oxygen ions with two types of interstices for the cations. In the unit cell which contains eight molecules, there are eight so-called A sites which are tetrahedrally coordinated by oxygen atoms and sixteen B sites which are octahedrally coordinated. In the simplest arrangement, or 'normal' spinel structure, the eight X atoms occupy the A sites with the sixteen Y atoms in the B sites. It was shown subsequently by Barth and Posnjak (1932) that an 'inverse' structure was also permissible, in which the A sites are occupied by eight of the Y atoms, with the X-atoms and eight remaining Y atoms distributed at random amongst the B sites. It was demonstrated by these authors that it was possible to distinguish between these two structures by measurement of the X-ray diffraction intensities in the case of spinel-type structures such as $MgFe_2O_4$ in which the metal atoms have appreciably different scattering amplitudes. In the case of spinel $MgAl_2O_4$, itself, however, no distinction can be made, for the scattering amplitudes of the Mg^{2+} and Al^{3+} ions are practically identical. For neutrons there is sufficient difference between the two values, which are $0{\cdot}52 \times \times 10^{-12}$ cm and $0{\cdot}35 \times 10^{-12}$ cm respectively, to differentiate between the 'normal' and 'inverse' structures.

Fig. 245 shows the experimental neutron diffraction intensities of the significant reflections alongside curves calculated theoretically for the 'normal' and 'inverse' arrangements in terms of the parameter u which describes the oxygen positions. For the actual positions in terms of u the reader is referred to Wyckoff (1931), it being remarked that in an ideal structure u would be equal to $\frac{3}{8}$. From the results shown in the Figure it is inferred that the structure is the 'normal' one and that $u = 0{\cdot}387$ with a probable error of little more than 0·001. The 'normal' and 'inverse' arrangements are, as pointed out by Verwey and Heilmann (1947), only the two extremes of a continuous range of cationic distributions which satisfy the spinel symmetry. Detailed consideration of the neutron diffraction intensities for $MgAl_2O_4$ suggests that the 'normal' structure may not be completely achieved and there is possibly a small movement towards the 'inverse' form, although the accuracy of intensity measurement did not permit a definite conclusion to be drawn on this point.

Distinction between the 'normal' and 'inverse' structures for many ferrites, which possess the spinel type of structure, has been made by X-rays, leading to the conclusion that ferromagnetism in ferrites is associated with the 'inverse' arrangement. In many cases,

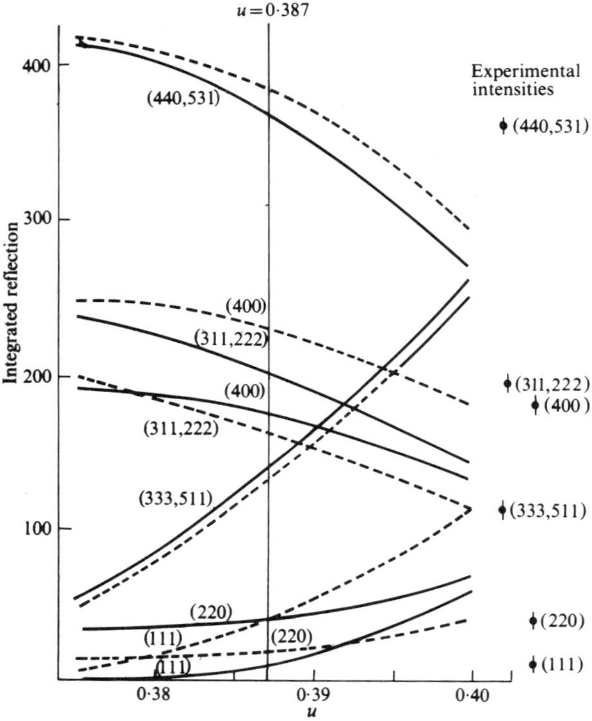

FIG. 245. $MgAl_2O_4$: variation of intensities with oxygen parameter u. ——— calculated intensities for 'normal' structure. - - - - calculated intensities for 'inverse' structure.

such as $MnFe_2O_4$, Fe_3O_4, $CoFe_2O_4$, $NiFe_2O_4$, and $ZnFe_2O_4$, X-ray measurements are inconclusive on account of the similarity of scattering amplitude between the divalent metal ions and the ferric ions, particular as precise interpretation of the diffraction intensities is made more difficult by the fact that the oxygen parameter has to be determined also. $ZnFe_2O_4$ was readily shown by Hastings and Corliss (1953) to have the normal structure, the distinction being simplified by the substantial differences between the neutron scattering lengths for iron and zinc which were listed in Table 25. In many cases, such as magnetite, Fe_3O_4, the deduction of the type of structure may be assisted by determination of the contributions to the diffraction pattern from coherent magnetic reflections and some further examples of this will be mentioned in

Chapter 14. In the case of $ZnFe_2O_4$, though, the material is paramagnetic and there are no coherent magnetic reflections so that the distinction between the normal and inverse structures is made from nuclear data alone.

An investigation of magnesium ferrite aluminate powders by Bacon and Roberts (1953) illustrates in two respects the application of neutron diffraction methods to ferrite studies, apart from deductions based on magnetic scattering. A distinction is drawn between the positions occupied by magnesium and aluminium atoms and it is also demonstrated that the oxygen parameter *u* can be determined more accurately by neutrons than X-rays. This latter advantage follows from the fact that the scattering amplitude of oxygen for neutrons is greater than that of magnesium and aluminium and is as much as 60 per cent of the value for iron, in contrast to its inferiority to that of the metals in the case of X-rays.

One of the complications in the interpretation of the diffraction patterns of ferrites is the large number of parameters which have to be determined. Quite apart from the orientation and magnitude of the magnetic moments it is always necessary to consider partial degrees of inversion as well as the departure of the oxygen parameter *u* from its ideal value. This is emphasized in the study of $MgFe_2O_4$ by Corliss, Hastings, and Brockman (1953), who show that the structure factors for various reflections are strongly dependent on either the degree of inversion or the value of *u*, but not on both. They also show that the sum $(F_{311}^2+F_{222}^2)$ is almost independent of the choice of either of these variables. For a number of reflections the ratio

$$F_{hkl}^2/(F_{311}^2+F_{222}^2)$$

is calculated as a function of *u* for various degrees of inversion and from the experimentally observed value of this intensity ratio it is then possible to establish a plot of *u* versus inversion ratio. Such a series of plots for four reflections is shown in Fig. 246, which makes clear how accurately both quantities can be determined as the intersection point of the four lines in the diagram.

A similar method of analysis has been used for the nuclear intensity data of $MnFe_2O_4$ for which Hastings and Corliss (1956*b*) deduced that $u = 0{\cdot}3846 \pm 0{\cdot}0003$ and that the fraction of A sites occupied by Mn^{2+} ions is $0{\cdot}81 \pm 0{\cdot}03$. Thus this ferrite has very nearly the 'normal' spinel structure.

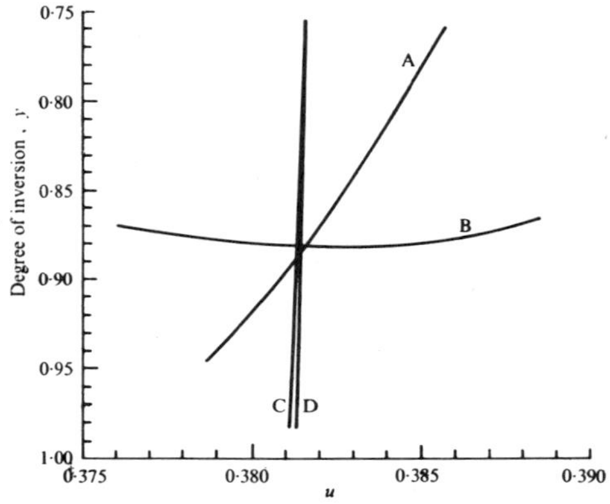

FIG. 246. The determination of the degree of inversion y and the oxygen parameter u for $MgFe_2O_4$. Curve A shows a calculated variation of y with u for the observed ratio of $F^2_{220}/(F^2_{311}+F^3_{222})$; curve B is for $F^2_{400}/(F^2_{311}+F^2_{222})$; curve C is for $(F^2_{422}+F^2_{511}+F^2_{333})/(F^2_{311}+F^2_{222})$ and curve D is for $(F^2_{511}+F^2_{333})/F^2_{311}$. The near-intersection point of these four curves determines the parameters y and u very accurately. (Corliss, Hastings, and Brockman 1953.)

Powder-analysis refinement

In the early years of neutron diffraction studies a great many investigations of the ordering and magnetic structures of ferrites were made and it is instructive to consider how such studies could be carried out at the present time, using the neutron beams of higher intensity which are now available. A characteristic of those studies of ferrite structures is that the interpretation is made in terms of a number of chosen parameters, and this work would now seem to be an admirable subject for the powder technique of profile refinement which we discussed in Chapter 4. This technique would seem equally applicable to other problems in which both the ordering and the precise positional coordinates of atoms of closely similar scattering power are under discussion. As an example we mention the β-sialons denoted by $Si_{6-0\cdot75x}Al_{0\cdot67x}O_xN_{8-x}$, where x varies from zero to about 6 and the α, β forms of silicon nitride, a series of ceramics which have been much studied and discussed by Jack and his colleagues (see e.g. Wild, Grieveson, and Jack 1972) and others. The α, β forms of Si_3N_4 are hexagonal structures containing four and two molecules per unit cell respectively and requiring the determination of 18 and 9 positional parameters in the two cases.

It is possible that a few per cent of the nitrogen atoms are replaced by oxygen, and therefore neutrons, for which the respective scattering lengths are $0{\cdot}94 \times 10^{-12}$ cm, $0{\cdot}58 \times 10^{-12}$ cm, are much more favourably placed to distinguish between those two atoms than are X-rays. Nevertheless, the intensity differentials are very small and of the same order (1 or 2 per cent) as the statistical accuracies in the counting rates, or inaccuracies caused by the $\lambda/2$, $\lambda/3$ contamination in the incident neutron beams and of errors due to preferred orientation of crystallites in the samples. Some preliminary neutron measurements have been made by Bacon and Plant (1974, unpublished), employing both profile refinement and conventional methods of powder analysis, but no firm conclusions regarding the structural details have yet been drawn. Indeed these materials may serve to indicate the approximate limit of complexity which may be handled adequately by powder-diffraction methods at the present time.

12.3. Inorganic cyanides

In quite a different field of application, several studies of inorganic cyanides take advantage of the ability of neutrons to distinguish between the neighbouring atoms carbon and nitrogen, which have scattering lengths of $0{\cdot}66 \times 10^{-12}$ cm and $0{\cdot}94 \times 10^{-12}$ cm respectively. The distinction is difficult to make with X-rays, particularly in the presence of heavy atoms. Hvoslef (1958) showed that in mercuric cyanide, $Hg(CN)_2$, it is the carbon atom which binds the cyanide group to the metal ion. Likewise in $K_3Co(CN)_6$ Curry and Runciman (1959) showed that it is the carbon atom, rather than the nitrogen, which is adjacent to the cobalt atom in the octahedral $[Co(CN)_6]^{3-}$ groups.

Several investigations have been made of NaCN and KCN in the regions of temperature where these substances are cubic. It is necessary to assume that the linear CN^- ions achieve effective cubic symmetry either by rotation or by disorder of their angular orientation. It is to be noted that both sodium and potassium have small scattering lengths, so that the diffraction intensities are very largely determined by the orientation and motion of the cyanide group. A study of KCN by Elliott and Hastings (1961) suggested free rotation, although this was not supported by later thermodynamic information and nuclear magnetic resonance experiments.

A later study of single crystals of NaCN by Row, Hinks, Price, and Susman (1973) and a re-interpretation of the data of Prince, Rowe, Rush, Prince, Hinks, and Susman (1972) for KCN leads to a model of preferred orientation of the axis of the CN ion, in contrast to either complete rotation or exclusive orientation in a particular direction. The conclusions for NaCN at 295 K and KCN at 180 K and 295 K are summarized in Fig. 247, which is a plot of the density

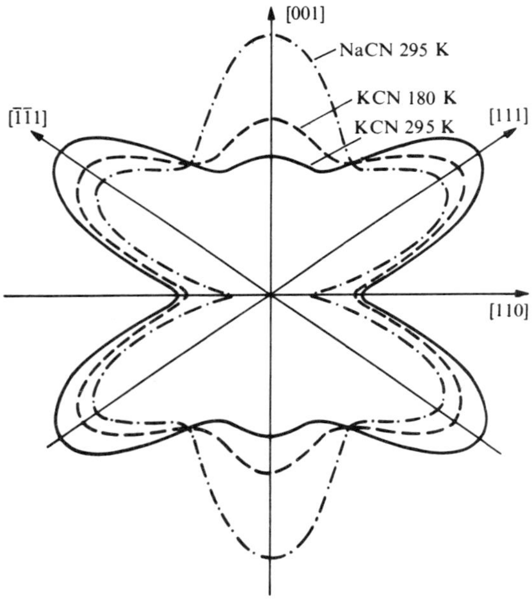

FIG. 247. A plot in the (110) plane of the density of C and N atoms at the surface of a sphere whose radius is equal to half the C≡N bond length. Curves are shown for NaCN at 295 K and KCN at 180 K and 295 K. In the first case there is a marked excess of CN groups oriented along the [001] axis compared with the [111] axis; in the second and third cases the converse result is found.(Rowe *et al.* 1973.)

of C and N atoms in the surface of a sphere of radius equal to half of the C≡N bond length. It will be seen that KCN has a maximum probability for orientation along the [111] directions, whereas NaCN has a maximum probability along [001] directions. There also appear to be substantial random displacements of both the metal and CN ions from their ideal positions in the rock-salt type of

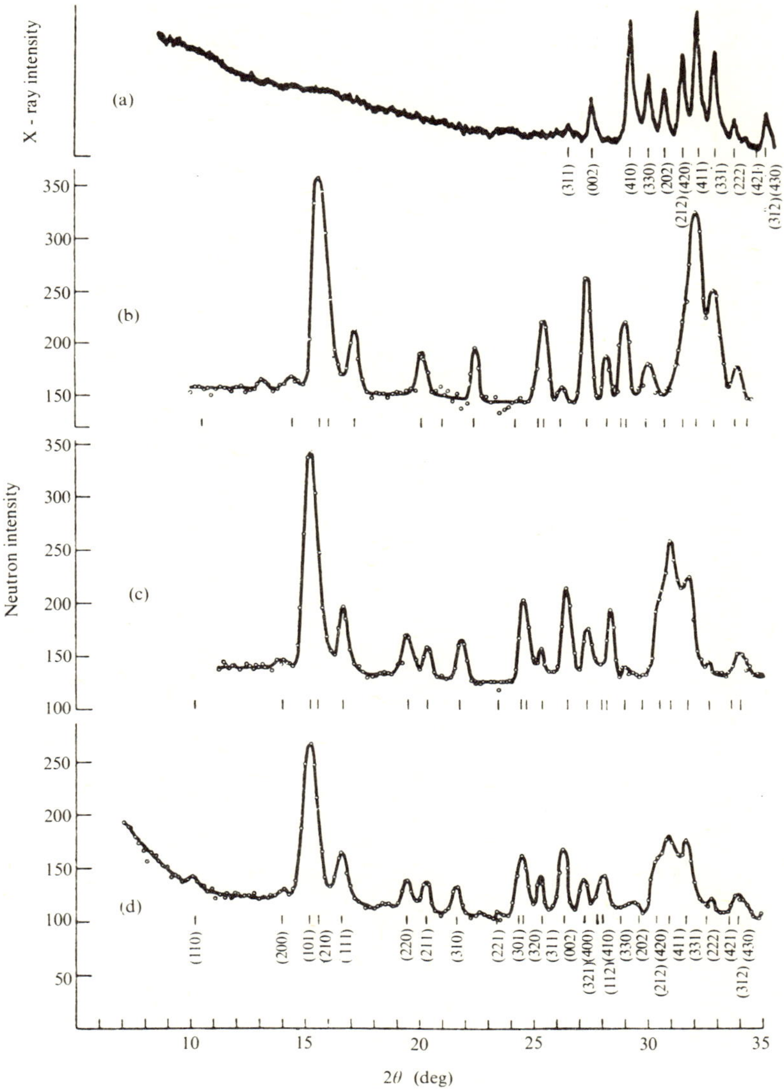

FIG. 248. Curve (a) shows an X-ray diffraction pattern of a Mn–Cr σ phase. Curves (b), (c), (d) are neutron diffraction patterns of various Ni–V σ phases, and the marked differences from (a) indicate that there is ordering of the nickel and vanadium atoms among the crystallographically different sites of the σ phase. The alloys shown are (b) $Ni_{13}V_{17}$, (c) $Ni_{11}V_{19}$, and (d) Ni_9V_{21}. (Kasper and Waterstrat 1956.)

structure, resulting in large Debye–Waller factors. Interpretation according to these models results in discrepancy factors R which range between 2 per cent and 3 per cent for the three sets of data.

12.4. σ-phases

Kasper and Waterstrat (1956) have studied the ordering of the component atoms in σ-phase alloys of the transition elements where there are five crystallographically different types of atom and among which the component distribution is very difficult to distinguish with X-rays because of the closeness of the X-ray scattering factors. Definite evidence of ordering was found for Ni–V, Fe–V, and Mn–Cr alloys, all of which are favourable examples to study with neutrons because of the very different values of b for the pairs of atoms. The existence of ordering is very clearly demonstrated in Fig. 248 which shows diffraction patterns for a number of Ni–V alloys. If no ordering occurred and the Ni and V atoms were distributed at random among the different sites, then the neutron pattern would be identical with the X-ray pattern which is shown at the top of the Figure; in fact the neutron and X-ray patterns are quite different from each other.

A study of the atomic positions in the σ phase of Fe–Cr has been reported by Bykov (1961, unpublished), and some preliminary measurements have been made of the ternary σ phases in the Fe–Cr–Mn system, making use of the null matrix method.

A sample of composition 20 % Fe, 25 % V, 55 % Mn, which entails an average scattering amplitude of zero, was found to give no coherent neutron reflections, although X-rays showed a homogeneous σ phase. This would suggest that there was no ordering of the constituent metals in this particular sample.

13

INVESTIGATIONS OF MAGNETIC MATERIALS: I. METALS AND ALLOYS

THE most significant contribution to the study of the solid state made by neutron diffraction has probably been the knowledge gained from the examination of magnetic materials. Historically the first materials to be investigated were the antiferromagnetic oxides, beginning with manganous oxide, MnO, in 1949, but we shall find it more convenient to consider first the metallic elements themselves. In the years which have elapsed since the first powder samples were examined with low-neutron fluxes, very many elements and compounds have been studied, and it has long since been impracticable to refer to them all in a book of this kind. We shall therefore consider only particular examples, chosen to indicate the range of magnetic structures which are found and the general development of the techniques for measuring and interpreting the diffraction patterns. There are several detailed compilations of structural information to which a reader in search of knowledge about a specific substance may refer. A series of five volumes and two supplements of *Tables of magnetic structures determined by neutron diffraction* issued by the Institute of Nuclear Techniques, Cracow, Poland covers most work up to 1971 and a loose-leaf set of *Magnetic structure data sheets* for later work is sponsored by the Neutron Diffraction Commission of the International Union of Crystallography and produced by the Brookhaven National Laboratory, U.S.A. Examples of the form of information which has been provided in this way are shown in Figs. 249 and 250.

13.1. The transition metals of the 3d group

The ferromagnetism of iron, cobalt, and nickel and the magnetic properties of the neighbouring elements constitute the best known of the basic phenomena which must be explained by any theory of magnetism. We shall show how neutron diffraction studies provide the basic structural information on an atomic scale in terms of which the macroscopic magnetic behaviour must be interpreted.

MuTe	a = 6.597 Below T_N there is a contraction of the cubic cell edge; there is no detected trigonal component to the distortion 1965/24	$T_N \approx 10$ 1965/24 $T_N = 9.8$ 1963/32	Antiferromagnetic compound with ordering of the second kind. Fig. 2. 1965/24 MSG - C_{2c} 2/m' /B^{90}_{15}/ 1963/32		1962/6 1962/12 1962/14
Microcrystalline particles of γ-Fe in a single-crystal copper matrix	a = 3.588 1957/6 The iron particles in the 1962/1 paper have lattice constants that are reasonable for bulk γ-iron and hence are probably negligibly strained.	$T_N = 8$ $-13°K < \Theta < -8°K$ 1962/1	Antiferromagnetic - microcrystalline particles of γ-Fe of nearly identical orientation in a single - crystal copper matrix give a neutron diffraction pattern, consistent with the development of antiferromagnetism. The spin structure resembles that of γ- manganese, but with the spin vectors inclined at about 19° from the normal to the ferromagnetic sheets. The spin directions in alternate sheets are antiparallel each other. Fig. 13. 1962/1	μ(Fe) = 0.7 1962/1	1938/3 1957/6 1962/1
Fe_3Al phase	a = 2.8961 1932/1, 1958/17 Fe_A occupies simple-cubic sublattices Fe_B Al rock-salt-type sublattice.	$T_c = 773$ 1958/14 1967/15	Ferromagnetic. Order - disorder transition change the magnetic properties. 1958/14 1959/8 MSG - I 4/m m'm' /B^{537}_{139}/	$\mu(Fe_A) = 1.46 \pm 0.1$ $\mu(Fe_B) = 2.14 \pm 0.1$ 1958/14	1959/1 1960/11 1961/17 1962/26 1967/15

FIG. 249. Facsimile of page from *Tables of magnetic structures determined by neutron diffraction* compiled by the Institute of Nuclear Techniques, Cracow, Poland.

NEUTRON DIFFRACTION COMMISSION – MAGNETIC STRUCTURE DATA SHEETS

AUTHOR(S)	B. Lambert-Andron, G. Berodias, and D. Babot		SUBMITTED TO PUBLISHED IN J. Phys. Chem. Solids 33, 87 (1972)
TITLE	Magnetic Structures of the Compounds Fe_2MSe_4 (M = Ti, V, Cr, Fe, Co, Ni)		
ADDRESS OF FIRST AUTHOR	Centre d'Etudes Nucleaires, Rue des Martyrs, 38 Grenoble, France		
MATERIAL	Fe_2 Ni Se_4		FORM Powder
CRYSTAL STRUCTURE	SPACE GROUP I2/m (C^3_{2h})	TYPE Cr_3 S_4	POSITIONS OF MAGNETIC ATOMS Fe in 2(c): 0,0,½ Fe, Ni in 4(i): x,0,z
MAGNETIC STRUCTURE	Chemical unit cell, a = 6.18Å, b = 3.56Å, c = 10.94Å, β = 91.7° Atomic positional parameters, x = 0.047, z = 0.231 Magnetic cell same size Collinear ferrimagnet, T_c = 67°K Ferromagnetic (001) layers coupled antiparallel	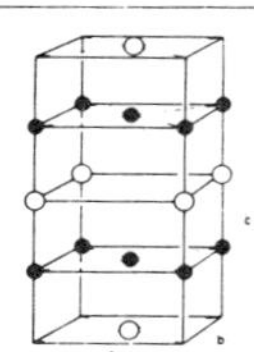 Fe,Ni in 4(i) ● Fe in 2(c) ○	
MAGNITUDE AND DIRECTION OF MOMENTS	Moments at 4.2°K Fe in 2(c) sites 1.9 μ_B; mean moments in 4(i) sites 0.9 μ_B		
ADDITIONAL INFORMATION	Crystal structure refinement at 298°K Magnetization measurements between 4.2°K and 300°K Mössbauer measurements Electrical resistivity measurements between 4.2°K and 300°K		
REFERENCES	1) A.F. Andresen, Acta Chem. Scand. 22, 827 (1968) 2) B. Lambert-Andron and G. Berodias, Solid State Commun. 7, 623 (1969)		

FIG. 250. Example of *Magnetic structure data sheet* published by the Neutron Diffraction Commission of the International Union of Crystallography.

Iron, cobalt, and nickel

The powder diffraction pattern of polycrystalline iron, as recorded by Shull, Wollan, and Koehler (1951), with the addition of a lower angle section reported later by Shull and Wilkinson (1953), is illustrated in Fig. 251. The pattern can be interpreted as being due to a superposition of the effects of the isotropic nuclear scattering by iron atoms, with their relatively large scattering amplitude of $0{\cdot}95 \times 10^{-12}$ cm, and the magnetic scattering resulting from the

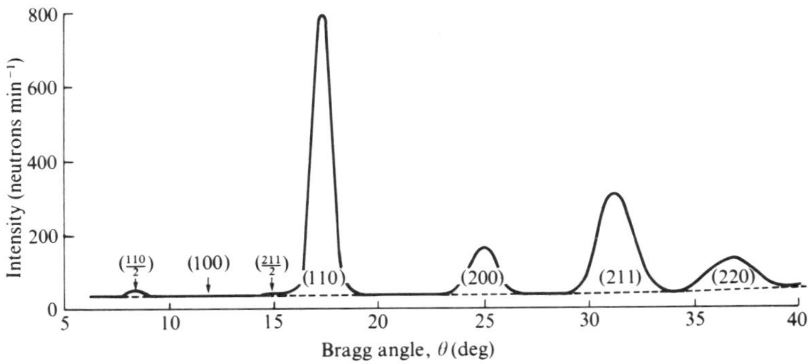

FIG. 251. The neutron diffraction pattern of polycrystalline iron. (After Shull, Wollan, and Koehler (1951) and Shull and Wilkinson (1953).)

magnetic moment of the atoms. Fig. 252 indicates how the resultant differential scattering cross-section varies with the Bragg angle θ and shows how the magnetic contribution can be assessed by subtraction of the isotropic nuclear component which greatly predominates. The magnitude of this magnetic contribution is in agreement with that calculated from eqn (6.12) if it is assumed that the effective value of the quantum number S is 1·11. This is what would be expected from measurements of the saturation magnetization of iron which suggest a magnetic moment of $2{\cdot}22\ \mu_B$ for each iron atom. Consequently the magnetic scattering amplitude will be $0{\cdot}598f \times 10^{-12}$ cm, where f is the amplitude form factor discussed in Chapters 6 and 7, and calculation gives the result which is shown in Fig. 252 in relation to the experimental observations. In carrying out this calculation it is assumed that the factor q^2 is equal to $\frac{2}{3}$. This will be the case both if the unique direction of the magnetic

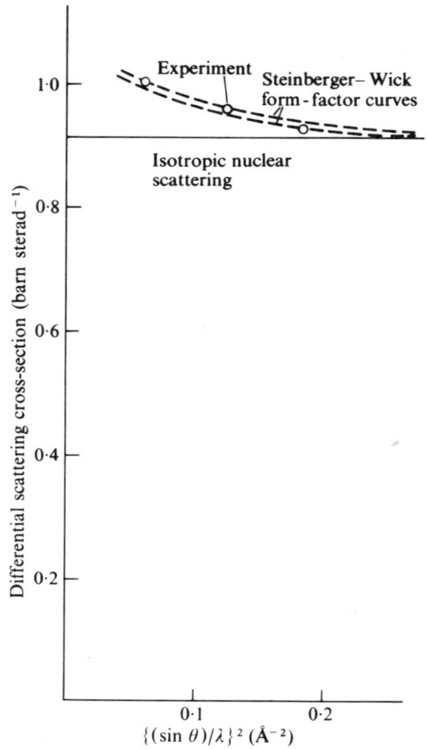

FIG. 252. The differential scattering cross-section of iron in relation to the known nuclear scattering and the extreme possibilities of the Steinberger–Wick calculation of the magnetic form factor. (Shull, Wollan, and Koehler 1951.)

moments within a domain may take any orientation relative to the crystal axes with equal probability and also if the magnetic moments may only be aligned along, say, any one of the cube edges ⟨100⟩, as is considered to be the case from magnetic studies. It is not possible from studies of polcrystalline iron samples to draw any deductions concerning the directions of domain magnetization relative to the crystallographic axes.

It is of interest to examine the diffraction pattern in greater detail in relation to the predictions made from certain theories which have been put forward to account for the particular value of 2·22 μ_B found for the magnetic moment of the iron atom. It was suggested by Hume-Rothery, Irving, and Williams (1951) that this was a mean value due to an ordered arrangement of two types of iron atom, namely Fe^{3+} ions with moment 5·0 μ_B at the cube corners and atoms with zero moment at the body centres. This would give a magnetic moment of 5 μ_B per unit cell of two atoms, or 2·5 μ_B per atom. Similarly, according to Zener's theory (1952) there would also be two types of iron atoms, with in this case magnetic moments of 5·0 μ_B and 1·0 μ_B respectively arranged anti-ferromagnetically on the two types of site. On this theory the magnetic moment per unit cell would equal $(5\,\mu_B - \mu_B)$, i.e. 2 μ_B per atom. In either instance a (100) superlattice line should be observed. A careful search for this was made by Shull and Wilkinson (1953) who found no intensity in this position apart from a very small contribution which could be accounted for by the known second-order contaminant from the (200) reflection. The low-angle pattern observed by these authors is that shown to the left in Fig. 251. Allowing for experimental error they conclude that F_{100}^2, the differential magnetic scattering cross-section per atom at the (100) position, is not greater than $0{\cdot}0010 \times 10^{-24}$ cm^2. It follows that if there is an ordered array of two types of iron atom at the corners and centres of the unit cell then their magnetic moments cannot differ algebraically by more than 0·4 μ_B, in contrast to the values of 6·0 μ_B, 5·0 μ_B respectively which are expected on the two theories mentioned above.

If, on the other hand, there was a random distribution of the two types of magnetic moment then no superlattice lines would be expected but there would be incoherent magnetic diffuse scattering of the kind shown by a paramagnetic material. The intensity would depend on $(\mu_1 - \mu_2)$, where μ_1, μ_2 are the two magnetic moments, and it would fall off with increasing angle according to a form factor such as was illustrated in Fig. 110. In fact no such diffuse scattering was observed. Fig. 253 shows the results of Shull and Wilkinson's measurements. After correction for ordinary thermal inelastic scattering the background intensity is found to be practically isotropic, in contrast with the marked fall-off with angle which would be expected if there were a random distribution of d^5- and d^{10}-type iron atoms. The diffuse scattering was also studied when the sample was magnetized along the scattering vector, that is with the magnetic field normal to the reflecting planes, under which conditions all magnetic scattering should disappear. The diffuse scattering did not change by more than 1 per cent and from this observation it was deduced that the difference $(\mu_1 - \mu_2)$ could not be greater than 0·6 μ_B.

The above observations show that, so far as the neutrons are concerned, there is neither an ordered nor disordered arrangement of two types of iron atom having appreciably different magnetic moments. The only possibility remaining, in order to account for the odd moment of 2·22 μ_B, seemed to be that there is a rapid transition from one state to another in a time which is much shorter than the neutron passage time, which is about 10^{-13} s. To the neutrons all the atoms would then appear to be identical.

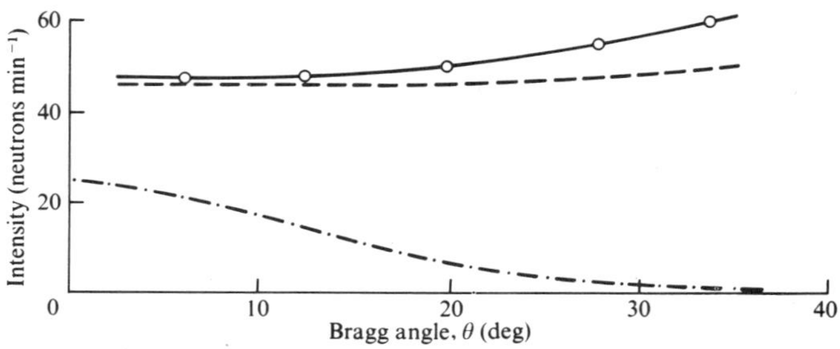

FIG. 253. The diffuse scattering from polycrystalline iron. -O-O- Experimentally observed background scattering. ---- Background after correction for thermal diffuse scattering. -·-·-· The calculated magnetic diffuse scattering, showing form factor variation, which would be expected for a random distribution of Fe(d^5) and Fe(d^{10}) atoms. (Shull and Wilkinson 1953.)

The diffraction pattern of face-centred cubic cobalt shows the effect of magnetic scattering more strikingly than for iron, since in the forward direction when $\theta = 0°$ the magnetic scattering amplitude of the atom is greater than the nuclear scattering. The two amplitude values are $0{\cdot}47 \times 10^{-12}$ cm and $0{\cdot}25 \times 10^{-12}$ cm respectively. The relative contributions of the magnetic and nuclear scattering to the experimentally observed values of the differential scattering cross-sections for the diffraction peaks are indicated in Fig. 254. The experimental curve extrapolates satisfactorily at $\theta = 0°$ to a value of total differential cross-section appropriate to a Bohr magneton value of 1·74 per cobalt atom. This is the value obtained from saturation magnetization studies on cubic cobalt and leads, by application of eqn (6.12), to a magnetic scattering amplitude in the forward direction of $0{\cdot}47 \times 10^{-12}$ cm, as mentioned above.

In the case of metallic nickel it can be seen from Table 16 (p. 211) that the magnetic scattering amplitude is only 15 per cent of the nuclear value, even in the forward direction. As a result, the intensity term for the magnetic scattering in eqn (6.18) for an unpolarized neutron beam is less than 2 per cent of the nuclear term, so that the magnetic scattering is barely detectable in practice.

We refer back to Chapter 7 for an account of the later, and much more accurate, measurements of Fe, Co, and Ni using polarized neutrons which have provided detailed information about the form factor f and which have led to accurate knowledge of the distribution of the unpaired electron spins within the individual atoms.

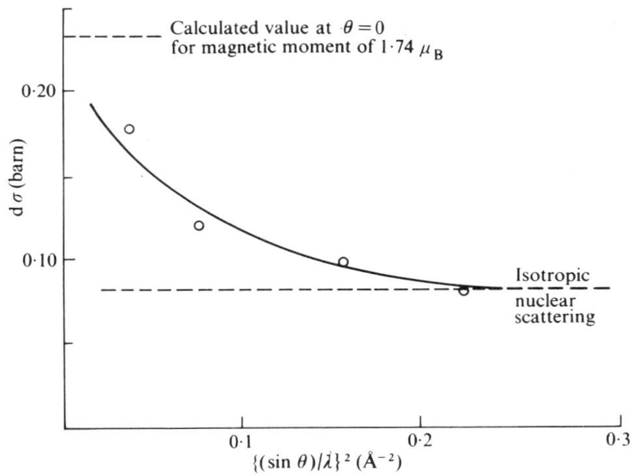

FIG. 254. Variation with angle θ of the differential scattering cross-section of the cobalt atom in α-Co. (Shull, Wollan, and Koehler 1951.)

Chromium and manganese

The magnetic structures of chromium and manganese, both of which are antiferromagnetic, have proved much more difficult to determine in detail and have led to many investigations. Neither of the two metals is yet fully understood.

Chromium was first studied with neutrons by Shull and Wilkinson (1953), who demonstrated that the metal has an antiferromagnetic structure in which the atoms at the corners and centres of the body-centred unit cell have their magnetic moments arranged antiparallel.

From the intensity of the resulting (100) reflection it was concluded that the magnetic moment was about 0·40 μ_B per atom. Rather surprisingly it was found that the Néel temperature was 473 K, in contrast with 313 K for the anomaly in various physical properties such as resistivity (Fine, Greiner, and Ellis 1951) and specific heat (Beaumont, Chihara, and Morrison 1960). These early neutron measurements were made with powdered samples and were not supported by later studies with single-crystals from which Corliss, Hastings, and Weiss (1959) found a Néel temperature of about 311 K, in line with expectations from the measurements of physical properties. For some time it was suggested that sample purity was the cause of these discrepancies, particularly in view of the work of de Vries (1959), who showed that the addition of 1 per cent of impurity metals made vast differences to the temperature at which the anomaly in electrical resistivity was observed. The measurements of the writer (Bacon 1961) establish that purity is not the primary cause of the discrepancy and leave the possibility that perhaps the Néel temperature is sensitive to differences of domain size and strains brought about by different methods of crystal growth and subsequent treatment. Fig. 255 shows curves of

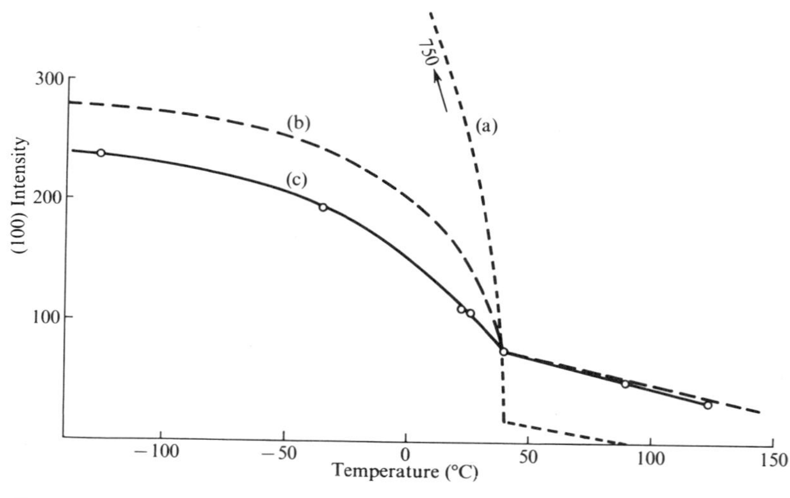

FIG. 255. The variation of the neutron diffraction intensity with temperature for the (100) antiferromagnetic reflection of very pure chromium. The curves are for (a) a single crystal, (b) a large-grain polycrystalline sample, and (c) a small-grain polycrystalline sample.

neutron intensity versus temperature for both a single-crystal and polycrystalline fragments of very pure Australian metal. A second complication in the magnetic structure is the occurrence of a long-range modulation which gives satellites in the diffraction pattern. These were first reported by Corliss, Hastings, and Weiss (1959), who suggested the existence of antiphase domains, giving a modulation distance of about 26 unit cells. This model was used by Hastings (1960) to explain a further transition at 120 K, noted by Bykov, Golovkin, Ageef, Levdik, and Vinogrador (1960), as due to rotation of the magnetic moments from positions parallel to the domain walls to new positions perpendicular to them. However, the model, which we illustrated earlier in Fig. 166 (p. 272), cannot be correct because it predicts third-order satellites having an intensity of about 11 per cent of the first-order pair, as we have discussed in Chapter 8 (p. 273), and these third-order satellites have never been observed. The alternative suggestion of a spin-spiral structure is also unlikely to be correct, as it becomes very difficult to account for the rotation of the moments at 120 K. Further, the spiral model has been ruled out by an experiment with polarized neutrons by Brown, Wilkinson, Forsyth, and Nathans (1965). It is believed that the correct structure, suggested by Shirane and Takei (1962), is one in which the moment varies sinusoidally along the modulation axis, as in Fig. 168 (p. 274), with a period of about 26 unit cells. At temperatures below the change point of 120 K the moments lie along the modulation axis but switch to a perpendicular direction above this temperature. A normal single crystal of chromium is believed to consist of three types of domain having their modulation directions parallel respectively to the three cubic axes. Each modulation direction can be associated with either of two spin directions. Thus the [100] modulation direction can be associated with spins which are parallel to either the y- or the z-axis. The normal equality of distribution among the various possibilities can be affected by cooling the crystal through the Néel point of 311 K in a strong magnetic field. For such a model, in which the magnitude of the magnetic moments varies sinusoidally, the maximum value of the moment is deduced to be 0·59 μ_B. This is to be contrasted with values of 0·41 μ_B and 0·45 μ_B, respectively for the spiral and the antiphase-domain models.

The earlier suggestion that the behaviour of chromium is affected by stresses in the material is supported by later work by Bacon and

Cowlam (1969), who have examined heavily crushed powder. They find that long-range magnetic order exists up to about 450 K, which is close to Shull and Wilkinson's original Néel temperature, and it is of the simple antiferromagnetic type first postulated. With reduction of temperature this structure gradually transforms to the incommensurate sinusoidal structure which we have just discussed. The low-temperature change point, at which the change in direction of the spins takes place, no longer occurs at a precise temperature and the change is incomplete even at 4·2 K. When these cold-worked samples are annealed the simple antiferromagnetic structure becomes less prominent and the temperatures of the transformations become more sharply defined until the normal, single-crystal type of behaviour is reached. The sequence of changes in behaviour of the sample is indicated in Fig. 256.

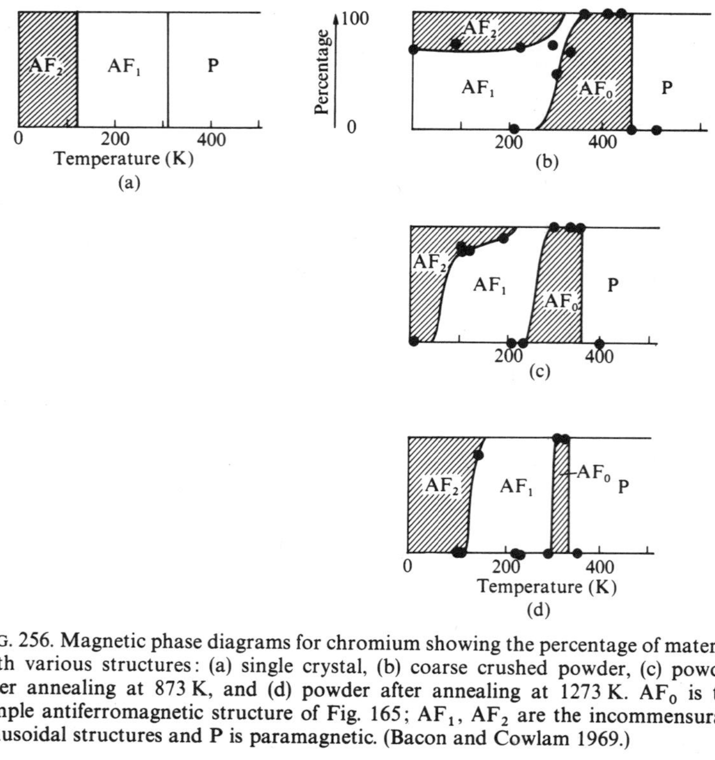

FIG. 256. Magnetic phase diagrams for chromium showing the percentage of material with various structures: (a) single crystal, (b) coarse crushed powder, (c) powder after annealing at 873 K, and (d) powder after annealing at 1273 K. AF_0 is the simple antiferromagnetic structure of Fig. 165; AF_1, AF_2 are the incommensurate sinusoidal structures and P is paramagnetic. (Bacon and Cowlam 1969.)

α-manganese, which is the room-temperature form of metallic manganese, has an unusual body-centred structure which contains 58 atoms in a cubic unit cell. The atoms are of four different types I–IV, comprising 2, 8, 24, and 24 atoms respectively. An early measurement by Shull and Wilkinson (1953) showed that it was antiferromagnetic below 100 K. The diffraction patterns at room temperature and at 20 K are contrasted in Fig. 257. The important

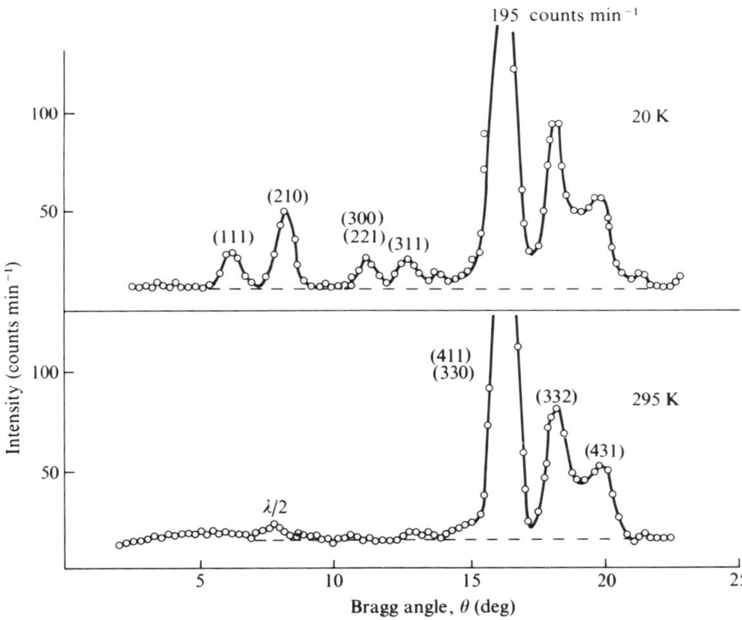

FIG. 257. Neutron diffraction patterns for α-manganese at 20 K and 295 K, showing antiferromagnetic reflections at the low temperature. (Shull and Wilkinson 1953.)

differences between the two patterns are that at low temperature there is a significant reduction in the angularly-dependent background of diffuse scattering and there are additional coherent peaks, corresponding to indices (111), (210), (300, 221), and (311). It is noteworthy that for each of these reflections the sum $(h+k+l)$ is an odd number. Later work by Kasper and Roberts (1956), again with a powder sample, suggested two possible detailed structures which would account for the experimental data. In each of these structures it was concluded that there was no magnetic moment on atoms of type IV but there were two different moment values among types I, II, and III. A new study using both powder and single crystals was made by Oberteuffer, Marcus, Schwartz, and Felcher (1970) and shows several interesting features. The powder measurement used a time-of-flight spectrometer in which the neutrons scattered at a constant angle of 90° are energy-analysed in terms of their flight-time. A picture of part of the record is given in Fig. 258, which illustrates particularly well the high resolution

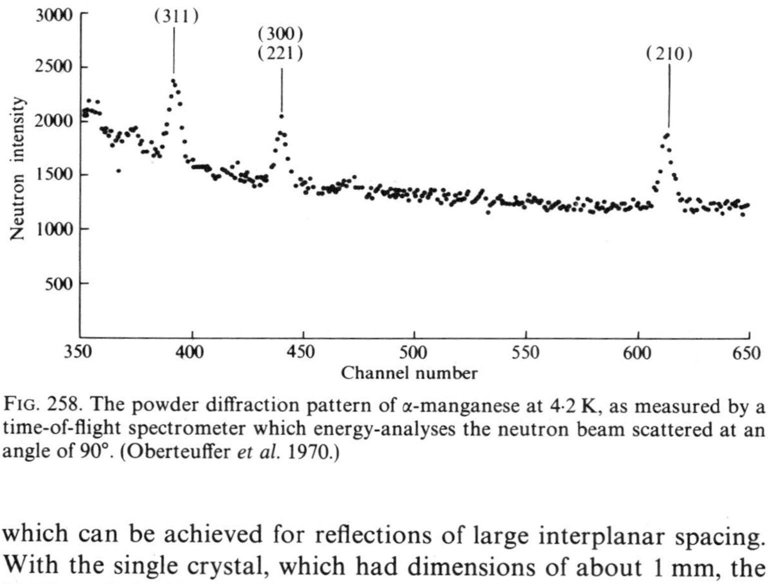

FIG. 258. The powder diffraction pattern of α-manganese at 4·2 K, as measured by a time-of-flight spectrometer which energy-analyses the neutron beam scattered at an angle of 90°. (Oberteuffer *et al.* 1970.)

which can be achieved for reflections of large interplanar spacing. With the single crystal, which had dimensions of about 1 mm, the [110] and [100] zones were examined, and it was demonstrated that the crystal consisted of a random distribution of magnetic domains by showing that equivalent magnetic reflections in different quadrants of space had equal intensities. Measurements were made at 77 K, 50 K, and 4 K, and it was concluded that the four types of manganese atom bore, respectively, magnetic moments of 1·72 μ_B, 1·47 μ_B, $-1·11$ μ_B, and 0·02 μ_B, the latter moment being insignificantly different from zero, as in the earlier work of Kasper and Roberts. The occurrence of only odd combinations of the sum of indices $(h+k+l)$ confirms that there is an antiferromagnetic reversal of spins, for each type of atom, for those positions which are related by body-centring. However, the so-called 'reliability factor'—or, more correctly named, discrepancy factor—which is defined here as $\sum |F_0^2 - F_c^2|/\sum F_0^2$, where F_0, F_c are the observed and calculated values of the structure factors, could not be reduced below 18 per cent for the magnetic reflections in terms of this model, in comparison with a much lower value of 4 per cent for the nuclear reflections. It is doubtful therefore that the model is correct, and Kunitomi, Yamada, Nakai, and Fujii (1969) have suggested a non-collinear model. This has been explored by Yamada *et al.* (1970), using the high-flux reactor at Brookhaven and a much larger single crystal which measured 2 mm × 3 mm × 9 mm. Thirty magnetic reflections were measured and interpreted in terms of Yamada's model which requires 13 parameters. Much better agreement was obtained than when the collinear model was used. In order to determine the moment values it is necessary to make precise assumptions concerning the form factor of manganese which will be closely dependent on the relative proportions of 3d and 4s electrons. The influence of these assumptions on the final values is indicated in Fig. 259 in which the values of the moments are shown as a function of the fractional contribution made by the 4s electrons. With the most plausible assumptions and using the same form factor for all the sites it was concluded that the values of the magnetic moments for the four types of atom are 1·9 μ_B, 1·7 μ_B, 0·6 μ_B, and 0·2 μ_B at 4·4 K.

Vanadium

The diffraction pattern of vanadium, which has a body-centred cubic structure, is particularly noteworthy for the almost complete absence of coherent diffraction peaks as seen in Fig. 260. This is a consequence of the extremely small coherent nuclear scattering amplitude of $-0{\cdot}05 \times 10^{-12}$ cm, as referred to earlier in Table 2 (p. 38). Practically the whole of the nuclear scattering is isotropic spin-disorder diffuse scattering, and a weak (110) reflection is the only visible coherent peak. The pattern at 20 K shows no (100) reflection and no angularly-dependent diffuse scattering. It is concluded that any magnetic moment must be less than 0·1 μ_B.

13.2. Rare-earth metals

The most spectacular example of the contribution of neutron diffraction technique to our knowledge of magnetic structures is

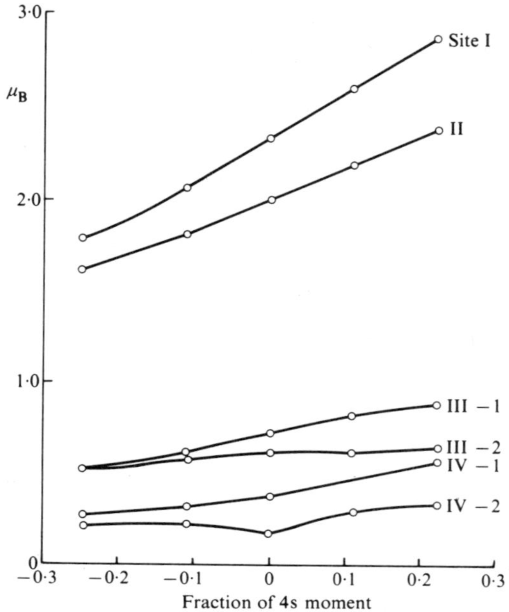

FIG. 259. The magnetic moments of the different types of Mn atom in α-manganese in terms of the fraction of the moment which is judged to be contributed by the 4s electrons. For atoms on sites III, IV there are two sets of positions which are *not* equivalent magnetically. (Yamada *et al.* 1970.)

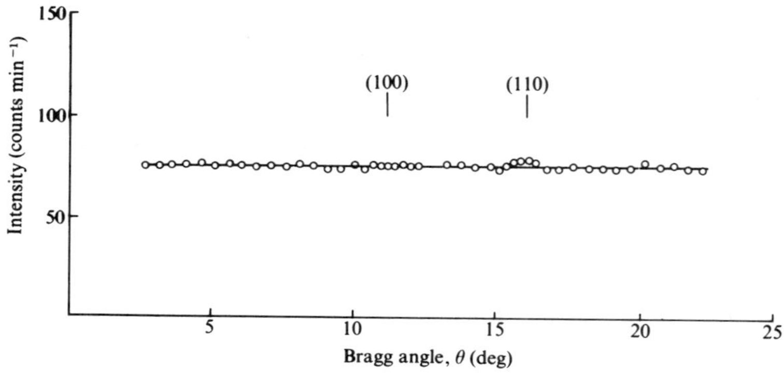

FIG. 260. The neutron diffraction pattern of vanadium at 20 K, showing the extremely weak coherent nuclear diffraction, and no detectable antiferromagnetic reflection at the (100) position. (Shull and Wilkinson 1953.)

provided by the metals of the rare-earth group. The structures have proved to be very intricate and varied and the experimental task of elucidating them has been difficult because of the very large coefficients of absorption which most of the elements show, as listed earlier in Table 6 (p. 72). The fact that success has been achieved is due mainly to two factors—the technology for producing pure single crystals developed by Spedding at Ames, Iowa, and the ingenuity of Koehler and his colleagues at Oak Ridge National Laboratory, who have interpreted the diffraction patterns.

As we have already discussed in Chapters 6 and 7 the magnetic moments of the rare-earth elements arise from unpaired electrons in an inner shell and contributions to the moment arise from both the orbital and spin momenta. The elements which have been studied most are those from gadolinium to thulium, in the second half of the rare-earth series. In each case the three valence electrons $5d^1 6s^2$ are, in the metals, lost to conduction bands, so that the atomic sites are occupied by tripositive ions, arranged in each case in a hexagonal close-packed structure. At room temperature all the metals are paramagnetic. On cooling, gadolinium becomes ferromagnetic immediately below room temperature, and maintains this state down to the lowest temperatures. All the other elements show successively two or more ordered magnetic arrangements as they are progressively cooled and the temperatures at which changes occur also display anomalies and discontinuities in electrical and thermal, as well as the macroscopic magnetic, properties. Table 26 summarizes the magnetic structures which are displayed by Gd, Tb, Dy, Ho, Er, and Tm, and we shall show how this information has been provided by the neutron diffraction observations. The table also lists the cross-sections for absorption for neutrons of wavelength 1·08 Å. In comparison with normal elements these are very large values, thus giving rise to the experimental difficulties mentioned earlier. For gadolinium, with the enormous σ value of 20 000 barns, significant progress has only really been made since single crystals constituted of the isotope ^{160}Gd became available. For this isotope the absorption coefficient is only 20 barns.

Inspection of Table 26 will show that a variety of structures are displayed by the elements listed in the table and these are identified by the letters (a) → (g), which refer to the diagrams in Fig. 261, where the moment directions in successive layers of the structure are shown. In order to make clear how these structures have been

TABLE 26

Magnetic structures of some rare-earth metals

Elements	Gadolinium	Terbium	Dysprosium	Holmium	Erbium	Thulium	
	$_{64}$Gd	$_{65}$Tb	$_{66}$Dy	$_{67}$Ho	$_{68}$Er	$_{69}$Tm	
σ_a	20 000	26	535	40	100	71	
	293 K	Paramagnetic 230 K Helical $\alpha = 18-20°$(*e*) 220 K	Paramagnetic	Paramagnetic	Paramagnetic	Paramagnetic	300 K 0 °C
	Ferromagnetic with changing angle between moment direction and *c*-axis (*g*)	Ferromagnetic with moments in basal plane $9{\cdot}0\,\mu_B$ (*f*)	176 K Helical antiferromagnetic $\alpha = 26$–$43°$ (*e*)	130 K			200 K
			88 K Ferromagnetic with moments in basal plane $9{\cdot}5\mu_B$ (*f*)	Helical antiferromagnetic $\alpha = 30$–$50°$(*e*) 20 K	85 K *c*-axis modulated antiferromagnetic 7*c*/2 $7{\cdot}6\mu_B$ (*b*) 52 K Combination of spiral with antiphase domain (*c*) 20 K	57 K *c*-axis modulated antiferromagnetic (*b*) $7{\cdot}0\mu_B$ 32 K Unbalanced antiphase system giving ferromagnetism (*a*)	100 K
				Conical spiral $\mu_{\parallel} = 1{\cdot}7\,\mu_B$ $\mu_{\perp} = 9{\cdot}5\,\mu_B$(*d*)	Conical spiral $\mu_{\parallel} = 7{\cdot}6\,\mu_B$ $\mu_{\perp} = 4{\cdot}3\,\mu_B$(*d*)		0 K

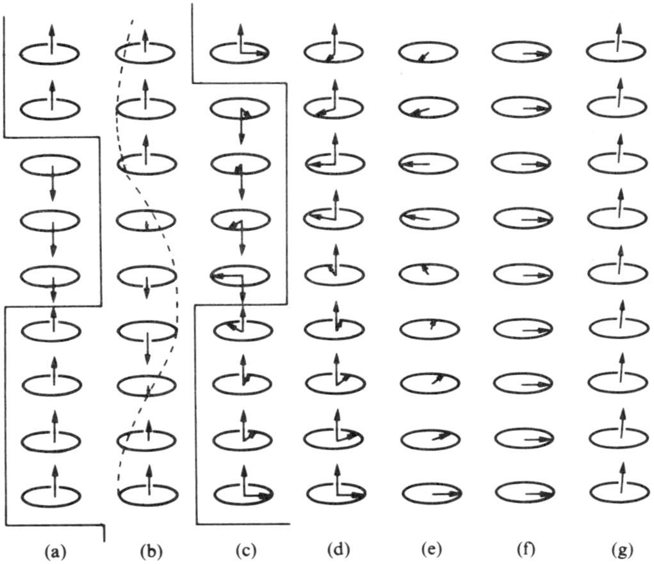

FIG. 261. Summary of types (a)–(g) of magnetic structure found amongst the rare-earth metals and their alloys. The circles indicate successive horizontal layers of atoms, spaced along the c-axis of the structures, and the arrows indicate the magnitude and direction of the magnetic moments, particularly in relation to their horizontal and vertical components. (After Koehler 1965.)

deduced from the diffraction patterns it will be instructive to consider some particular examples. Possibly the best example is erbium, which shows three different magnetic structures as it is progressively cooled below room temperature. Following Koehler (1972), we show diagrammatically in Fig. 262 the way in which the diffracted neutron intensity is distributed in reciprocal space for a single crystal of erbium at various temperatures. In Fig. 262 (a), at temperatures above 85 K, intensity is observed only at reciprocal lattice points and is due solely to nuclear scattering. Thus the diagram looks like a section of an X-ray rotation photograph for a crystal mounted to rotate about its c-axis; there is no intensity at the (0001) position because of cancellation of the contributions from planes at heights of 0 and $\frac{1}{2}c$ in the close-packed hexagonal structure. The intensity shown in this diagram (a) appears at *all* temperatures. At temperatures below 85 K other intensity also appears, as indicated in (b), (c), and (d) of Fig. 262, and it is magnetic in origin. In (b),

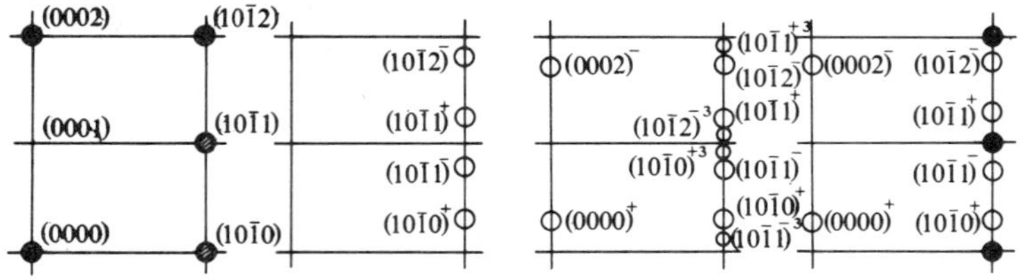

(a) (b) (c) (d)

FIG. 262. The intensity distributions in the reciprocal lattice of erbium at various temperatures. (a) shows the purely nuclear scattering above 85 K. The other diagrams show the additional, *magnetic*, scattering: at (b) for 85–52 K, at (c) for 52–20 K, and at (d) below 20 K. In the latter case magnetic intensity (shown black) appears at some of the lattice points themselves. (After Koehler 1972.)

which is applicable in the region from 85 K to 52 K, there is a pair of satellite reflections, such as $(10\bar{1}1)^+$ and $(10\bar{1}1)^-$, associated with each reciprocal lattice point, *except* for the points (000*l*). There is no extra intensity at the lattice points themselves, and the displacement of the satellites from these points is always in the direction of the *c*-axis. This distribution of intensity arises from a structure of the type shown in Fig. 261 (b). There are four important features. First, the moment directions are necessarily parallel or antiparallel to the *c*-axis, to account for absence of (000*l*) reflections. Secondly, the sinusoidal variation of the magnitude of the moments is deduced from the fact that only first-order satellites are observed. Thirdly, the fact that the satellites are displaced in the direction of the *c*-axis shows that it is along the *c*-axis that the sinusoidal variation of moment takes place. Fourthly, from the measured value of the displacement of the satellites it is possible to deduce the wavelength of the sinusoidal variation. As Fig. 262 (b) suggests, the vertical displacement of the satellites is about $\frac{2}{7}$ths of the distance between successive lattice points and this leads to the deduction, following the discussion of Chapter 8, that the wavelength must be $\frac{7}{2}$ times as great as the *c*-dimension of the unit cell. It is observed experimentally that this distance remains constant throughout the range of existence of this magnetic structure from 52 K to 85 K. This is indicated more directly in Fig. 263, where the repetition distance remains constant at 3·5 units for values of T/T_N greater than about 0·6. Below 52 K a change in the intensity pattern takes place, and this is shown at (c) in Fig. 262. First-order satellites are now associated with the (000*l*) reflections also, and indicate that there is now a component of magnetic moment within the basal planes which is taking part in the ordering. At the same time it is observed that the other reciprocal lattice points display third-order satellites (such as $(10\bar{1}1)^{+3}$) in addition to the first-order ones. This can be explained by a structure indicated in Fig. 261 (c), where a basal plane component, of about 3·6 μ_B, rotates around with a turn angle of about 45° per layer, accompanied by a more intense

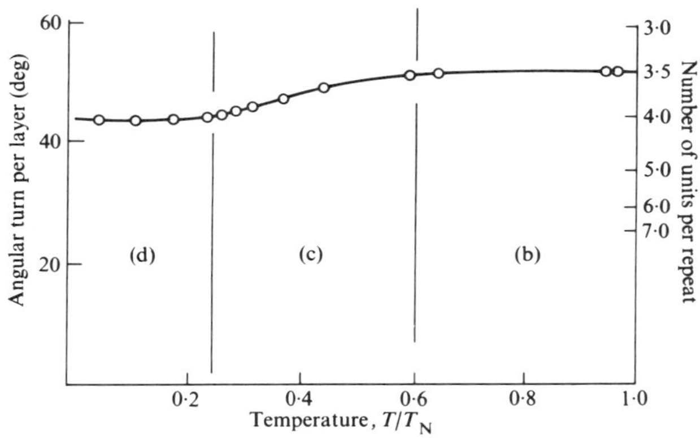

FIG. 263. The change in the vertical repeat distance of erbium with temperature, expressed as a fraction of the Neél temperature of 85 K. When $T/T_N > 0{\cdot}6$ structure (b) of Fig. 261 exists, with a repeat distance of seven layers, i.e. $\frac{7}{2}c$. When $0{\cdot}24 < T/T_N < 0{\cdot}6$ structure (c) exists with a steadily increasing turn-angle for the horizontal component of moment; when $T/T_N < 0{\cdot}24$ the conical spiral (d) exists with a constant turn-angle of about 45°.

component of moment parallel to the *c*-axis and equal to about 8 μ_B. The magnitude of this latter moment is effectively constant, but its sign changes and a sequence of four upward-pointing moments is succeeded by four downward-pointing moments, and so on. This complicated arrangement is succeeded below 20 K by the structure in Fig. 261 (d), which causes the crystal to become ferromagnetic. This structure is termed a 'conical spiral', in which a ferromagnetic component of 7·6 μ_B along the *c*-axis is accompanied by a helical spiralling component within the basal plane, and equal to 4·3 μ_B. The effect of this combination on the diffraction pattern is indicated in Fig. 262 (d). First-order satellites accompany the (*hkil*) reflections and the even (000*l*) reflections. At the same time the ferromagnetic component gives magnetic intensity (shown in black) at the reciprocal lattice points themselves, but not at any (000*l*) points, because the ferromagnetic moments are perpendicular to the (000*l*) planes. The rotation of the basal plane moments (from layer to layer) remains at about 45° throughout the range of temperature for which this form of structure exists.

As a further example we discuss the case of the metal thulium which, as Table 26 shows, displays two different ordered modifications. Over the region of temperature from 57 K to 32 K there is a *c*-axis sinusoidally modulated antiferromagnetic structure similar to that which we have already met with erbium and which is represented by Fig. 261 (b). Accordingly the diffraction pattern incorporates (Fig. 264 (b)) a single pair of magnetic satellites at each allowed

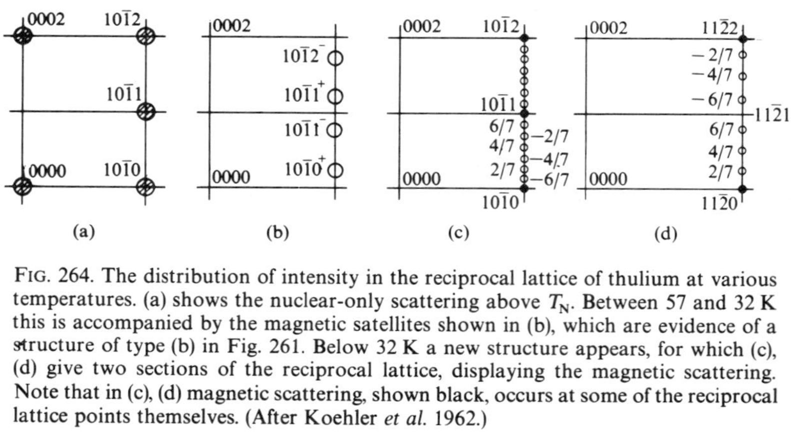

FIG. 264. The distribution of intensity in the reciprocal lattice of thulium at various temperatures. (a) shows the nuclear-only scattering above T_N. Between 57 and 32 K this is accompanied by the magnetic satellites shown in (b), which are evidence of a structure of type (b) in Fig. 261. Below 32 K a new structure appears, for which (c), (d) give two sections of the reciprocal lattice, displaying the magnetic scattering. Note that in (c), (d) magnetic scattering, shown black, occurs at some of the reciprocal lattice points themselves. (After Koehler *et al.* 1962.)

nuclear reflection position, with the exception of the (000*l*) reflections, just as for the corresponding form of erbium. It is again found from the separation of the satellites that the modulating wave has a length of about $\frac{7}{2}c$. Below 32 K a new structure exists, for which the magnetic intensity found in two sections through the reciprocal lattice is shown in (c) and (d) of Fig. 264. Again there are no magnetic contributions to (000*l*) reflections, indicating that the magnetic moments are parallel to the *c*-axis. There is a small magnetic contribution at some of the reciprocal lattice points themselves, such as (11$\bar{2}$0) and (11$\bar{2}$2), but not (11$\bar{2}$1), and this is consistent with a mean ferromagnetic moment per atom of 1 μ_B, in contrast with a measured moment of 7 μ_B per atom for polycrystalline thulium in the paramagnetic state. The intensity distribution among the satellites along a vertical row in the reciprocal lattice, as in Fig. 264 (c) and (d), is the Fourier transform of the distribution of moment along the *c*-axis of the structure, and it is deduced that this is of the form shown in Fig. 265, where succeeding columns of atoms are

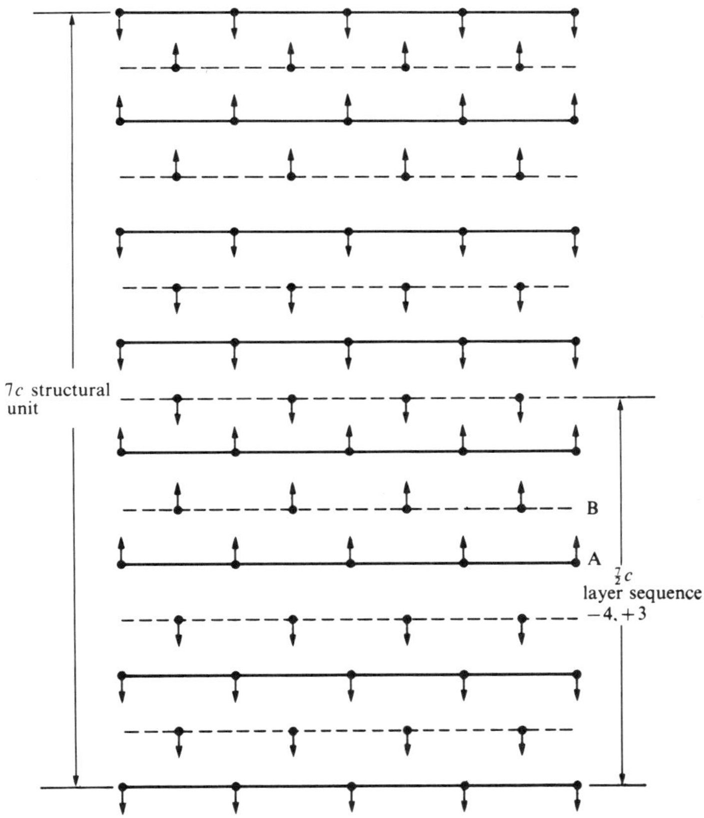

FIG. 265. The magnetic structure postulated for thulium at 4·2 K. The full and broken horizontal lines represent the two types of layer A, B in the hexagonal close-packed structure. The layer sequence gives four layers of downward-pointing moments followed by three layers of upward-pointing moments, producing a repeat unit of fourteen layers, i.e. 7*c*, when the two types of layer A, B are taken into account. (After Koehler *et al.* 1962.)

drawn in turn through (000) and $(\frac{1}{3}\frac{2}{3}\frac{1}{2})$ types of atom in the close-packed hexagonal structure. For further details of how the Fourier coefficients are found from the neutron intensities, and used to assess the distribution of magnetic moment on the atoms, the reader is referred to the original paper by Koehler, Cable, Wollan, and Wilkinson (1962). It will be seen that the structure repeats at intervals of 7 unit cells, within which the moment directions in successive layers follow a sequence -4, $+3$, -4, $+3$, ... This is the sequence which is indicated in Fig. 261 (a) and gives rise to a *mean* moment per atom of 1 μ_B if the numerical moment of each atom is 7 μ_B. In considering the general change in the magnetic contributions to the diffraction pattern when the crystal is cooled, it is the onset of the second and third harmonics of the satellites, below about 32 K, which marks the change from the *c*-axis-modulated anti-ferromagnetic structure of (b) in Fig. 261 to the ferrimagnetic arrangement in Fig. 261 (a).

The lighter rare-earth metals, from cerium to europium in the first half of the series, have more complex crystal structures than the heavier elements and their magnetic structures have been less thoroughly studied. However, we mention in particular the metal samarium, which has yielded to investigation in the form of single crystals of the isotope ^{154}Sm. This isotope has a cross-section of only 60 barns, in contrast to the naturally occurring element which has the prohibitively large value of about 12 000 barns. We have already mentioned the special interest of samarium because of its unusual form factor which peaks at a value of $(\sin\theta)/\lambda$ which is substantially different from zero. Samarium has an unusual layer structure which consists of a nine-layer sequence ABABCBCAC ... of close-packed layers; one-third of these layers have an immediate environment which is characteristic of cubic close-packing, in contrast to the hexagonal close-packing of the majority. Two magnetic structures have been observed by Koehler and Moon (1972). Below 106 K each plane of hexagonal sites forms a ferromagnetic sheet in which the moments are parallel to the *c*-axis and the sheets order in sequence $0++0--0++\ldots$ along the *c*-axis, as shown in Fig. 266 (a). The zeros in this sequence indicate the cubic-type planes, whose moments are disordered, and the nine planes together constitute only one-half of the magnetic unit cell. In the succeeding half of the cell the moment orientations are reversed, thus making a complete cell of 18 layers, extending over 52 Å. On cooling below

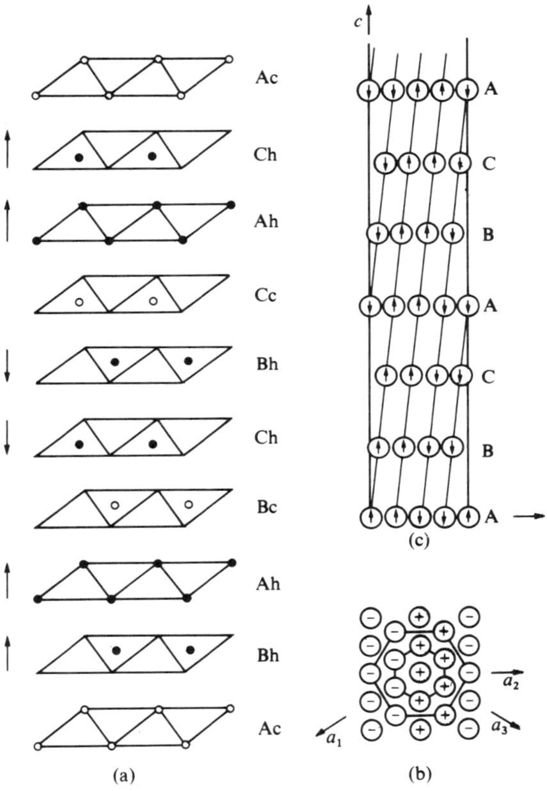

FIG. 266. The magnetic structures of samarium. The nine-layer sequence ABABC-BCAC illustrated at (a) includes layers with cubic or hexagonal environment indicated by c or h respectively. Between 106 K and 13·8 K the moments on the samarium atoms in the h layers order in pairs, giving a sequence ↑↑↓↓ and a unit of 18 layers, only half of which is shown in (a). Below 13·8 K the moments in the c layers order in the manner shown at (b), which is a plan of a layer and indicates that *lines* of atoms order perpendicular to the axis a_2. The outcome, indicated in (c) which is a vertical section through the cell, is that ferromagnetic sheets have appeared, parallel to the $(10\bar{1}1)$ planes. (Koehler and Moon 1972.)

13·8 K additional magnetic reflections appear in the diffraction pattern. These can be interpreted by assuming that the ordering of the atoms in the hexagonally-packed planes does not change, but that order sets in amongst the cubically-packed planes, in spite of the fact that these are almost 9 Å apart. Their moments become parallel or antiparallel to the c-axis in such a way that the layers are constituted only by ferromagnetic *lines* of atoms perpendicular to the axis a_2 and they do not form uniform sheets. These lines of atoms are indicated in Fig. 266 (b), which shows a view of an individual layer. In Fig. 266 (c) a section is drawn perpendicular to the layers, which shows how the moment directions for the ferromagnetic lines vary from layer to layer. The arrangement is such that ferromagnetic sheets have appeared parallel to the $(10\bar{1}1)$ planes. The resulting magnetic cell extends for 104 Å in the c-direction.

13.3. Rare-earth alloys

It is convenient to discuss first the alloys between the heavy rare-earth elements which we have examined above. These elements form a continuous series of solid solutions among themselves. An extended series of investigations of these has been carried out by Koehler and his co-workers, and they have also studied the alloys of these elements with the non-magnetic element yttrium. The yttrium alloys are of particular interest because yttrium has a similar configuration of outer electrons to the rare-earth metals and the same crystal structure, with a unit cell which is very close in size to that of gadolinium. Thus yttrium serves as an ideal non-magnetic diluent for the rare-earth metals. Some simple relationships hold for the metal and for both their alloys among themselves and those with yttrium. On cooling from room temperature it is found that the temperature at which magnetic order first sets in falls on a single universal curve when plotted against $(g-1)^2J(J+1)$ for a single metal, or against the average value of this quantity, when the average is taken over the atoms of an alloy. Here J is the angular momentum quantum number and g is the Landé splitting factor, and reference back to p. 222 will show that this expression signifies the square of the projection of the spin momentum on the direction of the total angular momentum vector $\mathbf{J}$. Moreover, the Curie points for the ferromagnetic gadolinium–yttrium alloys lie on the same curve as the Néel temperatures for the paramagnetic to antiferromagnetic change which occurs for the alloys of the other elements with yttrium. This curve is shown in Fig. 267 and includes not only a wide range of yttrium alloys but also a number of equiatomic alloys among the rare-earths themselves, such as ErDy, ErTb, and HoTb. The similar behaviour for both the binary AB

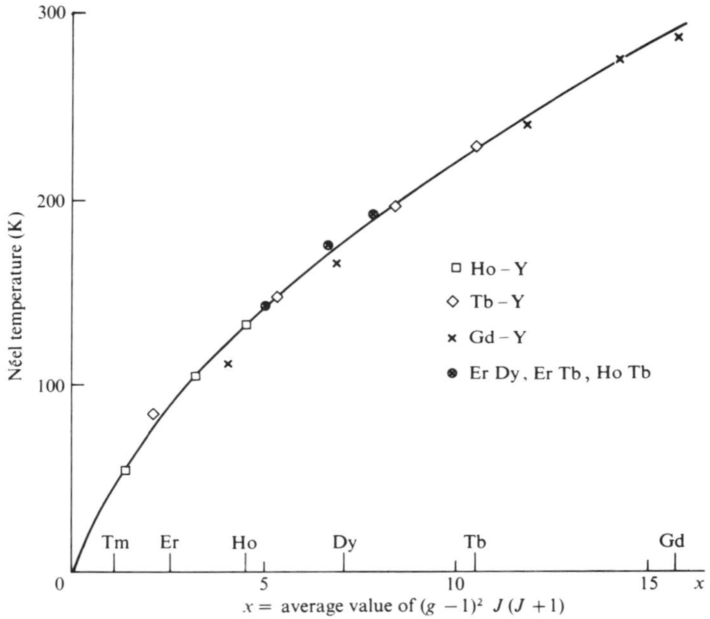

FIG. 267. A single curve is followed by the Neél temperatures of the rare-earth metals, their binary alloys, and alloys with yttrium when plotted against the quantity $(g-1)^2J(J+1)$ averaged over the constituent atoms. g is the Landé factor and J is the resultant angular momentum quantum number.

alloys and the yttrium alloys suggests that in the AB alloys both the A and B moments begin to order at the same temperature. It is also found that there is a common curve for the interlayer turn-angles in the helimagnetic structures which are first formed. This is illustrated in Fig. 268.

13.4. Transition-metal alloys

The majority of neutron studies of the magnetic structures of alloys have involved the transition metals of the iron group. In many cases it has been necessary to determine both the magnetic order and the chemical order among the constituent elements for, as we have seen in Chapter 12, neutrons are particularly valuable for distinguishing between many of the members of the iron group of elements. There are many different kinds of information which we may seek within this group of alloys. In some cases, such as the

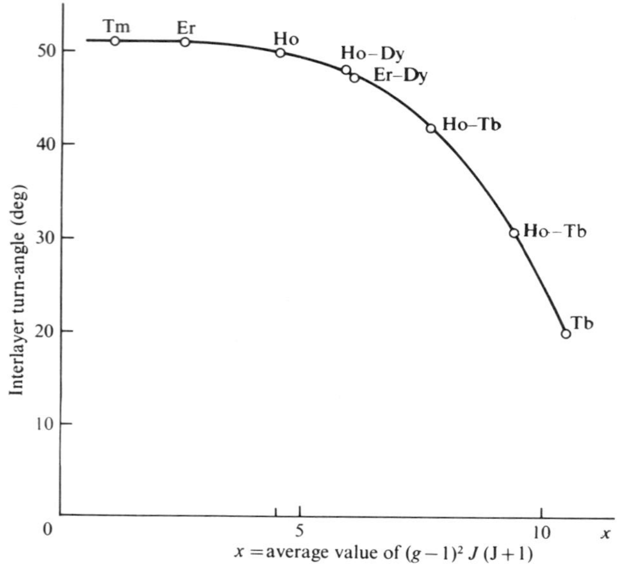

FIG. 268. Relation between interlayer turn-angle and the average value of $(g-1)^2 \times$ $\times J(J+1)$ for helimagnetic structures of rare-earth metals and binary alloys just below the Néel temperature.

Au–Mn system, we may observe several quite different types of magnetic structure as we alter the proportions of the constituents and we shall need to determine the details of these. In some alloys more than one structure may co-exist, not necessarily as two separate phases but possibly as a series of regions of short-range order which are themselves determined by local inhomogeneities of composition and which are coherently linked over large distances. We may also examine disordered ferromagnetic alloys, for which a study of the diffuse scattering in the background of the diffraction pattern permits us to assess the magnetic moments of the individual constituents, in contrast to obtaining the mean value of the moment which is given by macroscopic magnetic measurements. We can observe also the effect of inserting foreign atoms in a magnetic matrix, measuring both the magnitude and the range of the magnetic disturbance which is introduced into the magnetic material.

We shall discuss examples of the different kinds of problem in order to indicate the wide range of information which can be obtained.

One of the first alloy studies was made by Shull and Wilkinson (1955) to determine the individual magnetic moments of the two constituents in ordered ferromagnetic alloys such as Ni_3Fe. By using neutrons it is possible to determine separately the values for Ni and Fe, whereas macroscopic magnetic measurements only yield the mean moment per atom. The diffraction pattern shows weak superlattice reflections at the (100) and (110) positions, illustrated in Fig. 269, which are due to both nuclear and magnetic

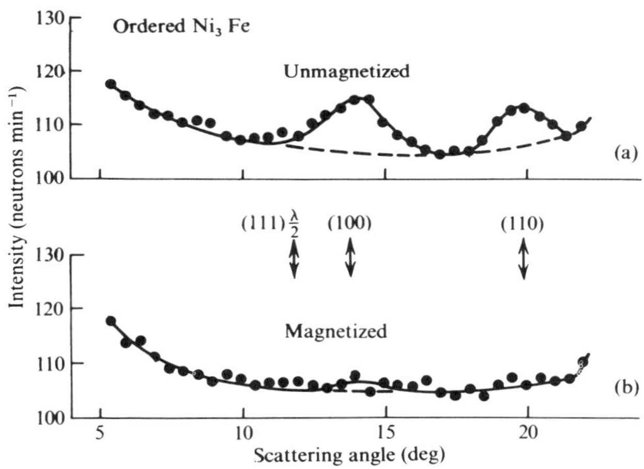

FIG. 269. The (100) and (110) superlattice reflections in the diffraction pattern of ordered Ni_3Fe for the case of (a) unmagnetized and (b) magnetized samples. (Shull and Wilkinson 1955.)

contributions. However, the nuclear contribution is very small because the nuclear scattering amplitudes of iron and nickel are rather similar. The magnetic contribution can be assessed directly by applying a saturating magnetic field in a direction parallel to the scattering vector, thus reducing the scattered intensity to that shown in Fig. 269 (b). Under these circumstances q^2 in eqn (6.16) is zero and the magnetic intensity is zero. Hence from the *reduction* of intensity we can obtain the value of $(\mu_{Ni} - \mu_{Fe})^2$. By a similar kind of procedure we can measure the magnetic contribution to a normal face-centred cubic reflection, such as (111), and this yields the value of $(\frac{3}{4}\mu_{Ni} + \frac{1}{4}\mu_{Fe})$. This same quantity can be obtained from saturation magnetization data. In this way, we can determine the values of μ_{Ni} and μ_{Fe} individually, but there are two possible solutions because of uncertainty in the sign of the quantity $(\mu_{Ni}-\mu_{Fe})$. This is illustrated graphically in Fig. 270.

Iron–aluminium, Fe_3Al, is a good example of an alloy for which the chemical order is readily found with X-rays but neutrons are needed to show the magnetic order. In the X-ray structure of Bradley and Jay (1932), shown in Fig. 271, there are two different

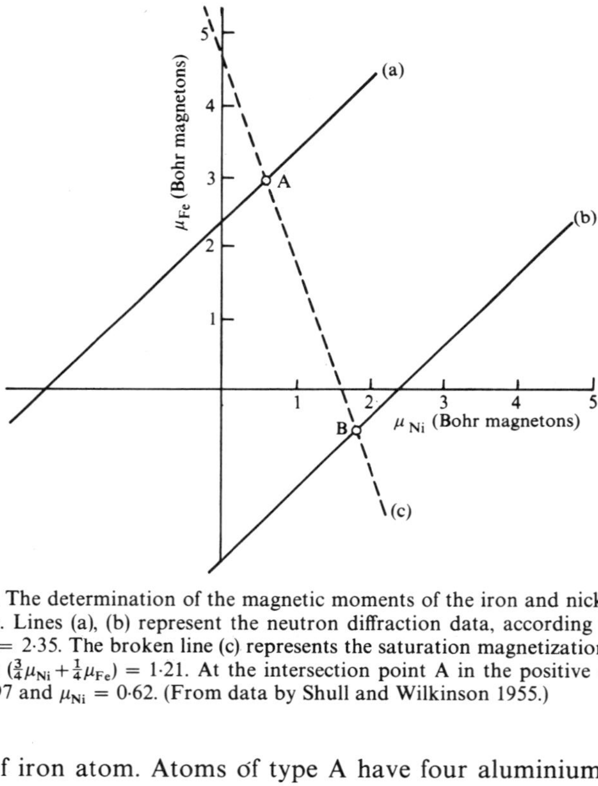

FIG. 270. The determination of the magnetic moments of the iron and nickel atoms in Ni_3Fe. Lines (a), (b) represent the neutron diffraction data, according to which $|\mu_{Fe}-\mu_{Ni}| = 2{\cdot}35$. The broken line (c) represents the saturation magnetization conclusion that $(\frac{3}{4}\mu_{Ni}+\frac{1}{4}\mu_{Fe}) = 1{\cdot}21$. At the intersection point A in the positive quadrant $\mu_{Fe} = 2{\cdot}97$ and $\mu_{Ni} = 0{\cdot}62$. (From data by Shull and Wilkinson 1955.)

types of iron atom. Atoms of type A have four aluminium atoms and four iron atoms as their nearest neighbours, with six further iron atoms as next-nearest neighbours. For atoms of type D, on the other hand, all the eight nearest-neighbours are iron atoms and next-nearest neighbours are all aluminium atoms. It was shown by Nathans, Pigott, and Shull (1958), using polarized neutrons, that the two types of iron atom had substantially different magnetic

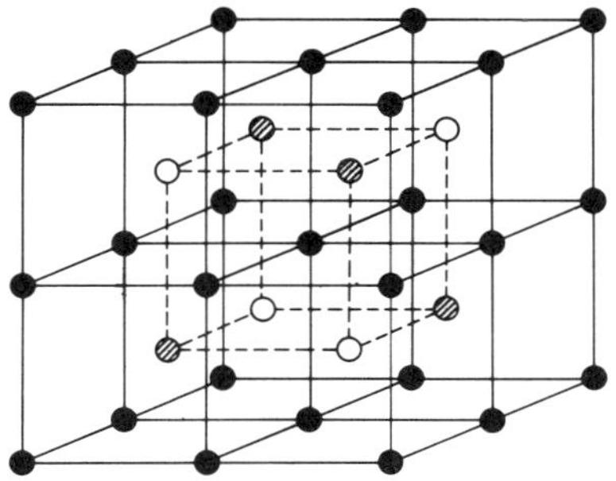

FIG. 271. The ordered arrangement of aluminium and iron atoms in Fe_3Al as deduced by Bradley and Jay (1932) from X-ray measurement.

moments, amounting to $(1{\cdot}46 \pm 0{\cdot}1)\,\mu_B$ and $(2{\cdot}14 \pm 0{\cdot}1)\,\mu_B$ respectively. The effective nuclear and magnetic scattering amplitudes for the different reflections are

$$b_{111} = b_D - b_B \qquad \text{and} \qquad p_{111} = p_D$$

$$b_{200} = 2b_A - b_B - b_D \qquad\qquad p_{200} = 2p_A - p_D$$

$$b_{220} = 2b_A + b_B + b_D \qquad\qquad p_{220} = 2p_A + p_D,$$

where the subscripts relate to the iron atoms on the A and D sites and the aluminium atoms on the B sites, which bear no magnetic moments. The accuracy of the measurement of magnetic moment depends on a precise determination of the ratios b_{hkl}/p_{hkl}, which can be done particularly accurately with polarized neutrons even for magnetic reflections of low intensity.

The Au–Mn system emphasizes how magnetic structure in an alloy series may change with the proportions of the constituents. In the neighbourhood of the equi-atomic composition, AuMn, the structure (Bacon 1962) is based on the very simple antiferromagnetic arrangement shown in Fig. 272. Above and below a transformation temperature of about 410 K the two variants shown in (a) and (b) in the Figure are found. Both structures are actually of tetragonal symmetry, and c/a is slightly less than, or slightly greater than, unity in the two cases respectively. Later and more accurate work

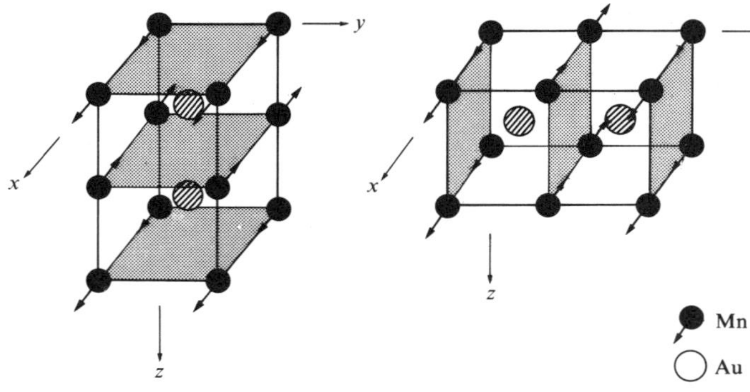

FIG. 272. Structures of Au–Mn alloys near the composition AuMn. (a) is the structure of an alloy with 49·2 per cent of Au at 453 K, where $c/a = 0{\cdot}97$. Below 403 K there is a transformation to structure (b) and a c/a ratio of 1·04. In each case the magnetic sheets are perpendicular to a *short* axis of the tetragonal cell.

by Bacon and Plant (1973) shows that the magnetic moments do not lie precisely in the magnetic sheets, shown shaded in the Figure, but at an angle of 20° to them. From these measurements it was concluded that the magnetic moment on the manganese atoms was $(4{\cdot}6 \pm 0{\cdot}06)\,\mu_B$.

Quite a different structure is found for alloys near the composition Au_2Mn. In fact this structure is the prototype structure for helimagnetism which was examined by Herpin, Meriel, and Villain (1959) and which we illustrated earlier in Fig. 158 (p. 263). Within each plane of manganese atoms the magnetic moments are all aligned parallel to each other but there is a rotation of 51°, about the c-axis, between the moment directions in successive planes spaced $\frac{1}{2}c$ apart. This results in the appearance of the characteristic satellite reflections which require non-integral indices for their description.

At the manganese-rich end of the composition range of gold–manganese alloys a third magnetic structure is found. This structure, illustrated in Fig. 273, was first observed by Meneghetti and Sidhu (1957) in the Cu–Mn system. It is regarded as characteristic of the γ-form of manganese, although the latter does not exist in a pure form and requires the presence of a few per cent of copper or gold atoms to stabilize it. The diffraction pattern of this structure is distinguished by the purely magnetic reflections (110) and (201). Successive (110) planes drawn through A, B, and C are shown shaded in Fig. 273, and it will be noted that these planes constitute ferromagnetic sheets and that they are interleaved by similar planes through P and Q in which the moments are oppositely directed. Thus the neutron contribution from the interleaving planes reinforces that from the main (110) planes and gives a

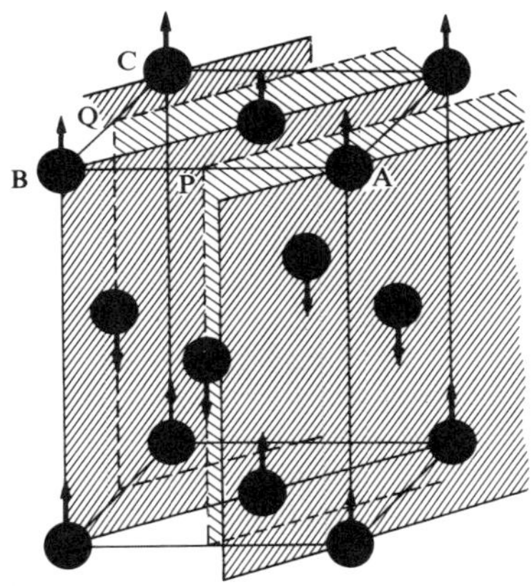

FIG. 273. The face-centred γ-manganese structure observed in manganese-rich Cu–Mn and Au–Mn alloys. Successive ferromagnetic (110) sheets through A, B, and C are interleaved by sheets of oppositely directed moments through P and Q, giving an antiferromagnetic arrangement.

magnetic reflection. Some typical diffraction patterns for manganese-rich alloys in the Au, Mn series are shown in Fig. 274, and we emphasize two interesting features. With increase of the gold content the magnetic (110) reflection disappears and the nuclear reflections become very weak. The latter happens because the nuclear scattering amplitude of manganese is negative and we are approaching a nul-matrix alloy, such as we described in Chapter 4, in which there is mutual cancellation of the nuclear scattering contributions from a random arrangement of gold and manganese atoms. At the same time we note the appearance of a broad intense peak near the position of the absent (100) Bragg reflection. This diffuse peak has been studied by many workers, and several different interpretations of its existence have been put forward. For the samples shown in Fig. 274 (Bacon and Cowlam 1970) it

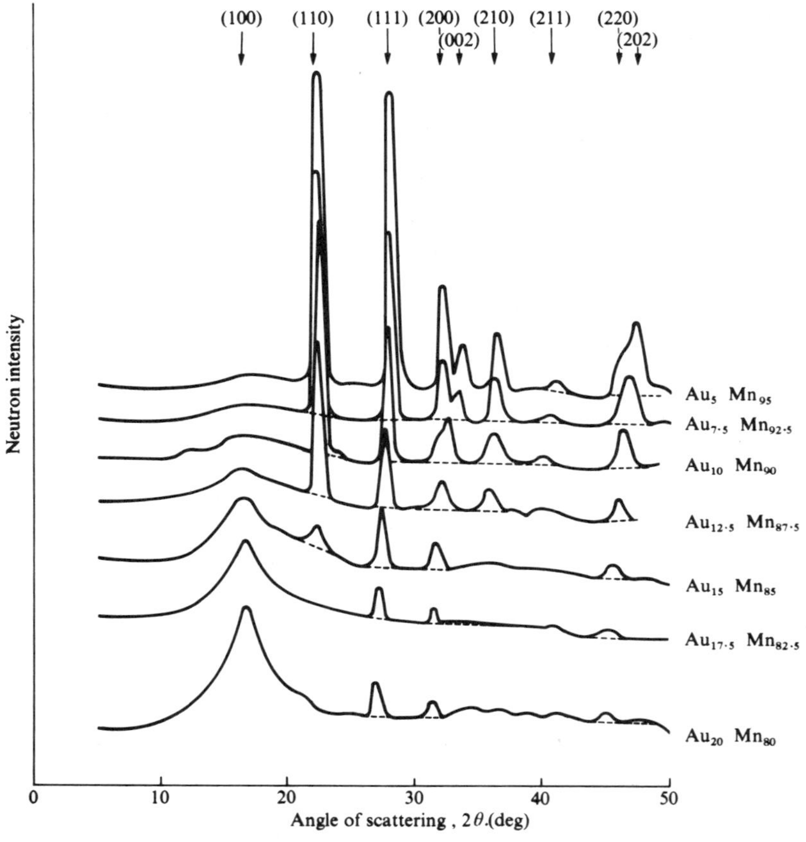

FIG. 274. Neutron diffraction patterns of manganese-rich Au–Mn alloys of various compositions. Note the intense antiferromagnetic (110) reflection for the higher contents of manganese. (Bacon and Cowlam 1970.)

is probably due to short-range order among the gold and manganese atoms.

Our discussion of the Au–Mn alloys leads quite naturally to an examination of the contribution of neutron diffraction to an understanding of the Heusler alloys. These were so-named in 1903 when Heusler first discovered that it was possible to make ferromagnetic alloys entirely from non-ferromagnetic components such as copper, manganese, tin, and aluminium. Typical examples were Cu_2MnAl and Cu_2MnSn. Later research has shown that both ferromagnetic and antiferromagnetic alloys occur among the Heusler series, and the crucial factor is the presence of manganese atoms which couple either ferromagnetically or antiferromagnetically among themselves according to their environment. Observations from neutron diffraction have permitted the determination of both the chemical order among the constituent atoms and also the nature of the alignment of the magnetic moments of the manganese atoms. For a general account of the Heusler alloys the reader is referred to an article by Webster (1969). We ourselves will simply comment on the extension of the binary Au–Mn alloy system to the ternary system $Au_2(Mn, Z)$. Here the element Z may be either Al, Zn, In, Ga, or Cu, and the system includes both ferromagnetics such as Au_2MnAl and antiferromagnetics such as Au_2MnZn. The magnetic arrangements in powdered samples of alloys in many of these ternary systems and also the chemical order between the manganese and Z atoms have been examined with neutrons by Bacon and Plant (1973). There is a wide variety of behaviour. Whereas Al, Ga, and In each orders chemically with the manganese, thus producing the ordered arrangement of Fig. 275 (b) with its enlarged unit cell,

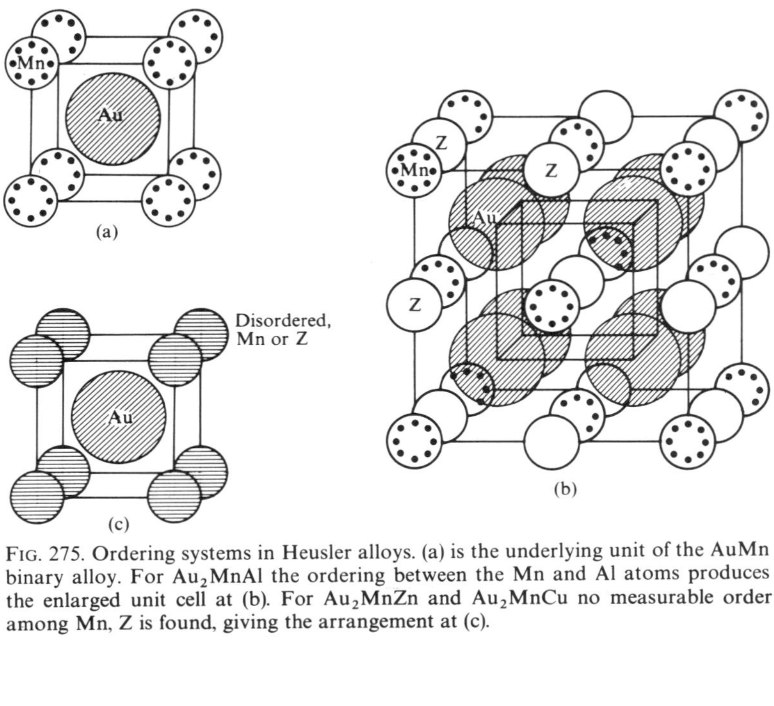

FIG. 275. Ordering systems in Heusler alloys. (a) is the underlying unit of the AuMn binary alloy. For Au_2MnAl the ordering between the Mn and Al atoms produces the enlarged unit cell at (b). For Au_2MnZn and Au_2MnCu no measurable order among Mn, Z is found, giving the arrangement at (c).

there is no measureable order for zinc or copper, where the arrangement is that shown at (c) in this Figure. The measurements of the Néel temperatures for the alloys can all be correlated by making the assumption that the dominant interaction is between manganese spins in odd coordination shells of neighbours. Only the copper derivatives are anomalous and seem to require the assumption that the copper atoms actually modify the strength of the coupling between manganese atoms.

In the alloy series $Au_2(Mn, Al)_2$ there is a continuous change from AuMn, which is antiferromagnetic, with the structure already shown in Fig. 275 (a), to Au_2MnAl, which is ferromagnetic. Between the two extremes it is found that alloys show both antiferromagnetic and ferromagnetic neutron reflections, although only a single crystallographic phase is detectable with X-rays. The susceptibility–temperature curve for these alloys is found to show two peaks. Inevitably there will be local inhomogeneities of chemical order, since the exact chemical ratio of atoms cannot be matched by *integral* numbers of the two constituents, and these inhomogeneities will determine a local allegiance to either an antiferromagnetic or a ferromagnetic structure. We suppose, in particular, that where an antiferromagnetic region of a few atoms occurs it will be phased coherently with similar neighbouring regions. In this way coherent regions of both antiferromagnetism and ferromagnetism can be built up, with the relative strengths which are indicated by the two curves in Fig. 276.

A more striking example of the co-existence of two magnetic structures was found by Bacon and Crangle (1963) in Pt–Fe alloys in the neighbourhood of the composition Pt_3Fe. The latter alloy shows complete chemical ordering of the iron and platinum atoms with the face-centred cubic arrangement which is characteristic of Cu_3Au. At this 3:1 ratio of atoms, and also for lower concentrations of iron, the magnetic structure consists of ferromagnetic (110) sheets arranged with their moments lying within the sheets but directed alternately + and − from sheet to sheet, to produce the antiferromagnetic arrangement illustrated in Fig. 277 (a). For alloys which contain more iron atoms than in Pt_3Fe there will inevitably be some near-neighbour iron atoms, such as the atom A in Fig. 277 (b), and under these circumstances a second structure appears, as in Fig. 277 (b), in which the ferromagnetic sheets are parallel to the (100) planes. Over the composition range of 25–30

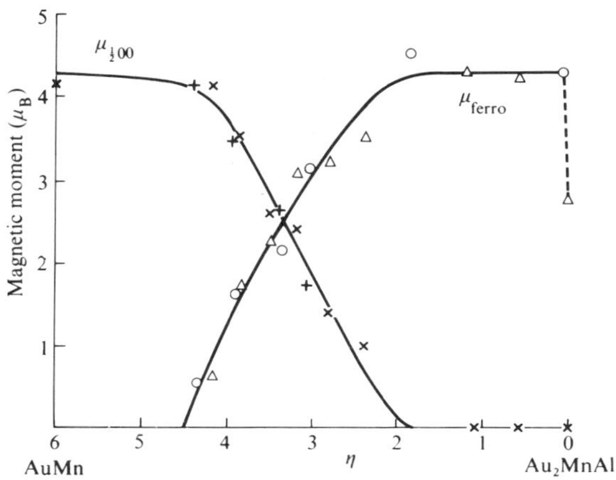

FIG. 276. The variation with composition in the alloy series $Au_2(Mn, Al)_2$ of the components of magnetic moment which are contributed separately to the antiferromagnetic and ferromagnetic structures. The abscissa is the parameter η, which is the average number of manganese atoms which are cube-edge near neighbours to each manganese atom. η can be computed from the composition, and the experimentally-measured order parameter x. For $\mu_{\frac{1}{2}00}$, $\times$ represents slow-cooled alloys and $+$ represents quenched alloys. For μ_{ferro}, $\triangle$ represents slow-cooled alloys and $\bigcirc$ represents quenched alloys. (Bacon and Mason 1967.)

per cent of iron the two types of antiferromagnetic structure are co-existent. The magnetic diffraction lines are sharp, and it is necessary to assume that there is coherence in phase for the scattered neutrons, and hence of the moment direction, over many unit cells. With further introduction of iron atoms there is a transition to ferromagnetism. The over-all allocation of magnetic moments among the three magnetic states varies with composition as shown by the curves in Fig. 277 (c).

A series of manganese-rich alloys with the cubic Cu_3Au structure show a quite different kind of behaviour. Kren *et al.* (1966) have studied Mn_3Pt, Mn_3Rh, and $Mn_3Pt_{0\cdot5}Rh_{0\cdot5}$. Each of these alloys shows an antiferromagnetic structure with the triangular arrangement of manganese moments, which is illustrated in Fig. 278 (a). The magnetic unit cell is of the same size as the chemical cell and the magnetic moments lie in (111) sheets. However, in the case of Mn_3Pt this triangular structure does not persist up to the paramagnetic transformation, and there is an intervening collinear structure with a unit cell which is doubled in one direction as

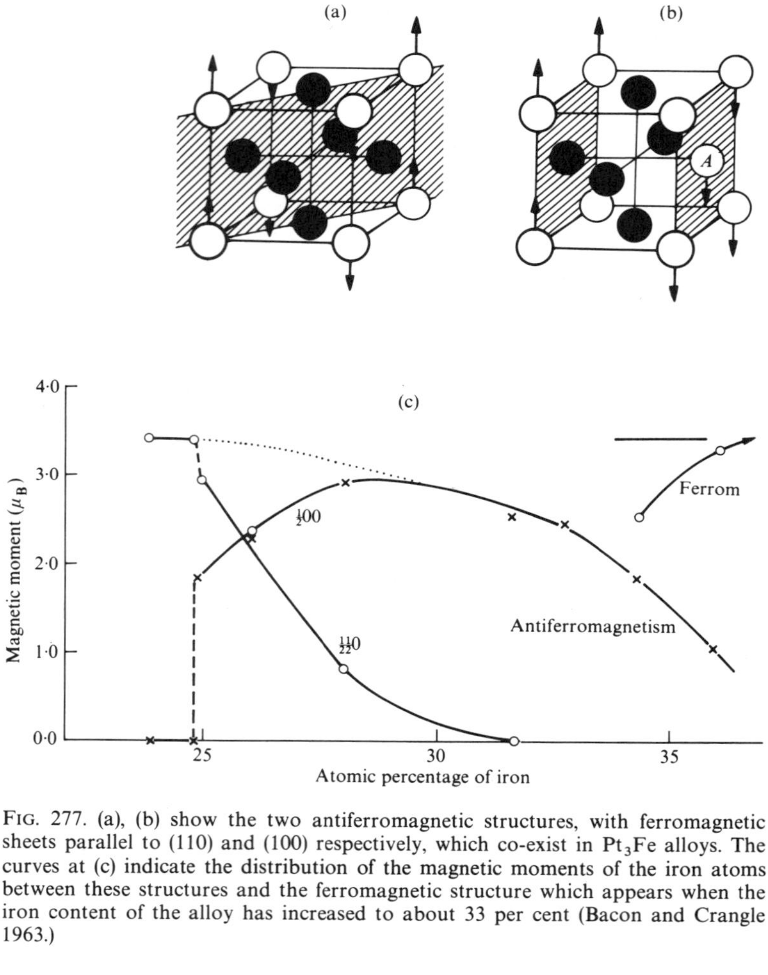

FIG. 277. (a), (b) show the two antiferromagnetic structures, with ferromagnetic sheets parallel to (110) and (100) respectively, which co-exist in Pt_3Fe alloys. The curves at (c) indicate the distribution of the magnetic moments of the iron atoms between these structures and the ferromagnetic structure which appears when the iron content of the alloy has increased to about 33 per cent (Bacon and Crangle 1963.)

illustrated in Fig. 278 (b). It is found that this collinear arrangement also occurs for small replacements of platinum by rhodium, as indicated by the phase diagram in Fig. 278 (c), and also for small excesses of platinum in alloys between Mn_3Pt and $Mn_{2\cdot 93}Pt_{1\cdot 07}$. The distinctive triangular and collinear structures are characterized quite clearly in the neutron diffraction pattern by the (110) and $(10\frac{1}{2})$ reflections respectively.

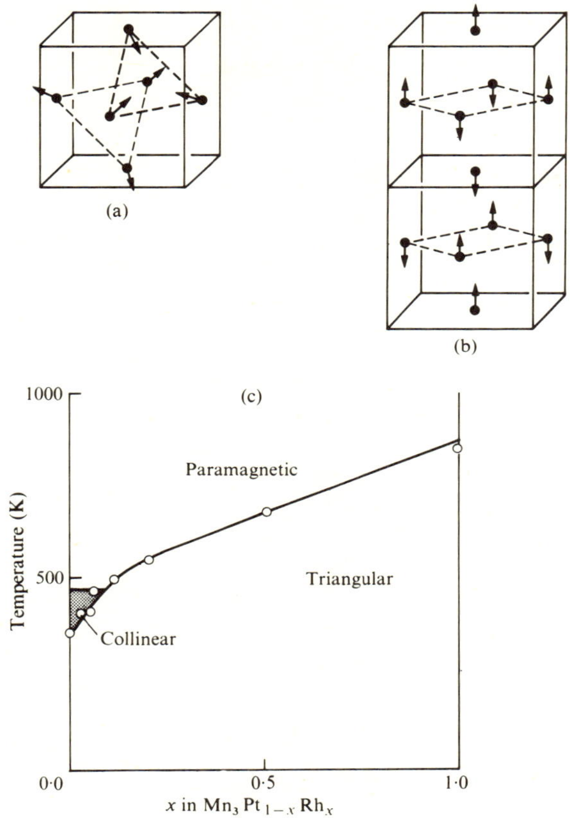

FIG. 278. Magnetic structures in the alloy series Mn_3(Pt, Rh), indicating only the manganese atoms. The usual non-collinear structure shown at (a) is replaced in a small region of the phase diagram (indicated at (c)) by a collinear structure (b) with a doubled unit cell. (Kren *et al.* 1966.)

Mn_3Ga, Mn_3Sn, and Mn_3Ge have a hexagonal crystal structure which is closely related to the cubic compounds that we have just discussed, and Fig. 279 indicates the ordered superlattice which is formed. In Mn_3Ga (Kren and Kadar 1970) there is again a triangular arrangement of spins, and these lie in the basal plane of the hexagonal cell, which is the equivalent of the (111) plane in the cubic compounds. The moments each make an angle of 45°

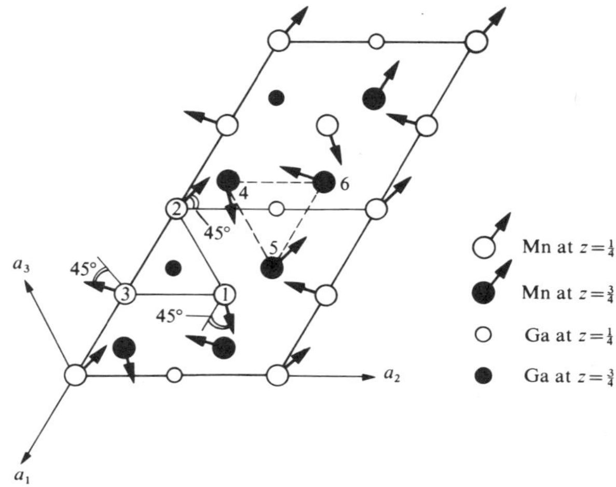

FIG. 279. A plan of the ordered magnetic superlattice for the alloy Mn_3Ga. There is a triangular non-collinear arrangement of the spins on the manganese atoms, lying in the basal plane of the hexagonal unit cell. The moments make an angle of 45° with the three horizontal hexagonal axes a_1, a_2, and a_3, and the figure distinguishes the moment directions for the layers of atoms at $z = \frac{1}{4}, \frac{3}{4}$ respectively. (Kren and Kadar 1970.)

with the horizontal hexagonal axes, as shown in the projection of Fig. 279, which distinguishes the manganese atoms in successive layers at levels of $z = \frac{1}{4}$ and $z = \frac{3}{4}$ in the unit cell. On the other hand, for Mn_3Sn and Mn_3Ge (Kouvel and Kasper 1965) the spins, although still forming a triangular arrangement, lie in a plane which is perpendicular to the basal plane. Moreover, a separate phase of Mn_3Sn has been identified below 270 K, and Mn_3Ga undergoes a change to a collinear ferrimagnetic structure of tetragonal symmetry above 770 K.

13.5. Disorder and defects in ferromagnetic alloys

We have seen already how it is possible to determine individually the magnetic moments of the two constituents in an ordered ferromagnetic alloy such as Ni_3Fe. We shall show that it is possible to make a similar measurement in a disordered alloy by studying the background scattering. We shall also find that the same kind of study will enable us to determine the nature of the disturbance produced by a foreign atom in a magnetic matrix. For example, we can find out what is the effect of introducing a small proportion of nickel atoms to a sample of iron.

We will first examine the scattering from a disordered binary alloy by extending our discussion in Chapter 2, which led to eqn (2.23). In our present problem the two types of magnetic moment are distributed at random amongst the available atomic sites, in a similar way to that in which the various possible isotopes were distributed in our earlier discussion. In eqn (2.23) we replace the nuclear scattering amplitude b_r by qp_r, where p_r is the magnetic scattering amplitude of the atom concerned and $\mathbf{q}$ is the magnetic interaction vector. Accordingly, the differential magnetic scattering cross-section for any angle of scattering 2θ will be

$$\frac{d\sigma}{d\Omega} = q^2\{\overline{p_r^2} - (\bar{p}_r)^2\}, \tag{13.1}$$

where the value of p_r will incorporate the value of the magnetic form factor of the atom which is appropriate to the value of θ. If the fractional amounts of the two components are c_1, c_2, so that $c_1 + c_2 = 1$, then this equation becomes

$$\begin{aligned}\frac{d\sigma}{d\Omega} &= q^2\{c_1p_1^2 + c_2p_2^2 - (c_1p_1 + c_2p_2)^2\}\\ &= q^2\{c_1c_2p_1^2 + c_1c_2p_2^2 - 2c_1c_2p_1p_2\}\\ &= q^2c_1c_2(p_1 - p_2)^2. \end{aligned} \tag{13.2}$$

We can determine experimentally how much of the background of the diffraction pattern is due to this 'ferromagnetic disorder' scattering by applying a magnetic field in the direction of the scattering vector and observing the reduction in the background count. Since the amplitudes p_1, p_2 contain the magnetic form factors f_1, f_2, which fall off rapidly as θ increases, the contribution to the background will only be important at small angles of scattering. By making measurements at several different angles it is possible to deduce the value of $(\mu_1 - \mu_2)^2$ by extrapolating to $\theta = 0°$. Some typical measurements by Shull and Wilkinson (1955) for a series of Fe–Cr alloys are shown in Fig. 280. From magnetization data for the same sample the mean magnetic moment $c_1\mu_1 + c_2\mu_2$ can be found. Combination of this value with that of $(\mu_1 - \mu_2)^2$ results in two possible sets of values for μ_1, μ_2, just as for the ordered alloys

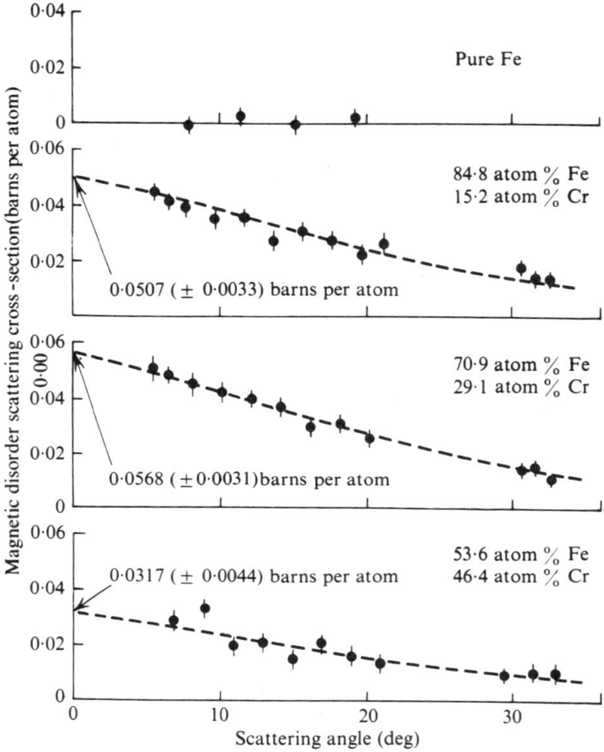

FIG. 280. The angular variation of the ferromagnetic disorder scattering for a series of disordered Fe–Cr alloys. (Shull and Wilkinson 1955.)

which we mentioned earlier. These two possible conclusions are indicated in Fig. 281, which shows how the individual magnetic moments of iron and chromium vary with the composition of the alloy. In practice these measurements are quite difficult to make accurately because they often involve the subtraction of two relatively large quantities, namely, the background count from all sources with and without the applied field. It is also necessary to make a correction for the change in absorption coefficient when the sample is magnetized; this effect, which is often known as the 'single transmission effect' will be discussed in Chapter 15. We shall also show later how it is possible to make an unambiguous choice between the two possible solutions given in Fig. 281 by making measurements with polarized neutrons.

The studies of alloys which we have just considered show, for certain elements at least, substantial variations of the moments with composition. In a given disordered alloy there will inevitably be substantial statistical variations of local environment, and the observed values of μ_1, μ_2 will be no more than representative of the general statistical average. Much more valuable information

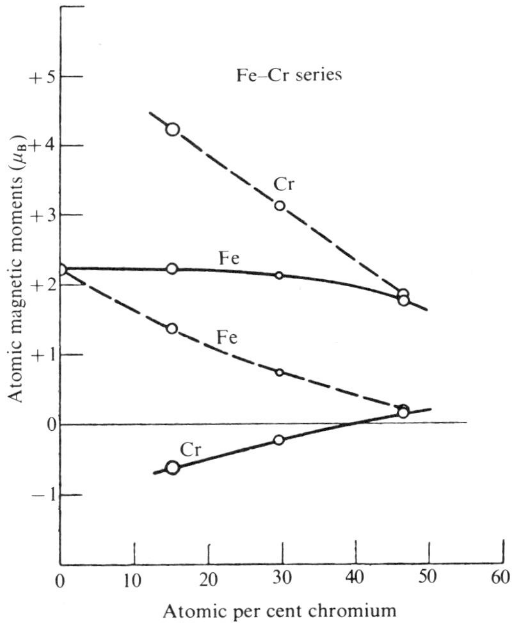

FIG. 281. The variation of magnetic moments with composition in Fe–Cr alloys. The two alternative solutions are represented by the full and dashed lines respectively. (Shull and Wilkinson 1955.)

could be forthcoming from dilute alloys, containing only 1 or 2 per cent of a second constituent, so that these 'foreign' atoms will be sufficiently far apart to have no influence on each other. We should then be able to assess the true effect on the matrix of such a single atom. The contribution to the background scattering will be much smaller for a dilute alloy of this kind, and an improved technique of measurement has been developed for this purpose by Low and Collins (1963). Before discussing this technique we shall derive an expression for the magnitude and distribution of the diffuse scattering for a single foreign atom or defect.

It will prove to be convenient to do this by an extension of eqn (2.20), in order to sum up the magnetic scattering contributions from the atoms in an alloy sample. For atoms at O and P in Fig. 282 (a) it follows that the phase difference between the scattered contributions is $(2\pi/\lambda)2x \sin \theta$, where 2θ is the angle of scattering. This can be written as $(4\pi \sin \theta/\lambda)r \cos \phi$ where ϕ is the angle between ON and OP or, alternatively, $\mathbf{Q} \cdot \mathbf{r}$, where $\mathbf{Q}$ is the vector $(\boldsymbol{\kappa} - \boldsymbol{\kappa}_0)$ which we discussed in Chapter 6, i.e. the vector which is in the direction ON and of magnitude $(4\pi \sin \theta)/\lambda$.

Let us now consider a sample of material made up of unit cells as in Fig. 282 (b). Taking an arbitrary point O as origin the differential

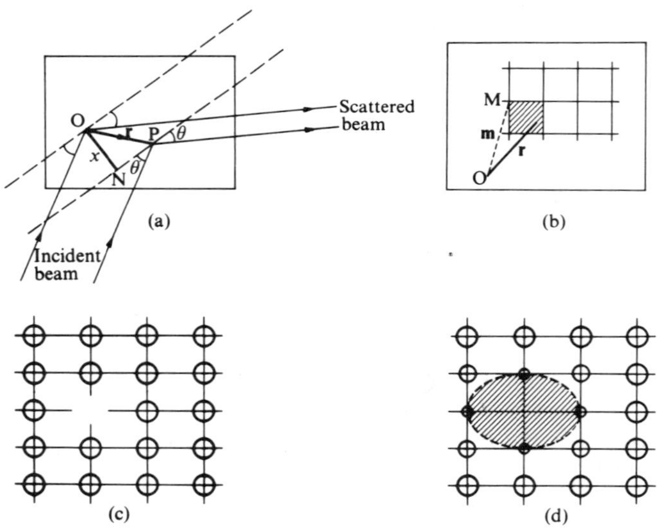

FIG. 282. Calculation of defect scattering by foreign atoms in dilute alloys.

magnetic scattering cross-section at an angle 2θ will be, utilizing eqn (6.8),

$$\frac{d\sigma}{d\Omega} = \left(\frac{e^2\gamma}{mc^2}\right)^2 q^2 \left| \int_{\text{sample}} dV \, \rho(r) \exp(i\mathbf{Q} \cdot \mathbf{r}) \right|^2, \qquad (13.3)$$

where $\rho(r)$ is the spin density per unit volume at the position r and the integration is carried out over the whole volume V of the sample. The value of Q, and its direction, will be dependent on the angle of scattering for which the measurements are being made. Alternatively, we can make the summation in turn over each of the unit cells in the sample, relating the phase of the contribution from each cell to the phase applicable to the corner of the cell, such as the point M identified by the vector **m**. In this way the cross-section becomes proportional to

$$\left|\sum_m \exp(\mathrm{i}\mathbf{Q}\,.\,\mathbf{m}) \int_{\text{cell}} \mathrm{d}V\, \rho(r) \exp\{\mathrm{i}\mathbf{Q}\,.\,(\mathbf{r}-\mathbf{m})\}\right|^2, \qquad (13.4)$$

where the summation is taken over all the unit cells in the sample and the integral is taken, in turn, over each cell.

In our dilute alloy, or indeed for any matrix which possesses defects, any cell which includes a defect will have a distinctive value of $\rho(r)$, and it is convenient to define

$$I_m = \int_{\text{cell}} \mathrm{d}V\, \rho(r) \exp\{\mathrm{i}\mathbf{Q}\,.\,(\mathbf{r}-\mathbf{m})\}$$

as a quantity unique to a particular cell m, in contrast to the average value of this quantity over all cells which we write as $\bar{I}$; thus $\bar{I} = (1/N)\sum_m I_m$, where N is the total number of unit cells in the sample.

We can then separate the expression (13.4) into two terms as

$$\left|\sum_m \exp(\mathrm{i}\mathbf{Q}\,.\,\mathbf{m})(I_m-\bar{I})+\bar{I}\sum_m \exp(\mathrm{i}\mathbf{Q}\,.\,\mathbf{m})\right|^2.$$

Because $\bar{I}$ is the average value, the cross-terms in the expression of this squared series will reduce to zero and the expression becomes

$$\left|\sum_m \exp(\mathrm{i}\mathbf{Q}\,.\,\mathbf{m})(I_m-\bar{I})\right|^2+\bar{I}^2\left|\sum_m \exp(\mathrm{i}\mathbf{Q}\,.\,\mathbf{m})\right|^2. \qquad (13.5)$$

The first term in this expression is now the diffuse magnetic scattering, which is distributed widely in the background of the diffraction pattern, whereas the second term is the coherent scattering, which is concentrated very sharply into the Bragg reflections. It is evident that the first term depends on the *departures* from perfection in the otherwise perfect periodic array. There will be no correlation or interference between contributions from individual defects positioned at random.

Let us first calculate the contribution to the diffuse scattering from a single defect. The contribution to the first term in (13.5) will be

$$\left|\sum_m \exp(\mathrm{i}\mathbf{Q}\,.\,\mathbf{m})\int \mathrm{d}V\,\{\rho(r)-\rho_{\mathrm{av}}.(r)\}\exp\{\mathrm{i}\mathbf{Q}\,.\,(\mathbf{r}-\mathbf{m})\}\right|^2$$

$$=\left|\int_{\text{sample}} \mathrm{d}V\,\{\rho(r)-\rho_{\mathrm{av}}(r)\}\exp(\mathrm{i}\mathbf{Q}\,.\,\mathbf{r})\right|^2,$$

which permits, if necessary, the effects of a defect to spread to neighbouring cells.

If there is a concentration c of such defects and a total N of unit cells, then the differential diffuse scattering will be proportional to

$$Nc\left|\int_{\text{sample}} \mathrm{d}V\,\{\rho_m(r)-\rho_{\mathrm{av}}(r)\}\exp(\mathrm{i}\mathbf{Q}\,.\,\mathbf{r})\right|^2, \qquad (13.6)$$

assuming that there is no interaction between the defects and that their effects are additive.

For a dilute alloy it is more useful to express the scattering in terms of the departure from ρ_0, the spin density in the undisturbed matrix, rather than from ρ_{av}. When this is done expression (13.6) becomes, after restoring the numerical constants,

$$\left(\frac{\mathrm{d}\sigma}{\mathrm{d}\Omega}\right)_{\text{diffuse}}=\left(\frac{e^2\gamma}{mc^2}\right)^2 q^2Nc(1-c)\left|\int_{\text{sample}} \mathrm{d}V\,\rho'(r)\exp(\mathrm{i}\mathbf{Q}\,.\,\mathbf{r})\right|^2,$$

where $\rho'(r)$ is the departure from the spin density in the undisturbed matrix, to be found at distance r from a defect.

If the magnetic disturbance is strictly limited to the removal of a magnetic atom, as in Fig. 282 (c), with no associated disturbance on the neigbouring atoms, then the above expression will vary with scattering angle in just the same way as an atomic form factor, since this is what the quantity $\int\rho'(r)\exp(\mathrm{i}\mathbf{Q}\,.\,\mathbf{r})\,\mathrm{d}V$ represents. However, if the effect of the disturbance is more widespread, as in Fig. 282 (d), where it affects the neighbouring atoms in the surrounding matrix, then $\rho'(r)$ will retain a value out to larger values of r and the diffuse scattering will be more sharply defined. Quantitatively, the magnitude and distribution of the disturbance is the Fourier transform of the diffuse neutron intensity. Accordingly we can use the measurements of the background neutrons to determine the details of the magnetic disturbance, and we will now describe the modified technique of Low and Collins (1963), which permits this to be done accurately for very dilute alloys.

The aim of the experimental arrangement is to increase the intensity of the scattered neutron beams and also to reduce the effects of other types of contribution to the background. This is achieved by using long-wavelength neutrons, greater than 4 Å, which are too long for Bragg reflection and therefore avoid double Bragg scattering, which is normally a contribution to the background which it is difficult to calculate very accurately. In the experimental arrangement of Fig. 283 the long-wavelength neutrons are selected

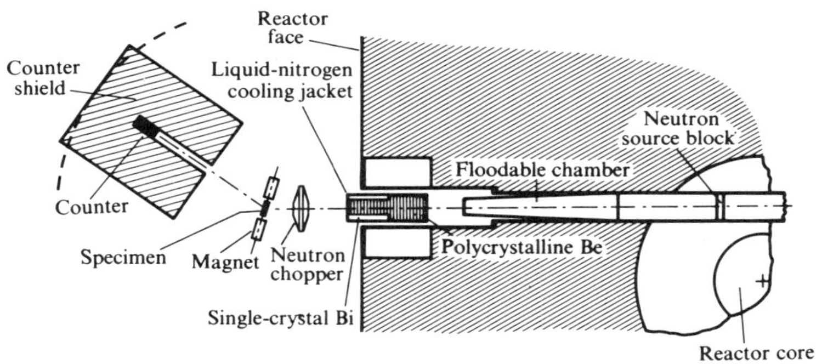

FIG. 283. Diagram of apparatus for measuring the defect scattering in magnetic alloys using a loosely collimated beam of neutrons with $\lambda > 3{\cdot}95$ Å, produced by transmission through a filter of polycrystalline beryllium. (G.G.E. Low.)

by passing the incident beam through a block of polycrystalline beryllium, which transmits no neutrons which are shorter in wavelength than the cut-off value of 3·95 Å. A further advantage of the long wavelength is that it is possible to relax the angular collimation of the incident beam to as much as 2° without impairing the definition of the measured pattern, which depends on $(\sin\theta)/\lambda$. It will be seen from the Figure that the apparatus also incorporates a simple form of chopper. This is in order to exclude any background contribution which is due to inelastic scattering and operates by including a gating delay in the detecting circuits to correspond to the time of flight of neutrons of wavelength about 5 Å. The results of these experiments show some very interesting differences in the behaviour of different solute atoms. We summarize the results of the experiments with iron, for which at least 15 different metallic impurities have been tried, usually in quantities between 1 per cent and 2 per cent. In all cases there is a reduction of the magnetic moment on the impurity site to a value which is lower than that of an iron atom, but the extent of this reduction varies widely between different elements. For cobalt the moment falls only from 2·2 μ_B, the value for iron, to 2·1 μ_B, but at the other extreme titanium and chromium produce an oppositely directed moment, i.e. in anti-ferromagnetic alignment, of $-0{\cdot}7\ \mu_B$. Equally wide variations are found amongst the various impurity metals for the extent to which their disturbing effect spreads outwards to the surrounding matrix of iron atoms. Some contrasting plots of the diffuseness of the scattering are given in Fig. 284. Thus manganese and titanium give broad

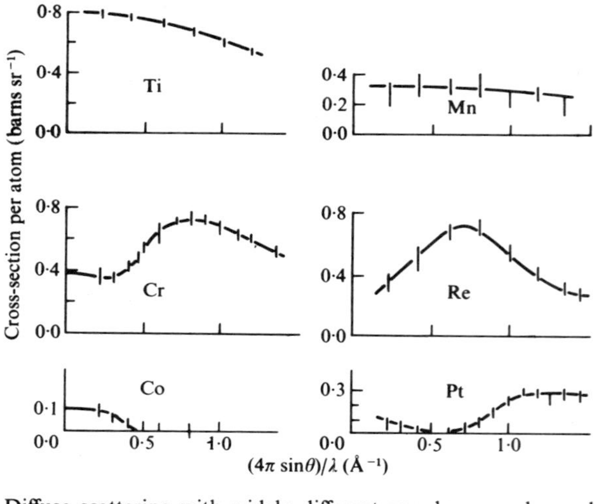

FIG. 284. Diffuse scattering with widely different angular spreads produced from dilute alloys of Ti, Mn, Cr, Re, Co, and Pt in iron. For Mn and Ti the disturbance scarcely extends beyond the impurity site itself. (Low and Collins 1963.)

distributions which indicate that the disturbance scarcely extends beyond the impurity site itself. On the other hand, all other impurities give much sharpened patterns, which are interpreted to show that there is a nett increase of moment which is distributed amongst the iron atoms which surround the impurity. Fig. 285 indicates how this distribution of moment may vary. For V, Cr, Mo, Ru, W, Re, and Os, all of which are elements which lie to the left or beneath Fe in the periodic table, the increase of moment is concentrated at a distance of about 5 Å from the impurity, as in the lower curve in the Figure. This distance corresponds roughly to the fourth and fifth neighbours of the impurity, and the magnitude of $\rho'(r)$ suggests

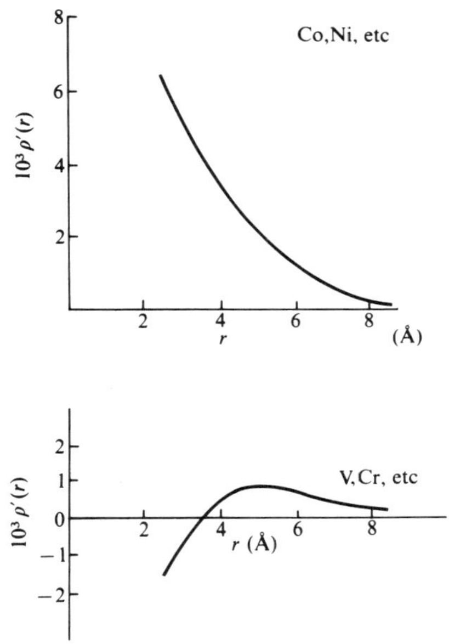

FIG. 285. Transfer of magnetic moment to the iron neighbours of an impurity atom in dilute iron alloys. The upper curve relates to atoms such as Co, Ni, and Pt which lie to the right of iron in the periodic table; the lower curve is for atoms such as V, Cr and Ru which lie below or to the left of iron. The ordinate is 10^3 times $\rho'(r)$, the fractional disturbance of the magnetic moment density in the matrix. (Collins and Low 1965.)

that these atoms have their moments increased by about 0·5 per cent. This may seem a small increase but it must be remembered that there are about 50 atoms within a spherical shell extending from 4 Å to 6 Å so that, over all, there is quite a significant redistribution of magnetic moment about the impurity. Elements which lie to the right of Fe in the periodic table, such as Co, Ni, Rh, Pd, Ir, and Pt, behave differently, as indicated by the upper curve in Fig. 285. For them the disturbance of moment is larger for the nearer neighbours, and it falls off steadily as we go outwards beyond a distance of 3 Å, where it amounts to about 3 per cent. It is emphasized that these quite small changes in magnetic moment which are deduced from these experiments indicate the high accuracy that has been attained in the identification of the diffuse magnetic scattering.

Let us return to our calculation of the diffuse scattering intensity for a disordered or defective material. It is easy to show that eqn (13.6) which we derived above is equivalent to eqn (13.2) which we used earlier. For, if we have an alloy with concentrations c_1, c_2 of atoms with magnetic scattering amplitudes p_1, p_2, there will be c_1 cells for which

$$\left(\frac{e^2\gamma}{mc^2}\right)\int \mathrm{d}V\ \rho(r)\exp\{\mathrm{i}\mathbf{Q}\,.\,(\mathbf{r}-\mathbf{m})\} = p_1,$$

which includes the appropriate form factor, and c_2 cells for which

$$\left(\frac{e^2\gamma}{mc^2}\right)\int \mathrm{d}V\ \rho(r)\exp\{\mathrm{i}\mathbf{Q}\,.\,(\mathbf{r}-\mathbf{m})\} = p_2,$$

and accordingly the average value will be $(c_1p_1+c_2p_2)$.

Hence for c_1 cells the $(\rho(r)-\rho_{\mathrm{av}})$ term will be given by p_1- $-(c_1p_1-c_2p_2)$, i.e. $c_2(p_1-p_2)$, and for c_2 cells this term will be $c_1(p_1-p_2)$. Accordingly, in terms of expression (13.6), the resulting intensity for a random distribution will be

$$q^2\{c_1c_2^2(p_1-p_2)^2+c_2c_1^2(p_1-p_2)^2\}$$
$$=q^2c_1c_2(p_1-p_2)^2$$

for each unit cell, which agrees with eqn (13.2).

This expression, we emphasize, is the magnetic contribution to the intensity of the diffuse scattering. There will also be a nuclear contribution, and for a beam of unpolarized neutrons these two contributions are additive, giving a resultant intensity equal to

$$c_1c_2\{(b_1-b_2)^2+q^2(p_1-p_2)^2\}. \tag{13.7}$$

If a beam of *polarized* neutrons is used then there is coherence between the nuclear and magnetic amplitudes, with an effective amplitude equal to $b+(\boldsymbol{\lambda}\,.\,\mathbf{q})p$, which will equal $b\pm p$. Eqn (13.7) then becomes

$$c_1c_2\{(b_1-b_2)^2+q^2(p_1-p_2)^2+2(\boldsymbol{\lambda}\cdot\mathbf{q})(b_1-b_2)(p_1-p_2)\}. \quad (13.8)$$

We note that this expression reduces to the form of eqn (13.7) in the case of unpolarized neutrons because the term $\boldsymbol{\lambda}\cdot\mathbf{q}$ then averages to zero. We can, however, use eqn (13.8) for a polarized beam to determine the actual sign of (p_1-p_2) and thus avoid any ambiguity in the measurements which we discussed earlier.

In the experiments of Collins and Forsyth (1963) the differential diffuse scattering cross-sections were measured

(1) (I_1) for neutrons having their spin direction parallel to the direction of magnetization of the sample (i.e. antiparallel to the spin direction in the sample) when the magnetic field is applied perpendicular to the scattering vector—thus $\boldsymbol{\lambda}\cdot\mathbf{K} = -1$ and $\boldsymbol{\lambda}\cdot\mathbf{q} = +1$ (see Fig. 117, p. 214); and

(2) (I_2) for neutrons having their spin direction antiparallel to the direction of magnetization (i.e. parallel to the spin direction in the sample)—thus $\boldsymbol{\lambda}\cdot\mathbf{K} = +1$ and $\boldsymbol{\lambda}\cdot\mathbf{q} = -1$.

In each case $q_\bullet$ is unity. We can therefore compute from eqn (13.8) the ratio of the difference to the sum of these two measurements, obtaining

$$\frac{I_1-I_2}{I_1+I_2} = \frac{2(b_1-b_2)(p_1-p_2)}{(b_1-b_2)^2+(p_1-p_2)^2}. \quad (13.9)$$

It follows from this equation that if $I_1 > I_2$ then (b_1-b_2) and (p_1-p_2) will have the same sign. Thus from a knowledge of b_1 and b_2 we can determine the sign of (p_1-p_2) and, hence, from our previous measurement we can determine p_1, p_2 without the ambiguities which are inherent in Figs. 270 and 281. The use of this method of discrimination with polarized neutrons was first applied by Collins and Forsyth (1963) to Fe–Co and Fe–Ni alloys.

14

INVESTIGATIONS OF MAGNETIC MATERIALS: II. OXIDES, SULPHIDES, HALIDES, ETC.

APART from the transition elements themselves, the materials whose magnetic structures have been studied most extensively have been the compounds of the iron group of transition metals and, subsequently, the rare earths. The first magnetic study was of manganous oxide, MnO, by Shull and Smart (1949). Many other investigations have also examined MnO and the related oxides FeO, CoO, and FeO, and this group as a whole serves as a good example of the way in which information on magnetic structure has been progressively deduced from the neutron measurements.

14.1. Oxides, sulphides of types MX, MX_2

MnO, CoO, NiO, FeO

It was inferred by Bizette, Squire, and Tsai (1938) and Squire (1939) from the variation with temperature of the magnetic susceptibility of MnO that this oxide became antiferromagnetic below 120 K. It was postulated that the manganous ions were aligned in equal numbers parallel and antiparallel to some particular direction, thus giving zero net magnetization. Direct evidence for this was obtained by Shull and Smart (1949) who measured the neutron diffraction patterns at 80 K and 293 K. These temperatures are respectively below and above the transition temperature deduced from the magnetic measurements. The two patterns are shown in Fig. 286, and their most noteworthy feature is the appearance at low temperature of additional diffraction peaks. In particular there is an intense peak at a Bragg angle of about 6° which cannot be explained in terms of the conventional chemical unit cell for which $a_0 = 4{\cdot}426$ Å, as established by X-ray measurements which show MnO to have the NaCl-type of face-centred cubic structure. It is, however, possible to account for this low-angle reflection, and for the other additional peaks, in terms of a unit cell with a side twice as great as the above, namely 8·85 Å. On this basis the intense

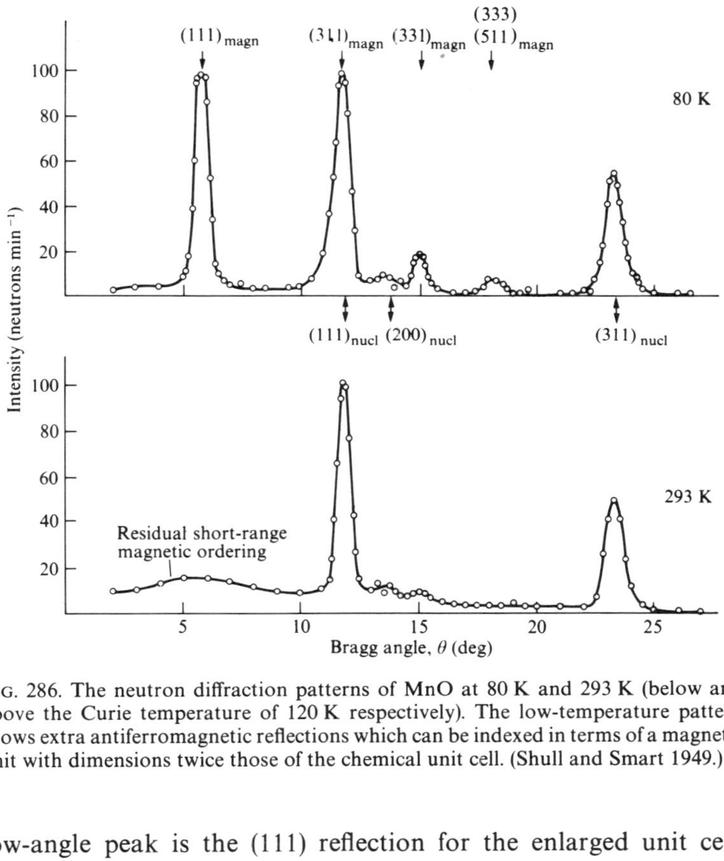

FIG. 286. The neutron diffraction patterns of MnO at 80 K and 293 K (below and above the Curie temperature of 120 K respectively). The low-temperature pattern shows extra antiferromagnetic reflections which can be indexed in terms of a magnetic unit with dimensions twice those of the chemical unit cell. (Shull and Smart 1949.)

low-angle peak is the (111) reflection for the enlarged unit cell. Observations at intermediate temperatures showed that the intensity of this reflection falls off with increase of temperature as shown in Fig. 287, by Shull, Strauser, and Wollan (1951), becoming zero in the neighbourhood of the transition temperature as inferred from susceptibility and specific heat measurements. It is clear therefore that this is a coherent reflection due to magnetic scattering and that the magnetic unit cell, i.e. the unit of pattern so far as magnetic moments are concerned, has a side $2a_0$ which is twice as great as for the chemical unit cell. The precise intensities of the magnetic lines observed, namely, (111), (311), (331), and (511), will depend on the actual arrangement of the magnetic moments within the magnetic unit cell. The observed intensities have been compared with those calculated for various magnetic structures suggested by Néel (1948) assuming the truth of eqn (6.12) for the magnetic scattering amplitude appropriate to Mn^{2+} atoms arranged in oriented fashion and inserting the variation of form factor f deduced from the paramagnetic scattering studies.

There has been much discussion about the detailed magnetic structure in this case, and in fact from powder data it is not possible

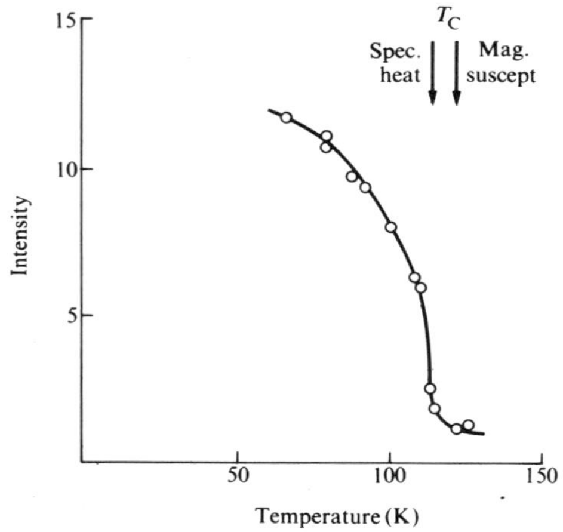

FIG. 287. The variation with temperature of the intensity of the (111) magnetic reflection of MnO, shown in relation to the values of the Néel temperatures deduced from magnetic susceptibility and specific-heat measurements. (Shull, Strauser, and Wollan 1951.)

to come to an unambiguous solution. A conclusive answer can only be obtained from measurements with a single-crystal, single-domain sample. However, the doubling of the unit cell which is observed in the powder pattern certainly means that the spin directions on second-nearest neighbours, such as atoms A, D in Fig. 288, are oppositely directed. The magnetic coupling between them must be of an indirect kind, via the oxygen ions which separate them. The structure model which was first suggested by Shull, Strauser, and Wollan was that shown at (a) in Fig. 288, with the magnetic moments directed along the cube edges in order to give quantitative agreement with the magnetic intensities. A similar conclusion was reached for CoO, and it was rather surprising that the two substances should have the same magnetic structure since,

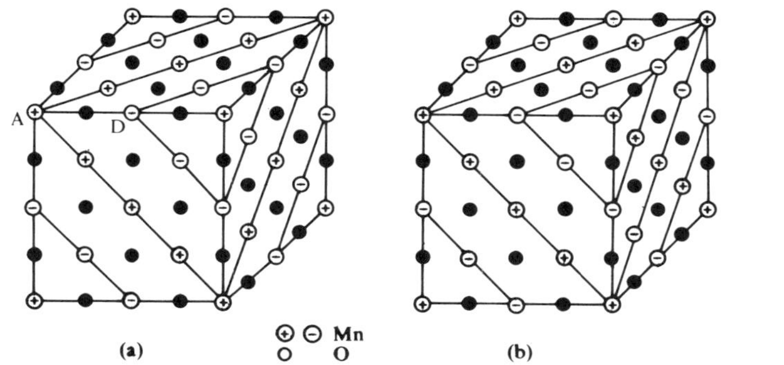

FIG. 288. Possible structural models for MnO, with oppositely directed magnetic moments on next-nearest-neighbour manganese atoms, such as A and D. Model (a) is the one which was originally chosen by Shull, Strauser, and Wollan and consists of ferromagnetic sheets of atoms parallel to one of the (111) planes. Model (b) is an alternative which was suggested by Li. Later work established that for MnO a model of type (a) was correct, with the magnetic moments lying in the (111) sheets.

whereas MnO becomes rhombohedral below the Neél temperature, CoO becomes tetragonal with c/a less than unity. Li (1955) showed that model (b) in the Figure would give the same magnetic intensities as model (a) with its moments aligned along [001]. Moreover, because in a powder pattern one can only observe the mean intensity over the different faces of a form such as {111}, it turns out that for model (b) the intensities will be independent of the magnetic moment direction. This direction can therefore be chosen to suit the evidence from the crystallographic distortion below the Néel temperature, and accordingly Li suggested that for MnO (and NiO) the moments lay along [111], whereas for CoO the moments were along [100], and the structure is then compatible with a tetragonal distortion. However, Roth (1958*a*) was able to show, from more exact experimental data, that neither Li's conclusions nor the original suggestions of Shull, Strauser, and Wollan were correct. Roth's measurements were carried out down to 4·2 K, where the magnetic moment alignment is complete, and his diffraction patterns had much improved angular resolution, so that not only was the important magnetic line (311) well resolved from the neighbouring nuclear line (222) but also several more magnetic reflections could be measured accurately. The improvement can be seen in Fig. 289, which should be compared with Fig. 286. It became

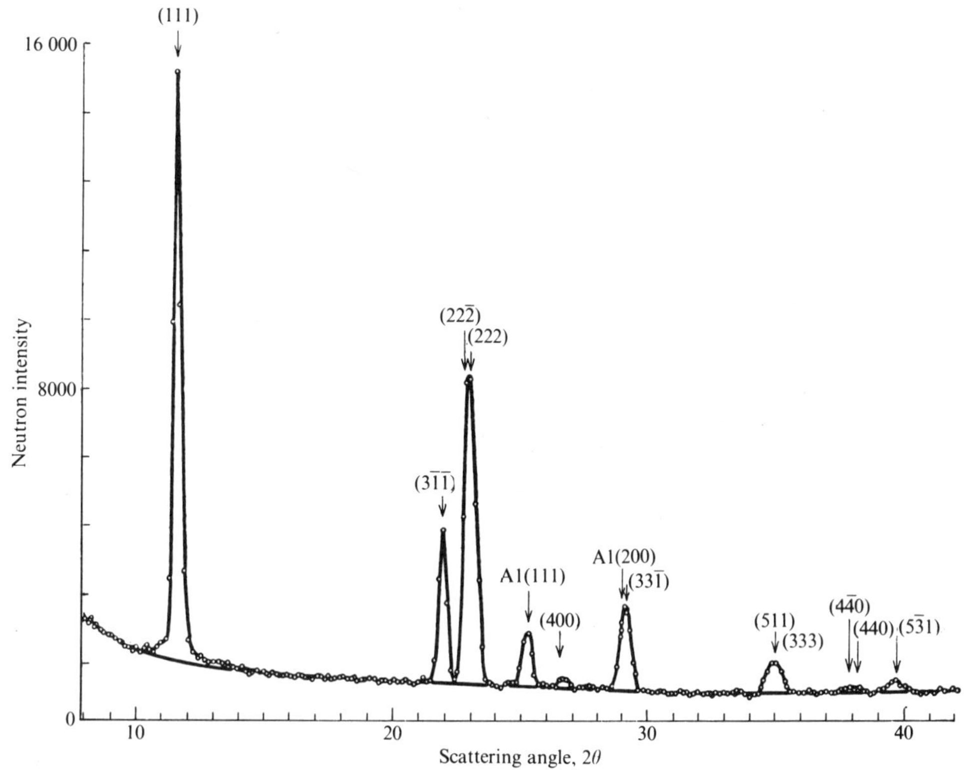

FIG. 289. A diffraction pattern for MnO at 4·2 K with much improved angular resolution. Both magnetic and nuclear lines are indexed in terms of the enlarged magnetic unit cell. The magnetic reflections have all-odd indices and the nuclear indices are all even. The latter fall into two classes. Those such as (222) for which (h, k, l) are odd $\times 2$ are intense; those such as (400), (440) for which (h, k, l) are even $\times 2$ are weak. This is because the lattice is face-centered cubic and the scattering lengths of Mn, O are of opposite sign (Roth 1958a).

clear that both the intensity calculated for (111) and the ratio of intensities for (111)/(113), on either the original [00′1] model or the Li model, were too small to satisfy the experimental observations. Good agreement could be found, however, for a model of type (a) if it was assumed that the magnetic moments lay *in* the (111) sheets. It is not possible from the powder data to say in what particular direction within the sheets the moments lie. Roth's data have been confirmed by Corliss, Elliott, and Hastings (1956). At 4·2 K the distortion from cubic symmetry is sufficient to identify positively the magnetic reflection as (111), rather than $(11\bar{1})$, and thus confirm that the structure is based on ferromagnetic sheets perpendicular to the rhombohedral axis.

Similar results were obtained by Roth for NiO, so that it seems likely that the structure is again that shown in Fig. 288 (a), where the magnetic moments lie in ferromagnetic (111) sheets and are directed alternatively + and − for successive sheets. The similarity to MnO thus agrees with the observation by Tombs and Rooksby (1950) that both MnO and NiO become rhombohedral below the antiferromagnetic transition temperature.

On the other hand, Roth showed that CoO, which becomes tetragonal, gives neutron intensities which are not compatible with model (a) for moments either within the (111) plane or along the [100] axes, nor for model (b). The best agreement was secured by assuming a variation of model (a) in which the spins lie in a direction close to the [117] axis, which is inclined at about 11° to the cube edge. However, this conclusion has been modified by van Laar (1965), using measurements with a diffractometer having much improved angular resolution. van Laar's measurements were made at the temperature of liquid nitrogen and were able to separate the individual tetragonal peaks, such as (113), (311) and (331), (313), as indicated in Fig. 290. The ratio of the intensities of these peaks is a sensitive index of the direction of the moments and it was concluded that the spin axis lay in the $(1\bar{1}0)$ plane making an angle with the *c*-axis of $(27{\cdot}4 \pm 0{\cdot}5)°$. This gives the observed ratio of 0·37 for the (331), (113) intensities compared with 0·64 for 11·5°. However, such an orientation of the spins means that they are only tipped out of the (111) planes by 8°, and there would thus be very little difference between the structure of MnO and CoO. This would be surprising when the distortion from cubic symmetry below the Néel temperature is rhombohedral for MnO but tetragonal

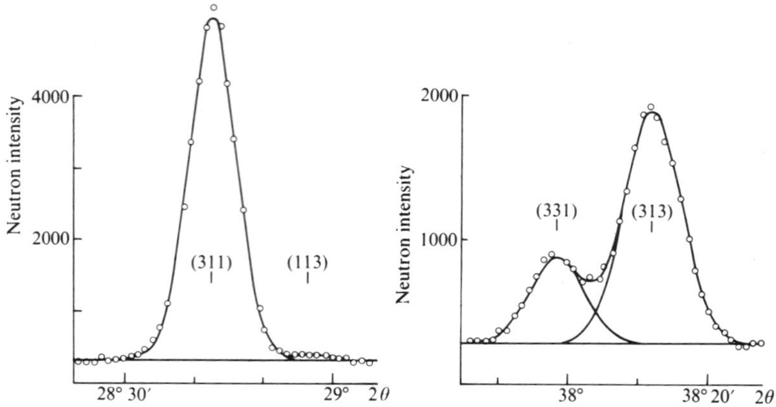

FIG. 290. Separation of the pairs of diffraction peaks (311), (113) and (331), (313) of CoO using a powder diffractometer of high angular resolution. (van Laar 1965.)

for CoO. Accordingly van Laar proposed a multi-spin-axis structure in which there are four pairs of spin directions, each pair comprising a direction and its antiparallel. Each direction is inclined at 27·4° to the *c*-axis (the same angle as for the collinear structure considered above), and the structure as a whole is the sum of the four substructures illustrated in Fig. 291, based on atoms at positions (000), ($\frac{1}{4}0\frac{1}{4}$), ($\frac{1}{4}\frac{1}{4}0$), and ($0\frac{1}{4}\frac{1}{4}$). This structure produces the same powder intensities as the collinear structure, but with these intensities distributed equally between all faces of a form $\{hkl\}$. The important distinction is that the multi-spin-axis structure would be expected to be tetragonal, not rhombohedral, in agreement with what is observed experimentally. Hence it is judged more probable that this structure is correct. Further support for this inclusion is provided by some measurements with single crystals of CoO by van Laar, Schweizer, and Lemaire (1966). The crystal was cooled through the Néel point under a small pressure, resulting in a practically untwinned antiferromagnetic tetragonal crystal. It was found that the intensities of different faces of all crystallographic forms were accurately the same. This could only occur for the collinear structure in the unlikely event that the four possible antiferromagnetic domains were present in each twin with exactly the same volume. Further support for the multi-axis model comes from Bertaut (1969), who shows how suitable interpretations of the tetragonal symmetry

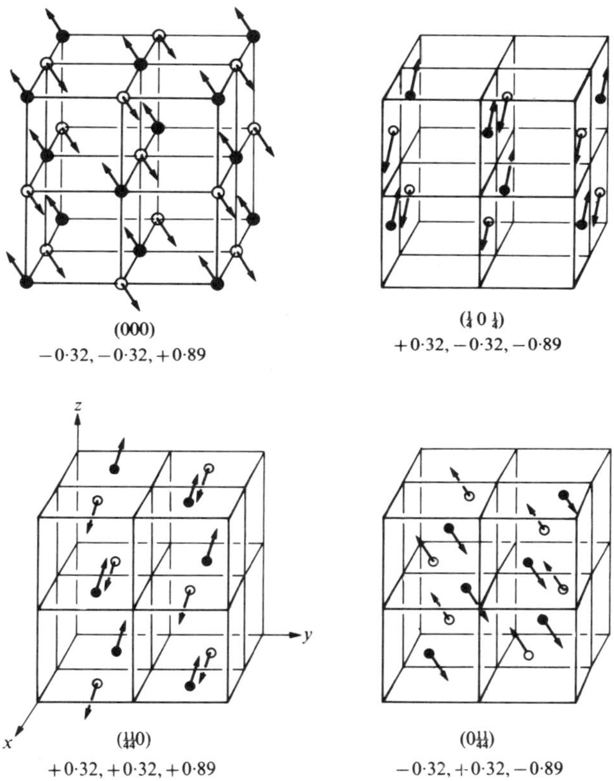

FIG. 291. The multi-spin-axis structure of CoO showing the magnetic moment directions for each of the four component substructures. All the moments are inclined at $\pm 27°$ to the *c*-axis and are symmetrically disposed between a pair of $\pm x$-, $\pm y$-axes. The decimal figures indicate the direction cosines of the moment directions, at the atomic positions specified.

may be provided in terms of either representational analysis or the Shubnikov groups.

The last member of this series of oxides, FeO, has yet a different structure. The low-temperature form is rhombohedral with an elongation along [111] but the diffraction pattern is distinguished by the absence of a (111) magnetic reflection, apart from some diffuse scattering, although it shows the other reflections such as (311) and (331) which are characteristic of the doubled unit cell.

The absence of the (111) peak indicates that the magnetic moments must lie along the scattering vector for this reflection. Accordingly Shull, Strauser and Wollan (1951) proposed an antiferromagnetic arrangement of (111) sheets of iron atoms with their moments perpendicular to these planes.

Further work has shown that the detailed structure of FeO is considerably more complicated than the previous paragraph suggests. The material is deficient in iron and samples usually lie between the compositions $Fe_{0\cdot91}O$ and $Fe_{0\cdot95}O$. In later neutron measurements, both above and below the transformation temperature of 198 K, Roth (1960*b*) was able to show that the occupancy of the normal octahedral cation sites in the NaCl-type structure was only about 75 per cent, and that a substantial number of iron atoms were present in interstitial tetrahedral sites. At the same time the antiferromagnetic neutron intensities were much weaker than would be expected, even if it were assumed that the orbital momentum of the Fe^{2+} ions was completely quenched, which would lead to a magnetic moment of $4\mu_B$. Roth concluded therefore that many of the magnetic spins which were present did not take part in the antiferromagnetic ordering and he suggested that the material contained paramagnetic islands in which two octahedral vacancies and one iron atom on a tetrahedral interstitial site were associated. The nature of the defects has been further investigated by Cheetham, Fender, and Taylor (1971) who carried out neutron measurements at temperatures between 1070 K and 1470 K, in order to take account of the fact that FeO is only in stable equilibrium above 840 K and almost all of the previous work had been done with quenched samples. Only the nuclear scattering is investigated under these circumstances, and no information is obtained about the magnetic properties of the clusters, but it is to be noted that neutrons nevertheless have two advantages compared with X-rays. The absorption effects in the heating furnace are much smaller, leading to more accurate measurement of intensities, and the absence of any nuclear form factor means that reflections can be explored out to higher values of the scattering angle. In each experiment at least nine diffraction peaks could be measured and their intensities were interpreted in terms of an occupation number for the tetrahedral iron atoms which amounted to about 0·03. It was concluded that clustering certainly persists into the equilibrium region and the ratio of the number of vacancies and interstitial atoms at different temperatures and compositions lay between 3 and 4. The ratio was certainly much greater than the value of 2 suggested by Roth and possibly larger than the 13:4 ratio for the type of cluster proposed by Koch and Cohen (1969) in an X-ray investigation.

EuO, EuS, EuTe

The europium compounds, EuO, EuS, and EuTe have the same NaCl-type structure as the iron-group oxides, and a study of their magnetic properties indicates how these depend on the relative strengths of the different magnetic interactions. The size of the unit

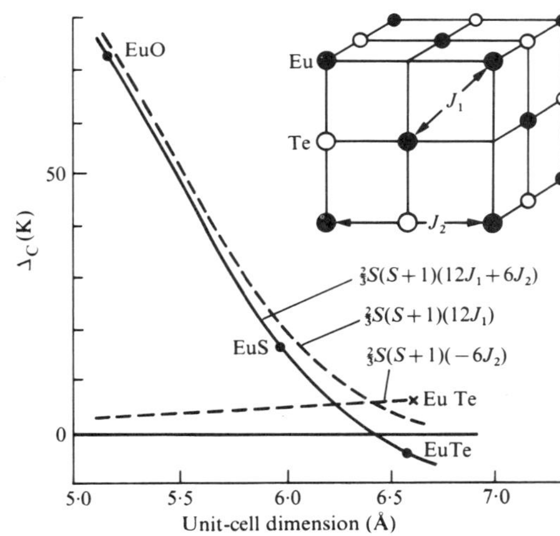

FIG. 292. The full-line curve shows T_C for EuO, EuS and the Curie–Weiss constant Δ_C for EuTe, plotted against unit-cell dimension, and the cross × indicates the Néel temperature of EuTe. The two broken lines then indicate how J_1, $-J_2$ vary with the size of the unit cell. Thus for EuTe the negative J_2 is numerically about 5 times as great as J_1. (After McGuire *et al.* 1963.)

cell increases from about 5·2 Å in EuO to 6·6 Å in EuTe, corresponding to a change in separation of nearest-neighbour europium atoms from 3·6 Å to 4·6 Å. There is a corresponding change from ferromagnetism in EuO (and in EuS) to antiferromagnetism (Will, Pickart, Alperin, and Nathans 1963) in EuTe. The full-line curve in Fig. 292 shows how the ferromagnetic Curie temperature

T_C and the paramagnetic Curie–Weiss constant Δ_C for the anti ferromagnetic EuTe vary with the unit-cell dimension. It can be shown from molecular field theory (Smart 1952) that $\Delta_C = T_C = \frac{2}{3}S(S+1)(12J_1+6J_2)$, where J_1 is the direct exchange interaction between nearest-neighbour europium atoms and J_2 is the super-exchange interaction between next-nearest-neighbour metal atoms, which are separated by the anions. It can also be shown that the Néel temperature for the antiferromagnetic material is given by

$$T_N = \tfrac{2}{3}S(S+1)(-6J_2),$$

and this point, for EuTe, is also shown in Fig. 292. The two broken lines in the Figure then indicate the postulated variations (McGuire *et al.* 1963) of J_1 and $-J_2$ with the size of the unit cell. For EuTe the negative next-nearest superexchange interaction J_2 is about 5 times as great as the positive direct exchange interaction, thus accounting for the change to antiferromagnetism in this compound.

Calculation of intensities

In our above discussion of the interpretation of the neutron diffraction data of the antiferromagnetic oxides we have said very little about the magnetic quantum states of the metallic atoms. In fact, of course, the interpretation consists not only in choosing a magnetic structure and moment direction which will give the correct relative intensities for the various reflections but also in correlating the absolute intensities of the magnetic diffraction peaks with the value of S in eqn (6.12). The case of MnO is simple since the Mn^{2+} ion is in a spectroscopic S state, and its magnetic moment is due to spin only. Agreement with the absolute neutron intensities, determined most easily by using the nuclear intensities for calibration, is secured by writing $S = \frac{5}{2}$, corresponding to a magnetic moment of 5 μ_B in eqn (6.12) for the scattering amplitude p. EuTe is also a straightforward example, the seven 4f electrons which provide the magnetic moment being in a pure S state. In NiO the nickel ion Ni^{2+} has orbital momentum as well as spin momentum, as listed in Table 13 (p. 190), but the neutron intensities are satisfied if it is assumed that the orbital contribution to the magnetic moment is completely quenched by the crystalline field, thus supporting the early conclusions of Van Vleck (1932, pp. 282–310) and of Schlapp and Penney (1932); the effective value of S in eqn (6.12) is therefore unity. On the other hand, it seems evident that certainly for CoO only partial quenching of the orbital moment takes place. For CoO the effective value of S is 1·9, corresponding to 3·8 μ_B compared with a 'spin-only' value of 3 μ_B. Yet a further complication which must be considered in interpreting these magnetic intensities is the uncertainty which exists in our knowledge of the angular variation of the magnetic form factor. It should be pointed out that the re-determination of the inclination of the magnetic moments in CoO by van Laar, which we have discussed above, depended not only on the much improved angular resolution of the diffractometer but also on the knowledge of an improved form factor (Scatturnin, Corliss, Elliott, and Hastings 1961) for the Co^{2+} ion. The accuracy of the form factor is of great importance in the case of FeO for which, in the absence of a low-angle (111) reflection, the interpretation has to be based solely on the quite weak (311), (331) magnetic reflections. The intensity of the latter is reduced by about 4 times because of the fall-off in the value of the form factor at this angle of scattering. Knowledge of the form factor also presupposes an awareness of its dependence on any covalency, such as we discussed for NiO in Chapter 8.

MnS

Not only does MnS occur in a form which has the same rock-salt structure as MnO, but there are also two other simple polymorphic forms, having respectively the zinc-blende and wurtzite structures. In each of these structures there is a close-packing of metal ions and the Mn^{2+} ions have twelve nearest Mn^{2+} neighbours, but there are different numbers of S^{2-} neighbours, namely six in the rock-salt structure but only four in the zinc-blende and wurtzite structures. All three forms have been examined by Corliss, Elliott, and Hastings (1956) with the aim of assessing the influence of the non-magnetic sulphur ion and the metal–sulphur bonds in determining the magnetic structure. For the rock-salt form the structure was found to be identical with that finally assigned by Roth to MnO, which we have just discussed above, with the magnetic moments lying at some undetermined angle within the (111) planes. As we have already pointed out, this structure is favourable to indirect exchange between next-nearest neighbour Mn^{2+} ions via the intermediate anion since this anion lies on a straight line between the two cations. On the other hand, in the zinc-blende form the oxygen atoms now connect *nearest* neighbours through a tetrahedral angle and in fact the coupling scheme which is found does have two-thirds of the *nearest* neighbours coupled antiferromagnetically. Both the rock-salt and zinc-blende structures have cubic close-packing of the metal ions. In contrast, although the wurtzite structure has hexagonal close-packing of the metal ions it shows similar tetrahedral Mn—S—Mn bonds as in the zinc blende form. In conformity with the influence of the indirect exchange mechanism it is found to show the same spin correlations between nearest and next-nearest neighbours as does the zinc-blende form. Fig. 293 indicates the different types of metal packing

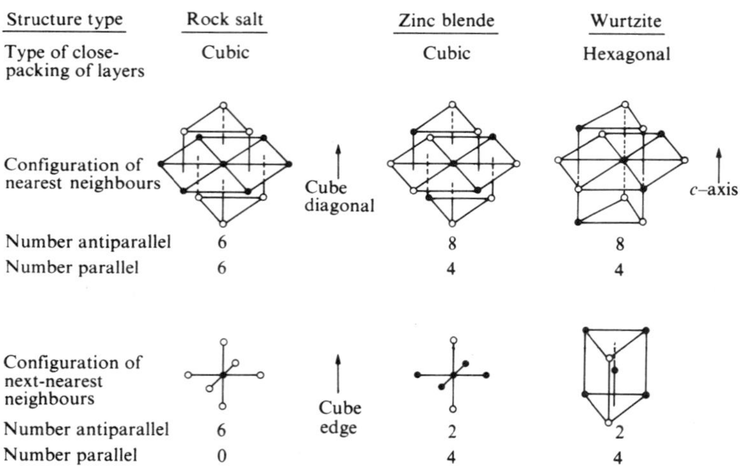

FIG. 293. Metal–metal nearest neighbours and next-nearest neighbours for the three forms of MnS. (After Corliss, Elliott, and Hastings 1956.)

and the nearest and next-nearest-neighbour spin relationships in each case.

In the initial stages of these studies of the three forms of MnS it was assumed that the form factor for the Mn^{2+} ions was that arrived at from earlier studies of iron-group compounds. As the structures were refined the reflection intensities were used to derive more accurate form-factor values for each of the polymorphic forms. Fig. 294 indicates the accuracy of the agreement between the three different sets of values and also with the data derived from the authors' reinvestigation of MnO. The mean curve drawn in this Figure is significantly lower than the later curves for Mn^{2+} which we showed in Fig. 139 (p. 241).

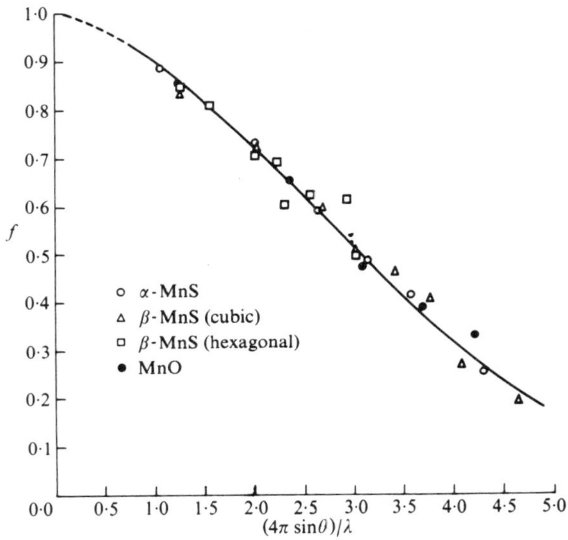

FIG. 294. The magnetic form factor for the Mn^{2+} ion, indicating the agreement between the experimental data from the three forms of MnS and from MnO. (Corliss, Elliott, and Hastings 1956.)

MnSe

This compound has the same rock-salt type of structure as MnO and the α-form of MnS and was found by Shull, Strauser, and Wollan (1951) to have the same antiferromagnetic structure, with anti-parallel alignment of the moments of the second-nearest neighbours.

CrSe: NiAs-type structures

Many sulphides, selenides, and tellurides have the nickel arsenide, NiAs, type of structure which is shown in Fig. 295 (a). This is a hexagonal structure in which the two kinds of atom lie in separate layers. The same structure is shown by some compounds of chromium and manganese with the metalloids As, Sb, and Bi. Among these there is a wide variety of magnetic behaviour. Thus CrSb

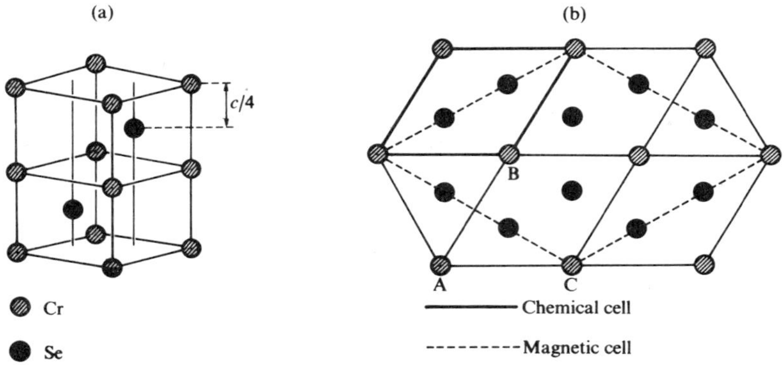

FIG. 295. (a) shows the atomic structure of CrSe. Below 280 K the material becomes antiferromagnetic and the diffraction pattern can be indexed in terms of the enlarged cell, of 3 times the volume, shown in projection in (b) and bordered by the broken lines.

(Snow 1952) is antiferromagnetic and MnAs, MnBi, (Andresen, Hälg, Fischer, and Stoll 1967) are ferromagnetic. Perhaps the most interesting compound of all, from the point of view of the magnetic structure, is chromium selenide, CrSe, which was the first unambiguous example of an antiferromagnetic arrangement of spins which were not collinear. This was demonstrated by Corliss, Elliott, Hastings, and Sass (1961), and a general discussion of spin configurations in NiAs structures was given by Hirone and Adachi (1957). In the powder diffraction pattern of CrSe intense magnetic superstructure lines develop below the Néel temperature of about 280 K. These lines can be indexed in terms of an enlarged hexagonal unit cell, which is shown in plan in Fig. 295 (b). The *a*-axis of the cell is increased by a factor of $\sqrt{3}$ compared with the chemical cell but the *c*-axis is unchanged, thus giving a three-fold increase of volume. A likely magnetic structure would be one in which the spins in the plane $z = \frac{1}{2}$ were antiparallel to those at $z = 0$ and with signs of $++-$ and $--+$ in these two planes, in order to account for the enlargement of the unit cell. Further, there are no $(000l)$ reflections when l is odd, so that the magnetic moments must be directed along the *c*-axis. However, the calculated intensities from a collinear model of this kind do not agree with what is observed experimentally. This is evident from Table 27, where calculated intensities are also shown for the so-called 'umbrella' model, which has the non-collinear arrangement of spins of Fig. 296 (a). This Figure shows

TABLE 27

CrSe: Comparison of observed and calculated intensities for collinear and non-collinear models

	I_{calc}		
hkil	Collinear model	Umbrella model	I_{obs}
$10\bar{1}1$	1858	1858	1858
$20\bar{2}1$	639	337	342
$21\bar{3}1$	514	235	216
$10\bar{1}3$	26	160	164

only the top half of the unit cell, including the layer of selenium atoms which lie in the plane at a distance $\frac{1}{4}c$ below the basal plane at the top of the cell in Fig. 295 (a). The umbrella motif is emphasized in diagram (b) of Fig. 296, which shows the three atoms A, B, C at $z = 0$ and three atoms A′, B′, C′ which have their moments directed antiparallel, in the plane $z = \frac{1}{2}$. The calculation for the umbrella model is made by a direct application of the summation equation (p. 206), bearing in mind that the vector **q** is different

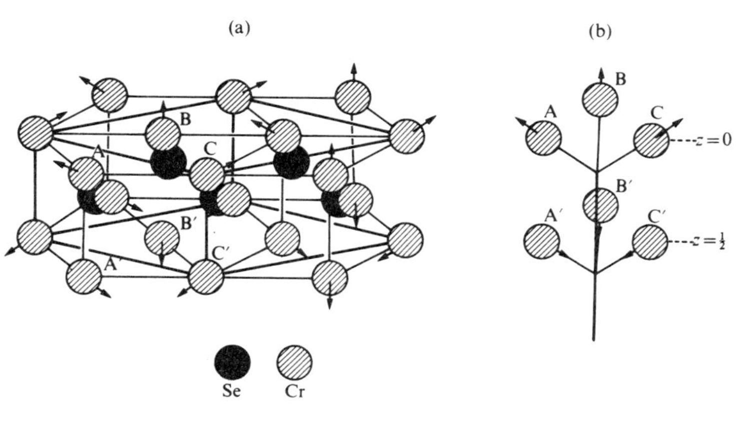

FIG. 296. Magnetic structure of CrSe constituting a non-collinear umbrella arrangement. (a) is the perspective view of the top half of the magnetic cell. The umbrella motif is emphasized at (b), which shows the three chromium ions A, B, C at $z = 0$ and the atoms A′, B′, C′ which lie beneath them at $z = \frac{1}{2}$.

for the three types of atom A, B, C and can be expressed, for each of them, in terms of eqn (6.9), whereby

$$\mathbf{q}_n = \boldsymbol{\varepsilon}(\boldsymbol{\varepsilon}\,.\,\mathbf{K}_n) - \mathbf{K}_n.$$

Here, as usual, $\boldsymbol{\varepsilon}$ is the scattering vector which is normal to the reflecting plane under consideration and $\mathbf{K}_n$ is the direction of the magnetic moment for each atom A, B, C in turn. When the summation is carried out it is found that the squares of the structure factors F^2 for the magnetic reflections fall into three classes:

for l even	$F^2 = 0$,	
for l odd and	$2h+k = 3n \pm 1$	$F^2 = 0{\cdot}654\,\mu_\perp^2(1+\varepsilon_z^2)$,
for l odd and	$2h+k = 3n$	$F^2 = 2{\cdot}61\,\mu_\parallel^2(1-\varepsilon_z^2)$.

In these expressions $\mu_\parallel$, $\mu_\perp$ are the components of magnetic spin, expressed in Bohr magnetons, parallel and perpendicular to the c-axis respectively, i.e. $\mu_\parallel = \mu \cos\phi$ and $\mu_\perp = \mu \sin\phi$, where μ is the resultant moment and ϕ is its inclination to the c-axis. ε_z is the component of the scattering vector $\boldsymbol{\varepsilon}$ in the z-direction. In principle, it is possible to determine both the magnitude of the magnetic moment and its inclination to the hexagonal axis from the above expressions for F^2. However, the reflections for which $2h+k = 3n$ are mainly nuclear, and the assessment of their magnetic contribution is rather inaccurate. Those reflections for which $2h+k = 3n \pm 1$, on the other hand, are purely magnetic and give an accurate value for $\mu_\perp$. This was found to equal $2{\cdot}9\,\mu_B$. The theoretical value of the 'spin-only' moment of chromium is $4{\cdot}0\,\mu_B$, so that the moments in the umbrella structure must be inclined to the c-axis at approximately 45°.

MnS_2, $MnSe_2$, and $MnTe_2$

The influence of the crystal structure and atomic environment on the mechanism of indirect exchange coupling which we discussed for the three forms of MnS has been explored further by Hastings, Elliott, and Corliss (1959) for the homologous series MnS_2, $MnSe_2$, and $MnTe_2$. These compounds are ionic combinations of Mn^{2+} ions with $(X_2)^{2-}$ groups in which the Mn^{2+} ions have the same face-centred arrangement as in MnO. The X_2 groups have their centres at the mid-points of the cube edges and lie with their axes parallel to the various body diagonals to give the pyrite structure.

As a result there are nearly tetrahedral Mn—X—Mn linkages between manganese atoms, via the intervening anions, which might be expected to lead to nearest-neighbour ordering as in β-MnS. On the other hand the occurrence of the face-centred arrangement of manganese atoms might lead to next-nearest-neighbour coupling as in MnO and α-MnS. In fact the first expectation is fulfilled, suggesting that it is the tetrahedral Mn—X—Mn linkage which is of particular importance.

The magnetic structures which are deduced from the neutron diffraction patterns are different in all three compounds, as the diagrams in Fig. 292 indicate, but the same relation between *nearest* neighbours occurs in each case. In all three compounds the Mn^{2+} ions have eight nearest neighbours with antiparallel

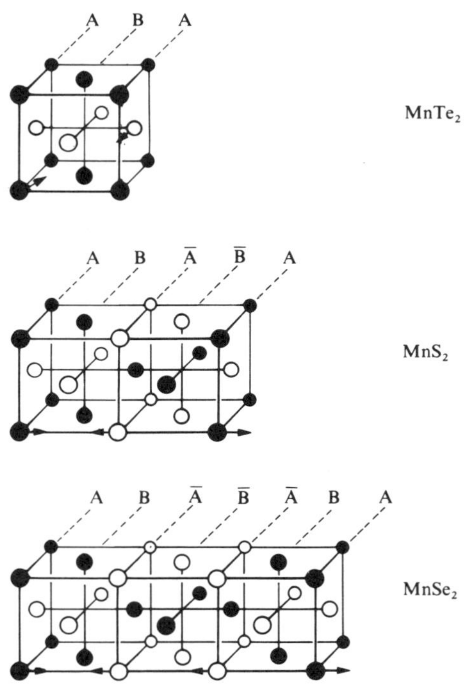

FIG. 297. The magnetic structures of $MnTe_2$, MnS_2, and $MnSe_2$, emphasizing the different packing of the A and B type layers in the three cases. The black and white atoms have their magnetic moments oppositely directed, parallel to the magnetic axis indicated by the arrows. (After Hastings, Elliott, and Corliss 1959.)

spins and four with parallel spins. On the other hand, the next-nearest-neighbour arrangements are quite different in the three cases. This interaction is always predominantly ferromagnetic and becomes increasingly so from MnS_2 to $MnSe_2$ and on to $MnTe_2$. For the first compound there are four parallel and two antiparallel spins, and for the last all six next-nearest neighbours have spins parallel to the central atom. In $MnSe_2$, which is an intermediate case, there are two types of manganese atoms; two-thirds of them have five parallel and one antiparallel next-nearest neighbours and the remaining third have four of them parallel and two antiparallel. Comparing the three structures in Fig. 297 we see that the constitution and arrangement of the first pair of planes, which we can characterize by A, B, are the same in all three cases. For $MnTe_2$ these are simply repeated successively to give the sequence ABAB... and a magnetic unit cell of the same size as the chemical one; the structure thus consists of horizontal ferromagnetic sheets. For MnS_2, however, the second chemical unit cell has magnetic moments with reversed signs so that the resulting sequence is $AB\bar{A}\bar{B}AB$... and the magnetic cell is twice as large as the chemical one. In the case of $MnSe_2$ the sequence is $AB\bar{A}\bar{B}\bar{A}BAB$, leading to a magnetic cell which is 3 times as large as the chemical cell.

For both MnS_2 and $MnSe_2$ the direction of the magnetic moments is along the unique, i.e. the lengthened, axis but for $MnTe_2$ the moments lie within the ferromagnetic sheets in a direction which cannot be defined from powder data.

14.2. Transition-group halides MX_2 and related compounds

MnF_2, FeF_2, CoF_2, and NiF_2

Investigations by Erickson (1953) have determined the magnetic structures of the above isomorphous fluorides which become antiferromagnetic at low temperatures. In each case it is found that the magnetic moments of the metal atoms sited at the body centres of the tetragonal unit cells, illustrated in Fig. 298, are antiparallel to those of the atoms at the corners of the unit cells. For MnF_2, FeF_2, and CoF_2 it is shown that the direction of orientation of the moments is parallel to the c_0-axis. This alignment accounts for the fact that there is no observable (001) reflection since q^2 is zero under these conditions, whereas for NiF_2 which

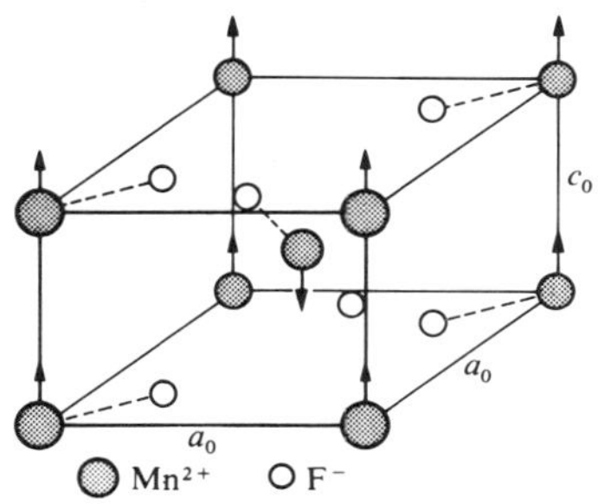

FIG. 298. The unit cell of MnF_2 showing the antiferromagnetic arrangement of the magnetic moments of the Mn^{2+} ions. (Erickson 1953.)

shows a finite (001) reflection it is deduced that the moments are inclined at about 10° to the c_0-axis.

The intensity data obtained by Erickson at various temperatures for these compounds demonstrate the applicability of Van Vleck's (1941) model of an antiferromagnetic material. According to this the magnetic moment should vary with temperature according to a Brillouin function $B_S(y)$ given by

$$B_S(y) = \frac{2S+1}{2S} \coth\left(\frac{2S+1}{2S}\right)y - \frac{1}{2S} \coth\left(\frac{1}{2S}\right)y, \tag{14.1}$$

where S is the effective quantum number of the atom and y is determined by T, the absolute temperature of measurement, and T_N, the Néel temperature, according to the relation

$$y = 3B_S(y)\frac{S}{S+1}\frac{T_N}{T}. \tag{14.2}$$

These two expressions define the variation of B_S with temperature T. The neutron intensities will vary with temperature according to the square of this function, and Fig. 299 illustrates how the experimental data for MnF_2 are in accord with a function $B^2_{\frac{5}{2}}(T)$ assuming a value of 75 K for the Néel temperature T_N in agreement with the value of 72 K found from susceptibility measurements.

With reduction of temperature below T_N the ordinary paramagnetic diffuse scattering disappears. Nevertheless it has been shown by Halpern and Johnson (1939) and confirmed by Tamor (private communication to Erickson (1952)) that inelastic diffuse magnetic scattering may occur due to collisions which involve a

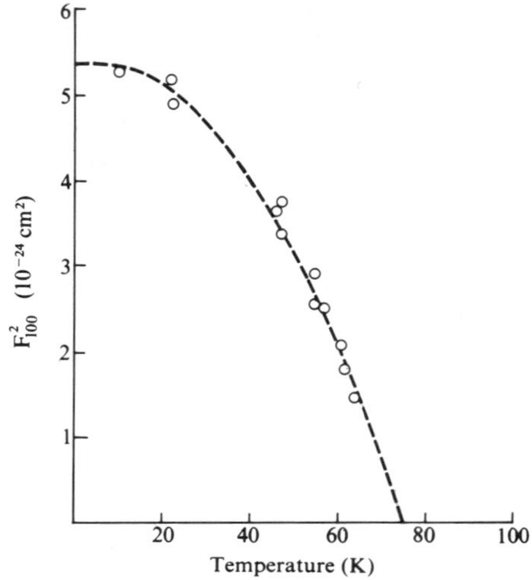

FIG. 299. Variation with temperature of the experimentally observed antiferromagnetic (100) reflection of MnF_2, in relation to the square of a Brillouin function calculated for a Néel temperature of 75 K and a spin quantum number of $\frac{5}{2}$. (Erickson 1953.)

change in the orientation of the magnetic spins. Tamor concludes that in the limiting case well below T_N the residual diffuse magnetic scattering should be a fraction $1/(S+1)$ of the paramagnetic scattering cross-section. This was confirmed by Erickson in the case of MnF_2 at 23 K for which the scattering cross-section in the forward direction was 0·48 barns per unit solid angle. This is equal to a fraction of $1/(\frac{5}{2}+1)$ of the differential paramagnetic scattering cross-section of 1·69 barns for the manganous ion Mn^{2+}.

The values found for the Néel temperature were as follows: MnF_2, 75 K; FeF_2, 90 K; CoF_2, 50 K; and NiF_2, 83 K. Fig. 300 provides a typical example of the observed change of appearance of a magnetic scattering peak as the temperature is reduced below the Néel temperature with progressive improvement in the perfection of the magnetic alignment as the disturbing effects of the thermal motion become less.

MnO_2

Manganese dioxide has the same rutile crystal structure as the fluorides which we have just discussed but Erickson's (1952) neutron diffraction measurements on a powdered sample suggested that it had a more complicated magnetic structure. In particular, the magnetic unit cell is larger than the chemical cell. Susceptibility measurements for powders and single crystals (Bizette 1951) led to a suggestion that above the Néel temperature of 84 K the spins on the corner and centre atoms of manganese form two mutually uncorrelated antiferromagnetic sublattices. This was supported by Erickson's data but no unambiguous conclusions could be

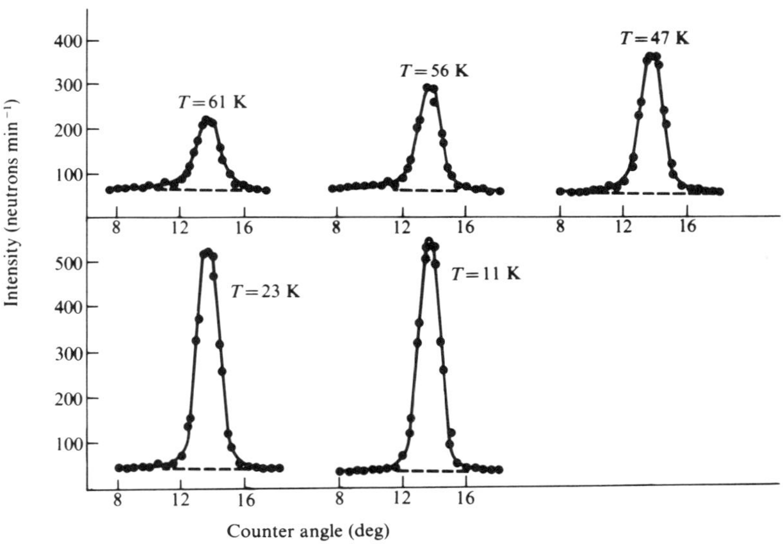

FIG. 300. The (100) magnetic reflection in the powder pattern of MnF_2, as recorded at various temperatures. (Erickson 1953.)

reached. Subsequently Erickson made some further unpublished measurements with a single crystal and found that the reflections observed could be indexed by assuming a magnetic unit cell having a c-axis 7 times as large as the chemical cell. This proved to be a very significant observation, for Erickson was providing the first experimental evidence for the helimagnetization which we discussed in Chapter 8. Yoshimori (1959) showed that the data could be interpreted quantitatively in terms of a spiral of pitch $\frac{7}{2}c$. The general conditions necessary for the occurrence of this type of magnetic structure in a rutile-type crystal are discussed in Yoshimori's paper in terms of the relative strengths of the different exchange interactions among the cations in the crystal. The pitch of the screw is found to depend on the ratio of the strength of the interaction J_2 between neighbouring cations along the c-axis to that of the interaction J_1 between neighbouring corner and body-centre cations.

Anhydrous halides: $MnCl_2$, $FeCl_2$, $CoCl_2$, etc.

The study of the anhydrous halides by Wollan and his co-workers at the Oak Ridge National Laboratory provides an excellent example of the power of neutron diffraction methods to unravel magnetic structure and magnetic behaviour. In these investigations the full range of techniques—using single crystals, low temperatures, and applied magnetic fields—has been employed to the utmost advantage.

The dihalides, particularly $FeCl_2$ and $CoCl_2$, have been known to have unusual magnetic properties for over sixty years, showing some of the characteristics of both ferromagnetism and antiferromagnetism in such a way as to have attracted to themselves a special name of 'metamagnetics'. Their crystals have hexagonal layer structures in which the metals are arranged in hexagonal nets separated by two intervening layers of halogen ions. More precisely they include two kinds of structure. The dibromides, and also MnI_2, crystallize with the hexagonal CdI_2 structure, whereas the dichlorides have the rhombohedral $CdBr_2$ structure—but these two types differ only in the sequence of stacking of the aggregated MX_2 layers. As an example, from which our description of the magnetic structures may be followed, we show in Fig. 301 the structure of $FeCl_2$ which has the second type of layer arrangement. The diagram emphasizes the sequence of layers rather than the underlying rhombohedral symmetry.

The magnetic structures have been described fully in papers by Wollan, Koehler, and Wilkinson (1958), Koehler, Wilkinson, Cable, and Wollan (1959), and Wilkinson *et al.* (1959), as the result of a long series of investigations using both powdered samples and (where available) single crystals, in which observations of the variation of neutron intensities in applied magnetic fields were of prime importance. $FeCl_2$ has also been studied in considerable detail by the Saclay workers, Ericson *et al.* (1958).

The basic magnetic structures of the iron and cobalt compounds are simple. Each layer of metal atoms is a ferromagnetic sheet

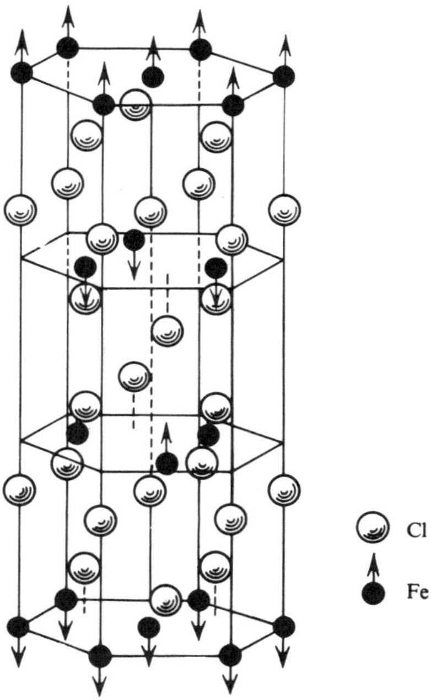

FIG. 301. The magnetic structure of $FeCl_2$. (After Ericson *et al.* 1958.)

but the magnetic moments are oppositely directed in successive sheets, as indicated in Fig. 301, so that the resulting structure is antiferromagnetic, with Néel temperatures ranging from 25 K to 1·8 K for the various compounds. For the iron compounds the magnetic moments are perpendicular to the layers, i.e. they are parallel to the unique crystallographic axis, whereas for the cobalt compounds the moments are parallel to the layers, and from observations with single crystals it was possible to show that they lay in the $[21\bar{3}0]$ directions. It was observed for $CoCl_2$ that equivalent (*hkil*) reflections were not of equal intensity, indicating that the crystal volume was not divided equally among the various possible domains with crystallographically equivalent axes. Moreover, when a magnetic field was applied a single domain direction could be favoured at the expense of the others and Fig. 302 shows how the intensity of the $(11\bar{2}3)$ antiferromagnetic reflection varies when a field is applied along the scattering vector. The moment orientation which is perpendicular to the field direction is the most favoured one and when the field has reached about 1200 Oe the entire crystal has become a single domain, and it remains largely in this state even if the field is subsequently removed. When more intense fields are applied the relatively weak antiferromagnetic coupling between successive layers of cobalt ions is broken down and all the moments rotate around into the field direction, thus giving a

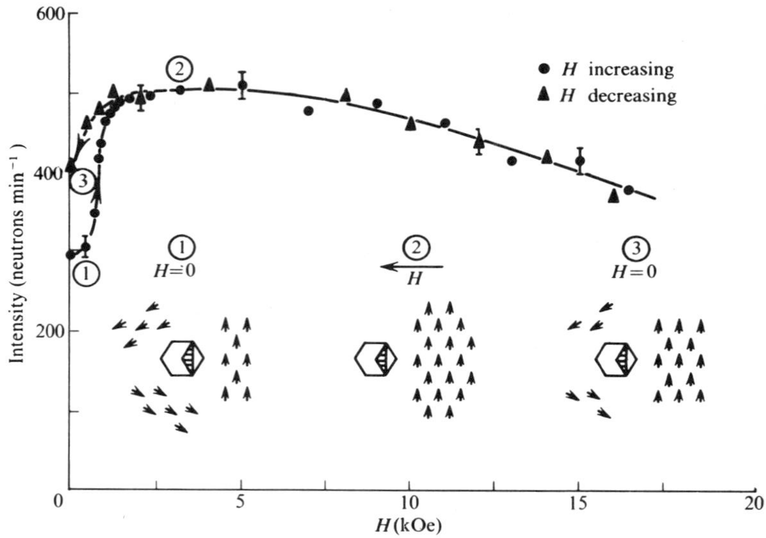

FIG. 302. The variation in intensity of the $(11\bar{2}3)$ antiferromagnetic reflection from a single crystal of $CoCl_2$ when an external magnetic field is applied along the scattering vector. In zero field, at (1) roughly equal numbers of moments point along the three equivalent $[21\bar{3}0]$ directions. On applying the field the moments rotate around until they are *perpendicular* to **H**, giving an increased intensity for $(11\bar{2}3)$, as at (2). In very large fields they are pulled into the field direction giving a net magnetization and a reduction of the antiferromagnetic intensity. On removal of the field much of the preferred orientation persists (3). (Wilkinson *et al.* 1959.)

net magnetization. By contrast, $FeCl_2$, which has a unique magnetic axis parallel to the crystallographic *c*-axis, cannot show such domain effects in weak fields but when moderate fields are applied parallel to the *c*-axis it shows the same breakdown of the interlayer coupling, producing a resultant magnetization. The breakdown seems to occur by direct reversal of the moments which are initially antiparallel to the field direction and not by any intermediate transition of *all* the moments into directions at right-angles to the field direction.

The magnetic structures of the manganese halides were found to be more complicated and could only be solved with the aid of single-crystal data and after much investigation of the behaviour of the crystals in applied magnetic fields. Fig. 303 shows the structure

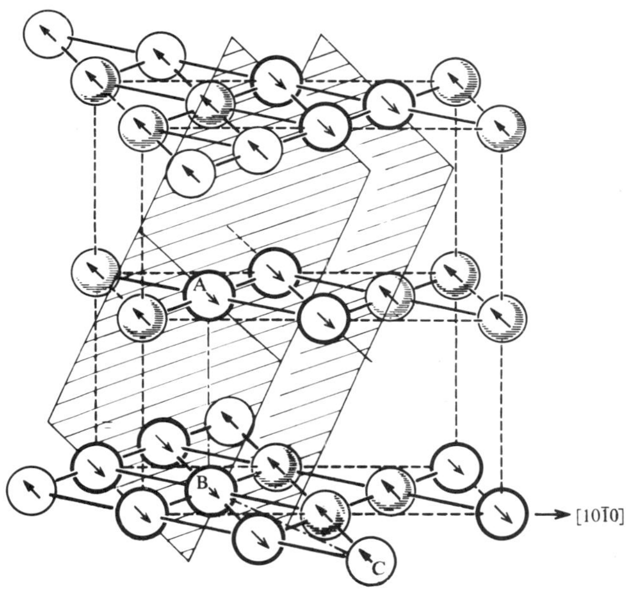

FIG. 303. The magnetic structure of $MnBr_2$ showing, dotted, one-half of the orthorhombic magnetic unit cell, which contains four layers of Mn atoms. The structure consists of ferromagnetic sheets of like spin parallel to (01$\bar{1}$1) planes and arranged in the sequence $++--$. A pair of sheets is shown with more heavily shaded atoms. The moment direction is parallel to the short sides of the cell. (After Wollan, Koehler, and Wilkinson 1958*a*.)

which was finally deduced. It may be considered as consisting of ferromagnetic sheets which are parallel to (01$\bar{1}$1) planes and which are arranged in a sequence $++--$. Here again, a domain structure is produced when such a crystal is cooled down below the Néel temperature in the absence of a magnetic field, because the spin direction characterizing an individual domain may take any one of three equivalent crystallographic directions. When an external magnetic field is applied one type of domain grows at the expense of the other two and a single-domain single crystal is produced and largely persists after the field is removed.

For all these halides it is found that the absolute intensities of the magnetic reflections lead to values of the atomic magnetic moments which are close to the 'spin only' values for the divalent metallic ions. It follows therefore that for the Fe^{2+} and Co^{2+} ions the orbital contribution must be completely quenched.

14.3. Haematite and ilmenite structures

α-Fe_2O_3 (haematite) was one of the first compounds to be studied by neutron diffraction and powder patterns at various temperatures between 80 K and 1000 K were examined by Shull, Strauser, and Wollan (1951). The reflections observed can all be accounted for in terms of the ordinary rhombohedral chemical unit cell. However, there are intense reflections corresponding to (111) and (100), which do not appear with X-rays owing to the mutual cancellation of the contributions from the various oxygen and iron atoms in the unit cell. With neutrons, intensity would be expected in these reflections for certain antiparallel arrangements of the magnetic moments on the iron atoms. This would be consistent with the magnetic behaviour of haematite. At room temperature very pure samples are very weakly ferromagnetic and this ferromagnetism disappears at 948 K which is very close to the Curie temperature of magnetite, leading Néel (1949) to the suggestion that the structure of haematite is antiferromagnetic but that there might be inclusions of ferromagnetic magnetite.

The sketch (a) at the left-hand side of Fig. 304 shows how the iron atoms in α-Fe_2O_3 are arranged along the body diagonal of the rhombohedral unit cell; the atoms belonging to some of the neighbouring cells are also shown. As we travel down the diagonal there could be three possible antiferromagnetic arrangements of the magnetic spins for the sequence of atoms AABB namely (a) $--++$, (b) $+--+$, and (c) $+-+-$. In assessing the neutron intensities which would be observed for each of these arrangements it must be remembered that the absolute direction of orientation of the spins relative to the crystal axes must also be considered, but for model (c) the intensities of the (111) and (100) reflections would be zero for *any* spin direction, so that only models (a) and (b) need be considered in interpreting the experimental

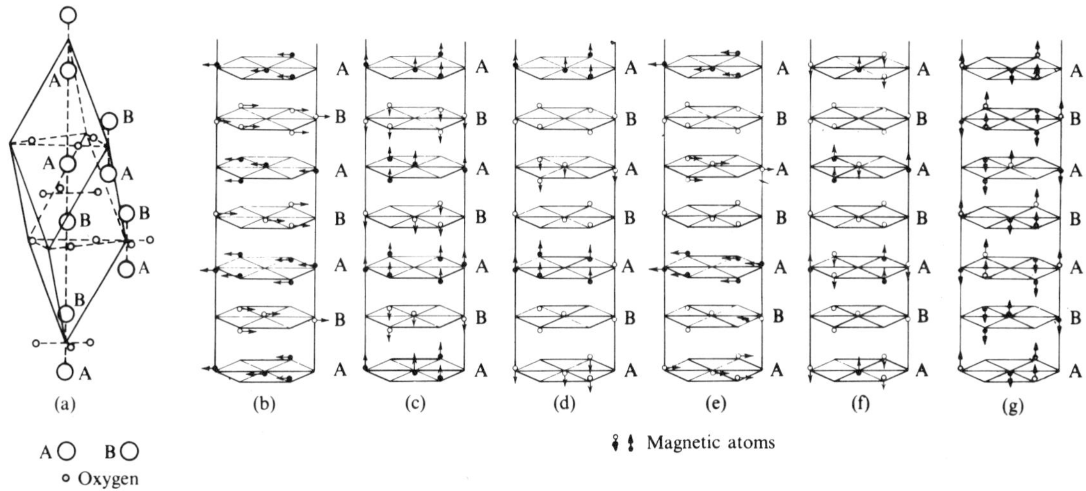

FIG. 304. The magnetic structures of various compounds of the haematite and ilmenite type. Diagram (a) shows the basic rhombohedral unit cell of haematite α-Fe_2O_3, which produces in the extended structure the hexagonal-type layers of A, B atoms shown at (b). At room temperature the magnetic moments lie in the plane of the layers, as (b) shows, but below 250 K the spins flip over and become perpendicular to the layers, as in (c).

In ilmenite, $FeTiO_3$, only alternate layers (the A layers) contain the magnetic iron atoms, the moments of which are perpendicular to the sheets as in diagram (d); this diagram shows only one-half of the magnetic unit cell, since the atoms in the top and bottom layers have their moments oppositely directed. For $NiTiO_3$ (e) the arrangement of atoms in the layers is the same but the moments now lie in the plane of the sheets. By contrast, $MnTiO_3$ has the structure (f) in which the magnetic unit cell is the same size as the chemical cell and the layers contain both + and − moments.

Diagram (g) shows Cr_2O_3, which is isomorphous with haematite. Both A and B layers now contain magnetic atoms, and each layer contains both + and − atoms.

results. Table 28 compares the observed neutron intensities at 80 K and 300 K with those calculated for three reasonable directions of orientation, respectively along the unit cell edges, along the body diagonal, and thirdly, perpendicular to the body diagonal. At room temperature the data are consistent with sequence (a) if the magnetic moments are perpendicular to the body diagonal. At low temperatures, below about 250 K, the magnetic moments flip over and become *parallel* to the body diagonal of the unit cell, but maintain the sequence $- - + +$.

TABLE 28

Deduction of the magnetic structure of α-Fe_2O_3

Magnetic reflection	Calculated intensity: Spin sequence $--++$, Orientation I	II	III	Calculated intensity: Spin sequence $+--+$, Orientation I	II	III	Experimental intensity 300 K	80 K
(111)	1·25	0	4·3	0·23	0	0·81	4·9	<0·05
(100)	1·40	1·59	0·96	2·32	2·64	1·59	0·91	1·37

Two alternative sequences of magnetic spin of the Fe^{3+} ions AABB along the axis of the cell are considered, each for three different orientations of magnetic moment: I, parallel to the unit cell edges; II, along the axis of the cell; III, in the (111) sheets and directed towards one of the three nearest neighbours.

The resulting extended structures in the material are indicated in (b) and (c) of Fig. 304, which show the hexagonal form of unit cell and emphasize how the iron atoms form ferromagnetic sheets with antiferromagnetic coupling between successive sheets. Between each pair of sheets is a layer of oxygen ions (which are not shown in the Figure), and the whole structure is only a slight distortion away from hexagonal close-packing. Thus the conclusion from the diffraction pattern means that at room temperature the magnetic moments lie *in* the ferromagnetic sheets but become perpendicular to these below 250 K.

Haematite is a special case of the more general ilmenite structure, named after the mineral $FeTiO_3$. Apart from slight displacements away from the ideal positions, the atomic sites in the ilmenite structure are the same as those for haematite. $FeTiO_3$ itself has been studied with neutrons by Shirane, Pickart, Nathans, and

Ishikawa (1959*a*), who were able to show (taking advantage of the significant difference between the nuclear scattering amplitudes of iron and titanium) that there is almost complete ordering of the iron and titanium atoms among the A, B metal positions. As a result alternate metal layers in $FeTiO_3$ are occupied by iron and titanium atoms. $FeTiO_3$ is antiferromagnetic and from a study of the powder pattern below the Néel temperature it was demonstrated that the magnetic unit cell is doubled in volume, by doubling along the original [111] rhombohedral axis, compared with the chemical unit cell. The structure is most easily considered in terms of the hexagonal units shown in Fig. 304. The detailed magnetic structure which was deduced is that shown at (d) in this Figure, with the magnetic moments directed perpendicular to the ferromagnetic sheets; only a half of the unit cell is shown in the Figure. The absolute neutron intensities and form factor variation indicated that there was some orbital contribution to the magnetic moment of the Fe^{2+} ion.

Two other oxides with ilmenite structures, $NiTiO_3$ and $MnTiO_3$, have been examined by Shirane, Pickart, and Ishikawa (1959*b*). The former, as indicated in Fig. 304(e), has a similar structure to ilmenite itself except for the fact that the magnetic moments lie *in* the planes of the ferromagnetic sheets, leading to reflections from the (111) planes. On the other hand, $MnTiO_3$ has the different structure shown in Fig. 304 (f) for which the magnetic unit cell is the same size as the chemical cell and in which the layers of manganese atoms contain both positive and negative moments, with these + and − spins lying parallel to the principal axis.

Shirane *et al.* (1959*a*) have also studied some solid solutions of ilmenite and haematite and have correlated the ferromagnetism which is observed in these solutions with ordering of the iron and titanium atoms.

The final illustration (g) in Fig. 304 is of Cr_2O_3 which is isomorphous with haematite. As in $MnTiO_3$ the layers of atoms contain spins of both kinds. Referring back to our initial discussion of haematite, it may be noticed that the arrangement of magnetic moments in the Cr_2O_3 structure corresponds to the sequence (c) (p. 470) which we considered, but rejected, for α-Fe_2O_3.

As seen in Fig. 304(g) the sequence of spins AABB along the rhombohedral axis is $+-+-$. The first neutron diffraction study by Brockhouse (1953*a*) was unable to decide the directions of the moments, but these were later recognized by Nathans, Riste, Shirane, and Shull (1958) as being along the *c*-axis as shown in Fig. 304 (g).

Cr_2O_3 and haematite are isomorphous and form a continuous series of solid solutions. Although their magnetic structures are quite different, for both the type of order and the direction of the moments, Cox *et al.* (1963) have shown that the two structures are connected via cone spiral structures and that there is a continuous change from one to the other in the mixed system $(1-x)Cr_2O_3 . xFe_2O_3$. The transition is summarized in Fig. 305. For pure Cr_2O_3 at the right-hand side of the diagram there are two sublattices of upward- and downward-pointing moments respectively. With introduction of Fe_2O_3 these become conical spirals and both the cone axis and the spiral wave vector are parallel to the hexagonal

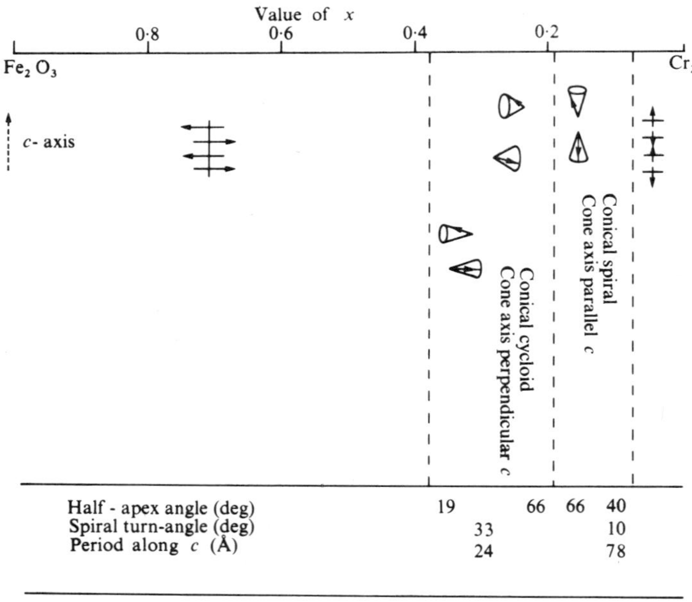

FIG. 305. Diagram to indicate the range of magnetic structures in the mixed system $(1-x)Cr_2O_3 . xFe_2O_3$. When $x = 0$ there are two sublattices with *vertical* upward- and downward-pointing moments respectively, becoming a conical spiral with increase of x. Initially the spiral axis is parallel to c, but it becomes perpendicular to c when $x > 0{\cdot}2$. As x further increases $>0{\cdot}35$ the conical system changes to the collinear arrangement, with *horizontal* moments, of Fe_2O_3 itself.

c-axis. The cone angle increases up to $x = 0{\cdot}15$ when it has reached about 66°, measured as the semi-apex angle. Meanwhile the spiral turn-angle has increased from about 10°, corresponding to a vertical period of about 78 Å, to about 33° which means a period of 24 Å. With further increase of x the axis of the cone switches to become perpendicular to c, although the wave vector remains parallel to c, which means that the spiral is of cycloidal type rather than a simple screw (see Fig. 306). Initially the semi-apex angle is again about 66°, but it falls rapidly with increase of x, having fallen to

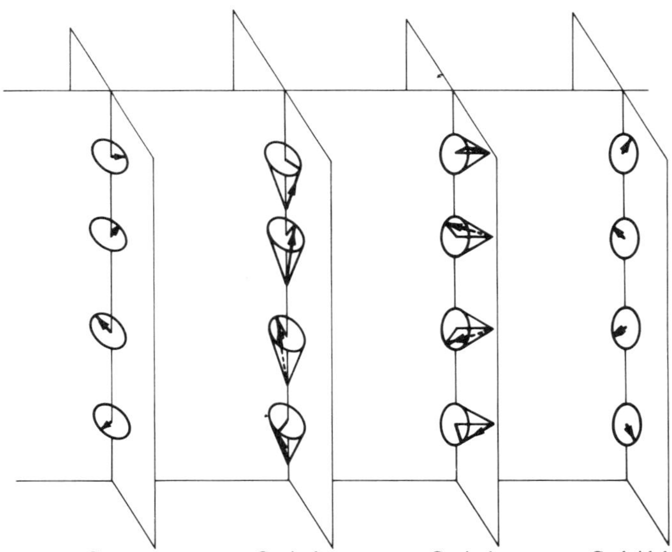

FIG. 306. Types of spiral.

19° when $x = 0{\cdot}35$ and the structure approaches the collinear arrangement which is characteristic of Fe_2O_3 itself. The two types of intermediate conical structures are illustrated in Fig. 307. In each case the spiral component is represented by a rotating radius in the base of the cone, the fundamental component of moment lies along the axis of the cone, and the resultant moment lies along a slant edge of the cone. The pure Fe_2O_3 structure is characterized by (0003) and $(10\bar{1}1)$ magnetic reflections, the Cr_2O_3 structure displays a magnetic ($01\bar{1}2$), and the intermediate conical structures show satellite magnetic reflections such as ($01\bar{1}2^-$), ($01\bar{1}2^+$) and ($10\bar{1}4^-$), ($10\bar{1}4^+$).

A study of Cr_2O_3 by Worlton *et al.* (1968) provides a good example of the use of neutron diffraction to examine the effect of pressure, which we mentioned in Chapter 4. The measurements were

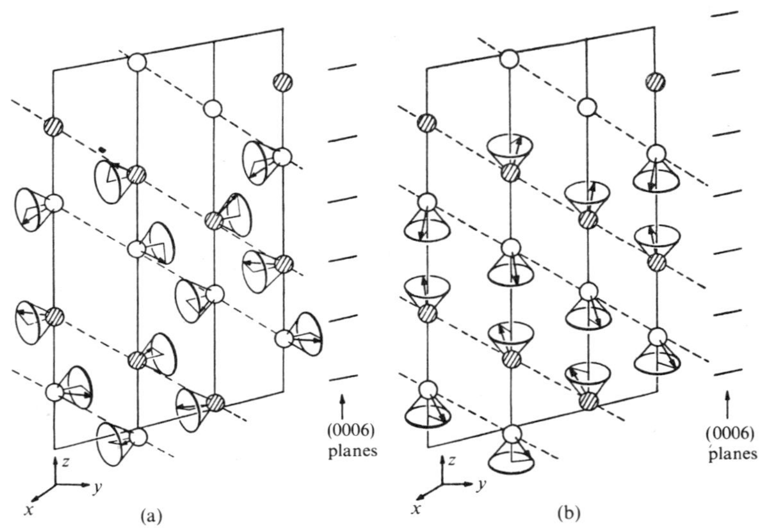

FIG. 307. The two intermediate conical structures which are found in the system $(1-x)Cr_2O_3 . xFe_2O_3$. In (a), which is based on the Fe_2O_3 structure, there is a conical cycloid with the cone axis perpendicular to c but the spiral wave vector is parallel to *c*. When $x < 0{\cdot}2$ the arrangement (b) exists; this is a conical spiral for which both the cone axis and the spiral wave vector are parallel to *c*. (Cox, Takei, and Shirane 1963.)

made by the time-of-flight method using a sample in which powdered Cr_2O_3 was mixed with liquid carbon disulphide, which transmits the pressure, and was then compressed. A determination of the Néel temperature was then made at pressures of 1 bar, 7·6 kbar and 13·3 kbar in turn, by observing the temperature dependence of the ($01\bar{1}2$), ($11\bar{2}3$) reflections. The former is mainly of magnetic origin, whereas the latter is a purely nuclear reflection and serves as a standard. Fig. 308 shows both the temperature-dependence of the scattering and the variation of Néel temperature with pressure.

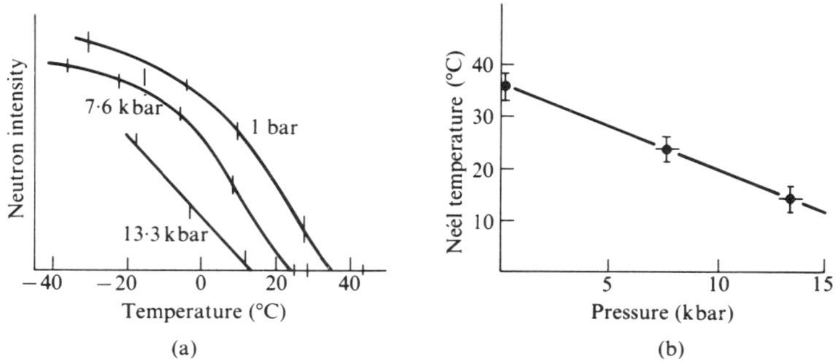

FIG. 308. (a) shows the variation of the intensity of the (01$\bar{1}$2) reflection with temperature at pressures of 1 bar, 7·6 kbar, and 13·3 kbar. (b) shows the change of Néel temperature with pressure deduced from these measurements. (Worlton, Brugger and Benion 1968.)

14.4. Perovskite structures

The ideal perovskite structure contains one molecule ABO_3 in a simple unit cell as shown in Fig. 309 (a). At the corners of the cell are the atoms B, which are small ions and which are separated by oxygen atoms at the mid-points of the cube edges. The atom A, which is a large ion, is placed at the body centre of the cell. Each B atom is therefore surrounded octahedrally by six other B atoms, separated in each case by an intervening oxygen atom. Thus compounds in which B, but not A, is a magnetic ion offer attractive subjects for the study of indirect exchange coupling between the magnetic ions via the intervening oxygen atoms. Ideally these perovskite structures are cubic but usually they are slightly distorted to a lower symmetry such as rhombohedral or monoclinic.

$LaBO_3$

Many of these substances have been studied by Wollan and his co-workers at the Oak Ridge National Laboratory. Among the simplest are the lanthanum compounds of the form $LaBO_3$, where B is one of the 3d transition elements Cr, Mn, Fe, Co, and Ni, which were examined by Koehler and Wollan (1957). Two simple types of antiferromagnetic structure have been found, as illustrated in Fig. 309 (b) and (c). In model (b) which is represented by $LaMnO_3$, the atoms lie in ferromagnetic sheets with the spin directions oppositely directed in successive sheets, giving a magnetic unit

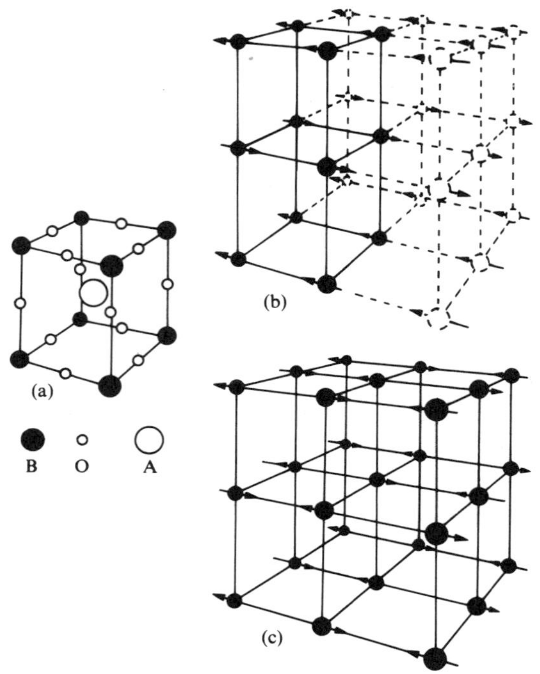

FIG. 309. (a) shows the ideal cubic structure of the perovskites ABO_3, containing one molecule per unit cell. (b) shows the magnetic structure of $LaMnO_3$, indicating only the manganese atoms. (c) shows the magnetic structure for $LaFeO_3$ and $LaCrO_3$, indicating the positions and moments of the iron or chromium atoms. (After Koehler and Wollan 1957.)

cell which is doubled in one direction only compared with the chemical cell. The nearest neighbours for each B ion are two ions with antiparallel spins and four with parallel spins. The moment directions lie within the ferromagnetic sheets but it is not possible from powder data to define the direction more specifically. A different structure, illustrated in Fig. 309 (c), is shown by $LaFeO_3$ and $LaCrO_3$. The sheets of atoms now contain both + and − spins and the magnetic cell is twice as large as the chemical cell in all three directions. Each B ion has six antiparallel nearest neighbours and the structure may be regarded as two interpenetrating face-centred lattices with the spin directions oppositely directed in the two cases. Again, the moments lie in some undetermined direction within the layer planes. With some of the compounds of low symmetry which show greater distortion from cubic symmetry, such as $ErFeO_3$ and $HoFeO_3$, it should be possible to define the moment directions from powder data taken with high angular resolution. The absolute intensities for $LaCrO_3$, $LaMnO_3$, and $LaFeO_3$, when interpreted according to the above models, are consistent with the scattering amplitudes calculated for the Cr^{3+}, Mn^{3+}, and Fe^{3+} ions if it is assumed that there is no contribution from the orbital moments. The Néel temperatures found for the three compounds are respectively 320 K, 100 K, and 750 K.

The compounds $LaCoO_3$ and $LaNiO_3$, also, have been examined at temperatures down to 4·2 K, but the diffraction patterns show no evidence of any magnetic ordering of the cobalt or nickel ions. Paramagnetic scattering is indeed observed, but it is considerably weaker than might be expected. Moreover, it falls off much less rapidly with increasing scattering angle than would be anticipated for 'spin-only' moments, suggesting that a significant proportion of the moment is due to an orbital contribution.

$LaMnO_3$, $CaMnO_3$, and the series $(La_{1-x}Ca_x)MnO_3$

A very full investigation of the properties of the series of perovskite compounds $(La_{1-x}Ca_x)MnO_3$ has been made by Wollan and Koehler (1955), in which the neutron diffraction data were supported by X-ray diffraction studies of the distortions of the structures from the ideal cubic symmetry and by measurements of magnetization. In $La^{3+}Mn^{3+}O_3$, which is one end member of the series, the manganese occurs entirely as the Mn^{3+} ion which, as we proceed through the series, is continuously replaced by Mn^{4+} until we arrive at the end member $Ca^{2+}Mn^{4+}O_3$, in which the manganese is entirely tetravalent.

We have already mentioned above that $LaMnO_3$ has the magnetic structure shown at (b) in Fig. 309 and which is designated as A in the series of structure elements given in Fig. 310. $CaMnO_3$, at the other end of the series, has the type G structure, similar to $LaFeO_3$ and $LaCrO_3$, which was shown in Fig. 309 (c). At intermediate compositions a number of different ferromagnetic and antiferromagnetic structures, including mixed phases, occur, and for a full account the reader is referred to the original paper. Some of the main conclusions can, however, be appreciated from Fig. 311, which indicates how the magnetic moments are distributed

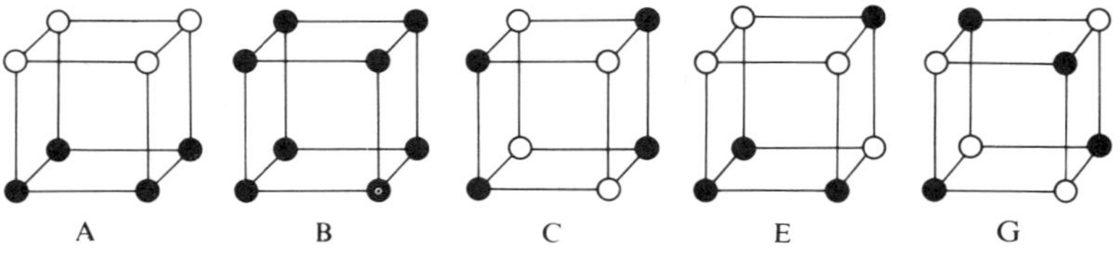

FIG. 310. Octants of the various magnetic unit cells found in the series of perovskite-type compounds $(La_{1-x}Ca_x)MnO_3$. The filled and open circles indicate oppositely-directed magnetic moments. (After Wollan and Koehler 1955.)

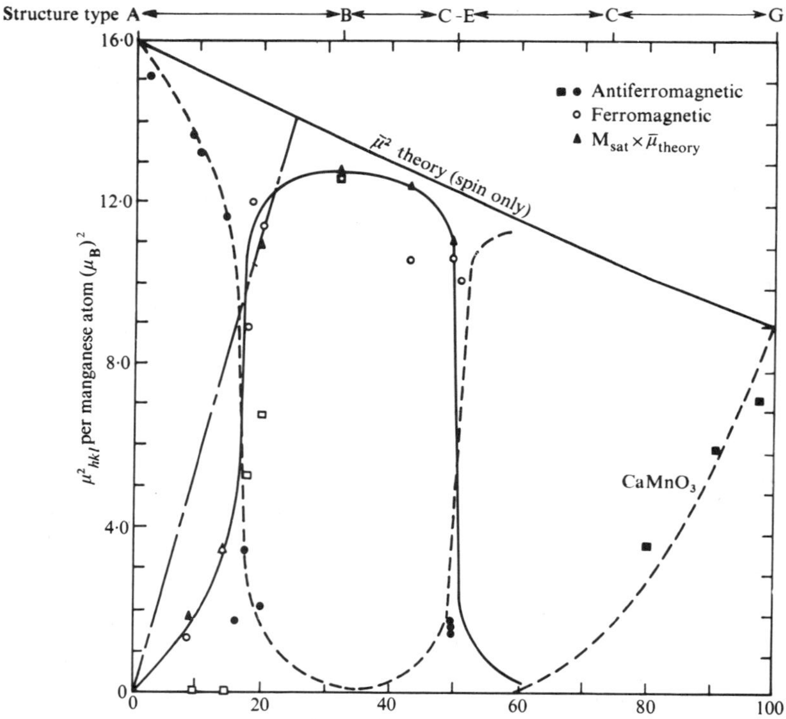

FIG. 311. The ferromagnetic and antiferromagnetic moments in the system $(La_{1-x}Ca_x)MnO_3$ as a function of the percentage of tetravalent manganese. The full-line curve is the contribution to μ^2 from ferromagnetic ordering (as measured by neutron diffraction or magnetic techniques), whereas the dashed line, up to 60 per cent Mn^{4+}, indicates the contribution from antiferromagnetism. The contribution from these two curves together is very close to the theoretical value of μ^2 for spin moments only. The composition range over which transition takes place from one structure type to the next, using the nomenclature of Fig. 310, is indicated along the top of the Figure. (From Wollan and Koehler 1955.)

between antiferromagnetic and ferromagnetic phases. With increase of Mn^{4+} content the A type of $LaMnO_3$ structure is gradually replaced by a simple ferromagnetic structure B (illustrated in Fig. 310) in which all the moments are parallel and which is the sole phase present for an Mn^{4+} content of 40 per cent. Proceeding from the Ca end of the system the type G structure of $CaMnO_3$ is replaced, at 80 per cent of Mn^{4+}, by the structure shown as type C, which requires a doubling of the unit cell in two directions as seen more clearly in Fig. 312 (a). In the central region of the diagram there was found a much more complicated antiferromagnetic phase which is illustrated in Fig. 312 (b). The unit cell, which multiplies the chemical cell by 4 times in two directions and by twice in the third direction, is obtained by stacking small cubes of types C and E (as defined in Fig. 310) in an ordered manner and constituted with an ordered arrangement of Mn^{3+}, Mn^{4+} ions.

Rare-earth–iron perovskites

The study of the magnetic structures of perovskites was extended by Koehler, Wollan, and Wilkinson (1960) to cases in which the

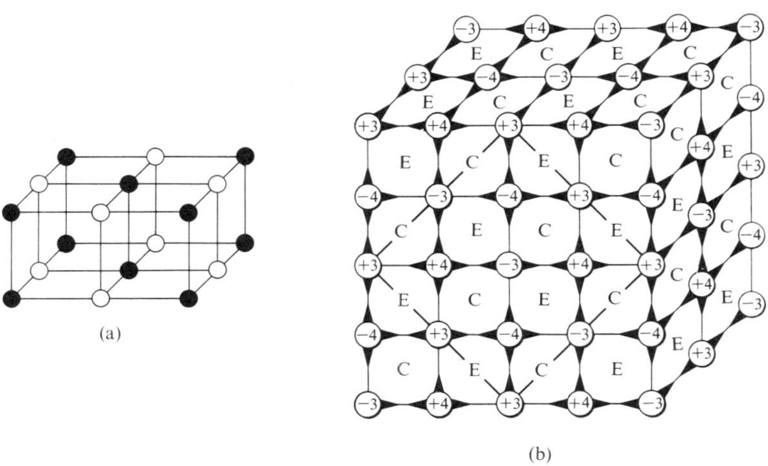

(a) (b)

FIG. 312. The magnetic structures of phases in the system $(La_{1-x}Ca_x)MnO_3$. Diagram (a) shows the full unit cell of the type-C phase, for which two of the axes are doubled compared with the chemical cell. (b) represents the complicated C–E type of arrangement which occurs in the middle of the composition range. It is built up by an ordered stacking of octants of types C, E. A small alternative cell has axes $\sqrt{2}c$, $\sqrt{2}c$, $2c$, outlined by the diagonal lines. There is evidence for ordering of the Mn^{3+}, Mn^{4+} ions as indicated in the Figure. (From Shull and Wollan 1956.)

atom A in the molecule ABO_3 is a magnetic ion. The compounds which have been studied are those of the rare earths $NdFeO_3$, $HoFeO_3$, and $ErFeO_3$.

At ordinary temperatures these compounds are antiferromagnetic due to ordering of the iron ions, in just the same way as we have seen to occur above for $LaFeO_3$. This ordering persists up to quite high temperatures, and the Néel temperatures determined for the three compounds are 760 K, 700 K, and 620 K respectively. In each case the type of ordering is the same as for $LaFeO_3$, in which each Fe^{3+} ion is surrounded octahedrally by six antiparallel nearest neighbours as represented by the structure element G in Fig. 310 and by the extended arrangement (c) in Fig. 309. In terms of this structure the values of the magnetic moments for the Fe^{3+} ion in the three compounds are estimated to be 4·57 μ_B, 4·60 μ_B, and 4·62 μ_B respectively. These values are significantly less than the expected values of 5 μ_B. The atomic structures of these perovskites are distorted from cubic to orthorhombic and in the case of $HoFeO_3$ and $ErFeO_3$ the distortion from cubic symmetry is sufficient to give partial resolution of some of the near-equivalent lines in the powder diffraction pattern. As a result it is possible to draw some approximate conclusions regarding the moment directions from powder data. It is concluded that at room temperature the magnetic moments are parallel to the [100] axis for both substances, but a rotation takes place with reduction of temperature and at 43 K the moments lie in a $(1\bar{1}0)$ plane. At 1·25 K it is possible to specify the directions more precisely as being parallel to [001] for the holmium compound but parallel to [110] for erbium.

As the temperature is lowered to the liquid-helium range order begins to develop among the magnetic moments of the rare-earth ions, which occupy the body-centre positions in the idealized cubic structures of Fig. 309 (a). A general impression of the growth of magnetic order with reduction of temperature for $ErFeO_3$ is provided by Fig. 313. The top three curves in this figure are the powder diffraction patterns at 955 K (above the Fe–Fe ordering temperature), 295 K, and 43 K, respectively. The bottom curve is a 'difference' pattern which shows the change in the diffraction effects between 43 K and 1·3 K. Not only do we observe the development of new intense coherent reflections but also the disappearance of the paramagnetic background scattering from the erbium ions, giving the negative background level in the difference pattern.

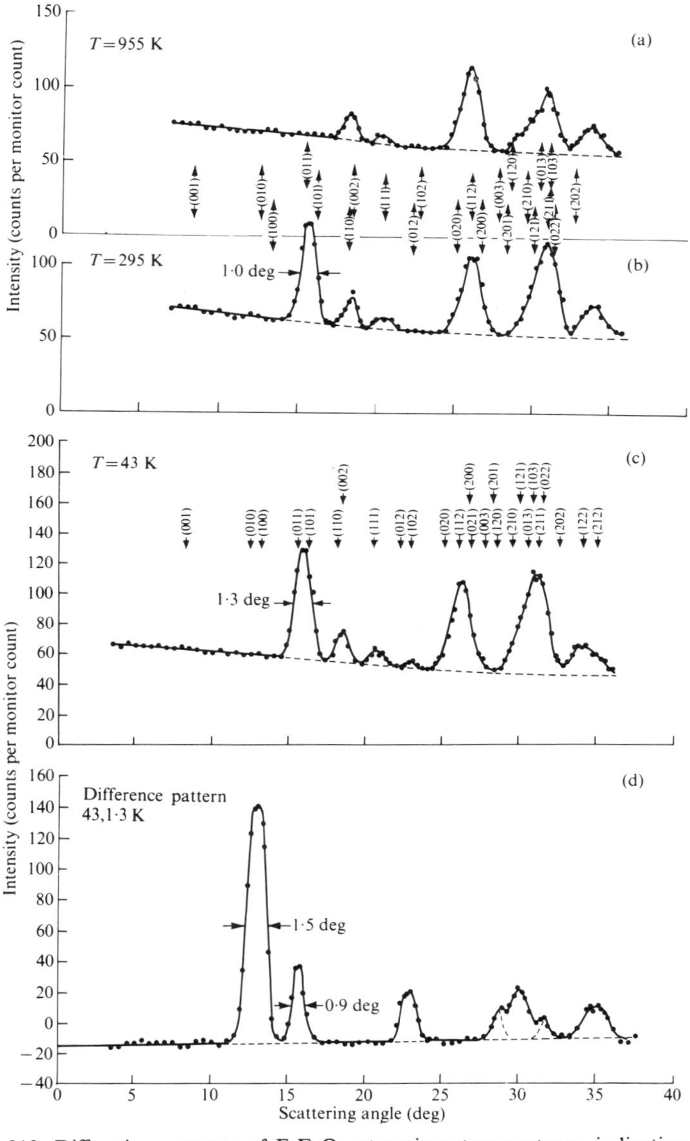

FIG. 313. Diffraction patterns of $ErFeO_3$ at various temperatures, indicating the growth of magnetic order. Curve (a) is taken at 955 K, which is above the temperature at which ordering of the Fe atoms takes place. The additional lines in curve (b) at 295 K indicate the magnetic reflections due to the Fe–Fe ordering; this ordering is slightly different at 43 K (curve (c)), corresponding to a reorientation of the moment directions. Curve (d) is a 'difference' curve showing the change of diffraction pattern between 43 K and 1·3 K. At the lower temperature the paramagnetic background scattering due to the erbium ions has disappeared, and additional coherent magnetic reflections appear as order develops among these ions. (From Koehler, Wollan, and Wilkinson 1960.)

Analysis of the data shows that the basic arrangement taken up by the holmium or erbium moments is the antiferromagnetic structure type C (in Fig. 310) in which a vertical line of positive moments is surrounded by four vertical lines of negative moments at nearest-neighbour distances. For erbium this arrangement seems to be accurately achieved with the moments parallel and antiparallel to the vertical [001] axis and having a magnitude of 5·8 μ_B at 1·25 K. This arrangement, together with the arrangement parallel to [110] taken up by the iron moments at this temperature, can be seen in model (a) of Fig. 314. For $HoFeO_3$, however, the neutron intensities can only be explained by assuming a more complicated arrangement for the precise orientation of the Ho^{3+} moments. It is concluded, as shown in model (b) of Fig. 314, that the structure is not strictly

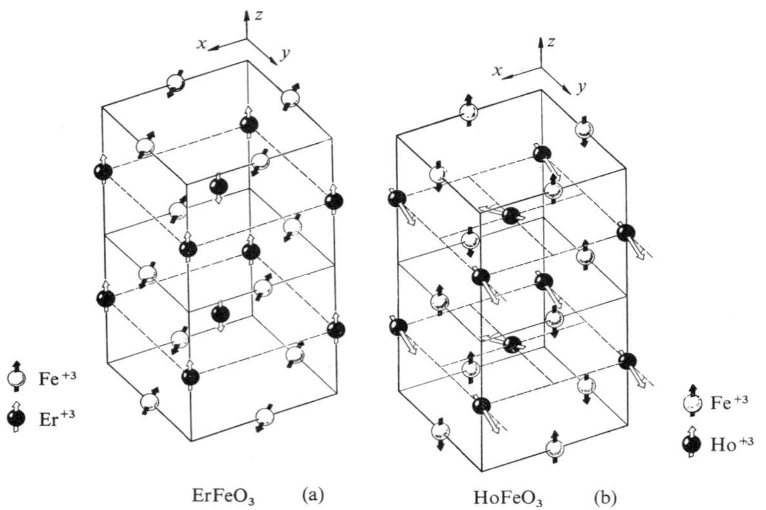

FIG. 314. The proposed magnetic structures for (a) $ErFeO_3$ and (b) $HoFeO_3$, at 1·3 K. The latter structure is not strictly antiferromagnetic, for the magnetic moments of the Ho^{3+} ions are directed slightly to one side of the *y*-axis, as shown in the Figure, resulting in a net ferromagnetic moment parallel to [100]. (After Koehler, Wollan, and Wilkinson 1960.)

antiferromagnetic. In fact the two sets of moments are not quite antiparallel but are directed at angles of 27° and (180°–27°) to the [010] *y*-axis. As a result the compound is ferromagnetic in this region of temperature, with a net ferromagnetic moment of 3·4 μ_B parallel to the [100] x-axis. The Néel temperatures for the ordering of the rare-earth moments are found to be 6·5 K and 4·3 K for the holmium and erbium compounds respectively. The magnitudes of the moments are 7·5 μ_B and 5·8 μ_B, which are substantially lower than the saturation moments of the free ions which are 10·0 μ_B and 9·0 μ_B for Ho^{3+} and Er^{3+} respectively.

By contrast to the compounds of holmium and erbium, $NdFeO_3$ showed no evidence for ordering of the neodymium ions at temperatures down to 1·3 K although, as we have mentioned above, it showed the normal type G ordering of the moments of the iron ions.

Mn_3AB compounds

A further series of perovskite compounds is known in which manganese ions occupy the sites of the oxygen atoms in perovskite itself. The simplest of these is Mn_3GaN which has been studied by Bertaut, Fruchart, Bouchard, and Fruchart (1968), and both this compound and the series as a whole are interesting for the triangular arrangements of spins which are found.

In examining this series it is convenient to redraw the perovskite unit cell of Fig. 309 (a) by placing the atom A at the corner of the cell so that the oxygen atoms will lie at the centres of the faces. We then have the arrangement of Fig. 315 (a) in which, for our present purpose, the face-centring atoms will be manganese. Bertaut *et al.* showed that Mn_3GaN was antiferromagnetic with a Néel temperature of 298 K, and it was deduced from the neutron

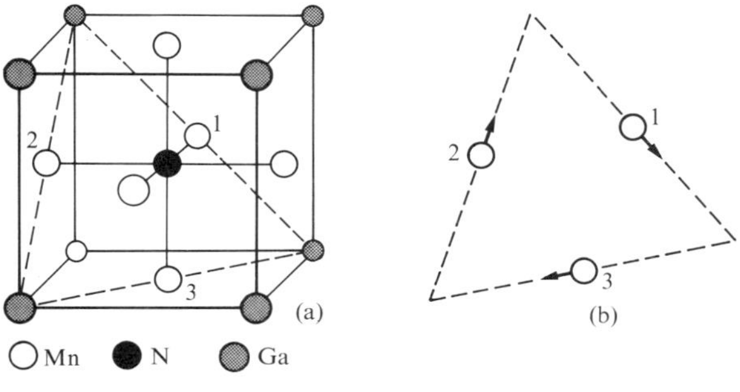

FIG. 315. Triangular arrangement of spins for the manganese atoms in Mn_3 GaN, in which the spins lie in the (111) plane at 120° to each other.

diffraction patterns at liquid-helium and liquid-nitrogen temperatures that there was a triangular arrangement of spins as drawn in Fig. 315 (b), where the spins from the three sublattices all lie in the (111) plane but with an inclination of 120° to each other. The magnetic moment of the manganese atoms is $1{\cdot}17\,\mu_B$ at 4·2 K. For Mn_3ZnN, however, (Fruchart *et al.* 1971*a*, *b*) there are two antiferromagnetic states. Between 183 K and 140 K there is a similar triangular arrangement of spins to that found in Mn_3GaN, but below 140 K there is a transition to the new arrangement illustrated in Fig. 316. The magnetic reflections can then be indexed by choosing a tetragonal cell which has a volume 4 times as large as that in Fig. 315. The vertical axis of the latter cell has been doubled, and new horizontal axes have been chosen in directions at 45° to the original *x*-, *y*-axes, as indicated in the plan of the structure in Fig. 316 (a). The new unit cell, indicating only the manganese atoms, is drawn in perspective in Fig. 316 (b). The manganese atoms can

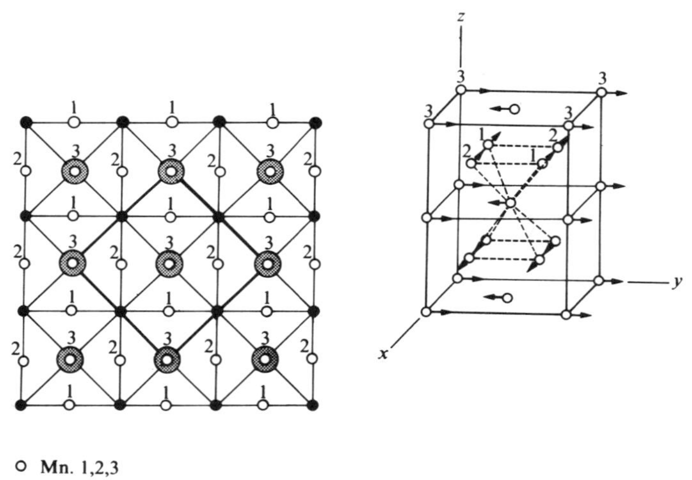

FIG. 316. Low-temperature magnetic structure of Mn_3ZnN, based on a tetragonal unit cell with a doubled *c*-axis and *a*-axes at 45° to those of the unit cell shown in Fig. 315. The magnetic manganese atoms are of two types, each forming an antiferromagnetic sublattice and with the two sets of moment directions orthogonal. The first set comprises the atoms labelled (1), (2) in Fig. 315 and has its moments along the new O*x*; the second set comprises the atoms (3) in Fig. 315 and has moments along the new O*y*.

be seen to be of two types and to lie on two antiferromagnetic lattices whose moment directions are orthogonal. For Mn_I, which comprises two-thirds of the manganese atoms (those originally identified as 1, 2 in Fig. 315), the moments lie along the new x-axis and μ is equal to 0·61 μ_B; for the remaining one-third of the manganese atoms, denoted by Mn_{II} and consisting of the atoms originally identified as 3, the moments are parallel to the y-axis and equal to 1·03 μ_B.

Mn_3NiN, studied by Fruchart *et al.* (1971*a*, *b*), shows a third type of behaviour. Below a temperature of 184 K the triangular arrangement characteristic of Mn_3GaN and shown in Fig. 315 (b) is found, but at higher temperatures the moments steadily rotate about the [111] axis until at 266 K, above which temperature the material has become paramagnetic, they have rotated through 90°. This behaviour is illustrated in Fig. 317.

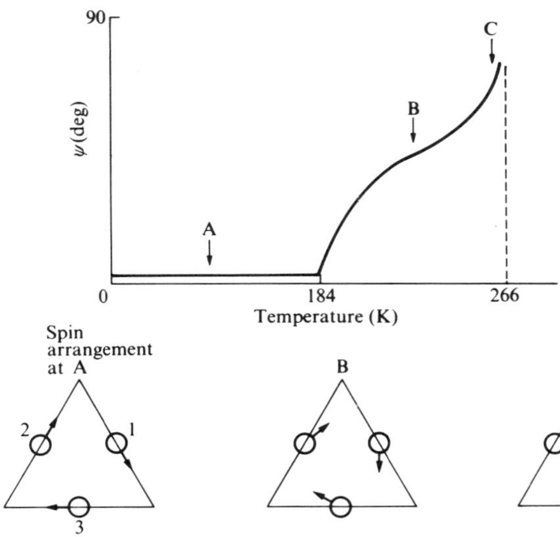

FIG. 317. Magnetic phase transition in Mn_3NiN. The structure is based on that of Mn_3GaN in Fig. 315. Below 184 K the spins lie along [110] directions in the (111) plane, as at A, but at higher temperatures an increasing rotation occurs, as at B; the rotation reaches 90°, as at C, at the Néel temperature of 266 K.

14.5. Manganites $RMnO_3$: non-collinear structures

It is convenient to comment here on the rare-earth manganites, $RMnO_3$, which have a quite different structure from the rare-earth ferrites such as $ErFeO_3$. These compounds do not possess the perovskite structure and they are of interest for their non-collinear arrangement of magnetic spins. The best-known example is $YMnO_3$ which has been studied in powder form by Bertaut and Mercier (1963), by Koehler *et al.* (1964), and finally, more conclusively, as a single crystal by Bertaut, Pauthenet, and Mercier (1967). The compounds (Fig. 318) have a hexagonal layer structure with a unit cell built up of six layers of oxygen atoms. The manganese atoms lie

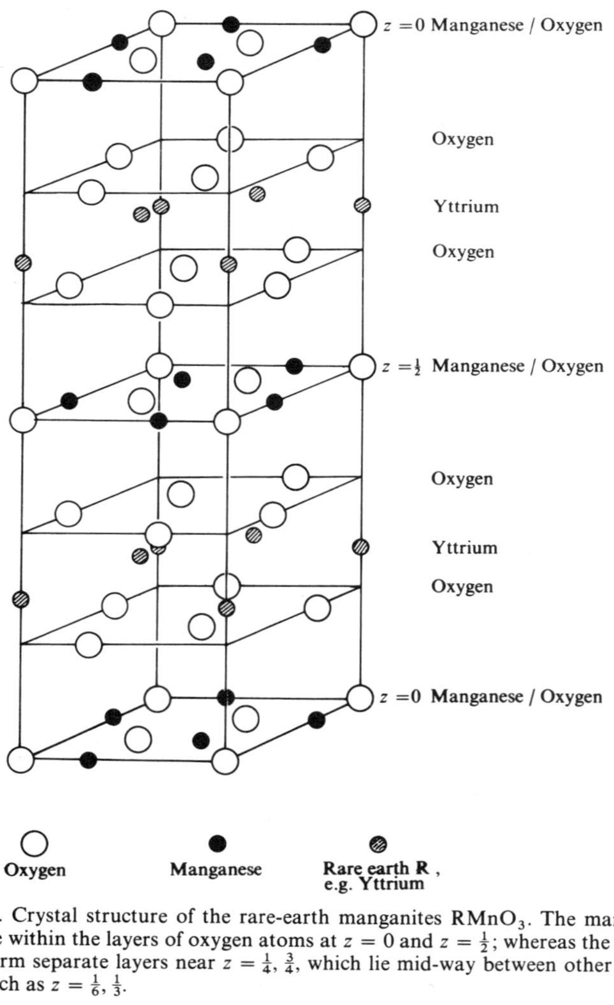

FIG. 318. Crystal structure of the rare-earth manganites $RMnO_3$. The manganese atoms lie within the layers of oxygen atoms at $z = 0$ and $z = \frac{1}{2}$; whereas the yttrium atoms form separate layers near $z = \frac{1}{4}, \frac{3}{4}$, which lie mid-way between other oxygen layers such as $z = \frac{1}{6}, \frac{1}{3}$.

within the oxygen layers at $z = 0$ and $z = \frac{1}{2}$, whereas the yttrium atoms constitute separate layers at $z = \frac{1}{4}$ and $z = \frac{3}{4}$, which interleave layers of oxygen. Two possible arrangements have been suggested for their magnetic structure, as indicated in Fig. 319, where, in principle, the angle γ can take any value. In each case there is a triangular arrangement of directions for the magnetic moments and the distinction between the two arrangements is seen by comparing the relative directions for the moments in the layers at $z = 0$, $z = \frac{1}{2}$ respectively. For the α model the layers at 0, $\frac{1}{2}$ have identical directions for the corresponding atoms, whereas for the

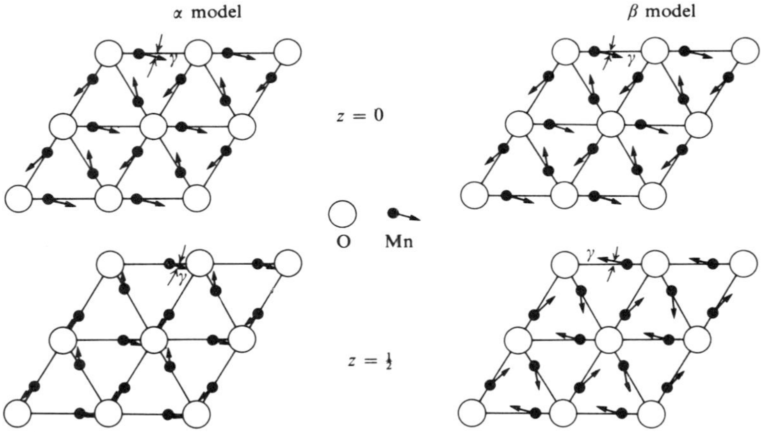

FIG. 319. Two possible arrangements of magnetic moments in the layers of manganese atoms at $z = 0$ and $z = \frac{1}{2}$ for $YMnO_3$. In principle the angle γ can take any value. For the α model the layers at 0 and $\frac{1}{2}$ have identical directions for the corresponding atoms, whereas for the β model the two layers show opposite directions.

β model the two layers have opposite directions. It can be shown that the experimental data for one of these compounds can always be interpreted in one of two ways, corresponding to a defined value of the angle γ in the model α or to a model of type β for which the value of γ is increased by 90°. We shall show that the relative intensities of individual reflections are strongly dependent on the value of γ, so that we can determine the value of γ but will not be able to say whether there is a type α or β structure. For $YMnO_3$ itself it is found that γ is zero for model α, so that the moments lie along the hexagonal axes; the alternative β solution requires the moments to be at right-angles to the axes. For $ErMnO_3$ the value of γ on the α model is 70°, and in the case of $HoMnO_3$

there is a phase change at a temperature of 50 K, with a rotation of the moments from directions along the axes (on the α model) to directions perpendicular to these. At the change point there is a very marked alteration in the relative intensities of the $(10\bar{1}0)$ and $(10\bar{1}1)$ reflections, as indicated in the curves of Fig. 320 for the powdered materials. When the moments lie along the axes $(10\bar{1}0)$ is large and $(10\bar{1}1)$ is small; when the moments are at right-angles to the axes the intensity of $(10\bar{1}0)$ becomes zero and $(10\bar{1}1)$ increases. This shows very plainly for the contrasting diffraction patterns of $HoMnO_3$ and $ErMnO_3$ at 4·2 K in Fig. 320.

These structures provide a suitable example for us to illustrate the method of calculation of the magnetic structure factor for a non-collinear structure, and we shall demonstrate that the variations of intensity shown in Fig. 320 are consistent with calculation. We have to remember in particular that the structure factor is a vector quantity, as indicated in eqn (8.4), and it will often be necessary to calculate separately its components in two directions at right-angles. Fig. 321 (a) shows the magnetic moment positions and directions for the layers at $z = 0$ and $z = \frac{1}{2}$ for the α-manganite structure which we have already illustrated more fully in Figs 318 and 319. We shall proceed to calculate the magnetic structure factor for the $(10\bar{1}0)$ reflection. For atom (1) in the layer at $z = 0$ we can resolve the spin vector, as in Fig. 321 (b), into a component $-\cos(60°-\gamma)$ along Oy and $\sin(60°-\gamma)$ along the orthogonal axis Ow; for atom (2) there is a component $\cos\gamma$ along Oy and $\sin\gamma$ along Ow, and for atom (3) the components are $-\cos(60°+\gamma)$ along Oy and $-\sin(60°+\gamma)$ along Ow. These pairs of components are, respectively, parallel and perpendicular to the $(10\bar{1}0)$ plane whose intersection with the basal plane is shown in diagram (b). The same components will apply to the three manganese atoms in the layer at $z = \frac{1}{2}$. We can, therefore, make the summation required for eqn (8.4) by listing, for each of the six atoms, as in the Table below, the resolved components of the spin vector *parallel* to the reflection plane and also the real and imaginary parts of the exponential term. We emphasize that our particular example is simplified by the fact that for $(10\bar{1}0)$ all six atoms have their moments lying in a plane which is perpendicular to the reflection plane, with the result that the component $\mathbf{S}_p$ in the reflection plane lies purely in the plane of the paper in Fig. 321 (b): more generally there would be a second component perpendicular to the paper.

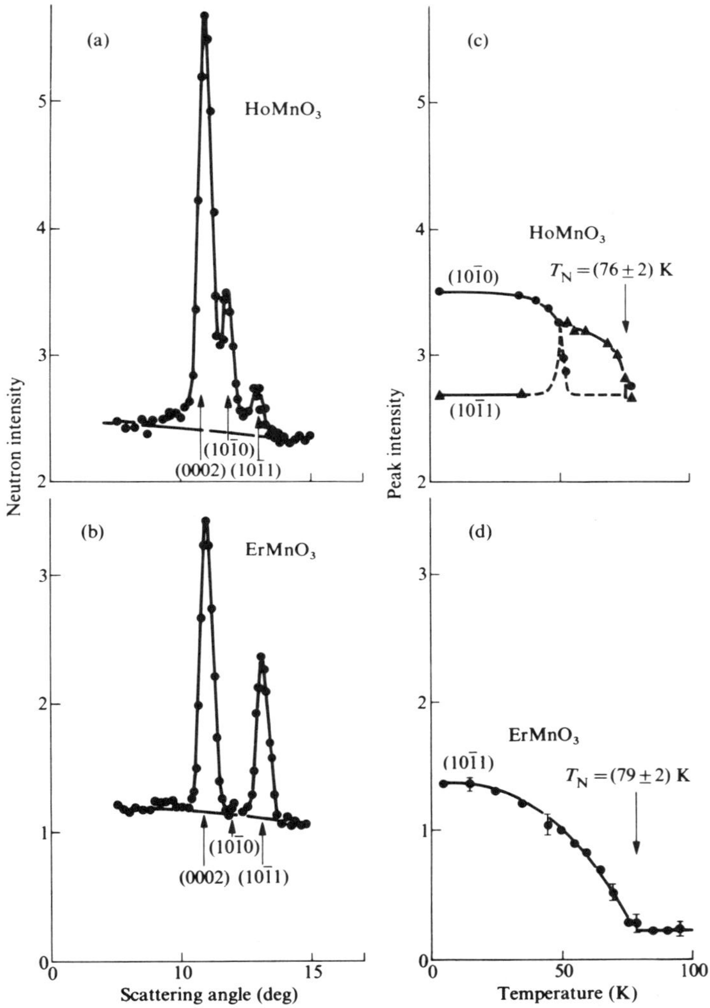

FIG. 320. (a), (b) show the low-angle portion of the diffraction patterns of $HoMnO_3$ and $ErMnO_3$ at 4·2 K, contrasting the ratio of the (10$\bar{1}$0) and (10$\bar{1}$1) reflections in the two cases. For $HoMnO_3$ diagram (c) indicates a marked change in this ratio at 50 K which is ascribed to a rotation of the moments through 90°. No change is evident in (d) for $ErMnO_3$. (After Koehler *et al.* 1964.)

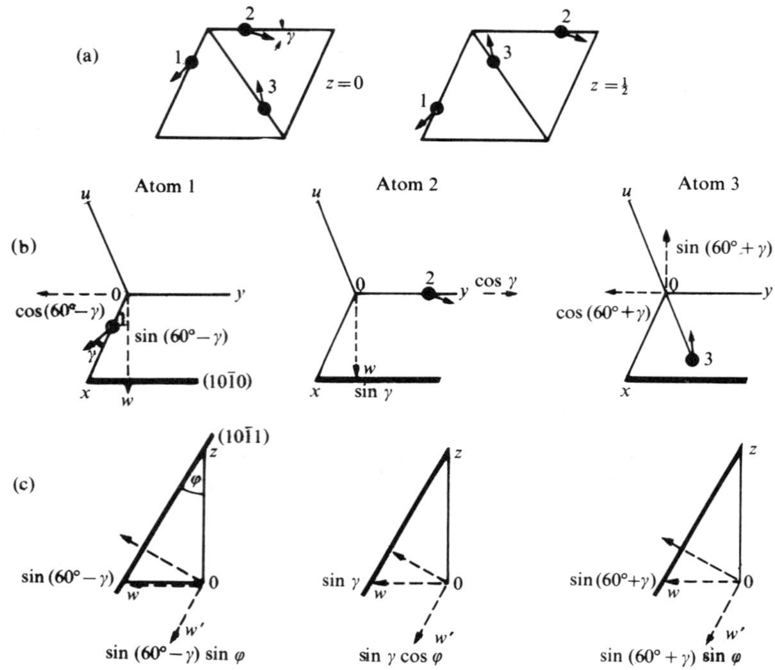

FIG. 321. Diagrams to assist in the calculations of the intensities of the magnetic reflections from the α structure of $RMnO_3$ described in the text.

(10$\bar{1}$0) reflection.

Atomic position	Spin component in reflection plane, parallel Oy	$\exp\{2\pi i(hx/a+ky/b+lz/c)\}$
$\frac{1}{3}00$	$-\frac{1}{2}\cos\gamma-\frac{\surd 3}{2}\sin\gamma$	$-\frac{1}{2}+i\frac{\surd 3}{2}$
$0\frac{1}{3}0$	$\cos\gamma$	1
$\frac{2}{3}\frac{2}{3}0$	$-\frac{1}{2}\cos\gamma+\frac{\surd 3}{2}\sin\gamma$	$-\frac{1}{2}-i\frac{\surd 3}{2}$
$\frac{2}{3}0\frac{1}{2}$	$-\frac{1}{2}\cos\gamma-\frac{\surd 3}{2}\sin\gamma$	$-\frac{1}{2}-i\frac{\surd 3}{2}$
$0\frac{2}{3}\frac{1}{2}$	$\cos\gamma$	1
$\frac{1}{3}\frac{1}{3}\frac{1}{2}$	$-\frac{1}{2}\cos\gamma+\frac{\surd 3}{2}\sin\gamma$	$-\frac{1}{2}+i\frac{\surd 3}{2}$

therefore

$$\sum \mathbf{S}_p f \exp\{2\pi i(hx/a+ky/b+lz/c)\}$$

$$= Sf\,\mathbf{y}\begin{pmatrix} \tfrac{1}{4}\cos\gamma+\tfrac{\sqrt{3}}{4}\sin\gamma+\cos\gamma+ \\ +\tfrac{1}{4}\cos\gamma-\tfrac{\sqrt{3}}{4}\sin\gamma+\tfrac{1}{4}\cos\gamma+ \\ +\tfrac{\sqrt{3}}{4}\sin\gamma+\cos\gamma+\tfrac{1}{4}\cos\gamma- \\ -\tfrac{\sqrt{3}}{4}\sin\gamma \end{pmatrix}+$$

$$+iSf\,\mathbf{y}\begin{pmatrix} -\tfrac{\sqrt{3}}{4}\cos\gamma-\tfrac{3}{4}\sin\gamma+\tfrac{\sqrt{3}}{4}\cos\gamma- \\ -\tfrac{3}{4}\sin\gamma+\tfrac{\sqrt{3}}{4}\cos\gamma+\tfrac{3}{4}\sin\gamma- \\ -\tfrac{\sqrt{3}}{4}\cos\gamma+\tfrac{3}{4}\sin\gamma \end{pmatrix}$$

$$= Sf(3\cos\gamma)\mathbf{y},$$

where S and f are the numerical value of the spin quantum number and the form factor of the atom respectively. Accordingly the intensity of the $(10\bar{1}0)$ reflection will be proportional to $9\cos^2\gamma(Sf)^2$. Thus

$$|\mathbf{F}_{\text{magn}}|^2_{10\bar{1}0} = 9(Sf)^2\cos^2\gamma. \tag{14.3}$$

The same result would be arrived at for the intensities of the $(0\bar{1}10)$ and $(\bar{1}100)$ reflections.

We can take a more general example by calculating the magnetic structure factor for the $(10\bar{1}1)$ reflection. This is a plane which is not parallel to the plane in which lie the magnetic spins. Fig. 321 (c) shows a section in the plane of the axes Oz, Ow, i.e. perpendicular to the plane $(10\bar{1}1)$. For atom (1) the component $\sin(60°-\gamma)$ along Ow, which we drew in Fig. 321 (b), now is itself resolved parallel and perpendicular to the reflection plane. The parallel component will equal $\sin(60°-\gamma)\sin\phi$, so that we now have two components for $\mathbf{S}_p$, namely, a component $-\cos(60°-\gamma)$ which is parallel to Oy and a component $\sin(60°-\gamma)\sin\phi$ in the direction Ow', which is the projection of the direction Ow on to the reflection plane. The table for summation of the contributions from the six atoms accordingly becomes, after calculation also of the new values of the exponential term,

$(10\bar{1}1)$ reflection.

Atomic position	Spin components in reflection plane parallel Oy	parallel Ow'	$\exp\left\{2\pi i\left(\frac{hx}{a}+\frac{ky}{b}+\frac{lz}{c}\right)\right\}$
$\frac{1}{3}00$	$-\frac{1}{2}\cos\gamma - {}$ $-\frac{\sqrt{3}}{2}\sin\gamma$	$(\frac{\sqrt{3}}{2}\cos\gamma - {}$ $-\frac{1}{2}\sin\gamma)\sin\phi$	$-\frac{1}{2}+i\frac{\sqrt{3}}{2}$
$0\frac{1}{3}0$	$\cos\gamma$	$\sin\gamma\sin\phi$	1
$\frac{2}{3}\frac{2}{3}0$	$-\frac{1}{2}\cos\gamma + {}$ $+\frac{\sqrt{3}}{2}\sin\gamma$	$(-\frac{\sqrt{3}}{2}\cos\gamma - {}$ $-\frac{1}{2}\sin\gamma)\sin\phi$	$-\frac{1}{2}-i\frac{\sqrt{3}}{2}$
$\frac{2}{3}0\frac{1}{2}$	$-\frac{1}{2}\cos\gamma - {}$ $-\frac{\sqrt{3}}{2}\sin\gamma$	$(\frac{\sqrt{3}}{2}\cos\gamma - {}$ $-\frac{1}{2}\sin\gamma)\sin\phi$	$\frac{1}{2}+i\frac{\sqrt{3}}{2}$
$0\frac{2}{3}\frac{1}{2}$	$\cos\gamma$	$\sin\gamma\sin\phi$	-1
$\frac{1}{3}\frac{1}{3}\frac{1}{2}$	$-\frac{1}{2}\cos\gamma + {}$ $+\frac{\sqrt{3}}{2}\sin\gamma$	$(-\frac{\sqrt{3}}{2}\cos\gamma - {}$ $-\frac{1}{2}\sin\gamma)\sin\phi$	$\frac{1}{2}-i\frac{\sqrt{3}}{2}$

therefore

$$\sum \mathbf{S}_p f \exp\left\{2\pi i\left(\frac{hx}{a}+\frac{ky}{b}+\frac{lz}{c}\right)\right\}$$

$$= Sf\,\mathbf{y}\begin{pmatrix} \frac{1}{4}\cos\gamma+\frac{\sqrt{3}}{4}\sin\gamma+\cos\gamma+ \\ +\frac{1}{4}\cos\gamma-\frac{\sqrt{3}}{4}\sin\gamma-\frac{1}{4}\cos\gamma- \\ -\frac{\sqrt{3}}{4}\sin\gamma-\cos\gamma-\frac{1}{4}\cos\gamma+ \\ +\frac{\sqrt{3}}{4}\sin\gamma \end{pmatrix}+$$

$$+iSf\,\mathbf{y}\begin{pmatrix} -\frac{\sqrt{3}}{4}\cos\gamma-\frac{3}{4}\sin\gamma+\frac{\sqrt{3}}{4}\cos\gamma- \\ -\frac{3}{4}\sin\gamma-\frac{\sqrt{3}}{4}\cos\gamma-\frac{3}{4}\sin\gamma+ \\ +\frac{\sqrt{3}}{4}\cos\gamma-\frac{3}{4}\sin\gamma \end{pmatrix}+$$

$$+Sf\sin\phi\,\mathbf{w}'\begin{pmatrix}-\frac{\sqrt{3}}{4}\cos\gamma+\frac{1}{4}\sin\gamma+\sin\gamma+\\ +\frac{\sqrt{3}}{4}\cos\gamma+\frac{1}{4}\sin\gamma+\frac{\sqrt{3}}{4}\cos\gamma-\\ -\frac{1}{4}\sin\gamma-\sin\gamma-\frac{\sqrt{3}}{4}\cos\gamma-\\ -\frac{1}{4}\sin\gamma\end{pmatrix}+$$

$$+\mathrm{i}Sf\sin\phi\,\mathbf{w}'\begin{pmatrix}\frac{3}{4}\cos\gamma-\frac{\sqrt{3}}{4}\sin\gamma+\frac{3}{4}\cos\gamma+\\ +\frac{\sqrt{3}}{4}\sin\gamma+\frac{3}{4}\cos\gamma-\frac{\sqrt{3}}{4}\sin\gamma+\\ +\frac{3}{4}\cos\gamma+\frac{\sqrt{3}}{4}\sin\gamma\end{pmatrix}$$

$$=\mathrm{i}Sf(-3\sin\gamma\,\mathbf{y}+3\cos\gamma\sin\phi\,\mathbf{w}').$$

Hence the intensity of the $(10\bar{1}1)$ reflection will be proportional to

$$|\mathbf{F}_{\text{magn}}|^2 = 9(\sin^2\gamma+\cos^2\gamma\sin^2\phi)(Sf)^2$$
$$= 9(1-\cos^2\phi\cos^2\gamma)(Sf)^2.$$

The value of $\cos^2\phi$ for the $(10\bar{1}1)$ plane can be calculated in terms of the unit cell lengths c, a and is found to be equal to $c^2/\{c^2+(\sqrt{3}a/2)^2\}$, therefore

$$|\mathbf{F}_{\text{magn}}|^2_{10\bar{1}1} = 9(Sf)^2\left\{1-\frac{c^2}{c^2+(\sqrt{3}a/2)^2}\cos^2\gamma\right\}. \tag{14.4}$$

If we carry out similar calculations for the type-β structure we find that

$$|\mathbf{F}_{\text{magn}}|^2_{\substack{10\bar{1}0\\ \beta}} = 9(Sf)^2\sin^2\gamma \tag{14.5}$$

and

$$|\mathbf{F}_{\text{magn}}|^2_{\substack{10\bar{1}1\\ \beta}} = 9(Sf)^2\left\{1-\frac{c^2}{c^2+(\sqrt{3}a/2)^2}\sin^2\gamma\right\}. \tag{14.6}$$

Inspection of eqns (14.3), (14.4), (14.5), and (14.6), and the corresponding pairs for other reflecting planes, shows that the relative intensities of the various reflections are sensitive to the value of the angle γ. Moreover, if a value of γ can be found to satisfy experimental data in terms of the α model then a solution is also possible for the β model if the value of γ is increased by 90°. Thus it is not possible to distinguish between the two models from intensity data but, assuming either one of them, the angle γ for the magnetic moments can be specified.

14.6. Iron-group trifluorides

The trifluorides of the iron-group elements crystallize in a modification of a cubic structure with the metal atoms at the cube corners and the fluorines near the centres of the cube edges. The interatomic relationships are therefore similar to those of 3d metal ions and the oxygen ions in the perovskites; the modified structures are rhombohedral and the relation between the rhombohedral cell and the underlying distorted cube can be seen in Fig. 322. (For

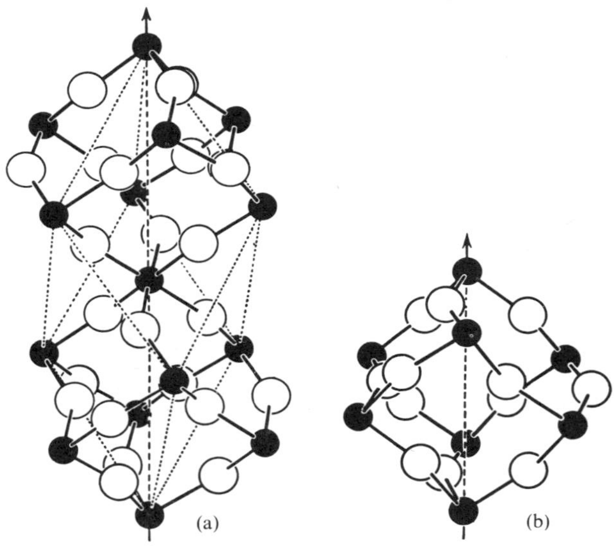

FIG. 322. The bimolecular rhombohedral unit cell (a) of the iron-group trifluorides which is built up of two of the distorted cubes, shown at (b), that have metal atoms at their corners and fluorine atoms displaced from the mid-point of the pseudo-cell edges. (Hepworth *et al.* 1957.)

a more complete account of the crystallography of these compounds the reader is referred to a paper by Hepworth, *et al.* (1957).) It is not surprising that the magnetic structures of the trifluorides, as determined by Wollan, Child, Koehler, and Wilkinson (1958), are similar to those of the corresponding perovskite type compounds. The relationships are summarized in Table 29. Thus in CrF_3, FeF_3, and CoF_3, which have the G-type structure, each

TABLE 29

Antiferromagnetic structures of perovskites and trifluorides

	T_N (K)	Structure type in Fig. 310		T_N (K)	
CrF_3	80	G	G	320	$LaCrO_3$
MnF_3	43	A	A	100	$LaMnO_3$
FeF_3	394	G	G	750	$LaFeO_3$
CoF_3	460	G	no order		$LaCoO_3$
VF_3	no order		no order		$LaVO_3$

magnetic ion is coupled via the intervening anion to six antiferromagnetic nearest neighbours. We found earlier for the perovskites that the distortions from cubic symmetry were not sufficient to permit the determination of the moment directions from powder data. However, for the fluorides the distortion from cubic symmetry is sufficiently great to resolve the magnetic reflections adequately enough to give an indication of the inclination of the moments relative to the rhombohedral axis. Fig. 323 illustrates the splitting of what would be a cubic type (111) reflection into its (111) and (100) rhombohedral components. For CrF_3 and FeF_3 the intensity ratio of (111) to (100) is in good agreement with what would be expected if the magnetic spins were parallel to the (111) rhombohedral plane. It was possible to demonstrate with CrF_3 that the application of a magnetic field could influence the orientation of the spins within the (111) planes. Thus when the field was applied along the scattering vector of the (100) planes the intensity of this reflection was increased by 50 per cent suggesting that the spins tended to be rotated in the (111) planes until they lay in a direction at right-angles to the applied field. The experimental data illustrating the intensity increase are shown in Fig. 324. By contrast to the effect on the (100) reflection no change is noted when the field is applied along the scattering vector of the (111) reflection, since all the spins are already at right angles to this direction. CoF_3, however, has a different moment direction from CrF_3, FeF_3. The (111) intensity is very small, as can be seen in Fig. 323(c), and this suggests that the spin direction is perpendicular to (111) planes. For both the iron and chromium compounds the absolute intensities are in agreement with those calculated for spin-only moments, but in CoF_3 there appears to be an orbital contribution to the magnetic moment.

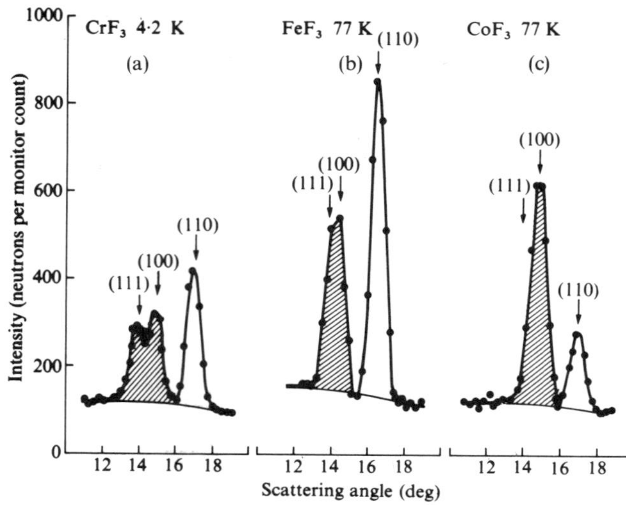

FIG. 323. Portions of the low-angle diffraction patterns of the iron-group trifluorides. Because of the distortion from cubic symmetry the (111) reflection is replaced by two reflections which index as (111) and (100) according to the rhombohedral unit cell. The ratio of these two intensities is very sensitive to the orientation of the magnetic moments. For CrF_3, where they are partially resolved, the ratio is close to 0·71, which is the calculated value if the spins lie parallel to the (111) planes. For FeF_3 the orientation is probably the same but for CoF_3 the intensity of (111) is practically zero, suggesting that the moments are perpendicular to the (111) planes. (Wollan *et al.* 1958*b*.)

MnF_3 differs from the Cr, Fe, and Co compounds and, like its counterpart $LaMnO_3$, has the A-type antiferromagnetic structure consisting of ferromagnetic layers arranged so that the spins in alternate layers are oppositely directed. The structure is actually monoclinic and it can be shown that the spins must lie in or nearly in the ferromagnetic sheets. Within experimental error the magnetic moment of Mn^{3+} comes to the spin-only value of 2 μ_B.

It will be noted from Table 29 that for neither the trifluoride nor the perovskite compound does vanadium show any magnetic ordering. Only for cobalt does there appear to be a different type of ordering in the two types of compound. However, there appears to be some evidence that the spin state of cobalt in $LaCoO_3$, in which no antiferromagnetic order was found, is very sensitive to the presence of Co^{4+} ions, which are difficult to eliminate entirely.

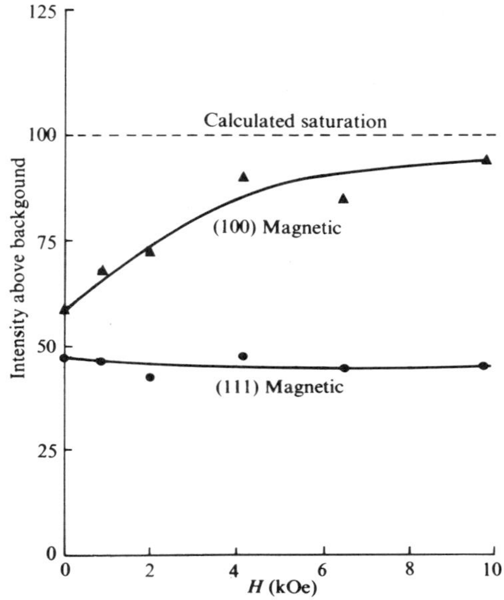

FIG. 324. The increase of intensity of the (100) magnetic reflection of CrF_3 when a magnetic field is applied along the scattering vector. It is concluded that the spins are rotated around within the (111) sheets so that a large proportion of them come parallel to (100) planes. By contrast, no change is found in the intensity of the (111) reflection for a field applied along its scattering vector. (Wollan *et al.* 1958*b*.)

This may account for the apparent difference of magnetic structure type between CoF_3 and $LaCoO_3$.

14.7. Magnetite, ferrites, and chromites

The study of magnetite by Shull, Wollan, and Koehler (1951) provides one of the earliest and best examples of the value of neutron diffraction for investigations of magnetic structure. As described by Verwey and Heilmann (1947) the crystallographic structure of Fe_3O_4 is that of an 'inverted' spinel, with the tetrahedral A sites occupied solely by trivalent Fe^{3+} ions and the octahedral B sites occupied both by the Fe^{2+} ions and the remaining Fe^{3+} ions in random distribution. Néel (1948) accounted for the observed magnitude of the saturation magnetization by postulating that the ions on the A, B sites were coupled antiferromagnetically. Such a structure in which more magnetic moments point in one direction than in the antiparallel direction will be ferromagnetic. In order to call attention both to this fact and to its underlying *antiferromagnetic* nature the term *ferrimagnetic* is applied to it.

The neutron intensities in the powder diffraction pattern of powdered magnetite at room temperature, obtained by Shull, Wollan, and Koehler (1951), are in very good agreement with what is expected for the Néel structure, assuming that the Fe^{2+}, Fe^{3+} ions have spin quantum numbers of 2 and $\frac{5}{2}$ respectively. A comparison of the experimental and calculated intensities is shown in Table 30. In arriving at the calculated intensities it was

TABLE 30

Comparison of the neutron intensities (in neutrons min^{-1}) observed for Fe_3O_4 at room temperature with those calculated for the Néel ferrimagnetic structure

Reflection	Calculated intensities: Magnetic	Nuclear	Total	Observed intensity
(111)	902	32	934	860
(220)	125	218	343	360
(311) (222)	112	948	1060	1070
(400)	116	649	765	780
(331)	94	16	110	135
(422) (333, 511)	20	670	690	700
(440) (531)	32	1692	1724	1730

The angular resolution of the diffraction pattern was insufficient to resolve the pairs of lines which are bracketed together.

assumed that the oxygen parameter u was exactly $\frac{3}{8}$, as deduced from X-ray results. It would be possible to examine this conclusion more closely, in the way described already in the measurement of the parameters of $MgFe_2O_4$ and $MnFe_2O_4$ in Chapter 12, by careful quantitative examination of the nuclear scattering.

Intensity calculations were made for other magnetic models but only for Néel's ferrimagnetic structure was agreement obtained with the observed neutron intensities. It is to be noted that the magnetic unit cell is the same size as the chemical unit cell so that there are no peaks in new positions to be attributed to magnetic scattering. This can be realized from Fig. 325 which compares the neutron and X-ray diffraction patterns. The magnetic structure is such that there is no reduction in multiplicity, i.e. in the number of planes which contribute to a powder reflection, and it is not possible to deduce from powder data the absolute orientation of the magnetic

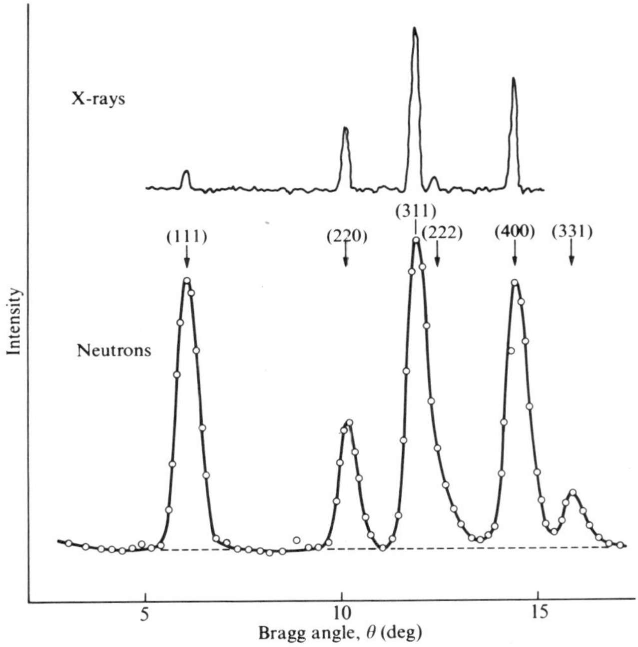

FIG. 325. The neutron diffraction pattern of magnetite at room temperature, with a corresponding X-ray pattern for comparison. No new lines appear in the neutron pattern and the magnetic unit cell is the same size as the chemical cell. Certain of the reflections, however, have an appreciable magnetic contribution; the (111) reflection, for example, is almost entirely magnetic in origin. (Shull, Wollan, and Koehler 1951.)

moments relative to the crystallographic axes. The orientation has been determined, however, from later measurements with single crystals by Hamilton (1958*b*) who also investigated the phase change which magnetite shows at about 119 K. Such a change had been recognized for many years from thermal, magnetic, and electrical measurements and was postulated by Verwey, Haayman, and

Romeijn (1947) to be due to an ordering among the Fe^{2+} and Fe^{3+} ions on the octahedral sites. Hamilton's measurements showed that the proposed ordering scheme, as illustrated in Fig. 326, was correct. Below 119 K the structure is orthorhombic with the octahedral Fe^{3+} ions lying in rows parallel to the short *a*-axis and the octahedral Fe^{2+} ions lying in rows which are parallel to the slightly longer *b*-axis. The direction of the magnetic moments

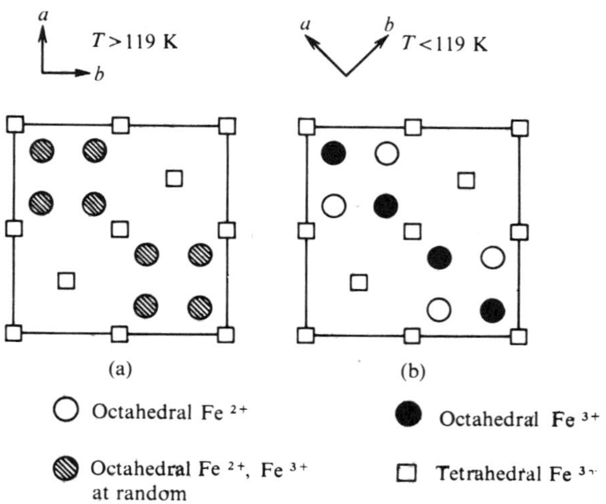

FIG. 326. Projections on the cube plane of the metal atoms in Fe_3O_4. (a) Above 119 K and (b) below 119 K. Below this temperature there is an ordered arrangement of the Fe^{2+} and Fe^{3+} ions among the octahedral sites, as shown in (b). (Shull 1959.)

is parallel to the orthorhombic *c*-axis so that in ordinary circumstances there is no (002) reflection. However, a measurable reflection can be produced, as expected, if the moments are forced away from the *c*-axis by applying a magnetic field in a direction inclined to *c*. It is indeed necessary to maintain a field along one of the original cubic axes before the crystal is cooled below the transition temperature in order to ensure that a single-domain untwinned orthorhombic crystal will be produced.

Since the original investigation of magnetite many ferrites of the general form $A^{2+}B_2^{3+}O_4$ have been studied with neutrons. We have already referred in Chapter 12 to the information on ferrite structures which could be obtained from observations of solely nuclear scattering by, for example, making observations above the Curie temperature. As in the case of magnetite it is possible to obtain much additional information by studying the magnetic intensities. The first observations were made on magnesium ferrite both by Bacon and Roberts (1953), who assessed the magnetic contributions by making observations above and below the Curie temperature, and by Corliss, Hastings, and Brockman (1953), who made measurements only at temperatures for which the structure was ferrimagnetic and deduced the magnetic contributions by the loss of intensity when their samples were magnetized normal to the reflecting planes. Both investigations demonstrated the incomplete inversion of $MgFe_2O_4$ and deduced values for the oxygen parameter u in good agreement with one another. Subsequently many other ferrites have been investigated, notably by Corliss and Hastings and their colleagues. For example, Hastings and Corliss (1953) have studied nickel ferrite, $NiFe_2O_4$, which they showed to be at least 80 per cent inverted, with an antiferromagnetic alignment of the moments on the A, B sites and, also, zinc ferrite, $ZnFe_2O_4$. The latter was shown to have the 'normal' spinel structure, with the diamagnetic zinc ions occupying the A sites. At room temperature no coherent magnetic reflections were observed, indicating that any magnetic interaction between the Fe^{3+} ions on the B sites must be weak. Later measurements by the same workers (Hastings and Corliss 1956*a*), suggested that there was a transition to an antiferromagnetic structure below 9 K as evidenced by the disappearance of the paramagnetic background scattering from the iron atoms and the appearance of a group of overlapping superlattice lines. An enlarged magnetic cell, which is unusual for ferrite structures, was necessary to explain these superlattice lines. A tentative model structure was proposed in which ferromagnetic parallel bands, two or three atoms wide, cross the crystal; successive bands have alternately + and − spins.

Hastings and Corliss (1956*b*) have studied manganese ferrite, $MnFe_2O_4$, which is a very favourable case for determining the distribution of the cations among the A, B sites because of the negative scattering amplitude of manganese. The structure is found to be largely of the 'normal' spinel type with a fraction $0{\cdot}81 \pm 0{\cdot}03$ of the Mn^{2+} ions in the A sites. The mean moments on the A and B sites were shown to be $4{\cdot}33\ \mu_B$ and $3{\cdot}78\ \mu_B$ respectively at room-temperature and $4{\cdot}60\ \mu_B$ and $4{\cdot}60\ \mu_B$ at $4{\cdot}2$ K, indicating that the A, B sites are saturated to unequal extents at room temperature. The net saturation moment of 4·60 μ_B agrees with the results of magnetization measurements but leaves unexplained why this, and the values for Mn^{2+}, Fe^{3+}, is less than the theoretical value of 5 μ_B. Some limitations to the accuracy of the measurements are set by uncertainty in form-factor information and preclude more precise conclusions from powder data.

Prince and Treuting (1956) have studied copper ferrite, $CuFe_2O_4$, which is a distorted form of an inverse spinel and has tetragonal symmetry. They find that although the oxygen octahedra are elongated parallel to the *c*-axis, nevertheless the FeO_4 tetrahedra are completely undistorted and have the same Fe—O distance as is found in the ordinary cubic inverse ferrites. Prince (1957) has studied the related compound copper *chromite*, $CuCr_2O_4$, which is a tetragonally distorted *normal* spinel. The particular interest of this compound is that the magnetic moment per molecule is only about 0·5 μ_B in contrast with a value of 5 μ_B which would be expected for two Cr^{3+} ions (3 μ_B) with a Cu^{2+} ion (1 μ_B) in antiparallel alignment. The neutron data are consistent with a model proposed by Yafet and Kittel (1952) in which the chromium atoms in an individual sheet are all in parallel alignment but the moment directions in successive sheets are *almost* oppositely directed giving a net small moment due to chromium which is perpendicular to the sheets of atoms. This resultant, represented by PQ in Fig. 327, is antiparallel to the Cu^{2+} moment QR, thus yielding a small net amount RP of 0·5 μ_B. Two possible angular combinations which are satisfactory are indicated in the figure but the neutron results, from powder data, are not sufficiently precise to distinguish between them. Measurements with polarized neutrons would be significant for the net magnetization is in the same direction as the copper moment QR in the lower model of Fig. 327, but it is oppositely directed for the upper model.

A more complicated non-collinear structure has been found by Hastings and Corliss (1962) in $MnCr_2O_4$. Observations at room temperature establish that $MnCr_2O_4$ is a normal spinel and the diffraction pattern remains unchanged until 43 K, when magnetic ordering sets in, giving a magnetic contribution to the fundamental spinel reflections. At 18 K a further ordering transition occurs and additional lines appear which cannot be indexed on either the original unit cell or on any reasonably enlarged cell. This pattern cannot be interpreted in terms of a Néel model and a ferrimagnetic spiral model is proposed, in line with the predictions of Lyons, Kaplan, Dwight, and Menyuk (1962). According to this model there are three magnetic substructures, one composed of the manganese atoms on the tetrahedral A sites and two others consisting of different types of chromium atom on the octahedral B sites; the division of the B sites is the same as that postulated in Fig. 326 (b) for the low-temperature form of magnetite. Each of the three types of site is associated with its own spin cone, as indicated in Fig. 328.

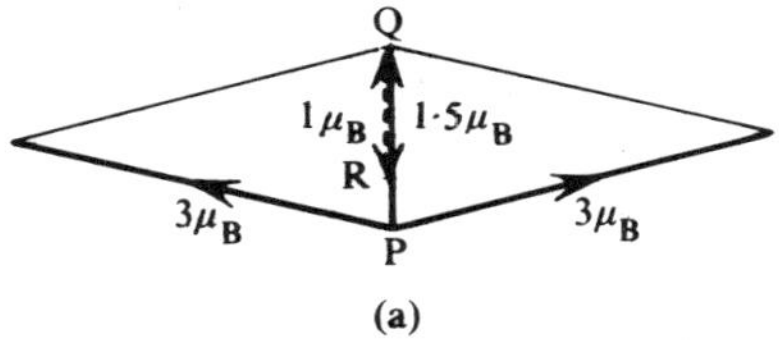

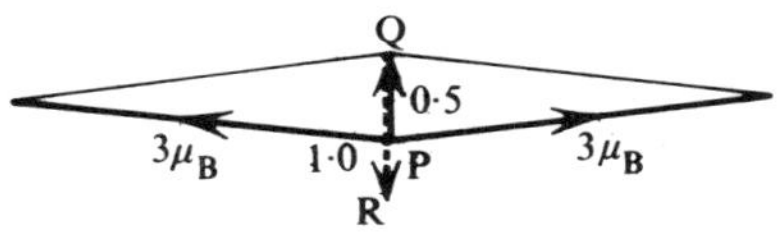

FIG. 327. Two possible arrangements of the magnetic spins in copper chromite $CuCr_2O_4$, in order to produce a resultant of 0·5 μ_B from two Cr^{3+} (3 μ_B) ions and one Cu^{2+} (1 μ_B). In (a) the resultant moment PR is oppositely directed to the copper moment QR but in (b) it is in the same direction. (After Prince 1957.)

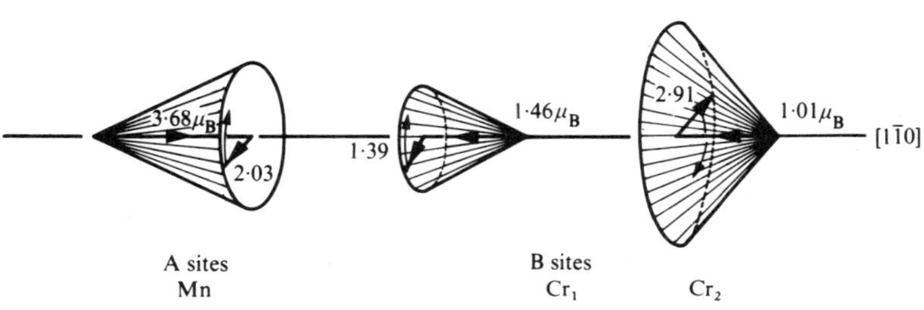

FIG. 328. Diagrams of the magnetic-spin cones in $MnCr_2O_4$ for the manganese atoms on the tetrahedral A sites and the two different types of chromium atom on the octahedral B sites. In each case there is an axial component of spin along the [1$\bar{1}$0] direction, which contributes to fundamental diffraction peaks, and a rotating normal component which yields the satellite reflections.

The axial components of the spin vectors contribute to the fundamental peaks, in the ordinary nuclear positions, and the rotating components contribute to the satellite reflections. Detailed comparison with the experimental data suggests that the axis of the cones is the $[1\bar{1}0]$ direction, that the magnetic moments are as shown in Fig. 328, and that a rotation of about $2\pi/10$ occurs between successive atoms for the transverse spin moment.

14.8. The sulphates RSO_4 and $CrVO_4$

We complete our examples of magnetic structures by considering the anhydrous sulphates, RSO_4, and chromium vanadate, $CrVO_4$. Of these, $CrVO_4$, $FeSO_4$, $NiSO_4$, and the α-form of $CoSO_4$ all have a common orthorhombic crystal structure (Frazer and Brown 1962) of space group *Cmcm*, in which the magnetic metal atom occupies the special positions (000), $(00\frac{1}{2})$, $(\frac{1}{2}\frac{1}{2}0)$, and $(\frac{1}{2}\frac{1}{2}\frac{1}{2})$. They become antiferromagnetic at low temperatures, and in all cases the powder diffraction patterns of the ordered magnetic state can be indexed in terms of the ordinary chemical unit cell. Three collinear models are possible for the arrangement of the magnetic moments and they are illustrated in Fig. 329. $CrVO_4$ is found to have the structure (a), designated as M_1, in which there are ferromagnetic sheets of chromium atoms, parallel to (001), which are stacked antiferromagnetically. The single spin direction is found to make angles of 27°, 64°, and 81° with the crystallographic axes, but there are probable errors of 15° on these values, and an orientation along the x-axis is scarcely outside experimental error. $FeSO_4$ and $NiSO_4$ have the structure M_2 (b), in which the sheets are themselves antiferromagnetic but they are stacked in a ferromagnetic manner. In each case the moment directions lie very close to the y-axis. However, α-$CoSO_4$ shows magnetic reflections which are characteristic of both structures (b) and (c), and it is not possible to account for all these reflections with a collinear model. A solution can be found in terms of the coplanar model of Fig. 330, in which all the spins lie in the (100) plane but are variously tilted above and below the direction of the y-axis by 25° in the manner shown in the Figure.

A second form of $CoSO_4$, designated as β-$CoSO_4$, (Brown and Frazer 1963) is stable above 873 K but can be kept for some weeks at room temperature and below. The crystal structure is orthorhombic but with a different space group (*Pnma*), from the

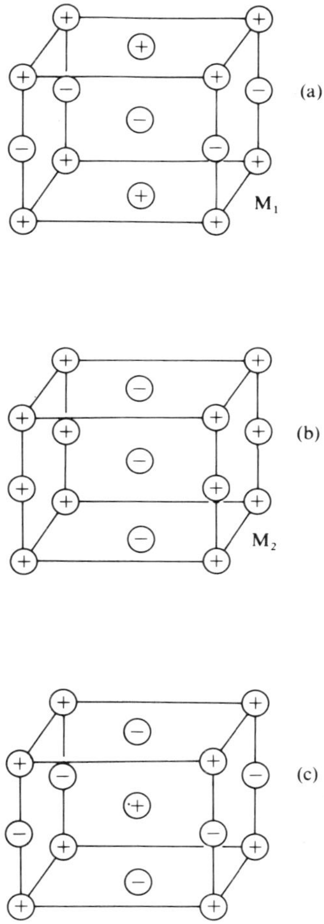

FIG. 329. Collinear magnetic structures found in the anhydrous sulphates and vanadates.

α form and with the cobalt atoms at (000), ($0\frac{1}{2}0$), ($\frac{1}{2}0\frac{1}{2}$), and ($\frac{1}{2}\frac{1}{2}\frac{1}{2}$). This material becomes antiferromagnetic at low temperatures, and although it is possible to index the powder pattern on the ordinary unit cell the extra reflections which are observed are too diverse to be accounted for by a collinear model. The four cobalt atoms in the unit cell have their magnetic moments differently oriented, though necessarily in a restricted manner since the vector sum

$\mathbf{S}_1+\mathbf{S}_2+\mathbf{S}_3+\mathbf{S}_4$ of the four spins must be zero. Indeed the four moments can be correlated and defined in terms of the three orthogonal vectors $\mathbf{L}_1$, $\mathbf{L}_2$ and $\mathbf{L}_3$ of Fig. 331, and it can be shown that

$$4\mathbf{S}_1 = \mathbf{L}_1+\mathbf{L}_2+\mathbf{L}_3,$$

$$4\mathbf{S}_2 = \mathbf{L}_3-\mathbf{L}_1-\mathbf{L}_2,$$

$$4\mathbf{S}_3 = \mathbf{L}_2-\mathbf{L}_1-\mathbf{L}_3,$$

and

$$4\mathbf{S}_4 = \mathbf{L}_1-\mathbf{L}_2-\mathbf{L}_3,$$

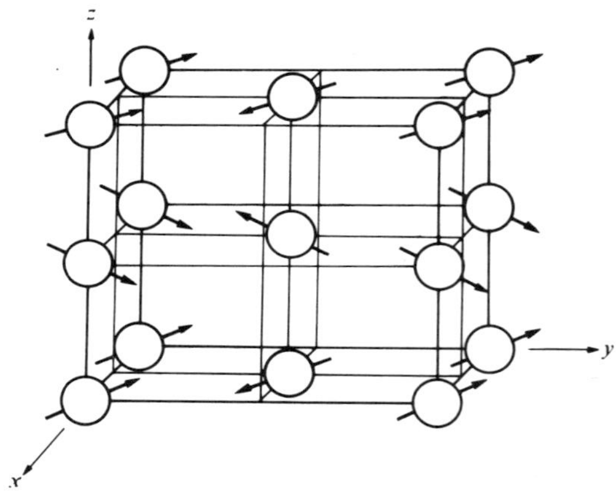

FIG. 330. The non-collinear magnetic structure for α-$CoSO_4$. All the cobalt spins lie in the (100) plane, but they are tilted by 25° either above or below the direction of the y-axis, as shown.

where the magnitudes of $\mathbf{L}_1$, $\mathbf{L}_2$, and $\mathbf{L}_3$ are $3{\cdot}0\,\mu_B$, $4{\cdot}6\,\mu_B$, and $3{\cdot}3\,\mu_B$ respectively, amounting in each case to a numerical value of $3{\cdot}2\,\mu_B$ for the moment on the cobalt atom. Addition of the above four equations shows, as expected, that the vector sum $\sum \mathbf{S}$ is equal to zero.

$MnSO_4$ (Will, Frazer, Shirane, Cox, and Brown 1965) offers an interesting development of the structure M_1 of Fig. 329, which we have associated with $CrVO_4$; the additional peaks which appear in the diffraction pattern of $MnSO_4$ at liquid–helium temperature, due to magnetic scattering, are found to be close to or to coincide with the positions calculated from the ordinary unit cell. However, more precise measurement shows small displacements, which are credited to the occurrence of satellite peaks from a helical spin arrangement with a propagation vector in the [100] direction. Fig. 332 shows how the reciprocal lattice vectors of the satellite

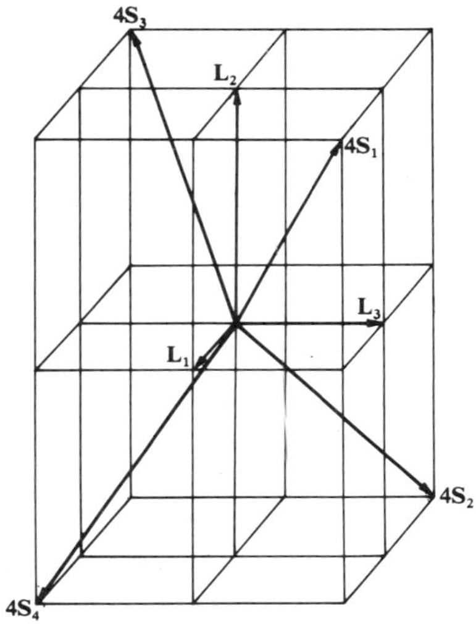

FIG. 331. The different directions of the magnetic moment vectors S_1, S_2, S_3, and S_4 for the four cobalt atoms in the unit cell of β-$CoSO_4$. The three fundamental orthogonal vectors $\mathbf{L}_1$, $\mathbf{L}_2$, and $\mathbf{L}_3$ in the diagram have magnitudes 3·0 μ_B, 4·6 μ_B, and 3·3 μ_B respectively; each of the S vectors equals 3·2 μ_B. (Brown and Frazer 1963.)

spots are all derived by adding a vector in the [100] direction, equal to about one-sixth of the fundamental spacing, to lattice points in fundamental positions. This is an example of the principle which we discussed in Chapter 8. At the same time there are contributions to the magnetic reflections which are associated with the M_1 structure. The pattern as a whole can be interpreted in terms of a conical spiral of the cycloidal kind, which we differentiated from the simple conical spiral on p. 475. The structure is illustrated in Fig. 333, in which the orientation of the x, y-axes is the reverse of the normal practice in order to indicate more clearly how the azimuth of the magnetic moments varies in the x-direction, rotating through 360° in the course of six unit cells. It will be seen by comparison

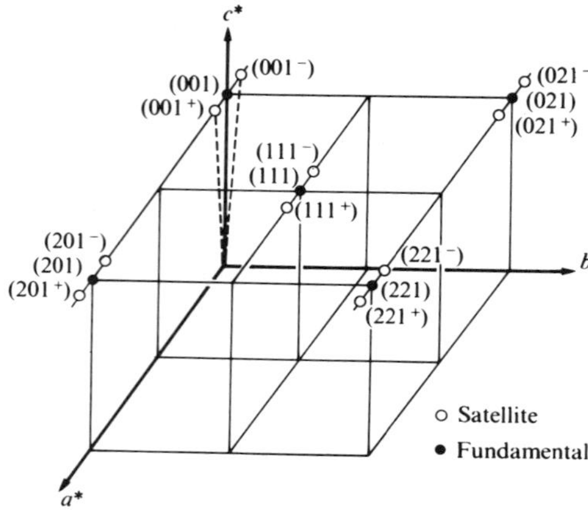

FIG. 332. Reciprocal lattice diagram indicating the position of the satellite peaks in the diffraction pattern of $MnSO_4$ at 4 K. Each satellite peak is displaced by $\frac{1}{6}a^*$ from a fundamental lattice point. At the same time there are magnetic contributions at fundamental positions expected for the M_1 structure of Fig. 329. (Will *et al.* 1965.)

with diagram (a) of Fig. 329 that the simple up and down moments of a collinear model have been replaced by a very open cone, for which the half-angle is in fact 78°. We also emphasize the cycloidal nature of the spiral by commenting that the propagation direction

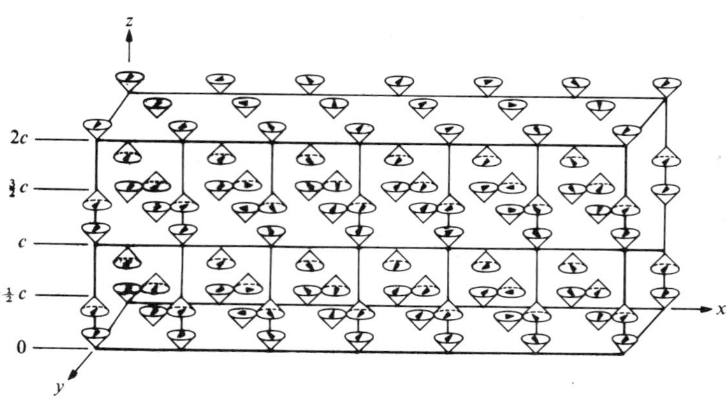

FIG. 333. A representation of the cycloidal spiral magnetic structure of $MnSO_4$, showing only the manganese atoms. The magnetic moments lie on the surface of cones which are oppositely directed in successive horizontal sheets. The spiral component of the moment rotates, cycloidally, with progression along 0*x*, leading to a repeat in the course of the six unit cells drawn in the Figure.

$0x$, of the spiral lies within the plane (001) which contains the spiralling component of the moment; for an ordinary spiral the propagation direction is perpendicular to the plane of the spiralling spins. Table 31 indicates the agreement between the observed and

TABLE 31

Observed and calculated intensities for fundamental and satellite reflections of magnetic origin in $MnSO_4$ at 4·2 K

hkl	$j\|F\|^2_{calc}$	$j\|F\|^2_{obs}$
$(001)^0$	0	40
$(001)^{\pm}$	38	
$(111)^-$	41	40
$(111)^0$	11	11
$(021)^0$	5	
$(021)^{\pm}$	34	61
$(111)^+$	34	
$(201)^-$	11	5
$(201)^0$	4	5

calculated magnetic intensities for this structure, the 'observed' data being taken as the difference between patterns measured at 77 K and 4·2 K. It will be noted that the planes (001) and (021) produce only single satellites, which are labelled $(001)^{\pm}$ and $(021)^{\pm}$ in the table, whereas the planes (111) and (201) each produces a well-separated pair of satellites. This is because the propagation vector of the spiral is in the [100] direction so that, as can be inferred from Fig. 332, there is no difference in spacing between, for example, satellites $(021)^-$ and $(021)^+$.

15

SOME TOPICS IN NEUTRON POLARIZATION

15.1. Introduction

WE shall describe in this chapter a number of effects which depend on magnetic diffraction and show in particular how in recent years, with the availability of high neutron fluxes, these have led to a technique called 'neutron polarization analysis'. This technique involves a study of the manner in which the direction of the neutron spin may be changed as a result of the scattering process and seems likely to make an immense contribution to the study of materials, particularly because of the way in which it can distinguish between the many different types of scattering, such as nuclear and magnetic Bragg peaks, paramagnetic and other forms of incoherent scattering, which we have already encountered.

In Chapter 6 we showed how a beam of neutrons may be partially, or practically completely, polarized by reflection from a suitable magnetized single crystal, and we have described how polarized-neutron techniques are used in the study of ferromagnetic and antiferromagnetic materials. The polarization was made possible by the coherence between the nuclear and magnetically scattered waves under these conditions, resulting in different differential scattering cross-sections for neutrons having their spin directions parallel and antiparallel to the direction of magnetization in the sample. This same distinction underlies a much earlier demonstration (Powers 1938; Bloch, Hamermesh, and Staub 1943; Bloch, Condit, and Staub 1946) that partial polarization could be produced by passing a beam of neutrons through a highly magnetized block of polycrystalline iron. Referring back to Chapter 6 we recall eqn (6.15) given by Halpern and Johnson (1939) for the differential scattering cross-section of an atom, namely,

$$d\sigma = b^2 + 2bp\mathbf{q} \cdot \boldsymbol{\lambda} + p^2q^2, \tag{15.1}$$

where $\mathbf{q}$ is the magnetic interaction vector, defined by eqn (6.9), and $\boldsymbol{\lambda}$ is a unit vector in the direction of polarization of the neutron.

On account of the cross-product term, representing the interference between the nuclear and magnetically scattered waves, the value of $d\sigma$ will be different for the two spin states parallel and antiparallel to the direction of magnetization into which an ordinary unpolarized neutron beam may be resolved. If therefore we carry out a transmission experiment, of the kind described in Chapter 5, then the effective transmission cross-section σ_{eff} will be different for the two spin states. Consequently one spin state will be favoured relative to the other, giving a partial polarization of the beam. It was first shown by Bloch (1936) that the two cross-sections under magnetization $(\sigma_{\text{eff}})_{\text{magn}}$ could be expressed in terms of the normal transmission cross-section σ_{eff} in the demagnetized condition, and a polarization cross-section σ_{p}, according to the relation

$$(\sigma_{\text{eff}})_{\text{magn}} = \sigma_{\text{eff}} \pm \sigma_{\text{p}}. \tag{15.2}$$

The three cross-sections in this equation will each depend on the neutron wavelength in a complicated way because of crystal diffraction effects. We can calculate this variation from eqn (15.1) and an extension of eqn (2.30), in which we expressed the total integrated nuclear scattering. We can rewrite the latter equation in a form suitable for the present discussion by replacing F^2 by $d\sigma$ as given by eqn (15.1). The additional cross-section for the magnetized sample will therefore be

$$\sigma_{\text{p}} = N_{\text{c}}\lambda^2 \sum_{\substack{hkl \\ d \geqslant \lambda/2}} (2bp\mathbf{q} \cdot \boldsymbol{\lambda}) d_{hkl}. \tag{15.3}$$

Compared with eqn (2.30) a factor of $\frac{1}{2}$ has been inserted in order to give the cross-section per *atom* instead of per unit cell, it being remembered that the body-centred unit cell of iron contains two atoms. Replacing N_{c}, the number of unit cells cm^{-3}, by $1/a_0^3$, where a_0 is the linear dimension of the unit cell, we have

$$\sigma_{\text{p}} = \frac{\lambda^2}{a_0^3} \sum_{\substack{hkl \\ d \geqslant \lambda/2}} (2bp\mathbf{q} \cdot \boldsymbol{\lambda}) d_{hkl}. \tag{15.4}$$

From the definition of $\mathbf{q}$ provided by eqn (6.9), it follows that $\mathbf{q} \cdot \boldsymbol{\lambda}$ is equal to $\pm q^2$ if $\boldsymbol{\lambda}$ is parallel or antiparallel to the magnetization direction $\mathbf{K}$. The average value of q^2 around the powder diffraction ring at a Bragg angle θ will, by eqn (6.10), be equal to the average value of $\sin^2 \alpha$, which will be $\frac{1}{2}(1 + \sin^2 \theta)$ since α varies around the ring from θ up to 90°. Our expression for σ_p therefore becomes

$$\sigma_p = \frac{\lambda^2}{a_0^3} \sum_{\substack{hkl \\ d \geqslant \lambda/2}} bp(1+\sin^2\theta)d_{hkl}. \tag{15.5}$$

The magnetic scattering amplitude p can be evaluated from eqn (6.12) if the variation with θ of the amplitude form factor f is known, thus leading to the value of σ_p for any particular wavelength. In this way Steinberger and Wick (1949) arrived at the curve shown in Fig. 334 for the variation of σ_p with neutron wavelength. As would be expected the curve shows discontinuities corresponding to the onset of diffraction from the various (hkl) planes.

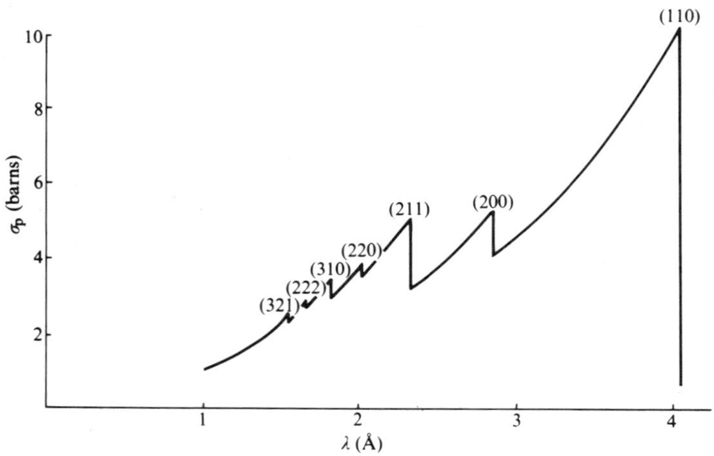

FIG. 334. The variation with wavelength of the polarization cross-section of iron, as calculated theoretically by Steinberger and Wick. The discontinuities in the curve correspond to the onset of reflection from the crystallographic planes indicated. (Steinberger and Wick 1949.)

15.2. Single transmission effect

The value of σ_p can be readily determined experimentally by measuring the transmission cross-section of a block of iron before and after magnetization. If an incident beam of intensity $\mathscr{I}_0$ is considered to be resolved into two equal components, each of intensity $\frac{1}{2}\mathscr{I}_0$ and of positive and negative spin, it follows from eqn (15.2) that the transmitted intensity under magnetization will be

$\mathscr{I}_{\text{magn}}$ such that

$$\mathscr{I}_{\text{magn}} = \tfrac{1}{2}\mathscr{I}_0 \exp\{-Nt(\sigma_{\text{eff}}+\sigma_{\text{p}})\} + \tfrac{1}{2}\mathscr{I}_0 \exp\{-Nt(\sigma_{\text{eff}}-\sigma_{\text{p}})\}$$
$$= \mathscr{I}_0 \exp(-Nt\sigma_{\text{eff}}) \cosh(Nt\sigma_{\text{p}}), \qquad (15.6)$$

where t is the thickness of the iron block and N is the number of atoms cm^{-3}.

In the absence of magnetization the transmitted intensity would be

$$\mathscr{I}_{\text{demagn}} = \mathscr{I}_0 \exp(-Nt\sigma_{\text{eff}}).$$

Thus the effect of the magnetic field is to produce an *increase* $\Delta\mathscr{I}$ in the transmitted intensity such that

$$\Delta\mathscr{I}/\mathscr{I}_{\text{demagn}} = \cosh(Nt\sigma_{\text{p}}) - 1 \doteqdot \tfrac{1}{2}N^2t^2\sigma_{\text{p}}^2. \qquad (15.7)$$

This fractional increase of transmitted intensity is normally referred to as the 'single transmission effect' τ_{s}. In this way values of the polarization cross-section σ_{p} for monochromatic neutrons of various wavelengths were determined by Hughes, Wallace, and Holtzmann (1948) and by Fleeman, Nicodemus, and Staub (1949). Both the form of dependence upon λ and the absolute value are in good agreement with the calculation of Steinberger and Wick, whose curve was illustrated in Fig. 334. For a non-monochromatized beam from the thermal column of a reactor a value of about 3 barns is found for σ_{p}, but the precise value varies according to the actual spectrum of neutron wavelengths.

It has been assumed in our discussion so far that the magnetic moments of all the domains in the polycrystalline block of iron are parallel, that is that the iron is completely saturated by the magnetic field. In practice this will be only approximately true and depolarization will be produced by the departure from perfect alignment of the domains. As a result of this the increase in transmission produced by magnetization is reduced and eqn (15.7) is modified to give

$$\Delta\mathscr{I}/\mathscr{I}_{\text{demagn}} = \tfrac{1}{2}N^2t^2\sigma_{\text{p}}^2 f(\eta_{\text{d}}/\varepsilon t), \qquad (15.8)$$

where the function $f(x)$ is defined by

$$f(x) = 2x^2\{\text{e}^{-1/x} + (1/x) - 1\}. \qquad (15.9)$$

The quantity η_{d} in the depolarization term is related to the size δ of the domains in the iron and also to the distance travelled by the neutron during one full Larmor precession about the internal field. The value of η_d in the general case has been deduced by Halpern and Holstein (1941), but for neutron wavelengths in the thermal region and large crystallites it is approximately equal to $\frac{1}{2}\delta$. The quantity ε is the fractional reduction of magnetization below saturation, namely, $(M_\infty - M)/M_\infty$, where M is the actual magnetization and M_∞ is its saturation value.

The importance of the depolarization term can be appreciated from a typical practical case. If the magnetization is 1 per cent below the saturation value for a sample of thickness $t = 1$ cm and domains of dimension $\delta = 2 \times 10^{-3}$ cm, then the 'single transmission effect' τ_s will fall to 30 per cent of its maximum value. The presence of depolarization is most easily detected experimentally by observing the variation of the single transmission effect with the specimen thickness. For saturation magnetization τ_s will be proportional to t^2, as expressed by eqn (15.7), but under the influence of depolarization it will increase more slowly. This is illustrated in Fig. 335, which shows the experimental values of τ_s obtained by

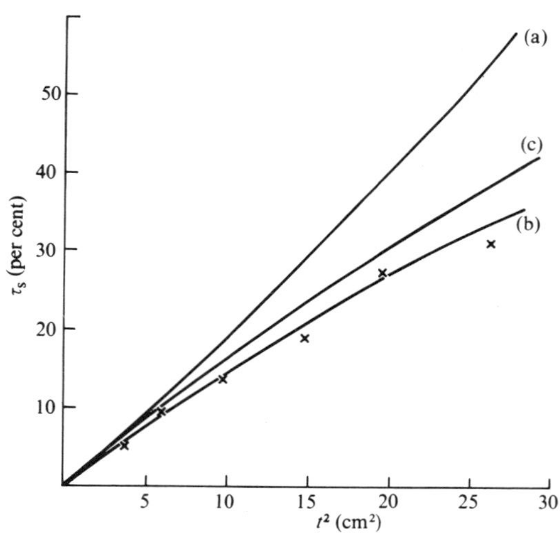

FIG. 335. The variation of the single transmission effect τ_s for iron with specimen thickness showing the effect of depolarization with incomplete magnetic saturation. Experimental points are shown in relation to the following calculated curves, assuming $\sigma_p = 2{\cdot}35 \times 10^{-24}$ cm^2. (a) $\tau_s = \cosh(Nt\sigma_p) - 1$; (b) corrected for depolarization, with $\eta_d/\varepsilon = 10$ cm, and for beam-hardening; (c) corrected for beam-hardening only. (After Fleeman, Nicodemus, and Staub 1949.)

Fleeman, Nicodemus, and Staub (1949) in relation to both the curve appropriate to saturation and also a curve modified according to the value of $f(\eta_d/\varepsilon t)$ obtained from the known values of η_d, and ε. These measurements were made with a non-monochromatized beam of thermal neutrons, and under these circumstances it is necessary to allow for the 'hardening' of the beam, by preferential absorption of the longer wavelengths, as it passes through the block of iron. A correction for this effect has been included in deducing curve (b) of Fig. 335. The effect of hardening alone is illustrated by comparison with curve (c). In these experiments a very high magnetizing field of 11 200 G was used, so that the depolarizing effect was relatively small. The manner in which it increases rapidly as the field is reduced can be seen in Fig. 336,

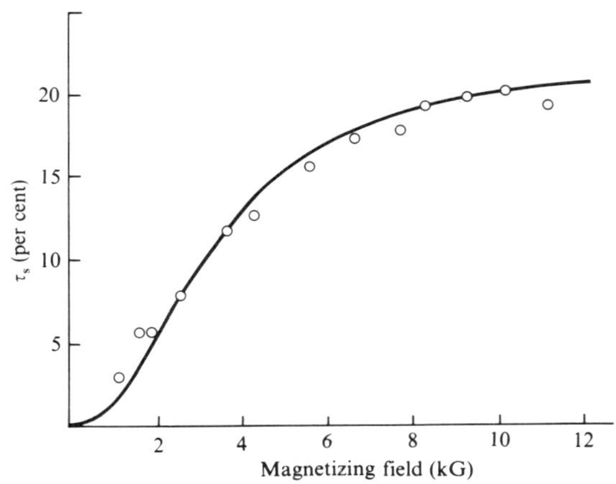

FIG. 336. The variation of the single transmission effect with magnetizing field, showing the experimental results of Fleeman. Nicodemus, and Staub in relation to the theoretical curve for $\sigma_p = 2{\cdot}35 \times 10^{-24}$ cm^2 and allowing for hardening of the slow neutron beam in passing through the iron sample, of thickness 3·81 cm. (After Fleeman, Nicodemus, and Staub 1949.)

which shows both the experimental points and a theoretical curve for a block of iron of thickness 3·81 cm.

With a field of about 12 000 G a degree of polarization of about 40 per cent is obtained with a thickness of 2 cm of iron.

15.3. Double transmission effect

If the partially polarized neutron beam produced by transmission through a magnetized block of iron is caused to pass through a second block, then the resultant transmitted beam will vary in intensity depending on the relative orientations of the magnetic fields in the two blocks. In effect the second block becomes an 'analyser' of the polarization produced by the first, or 'polarizer' block. The behaviour can be seen most simply by comparing the results for parallel and antiparallel fields. For parallel fields the intensity on emergence from the second block, also of thickness t, will be

$$\begin{aligned}\mathscr{I}_{\mathrm{par}} &= \tfrac{1}{2}\mathscr{I}_0 \exp\{-2Nt(\sigma_{\mathrm{eff}}-\sigma_{\mathrm{p}})\}+\tfrac{1}{2}\mathscr{I}_0 \exp\{-2Nt(\sigma_{\mathrm{eff}}+\sigma_{\mathrm{p}})\}\\ &= \mathscr{I}_0 \exp(-2Nt\sigma_{\mathrm{eff}})\{1+2(Nt\sigma_{\mathrm{p}})^2\},\end{aligned} \tag{15.10}$$

whereas

$$\begin{aligned}\mathscr{I}_{\mathrm{antip}} &= \mathscr{I}_0 \exp\{-Nt(\sigma_{\mathrm{eff}}-\sigma_{\mathrm{p}})\}\,.\,\exp\{-Nt(\sigma_{\mathrm{eff}}+\sigma_{\mathrm{p}})\}\\ &= \mathscr{I}_0 \exp(-2Nt\sigma_{\mathrm{eff}}),\end{aligned} \tag{15.11}$$

which is in fact the same intensity as would be expected if the blocks were unmagnetized.

We can define a 'double transmission effect' τ_{D} by the ratio

$$\tau_{\mathrm{D}} = (\mathscr{I}_{\mathrm{par}}-\mathscr{I}_{\mathrm{antip}})/\mathscr{I}_{\mathrm{av}} = 2(Nt\sigma_{\mathrm{p}})^2. \tag{15.12}$$

In the early measurements of the double transmission effect such an effect was indeed found, but it was small and the antiparallel intensities were higher than expected. This was to be accounted for by two reasons. First, depolarization within the blocks, due to incomplete saturation, is much more important than for the single transmission case. With two blocks of thickness 1 cm and $\delta = 2\times\times 10^{-3}$ cm the double transmission effect falls to $2\frac{1}{2}$ per cent of its maximum value when the magnetizing field is 1 per cent below saturation. Secondly, depolarization occurring in the intervening region of space between the polarizer and analyser is very important when the two fields are antiparallel. Burgy *et al.* (1950) have shown that great care must be taken to avoid spin transitions at discontinuities in the magnetic field. Unless this is achieved the neutron spin will change its alignment in space when the field changes direction, as we have already noted in our discussion of the polarized-neutron spectrometer in Chapter 6. By suitable disposition of subsidiary magnets to ensure that there was no intervening region over which the field varied irregularly in direction these workers were able to show that the expected magnitude of the double transmission effect was obtained. Fig. 337 shows some experimental results for the variation of τ_D with block thickness, for polarizer and analyser of equal thickness, together with a calculated theoretical curve taking into account hardening and depolarization within the iron.

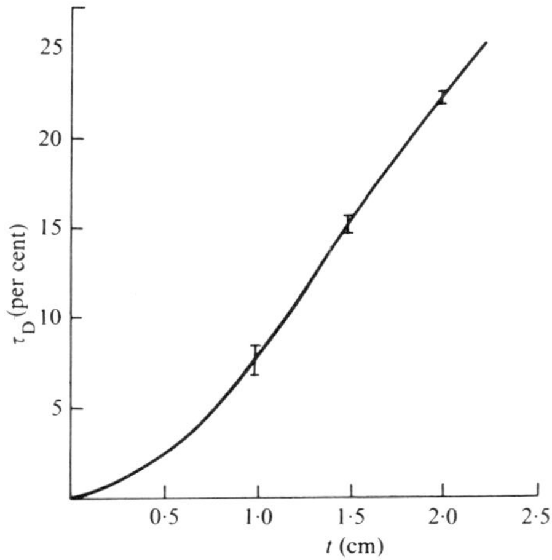

FIG. 337. The variation of the 'double transmission effect' τ_D with block thickness, for polarizer and analyser of equal thickness. The experimental points are shown in relation to a theoretical curve which allows for depolarization and beam hardening, using a non-monochromatic thermal-neutron beam. (Burgy *et al.* 1950.)

15.4. Measurement of domain size

The techniques developed by Burgy *et al.* (1950) for the true measurement of the double transmission effect have been used by them to determine domain sizes in iron. The method depends on a determination of the depolarizing effect, as evidenced by the change in intensity of the transmitted beam, when an iron foil, about 50 μm thick, is inserted in a magnetic shield situated between the polarizing and analysing blocks. The degree of depolarization will depend on the extent to which the neutron spin is reoriented by the variously directed fields in the individual domains of the unmagnetized iron. For a given neutron velocity and domain magnetization this will depend on the size of the domains. The experimental arrangement is indicated in Fig. 338, and the experiment consists in measuring the ratio $E_p(t)$ of the transmitted intensities with and without parallel fields on the polarizer and analyser when the depolarizing foil is in position. The measurement is then repeated with a thicker piece of iron, about 0.25 mm thick, which is sufficient to cause complete depolarization, giving a ratio $E_p(\infty)$. For a given type of iron the value of $(E_p(t)-E_p(\infty))$ is plotted as a function of the foil thickness, giving results of the type shown in Fig. 339 for two

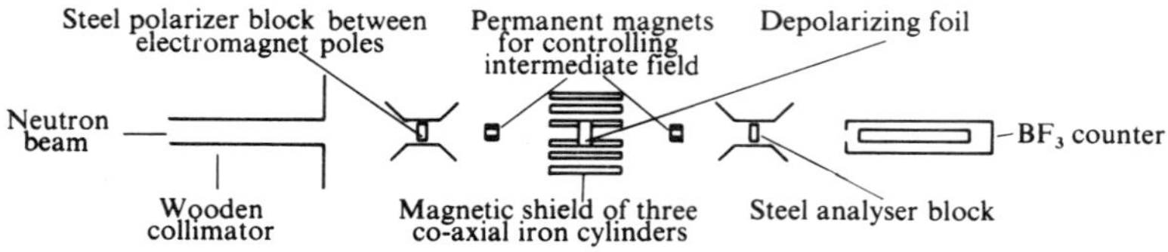

FIG. 338. Experimental arrangement for measurement of the 'double transmission effect' for iron and for the determination of the domain size from the depolarizing effect of an iron foil. (Burgy *et al.* 1950.)

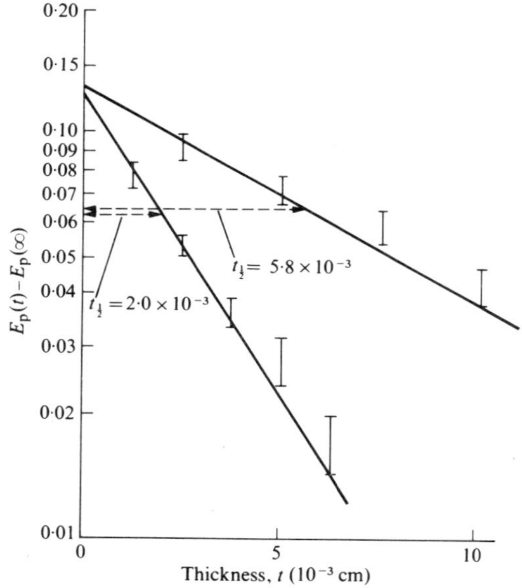

FIG. 339. Variation of neutron depolarization with foil thickness for two varieties of Armco iron. The depolarization is approximately an exponential function of t, giving a linear relation when the ordinate scale is logarithmic as above. From the indicated values of the half-thickness $t_{\frac{1}{2}}$ the domain sizes in the two cases are deduced to be $1{\cdot}5\times10^{-4}$ cm and $0{\cdot}52\times10^{-4}$ cm respectively. (Burgy *et al.* 1950.)

samples of Armco iron which have had different metallurgical treatments. The average domain size $\bar{\delta}$ can be inferred from the thickness $t_{\frac{1}{2}}$ which reduces the ordinate to a half of its maximum value, according to formulae obtained by Halpern and Holstein (1941) and discussed by Burgy *et al.* (1950). For the two samples of iron used in Fig. 339 values of $0{\cdot}52 \times 10^{-4}$ cm and $1{\cdot}5 \times 10^{-4}$ cm respectively were obtained for the domain sizes. From a series of determinations made with different samples it was found that the value of the domain size $\bar{\delta}$ obtained in this manner was about 75 per cent of the grain size as given by standard metallographic techniques.

15.5. Magnetic refraction at domain boundaries

The study of the single transmission effect for magnetized samples of iron by Hughes, Wallace, and Holtzmann (1948) revealed an effect due to the diffraction of neutrons at domain boundaries in iron. These workers found that when very narrow beams, of the order of minutes of arc, were used then there was a very great increase in the single transmission effect. For example, the normal value of about 3 per cent for the coefficient τ_s in a 1 cm thickness of iron could be increased to 13 per cent when very narrow beams were used. It was also found that this spurious single transmission effect could be obtained with quite small magnetizing fields and, in fact, did not occur for fields higher than a few hundred gauss.

This effect has been investigated fully by Hughes *et al.* (1949), who explained it as being due to magnetic diffraction at domain boundaries occurring at low fields when the directions of domain magnetization are being aligned in the most favoured crystallographic direction. By contrast the polarization effects which we have discussed previously occur at the much higher fields under which the domain magnetization is being rotated away from the crystallographic axis and into the direction of the applied field. To understand the effect we first note that by an extension of eqn (5.1) the index of refraction for neutrons in iron is given by

$$n^2 = 1 - \frac{N\lambda^2}{\pi}(b \pm p), \tag{15.13}$$

the positive or negative sign depending on the orientation of the neutron spin relative to the direction of domain magnetization.

In the present case, where the scattering is measured close to $\theta = 0°$, the form factor for magnetic scattering is unity and this equation reduces to the simple form

$$n^2 = 1 - \left(\frac{N\lambda^2}{\pi}b \pm \mu_N \frac{B}{E}\right), \tag{15.14}$$

where μ_N is the magnetic moment of the neutron, E is its energy, and B is the magnetic induction within a domain. The implications of this equation will be considered in more detail later in this chapter. As a neutron proceeds through the iron from domain to domain it will undergo refraction at the intervening boundaries, suffering deviations which will depend on its spin orientation and on the orientation of the boundary and the directions of magnetization on the two sides of it. The maximum deviation which it can suffer will be twice the critical angle θ_c, amounting to about 20′ for thermal neutrons, but the average deviation will be much less than this. If the domain boundaries and their magnetizations are assumed for simplicity to be oriented at random, then the average deviation ζ is calculated, by Hughes *et al.* (1949), to be only 0·029′. In traversing a block of iron successive deviations will occur, and it can be shown that an initially parallel beam would emerge with an angular distribution of Gaussian shape and a width Ω given by

$$\Omega = \zeta(t/\delta)^{\frac{1}{2}}, \tag{15.15}$$

where t is the block thickness and δ is the domain size. Each angular component of the incident beam will be broadened according to this equation, and in the case of experimental arrangements in which the beam is defined by very narrow slits there will be marked scattering outside the beam when it passes through the unmagnetized iron. When, however, the block is magnetized sufficiently to destroy the domain boundaries the effect will disappear. Thus the previously refracted neutrons will be restored to the emergent beam, giving increased transmission and an anomalously large value of τ_s. Fig. 340 shows the contours of the beam obtained by the above authors after passing through a block of iron 0·57 cm in thickness in the magnetized and unmagnetized conditions respectively. Neglecting the small-angle refraction effect it would have been inferred that τ_s, defined as $\Delta\mathscr{I}/\mathscr{I}_{\text{demagn}}$, was 100 per cent, whereas the true single transmission effect due to polarization amounts to only

1 per cent, as indeed might be deduced from Fig. 335. By obtaining beam contours for different thicknesses of iron Hughes *et al.* (1949) demonstrated the proportionality between Ω and $t^{\frac{1}{2}}$, which eqn (15.15) expresses, and from the observed values of Ω and a knowledge of δ which was equal to $3{\cdot}4 \times 10^{-3}$ cm for the particular iron used, they found that ζ was equal to 0·027′. This is in very adequate agreement with the calculated value of 0·029′, which was quoted above, to justify the given explanation of the phenomenon.

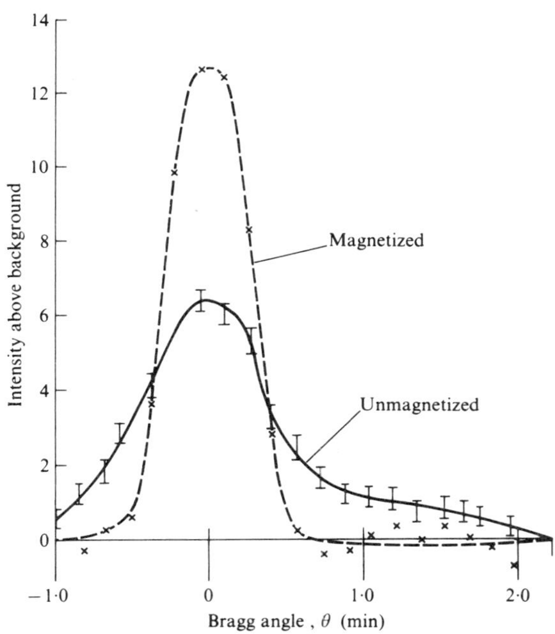

FIG. 340. The effect of small-angle scattering on the shape of a neutron beam transmitted through unmagnetized iron. The full-line curve shows the broadened beam, in contrast to the dotted curve which is obtained when the iron is magnetized and which is identical in shape with that calculated from the slit geometry. (Hughes, Burgy, Heller, and Wallace 1949.)

15.6. Reflection by magnetized mirrors

The double refraction of iron for neutrons, which we have just discussed in order to account for the small-angle scattering measurements, has been demonstrated more directly from observations of total reflection at the surface of magnetized mirrors. Just as a monochromatic neutron beam will have different refractive indices for the two spin orientations, so also will there be two different critical values of the glancing angle θ_c below which total reflection will take place.

We deduce from eqn (15.14) that the two values of θ_c will be given by

$$\theta_c = \left(\frac{N\lambda^2}{\pi}b \pm \mu_N \frac{B}{E}\right)^{\frac{1}{2}}. \tag{15.16}$$

The actual form of the magnetic term in this equation and in eqn (15.14) was initially the subject of much theoretical discussion (Bloch 1937; Halpern, Hamermesh, and Johnson 1941; Ekstein 1949, 1950; Halpern 1949). The particular points under question were whether **B**, the magnetic flux density, or **H**, the magnetic field, in the iron was the significant quantity, and also whether the magnetic term depended on the angle between **B** and the direction of the neutron motion. The experimental measurements with magnetized mirrors have indeed decided in favour of the magnetic term as given in eqn (15.16). The effective total coherent scattering amplitude in these equations can be expressed as the sum of the nuclear and magnetic components

$$b_{\text{total}} = b_{\text{nucl}} \pm \mu_N \frac{B}{E}\frac{\pi}{\lambda^2 N}. \tag{15.17}$$

It can be shown that the term $\mu_N B\pi/E\lambda^2 N$ is identical with the magnetic scattering amplitude p, defined by eqn (6.12) as $e^2\gamma S/mc^2$ for the case of $\theta = 0°$ for which f is equal to unity, by making the substitutions

$$\left.\begin{aligned} B &= 4\pi M = 4\pi N(S/\tfrac{1}{2})(eh/4\pi mc), \\ \mu_N &= \gamma eh/4\pi m_n c, \\ E &= \tfrac{1}{2}m_n v^2 = h^2/2m_n\lambda^2, \end{aligned}\right\} \tag{15.18}$$

where M is the intensity of magnetization, m, m_n are the masses of the electron and neutron respectively, and γ is the magnetic moment of the neutron expressed in nuclear Bohr magnetons.

In the experimental investigation of reflection at magnetized mirrors by Hughes and Burgy (1951) a beam of thermal neutrons filtered through a block of BeO was used in order to obtain a higher intensity than would be available from a crystal monochromator. The BeO does not transmit any neutrons of wavelength less than 4·4 Å, these being scattered from the beam, and the intensity falls off as $1/\lambda^4$ above this wavelength, as determined by the reactor spectrum. This filtered beam was collimated by fine slits and allowed to fall at grazing incidence θ on an iron mirror magnetized parallel to the neutron propagation direction. The variation of the reflected intensity with θ is shown in Fig. 341. As θ is increased from zero

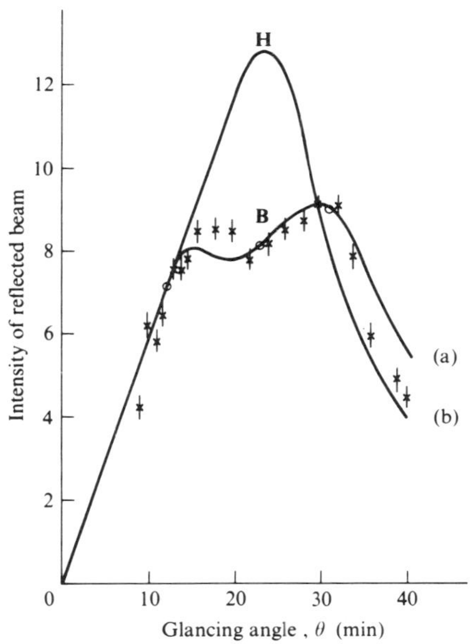

FIG. 341. The variation with glancing angle of the intensity of reflection from a magnetized iron mirror, for a beam of filtered neutrons. The experimental points are shown in relation to two theoretical curves which assume a dependence of the refractive index on **B** and **H** respectively. The magnetic field is in the plane of the mirror, and, except for the three circled points, along the neutron propagation direction. For these latter three points the field was perpendicular to the neutron direction. (Hughes and Burgy 1951.)

there is an initial linear rise in reflected intensity as more of the incident beam is intercepted by the mirror. When θ has reached the value of the critical angle θ_c for neutrons of wavelength 4·4 Å and the spin state corresponding to the *negative* sign in eqn (15.16) neutrons of this spin state will begin to disappear from the reflected beam, being, of course, transmitted. With further increase of θ a second peak occurs in the reflected intensity as the value of θ_c for the other spin state is exceeded. Curve (a) in the Figure is a theoretical curve based on eqn (15.14) and (15.16), making due correction for the finite reflectivity above the critical angle and for the angular resolution of the experimental arrangement. In deriving this curve it is assumed that the value of **B** in the iron domains is 22 500 G, which is the value appropriate to a magnetic scattering amplitude in the forward direction of $0{\cdot}60 \times 10^{-12}$, as given in Table 16 (p. 211) and which can be derived from the first expression given in eqns (15.18).

15.7. Polarization of the mirror-reflected beam

In the case of iron the nuclear and magnetic scattering amplitudes in eqn (15.17) are equal to $0{\cdot}95 \times 10^{-12}$ cm and $0{\cdot}60 \times 10^{-12}$ cm respectively. Consequently the value of b_{total} is positive for each of the two spin states, thus making each refractive index less than unity and giving total reflection at the iron mirror below certain critical values of $(\theta_c)_+$ and $(\theta_c)_-$ respectively. For a *monochromatic* neutron beam incident at an angle between $(\theta_c)_+$ and $(\theta_c)_-$ the reflected beam would be almost completely polarized, for one spin state would be totally reflected and the other almost completely transmitted (the reflectivity above the critical angle is shown in Fig. 88 (p. 158)). With a non-monochromatized beam *partial* polarization is produced by reflection from iron, for a greater range of wavelengths is reflected for one spin state than for the other.

For cobalt, however, the situation is rather different, for here the values of the nuclear and magnetic scattering amplitudes are $0{\cdot}25 \times 10^{-12}$ cm and $0{\cdot}47 \times 10^{-12}$ cm respectively, assuming that the saturation value of **B** is effective. Thus only for one of the two spin states is the refractive index less than unity, and only this state can undergo total reflection. It follows therefore that with a saturated cobalt mirror complete polarization can be obtained even when the incident beam has not been monochromatized. For any chosen value of the glancing angle θ the reflected neutrons, which will all be of the same spin state, will be those of wavelengths greater than the value of λ obtained by substituting this θ value in eqn (15.16). This conclusion has been tested by Hughes and Burgy (1951), who took special steps to ensure the production of a highly magnetized cobalt mirror. Polarization of 100 per cent was obtained within the experimental accuracy of 1 per cent. However, the intensity of the polarized beam which can be obtained in this way may be about 50 times less than for the beam obtained by Bragg reflection from a magnetized crystal (see Chapter 6) but the collimation is correspondingly better, being about $\frac{1}{40}^{\circ}$ in the magnetized mirror case and 1° for the crystal reflection. However, the crystal reflected beam is truly monochromatic, whereas the mirror reflected beam contains all wavelengths beyond the critical value. From reactors with increased flux the intensities of the beams of polarized neutrons are, of course, correspondingly increased in each case.

15.8. Polarization effects in Bragg-reflected beams

In discussing the significance of the magnetic interaction vector **q** in Chapter 6 we showed how the intensity of the (111) reflection for magnetized Fe_3O_4 varies with the field direction. There is a marked increase in the intensity when the field is aligned perpendicular to the scattering vector, making $q^2 = 1$, and a decrease when the field is along the scattering vector, making q^2 zero and giving no magnetic scattering.

When similar experiments were made with magnetized iron, using the (110), (220) reflections, Shull, Wollan, and Koehler (1951) found that there was a *decrease* in intensity even when the magnetic field was applied perpendicular to the scattering vector. This anomaly can be accounted for in terms of the polarization effects which we have been discussing in the present chapter. When an initially unpolarized beam is diffracted by a block of iron arranged in the symmetrical transmission position with the field perpendicular to the scattering vector, then the intensities $\mathscr{P}_+$, $\mathscr{P}_-$ in the (110) reflection for the two spin states will be

$$\begin{aligned}\mathscr{P}_+ &= \text{constant} \times t \exp\{-N_c t \sec\theta(\sigma_{\text{eff}} - \sigma_p)\}(b - p_{110})^2 \\ \mathscr{P}_- &= \text{constant} \times t \exp\{-N_c t \sec\theta(\sigma_{\text{eff}} + \sigma_p)\}(b + p_{110})^2\end{aligned},$$

where t is the thickness of the iron block and σ_p is the polarization cross-section as given by eqns (15.2) and (15.5). This follows from an extension of eqn (4.11).

For an unmagnetized sample the transmitted intensity will be

$$\mathscr{P} = \text{constant} \times 2t \exp(-N_c t \sec\theta\sigma_{\text{eff}})(b^2 + \tfrac{2}{3}p_{110}^2).$$

Hence the fractional increase $\mathscr{R}_s$ in the intensity of the reflection produced by applying the magnetic field will be

$$\mathscr{R}_s = \frac{(\mathscr{P}_+ + \mathscr{P}_-) - \mathscr{P}}{\mathscr{P}}$$

$$= [(b-p_{110})^2 \exp\{-(\sigma_{\text{eff}}-\sigma_p)\times N_c t \sec\theta\} + (b+p_{110})^2 \times \exp\{-(\sigma_{\text{eff}}+\sigma_p)N_c t \sec\theta\} - 2(b^2+\tfrac{2}{3}p_{110}^2)\times \exp\{-\sigma_{\text{eff}}N_c t \sec\theta\}] \Big/ 2(b^2+\tfrac{2}{3}p_{110}^2)\exp(-\sigma_{\text{eff}}N_c t \sec\theta) \tag{15.19}$$

whereas if polarization effects are neglected the value of $\mathscr{R}_s$ is given by the corresponding expression with σ_p equal to zero. At a wavelength of 1·2 Å the values of $\sigma_{\text{eff}}, \sigma_p$ are $11{\cdot}26\times10^{-24}$ cm, $1{\cdot}8\times10^{-24}$ cm^2 respectively. For a thickness of 1 cm it is found from eqn (15.19) that $\mathscr{R}_s$ is calculated to be $-0{\cdot}05$, corresponding to a 5 per cent *decrease* in intensity on magnetization, whereas if polarization is neglected an *increase* of 4·4 percent would be expected. This conclusion is in good agreement with the experimental result of Shull, Wollan, and Koehler.

The expected increase of 4·4 per cent in the absence of polarization effects is, of course, small because the contribution of magnetic scattering to the (110) and other reflections for iron is so much smaller than the nuclear scattering. The advantages of a reflection such as the (111) of magnetite, Fe_3O_4, for demonstrating the effect of magnetization are twofold. This reflection is almost entirely magnetic in origin, and this means not only that much greater relative changes are produced, in the absence of any large constant nuclear intensity, but also that polarization effects are very small since these depend on the cross-product term representing interference between the nuclear and magnetic scattering. In this case, therefore, the expected increase of intensity is observed on magnetization, as was illustrated in Fig. 116.

It will be realized from the previous discussion that, in general, the assessment of magnetic scattering by observing the intensity increase when a magnetic field is applied perpendicular to the scattering vector may be greatly complicated by polarization effects. This is the reason for the adoption of the experimental method of observing the net intensity *decrease* when a field is applied *along* the scattering vector. Under these conditions q becomes zero and

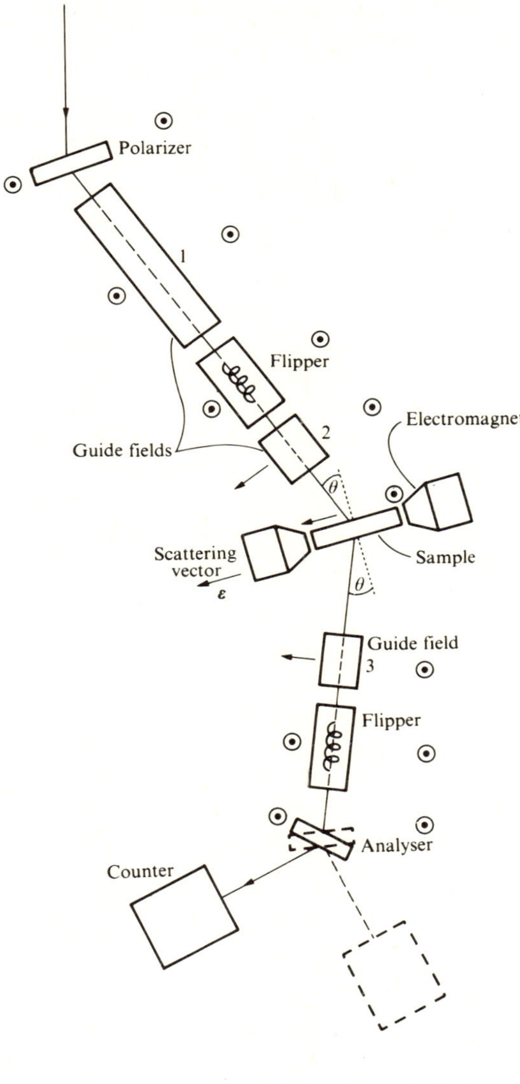

FIG. 342. Experimental arrangement for polarization analysis. Arrows adjacent to the guide fields on the left-hand side of the neutron path show the field directions when the magnetic field at the sample is arranged to be along $\boldsymbol{\varepsilon}$, the scattering vector. The arrows on the right-hand side (with the guide fields 2 and 3 rotated through 90°) are appropriate when the sample field is at right-angles to $\boldsymbol{\varepsilon}$, i.e. normal to the plane of the paper. (After Moon, Riste, and Koehler 1969.)

there is no magnetic scattering, and accordingly there are no complications from polarization effects.

15.9. Polarization analysis

In our foregoing discussion we have indicated how the total intensity of the neutrons scattered from various samples depends on polarization effects. We shall show that it is very rewarding to examine in more detail how these neutrons are distributed between the different polarizations and indeed to conduct a polarization analysis of the scattered beam. Such an analysis becomes a practical possibility for the intense neutron beams which are now available from high-flux reactors. The procedure and its practical possibilities can be seen best by considering the use of a triple-axis spectrometer modified in a manner described by Moon, Riste, and Koehler (1969) and illustrated in Fig. 342. On the first and third axes of this spectrometer are placed cobalt–iron crystals of the type which we have described in Chapter 6 for the production of polarized neutrons. The spectrometer also includes two radio-frequency coils which act as spin-flipping devices, as described on p. 220. The sample under investigation is mounted on the second axis of the instrument and is fixed between the poles of an electromagnet which can be rotated about a horizontal axis so that its field can easily be changed from horizontal to vertical, thus enabling the field to be either along or perpendicular to the scattering vector $\boldsymbol{\varepsilon}$; the guide fields on coils (2) and (3) can be similarly rotated. The polarizing and analysing crystals will act as devices which reflect unimpaired all neutrons of upward (+ve)-pointing spin but have zero reflectivity for downward (−ve)-pointing spins. When the electromagnetic field at the sample is vertical the neutron polarization remains vertical throughout. When neither of the spin-flippers is activated the neutrons incident on the sample will be of upward (+ve) spin and the only ones which reach the detector will also be of +ve spin. If the second flipper is activated, the incident neutrons will be +ve but the only ones which are detected at the counter will be those which have undergone a spin reversal in the sample and have emerged with −ve spin. The converse situation holds if it is simply the first flipper which is activated. Thus with this arrangement and by suitable manipulation of the flippers we can measure individual cross-sections for the spin transitions $\uparrow\uparrow$, $\uparrow\downarrow$, $\downarrow\downarrow$, and $\downarrow\uparrow$. Moreover, we can do this both for the case when the neutron spin direction in the sample is along the scattering vector and also for the case when the spin is perpendicular to the scattering vector. In the former case the field directions along the neutron path are indicated by the arrows at the left-hand side of the path in Fig. 342. In the latter case all the fields are vertical, as indicated at the right-hand side of the Figure. In the former case there will be four adiabatic rotations of spin: first, between the first flipper and guide field (2), then both before and after the sample, by angle θ in each case, and finally between the guide field (3) and the second flipper.

To take into account the polarization of both the incident and scattered beams we need to describe the scattering amplitude in terms of the Pauli spin matrices (see Halpern and Johnson 1939; Marshall and Lovesey 1971) as

$$U = (u'd')\{b + p\mathbf{q} \cdot \hat{\boldsymbol{\sigma}}\}\begin{pmatrix} u \\ d \end{pmatrix}, \tag{15.20}$$

where $\hat{\boldsymbol{\sigma}}$ is the Pauli spin operator and the matrix elements u, d and u', d' refer to the spin states of the incident and scattered beams as 'up' or 'down'. This equation is for a single isotope and a nucleus without spin, but can be extended to nuclei of spin I by writing

$$U = (u'd')\{b_{\mathscr{S}} + (p\mathbf{q} + b_s\, I\boldsymbol{\mathscr{I}}/\{I(I+1)\}^{\frac{1}{2}}) \cdot \hat{\boldsymbol{\sigma}}\}\begin{pmatrix} u \\ d \end{pmatrix}. \tag{15.21}$$

Here $b_{\mathscr{S}}$ is the coherent part of the nuclear scattering amplitude, equivalent to $(\mathscr{S}/4\pi)^{\frac{1}{2}}$ of eqn (2.15) and b_s is the incoherent part, equal to $(s/4\pi)^{\frac{1}{2}}$. $\boldsymbol{\mathscr{I}}$ is a unit vector in the direction of the nuclear spin.

For convenience we define a vector $\boldsymbol{\mathscr{A}}$ by

$$\boldsymbol{\mathscr{A}} = p\mathbf{q} + BI\boldsymbol{\mathscr{I}}, \tag{15.22}$$

where we have written B for the quantity $b_s/\{I(I+1)\}^{\frac{1}{2}}$, so that eqn (15.21) then becomes

$$U = (u'd')(b_{\mathscr{S}} + \boldsymbol{\mathscr{A}} \cdot \hat{\boldsymbol{\sigma}})\begin{pmatrix} u \\ d \end{pmatrix} \tag{15.23}$$

We expand this equation into the matrix components of the Pauli spin operator as

$$\begin{aligned} U &= (u'd')\left\{b_{\mathscr{S}} + \mathscr{A}_x\begin{pmatrix} 0 & 1 \\ 1 & 0 \end{pmatrix} + \mathscr{A}_y\begin{pmatrix} 0 & -\mathrm{i} \\ \mathrm{i} & 0 \end{pmatrix} + \mathscr{A}_z\begin{pmatrix} 1 & 0 \\ 0 & -1 \end{pmatrix}\right.\begin{pmatrix} u \\ d \end{pmatrix} \\ &= b_{\mathscr{S}}(u'u + d'd) + \mathscr{A}_x(u'd + ud') + \mathrm{i}\mathscr{A}_y(-u'd + ud') + \mathscr{A}_z(u'u - d'd), \end{aligned} \tag{15.24}$$

where z is the direction of the neutron polarization. We can then write down the scattering amplitudes for the four types of spin transition which we can examine in the experiment as

$$
\begin{array}{lllllll}
\uparrow\uparrow & u=1 & d=0 & u'=1 & d'=0 & U^{++} & = b_{\mathscr{S}}+\mathscr{A}_z \\
\downarrow\downarrow & u=0 & d=1 & u'=0 & d'=1 & U^{--} & = b_{\mathscr{S}}-\mathscr{A}_z \\
\uparrow\downarrow & u=1 & d=0 & u'=0 & d'=1 & U^{+-} & = \mathscr{A}_x+\mathrm{i}\mathscr{A}_y \\
\downarrow\uparrow & u=0 & d=1 & u'=1 & d'=0 & U^{-+} & = \mathscr{A}_x-\mathrm{i}\mathscr{A}_y
\end{array}
\tag{15.25}
$$

If we recall that $p\mathbf{q}$ is equal to $-e^2\gamma f\mathbf{S}_\mathrm{p}/mc^2$, where $\mathbf{S}_\mathrm{p}$ is the projection of the spin vector on to the reflection plane (see p. 264) then we can recast these equations into a form similar to that of Moon *et al.* (1969), dropping the subscript of $b_{\mathscr{S}}$, as

$$
\begin{aligned}
\uparrow\uparrow \quad U^{++} &= b-\frac{e^2\gamma}{mc^2}f(S_\mathrm{p})_z+BI_z, \\
\downarrow\downarrow \quad U^{--} &= b+\frac{e^2\gamma}{mc^2}f(S_\mathrm{p})_z-BI_z, \\
\uparrow\downarrow \quad U^{+-} &= -\frac{e^2\gamma}{mc^2}f\{(S_\mathrm{p})_x+\mathrm{i}(S_\mathrm{p})_y\}+B(I_x+\mathrm{i}I_y), \\
\downarrow\uparrow \quad U^{-+} &= -\frac{e^2\gamma}{mc^2}f\{(S_\mathrm{p})_x-\mathrm{i}(S_\mathrm{p})_y\}+B(I_x-\mathrm{i}I_y),
\end{aligned}
\tag{15.26}
$$

These expressions have assumed that the nuclear spins are oriented in space, and will have to be averaged over all directions to apply to their usual random orientation.

Before proceeding to the experimental application of the above analysis we will correlate it, for the simple case of no nuclear spin, with eqn (6.15). If for an 'upward' pointing incident spin we collect *both* scattered polarizations then we shall have an effective differential cross-section equal to

$$
\left|b-\frac{e^2\gamma}{mc^2}f(S_\mathrm{p})_z\right|^2+\left|-\frac{e^2\gamma}{mc^2}f\{(S_\mathrm{p})_x+\mathrm{i}(S_\mathrm{p})_y\}\right|^2
$$

$$
=b^2-2b_{\mathscr{S}}\frac{e^2\gamma}{mc^2}f(S_\mathrm{p})_z+\left(\frac{e^2\gamma}{mc^2}f\right)^2\{(S_\mathrm{p})_z^2+(S_\mathrm{p})_x^2+(S_\mathrm{p})_y^2\},
$$

which is $b^2+2bp\mathbf{q}\cdot\boldsymbol{\lambda}+p^2q^2$ of eqn (6.15), since $\mathbf{q}\cdot\boldsymbol{\lambda}=q_z$.

Returning to a discussion of eqn (15.26) in relation to the experimental arrangement in Fig. 342 we emphasise that z is the direction of the neutron polarization and show in Fig. 343 the nomenclature of the axes and the vectors for the two cases which we postulated in Fig. 342 in which the polarization direction of the neutrons is parallel and perpendicular, respectively, to the scattering vector.

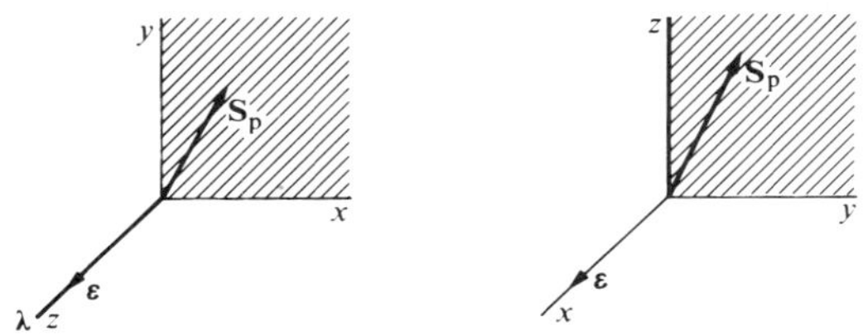

Polarization direction z is parallel to scattering vector $\boldsymbol{\varepsilon}$

$\lambda.\boldsymbol{\varepsilon}=1$

(a)

Polarization direction z is perpendicular to scattering vector $\boldsymbol{\varepsilon}$

$\lambda.\boldsymbol{\varepsilon}=0$

(b)

FIG. 343. Geometry and nomenclature of the neutron polarization and sample spin vectors in relation to the scattering vector $\boldsymbol{\varepsilon}$: the magnetic interaction vector $\mathbf{q}$ is parallel to $\mathbf{S}_\mathrm{p}$.

We can draw several important conclusions from an examination of eqn (15.26):

1. The coherent nuclear scattering of amplitude b, which gives rise to the Bragg peaks in the diffraction pattern, is always non-spin-flip scattering ($++$ or $--$).
2. Isotope-disorder scattering, which arises from randomness in the value of b, will also be non-spin-flip.
3. The nuclear spin-disorder scattering will be of two types. The component of nuclear spin I_z which is parallel to the polarization direction will produce non-spin-flip scattering; the components I_x, I_y which are perpendicular to the polarization direction will give spin-flip scattering.
4. For magnetic scattering, as we have already seen in Chapter 6, only $\mathbf{S}_\mathrm{p}$, the component of magnetic moment which is perpendicular to the scattering vector $\boldsymbol{\varepsilon}$, will be effective in scattering neutrons. If, as in Fig. 343 (a), the neutron polarization is along the scattering vector then the component $(\mathbf{S}_\mathrm{p})_z$ will be zero and all the magnetic scattering will be spin-flip scattering.

The above conclusions have been demonstrated experimentally by Moon *et al.* (1969), and their results are shown in Fig. 344. The curves (a) in Fig. 344 show the situation for nickel, which displays isotope incoherence but has zero nuclear spin. Moreover, if the experimental arrangement is such that $\boldsymbol{\lambda} \cdot \boldsymbol{\varepsilon} = 1$ there will be no magnetic scattering. Accordingly, when one of the flippers is activated no neutrons will be detected, as Fig. 344 (a) shows, whereas without activation a substantial number of incoherent scattered neutrons will be observed at a scattering angle which is not giving Bragg reflection. The curves (b) in Fig. 344 refer to the spin-incoherent

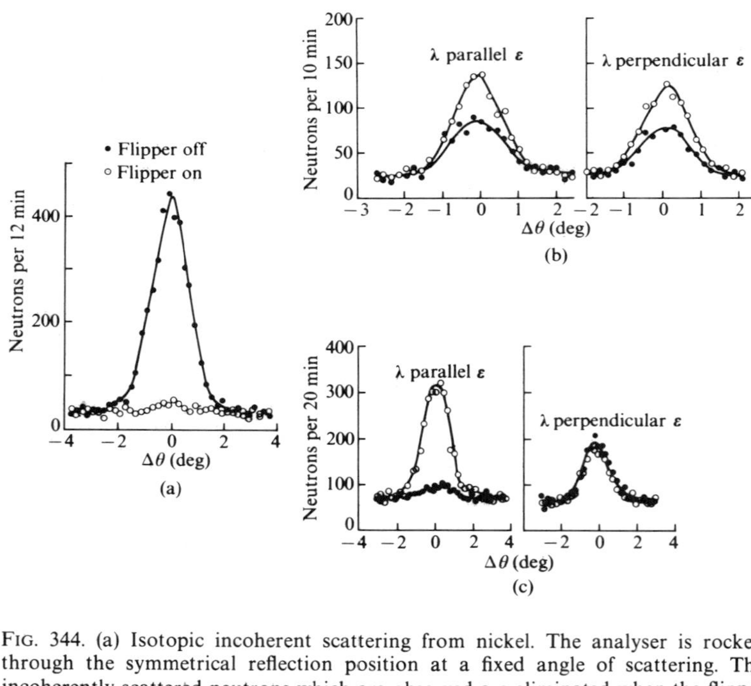

FIG. 344. (a) Isotopic incoherent scattering from nickel. The analyser is rocked through the symmetrical reflection position at a fixed angle of scattering. The incoherently scattered neutrons which are observed are eliminated when the flipper is operated, thus permitting only $+-$ neutrons to be counted. (b) Nuclear spin-incoherent scattering from vanadium; the detected neutrons are reduced by a factor of 2 when the flipper is operated, for either direction of polarization. (c) Paramagnetic scattering from MnF_2; when $\boldsymbol{\lambda}$ is parallel to $\boldsymbol{\varepsilon}$ the magnetic scattering is entirely spin-flip and disappears entirely when the flipper is off; when $\boldsymbol{\lambda}$ is perpendicular to $\boldsymbol{\varepsilon}$ the spin-flip and non-spin-flip portions of the scattering are equal and there is no change when the flipper is operated. (Moon, Riste, and Koehler 1969.)

scattering from vanadium. Vanadium consists of only a single isotope, for which b is very small, but which shows large spin-incoherent scattering. There is no significant amount of magnetic scattering, and there will be a random arrangement of the nuclear spins. From the form of eqn (15.26) it will be seen that the I_z components contribute to non-spin-flip scattering and the equal I_x, I_y components will give spin-flip scattering. Thus in Fig. 344 (b) the contribution with the flipper on will be twice as great as that with the flipper inoperative, and this result is the same for all directions of neutron polarization. Curves (c) of Fig. 344 show an example for paramagnetic scattering, from MnF_2 above its Néel point. If $\boldsymbol{\lambda} \cdot \boldsymbol{\varepsilon} = 1$ then, as seen in Fig. 343 (a), $(\mathbf{S}_p)_z$ is zero and the scattering is entirely spin-flip. On the other hand, when $\boldsymbol{\lambda} \cdot \boldsymbol{\varepsilon} = 0$ it is $(\mathbf{S}_p)_x$ which is zero, and the equality of $(\mathbf{S}_p)_z$ and $(\mathbf{S}_p)_y$ ensures that the non-spin-flip and the spin-flip portions of the scattering are equal, as Fig. 344 (c) demonstrates. It will be noted that the total cross-section, i.e. the sum of the non-spin-flip and the spin-flip contributions, is not dependent on the direction of polarization.

In our discussion of liquids and glasses we shall see that an important problem is to separate the coherent and incoherent parts of the scattering. In the case of hydrogenous substances this is effectively the separation of the spin-incoherent scattering from all other forms of nuclear scattering, since isotopic incoherence will be very small and there is no magnetic scattering to consider. This can readily be done if the non-spin-flip and spin-flip cross-sections are measured separately, for it can be shown from the eqns (15.26) that

$$\begin{pmatrix}\text{non-spin-flip}\\ \text{cross-section}\\ ++\cdot\end{pmatrix} = \frac{1}{3}\begin{pmatrix}\text{spin-}\\ \text{incoherent}\\ \text{cross-section}\end{pmatrix} + \begin{pmatrix}\text{other}\\ \text{nuclear}\\ \text{contributions}\end{pmatrix}$$

and

$$\begin{pmatrix}\text{spin-flip}\\ \text{cross-section}\\ +-\end{pmatrix} = \frac{2}{3}\begin{pmatrix}\text{spin-}\\ \text{incoherent}\\ \text{cross-section}\end{pmatrix}, \qquad (15.27)$$

where the cross-sections on the right-hand side refer to measurements with an unpolarized neutron beam. It follows therefore that, for hydrogenous substances,

$$\begin{aligned}\begin{matrix}\text{coherent}\\ \text{fraction}\end{matrix} &= \frac{\begin{pmatrix}\text{other nuclear}\\ \text{contributions}\end{pmatrix}}{\begin{pmatrix}\text{spin-}\\ \text{incoherent}\\ \text{cross-section}\end{pmatrix}+\begin{pmatrix}\text{other}\\ \text{nuclear}\\ \text{contributions}\end{pmatrix}}\\ &\equiv 1-\frac{\begin{pmatrix}\text{spin-}\\ \text{incoherent}\\ \text{cross-section}\end{pmatrix}}{\begin{pmatrix}\text{spin-}\\ \text{incoherent}\\ \text{cross-section}\end{pmatrix}+\begin{pmatrix}\text{other}\\ \text{nuclear}\\ \text{contributions}\end{pmatrix}}\\ &= 1-\frac{\frac{3}{2}\begin{pmatrix}\text{spin-flip}\\ \text{cross-section}\\ +\,-\end{pmatrix}}{\begin{pmatrix}\text{spin-flip}\\ \text{cross-section}\\ +\,-\end{pmatrix}+\begin{pmatrix}\text{non-spin-flip}\\ \text{cross-section}\\ +\,+\end{pmatrix}}.\end{aligned}$$

In this way we can separate the coherent and incoherent fractions in a normal unpolarized experiment by measuring the $(+\,-)$ and $(+\,+)$ contributions with the polarization-analysis spectrometer.

An equally valuable application can be seen if we consider the situation when spin-incoherent and paramagnetic scattering occur together, or indeed we may consider the general case when isotopic disorder, Bragg scattering, multiple Bragg scattering, and thermal diffuse scattering also contribute, as in a normal measurement of diffuse scattering. We note first that none of this latter group of varieties of scattering can contribute to the spin-flip scattering. In fact the latter is made up solely of the paramagnetic scattering and the spin-incoherent scattering, and the following two relations can be shown to arise from eqns (15.26); if $\boldsymbol{\lambda}\,.\,\boldsymbol{\varepsilon} = 1$, then

$$\begin{pmatrix}\text{spin-flip}\\ \text{scattering}\\ +\,-\end{pmatrix}=\begin{pmatrix}\text{total}\\ \text{paramagnetic}\\ \text{scattering}\end{pmatrix}+\tfrac{2}{3}\begin{pmatrix}\text{total incoherent}\\ \text{nuclear spin}\\ \text{scattering}\end{pmatrix}$$

whereas if $\boldsymbol{\lambda} \cdot \boldsymbol{\varepsilon} = 0$, then

$$\begin{pmatrix}\text{spin-flip}\\ \text{scattering}\\ + -\end{pmatrix} = \tfrac{1}{2}\begin{pmatrix}\text{total}\\ \text{paramagnetic}\\ \text{scattering}\end{pmatrix} + \tfrac{2}{3}\begin{pmatrix}\text{total incoherent}\\ \text{nuclear-spin}\\ \text{scattering}\end{pmatrix}.$$

Thus if the spin-flip scattering is measured for both of the two different geometrical arrangements of the sample field which we discussed for Fig. 342, then the paramagnetic and nuclear-spin scattering can be separated. Of course, if only one of these two types of scattering is present a single measurement of the spin-flip scattering for *one* geometrical arrangement will remove the isotopic, Bragg, multiple Bragg, and thermal diffuse scattering. This is evident from the two diffraction patterns for MnF_2 in Fig. 345 (Moon *et al.*)

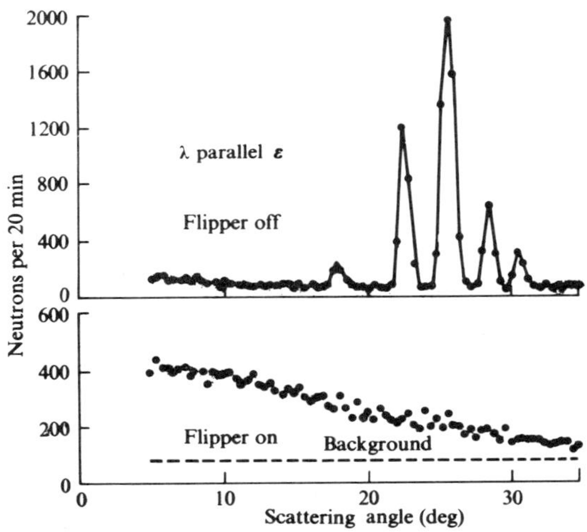

FIG. 345. The powder diffraction pattern of MnF_2 when $\boldsymbol{\lambda}$ is parallel to $\boldsymbol{\varepsilon}$. When the flipper is on only paramagnetic scattering (and the small spin-incoherent scattering of the manganese) can contribute to the diffuse background. (Moon, Riste, and Koehler 1969.)

for which $\boldsymbol{\lambda} \cdot \boldsymbol{\varepsilon} = 1$. These were recorded by setting the diffracting sample in turn to give different values of the scattering angle 2θ and then recording the count (for the standard inclination of the analyzer appropriate to the neutron wavelength) both with and without the flipper in operation. The appearance of the paramagnetic scattering in the former case is very striking; there is no contribution from it in the latter case because $(\mathbf{S}_p)_z$ is zero when $\boldsymbol{\lambda} \cdot \boldsymbol{\varepsilon} = 1$. A correction can be made for a small amount of spin-incoherent scattering from the manganese.

Perhaps an even more convincing example of the segregation of paramagnetic scattering is illustrated in Fig. 346 (Moon *et al.* 1972) which shows polarization analysis data for Gd_2O_3, containing the Gd isotope of mass number 160 in an extremely pure state and thus giving a sample of very small absorption. There is no nuclear spin-incoherent scattering, and when $\boldsymbol{\lambda} \cdot \boldsymbol{\varepsilon} = 1$ there is a very clear separation of the paramagnetic scattering when the spin-flipper is operated. From the variation of scattering with angle the atomic form-factor curve indicated by the open circles in Fig. 347 was obtained. This is in very good agreement with the Blume–Freeman–Watson (1962, 1964) calculation except at very small angles of scattering. On the other hand, there is a significant difference from

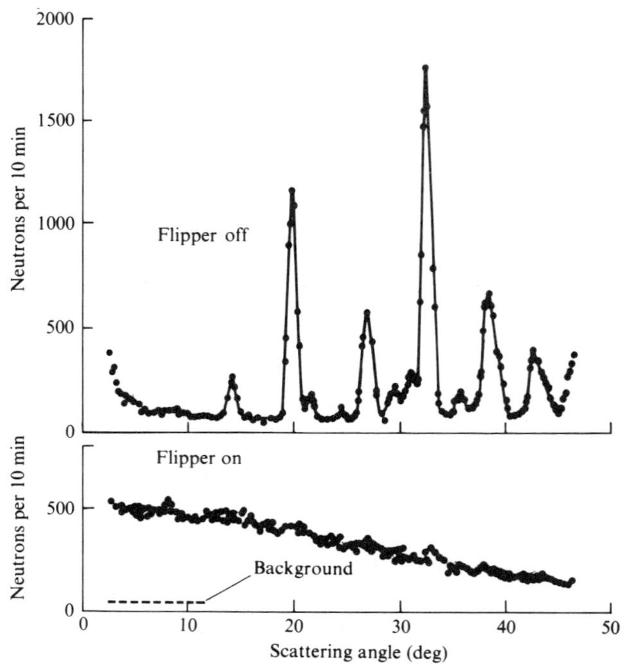

FIG. 346. Polarization-analysis pattern of $^{160}Gd_2O_3$. With the flipper off only the coherent nuclear scattering is detected, whereas when the flipper is operated only the paramagnetic scattering appears. In a conventional experiment the resulting pattern would be the sum of the two shown. (Moon *et al.* 1972.)

the experimental points for metallic gadolinium which were obtained by Moon *et al.* (1972); these are the closed circles in the Figure.

Examination of the coherent scattering which produces the *Bragg peaks* in the diffraction pattern will show that polarization analysis provides a direct means of separating the nuclear and magnetic contributions. This is especially valuable for antiferromagnetic materials since it makes it unnecessary to carry out duplicate experiments above and below the Néel temperature. We see from

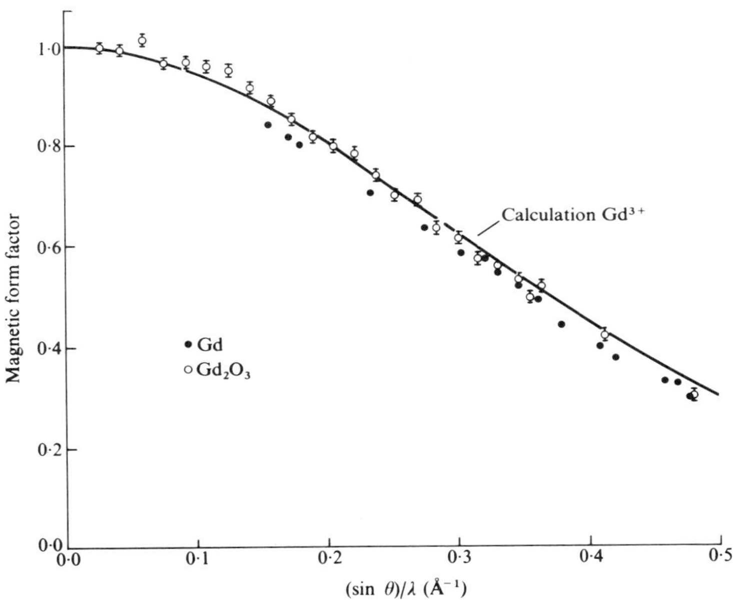

FIG. 347. The open circles show the magnetic form factor of gadolinium as determined from the paramagnetic scattering of $^{160}Gd_2O_3$ which was shown in Fig. 346. This is in good agreement with the full-line curve which represents the calculation of Blume–Freeman–Watson (1962, 1964). There is a significant difference from the closed circles which show the form factor measured for the metal. (Moon *et al.* 1972.)

eqns (15.26) that the coherent nuclear scattering amplitude b contributes only to the non-spin-flip scattering, whereas $(e^2\gamma/mc^2)$ Sf, which is our magnetic scattering amplitude p, affects both spin-flip and non-spin-flip scattering. However, if we arrange that $\boldsymbol{\lambda} \cdot \boldsymbol{\varepsilon} = 1$, i.e. that the neutron polarization is along the scattering vector, as in Fig. 343 (a), then there will be no magnetic contribution to the non-spin-flip peaks. Consequently the non-spin-flip peaks will be purely nuclear and the spin-flip peaks will be purely magnetic. Fig. 348 illustrates very clearly how the nuclear and magnetic contributions have been separated in this way in the powder pattern of haematite (Moon *et al.* 1969).

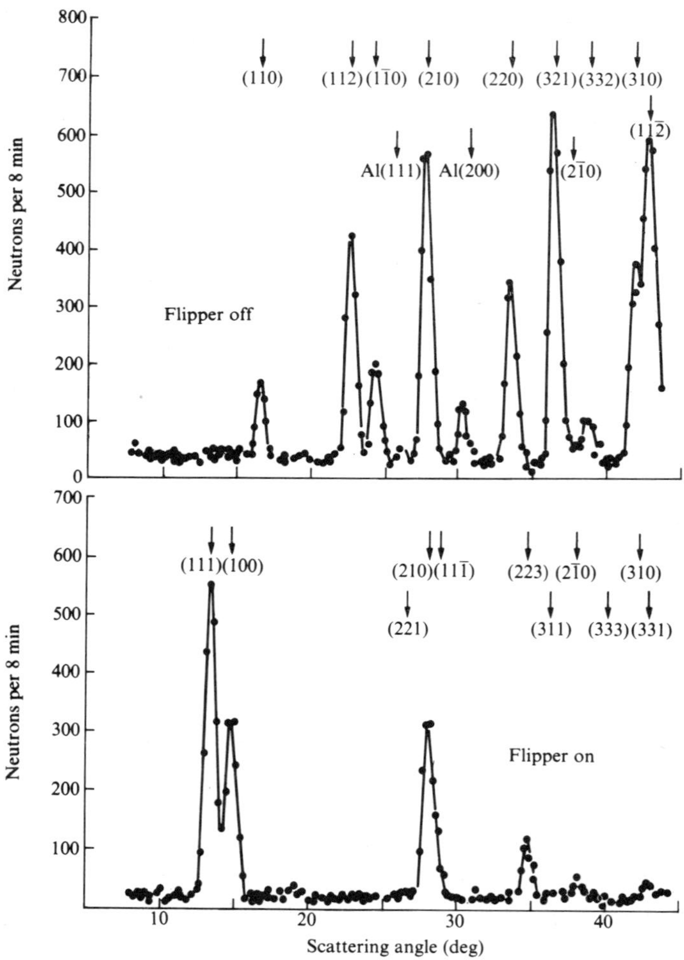

FIG. 348. Separation of the nuclear and magnetic peaks in the powder pattern of α-Fe_2O_3 by polarization analysis. When $\lambda . \varepsilon = 1$ the non-spin-flip peaks are purely nuclear and the spin-flip peaks are purely magnetic. (Moon *et al.* 1969.)

15.10 Polarizing filters

The above examples will indicate the truth of the general conclusion that polarization analysis is of immense value when different types of scattering have to be identified and evaluated separately. Many existing reactors have sufficient flux for the technique to be applied to coherent Bragg diffraction but high-flux reactors are needed to exploit the technique fully in the study of the various kinds of diffuse scattering. This is because the triple-axis spectrometer which we showed in Fig. 342 involves three separate crystal reflections and also because the reflectivity of the polarizing crystals used at present is low. There is a loss in observed intensity by a factor of about 50 in going from a conventional two-axis spectrometer to a three-axis instrument with Co–Fe polarizer and analyser. There is a prospect of remedying some of the loss of intensity by using a 'polarizing filter' (Williams 1973) in place of the analysing crystal of Co–Fe. Not only does such a filter give a direct increase in the neutron counting rate, but it is also not dependent on the precise angle of incidence of the neutrons. Thus in many experiments the permitted angular divergence of the beam in a horizontal plane may be relaxed. In a typical arrangement an angular resolution of 1·25° may be employed, compared with the limited acceptance angle of about 0·08° for a Co–Fe crystal, and this gives a further increase of intensity by about 15 times.

There are a number of possible types of polarization filter, and the magnetized block of iron which we described earlier in this chapter, based on the spin-directional dependence of *magnetic* scattering, is the simplest one. Such a filter has been proposed by Ahmed *et al.* and Campbell *et al.* (1974) and is useful at wavelengths greater than 3 Å but becomes inefficient at shorter wavelengths. A second type utilizes the spin-dependence of the nuclear scattering by nuclei such as protons for which the scattering lengths of the $\uparrow\uparrow$ and $\uparrow\downarrow$ spin combinations are widely different. At low temperatures, around 1 K, and under magnetic fields of about 20 kOe, it is possible (Abragam and Borghini 1964) to polarize protons with a beam of 4 mm microwaves. A suitable material is the hydrated nitrate $La_2Mg_3(NO_3)_{12} . 24H_2O$, and a polarized-proton filter of this material is under construction at the Institut Laue–Langevin at Grenoble. In this type of filter, where the nuclear polarization is achieved dynamically, the performance is somewhat impaired by drift in the microwave frequency and the magnetic field, and more promising performance is offered by a filter using a material in which the nuclear *capture*, rather than the nuclear scattering, is dependent on the direction of the neutron spin. Thus the polarization of the beam is produced by preferential absorption of one of the two neutron spin states. Williams's (1973) proposal employs single crystals of cerous magnesium nitrate in which 8 per cent of the Ce^{3+} ions are replaced by ions of $^{149}Sm^{3+}$, thus yielding $Ce_{1\cdot84}$ $^{149}Sm_{0\cdot16}Mg_3(NO_3)_{12}.24H_2O$. The samarium isotope ^{149}Sm has a strong resonance peak at a neutron wavelength of 0·95 Å, and

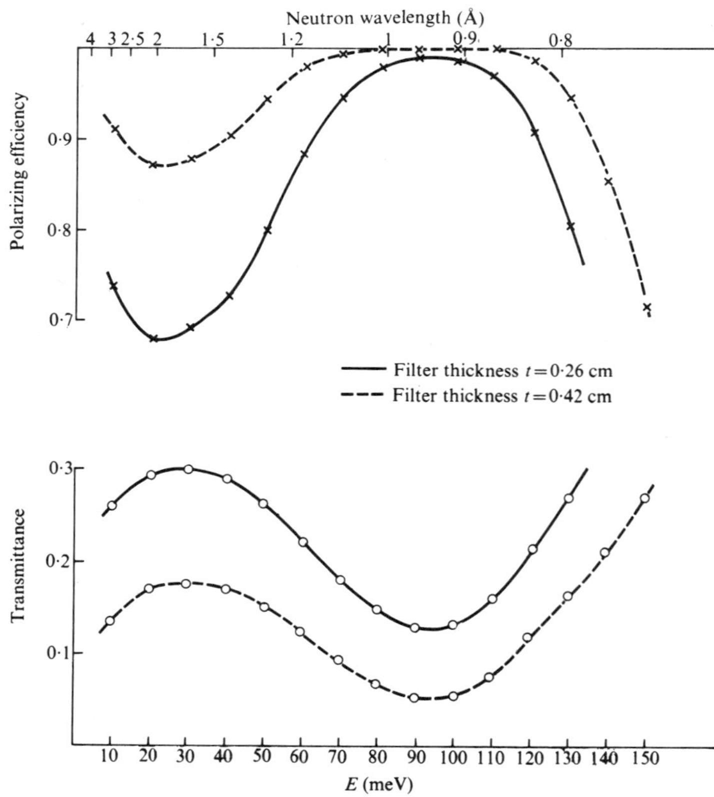

FIG. 349. The calculated variation with neutron energy and wavelength, for the polarizing efficiency and transmittance of a filter of $Ce_{1.84}$ $^{149}Sm_{0.16}Mg_3(NO_3)_{12}.24H_2O$ at a temperature of 0·015 K. The full-line curves are for a filter thickness t of 0·26 cm and the broken lines for t = 0·42 cm. (Williams 1973.)

its polarization within the paramagnetic salt can be achieved at a temperature below 0·1 K in a magnetic field of 10 kOe by the static method described by Rose (1949) and Gorter (1948). The intention is to cool the single crystal to 0·015 K, using a helium 3/helium 4 dilution refrigerator, with the prospect of securing the values of transmittance and polarization efficiency which are shown in Fig. 349. In this Figure calculated performances are shown for crystals of two different thicknesses. A crystal of thickness 0·26 cm gives a polarizing efficiency of 0·98 and a transmittance of 0·15 over a substantial region of wavelength close to 1 Å. The thicker crystal of 0·42 cm is proposed for the wavelength region between 1·5 Å and 4 Å. The broad range of the neutron spectrum over which this type of filter is satisfactory, in contrast with the sharp sensitivity of the Co–Fe crystal, is of additional value when considering its use in experiments on inelastic scattering. It should be possible to use such a filter for polarization analysis in a time-of-flight instrument which analyses a wide range of wavelengths at a single angle of scattering.

16

THE STUDY OF LIQUIDS, GLASSES, AND GASES

16.1. Introduction

THE atomic structure of a liquid is intermediate between the perfect disorder of atoms or molecules in a gas and the highly perfect three-dimensional order in a solid. The immediate environment of an atom in a liquid is not very different from that in a solid and the density of packing is only a few per cent lower, but the order is very local and no long-range order occurs. A good deal of information on the structures of liquids has been forthcoming from studies of the way in which they scatter neutrons and some of this information can be obtained in no other way. By its very nature, and because the atomic arrangement is continually changing, the structure of a liquid is difficult to define and to describe and the diffraction pattern for any radiation is correspondingly difficult to interpret. Fig. 350 shows the results of one of the earliest measurements of a diffraction pattern of a liquid using neutrons and is for liquid lead at 623 K (Sharrah and Smith 1953). It is immediately evident that the pattern is quite diffuse and the peaks are much broader than

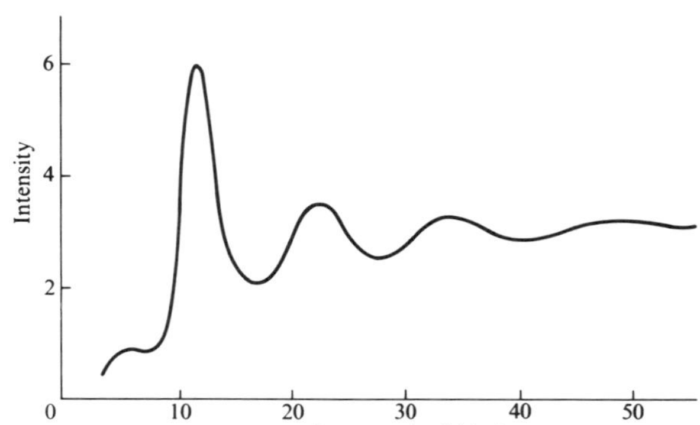

FIG. 350. An early diffraction pattern of a liquid. Lead at 623 K and a neutron wavelength $\lambda = 1{\cdot}16$ Å. (Sharrah and Smith 1953.)

those from a solid, as shown, for example, in Fig. 248. This feature results, of course, from the absence of any long-range order.

We shall see that the distinctions which we have drawn between coherent and incoherent and between elastic and inelastic scattering of neutrons lead to some important advantages in using neutrons rather than X-rays for studying liquids. It will be necessary, however, to examine the scattering process in some detail before being able to appreciate these advantages. At the same time there are other more practical advantages which can be seen immediately. Thus the low absorption coefficients of most materials for neutrons are of great value. It is almost always possible to use samples in the transmission mode rather than the reflection mode, whereas X-ray studies are restricted to the latter, with the result that conclusions really relate only to a surface layer of liquid, which is unrepresentative and often heavily contaminated with foreign atoms. At the same time the low absorption coefficients for neutrons permit the use of containers and holders which are strong mechanically, so that materials can be observed under pressure or at high temperatures. This readily permits the study of liquid metals, whose structure may offer easier interpretation than that of the associated molecular aggregates which comprise most of the more conventional liquids.

Our study of solids has been based on a well-defined three-dimensional structure in which atoms are allocated precise positions. Because of their thermal energies atoms engage in vibratory motion about these equilibrium positions, and we have studied these motions in terms of the Debye–Waller factor, which reduces the intensities of the Bragg reflections, and in terms of the inelastic scattering which views the motions as a spectrum of waves in the solid. However, this type of model and approach is quite unsuitable for liquids because of the diffusion which occurs and the fact that individual atoms cannot be ascribed to particular sites. A snapshot picture of a liquid, with its close similarity to a solid so far as local order is concerned, is very misleading, for the dynamical behaviour of the atoms in solids and liquids is quite different.

Of fundamental importance in providing a general picture of the dynamical behaviour of a liquid is the curve shown in Fig. 351, which indicates the wavelength distribution of scattered neutrons, for a given angle of scattering, when the liquid is irradiated by a monochromatic beam of long-wavelength neutrons. The pattern is of two parts. First, the incident distribution is somewhat broadened, giving the so-called 'quasi-elastic' scattering. This is brought about by processes which strictly are inelastic, but the energy transfers are small and are related to the diffusive motions of the atoms. Because of this diffusion, permitted by the fact that atoms in a liquid are *not* attached to specific sites, there is no strictly elastic scattering. Secondly, in the region of smaller λ in Fig. 351 intrinsic inelastic scattering has taken place with quantized exchange of energy in much more substantial amounts than for the diffusive motions. These exchanges will include molecular rotations and vibrational motions of atoms within the molecules.

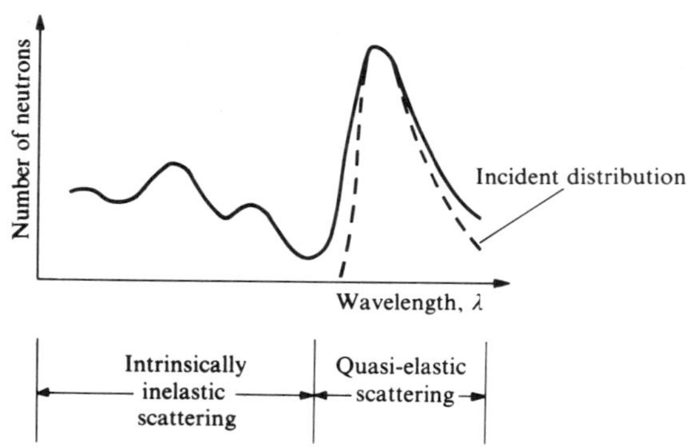

FIG. 351. The wavelength distribution of neutrons scattered from a liquid. The scattering of a beam of long-wavelength neutrons is observed at a fixed angle of scattering 2θ. The two parts of the pattern distinguish the quasi-elastic scattering, associated with the diffusion of the atoms, and the genuine inelastic scattering.

16.2. Space–time correlation functions

The most advantageous way of trying to interpret the scattering pattern of a liquid for neutrons, or for any other kind of radiation, is in terms of the space–time correlation function of van Hove (1954). This function of space and time describes what is basically the probability that we shall find an atom within a unit volume of space at distance r and time t, if we measure and count from an atom which we observed to be at a chosen origin, $r = 0$, at a starting time $t = 0$. We denote this function by $G(r, t)$, and it is a complicated function of r and t, which, in the present state of our knowledge, we are far from able to describe in detail; it is the aim of our neutron-beam studies to define it more accurately. It is convenient to divide the function into two parts. First, the atom which we observe at r at time t may be the *same* one as we saw at the origin at $t = 0$, and we call this part of the function the 'self-correlation' function $G_s(r, t)$; secondly, it may be a *different* atom which we observe at r, t, and this part of the probability is usually called the 'distinct-correlation' function $G_d(r, t)$. $G_s(r, t)$ will evidently be a simpler type of function than is $G_d(r, t)$, since the former describes the motion of a single atom. It will be clear that

$$G_s(r, t)+G_d(r, t) = G(r, t). \tag{16.1}$$

It is difficult to give simple diagrammatic representations of these two functions since they vary not only with distance but with time. We can, however, show curves for their variation with distance for times which are short and long, respectively, compared with the relaxation time t_0. This time t_0, which is of the order of 10^{-13} s, is the time which elapses before a disturbance brought about by displacement of an individual particle is effectively damped out. Fig. 352 shows two sets of curves. The curves (a) represent the

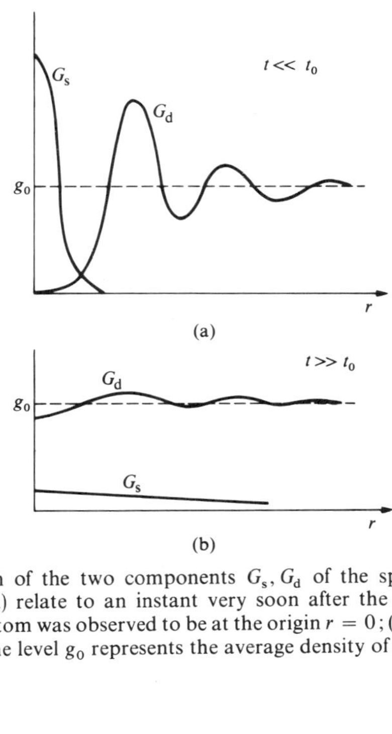

FIG. 352. The form of the two components G_s, G_d of the space–time correlation function. Curves (a) relate to an instant very soon after the 'starting time' $t = 0$, when a particular atom was observed to be at the origin $r = 0$; (b) shows the situation very much later. The level g_0 represents the average density of the liquid.

situation very shortly after an initial observation that a particular atom A was at the origin. Curve G_s here will indicate the new position of atom A, and it is naturally likely to be very close to the origin, with only a small probability that it has moved very far away. Likewise G_d indicates the position of *other* atoms, and evidently there is no likelihood that some other atom can yet have replaced A at the origin. On the other hand, the curves (b) in Fig. 352 show the situation long after our starting time. The curve for G_s indicates that atom A is likely to have diffused away from the origin, though still slightly more likely to be found at the origin than in any other particular position, whereas the curve for G_d shows that, so far as other atoms are concerned, diffusion has produced practically a uniform density. There is a minimum in G_d at the origin to compensate for the slight preference, in G_s, that the origin is still occupied by atom A.

The extreme case of $G_d(r, t)$ when $t = 0$ is of particular significance. This, by definition, will describe the situation at our starting time, when we have an atom A at the origin, and will be simply a *snapshot* picture of the atomic environment in the liquid and contains no dynamical information; this snapshot function is usually represented by the pair-distribution function $g(r)$ when suitably normalized. It is shown in the upper curve of Fig. 353 and, as expected, is not much different from the curve which we showed in Fig. 352 for $G_d(r, t)$ at a very small value of t. More interesting, however, is to compare our $g(r)$ curve for a liquid with the corresponding curves for solids and gases. Curves (b) and (c) in Fig. 353 do this. The curve (b) for the solid indicates the regular periodic arrangement of atoms over long distances. On the other hand, curve (c) for a gas shows that it is not possible to find a second atom within some specified short distance from the origin but beyond this distance the probability increases quite quickly and rapidly attains a constant value, indicating that the distribution of gas molecules is completely random.

16.3. Neutron scattering patterns

Let us now outline the relation between the neutron scattering pattern for the liquid and the distribution functions which we have just discussed. It will be useful to recall some of our results from Chapter 2 and to recast them in the form and nomenclature which

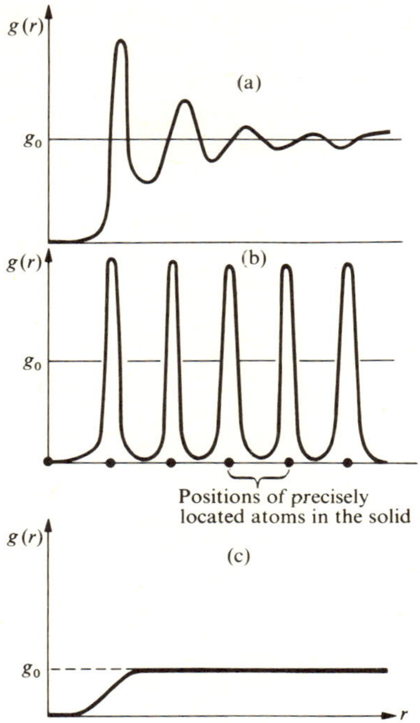

FIG. 353. The instantaneous distribution function $g(r)$ for (a) a liquid, (b) a solid, (c) a gas. The quantity $g(r)\delta V$ is the probability of finding a second atom within a volume δV at a distance r from a first atom which is taken as the origin.

is commonly used in studies of liquids. In eqn (2.20) we gave an expression for the 'differential scattering cross-section' for the scattering in a particular direction by an assembly of nuclei in a crystallite. We were then interested in those directions in which Bragg reflections occur, and we denoted the number of neutrons passing per unit solid angle in the direction of the (hkl) reflection by G_{hkl}. We could write a similar expression for *any* direction and denote it by the differential cross-section per unit solid angle $\mathrm{d}\sigma/\mathrm{d}\Omega$. For an assembly of N_0 nuclei we can rewrite eqn (2.20) as

$$\frac{\mathrm{d}\sigma}{\mathrm{d}\Omega} = \left| \sum_n b_n \exp(\mathrm{i}\mathbf{Q} \cdot \mathbf{R}_n) \right|^2, \tag{16.2}$$

where b_n, $\mathbf{R}_n$ represent the scattering length and position of the nth nucleus and $\mathbf{Q}$ is the vector $(\boldsymbol{\kappa} - \boldsymbol{\kappa}_0)$, which we discussed in

Chapter 9. We recall that κ, κ_0 are equal to $2\pi/\lambda$, $2\pi/\lambda_0$ and that for elastic scattering the incident and final wavelengths λ_0 and λ are equal. The equivalence of the expressions $\mathbf{Q}\,.\,\mathbf{R}_n$ and $2\pi(hx/a_0+$ $+ky/b_0+lz/c_0)$ is outlined in Fig. 354, which indicates that the difference of phase for the scattering by atoms at O and P is equal to $(\boldsymbol{\kappa}-\boldsymbol{\kappa}_0)\,.\,\mathbf{R}_n$.

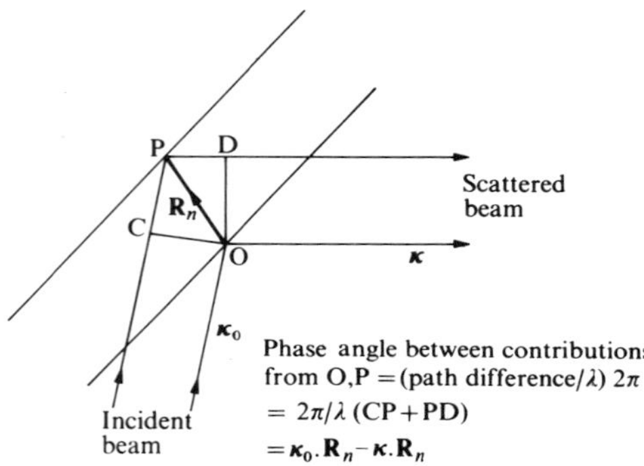

FIG. 354. Demonstration of equivalence of the expressions $2\pi(hx/a_0+ky/b_0+lz/c_0)$ and $\mathbf{Q}\,.\,\mathbf{R}_n$.

When the right-hand side of eqn (16.2) is expanded we find that

$$\frac{d\sigma}{d\Omega} = N_0\{\overline{b_r^2}-(\bar{b}_r)^2\}+(\bar{b}_r)^2|\sum \exp(i\mathbf{Q}\,.\,\mathbf{R}_n)|^2, \qquad (16.3)$$

which is equivalent to (2.22), and the two terms represent respectively the incoherent scattering, which is isotropic, and the coherent scattering which is directional. Indeed for a solid, because of the three-dimensional order, the coherent scattering is highly directional and appears most noticeably in the Bragg reflections which occur in directions corresponding to the reciprocal lattice points.

When we proceed to the study of inelastic scattering we can write an expression for a scattering amplitude, analogous to eqn (16.2), for a transition to any changed energy state of the system, energy having been gained or lost by interchange with the neutron. Any interchange will be governed by the conservation of energy and the magnitude of the amplitude expresses the probability that any particular change of state shall occur. For a given arrangement of the scattering sample and for a chosen angle of scattering the probability of scattering will depend on the magnitude of the energy interchange, and we have to consider the double differential cross-section $d^2\sigma/d\Omega\, dE$ which is the cross-section per unit solid angle *per unit range of energy interchange*. This quantity will be a function of both the angle of scattering and the interchange of energy. However there is correlation between the interchange of energy $\hbar\omega$, the momentum change $\hbar Q$, and the angle of scattering 2θ, according to the relation

$$\frac{\hbar^2 Q^2}{2m} = 2E_0 + \hbar\omega - 2(E_0^2 + \hbar\omega E_0)^{\frac{1}{2}} \cos 2\theta, \tag{16.4}$$

which is derived from eqns (9.1) and (9.2) and where E_0 is the incident neutron energy, so that $d^2\sigma/d\Omega\, dE$ is a function of Q and ω.

The energy and momentum changes which occur within a liquid will correlate with the changes in position and velocity and, therefore, with the correlation functions $G_s(r, t)$, $G_d(r, t)$, $G(r, t)$ which describe the liquid and its atomic behaviour. The detailed derivation of this relation, first done by van Hove (1954), is beyond the scope of this book and we shall simply quote the relationship which is, for a group of N nuclei,

$$\frac{d^2\sigma}{d\Omega\, dE} = N\frac{\kappa}{\kappa_0}\frac{1}{2\pi\hbar}\int_{-\infty}^{\infty} dt\, e^{-i\omega t} \sum e^{i\mathbf{Q}.\mathbf{r}}[(\bar{b})^2 G(r, t) + \\ + \{(\overline{b^2}) - (\bar{b})^2\} G_s(r, t)]. \tag{16.5}$$

This expression for $d^2\sigma/d\Omega\, dE$ separates into two parts and, by analogy with the expression for $d\sigma/d\Omega$ in eqn (16.3) the two contributions are called the coherent and incoherent scattering cross-sections. Thus

$$\frac{d^2\sigma_{coh}}{d\Omega\, dE} = N\frac{\kappa}{\kappa_0}\frac{1}{2\pi\hbar}(\bar{b})^2 \int_{-\infty}^{\infty} dt\, e^{-i\omega t} \sum_r e^{i\mathbf{Q}.\mathbf{r}} G(r, t) \tag{16.6}$$

and

$$\frac{d^2\sigma_{incoh}}{d\Omega\, dE} = N\frac{\kappa}{\kappa_0}\frac{1}{2\pi\hbar}\{\overline{b^2} - (\bar{b})^2\} \int_{-\infty}^{\infty} dt\, e^{-i\omega t} \sum_r e^{i\mathbf{Q}.\mathbf{r}} G_s(r, t) \tag{16.7}$$

The two integrals which appear in these expressions are known as 'scattering laws' and are written as $S(Q, \omega)$ and $S_s(Q, \omega)$, being functions of Q, ω. Thus we have

$$S(Q, \omega) = \frac{1}{2\pi} \iint G(r, t)\, e^{i(\mathbf{Q}.\mathbf{r} - \omega t)}\, d\mathbf{r}\, dt \tag{16.8}$$

and

$$S_s(Q, \omega) = \frac{1}{2\pi} \iint G_s(r, t)\, e^{i(\mathbf{Q}.\mathbf{r} - \omega t)}\, d\mathbf{r}\, dt. \tag{16.9}$$

These functions depend on the dynamics of the liquid system and not on the properties of the incident radiation. Nevertheless, different radiations such as X-rays, neutrons, and infrared rays can investigate different portions of the momentum and energy range.

16.4. Experimental measurements for simple liquids

In proceeding to investigate the extent to which experimental observations of neutron scattering will provide us with details of the structure of the liquid it is most convenient to combine eqns (16.6) and (16.8) to yield a relation between $d^2\sigma_{coh}/d\Omega\, d\omega$ and the coherent scattering law $S(Q, \omega)$. Thus we have

$$\frac{d^2\sigma_{coh}}{d\Omega\, d\omega} = N\frac{\kappa}{\kappa_0}(\bar{b})^2 S(Q, \omega), \tag{16.10}$$

remembering that the energy change is equal to $\hbar\omega$.

The geometry of a conventional diffraction experiment with a solid or liquid is illustrated in Fig. 355. The detecting counter will collect, at a given value of scattering angle 2θ, scattered neutrons of all energies. In general the change of momentum $\hbar\mathbf{Q}$ which they

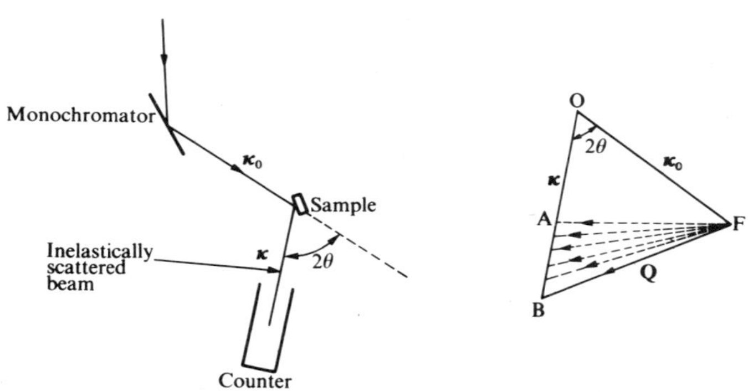

FIG. 355. For a counter which accepts neutrons in a wide range of energy, indicated by the range of OA → OB in the value of κ, there is a considerable range for the momentum transfer which has taken place, shown by the range of $\mathbf{Q}$ from FA to FB.

have undergone will vary considerably, as indicated by the variation of length of **Q** from FA to FB in Fig. 355. However, in the special case of X-ray diffraction the energy change $\hbar\omega$ is negligible in comparison with the initial photon energy E and the magnitude κ of the scattered wave vector is indistinguishable from κ_0. Consequently the range of variation of Q is negligible and the detector will record a count which is proportional to

$$\frac{\mathrm{d}\sigma_{\mathrm{coh}}}{\mathrm{d}\Omega} = \int \frac{\mathrm{d}^2\sigma_{\mathrm{coh}}}{\mathrm{d}\Omega\,\mathrm{d}\omega}\,\mathrm{d}\omega = N(\bar{b})^2 \int_{-\infty}^{\infty} S(Q,\omega)\,\mathrm{d}\omega, \qquad (16.11)$$

since $\kappa = \kappa_0$.

The integral on the right will be a function of Q, and we write it as $S(Q)$ and designate it the 'structure factor' of the liquid. Thus

$$S(Q) = \int_{-\infty}^{\infty} S(Q,\omega)\,\mathrm{d}\omega \qquad (16.12)$$

$$\mathrm{d}\sigma_{\mathrm{coh}}/\mathrm{d}\Omega_{\mathrm{X\text{-}rays}} = N(\bar{b})^2 S(Q). \qquad (16.13)$$

Fig. 356, taken from Egelstaff (1967), shows qualitatively the over-all variation of the structure factor $S(Q)$ with Q and for several values of Q indicates the form of variation of the scattering law $S(Q, \omega)$ as a function of ω. At low values of Q, $S(Q, \omega)$ shows sharp peaks at values of ω given by $\omega = cQ$, where c is the velocity of sound, together with a central peak. As Q is increased the width of the lines increase and they overlap, progressing at high Q values to a broad single-peaked function. We may regard the determination of $S(Q)$ as the initial aim of diffraction experiments with liquids and, from the foregoing, it is evident that it is obtainable directly in principle from an X-ray experiment—though not without practical difficulties because of high X-ray absorption and surface contamination of the liquid which may invalidate the results.

Evaluation of $\mathrm{d}\sigma_{\mathrm{coh}}/\mathrm{d}\Omega$ for X-rays from (16.8) and (16.11) gives

$$\begin{aligned}\frac{\mathrm{d}\sigma_{\mathrm{coh}}}{\mathrm{d}\Omega} &= \frac{N(\bar{b})^2}{2\pi} \iint G(r,t)\,\mathrm{e}^{\mathrm{i}\mathbf{Q}.\mathbf{r}}\,\mathbf{dr}\,\mathrm{d}t \int \mathrm{e}^{-\mathrm{i}\omega t}\,\mathrm{d}\omega. \\ &= \frac{N(\bar{b})^2}{2\pi} \int_{-\infty}^{\infty} \mathrm{e}^{-\mathrm{i}\omega t}\,\mathrm{d}\omega \int \mathrm{d}t \int G(\mathbf{r},t)\,\mathrm{e}^{\mathrm{i}\mathbf{Q}.\mathbf{r}}\,\mathbf{dr} \\ &= N(\bar{b})^2 \int G(r,0)\,\mathrm{e}^{\mathrm{i}\mathbf{Q}.\mathbf{r}}\,\mathbf{dr}, \end{aligned} \qquad (16.14)$$

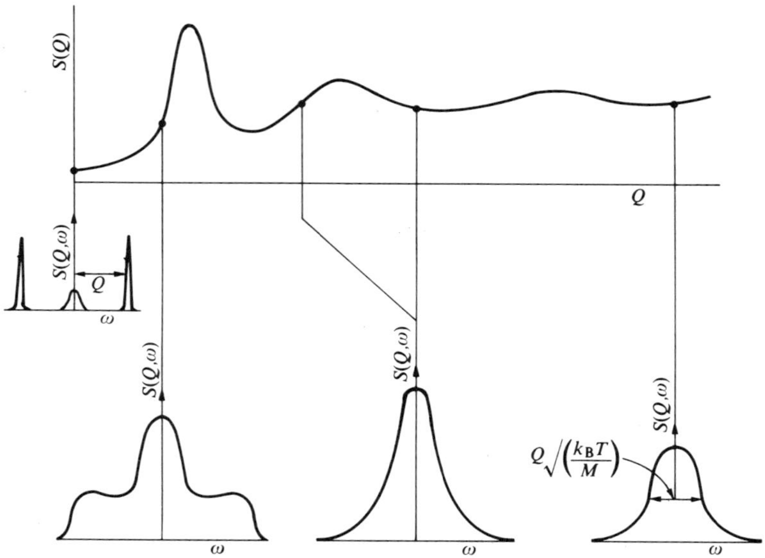

FIG. 356. The upper curve shows the variation of the function $S(Q)$ with the value of Q. The ordinate has been obtained at each point by integration of the area under the curve of $S(Q, \omega)$ versus ω, of which representative samples are shown below. (After Egelstaff 1967.)

since $(1/2\pi)\int_{-\infty}^{\infty} e^{-i\omega t}\, d\omega$ is the Dirac δ-function $\delta(t)$ which has a value only in the neighbourhood of $t = 0$ and for which $\int_{-\infty}^{\infty} \delta(t)\, dt = 1$.

But

$$\begin{aligned} G(r, 0) &= G_s(r, 0) + G_d(r, 0) \\ &= \delta(r) + G_d(r, 0) \\ &= \delta(r) + \rho g(r) \end{aligned}$$

where ρ is the number density of atoms, i.e. total number per unit volume, and we have followed the normal practice of normalizing $g(r)$ to unity at large r. Therefore eqn (16.14) becomes

$$\frac{d\sigma_{coh}}{d\Omega} = N(\bar{b})^2 \left(1 + \int \rho g(r)\, e^{i\mathbf{Q}.\mathbf{r}}\, d\mathbf{r}\right) \tag{16.15}$$

so that inversion of the scattered intensity function $\mathrm{d}\sigma_{\mathrm{coh}}/\mathrm{d}\Omega$ will yield $g(r)$, i.e. $G_{\mathrm{d}}(r, 0)$. We emphasize that it is the 'snapshot' function $g(r)$ which is obtained directly from an X-ray pattern in this way. Accordingly, since dynamical information is absent, the X-ray measurement is normally described as the 'static approximation'.

The term $\exp(\mathrm{i}\mathbf{Q}\cdot\mathbf{r})$ in eqn (16.15) has to take into account all possible orientations of $\mathbf{Q}$ and $\mathbf{r}$. When the necessary averaging is done we arrive at the expression

$$\frac{\mathrm{d}\sigma_{\mathrm{coh}}}{\mathrm{d}\Omega} = N(\bar{b})^2\left[1+\frac{4\pi\rho}{Q}\int_0^\infty \{g(r)-1\}r\sin Qr\,\mathrm{d}r\right], \qquad (16.16)$$

where $g(r)$ is normalized to unity as $r \to \infty$, in accordance with the fact that there is no long-range order and at large distances from the origin there is uniform probability of finding an atom. Alternatively, from (16.13) we may write an expression for the structure factor $S(Q)$ as

$$S(Q) = 1+\frac{4\pi\rho}{Q}\int_0^\infty \{g(r)-1\}r\sin Qr\,\mathrm{d}r, \qquad (16.17)$$

where we reiterate that $g(r)$ indicates the probability of finding an atom at the position r at the same time as we observe an atom at the origin, and ρ is the number of atoms per unit volume.

For neutrons, in spite of many advantages, the interpretation of the experimental situation is more difficult and considerable corrections must be applied in order to produce $S(Q)$ and, subsequently, $g(r)$. The difficulty arises because, as we have already pointed out in relation to Fig. 355, the neutron counter in a given angular position will receive neutrons having a substantial range of value of Q. A correction must therefore be applied if the angular variation of $\mathrm{d}\sigma_{\mathrm{coh}}/\mathrm{d}\Omega$ is to be correlated with the structure factor $S(Q)$ of eqn (16.12). The correction has been studied by a number of workers, notably by Placzek (1952) and by Ascarelli and Caglioti (1966), utilizing expansions of approximate functions for $S(Q, \omega)$. Not only is it necessary to take into account the range of Q values which is received at each angular setting but also to allow for the fact that the efficiency of the neutron detector varies with energy. If it is assumed that the detector efficiency varies as the reciprocal of the neutron velocity, i.e. as $1/\kappa$, then the Placzek correction to

$S(Q)$ is a function $f_P(Q)$ of Q such that

$$f_P(Q) = \frac{\bar{K}_0}{3\varepsilon\mu} - \frac{\alpha}{2\varepsilon}$$

$$= \frac{\bar{K}_0}{3\varepsilon\mu} - \frac{1}{\mu}(1-\cos 2\theta), \quad (16.18)$$

where $\bar{K}_0$ = average kinetic energy of the atoms of the liquid in units of kT, ε = incident neutron energy in units of kT, μ = nuclear mass/neutron mass, i.e. M/m, and $\alpha = \hbar^2Q^2/2MkT$.

Table 32 shows some values of the Placzek correction calculated at various values of Q for a number of different liquid metals by

TABLE 32

Placzek corrections†

Element	T (K)	$\frac{\bar{K}_0}{3\varepsilon\mu}$	$f_P(Q)$ $Q = 0$	$Q = 2{\cdot}15$	$Q = 4{\cdot}23$	$Q = 6{\cdot}19$	$Q = 8{\cdot}75$ (Å^{-1})
Li	459	0·0355	0·0355	0·0270	0·0022	−0·0358	−0·1072
V	293	0·0031	0·0031	0·0019	−0·0014	−0·0066	−0·0164
Zn	693	0·0057	0·0057	0·0048	0·0022	−0·0019	−0·0095
Sn	505	0·0023	0·0023	0·0017	0·0002	−0·0016	−0·0061
Pb	600	0·0016	0·0016	0·0013	0·0005	−0·0008	−0·003

† Evaluated for a $1/v$ detector with λ_0 = 1·015 Å and $\bar{K}_0$ = 1·5.

North, Enderby, and Egelstaff (1968*a*). Except for the light element lithium the correction terms seldom reach 1 per cent; the larger correction for lithium is accounted for by the appearance of the nuclear mass in the above expression for $f_P(Q)$.

We emphasize that eqns (16.11) and (16.15) relate the structure factor $S(Q)$ to the coherent scattering. Consequently, in order to determine $S(Q)$ it is essential to study liquids for which the neutron scattering is almost completely coherent. This condition is well satisfied by the liquid metals which were used in the above work, and for which the cross-sections are given in Table 33. On the other hand, the condition would *not* be satisfied by liquid sodium, for which the coherent and incoherent scattering cross-sections are $1{\cdot}55\times10^{-24}$ cm^2 and $1{\cdot}85\times10^{-24}$ cm^2 respectively, nor by hydrogeneous liquids in which the incoherent scattering by hydrogen atoms is usually heavily predominant.

TABLE 33

Element	σ_{coh}	σ_{incoh}
Sn	$4{\cdot}6 \times 10^{-24}$ cm^2	$0{\cdot}04 \times 10^{-24}$ cm^2
Bi	9·35	0·02
Pb	11·5	0·05
Tl	10·0	0·1

Only in recent years have accurate results, for which all necessary corrections have been made, become available. Fig. 357 shows three independent sets of data for liquid zinc (Egelstaff 1970) which are consistent to 5 per cent. North *et al.* (1968*a*) estimate that the experimental errors in their determinations of $S(Q)$ for liquid metals range from $\pm 0{\cdot}03$ for Pb to $\pm 0{\cdot}05$ for Tl in the region where $Q < 2$ Å^{-1} and rather larger than this for higher values of Q.

Any inadequacies in the determination of $S(Q)$ are likely to produce large errors in the distribution function $g(r)$ which, by

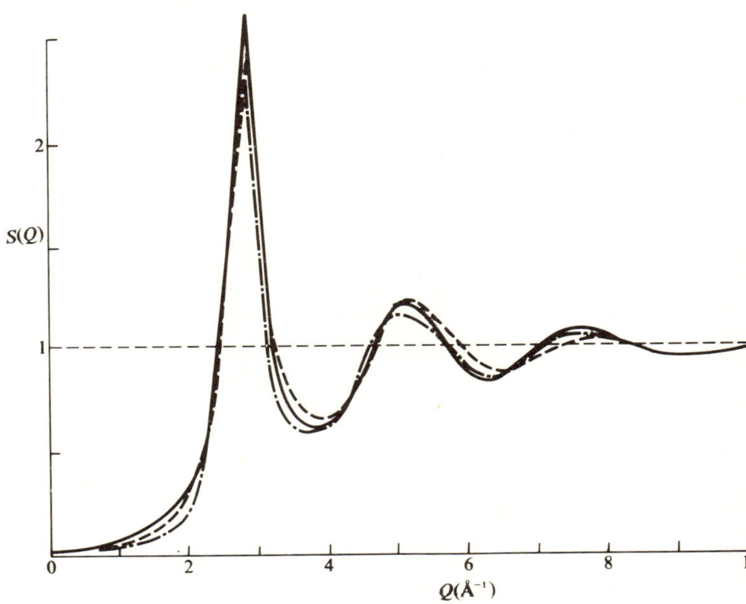

FIG. 357. Comparison of neutron diffraction data for liquid zinc from three independent investigations. (After Egelstaff 1970.)

eqn (16.15), is obtained by Fourier inversion of $S(Q)$. Thus

$$g(r) = 1 + \frac{1}{2\pi^2 nr} \int_0^\infty Q\{S(Q) - 1\} \sin(Qr) \, dQ, \qquad (16.19)$$

where n is the number of atoms per unit volume. Inevitably errors in $g(r)$ are caused by the cut-off in $S(Q)$ at both large and small values of Q. The interpretation of data for five liquid metals (Zn, Pb, Tl, Sn, and Bi) is described by North *et al.* (1968*b*). Fig. 358

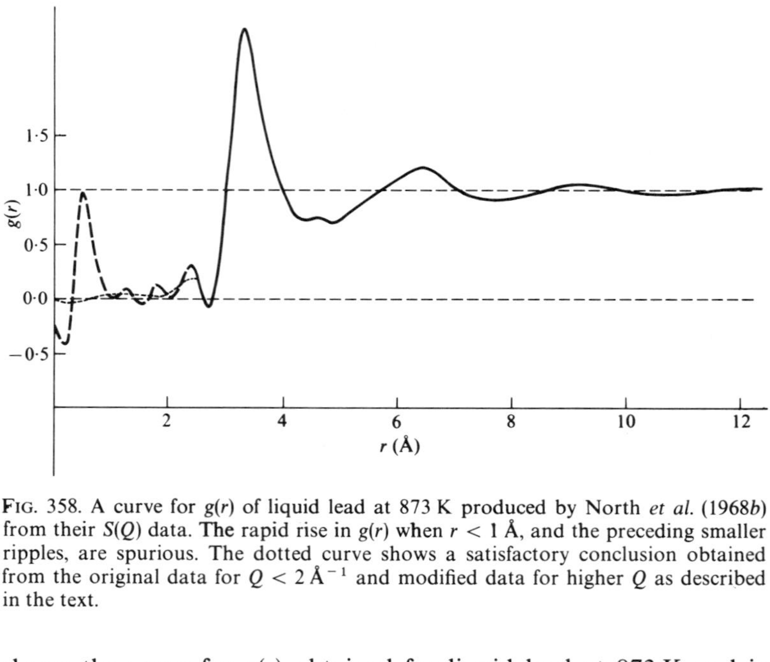

FIG. 358. A curve for $g(r)$ of liquid lead at 873 K produced by North *et al.* (1968*b*) from their $S(Q)$ data. The rapid rise in $g(r)$ when $r < 1$ Å, and the preceding smaller ripples, are spurious. The dotted curve shows a satisfactory conclusion obtained from the original data for $Q < 2\ \text{Å}^{-1}$ and modified data for higher Q as described in the text.

shows the curve for $g(r)$ obtained for liquid lead at 873 K and is noteworthy (in the broken curve) for the spurious rapid increase of $g(r)$ at values of r less than about 1 Å, together with smaller ripples due to the truncation errors. If these spurious effects are ignored and the data for $g(r)$ are transformed back to $S(Q)$, as shown in Fig. 359, it will be found that the original data are well reproduced except for low Q, where the modified curve goes negative. If we then use the original experimental data for $Q < 2\ \text{Å}^{-1}$ but the modified curve of Fig. 359 for $Q > 2\ \text{Å}^{-1}$ then we find, in the *dotted* curve in Fig. 358, that a satisfactory curve for $g(r)$ is now obtained at small values of r. North *et al.* conclude that their data for $g(r)$ is correct to ± 5 per cent for Pb in the region where $r > 2$ Å.

A further way of assessing the experimental results is to compare the curve for $g(r)$ deduced by experiment with curves deduced theoretically for various models. Fig. 360 shows the experimental results for argon and rubidium in comparison with a calculation

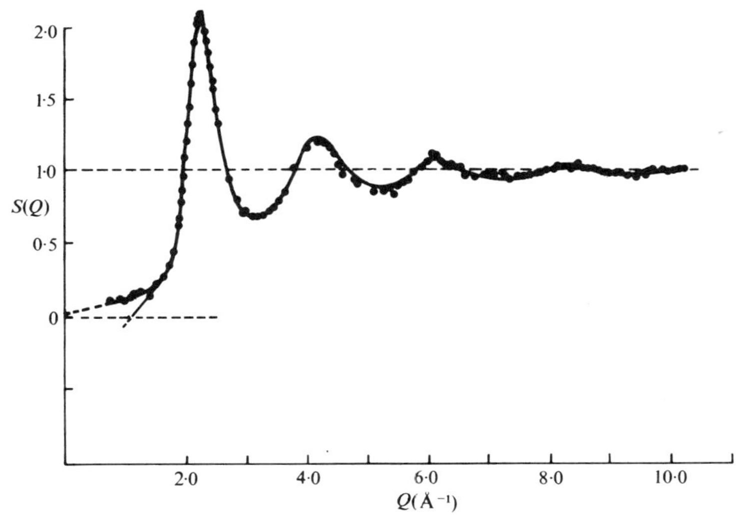

FIG. 359. The points show the experimental $S(Q)$ data of North *et al.* (1968*b*) for liquid lead at 873 K and the full line is derived by re-transformation of the derived $g(r)$ data of Fig. 358 when the spurious oscillations at small r are removed. This line is satisfactory except at small Q, where it goes negative and where the unmodified experimental curve (broken line) is preferred.

for a 'hard-sphere' model, according to which the pair potential between two atoms is zero for separations greater than some fixed value ρ and is infinite for any separation which is less than this value. This is one of three models which are widely used and which are represented in Fig. 361 as (a) the hard sphere, (b) the Lennard-Jones model, and (c) the modified Buckingham potential. Fig. 360 indicates how for argon the oscillations in $S(Q)$ at larger values of Q die out faster than would be expected for the hard-sphere model and the effect for rubidium is even more marked. This means that when the atoms in the liquid collide with each

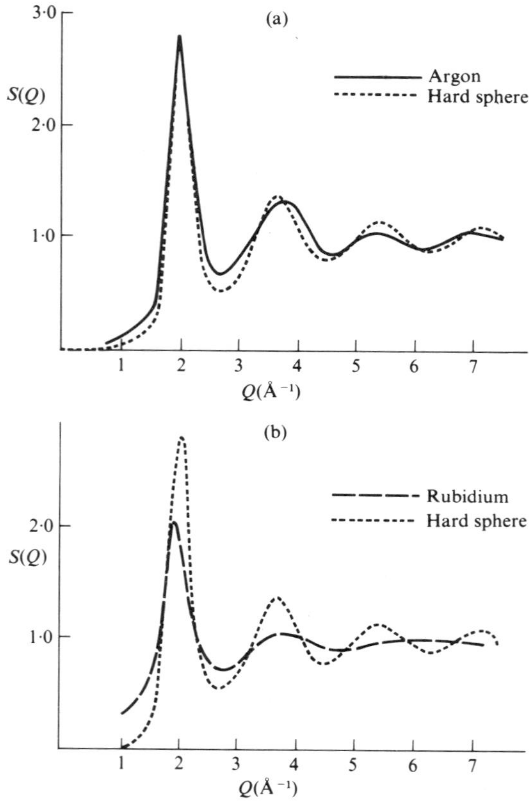

FIG. 360. Comparison of the experimentally determined liquid structure factors for argon (a) and rubidium (b) with the calculated curve for a hard-sphere model. (After Page *et al.* 1969.)

other then partial overlap occurs and the atoms are not infinitely rigid as the simple hard-sphere model assumes.

Most scattering measurements with liquids have used the ordinary neutron wavelength of about 1 Å which is employed in diffraction studies, but there are certain advantages which favour longer and shorter wavelengths. 'Cold' and 'hot' moderators may be used (see p. 138) in order to increase the output of neutrons of, respectively, longer or shorter wavelength than average. An advantage of a short wavelength is that the approach to the static approximation

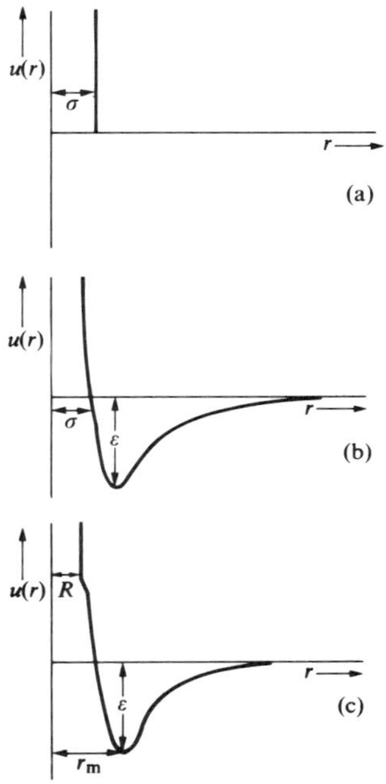

FIG. 361. Illustrations of spherically symmetrical potentials: (a) hard-sphere; (b) Lennard-Jones; (c) modified Buckingham. (Page 1973.)

is improved. Samples in the form of a thin slab, containing, for example, a liquid metal in an ordinary metal box are generally used, with a cross-sectional area of a few square centimeters and of such a thickness that the transmission is not less than 90 per cent. This is important if the correction for multiple scattering is to be made satisfactorily. There are many advantages in constructing the sample holder of vanadium, which contributes no Bragg peaks to confuse the scattering pattern. The low absorption coefficients of materials for neutrons, compared with X-rays, are of particular importance when long wavelengths are being used.

16.5. Binary alloys

When we pass from the study of simple monatomic liquids to liquids which contain two different atomic species we find that neutrons have a very distinctive advantage. If we consider as a simple case a binary alloy which contains two elements A and B we have to be concerned with three different interatomic potentials, related to the interactions A–A, B–B, and A–B respectively, and we have to consider three separate correlation functions. Instead of a single function $g(r)$ we have to consider $g_{AA}(r)$, $g_{BB}(r)$, and $g_{AB}(r)$, where, for example, $g_{AB}(r)$ is the probability that we shall find an atom of type B at position **r** when there is an atom of type A at the origin. In order to provide a suitable description we extend our expression (16.17) for the structure factor $S(Q)$ to provide a definition of three partial structure factors S_{AB}, S_{AA}, and S_{BB}. Thus

$$S_{AB} \equiv 1+\frac{4\pi\rho}{Q}\int_0^\infty \{g_{AB}(r)-1\}r\sin Qr\,dr, \tag{16.20}$$

where ρ remains equal to the *total* number of scattering atoms per unit volume.

When the three types of distribution are taken into account eqn (16.16) then becomes

$$\begin{aligned}\frac{d\sigma_{coh}}{d\Omega} &= Nc_A b_A^2\{1+c_A(S_{AA}-1)\}+Nc_B b_B^2\{1+c_B(S_{BB}-1)\}+\\ &\quad +2Nc_A c_B b_A b_B(S_{AB}-1)\\ &= N(c_A b_A^2+c_B b_B^2)+N\{c_A^2 b_A^2(S_{AA}-1)+c_B^2 b_B^2(S_{BB}-1)+\\ &\quad +2c_A c_B b_A b_B(S_{AB}-1)\}.\end{aligned} \tag{16.21}$$

where c_A, c_B are the atomic fractions of the two components A, B and b_A, b_B are their coherent scattering lengths. N is the total number of atoms in the sample.

Thus the scattering consists of a constant term, to which will be added isotropic contributions from incoherent and multiple scattering processes, and an angularly varying function which can be separated from it. From the latter function we wish to deduce separately the functions S_{AA}, S_{BB}, and S_{AB}. This can only be done if it is somehow possible to vary the scattering lengths b_A, b_B in three separate measurements, thus providing, for each value of Q, three equations from which the individual values of S_{AA}, S_{BB}, and S_{AB} can be determined. In principle this could be done by combining scattering data from X-rays, neutrons, and electrons, but it would be very difficult to correlate these on a single intensity scale. However, for certain alloys neutron diffraction offers a means of providing three separate equations if different isotopes, having different scattering lengths, are available. An example of such an application is provided by Enderby, North, and Egelstaff (1966), who studied the alloy Cu_6Sn_5. Three samples were used, consisting of pure tin alloyed respectively with ^{63}Cu, ^{65}Cu, and natural copper. The scattering lengths for copper in the three cases are $0{\cdot}67 \times 10^{-12}$ cm, $0{\cdot}11 \times 10^{-12}$ cm, and $0{\cdot}76 \times 10^{-12}$ cm. The accuracy of the determination is limited by the rather small difference between the first and third samples but, nevertheless it was demonstrated that there were evident differences between the three partial structure factors S_{AA}, S_{AB}, and S_{BB}. The three $S(Q)$ curves are shown in Fig. 362, and it will be noticed in particular that the first peak in the Cu–Sn curve is close to the position of the Cu–Cu peak and is *not* near to the mean position for Cu–Cu and Sn–Sn. Some later and more striking results are given by Page and Mika (1971), who studied four samples of molten CuCl which consisted, in turn, of ^{63}Cu ^{35}Cl, ^{nat}Cu ^{nat}Cl, ^{65}Cu ^{nat}Cl, and ^{65}Cu ^{37}Cl. The ratio of the scattering lengths of copper:chlorine were 0·58, 0·82, 1·16, and 2·41 respectively, thus covering a fourfold range. Fig. 363 shows the experimentally determined structure factors for the four samples and emphasizes how the size of the first peak decreases as the contribution from the chlorine atoms decreases. Fig. 364 contrasts the radial distribution functions determined by these experiments with the results of calculations which assume complete dissociation into Cu^+ and Cl^- ions. Agreement is good for the chlorine ions but not for the distribution of the copper ions.

16.6. Molecular liquids

The liquids which we have discussed so far, and particularly the liquid metals, have been chosen because the interpretation of their structure might be expected to be relatively simple, but they are certainly not typical liquids. More commonly thought of as a typical liquid would be water or an organic solvent, but detailed interpretation of the structures of these is much more complicated, for several reasons. These liquids are molecular liquids, and the molecules have a well-defined structure in which the atoms are separated by precise and constant distances. This results in well-defined peaks in the diffraction pattern, from which the more diffuse structure of the intermolecular arrangement has to be disentangled. At the same time we have to note that molecules may have complicated shapes which are far removed from the simple spheres in

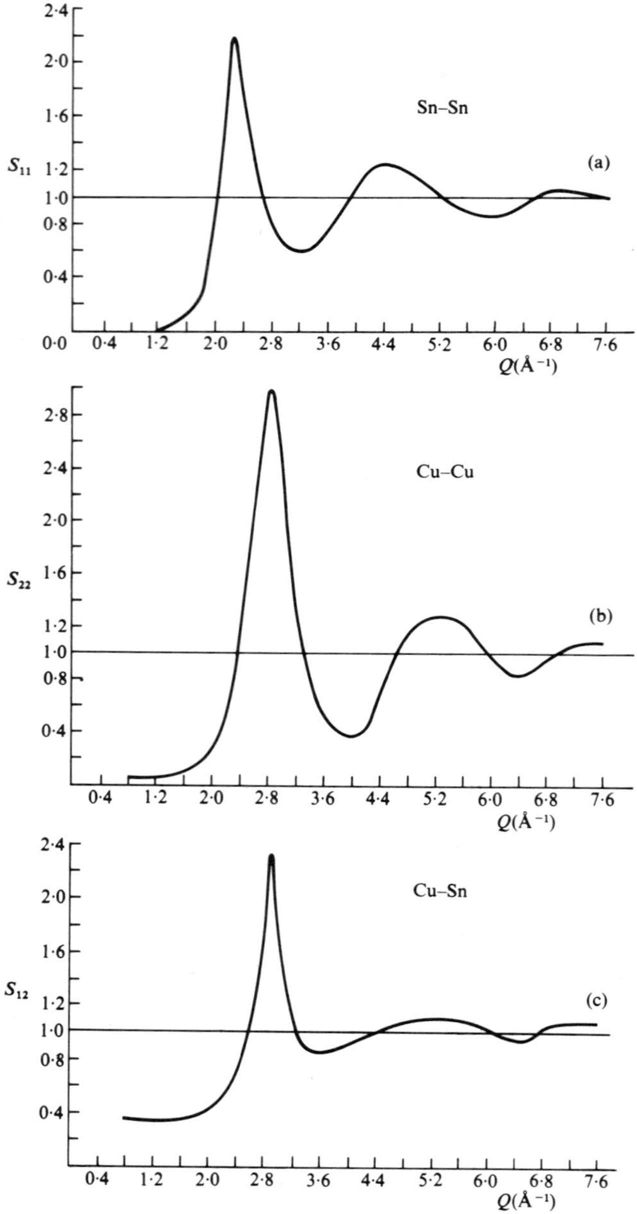

FIG. 362. Partial structure factors for (a) Sn–Sn, (b) Cu–Cu, and (c) Cu–Sn. (From Enderby *et al.* 1966.)

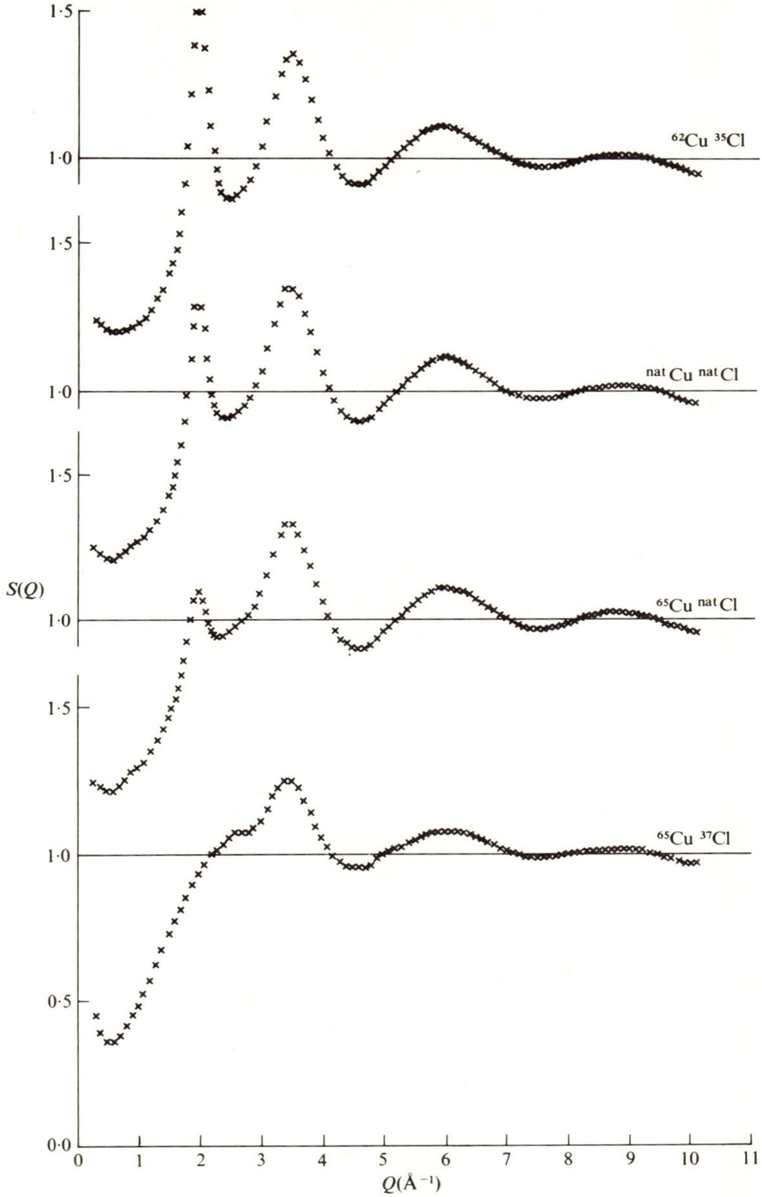

FIG. 363. Observed structure factors for molten CuCl at 750 K, for various isotopic compositions. (After Page and Mika 1971.)

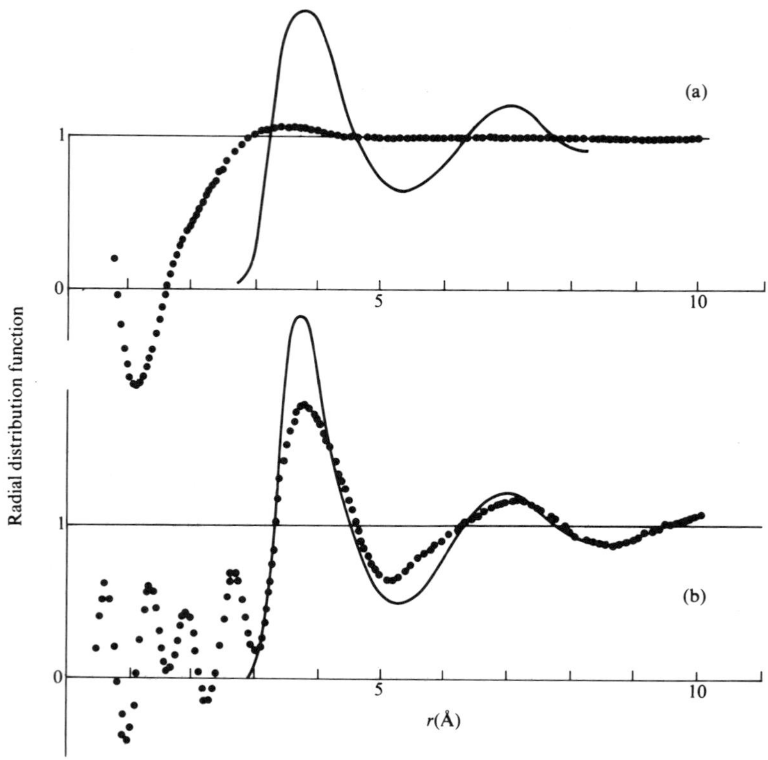

FIG. 364. Comparisons of the radial distribution functions for CuCl with the results of Monte Carlo calculations; (a) for the Cu–Cu distribution; (b) for the Cl–Cl distribution. In each case the closed circles represent the experimental data and the full-line curves are calculated. (After Page and Mika 1971.)

terms of which we have been able to regard the atoms of a liquid metal, for example. As a result the forces between molecules are directional in nature, and intermolecular distances will depend on the relative orientation of neighbouring molecules. In addition, as we have already noted, the scattering of ordinary water, but not of heavy water, and of many organic liquids will be predominantly the incoherent scattering from hydrogen, and this does not yield information on the structure factor $S(Q)$.

Consequent on the above considerations, study of the structures of molecular liquids is not yet in a very advanced state. The growing availability of high neutron fluxes is encouraging work in this field where, just as for simple liquids, the possibility of using different isotopes provides in principle the means of determining the several partial structure factors which are needed to define the structure of a molecular liquid. At the same time, as we think of larger and more complicated molecules, it is possible to use atomic, rather than isotopic, substitution, where it may be possible to retain a molecular shape by substituting, for example, a fluorine atom in place of a hydrogen atom.

16.7. Atomic motion

In our discussion of liquids so far we have emphasized the atomic environment of the atoms in the liquid and have drawn few conclusions regarding the actual motion of the liquid atoms. We have emphasized that it is the coherent neutron scattering which leads to our knowledge of $S(Q)$ which we have regarded as an essential first step in gaining a picture of the liquid. We return now to a study of atomic motion in liquids, and we shall see that information on the incoherent contribution to the scattering becomes of particular importance. This will mean that liquid water, with its predominance of incoherent scattering, becomes an attractive object for study. Ideally we require a simple monatomic liquid for this study, but there is no liquid which really satisfies this requirement. At first sight liquid vanadium, whose scattering is almost entirely incoherent, might be thought to qualify but unfortunately it both has a high melting point (1983 K) and is an extremely corrosive substance. We emphasize again, in relation to eqn (16.1), (16.5), and (16.6), that the incoherent scattering is determined by $G_s(r, t)$ which describes the motions of an individual atom whereas the coherent scattering depends on both $G_s(r, t)$ and $G_d(r, t)$ and is essentially more complicated.

On a macroscopic scale we can visualize motion within a liquid in terms of Fick's law, which describes the net motion of solute atoms in a liquid solution in which there exists a concentration gradient. The rate of transfer is proportional to the gradient, with a constant of proportionality which is called the diffusion coefficient and which is of the order of 10^{-9} m^2 s^{-1}. Ideally we wish to observe on an atomic scale and to follow the motion of an individual atom. This we are not able to do, but we may hope to define and measure the *average* behaviour of an atom. In particular we may think of the mean-square distance $\overline{r_t^2}$, travelled by an atom in time t, and it is possible to relate this quantity both to our diffusion coefficient D and to the self-correlation function $G_s(r, t)$ which we defined earlier in this chapter; We recall that this function gives the probability of finding an atom at a distance r and time t when that same atom was observed to be at the origin of coordinates when $t = 0$.

The relevant equations are

$$\overline{r_t^2} = \int_0^\infty r^2 G_s(r, t) 4\pi r^2 \, dr \tag{16.22}$$

and, if r, t are large,

$$G_s(r, t) = \frac{1}{(4\pi D t)^{\frac{3}{2}}} \exp\left(\frac{-r^2}{4Dt}\right). \tag{16.23}$$

The latter equation leads to the expression

$$\overline{r_t^2} = 6Dt. \tag{16.24}$$

For very short times, on the other hand, we can simply write $r = vt$, where v is the atomic velocity and $\frac{1}{2}Mv^2 = \frac{3}{2}k_B T$, and these two equations together lead to

$$\overline{r_t^2} = \left(\frac{3k_B T}{M}\right) t^2. \tag{16.25}$$

At intermediate times there will be a change-over from behaviour represented by eqn (16.24) to eqn (16.25), and under these circumstances the precise details of the atomic forces and positions will determine the exact motion. We can get an idea of the ranges of applicability of the two equations by considering that an atom will have to have moved several atomic distances before the macroscopic parameter D can be significant. At room temperature a typical atomic velocity is 3×10^2 m s^{-1}, so that an atom will traverse an interatomic distance of 3 Å in 10^{-12} s. This conclusion is substantiated by Fig. 365, which shows some neutron data for water by Sakamoto *et al.* (1962) and indicates that $\overline{r_t^2}$ has become proportional to t, as predicted by eqn (16.23), after a time equal to a few times 10^{-12} s. We shall make clearer shortly the manner in which the experimental neutron data have to be interpreted in order to yield the information in Fig. 365. For the moment we will justify the particular potential advantage of thermal neutrons for this kind of investigation. This becomes clear when we remark that if a neutron is to provide a picture of circumstances and conditions in a liquid then it will have to spend a time of the order of the few times 10^{-12} s, which we have computed above, in traversing an interatomic separation. This demands a wavelength of a few ångströms, which is very appropriate to a thermal neutron.

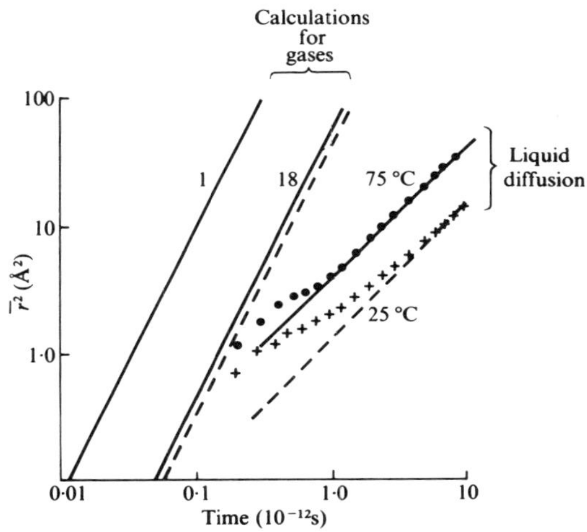

FIG. 365. Interpretation of the neutron data which show how the molecules in water move away from an initial position. The points • and × represent experimental observations at 75 °C and 25 °C respectively. The straight lines (on this logarithmic plot) indicate the results of calculations which assume, in turn, that the molecules behave as in a gas, with masses of 1 or 18, or according to a macroscopic view of diffusion. In each case full lines ——— are for 75 °C and the broken lines - - - - - are for 25 °C. (From Sakamoto *et al.* 1962.)

We return to a more detailed discussion of the interpretation of the experimental measurements which can be made with neutrons. If we are examining a liquid, such as water, in which the neutron scattering is almost entirely incoherent then eqns (16.7) and (16.9) show the relation between the neutron scattering and the self-correlation function $G_s(r, t)$. Fig. 366 shows a typical measurement of the magnitude of the scattering in terms of energy of the scattered neutron for a given angle of scattering from liquid water at 297·5 K, i.e. a curve of $S(Q, \omega)$ versus ω. This is an example of the quasi-elastic peak to which we referred earlier in Fig. 351 (see p. 546). A collection of such distribution curves, over the region of small values of ω, is shown in Fig. 367 for various values of Q, and from these distributions we can compute the quantity $\int S(Q, \omega) \cos \omega t \, \mathrm{d}\omega$ for given values of Q, t. This quantity is often named the intermediate scattering function $I(Q, t)$, i.e.

$$I(Q, t) = \int_{-\infty}^{+\infty} S(Q, \omega) \cos \omega t \, \mathrm{d}\omega. \tag{16.26}$$

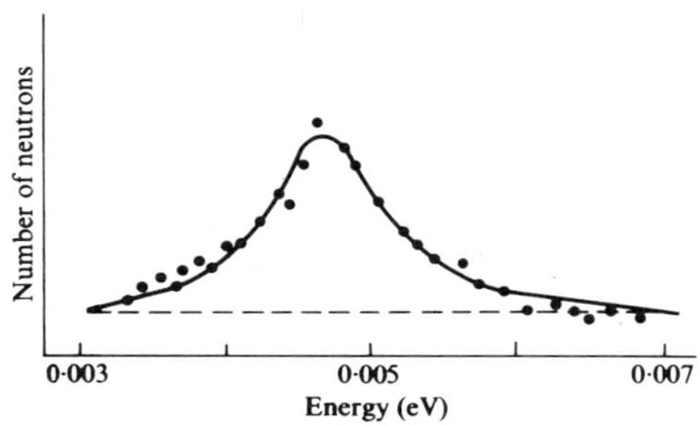

FIG. 366. Scattered spectrum of neutrons from liquid water at 297·5 K. (Brockhouse *et al.* 1963.)

It follows from eqn (16.9) that

$$G_s(r, t) = \frac{1}{2\pi} \iint S_s(Q, \omega) \, \mathrm{e}^{\mathrm{i}(\mathbf{Q}.\mathbf{r} - \omega t)} \, \mathrm{d}Q \, \mathrm{d}\omega$$

$$= \frac{1}{2\pi} \int_{-\infty}^{\infty} I(Q, t) \, \mathrm{e}^{\mathrm{i}(\mathbf{Q}.\mathbf{r})} \, \mathrm{d}Q. \tag{16.27}$$

$I(Q, t)$ is not dependent on the direction of $\mathbf{Q}$ but in the exponential term $\exp(\mathrm{i}\mathbf{Q} \,.\, \mathbf{r})$ the direction of $\mathbf{Q}$ must be taken into account. This can effectively be done by averaging $\exp(\mathrm{i}Qr \cos \theta)$ over all directions in $\mathbf{Q}$ space. If this is done, conveniently in polar co-ordinates, it is found that eqn (16.27) becomes

$$G_s(r, t) = \frac{1}{2\pi^2} \int_0^{\infty} I(Q, t) \frac{\sin Qr}{Qr} Q^2 \, \mathrm{d}Q, \tag{16.28}$$

in which only the magnitude, and not the direction, of $\mathbf{Q}$ is significant. For the results from water in Fig. 367 the curves in Fig. 368 show how the value of $G_s(r, t)$ varies with r for various values of t, and these curves, in turn, are used to compute the value of $\overline{r_t^2}$ by means of eqn (16.22). This results in the curves in Fig. 365 showing the

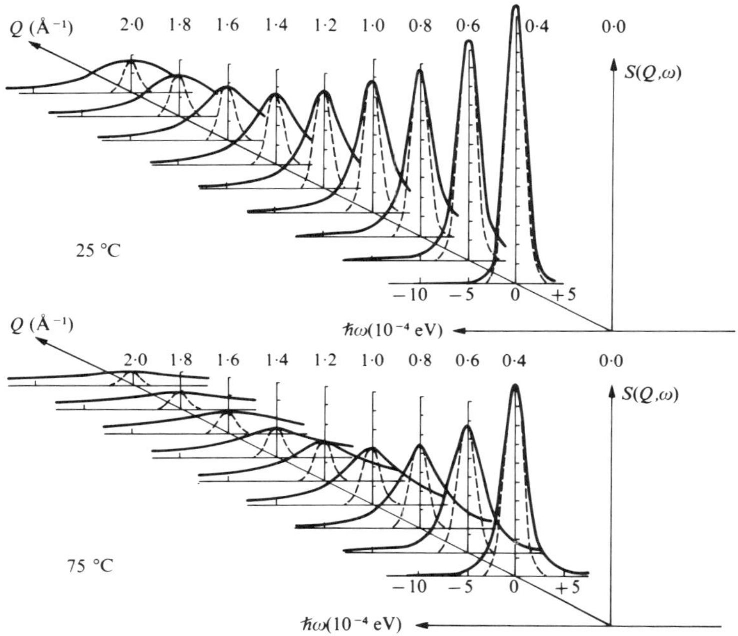

FIG. 367. A collection of $S(Q, \omega)$ versus ω curves for a variety of values of Q for liquid water at 25 °C and 75 °C. In the perspective drawing the three axes are $S(Q, \omega)$ vertical, ω left to right and Q from front to back. (Sakamoto *et al.* 1963.)

variation of $\overline{r_t^2}$ with t, which we have already discussed. In practice the accuracy to which r^2 can be determined by the above method is limited by the restricted range of variables over which the above integrations can be carried out, and a different mathematical approach was used by Sakamoto *et al.* in order to improve the accuracy of the experimental points shown in the Figure.

An alternative approach is to examine the width in ω, the energy change, of the Q–ω curve, and Fig. 369 presents some results for water, due to Larsson *et al.* (1960). For small values of Q, which correspond to large times and averaging over large distances, the width $\Delta\omega$ is proportional to Q^2, in agreement with expectation from a simple diffusion model, for which

$$\Delta\omega = 2DQ^2,$$

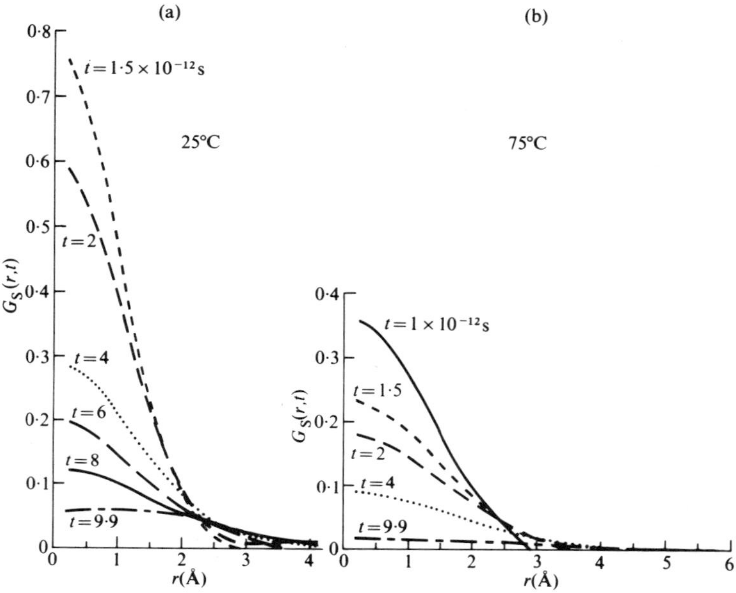

FIG. 368. The value of the self-correlation function $G_s(r, t)$ as a function of r for various values of t, for liquid water at (a) 25 °C, (b) 75 °C. (Sakamoto *et al.* 1962.)

where D is the diffusion coefficient. At large values of Q, which correspond to the local order over short distances, this model does not hold. On the perfect-gas model $\Delta\omega$ should become proportional to Q, which is consistent with Fig. 369. Equally compatible, however, are other models, such as the model of jump diffusion, which supposes that progression of an atom is by a succession of finite jumps of average length l_0 at intervals of time τ_0. From the value of D, given by the initial slope of the curve in Fig. 369 and of that value to which $\Delta\omega$ is tending at large values of Q, it is possible to deduce both l_0 and τ_0. The values are close to 2 Å and 10^{-12} s respectively for a wide variety of liquids (Egelstaff 1967, p. 131).

Many other liquids have been studied, and probably liquid sodium (Cocking 1963) and liquid argon (Dasannacharya and Rao 1965) are worthy of particular mention. Both are simple liquids, but in each case the cross-section is coherent and incoherent in roughly equal parts. For sodium the components are 1·55 barns and 1·85 barns respectively, and for argon they are 0·5 barns and 0·4 barns.

Consequently it is necessary to postulate adequate models of both the atomic structure and the individual atomic motions in order to interpret the scattering, and it is correspondingly difficult to deduce the details of either process from the experimental measurements.

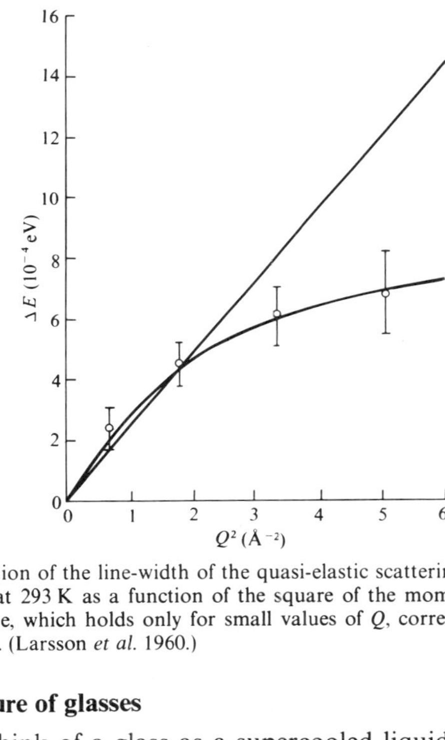

FIG. 369. Variation of the line-width of the quasi-elastic scattering curve (as in Fig. 366) for water at 293 K as a function of the square of the momentum transferred. The straight line, which holds only for small values of Q, corresponds to a simple diffusion model. (Larsson *et al.* 1960.)

16.8. Structure of glasses

We may think of a glass as a supercooled liquid or as a liquid with an extremely high viscosity which, with change of temperature, falls rapidly as the softening point is approached. When we think of the three-dimensional structure of a glass, and particularly of the way in which this can be investigated by studies of neutron scattering, it is the similarities and the differences between a glass and the extremes of solid and liquid which are important. Thus in a glass, as in a liquid, there is short-range atomic order but no long-range order. On the other hand, in a glass, as in a solid, the individual atoms have defined equilibrium positions, and they are not able to diffuse through the body of the material. In a liquid, as we have already seen, there is no strictly elastic scattering, but for a glass there is a meaningful distinction between elastic and inelastic scattering. As for a solid, the former tells us about the layout of the equilibrium atomic positions and the latter gives information concerning atomic vibratory motions. The vibratory motions of adjacent atoms are correlated and build up both longitudinal and transverse waves, but they are not simple plane waves, as they are for a normal solid, because the glass does not possess the long-range order which is characteristic of a crystal. We can still think in terms of the passage of phonons through a glass, but with increase of frequency the lifetime of the phonons becomes less and the concept becomes meaningless.

In an ordinary crystal the elastic component of scattering can be largely separated from the inelastic scattering in both X-ray and neutron experiments simply because the former gives rise to well-defined narrow peaks. For a glass the peaks are broadened, as shown in Fig. 370 for a neutron pattern, because of the absence of

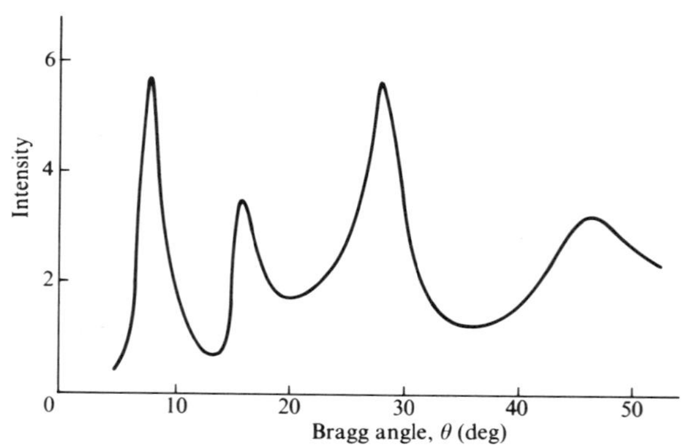

FIG. 370. A neutron diffraction pattern for vitreous silica at a wavelength of 1·13 Å. (Lorch 1970.)

long-range order, and the separation of the two types of scattering requires an energy-analysis. This analysis can be achieved with sufficient resolution only with neutrons. As with liquids the *total* scattering gives a picture of the instantaneous structure.

We recall that the most informative results about the liquid state have come from studies of simple monatomic liquids, particularly the liquid metals. Even for a diatomic liquid it is necessary to determine three partial structure factors, and the complication increases rapidly with the atomicity. By using different isotopes we saw that it was possible to isolate the partial structure factors for certain favourable diatomic liquids such as molten CuCl. Unfortunately the detailed circumstances are less favourable with glasses. Apart from selenium and vapour-deposited germanium and silicon, glasses are at least diatomic, and most of the simple glasses, such as SiO_2 and GeO_2, do not possess suitable isotopes of their constituent atoms. Nor do glasses consist of a large proportion of hydrogen, so it is not possible to assess the incoherent scattering separately and use it to give a frequency spectrum (but not details of the individual modes) as for the case of a polycrystalline aggregate. As a result of all these factors less progress has been made in the study of glasses than might at first sight have been expected, and we shall simply mention a few examples of some of the more recent work. For a more comprehensive survey the reader is referred particularly to articles by Leadbetter and his co-workers (Leadbetter and Wright 1972*a*, *b*; Leadbetter and Litchinsky 1970; Leadbetter 1973).

Some measurements of vitreous silica by Lorch (1970) are of particular interest in relation to our previous discussions in indicating the significance of the Placzek corrections, which take account of the departures from the 'static approximation' for ordinary neutron experiments. If we consider only the *elastic* part of the coherent scattering then eqn (16.14) will be replaced by

$$\left(\frac{\mathrm{d}\sigma_{\mathrm{coh}}}{\mathrm{d}\Omega}\right)_{\mathrm{elastic}} = N(\bar{b})^2 \int (G_{\mathrm{e}}(r)\,\mathrm{e}^{-2W}\,\mathrm{e}^{\mathrm{i}\mathbf{Q}.\mathbf{r}}\,\mathrm{d}r), \qquad (16.29)$$

without the necessity to invoke any approximation, since for the elastic scattering the magnitudes of $\boldsymbol{\kappa}$ and $\boldsymbol{\kappa}_0$ are equal. In this equation $\exp(-2W)$ is the Debye–Waller factor and $G_e(r)$ is the correlation function which relates the *equilibrium* positions of the atoms. Therefore if we correct for $\exp(-2W)$ we should be able to obtain from the elastic scattering an undistorted result for $G_e(r)$. Lorch's results for the variation of intensity with Q are shown in Fig. 371, both before and after correction for the Debye–Waller factor. It is believed that the small peak at 6·6 Å^{-1}, which had been noted in X-ray work but not in any previous neutron work, is significant; the peak does not appear, for example, in Fig. 370, where its expected position would be at $\theta = 35°$. Measurements of this kind, in which only the elastically scattered neutrons are recorded, are carried out on a triple-axis spectrometer in which the neutrons scattered by the sample fall on an analysing crystal set at such an inclination that the only neutrons which are reflected have a wavelength which is the same as that of the neutrons in the initial beam.

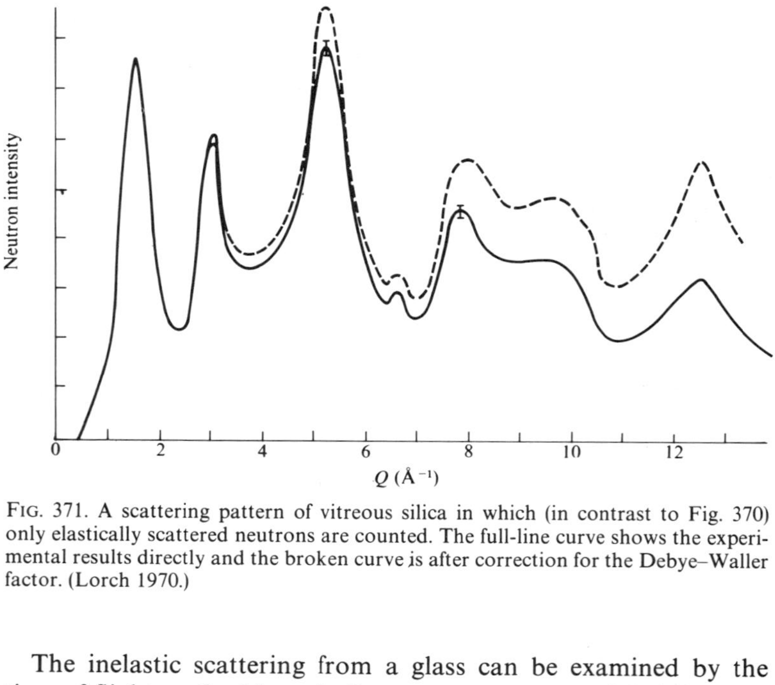

FIG. 371. A scattering pattern of vitreous silica in which (in contrast to Fig. 370) only elastically scattered neutrons are counted. The full-line curve shows the experimental results directly and the broken curve is after correction for the Debye–Waller factor. (Lorch 1970.)

The inelastic scattering from a glass can be examined by the time-of-flight method in a similar manner to a crystal. It is instructive to compare the results for polycrystalline and vitreous phases of the same material, as has been done by Leadbetter and Wright (1970) for BeF_2. The two sets of curves, measured for scattering angles of 30°, 60°, and 90°, are shown in Fig. 372, where the measured intensity of the neutrons is plotted as a function of energy. Below an energy of about 0·03 eV, corresponding to an optical wavenumber of 250 cm^{-1}, the positions and intensities of the peaks depend on the angle of scattering, indicating that coherent scattering effects are being observed. At higher values of energy there is little

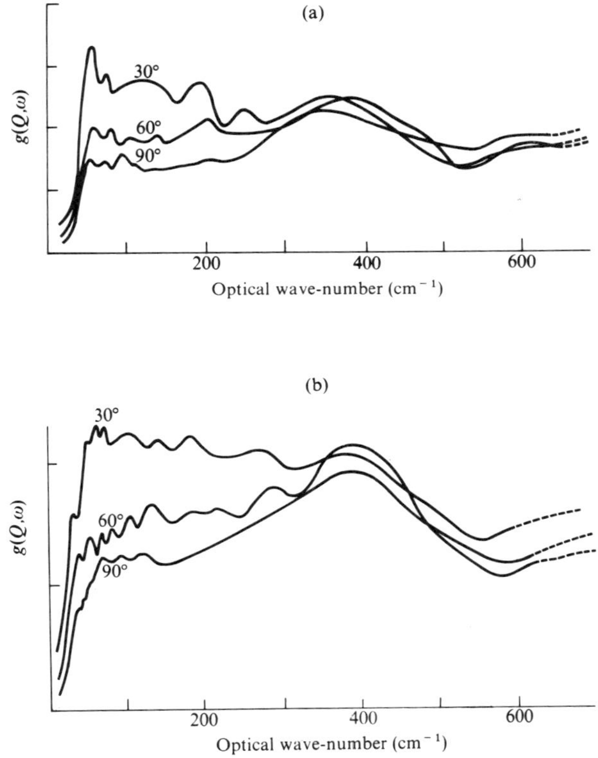

FIG. 372. A comparison of $g(Q, \omega)$ versus energy curves for (a) polycrystalline and (b) vitreous BeF_2 at scattering angles of 30°, 60°, and 90°. $g(Q, \omega)$ is defined as $Z(\omega) \sum (b_v)^2_{inc} u_v^2 \exp(-2W_v)/M_v$ and is derived, via eqn (9.6), from the experimentally measured $d^2\sigma/d\Omega\, dE$. (After Leadbetter and Wright 1970.)

dependence on scattering angle and little difference between the curves for the hexagonal crystalline form and the vitreous form. The scattering here is effectively incoherent and indicates a broad peak in the frequency distribution at about 380 cm^{-1}, in the same way as we were able to assess the frequency spectra of predominantly incoherent scatterers from polycrystalline data in Chapter 9. The conclusion from this type of experiment is that glasses, like crystals, possess quite well-defined phonon spectra but the dispersion curves broaden, relative to a crystal, with increasing q values because of the lack of long-range order in the glass.

16.9. Diffraction by gases

Very little work has been done to study the neutron scattering by gases since the original observations at Chalk River, Canada by Hurst and his co-workers (Alcock and Hurst 1949, 1951; Hurst and Alcock 1951) who studied deuterium, N_2, O_2, CH_4, and CF_4. For gases there is no correlation between the positions or orientations of neighbouring molecules, because of their larger distances of separation. As a result the scattering pattern is that of the molecule itself. The scattering is necessarily very weak and only became measureable by using pressures of about 30 atm and by making very careful correction for the scattering by the containing vessels.

Fig. 373 shows, as curve (a), the theoretical variation of scattering intensity with angle for a diatomic molecule, on the assumption

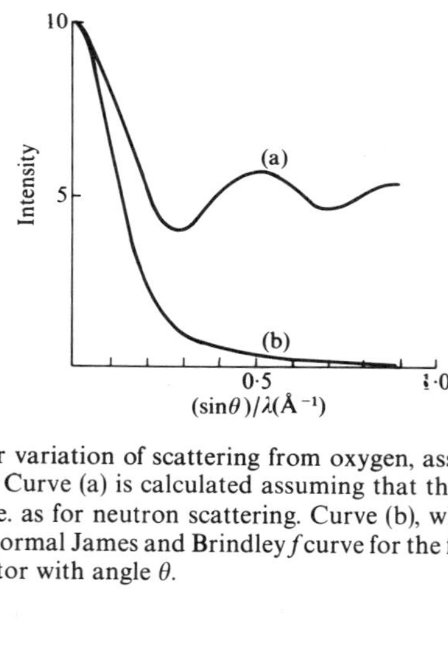

FIG. 373. The angular variation of scattering from oxygen, assuming an interatomic separation of 1·24 Å. Curve (a) is calculated assuming that the individual atoms act as point scatterers, i.e. as for neutron scattering. Curve (b), which is appropriate for X-rays, assumes the normal James and Brindley *f* curve for the fall-off of the electronic atomic scattering factor with angle θ.

that the two atoms which constitute the molecule act as point scatterers, that is they individually scatter isotropically. The curve is that appropriate to oxygen assuming an interatomic separation l_0 equal to 1·25 Å, and the first subsidiary peak occurs at an angle θ such that $(\sin\theta)/\lambda$ is equal to $0{\cdot}615/l_0$. We expect a curve of this form for the scattering of neutrons since the nuclear scattering is indeed isotropic. With X-rays, however, the scattering by each of the individual oxygen atoms falls off rapidly with $(\sin\theta)/\lambda$, since the electron cloud has dimensions of the same order as the wavelength, and when allowance is made for this the second curve in the figure is obtained. The subsidiary peaks, from which the interatomic separation could have been deduced, are smeared out.

These conclusions are borne out by experiment. Fig. 374 shows the experimental results for the neutron scattering by oxygen, given by Alcock and Hurst (1949), together with the much earlier X-ray

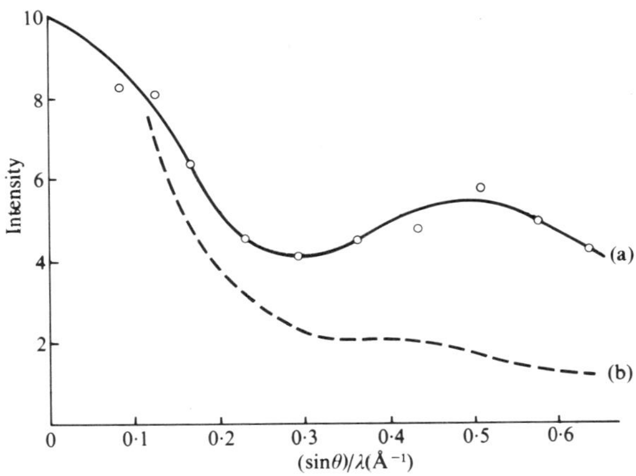

FIG. 374. The experimentally observed scattering by diatomic oxygen. Curve (a) is the neutron data obtained by Alcock and Hurst and shows a marked peak of intensity corresponding to an atomic separation of about 1·2 Å. In curve (b), which shows the X-ray results of Gajewski, the O—O peak is almost completely obliterated by the rapid fall-off of the atomic scattering factor.

results of Gajewski (1932). In the latter the peak characteristic of the l_0 value is scarcely discernible. A further example of how information may be obtained from the neutron measurements is provided by Fig. 375 which shows the results for CF_4. For a neutron wavelength of 1·06 Å two peaks at θ values of 19° and 32° respectively are observed, corresponding to a tetrahedral molecule with an edge of length 2·17 Å equal to the F—F bond and a radius of 1·33 Å equal to the C—F bond.

At the larger angles of scattering the observed intensity is slightly lower than that calculated by the usual semi-classical theory. The difference is ascribed to vibrations within the molecule and the amplitude of vibration can be deduced from the data, amounting to 0·045 Å in the case of CF_4 and giving a reduction in intensity of about 7 per cent. The modification of the theory is discussed by Alcock and Hurst (1951) and described in more detail by Pope (1952).

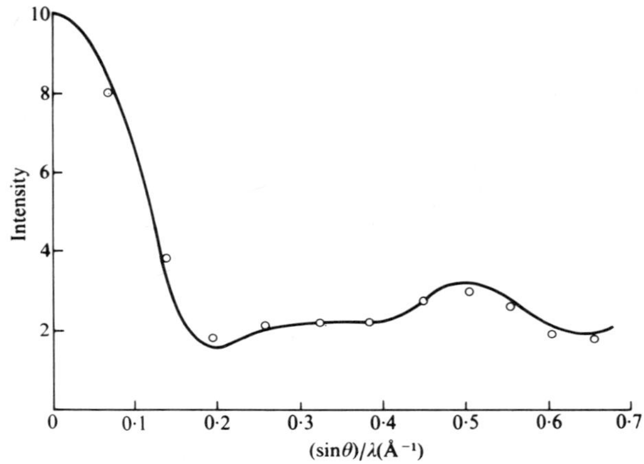

FIG. 375. The experimental data for the neutron scattering by carbon tetrafluoride in relation to a theoretical curve for a tetrahedral molecule with C—F bond equal to 1·33 Å. (Alcock and Hurst 1951.)

17
DEFECTS AND SMALL-ANGLE SCATTERING

WHEN neutrons are scattered elastically by a perfect three-dimensional arrangement of atoms the diffraction pattern will consist solely of Bragg peaks, assuming that the atoms do not give rise to either isotope or spin incoherence. If there are defects in the structure, consisting either of vacancies or of foreign atoms in normal or interstitial sites or of positional changes of atoms due to deformation, then diffuse scattering will occur. In the more general case this diffuse defect-scattering will be superimposed on the effects due to isotope- and spin-incoherence, which we discussed in Chapter 2, and also on effects due to multiple scattering, in particular double Bragg scattering. If the defect scattering can be assessed separately then its magnitude and spatial distribution will provide information about the nature and distribution of the defects.

17.1. Use of long-wavelength neutrons

The interpretation of the scattering is much simplified if neutrons of long wavelength are used, the wavelength being greater than twice the value of the maximum interplanar spacing in the diffracting sample. Under these circumstances no Bragg scattering can occur and the diffuse scattering can be identified much more confidently. With a sample of beryllium the neutron wavelength would have to be greater than 4 Å; with graphite it would need to be greater than 6·7 Å. It is of course true that X-rays of these same wavelengths would likewise be free from Bragg scattering but their absorption coefficients would be so large that accurate measurements of the diffuse scattering would not be possible. In this region of wavelength the linear absorption coefficient of copper, for example, is of the order of 20 000 cm^{-1}. Neutrons thus have a unique advantage for this kind of experiment and with many materials it is practicable to use wavelengths as great as 10 Å or 15 Å. In a very early application of the neutron technique Atkinson (1959) was able to show that the density of dislocations in cold-worked and fatigued samples of aluminium and copper was only about 1 per cent of what had previously been thought. This was because practically the whole of the small angle scattering previously measured with X-rays was due to double Bragg reflections.

17.2. Transmission measurements of defects

In practice two alternative procedures have been used. In the first of these, which is a simple but less informative method, the total defect cross-section is determined by measuring the transmission factor of the sample for a neutron beam of long wavelength. For a sample of thickness t the beam will be attenuated in intensity from $\mathscr{I}_0$ to $\mathscr{I}$, such that for λ greater than the Bragg cut-off

$$\mathscr{I} = \mathscr{I}_0 \exp\{-Nt(\sigma_a+\sigma_i+\sigma_{dis})-n_d t\sigma\}, \tag{17.1}$$

where σ_a, σ_i, and σ_{dis} are the cross-sections for true absorption, inelastic scattering, and isotopic and spin-disorder scattering and σ is the bound atom scattering cross-section for the atoms which constitute the material. N is the number of atoms per unit volume and n_d is the number of defects in the form of vacancies or interstitial atoms. As would be expected from Babinet's principle, both vacancies and interstitial atoms contribute to the scattering in exactly the same manner. A typical early measurement was made by Antal, Weiss, and Dienes (1955) to determine the number of defects produced in graphite after irradiation in a nuclear reactor. Their experimental procedure was to compare the transmission $\mathscr{I}_1$ for a sample of normal graphite with that $\mathscr{I}_2$ for an otherwise identical sample which had been irradiated. Thus

$$\mathscr{I}_2/\mathscr{I}_1 = \exp(-n_d t\sigma), \tag{17.2}$$

from which the value of n_d, the number of defects per unit volume, can be found. In a typical experiment the number of defects was found to be 2·6 per cent for an irradiation dose of $1{\cdot}1 \times 10^{20}$ neutrons cm^{-2}. Such a measurement is only accurate when σ is large compared with the cross-sections due to absorption, inelastic scattering, and isotope and spin disorder. Graphite is indeed a particularly favourable case, since the latter two cross-sections are zero and the sum of σ_a and σ_i is only 0·9 barns for a neutron wavelength of 8 Å, in comparison with a value of 4·7 barns for σ.

This simple interpretation of the defect scattering assumes that all the defects exist as isolated vacancies or interstitial atoms, whereas clustering of the defects will undoubtedly take place to some extent. At the same time distortions of the periodicity of the unit-cell structure will also take place, and more refined measurements aim to investigate these two features. Much more information is obtained by making the measurements over a wide range of wavelength and Fig. 376 shows an apparatus for doing this, employing a rotating mechanical monochromator bearing helical slots;

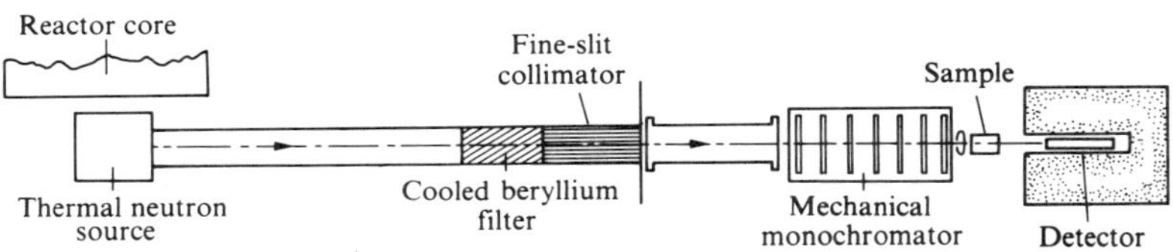

FIG. 376. Diagram of apparatus for measuring the transmission cross-section of defect samples as a function of wavelength. The wavelength is determined by the speed of rotation of the mechanical monochromator. (Courtesy A. L. Rodgers.)

the neutron wavelength can be varied by changing the speed of rotation. Variation of the scattering cross-section with wavelength is a general indication that the defects are not simply isolated vacancies or interstitial atoms. Scattering patterns can be computed for various possible models and compared with the experimental results. The method is useful for clusters up to about 30 Å in size; for larger defects than this it is possible to get more information by electron microscopy. Within this range of applicability it is possible to draw conclusions for samples in which there are at least 0·01 per cent of displaced atoms, using the high neutron fluxes which are now available. Some results are shown in Fig. 377 which indicates how the defect cross-section of a sample of germanium, irradiated by 10^{20} neutrons cm^{-2}, varies over a wavelength range from 6 Å to 14 Å (Clark, Mitchell, and Stewart 1971).

In principle the scattering from any form of vacancy, and its variation with neutron wavelength, can be calculated by integrating an expression of the type of eqn (2.20) over the atoms which form a cluster and then summing the cross-sections over all scattering angles. In practice this can be done with the aid of a computer for a wide range of postulated models, seeking eventual agreement with the curves such as Fig. 377, which are measured experimentally. It can be shown that the relaxation of the surrounding atomic

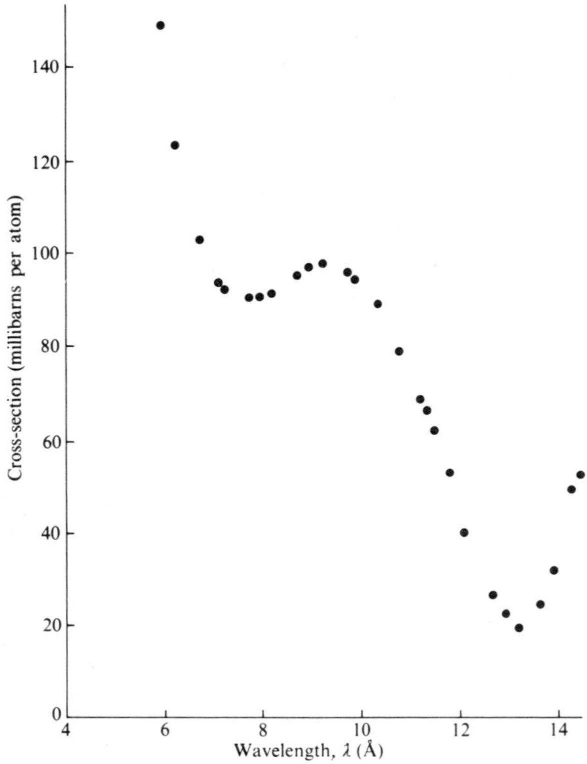

FIG. 377. Variation with wavelength of the defect cross-section of irradiated germanium allowing for transmutation. (After Clark, Mitchell, and Stewart 1971.)

positions is as important as the specific nature of the defects themselves and produces large effects on the cross-section. The importance of this was first pointed out by Mitchell and Wedepohl (1958) in relation to measurements on irradiated quartz and was considered quantitatively by Martin (1960), who calculated the effect on the cross-section of the inward movement of the nearest neighbours of a vacant site and their outward movement around an interstitial atom. Subsequently, Martin (1964) extended his calculations to more remote shells of neighbours, concluding that the general shape of the curve showing variation of total cross-section with wavelength could be found by considering the displacement of about 30 atoms. For a reasonable interpretation of the differential scattering cross-section it is probably necessary to consider the movements of 200 or 300 atoms.

Thus Fig. 378 (a) shows computed curves for a divacancy, i.e. two nearest-neighbour vacant sites, with and without relaxation. It is assumed for the latter curves that the nearest neighbours are relaxed by 5 per cent which amounts to 0·12 Å, towards the vacancies and in successive curves appropriate relaxations are extended respectively to second, third, fourth, and fifth shells. Most of the effect is dependent on the first and second shells. A second series of curves in Fig. 378 (b) assume a relaxation of 20 per cent, which equals 0·49 Å, for the nearest neighbours. It is evident that the enhanced relaxation assumed for the second set of curves produces a substantial difference in the variation of cross-section with wavelength. Calculations were made for a wide variety of defect models in germanium, including single interstitials, di-interstitials,

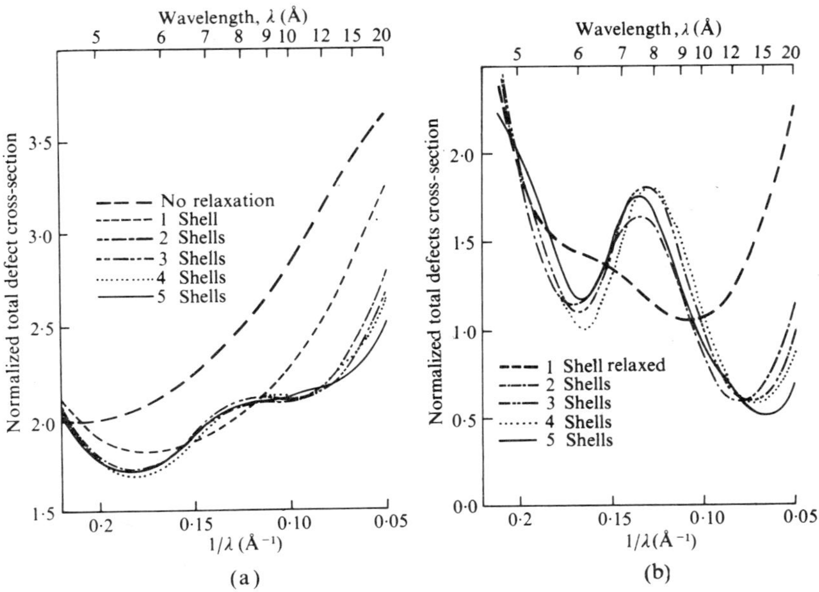

FIG. 378. Computed curves for the variation of cross-section with wavelength for a divacancy, i.e. two neighbouring vacant sites. The curves (a) show, successively, unrelaxed neighbours, first-shell relaxation by 5 per cent and appropriate extension of relaxation to second, third, fourth, and fifth shells. The curves (b) are for a first-shell relaxation by 20 per cent, with subsequent extension to second, third, fourth, and fifth shells. (Clark, Mitchell, and Stewart 1971.)

single vacancies, divacancies, multi-clusters, and amorphous regions, and the effect of a range of relaxations around the defects was included. By trial and error it is possible to attain optimum agreement with the experimental results; in all 160 models were considered. For the particular irradiated samples which were examined it was calculated that the defects were either divacancies or di-interstitials, corresponding to $8{\cdot}8 \times 10^{19}$ displaced atoms cm^{-3}, with quite an intricate lattice relaxation extending over the first few shells. The computed results are somewhat different for di-vacancies and di-interstitials, but the accuracy of the experimental measurements was not sufficient to distinguish between them. For most of the other likely models the calculations predicted precise features which were not observed experimentally. The accuracy of the measurements depends upon the magnitude of the cross-sections of the defects in relation to that of the unirradiated germanium. For the latter the cross-section is mainly due to true absorption and $\sigma = (2{\cdot}790 + 0{\cdot}904\,\lambda)$ barns when the wavelength λ is expressed in ångströms. By 10 Å, σ has risen to about 120 times as large as the defect cross-section of 90 millibarns which is evident in Fig. 377. A correction of about 10 millibarns has been applied to allow for the change of isotope-content by transmutation during the irradiation.

Similar measurements of gallium arsenide have been made by the same group of investigators, with some data being obtained out to a wavelength of 24 Å. In this case the interpretation of the variation with λ requires a quite different model from that postulated for germanium; the best fit is obtained with a vacancy–interstitial pair in which there is an inwards relaxation near the vacancy and an outwards relaxation near the interstitial. It is then deduced that the total number of displaced atoms is about $1{\cdot}5 \times 10^{21}$ cm^{-2}, which means 900 for each primary knock-on atom produced by irradiation. This value is in much closer agreement with what is predicted by radiation-damage theory than are the indirect values estimated from observations of changes in various physical properties.

17.3. Angular variation of scattering

A quite different, and more powerful, method of procedure is to study the variation of the defect scattering with the scattering angle 2θ, thus measuring the differential scattering cross-section.

This angular variation can be more directly interpreted in terms of a model of the defect than can the variation with wavelength of the *total* scattering. Again the early measurements were made with samples of graphite and were reported by Martin and Henson (1964). In order to eliminate errors due to inelastic scattering the scattered neutrons are energy-analysed by a time-of-flight measurement. Both the energy-analysis and the measurement of a differential, rather than the total cross-section, mean that the recorded rates of counting are low, and this is off-set to some extent by using a portion of moderator at low temperature, within the reactor, to increase the flux of long-wavelength neutrons. A diagram of this apparatus at A.W.R.E. Aldermaston, which is very similar to that originally employed by Martin and Henson, is shown in Fig. 379. Some measurements of irradiated graphite made with this apparatus are given in Fig. 380 in which a single smooth curve is found for

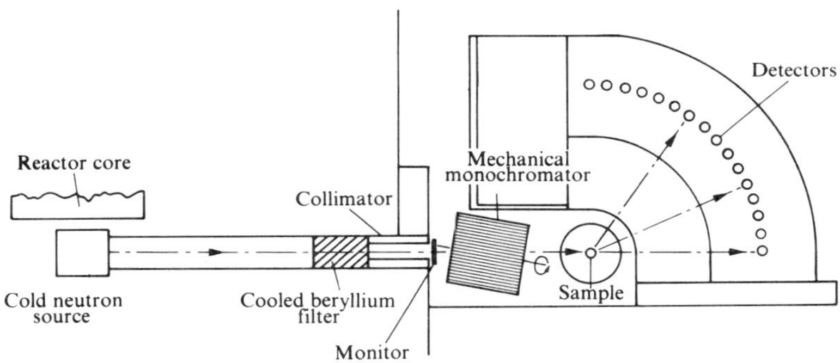

FIG. 379. Diagram of apparatus at A.W.R.E. Aldermaston for study of the angular variation of the defect scattering. (Courtesy of A. L. Rodgers.)

measurements made at wavelengths between 7·8 Å and 12 Å. From the average value of 30 mb sr^{-1} between Q values of 0·4 and 0·8 it is estimated that the defect concentration is about 7 per cent and that each primary carbon atom displaced by a neutron leads to about 120 secondary displaced carbon atoms. It will be noticed that for values of Q lower than 0·2 the neutron irradiation has produced a *decrease* in the scattering cross-section. This effect has been attributed to a decrease in the small-angle scattering brought about by the closure of small pores in the graphite by interstitial atoms produced during the irradiation.

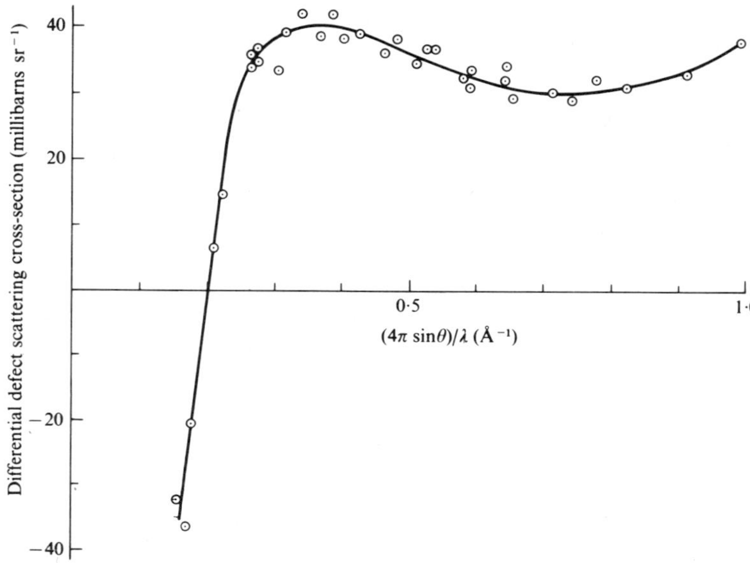

FIG. 380. Angular variation of defect scattering from irradiated graphite. (Mitchell and Stewart, unpublished.)

Ideally it should be possible to obtain directly a radial distribution function for the scattering by the defect by carrying out a Fourier inversion of the cross-section data. However, this is not very satisfactory in practice either because of the limited range of wavelength for which the total cross-section is measured or because of the limited range of values of $(\sin\theta)/\lambda$ for which the differential cross-section can be found. Consequently trial-and-error methods in terms of postulated defect models have been mainly used.

17.4. Non-stoichiometric compounds

Our examples of the use of neutrons for studying defects have been drawn from experiments with irradiated materials. It is also possible to study non-stoichiometric compounds in which the content of defects has been determined by chemical manipulation. In such materials, e.g. $Fe_{1-x}O$, there can be a shortage or excess of particular types of atom by several per cent. It is energetically unfavourable for the necessary structural adjustments to consist of random isolated point defects, and some kind of clustering invariably occurs. A determination of the nature of the clustering by neutron scattering observations may clarify the reasons for precise ranges of existence for particular structures among these non-stoichiometric compounds. A review of this kind of investigation has been given by Fender (1973), including details of the calculation of the cross-sections for different models. The problems are relatively more straightforward than for irradiated materials, since in the latter case interstitial atoms may be trapped in very metastable sites with much distortion of the positions of the neighbouring atoms. In the non-stoichiometric materials which we are now discussing the interstitial position can be predicted with greater confidence, and it is not necessary to explore such a wide range of models for the defect. A good example is provided by niobium carbide, which ideally has the rock-salt structure and the equiatomic composition NbC. However, up to 30 per cent of the carbon atoms may be absent. For a sample with the composition $NbC_{0\cdot86}$, Fig. 381 (Henfrey and Fender 1971) shows how the differential

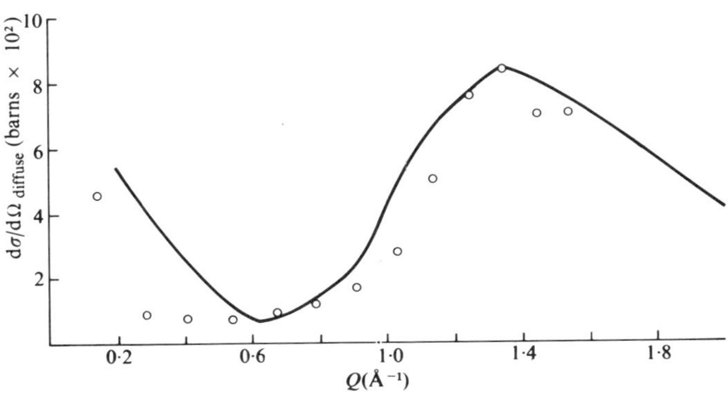

FIG. 381. Angular variation of diffuse scattering from $NbC_{0\cdot86}$ for long-wavelength neutrons. The open circles show the experimental points, and the curve is calculated for a model which postulates an excess of carbon atoms in the shells closest to the vacancies. (Henfrey and Fender 1971.)

scattering varies with $Q = (4\pi \sin \theta)/\lambda$ for long-wavelength neutrons. Supplementary measurements of the Bragg reflections give no evidence of atomic displacements but the differential scattering curve shows a marked peak for a value of Q equal to 1·4. This has been interpreted quantitatively for a model which assumes vacancy–

vacancy repulsion, i.e. it postulates that there is an excess of carbon atoms in the shells which are close to the vacancies. The curve in Fig. 381 is calculated for occupation probabilities of 0·96, 0·92, and 0·79 for the first, second, and third shells, respectively, around the vacancies. These figures are to be compared with 0·86, which would be the value if the distribution of carbon atoms was random.

Another interesting case, studied by Fender and Henfrey (1970), is that of niobium hydride, for which several per cent of hydrogen can be dissolved in the body-centred structure of the parent metal. It would be expected that the hydrogen atoms would be found in the tetrahedral sites of Fig. 382, and previous thermodynamic

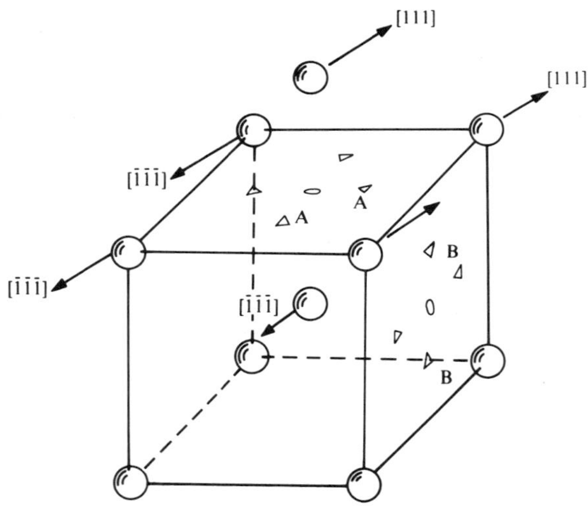

FIG. 382. The body-centred cubic structure of niobium hydride indicating the octohedral ○ and tetrahedral △ interstitial sites. A–A indicates two nearest-neighbour sites and B–B are next-nearest neighbours. The arrows indicate directions of displacement of niobium atoms, around the A–A positions, which have been investigated as a possible model of the defects. (Fender and Henfrey 1970.)

evidence suggested that they would form clusters. However, the variation of neutron scattering with angle, which was measured with deuterium to avoid the large incoherent scattering of hydrogen, suggests very strongly that the deuterium atoms are distributed singly and randomly. As Fig. 383 shows, the scattering is practically independent of angle, in contrast to the marked concentration in the forward direction which would be expected for clusters of various kinds.

A further type of material which may be studied is an alloy such as duralumin in which age-hardening occurs brought about by the segregation of the copper atoms on to particular planes, to give copper-rich patches known as Guinier–Preston zones, which may measure up to 50 Å in size and be separated from each other by a few hundred ångströms. The size and distribution of these regions can be inferred from the angular variation of the neutron scattering pattern, which can thus be used to follow their growth and to understand its dependence on temperature and time. Studies of this kind have been made by Raynal and Roth (unpublished) for Al–Mg alloys and by Clark and Meardon (1972) who have examined spinodal decomposition in Al–Zn alloys.

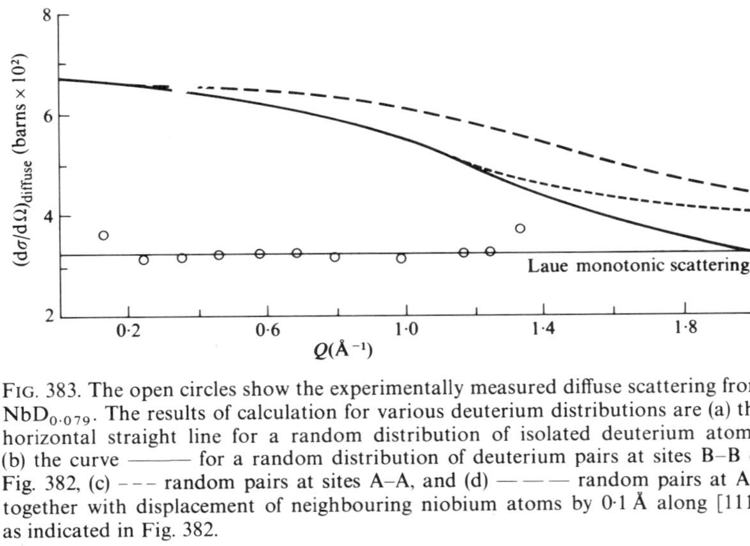

FIG. 383. The open circles show the experimentally measured diffuse scattering from $NbD_{0\cdot079}$. The results of calculation for various deuterium distributions are (a) the horizontal straight line for a random distribution of isolated deuterium atoms, (b) the curve ——— for a random distribution of deuterium pairs at sites B–B of Fig. 382, (c) - - - random pairs at sites A–A, and (d) — — — random pairs at AA together with displacement of neighbouring niobium atoms by 0·1 Å along [111], as indicated in Fig. 382.

17.5. Small-angle scattering

As we visualize the extension of the defect regions to larger sizes we eventually pass to the situation where the defects are distinct entities such as enclosed foreign crystallites or indeed the grains of a powder. In these cases also the pattern of scattered radiation will depend on the size and shape of these crystalline aggregates, and study of such materials with long-wavelength neutrons will often permit their size to be assessed free from complications due to double Bragg scattering. For particle dimensions which are large compared with interatomic distances the information will mainly be contained in the low-angle portion of the diffraction pattern, namely at scattering angles of a degree or even less. In normal circumstances this region of so-called small-angle scattering would not be capable of investigation because of the finite width of the undeviated beam with ordinary angles of collimation.

Measurements of Weiss

The small-angle scattering process for neutrons was first examined in detail by Weiss (1951), employing a well-collimated beam of neutrons, with an angular width of about 2′ of arc and which was then reflected by a glass mirror set at the critical angle of reflection for a wavelength of 3·0 Å. Thus all the reflected neutrons have wavelengths greater than 3·0 Å. In this region of the neutron spectrum the intensity is falling very rapidly with increase of λ, having a form

$$\nu_\lambda = \text{constant} \times \lambda^{-5}, \tag{17.3}$$

and the effective mean wavelength in the mirror-reflected beam is about 3·2 Å.

Weiss demonstrated that the process of small-angle scattering was a combination of diffraction and refraction, the predominant process depending on the differential phase change ϕ. ϕ is the difference in radians between the phase change for a neutron wave which traverses the particle diameter and that when the same distance is traversed in a vacuum, i.e.

$$\phi = (4\pi/\lambda)\ (1-n)R_{\mathrm{p}}, \tag{17.4}$$

where n is the refractive index of the particle and R_{p} is its radius. Interpretation is simple in two cases. When $\phi \gg 1$ refraction is predominant and the behaviour is described by the theory of von Nardroff (1926): when $\phi \ll 1$ only diffraction is important, as described by the theories of Rayleigh (1911) and Gans (1925). A full treatment, correlating these two extreme situations, was provided by van de Hulst (1946). Weiss studied the former case with a sample of powdered bismuth of 200–325 mesh for which the ϕ value was about 10, well within the refraction range. The result is indicated in Fig. 384 which displays a broadening of the beam in accordance with the equation

$$(\Omega^2 - \Omega_0^2)^{\frac{1}{2}} = \text{constant} \times (1-n)N_t^{\frac{1}{2}}. \tag{17.5}$$

Here Ω_0, Ω are the angular widths of the neutron beam before and after traversing the particles and N_t is the total number of particles traversed. If we substitute the value of $N\bar{b}\lambda^2/2\pi$ for $1-n$, from eqn (5.1) (p. 155), then the above equation becomes

$$(\Omega^2 - \Omega_0^2)^{\frac{1}{2}} = \text{constant} \times \frac{N\bar{b}\lambda^2}{2\pi} N_t^{\frac{1}{2}}. \tag{17.6}$$

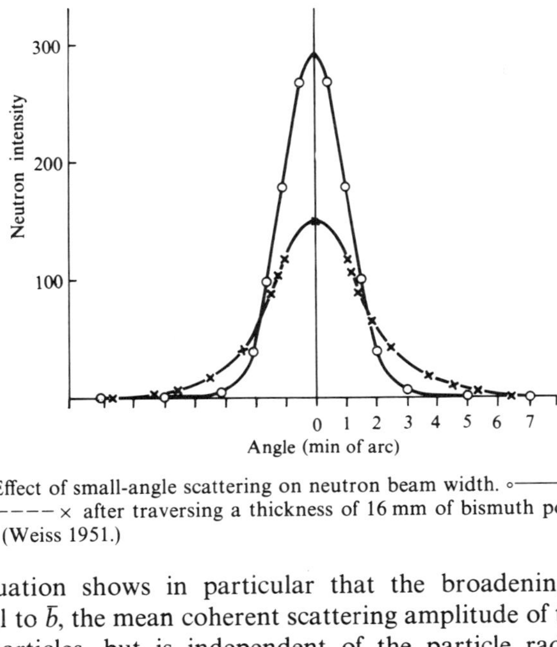

FIG. 384. Effect of small-angle scattering on neutron beam width. ○——○ incident beam ×-----× after traversing a thickness of 16 mm of bismuth powder (200–325 mesh). (Weiss 1951.)

This equation shows in particular that the broadening is proportional to $\bar{b}$, the mean coherent scattering amplitude of the atoms in the particles, but is independent of the particle radius. This analysis formed the basis of a method developed by Weiss, from a suggestion by Goldhaber, for determining the coherent scattering amplitude b. The principle of the method is to measure the angular broadening of the beam by a sample both before and after immersion in a liquid such as carbon disulphide. The powder is tightly packed for the experiment in a container made of very fine wire mesh, so that the liquid is able to flow in, and on account of its low viscosity flows freely around the individual particles but no powder is able to flow out. Thus N_t and λ in eqn (17.6) are the same in each experiment and $N\bar{b}$ for the particles is replaced by $N\bar{b}-(N\bar{b})_{CS_2}$, so that when $\bar{b}$ for the particles is positive, the same as it is for CS_2, the beam broadening is decreased. Similarly if $\bar{b}$ for the particles is negative the introduction of CS_2 will produce a broadening of the transmitted beam. We may write

$$\frac{(\Omega^2-\Omega_0^2)^{\frac{1}{2}}}{(\Omega^2-\Omega_0^2)^{\frac{1}{2}}_{CS_2}} = \frac{N\bar{b}}{N\bar{b}-(N\bar{b})_{CS_2}}, \tag{17.7}$$

and from a knowledge of N_{CS_2} and $\bar{b}_{CS_2}$ we can determine the magnitude of $\bar{b}$ for the particles, in addition to determining its sign. The determination will be most accurate if a liquid can be used with a refractive index close to that of the particles, thus producing a disappearance of the small-angle scattering. However, it is necessary to use a liquid with very small attenuation, and CS_2 and heavy water are among the few which are available. Table 34 shows

TABLE 34

Determination of coherent scattering amplitudes from small-angle scattering measurements

Element	Value of $\bar{b}$ (10^{-12} cm): From small-angle scattering	Value of $\bar{b}$ (10^{-12} cm): By Shull and Wollan
Nickel	$1{\cdot}17 \pm 0{\cdot}15$	1·03
Palladium	$0{\cdot}64 \pm 0{\cdot}03$	0·63
Antimony	$0{\cdot}56 \pm 0{\cdot}03$	0·54
Silicon	$0{\cdot}40 \pm 0{\cdot}02$	—
Arsenic	$0{\cdot}60 \pm 0{\cdot}03$	0·63

the values of $\bar{b}$ which were obtained by Weiss in this way, compared with those obtained earlier by Shull and Wollan (1951) using diffraction methods. It is interesting to note that the first determination of a value for silicon was made in this way. In addition to the above absolute determinations of $\bar{b}$, phase determinations were made for the first time for several elements.

As an example of small-angle scattering in which diffraction is predominant Weiss used particles of carbon black, with the reflecting mirror set at the critical angle for neutrons of wavelength 1·65 Å, yielding a beam with an effective mean wavelength of about 2·3 Å. Fig. 385 shows results for particles of diameter 600 Å and in this case the calculated value of the phase shift ϕ is of the order of 10^{-2} which is well within the diffraction range. Under these conditions it can be shown that the broadening of the beam by diffraction is given by

$$(\Omega^2 - \Omega_0^2)^{\frac{1}{2}} = \sqrt{3}\lambda_{\text{eff}}/\pi R_{\text{p}}. \qquad (17.8)$$

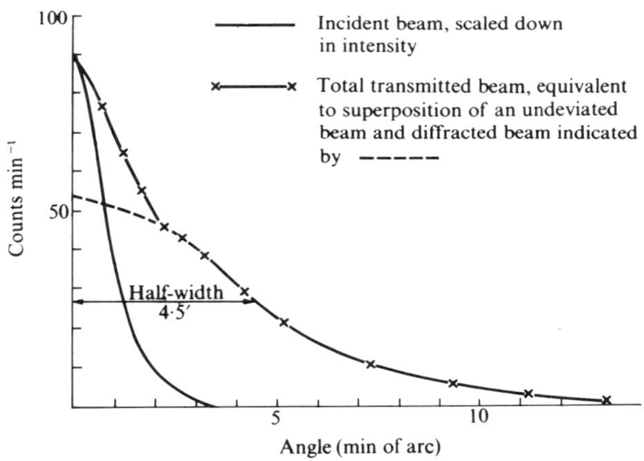

FIG. 385. Scattering of neutron beam by carbon-black particles of diameter 600 Å. ——— incident beam, scaled down in intensity. ×———× total transmitted beam, equivalent to superposition of an undeviated beam and diffracted beam indicated by ----. (Weiss 1951.)

Thus the broadening is inversely proportional to the radius of the particle, in contrast to the behaviour in the refraction region which was described by eqn (17.6).

As an example of small-angle magnetic scattering Weiss examined the study of unmagnetized iron by Hughes *et al.*, which we described in Chapter 15. For the domain size which was found, namely, $3{\cdot}4 \times 10^{-3}$ cm, and for a wavelength of 1·8 Å the value of the differential phase change ϕ is equal to about 6 rad. The analysis in terms of double refraction on a basis of geometrical optics is therefore justified.

Practical applications

Among classical examples of measurements of small-angle scattering we recall the study of the critical scattering of iron near the Curie point by Wilkinson and Shull (1956) which we examined in Chapter 9, with the conclusion that the regions of short-range order were about 12–25 Å. Later Lowde and Umakantha (1960) observed the scattering by spin waves in iron below an angle of about $\frac{1}{2}°$. In these cases, and of course for the measurements of Weiss, it was necessary for special steps to be taken to provide well-collimated incident beams in order that the angular region of interest would not be swamped by the incident beam itself. As a consequence the scattered intensities were very low and the precision of measurement was very limited. The advent of high-flux reactors, particularly when equipped with a cold source to provide neutrons of wavelength up to 20 Å and using guide tubes to ensure very low backgrounds from unwanted radiation, has completely altered the prospects and possibilities of measurements at very small angles and has led to the design of special instruments for this purpose. We have already mentioned that deviations from regular order, whether they be deviations of position or of composition, lead to diffuse scattering and from its study we have shown, for example, how relaxation around a vacancy extends to several shells of atoms and how an impurity in a magnetic alloy can effect the magnetic moments of atoms out to 10 Å. If we extend our measurements of diffuse scattering to smaller and smaller angles then we increase correspondingly the range of size of the defects, or other entities, which we can study. Not only may we measure the size and distribution of defects but we may also determine regularities of the same order of size. Thus Cribier *et al.* (1964) demonstrated that a Bragg peak found at an angle of 10′ in the diffraction pattern of niobium under a field of 1620 Oe, for neutrons of wavelength 4·7 Å, was due to the regular pattern of the lines of magnetic induction in the superconducting crystal. This angular position corresponds to a periodicity of 1500 Å and the periodicity was observed to vary with the magnitude of the applied field.

The Grenoble instrument for small-angle scattering, which is mounted on one of the assembly of guide tubes which we showed in Fig. 91 (see between pp. 10–11), has been described by Schmatz *et al.* (1974) and is capable of examining structural effects in the range from about 50–5000 Å. It operates over a wavelength range of 2–20 Å and obtains high angular resolution by using specimen–detector distances up to the immense value of 40 m. This means that values of **Q**, the momentum transfer vector of eqn (9.2) (p. 288), can be as small as 10^{-4} Å^{-1}, in contrast with, say, the minimum measured value of 0·2 Å^{-1} in the defect pattern shown in Fig. 380. Provision is also made in this instrument for the study of inelastic effects by making an energy analysis of the scattered neutrons. Experience with specialized instruments of this kind has revealed an immensely varied range of problems whose solution can be expedited by study of the small-angle scattering of neutrons and we shall mention some examples of these.

Ernst *et al.* (1971) have studied the small-angle scattering from ferromagnetic precipitates which can be grown within single crystals of copper containing 1 per cent of cobalt and conclude that they have a volume corresponding to spheres of about 150 Å radius. This work provides a good example from which to see how the effective dimensions of a defect or cluster can be inferred from the small-angle scattering pattern. Such particles as the above fall well within the diffraction range, in the terms we have considered, and Guinier (1963), following ordinary optical diffraction theory, has shown that the scattering falls off with angle 2θ according to the expression

$$\begin{aligned} I &= I_0 \exp(-16\pi^2 R^2 \theta^2 / 3\lambda^2) \\ &= I_0 \exp(-Q^2 R^2 / 3), \end{aligned} \tag{17.9}$$

where we have substituted Q for the quantity $(4\pi \sin\theta)/\lambda$, which for small angles is equal to $4\pi\theta/\lambda$. The important quantity in eqn (17.9) is R which, by analogy with classical dynamics, is called the radius of gyration of the particle for scattering. R^2 is defined as the mean value of the square of the distance of each scattering atom from an axis through the centre of gravity and parallel to the scattering vector, assuming that each atomic contribution to the summation of R^2 is weighted by the scattering amplitude of the atom. For a collection of identical particles which are oriented at random with respect to the incident beam the total intensity will be the sum of that from the individual particles but each of these, being differently oriented, will have a different value of R, unless the particles are spherical. In the expression given above R is averaged over all possible orientations of the particle and thus becomes given by

$$R^2 = \frac{\int \rho(r) r^2 \, \mathrm{d}V}{\int \rho(r) \, \mathrm{d}V}, \tag{17.10}$$

where $\rho(r)$ is the scattering amplitude per unit volume at any position $\mathbf{r}$ in the particle and the integrals are taken over the whole volume of the particle.

The value of $\rho(r)$ is measured relative to the surrounding medium, being the absolute value of ρ if the particles are in a vacuum but the differential value when they are immersed in a homogeneous medium. For magnetic particles in a non-magnetic matrix it will be the absolute magnetic scattering amplitude which is important.

It follows from eqn (17.9) that

$$\log I = \text{constant} - \frac{|Q^2| R^2}{3}. \tag{17.11}$$

so that the slope of a plot of the logarithm of the scattered intensity against $|Q|^2$ will be equal to $-R^2/3$. Fig. 386 shows such a plot for some of the measurements of Ernst *et al.* and from it a value of R of 104 Å can be deduced. Bearing in mind the well-known relation

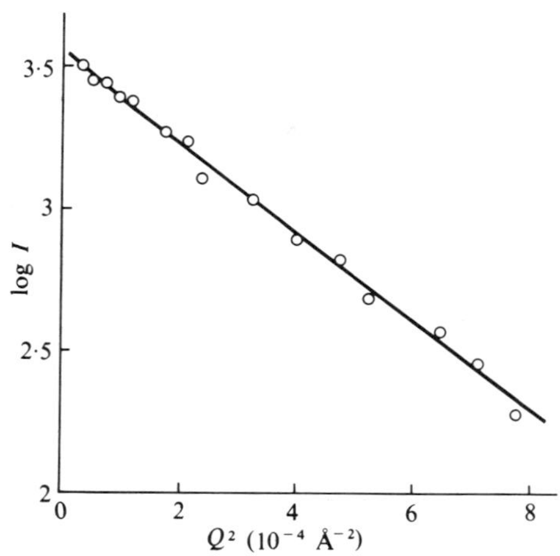

FIG. 386. The variation of the logarithm of the intensity of small-angle scattering with Q^2 for the measurements of ferromagnetic precipitates in copper with 1 per cent cobalt, by Ernst *et al.* (1971).

that the radius of gyration of a sphere is equal to $\sqrt{(\frac{3}{5})}$ of its radius we may regard the precipitates in the crystal as having the same volume as spheres of radius $104/(0{\cdot}6)^{\frac{1}{2}}$, i.e. about 140 Å.

Similarly, in studies of Ni_3Mn, Denkhaus *et al.* (1974) have deduced the existence of ferromagnetic clusters of about 1500 atoms, of elongated shape and with a major axis of about 40 Å.

In a very different field of research, these same ideas have been used to study the geometrical configurations of polymer molecules. Kirste *et al.* (1972) examined solutions of deuterated polystyrene in various solvents and also the solid hydrogeneous polymer. In such cases it is the scattering amplitude relative to the matrix which determines the small-angle scattering and in the solid or molten state of the bulk polymer this would normally be zero. However, because of the different scattering amplitudes of hydrogen and deuterium, the molecules in the solid can be revealed by deuterating a small fraction of them in what is mainly a hydrogeneous collection or, conversely, by examining a few hydrogeneous molecules in a largely deuterated matrix: the latter alternative has the advantage of resulting in a much-reduced incoherent background. In this way it was concluded for the glassy state of both polymethyl methacrylate (Kirste *et al.* 1974) and polystyrene (Wignall *et al.* 1974) that the structure of the molecules was that of the random coil, commonly found in solutions, rather than the compressed ball. The latter model would be expected to give a radius of gyration about 4 times smaller than that which was deduced from the experiments.

As a final example of these techniques we instance the study of biological macromolecules in solution by Stuhrmann (1974) and others, which will reveal the immense importance of hydrogen–deuterium substitutions. As in the case of polymers which we examined earlier, the scattering from the macromolecules depends on the differential scattering-length-density relative to the solvent. The intensity can be written as proportional to

$$\frac{\mathrm{d}\sigma}{\mathrm{d}\Omega}(Q) = \left| \sum_i b_i \exp(\mathrm{i}\mathbf{Q}\,.\,\mathbf{r}_i) - \rho_s \int_V \exp(\mathrm{i}\mathbf{Q}\,.\,\mathbf{r})\,\mathrm{d}r \right|^2, \qquad (17.12)$$

where the summation is taken over the atoms within the molecule and the integral is taken over the volume V of solvent which would otherwise occupy the molecular shape. ρ_s is the scattering-length-density of the solvent. We can of course vary the value of ρ_s by changing the hydrogen–deuterium ratio of the solvent. We have to remember however that when we change the hydrogen–deuterium ratio of the solvent we are also changing, albeit more slowly, the ratio for the macromolecules as well. In the case of haemoglobin, for example, there are 950 exchangeable protons in each molecule and it is estimated that equilibrium with the solvent has reached a degree of about 83 per cent after 12 hours. When the solvent contains 8 per cent of D_2O the value of ρ_s is zero and the small-angle scattering is then the same as would be given by the molecule *in vacuo*.

A further approach suggested by Stuhrmann is based on a reformulation of eqn (17.12) which divides the scattering into two components, first that due to the external shape of the molecule and represented by a form factor $F_F(\mathbf{Q})$ and, secondly, that due to the fluctuations of scattering within the molecule, i.e. to the departures of the scattering-length-density $\rho(r)$ from the average value ρ_m: this fluctuation scattering will have a separate form factor $F_S(\mathbf{Q})$. In these terms we can write eqn (17.12) in the form

$$\frac{d\sigma}{d\Omega}(Q) = (\rho_m V)^2\left(F_S(\mathbf{Q})+\frac{\rho_m-\rho_s}{\rho_m}F_F(\mathbf{Q})\right)^2, \qquad (17.13)$$

where $F_S(\mathbf{Q})$ and $F_F(\mathbf{Q})$ which we have just defined are given by

$$F_S(\mathbf{Q}) = \frac{1}{\rho_m V}\int_V [\rho(r)-\rho_m]\exp(i\mathbf{Q}\cdot\mathbf{r})\,dr \qquad (17.14)$$

and

$$F_F(\mathbf{Q}) = \frac{1}{V}\int_V \exp(i\mathbf{Q}\cdot\mathbf{r})\,dr. \qquad (17.15)$$

By making measurements in solvents containing different ratios of hydrogen to deuterium, thus varying the value of ρ_s, it is possible to determine separately the form factors F_S, F_F. It follows from eqn (17.12) that the scattered intensity extrapolated to $\mathbf{Q} = 0$ will be proportional to $|(\sum b_i)-\rho_s V|^2$. The square root of this extrapolated intensity will therefore be a linear function of ρ_s and will pass through a value of zero when $\sum b_i = \rho_s V$, i.e. when $\rho_m = \rho_s$. Some results for haemoglobin by Schelten *et al.* (1972) are shown in Fig. 387, where the intensity is zero for a heavy-water concentration of 40 per cent. For this concentration the value of

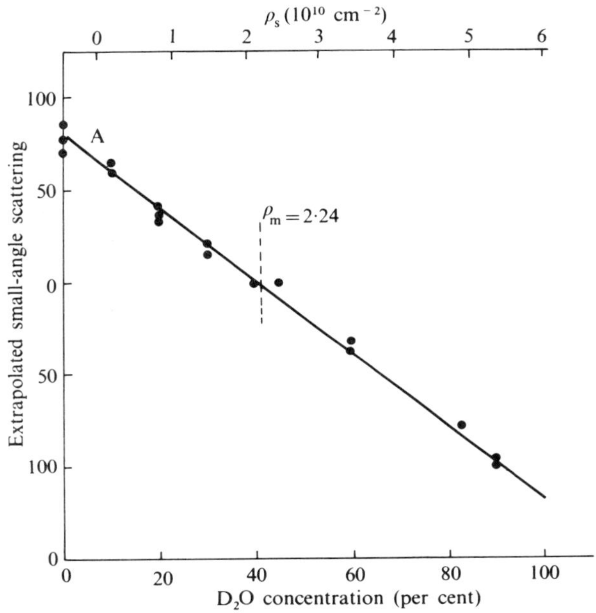

FIG. 387. The change of small-angle scattering from haemoglobin extrapolated to $Q = 0$, with the D_2O concentration in heavy–light water solvents. The scale at the top of the diagram gives the scattering-length density ρ_s of the solvent. A is the position at which $\rho_s = 0$. (From the results of Schelten *et al.* 1972.)

ρ_s is $2{\cdot}24 \times 10^{10}$ cm^{-2} and this, accordingly, is the value of ρ_m. We can utilize this value to compute the volume of a molecule as follows. Since the extrapolated small-angle scattering is zero

$$\sum b_i = \rho_s V = 2{\cdot}24 \times 10^{10}\ V. \tag{17.16}$$

For a molecule of haemoglobin in light water it is calculated that $\sum b_i$ is equal to 1543×10^{-12} cm. In a solvent containing 40 per cent of D_2O it is believed that 790 of the 950 exchangeable protons will have attained the 0·4 value of the deuterium–hydrogen ratio which is appropriate to the solvent and will thus have increased their scattering lengths by an average amount of $0{\cdot}4\ (0{\cdot}667 + 0{\cdot}374) \times 10^{-12}$ cm, where we recall the values of b for deuterium and hydrogen. Thus for a molecule of haemoglobin in a solvent containing 40 per cent D_2O

$$\sum b_i = 1543 + (790 \times 0{\cdot}4 \times 1{\cdot}041)\ \text{cm}$$
$$= 1872 \times 10^{-12}\ \text{cm}.$$

Hence from eqn (17.16) the value of V, the volume occupied by a molecule of haemoglobin, is 83 000 Å^3. We can also note in Fig. 387 that when the D_2O concentration is 8 per cent the value of ρ_s is zero and, hence, the point A in the figure indicates the scattering from a molecule in vacuo.

Having determined the value of ρ_m in the above manner we can proceed to evaluate the form factor F_F and the corresponding radius of gyration for the molecular shape. It follows from eqn (17.13) that when $(\rho_m - \rho_s)/\rho_m \to \infty$ the small-angle scattering will be determined solely by F_F. Hence if we determine, from the small-angle scattering curves, the value of the radius of gyration R for various solvent values ρ_s then the value R_g when $1/(\rho_m - \rho_s)$ is zero will correspond to the molecular shape F_F. Some results are shown in Fig. 388 for myoglobin from the work of Stuhrmann (1974), and it is found that R_g^2 is a linear function of $1/(\rho_m - \rho_s)$. It is deduced that the value of R_g^2 for the molecular shape is about 210 Å.

For a much more detailed survey of the applications of small-angle scattering which have become possible with the advent of very high-flux reactors the reader is referred to the article by Schmatz *et al.* (1974).

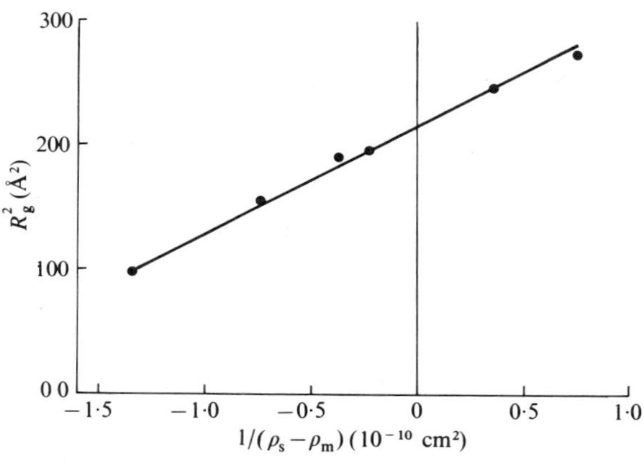

FIG. 388. Determination of the radius of gyration for the molecular shape of myoglobin from measurements in various D_2O–H_2O mixtures. (From the results of Stuhrmann (1974).)

LIST OF SYMBOLS

A COMPROMISE has been made between the conflicting aims of having a consistent set of symbols throughout the book and maintaining those employed in the original papers if this can be done without confusion. The symbols principally employed are given below.

A	mass number of nucleus
A_{hkl}	absorption factor
a	free scattering amplitude of a nucleus
a_0	side of unit cell
b	bound scattering amplitude
b_0	side of unit cell
c	velocity of light
c_0	side of unit cell
γ	magnetic moment of neutron in nuclear magnetons
d	interplanar spacing
e	electron charge
$\boldsymbol{\varepsilon}$	scattering vector
E	energy
F, F_{hkl}	structure amplitude factor for a unit cell
F_{magn}	magnetic contribution to F
F_{nucl}	nuclear contribution to F
f	form factor for magnetic scattering
f_L, f_S	individual form factors for orbital, spin contributions
η	standard deviation of mosaic blocks
g	acceleration due to gravity
g	Landé splitting factor
$g(\omega)$	frequency-distribution function
$g(r) = G_{\text{d}}(r, 0)$	pair-distribution function
$G_{\text{s}}(r, t)$	self-correlation function
$G_{\text{d}}(r, t)$	distinct-correlation function
$G(r, t)$	total-correlation function
θ	Bragg (glancing) angle
2θ	scattering angle
Θ	Debye temperature
h	crystallographic index
h	Planck's constant

$\hbar$	$h/2\pi$
$\mathbf{H}$	magnetic field
I	nuclear spin
J	quantum number of atom
J	exchange integral
k, k_B	Boltzmann's constant
k	Miller index
$\mathbf{K}$	magnetic spin vector
κ	wave number $2\pi/\lambda$
$\boldsymbol{\kappa}$	wave-number vector
λ	wavelength
$\boldsymbol{\lambda}$	neutron polarization vector
l	Miller index
L	quantum number
μ	linear absorption coefficient
μ_B	Bohr magneton
N_c	number of unit cells cm^{-3}
n	refractive index
p	magnetic scattering amplitude (including form factor)
$\mathbf{Q}$	momentum-transfer vector
Q	the crystallographic quantity $(N_c^2 F^2 \lambda^3/\sin 2\theta)$
$\mathbf{q}$	magnetic-interaction vector
$\mathbf{q}$	phonon wave vector
$\mathscr{R}^{\theta}$	integrated reflection (rotating crystal)
$\mathscr{R}^{\lambda}$	integrated reflection (white radiation–Laue)
S	spin quantum number
$\mathbf{S}_p$	projection of spin vector on reflecting plane
$S(Q, \omega)$	scattering law for liquid
$S(Q)$	structure factor for liquid
σ	total cross-section
$\mathscr{S}$	coherent scattering cross-section
s	incoherent scattering cross-section
$\hat{\boldsymbol{\sigma}}$	Pauli spin operator
$\boldsymbol{\tau}$	reciprocal lattice vector
W	Debye factor in e^{-2W} $(W = B\sin^2\theta/\lambda^2)$
ω	angular frequency of phonon

REFERENCES

PAGES on which work is cited appear in square brackets after each reference.

ABELN, O., DREXEL, W., GLASER, W., GOMPF, F., REICHARDT, W., and RIPFEL, H. (1968). In *Neutron inelastic scattering*, Vol. II, pp. 331–9. IAEA, Vienna. [140]

ABRAGAM, A. and BORGHINI, M. (1964). *Proc. low Temp. Phys.* **4**, 384. [541]

ABRAHAMS, S. C. (1963). *Phys. Rev.* **130**, 2230–7. [241]

—— WILLIAMS, H. J. (1963). *J. chem. Phys.* **39**, 2923–33. [278]

ADAIR, R. K. (1950*a*). *Phys. Rev.* **79**, 1018–9. [44]

—— (1950*b*). *Rev. mod. Phys.* **22**, 249–89. [163]

ADAMS, M. J., BLUNDELL, T. L., DODSON, E. J., DODSON, G. G., VIJAYAN, M., BAKER, E. N., HARDING, M. M., HODGKIN, D. C., RIMMER, B., and SHEAT, S. (1969). *Nature, Lond.* **224**, 491–5. [187]

AGERON, P. (1972). *Endeavour* **113**, 67–72. [8]

AHMED, N., CAMPBELL, S. J., and HICKS, T. J. (1974). *J. Phys. E.* **7**, 199–204. [541]

ALCOCK, N. Z. and HURST, D. G. (1949). *Phys. Rev.* **75**, 1609–10. [578]

—— —— (1951). *Phys. Rev.* **83**, 1100–5. [578]

ALDRED, B. K., EDEN, R. C., and WHITE, J. W. (1967). *Discuss. Faraday Soc.* **43**, 169. [375]

ALPERIN, H. A. (1961). *Phys. Rev. Lett.* **6**, 55–7. [227]

—— (1962). *J. phys. Soc., Japan* **17**, Suppl. B III, 12–15. [253]

—— BROWN, P. J., NATHANS, R., and PICKART, S. J. (1961). *Phys. Rev. Lett.* **8**, 237. [257]

ALVAREZ, L. W. and BLOCH, F. (1940). *Phys. Rev.* **57**, 111. [220]

ANDRESEN, A. F. (1958). *Acta crystallogr.* **11**, 612. [327]

—— HÄLG, W., FISCHER, P., and STOLL, E. (1967). *Acta Chem. scand.* **21**, 1543–54. [458]

ANTAL, J. J., WEISS, R. J., and DIENES, G. J. (1955). *Phys. Rev.* **99**, 1081–5. [582]

ANTONINI, B., FELCHER, G. P., MAZZONE, G., MENZINGER, F., and PAOLETTI, A. (1964). *Proc. Int. Conf. Magnetism, Nottingham*, pp. 288–90. Institute of Physics, London. [243]

ARNOLD, G. P., WENZEL, R. G., RABIDEAN, S. W., NERESEN, N. G., and BOWMAN, A. L. (1971). *J. chem. Phys.* **55**, 589–95. [326]

ASCARELLI, P. and CAGLIOTI, G. (1966). *Nuovo Cim.* **43B**, 375. [555]

ATKINSON, H. H. (1959). *J. appl. Phys.* **30**, 637. [581]

ATOJI, M. (1961). *J. chem. Phys.* **35**, 1950–60. [197, 259]

—— (1965). *Am. J. Phys.* **33**, 212–9. [281]

—— MEDRUD, R. C. (1959). *J. chem. Phys.* **31**, 332. [327]

—— RUNDLE, R. E. (1958). *J. chem. Phys.* **29**, 1306. [345]

BACON, G. E. (1951). *Proc. R. Soc.* **A209** 397–407. [76]

—— (1952). *Acta crystallogr.* **5**, 684–6. [386]

—— (1961). *Acta crystallogr.* **14**, 823–9. [402]

—— (1962). *Proc. phys. Soc.* **79**, 938–45. [423]

BACON, G. E. and COWLAM, N. (1969). *J. Phys.* C**2**, 238–51. [403]
—— —— (1970). *J. Phys.* C**3**, 675–86. [425]
—— CRANGLE, J. (1963). *Proc. R. Soc.* A**272**, 387–405. [428]
—— CURRY, N. A. (1956*a*). *Acta crystallogr.* **9**, 82–5. [344]
—— —— (1956*b*). *Proc. R. Soc.* A**235**, 552–9. [350]
—— —— (1962). *Proc. R. Soc.* A**266**, 95–108. [345]
—— —— WILSON, S. A. (1964). *Proc. R. Soc.* A**279**, 98–110. [349]
—— GARDNER, W. E. (1958). *Proc. R. Soc.* A**246**, 78–90. [344]
—— JUDE, R. J. (1973). *Z. Kristallogr.* **138**, 19–40. [88, 351, 366]
—— LOWDE, R. D. (1948). *Acta crystallogr.* **1**, 303–14. [69, 74]
—— MASON, E. W. (1967). *Proc. phys. Soc.* **92**, 713–25. [268, 429]
—— PEASE, R. S. (1953). *Proc. phys. Soc.* A**220**, 397–421. [333]
—— —— (1955). *Proc. phys. Soc.* A**230**, 359–81. [340]
—— PLANT, J. S. (1973). *J. Phys., F* **3**, 2003–20. [424, 427]
—— ROBERTS, F. F. (1953). *Acta crystallogr.* **6**, 57–62. [389, 503]
—— THEWLIS, J. (1949). *Proc. R. Soc.* A**196**, 50–64. [3]
—— TITTERTON, D. W. (1975). *Z. Kristallogr.* (In press). [347]
BARRETT, C. S. (1943). *Structure of metals.* McGraw-Hill, New York. [381]
——MUELLER, M. H., and HEATON, L. (1963). *Rev. scient. Instrum.* **34**, 847–8. [99]
BARTH, T. F. W. and POSNJAK, E. (1932). *Z. Kristallogr.* **82**, 325. [387]
BAUR, W. H. (1962). *Acta crystallogr.* **15**, 815. [347]
—— (1965). *Acta crystallogr.* **19**, 909–16. [347]
BEAUMONT, R. H., CHIHARA, H., and MORRISON, J. A. (1960). *Phil. Mag.* **5**, 188. [402]
BECKA, L. N. and CRUICKSHANK, D. W. J. (1963). *Proc. R. Soc.* A**273**, 435–54. [352]
BECKER, P. J. and COPPENS, P. (1974*a*). *Acta crystallogr.* A**30**, 129–47. [87]
—— —— (1974*b*). *Acta crystallogr.* A**30**, 148–53. [87]
BELOV, N. V., NERONOVA, N. N., and SMIRNOVA, T. S. (1957). *Sov. Phys. Cryst.* **2**, 311. [276, 281]
BENDT, P. J. and RUDERMAN, I. W. (1950). *Phys. Rev.* **77**, 575–9. [163]
BERTAUT, E. F. (1969). *J. Phys. Chem. Solids* **30**, 763–73. [450]
—— MERCIER, M. (1963). *Phys. Lett.* **5**, 27–9. [488]
—— FRUCHART, D., BOUCHARD, J. P., and FRUCHART, R. (1968). *Solid St. Commun.* **6**, 251–6. [485]
—— PAUTHENET, R. and MERCIER, M. (1967). *Phys. Lett.* **18**, 13. [488]
BETHE, H. (1937). *Rev. mod. Phys.* **9**, 79–244. [26]
—— (1949). *Phys. Rev.* **76**, 38. [168]
BIZETTE, H. (1951). *J. Phys. Radiat.* **12**, 161. [465]
—— SQUIRE, C. F., and TSAI, B. (1938). *C.r. hebd. Séanc. Acad. Sci. Paris* **207**, 449. [203, 444]
BLATT, J. M. (1948). *Phys. Rev.* **74**, 92. [168]
—— JACKSON, J. D. (1949). *Phys. Rev.* **76**, 18. [168]
BLECH, I. A. and AVERBACH, B. L. (1965). *Phys. Rev.* **137**A, 1113–16. [123]
BLOCH, F. (1936). *Phys. Rev.* **50**, 259. [191, 513]
—— (1937). *Phys. Rev.* **51**, 994. [524]
—— CONDIT, R. I. and STAUB, H. H. (1946). *Phys. Rev.* **70**, 972. [512]

BLOCH, F., HAMERMESH, M., and STAUB, H. (1943). *Phys. Rev.* **64**, 47. [512]
BLUME, M. (1963). *Phys. Rev.* **130**, 1670–6. [206]
—— FREEMAN, A. J., and WATSON, R. E. (1962). *J. chem. Phys.* **37**, 1245–53. [225, 245, 250, 538]
—— —— —— (1964). *J. chem. Phys.* **41**, 1874–8. [245, 538]
BOUTIN, H., SAFFORD, G. J., and BRAJOVIC, V. (1963). *J. chem. Phys.* **39**, 3135–40. [372]
BOZORTH, R. M. (1923). *J. Am. chem. Soc.* **45**, 2128. [328]
BRADLEY, A. J. (1935). *Proc. phys. Soc. Lond.* **47**, 879–99. [112]
—— JAY, A. H. (1932). *Proc. R. Soc.* A**136**, 210. [422]
BRAGG, W. H. (1915). *Phil. Mag.* (6) **30**, 305. [386]
BRAJOVIC, V., BOUTIN, H., SAFFORD, G. J., and PALEVSKY, H. (1963). *J. Phys. Chem. Solids* **24**, 617–24. [379]
BREIT, G. and WIGNER, E. (1936). *Phys. Rev.* **49**, 519. [25]
BROCKHOUSE, B. N. (1953*a*). *J. chem. Phys.* **21**, 961–2. [473]
—— (1953*b*). *Can. J. Phys.* **31**, 432–52. [183]
—— (1955). *Phys. Rev.* **99**, 601–2. [305]
—— (1958). *Bull. Am. phys. Soc.* **3**, 233 and AECL 1183. [292]
—— (1961). In *Inelastic scattering of neutrons in solids and liquids*, p. 147. IAEA, Vienna. [62]
BROCKHOUSE, B. N., BERGSMA, J., DASANNACHARYA, B. A., and POPE, N. K. (1963). In *Inelastic scattering of neutrons in solids and liquids*, Vol. I, p. 189–202. IAEA, Vienna. [570]
—— HURST, D. G. (1952). *Phys. Rev.* **88**, 542–7. [288]
BROWN, P. J. (1970). In *Thermal neutron diffraction* (ed. B. T. M. Willis), pp. 176–89. Clarendon Press, Oxford. [90, 206]
—— FRAZER, B. C. (1963). *Phys. Rev.* **129**, 1145–7. [506]
—— WILKINSON, C., FORSYTH, J. B., and NATHANS, R. (1965). *Proc. phys. Soc. Lond.* **85**, 1185–9. [403]
BRUGGER, R. M. (1965). In *Thermal neutron scattering* (ed. P. A. Egelstaff), pp. 53–96. Academic Press, London. [291]
—— BENNION, R. B., and WORLTON, T. G. (1967). *Phys. Lett.* **24A**, 714. [153]
BURAS, B. (1963). *Nukleonika* **8**, 259. [140]
—— (1967). *AEC–ENEA Seminar*, Santa Fe, New Mexico. [147]
—— LECIEJEWICZ, J. (1963). *Nukleonika* **8**, 75. [140]
—— —— (1964). *Phys. Stat. solidi* **4**, 349. [140]
BURAS, B. and O'CONNOR, D. (1959). *Nukleonika* **4**, 119. [140]
—— LECIEJEWICZ, J., NITO, W., SOSNOWSKA, I., SOSNOWSKI, J., and SHAPIRO, F. (1964). *Nukleonika* **9**, 523–37. [140]
BURGY, M. T., HUGHES, D. J., WALLACE, J. R., HELLER, R. B., and WOOLF, W. (1950). *Phys. Rev.* **80**, 953–60. [519]
—— RINGO, G. R., and HUGHES, D. J. (1951). *Phys. Rev.* **84**, 1160–4. [169]
BUSING, W. R. and LEVY, H. A. (1957). *J. chem. Phys.* **26**, 563–8. [364]
—— —— (1964). *Acta crystallogr.* **17**, 142–6. [364]
BYKOV, V. N., GOLOVKIN, V. S., AGEEF, N. V., LEVDIK, V. A., and VINOGRADOV, S. I. (1959). *Dokl. Akad. nauk. SSSR* **128**, 1153–6. [403]
CAGLIOTI, G. (1970). In *Thermal neutron diffraction* (ed. B. T. M. Willis), pp. 14–33. Clarendon Press, Oxford. [103]

CAGLIOTI, G. and RICCI, F. P. (1962). *Nucl. Instrum. Methods* **15**, 155–63. [103, 107]

—— PAOLETTI, A., and RICCI, F. P. (1958). *Nucl. Instrum. Methods* **3**, 223–8. [103]

—— —— —— (1960). *Nucl. Instrum. Methods* **9**, 195–8. [103]

CAMPBELL, S. J., AHMED, N., HICKS, T. J., EBDON, F. R. and WHEELER, D. A. (1974). *J. Phys. E.* **7**, 195–8. [541]

CARTER, R. S., PALEVSKY, H., and HUGHES, D. J. (1957). *Phys. Rev.* **106**, 1168–74. [295]

CASSELS, J. M. (1950). In *Progress in nuclear physics I* (ed. O. R. Frisch). Butterworth–Springer, London. [33, 51, 55]

—— (1951). *Proc. R. Soc.* A**208**, 527–34. [166, 289]

CHEETHAM, A. K., FENDER, B. F., and TAYLOR, R. I. (1971). *J. Phys. C.* **4**, 2160–5. [452]

CHIDAMBARAN, R., SEQUEIRA, A., and SIKKA, S. K. (1964). *J. chem. Phys.* **41**, 3616–22. [347]

CHILD, H. R., MOON, R. M., RAUBENHEIMER, L. J., and KOEHLER, W. C. (1967). *J. appl. Phys.* **38**, 1381–3. [250]

CHRISTIANSEN, C. (1884). *Weid. Ann.* **23**, 298. [175]

CHUMBLEY, L. C., DYER, R. F., and WALLIS, D. E. (1968). *J. Phys. E***1** 528–30. [Fig. 186]

CLAASSEN, A. (1930). *Phil. Mag.* (7) **9**, 57. [112]

CLARK, C. D., MITCHELL, E. W. J., and STEWART, R. J. (1971). *Cryst. Lattice Defects* **2**, 105–20. [583]

—— MEARDON, B. H. (1972). *Nature* (*Phys. Sci.*) **235**, 18–20. [591]

COCKING, S. J. (1963). In *Inelastic scattering of neutrons in solids and liquids*, pp. 227–36. IAEA, Vienna. [572]

—— WEBB, F. J. (1965). In *Thermal neutron scattering* (ed. P. A. Egelstaff), pp. 141–92. Academic Press, London. [14, 139]

COLLINS, M. F. and FORSYTH, J. B. (1963). *Phil. Mag.* **87**, 401–10. [443]

—— LOW, G. G. E. (1965). *Proc. phys. Soc. Lond.* **86**, 535–48. [441]

COMPTON, A. H. and ALLISON, S. K. (1935). *X-rays in theory and experiment*. Van Nostrand, New York. [101]

COOPER, M. J. (1970). In *Thermal neutron diffraction* (ed. B. T. M. Willis), pp. 51–67. Clarendon Press, Oxford. [90]

—— ROUSE, K. D. (1968). *Acta crystallogr.* A**24**, 405–10. [91]

—— —— (1970). *Acta crystallogr.* A**26**, 214–23. [85]

—— —— (1971). *Acta crystallogr.* A**27**, 622–8. [85]

—— —— WILLIS, B. T. M. (1968). *Acta crystallogr.* A**24**, 484–93. [85, 88]

COPPENS, P. (1967). *Science* **158**, 1577–9. [356]

—— (1970). In *Thermal neutron diffraction* (ed. B. T. M. Willis), pp. 82–100. Clarendon Press, Oxford. [356]

—— SABINE, T. M. (1969). *Acta crystallogr.* B**25**, 2442–51. [347, 360]

—— VOS, A. (1971). *Acta crystallogr.* B**27**, 146–58. [356]

—— SABINE, T. M., DELAPLANE, R. G., and IBERS, J. A. (1969). *Acta crystallog.* B**25**, 2451–8. [356]

CORLISS, L. M., ELLIOTT, N., and HASTINGS, J. M. (1956). *Phys. Rev.* **104**, 924–8. [449, 455]

—— —— —— SASS, R. L. (1961). *Phys. Rev.* **122**, 1402–6. [458].

CORLISS, L. M., ELLIOTT, N., and BROCKMAN, F. G. (1953). *Phys. Rev.* **90**, 1013–8. [389, 503]
—— —— WEISS, R. J. (1959). *Phys. Rev. Lett.* **3**, 211–2. [271, 402]
COULSON, C. A. (1961). *Valence* (2nd edn), p. 203. Oxford University Press. [348]
—— (1970). In *Thermal neutron diffraction* (ed. B. T. M. Willis), p. 78. Clarendon Press, Oxford. [354]
COX, D. E., TAKEI, W. J., and SHIRANE, G. (1963). *J. Phys. Chem. Solids* **24**, 405–23. [271, 474]
CRIBIER, D. and JACROT, B. (1960). *C.r. hebd Séanc. Acad. Sci. Paris* **250**, 2871–3. [304]
—— —— (1963). In *Inelastic scattering of neutrons in solids and liquids*, Vol. II, pp. 309–15. IAEA, Vienna. [305]
—— —— PARETTE, G. (1962). *J. phys. Soc., Japan* **17**, Suppl. BIII, 67–8. [314]
—— —— MADHAV, R. L., and FARNOUX, B. (1964). *Phys. Lett.* **9**, 106–7. [596]
—— ERICSON, M., JACROT, B., and GOBALAKICHENA, S. (1959). *C.r. hebd. Séanc. Acad. Sci. Paris* **248**, 1631–4. [305]
CURRY, N. A. (1965). *Proc. phys. Soc.* **86**, 1193–8. [251]
—— (1966). *Proc. phys. Soc.* **89**, 427–9. [251]
—— RUNCIMAN, W. A. (1959). *Acta crystallogr.* **12**, 674–8. [391]
DALE, D. H. and WILLIS, B. T. M. (1966). A.E.R.E. Report R 5195. [187]
DARWIN, C. G. (1914). *Phil. Mag.* **27**, 315 and **27**, 675. [68]
DASANNACHARYA, B. A. and RAO, K. R. (1965). *Phys. Rev.* **137**A, 417–27. [572]
DAVIES, F., RODGERS, A. L., TODD, M. C. J., ROSS, D. K., SANALAN, Y., WALKER, J., BELSON, J., CLARK, C. D., MITCHELL, E. W. J., and TUCKEY, G. S. G. (1968). In *Neutron inelastic scattering*, Vol. II, pp. 341–8. IAEA, Vienna. [139]
DEBYE, P. (1913). *Verh. dt. phys. Ges.* **15**, 678, 738, 857. [55]
—— (1914). *Annln. Phys.* **43**, 49. [55]
DENKHAUS, U., SCHELTEN, J., and SCHMATZ, W. (1974). *J. appl. Crystallogr.* **7**, 232. [599]
DIXON, A. E., WOODS, A. D. B., and BROCKHOUSE, B. N. (1963). *Proc. phys. Soc. Lond.* **81**, 973–4. [285]
DONNAY, G., CORLISS, L. M., DONNAY, J. D. H., ELLIOTT, N., and HASTINGS, J. M. (1958). *Phys. Rev.* **112**, 1917. [278]
DUCKWORTH, J. A. K., WILLIS, B. T. M., and PAWLEY, G. S. (1969). *Acta crystallogr.* A**25**, 482–4. [91, 352]
—— —— —— (1970). *Acta crystallogr.* A**26**, 263–71. [352]
DUNLOP, R. S. and SPEAKMAN, J. G. (1973). *Z. Kristallogr.* **138**, 100–12. [342]
EGELSTAFF P. A. (1951). *Nature, Lond.* **168**, 290. [58, 288]
—— [1967]. *Introduction to the liquid state.* Academic Press, London. [553, 572]
—— (1970). *Current problems in neutron scattering*, p. 56. C.N.E.N., Rome. [557]
—— HAYWOOD, B. C., and WEBB, F. J. (1967). *Proc. phys. Soc. Lond.* **90**, 681–96. [372]

Eisenhauer, C. M., Pelah, I., Hughes, D. J., and Palevsky, H. (1958). *Phys. Rev.* **109**, 1046–51. [300]
Ekstein, H. (1949). *Phys. Rev.* **76**, 1328–31. [524]
—— (1950). *Phys. Rev.* **78**, 731–2. [524]
Elliott, N. and Hastings, J. M. (1961). *Acta crystallogr.* **14**, 1018. [391]
Elliott, R. J. and Marshall, W. (1958). *Rev. mod. Phys.* **30**, 75–89. [314]
Ellis, W. C. and Greiner, E. S. (1941). *Trans. Am. Soc. Metals* **29**, 415. [381]
Ellison, R. D., Johnson, C. K., and Levy, H. A. (1971). *Acta Crystallogr.* **B27**, 333. [342]
Elsasser, W. M. (1936). *C.r. hebd. Séanc. Acad. Sci. Paris* **202**, 1029 [1]
Enderby, J. E., North, D. M., and Egelstaff, P. A. (1966). *Phil. Mag.* **14**, 961–70. [563]
Erickson, R. A. (1952). Thesis, Agricultural and Mechanical College of Texas. [463]
—— (1953). *Phys. Rev.* **90**, 779–85. [134, 212, 257, 462]
Ericson, M., de Gennes, P. G., Herpin, A., Jacrot, B., and Meriel, P. (1958). *J. Phys. Radiat.* **19**, 617. [466]
Ernst, M., Schelten, J., and Schmatz, W. (1971). *Phys. Stat. Solidi* (a) **7**, 469–76. [597]
Ewald, P. P. (1916). *Annln Phys.* **49**, 1 and **49**, 117. [68]
—— (1917). *Annln Phys.* **54**, 519. [68]
Fender, B. E. F. (1973). In *Chemical applications of thermal neutron scattering* (ed. B. T. M. Willis), pp. 250–69. Clarendon Press, Oxford. [589]
—— Henfrey, A. W. (1970). *Acta crystallogr.* **B26**, 1882–3. [590]
—— Jacobson, A. J., and Wedgwood, F. A. (1968). *J. chem. Phys.* **48**, 990–4. [254]
Fermi, E. (1950). *Nuclear physics*, University of Chicago Press. [156]
—— Marshall, L. (1947). *Phys. Rev.* **71**, 666–77. [78, 158]
—— Zinn, W. H. (1946). *Phys. Rev.* **70**, 103. [158]
Ferraris, G. and Franchini-Angela, M. (1972). *Acta crystallogr.* **B28**, 3572–83. [347]
Feshbach, H., Peaslee, D. C., and Weisskopf, V. F. (1947). *Phys. Rev.* **71**, 145–58. [25]
Fine, M. E., Greiner, E. S., and Ellis, W. C. (1951). *Trans. Am. Inst. Min. metall. Engrs* **191**, 56. [402]
Fleeman, J., Nicodemus, D. B., and Staub, H. H. (1949). *Phys. Rev.* **76**, 1774–81. [515]
Foldy, L. L. (1952). *Phys. Rev.* **87**, 693–6. [23]
—— (1958). *Rev. mod. Phys.* **30**, 471–81. [23]
Fowler, I. L. and Tunnicliffe, P. R. (1950). *Rev. scient. Instrum.* **21**, 734–40. [12]
Frazer, B. C. and Brown, P. J. (1962). *Phys. Rev.* **125**, 1283–91. [506]
—— Pepinsky, R. (1953). *Acta crystallogr.* **6**, 273–85. [333, 341]
—— Shirane, G., Cox, D. E., and Olsen, C. E. (1965). *Phys. Rev.* **140**A, 1448–52. [251]
Froman, P. O. (1952*a*). *Ark. Fys.* **4**, 191. [289]
—— (1952*b*). *Ark. Fys.* **5**, 53. [289]
Fruchart, D., Bertaut, E. F., Madar, R., and Fruchart, R. (1971*a*). *J. Phys.* (*Fr.*) C1 **32**, 876–7. [486]

FRUCHART, D., BERTAUT, E. F., MADAR, R., LORTHIOIR, G., and FRUCHART, R. (1971*b*). *Solid St. Commun.* **9**, 1793–7. [486]
GAJEWSKI, H. (1932). *Phys. Z.* **33**, 122. [579]
GANS, R. (1925). *Annln Phys.* **76**, 29. [592]
GARRETT, B. S. (1954). Oak Ridge National Lab. Rep. No. 1745. [345]
DE GENNES, P. G. (1958). *J. Phys. Chem. Solids* **4**, 223–6. [305, 313]
—— (1959). *J. Phys. Chem. Solids* **6**, 43–5. [314]
—— VILLAIN, J. (1960). *J. Phys. Chem. Solids* **13**, 10–27. [314]
GOLDBERGER, M. L. and SEITZ, F. (1947). *Phys. Rev.* **71**, 294–310. [162]
GOLDSMITH, H. H., IBSER, H. W., and FELD, B. T. (1947). *Rev. mod. Phys.* **19**, 259–97. [163]
GOMPF, F., REICHARDT, W., GLASER, W., and BECKURTS, K. H. (1968). In *Neutron inelastic scattering*, Vol. II, pp. 417–28. IAEA, Vienna. [153]
GORTER, C. J. (1948). *Physica* **14**, 504. [543]
GUINIER, A. (1963). *X-ray Diffraction*. Freeman and Co, London. [597]
GUREVICH, I. I. and TARASOV, L. V. (1968). *Low-energy neutron physics*. North Holland, Amsterdam. [59, 61]
HALBAN, H. and PREISWERK, P. (1936). *C.r. hebd. Séanc. Acad. Sci. Paris.* **203**, 73. [1]
HALLIDAY, D. (1950). *Introductory nuclear physics*. New York, Wiley. [47]
HALPERN, O. (1949). *Phys. Rev.* **76**, 1130–3. [524]
—— HAMERMESH, M., and JOHNSON, M. H. (1941). *Phys. Rev.* **59**, 981–96. [524]
—— HOLSTEIN, T. (1941). *Phys. Rev.* **59**, 960–81. [516, 521]
—— and JOHNSON, M. H. (1939). *Phys. Rev.* **55**, 898–923. [32, 191, 205, 463, 512, 531]
HAMERMESH, M., RINGO, G. R., and WATTENBERG, A. (1952). *Phys. Rev.* **85**, 483. [24]
—— SCHWINGER, J. (1946). *Phys. Rev.* **69**, 145. [47]
HAMILTON, W. C. (1958*b*). *Phys. Rev.* **110**, 1050–7. [501]
—— IBERS, J. A. (1968). *Hydrogen bonding in solids*. New York, Benjamin. [347]
—— LA PLACA, S. J. (1968). *Acta crystallogr.* **B24**, 1147–56. [355]
HASTINGS, J. M. (1960). *Bull. Am. Phys. Soc.* **5**, 455. [403]
—— CORLISS, L. M. (1953). *Rev. mod. Phys.* **25**, 114–21. [388, 503]
—— —— (1956*a*). *Phys. Rev.* **102**, 1460–3. [134, 503]
—— —— (1956*b*). *Phys. Rev.* **104**, 328–31. [389, 503]
—— —— (1962). *Phys. Rev.* **126**, 556–65. [504]
—— —— KUNUMAURA, W., and LA PLACA, S. (1967). *J. Phys. Chem. Solids* **28**, 1089–92. [281]
—— ELLIOTT, N., and CORLISS, L. M. *Phys. Rev.* **115**, 13–17. [460]
HAVENS, W. W., JR., RAINWATER, L. J., WU, C. S., and DUNNING, J. R. (1948). *Phys. Rev.* **73**, 963. [163]
HAYWOOD, B. C. G. (1974). *Acta crystallogr.* A **30**, 448–53. [129]
—— LEAKE, J. W. (1972). AERE Report R 7216. [150]
HEATON, L. R., MUELLER, M. H., ADAM, M. F., and HITTERMAN, R. L. (1970). *J. appl. Crystallogr.* **3**, 289–94. [138]
HELMHOLZ, L. and ROGERS, M. T. (1939). *J. Am. chem. Soc.* **61**, 2590. [328]
HENFREY, A. W. and FENDER, B. E. F. (1971). See Fender (1973). [589]

HEPWORTH, M. A., JACK, K. H., PEACOCK, R. D. and WESTLAND, G. J. (1957). *Acta crystallogr.* **10**, 63–9. [496]
HERPIN, A., MERIEL, P., and VILLAIN, J. (1959). *C.r. hebd. Séanc. Acad. Sci. Paris* **249**, 1334. [424]
HEWAT, A. (1973*a*). *J. Phys. C* **6**, 2559–72. [131]
—— (1973*b*). *Nature, Lond.* **246**, 90–1. [132]
HIBDON, C. T. and MUEHLHAUSE, C. O. (1950). *Phys. Rev.* **79**, 44–5. [44]
—— —— SELOVE, W., and WOOLF, W. (1950). *Phys. Rev.* **77**, 730. [44]
HIRONE, T. and ADACHI, K. (1957). *J. phys. Soc., Japan* **12**, 156. [458]
HUBBARD, C. R., QUICKSELL, C. O., and JACOBSON, R. A. (1972). *Acta crystallogr.* A**28**, 236–45. [143]
HUGHES, D. J. (1954). *Neutron optics.* Interscience, New York. [31]
—— BURGY, M. T. (1951). *Phys. Rev.* **81**, 498–506. [524]
—— —— HELLER, R. B., and WALLACE, J. W. (1949). *Phys. Rev.* **75**, 565–9. [521]
—— —— RINGO, G. R. (1950). *Phys. Rev.* **77**, 291–2. [169]
—— WALLACE, J. R., and HOLTZMANN, R. H. (1948). *Phys. Rev.* **73**, 1277. [515, 521]
VAN DE HULST, H. C. (1946). Thesis: *Optics of spherical particles and light scattering by small particles.* Wiley, New York. [592]
HUME-ROTHERY, W., IRVING, H. M., and WILLIAMS, R. J. P. (1951). *Proc. R. Soc.* A**208**, 431. [398]
HURST, D. G. and ALCOCK, N. Z. (1951). *Can. J. Phys.* **29**, 36–58. [578]
HVOSLEF, J. (1958). *Acta Chem. Scand.* **12**, 1568. [391]
IYENGAR, P. K. (1965). In *Thermal neutron scattering.* (ed. P. A. Egelstaff), pp. 97–140. Academic Press, London. [297]
IZYUMOV, Yu. A. and OZEROV, R. P. (1970). *Magnetic neutron diffraction*, pp. 533–44. Plenum Press, New York. [278]
JACOBSON, A. J., TOFIELD, B. C., and FENDER, B. E. F. (1973). *J. Phys.* C**6**, 1615–22. [240]
JACROT, B. (1962). In *Pile neutron research in physics*, pp. 393–408. IAEA, Vienna. [139]
—— KONSTANTINOVIC, J., PARETTE, G., and CRIBIER, D. (1963). In *Inelastic scattering of neutrons in solids and liquids*, Vol. II, pp. 317–26. IAEA, Vienna. [315]
JAMES, R. W. (1948). *The optical principles of the diffraction of X-rays.* Bell, London. [56, 66]
—— (1963). *Solid St. Phys.* **15**, 53–220. [179]
—— BRINDLEY, G. W. (1932). *Phil. Mag.* **12**, 81. [196]
JANIK, J. A., JANIK, J. M., MELLOR, J. and PALEVSKY, H. (1964). *J. Phys. Chem. Solids.* **25**, 1091–8. [302]
JAUHO, P. and PIRILA, P. (1970). *Phys. Stat. solidi* **42**, 757–66. [48]
JOHNSTON, D. F. (1966). *Proc. phys. Soc. Lond.* **88**, 37–52. [251]
JONES, F. W. and SYKES, C. (1937). *Proc. R. Soc. A***161**, 440. [381]
JUDE, R. J. (1971). Thesis. University of Sheffield. [325]
KAMB, B., HAMILTON, W. C., LA PLACA, S. J., and PRAKASH, A. (1971). *J. chem. Phys.* **55**, 1934–45. [326]
KASPER, J. S. and ROBERTS, B. W. (1956). *Phys. Rev.* **101**, 537–44. [405]
—— WATERSTRAT, R. M. (1956). *Acta crystallogr.* **9**, 289–95. [393]

KHAN, D. C. and ERICKSON, R. A. (1970). *Phys. Rev.* **B1**, 2243–9. [229]

KIMURA, M., SUGAWARA, M., OYAMADA, Y., TOMIYOSHI, S., SUZUKI, T., WATANABE, N., and TAKEDA, S. (1969). *Nucl. Instrum. Methods* **71**, 102–10. [2, 154]

KIRSTE, R. G., KRUSE, W. A., and SCHELTEN, J. (1972). *Makromolek. Chem.* **162**, 299–303. [599]

—— —— —— (1974). *J. appl. Crystallogr.* **7**, 188. [599]

KITTEL, C. (1971). *Introduction to solid state physics* (4th edn), Wiley, New York. [311]

KLEINMAN, D. and SNOW, G. (1951). *Phys. Rev.* **82**, 952. [156]

KOCH, F. and COHEN, J. B. (1969). *Acta Crystallogr.* B **25**, 275–87. [453]

KOEHLER, W. C. (1965). *J. appl. Phys.* **36**, 1078–87. [411]

—— (1972). In *Magnetic properties of rare-earth metals* (ed. R. J. Elliott), pp. 81–128. Plenum Press, London. [411]

—— CABLE, J. W., WOLLAN, E. O., and WILKINSON, M. K. (1962). *Phys. Rev.* **126**, 1672–8. [415]

—— MOON, R. M. (1972). *Phys. Rev. Lett.* **29**, 1468–72. [41, 247, 416]

—— WILKINSON, M. K., CABLE, J. W., and WOLLAN, E. O. (1959). *J. Phys. Radiat.* **20**, 180–4. [466]

—— WOLLAN, E. O. (1952). *Phys. Rev.* **85**, 491–2. [165]

—— —— (1953). *Phys. Rev.* **92**, 1380–6. [246]

—— —— (1957). *J. Phys. Chem. Solids* **2**, 100–106. [477]

—— —— WILKINSON, M. K. (1960). *Phys. Rev.* **118**, 58–70. [481]

—— YAKEL, H. L., WOLLAN, E. O., and CABLE, J. W. (1964). *Phys. Lett.* **9**, 93–5. [488]

KOESTER, L. (1965). *Z. Phys.* **182**, 328–36. [173]

—— (1967). *Z. Phys.* **198**, 187–200. [174]

—— KNOPF, K. (1971). *Z. Naturf.* **26**a, 391–9. [178]

—— —— (1972). *Z. Naturf.* **27**a, 901–5. [178]

—— UNGERER, H. (1969). *Z. Phys.* **219**, 300–10. [175]

KOUVEL, J. S. and KASPER, J. S. (1965). *Proc. Int. Conf. Magnetism, Nottingham*, pp. 169–70. Institute of Physics, London. [432]

KRANENDONK, J. VAN and VAN VLECK, J. H. (1958). *Rev. mod. Phys.* **30**, 1–23. [310]

KREN, E. and KADAR, G. (1970). *Solid St. Commun.* **8**, 1653–5. [431]

—— —— PAL, L., and SZABO, P. (1966). Central Research Institute for Physics, Budapest KFKI 2811. [429]

KRIEGER, T. J. and NELKIN, M. S. (1957). *Phys. Rev.* **106**, 290–5. [302]

KROHN, V. E. and RINGO, G. R. (1966). *Phys. Rev.* **148**, 1303–11. [24]

KUNITOMI, N., YAMADA, T., NAKAI, Y., and FUJII, Y. (1969). *J. appl. Phys.* **40**, 1265–9. [407]

KUZNIETZ, M. and WEDGWOOD, P. A. (1972). *Acta crystallogr.* **28**, 655. [41]

VAN LAAR, B. (1965). *Phys. Rev.* **138**A, 584–7. [449]

—— SCHWEIZER, J., and LEMAIRE, R. (1966). *Phys. Rev.* **141**, 538–40. [450]

LAMPI, E. E., FREIER, G. D., and WILLIAMS, J. H. (1950). *Phys. Rev.* **80**, 853–6. [325]

LARSSON, K. E., DAHLBORG, U. and HOLMRYD, S. (1960). *Ark. Fys.* **17**, 369. [571]

LAX, M. (1951). *Rev. mod. Phys.* **23**, 287–310. [162]

LEADBETTER, A. J. (1973). In *Chemical applications of thermal neutron scattering* (ed. B. T. M. Willis), Clarendon Press, Oxford, pp. 146–71. [575]
—— LITCHINSKY, D. (1970). *Discuss. Faraday Soc.* **50**, 62–73. [575]
—— WRIGHT, A. C. (1970). *J. non-cryst. Solids* **3**, 329. [576]
—— —— (1972*a*). *J. non-cryst. Solids* **7**, 23–36. [575]
—— —— (1972*b*). *J. non-cryst. Solids* **7**, 37–52. [575]
LEBECH, B. and MIKKE, K. (1967). Risö Report R164, Danish Atomic Energy Commission. [146]
—— PADLO, I. (1962). *Naturwissenschaften* **49**, 373. [327]
LI, Y.-Y. (1955). *Phys. Rev.* **100**, 627–31. [447]
LINDQVIST, O. and LEHMANN, M. S. (1973). *Acta Chem. scand.* **27**, 85–95. [41]
LOMER, W. M. and LOW, G. G. E. (1965). In *Thermal neutron scattering* (ed. P. A. Egelstaff), pp. 1–52. Academic Press, London. [59, 204]
LONSDALE, K. (1947). *Mineral. Mag.* **28**, 14. [70]
LOOPSTRA, B. O. (1966). *Nucl. Instrum. Methods* **44**, 181–7. [110]
LORCH, E. (1970). *J. Phys.* **C3**, 1314–22. [574]
LOW, G. G. E. and COLLINS, M. F. (1963). *J. appl. Phys.* **34**, 1195–9. [436]
—— and DYER, R. F. (1961). In *Inelastic scattering of neutrons in solids and liquids*, pp. 179–97. IAEA, Vienna. [291]
LOWDE, R. D. (1950). *Rev. scient. Instrum.* **21**, 835–42. [141]
—— (1951). *Nature, Lond.* **167**, 243. [141]
—— (1952). *Proc. phys. Soc. Lond.* **A65**, 857–8. [289]
—— (1954). *Proc. R. Soc.* **A221**, 206–23. [289]
—— (1960). *J. nucl. Energy* A (*React. Sci.*) **11**, 69–80. [291]
—— Umakantha, N. (1960). *Phys. Rev. Letters* **4**, 452–4. [596]
LUSHIKOV, V. I., POKOTILOVSKY, YU. N., STRELKOV, A. V., and SHAPIRO, F. L. (1969). *JETP Letters* **9**, 40. [162]
LYONS, D. H., KAPLAN, T. A., DWIGHT, K., and MENYUK, N. (1962). *Phys. Rev.* **126**, 540–55. [505]
MCGUIRE, T. R., ARGYLE, B. E., SHAFER, M. W., and SMART, J. S. (1963). *J. appl. Phys.* **34**, 1345–6. [453]
MCREYNOLDS, A. W. (1951). *Phys. Rev.* **84**, 969–72. [168]
—— WEISS, R. J. (1951). *Phys. Rev.* **83**, 171–2. [168]
MAIER-LEIBNITZ, H. (1962). *Z. angew. Phys.* **14**, 738. [170]
—— SPRINGER, T. (1963). *J. nucl. Energy.* A/B **17**, 217–25. [161]
MARCINKOWSKI, M. J. and BROWN, N. (1961). *J. appl. Phys.* **32**, 375–86. [382]
MARSHALL, W. and LOVESEY, S. W. (1971). *Theory of thermal neutron scattering*. Clarendon Press, Oxford. [206, 253, 301, 531]
MARTIN, D. G. (1960). *Phil. Mag.* **5**, 1235–46. [584]
—— (1964). *J. Phys. Chem. Solids* **25**, 1005–13. [584]
—— HENSON, R. W. (1964). *Phil. Mag.* **9**, 659–72. [587]
MENEGHETTI, D. and SIDHU, S. S. (1957). *Phys. Rev.* **105**, 130. [424]
MELKONIAN, E. (1949). *Phys. Rev.* **76**, 1744. [325]
MITCHELL, D. P. and POWERS, P. N. (1936). *Phys. Rev.* **50**. 486. [1]
MITCHELL, E. W. J. and WEDEPOHL, P. T. (1958). *Phil. Mag.* **3**, 1280–6. [584]
MODRZEJEWSKI, A. and KOBLA, J. (1969). *Krist. Tech.* **4**, 135–48. [98]
MOOK, H. A. (1966). *Phys. Rev.* **148**, 495–501. [227, 236]
—— SHULL, C. G. (1966). *J. appl. Phys.* **37**, 1034–5. [235]

MOON, R. M. (1964). *Phys. Rev.* **136**A, 195–202. [227, 237]
—— KOEHLER, W. C., CABLE, J. W., and CHILD, H. R. (1972). *Phys. Rev.* **B5**, 997–1016. [538]
—— —— TREGO, A. L. (1966). *J. appl. Phys.* **37**, 1036–7. [239]
—— RISTE, T., and KOEHLER, W. C. (1969). *Phys. Rev.* **181**, 920–31. [530]
MOORE, F. H., WILLIS, B. T. M., and HODGKIN, D. C. (1967). *Nature, Lond.* **214**, 130. [360]
MOORE, M. J., KASPER, J. S., and MANZEL, J. H. (1968). *Nature, Lond.* **219**, 848–9. [2, 154]
MOTT, N. F. and MASSEY, H. S. W. (1949). *The theory of atomic collisions* (2nd edn). Clarendon Press, Oxford. [25, 26]
MUELLER, M. H., LANDER, G. H., and REDDY, J. F. (1974). *Acta crystallogr.* A**30**, 667–71. [41]
NARDOFF, R. VON (1926). *Phys. Rev.* **28**, 240. [592]
NATHANS, R. and PAOLETTI, A. (1959). *Phys. Rev. Lett.* **2**, 254–6. [219, 227, 237]
—— ALPERIN, H. A., PICKART, S. J., and BROWN, P. J. (1963). *J. appl. Phys.* **34**, 1182–6. [252, 257]
—— PICKART, S. J., and ALPERIN, H. A. (1960). *Bull. Am. phys. Soc.* **5**, 455. [227]
—— PIGOTT, M. T. and SHULL, C. G. (1958). *J. Phys. Chem. Solids* **6**, 38–42. [422]
—— RISTE, T., SHIRANE, G., and SHULL, C. G. (1958). M.I.T. Structure of Solids Group. Technical Rep. No. 4. AFOSR TR 58–1058. [220, 282, 474]
—— SHULL, C. G., SHIRANE, G., and ANDRESEN, A. (1959). *J. Phys. Chem. Solids* **10**, 138–46. [216, 220]
NÉEL, L. (1948). *Annln Phys.* **3**, 137. [446, 499]
—— (1949). *Annln Phys.* **4**, 249. [470]
NILSSON, N. (1957). *Ark. Fys.* **12**, 247. [91]
NISHIKAWA, S. (1915). *Proc. Tokyo maths. phys. Soc.* **8**, 199. [386]
NIX, F. C. and SHOCKLEY, W. (1938). *Rev. mod. Phys.* **10**, 1–71. [381]
NORTH, D. M., ENDERBY, J. E., and EGELSTAFF, P. A. (1968*a*). *J. Phys.* **C1**, 784–94. [557]
—— —— —— (1968*b*). *J. Phys.* **C1**, 1075–87. [556]
OBERTEUFFER, J. A., MARCUS, J. A., SCHWARTZ, L. H., and FELCHER, G. P. (1970). *Phys. Rev.* **B2**, 670–7. [406]
PAGE, D. I. (1973). In *Chemical applications of thermal neutron scattering* (ed. B. T. M. Willis), pp. 173–200. Clarendon Press, Oxford. [561]
—— EGELSTAFF, P. A., ENDERBY, J. E., and WINGFIELD, B. R. (1969). *Phys. Lett.* A**29**, 296. [560]
—— MIKA, K. (1971). *J. Phys.* **C4**, 3034–44. [563]
PALEVSKY, H. and HUGHES, D. J. (1953). *Phys. Rev.* **92**, 202–3. [314]
PALMBERG, P. W., DEWAMES, R. E., and VREDEVOE, L. A. (1968). *Phys. Rev. Lett.* **21**, 682–5. [209]
PASSELL, L., BLINOWSKI, K., BRUN, T., and NIELSEN, P. (1964). *J. appl. Phys.* **35**, 933–4. [316]
—— —— —— —— (1965). *Phys. Rev.* **139**, A 1866–76. [316]

PAULING, L. (1940). *The nature of the chemical bond*. Cornell University Press. [195]
PETERSON, S. W. and LEVY, H. A. (1951). *J. chem. Phys.* **19**, 1416–8. [82]
—— —— (1952*a*). *J. chem. Phys.* **20**, 704–7. [328]
—— —— (1952*b*). *Phys. Rev.* **87**, 462–3. [48]
—— —— (1953). *Phys. Rev.* **92**, 1082. [326]
—— —— (1957). *J. chem. Phys.* **26**, 563. [344]
—— —— SIMONSEN, S. H. (1953). *J. chem. Phys.* **21**, 2084–5. [333]
—— —— —— (1954). *Phys. Rev.* **93**, 1120–1. [333]
PLACZEK, G. (1952). *Phys. Rev.* **86**, 377–88. [555]
POPE, N. K. (1952). *Can. J. Phys.* **30**, 597. [580]
POPOVICI, M. and GELBERG, D. (1966). *Nucl. Instrum. Methods* **40**, 77. [103]
POWERS, P. N. (1938). *Phys. Rev.* **83**, 641. [512]
PRINCE, D. R., ROWE, J. M., RUSH, J. J., PRINCE, E., HINKS, D. G., and SUSMAN, S. (1972). *J. chem. Phys.* **56**, 3697–702. [392]
PRINCE, E. (1957). *Acta crystallogr.* **10**, 554–6. [504]
—— and TREUTING, R. G. (1956). *Acta crystallogr.* **9**, 1025–8. [504]
RABI, I. I., ZACHARIAS, J. R., MILLMAN, S., and KUSCH, P. (1936). *Phys. Rev.* **53**, 318. [220]
RAINWATER, L. J., HAVENS, JR., W. W., DUNNING, J. R., and WU, C. S. (1948). *Phys. Rev.* **73**, 733. [166]
RAYLEIGH, LORD (1911). *Proc. R. Soc.* A. **84**, 25. [592]
RIETVELD, H. M. (1969). *J. appl. Crystallogr.* **2**, 65–71. [93, 130]
RIMMER, D. E. (1970). In *Thermal neutron diffraction*, pp. 211–20. (ed. B. T. M. Willis) Clarendon Press, Oxford. [256]
ROBERTSON, J. M. (1936). *Proc. R. Soc.* **A157**, 79. [350]
ROSE, M. E. (1949). *Phys. Rev.* **75**, 213. [47, 543]
ROTH, W. L. (1958*a*). *Phys. Rev.* **110**, 1333–41. [447]
—— (1958*b*). *Phys. Rev.* **111**, 772–81. [449]
—— (1960*a*). *J. appl. Phys.* **31**, 2000–11. [281]
—— (1960*b*). *Acta crystallogr.* **13**, 140–9. [452]
ROUSE, K. D. and COOPER, M. J. (1970). *Acta crystallogr.* **A26**, 682–91. [113]
ROWE, J. H., HINKS, D. G., PRICE, D. L., and SUSMAN, S. (1973). *J. chem. Phys.* **58**, 2039–42. [392]
RUDERMAN, I. W. (1949). *Phys. Rev.* **76**, 1572–84. [193]
RUNDLE, R. E. (1951). *J. Am. chem. Soc.* **73**, 4172–4. [327]
—— SHULL, C. G., and WOLLAN, E. O. (1952). *Acta crystallogr.* **5**, 22. [327]
RUSH, J. J. (1966). *J. chem. Phys.* **44**, 1722. [303]
—— and TAYLOR, T. I. (1965). In *Inelastic scattering of neutrons*, Vol. II, pp. 333–45. IAEA, Vienna. [302]
RUSSELL, H. N. and SAUNDERS, F. A. (1925). *Astrophys. J.* **61**, 38. [199]
SAILOR, V. L., FOOTE, H. L., LANDON, H. H., and WOOD, R. E. (1956). *Rev. scient. Instrum.* **27**, 26–34. [103]
SAKAMOTO, M., BROCKHOUSE, B. N., JOHNSON, R. G., and POPE, N. K. (1962). *J. phys. Soc., Japan*, **17**, Suppl. BII, 370–3. [568]
—— KUNITOMI, N., MOTOHASHI, H., and MINAKAWA, N. (1965). *Jap. J. appl. Phys.* **4**, 911. [103]
SATYA MURTHY, N. S., SAMANATHAN, C. S., BEGUM, R. J., SRINIVASAN, B. S., and MURTHY, M. R. L. N. (1969). *Ind. J. pure appl. Phys.* **7**, 546–52. [221]

SCATTURIN, V., CORLISS, L. M., ELLIOTT, N., and HASTINGS, J. M. (1961). *Acta crystallogr.* **14**, 19–26. [455]
SCHELTEN, J., SCHLECHT, P., SCHMATZ, W., and MAYER, A. (1972). *J. biol. Chem.* **247**, 5436–41. [601]
SCHERMER, R. I. and BLUME, M. (1968). *Phys. Rev.* **166**, 554–61. [48]
SCHLAPP, R. and PENNEY, W. G. (1932). *Phys. Rev.* **42**, 666. [455]
SCHMATZ, W., SPRINGER, T., SCHELTEN, J., and IBEL, K. (1974). *J. appl. Crystallogr.* **7**, 96–116. [596]
SCHOBINGER-PAPAMENTELLOS, P., FISCHER, P., VOGT, O., and KALDIS, E. (1973). *J. Phys.* **C6**, 725–37. [41]
SCHWINGER, J. (1948). *Phys. Rev.* **73**, 407. [23]
—— TELLER, E. (1937). *Phys. Rev.* **52**, 286–95. [47, 168]
SHARRAH, P. C. and SMITH, G. P. (1953). *J. chem. Phys.* **21**, 228–32. [544]
SHIRANE, G. (1959). *Acta crystallogr.* **12**, 282–5. [212, 282]
—— NATHANS, R., PICKART, S. J., and ALPERIN, H. A. (1964). *Proc. Int. Conf. Magnetism, Nottingham*, p. 223–7. Inst. Phys., London. [241]
—— PICKART, S. J., NATHANS, R., and ISHIKAWA, Y. (1959*a*). *J. Phys. Chem. Solids* **10**, 35. [472]
—— —— and ISHIKAWA, Y. (1959*b*). *J. phys. Soc., Japan* **14**, 1352. [472]
—— TAKEI, W. J. (1962). *J. phys. Soc. Japan* **17**, Suppl. BIII, 35–8. [274, 403]
SHULL, C. G. (1959). *J. Phys. Radiat.* **20**, 169–74. [502]
—— (1960). AFOSR TR 60–111. [103, 106]
—— (1967). *Trans. Am. cryst. Assoc.* **3**, 1–16. [23, 34]
—— (1968). *Phys. Rev. Lett.* **21**, 1585–9. [178]
—— FERRIER, R. P. (1963). *Phys. Rev. Lett.* **10**, 295–7. [24, 239]
—— MOOK, H. A. (1966). *Phys. Rev. Lett.* **16**, 184–6. [234]
—— SIEGEL, S. (1949). *Phys. Rev.* **75**, 1008–10. [381, 386]
—— SMART, J. S. (1949). *Phys. Rev.* **76**, 1256. [444]
——WILKINSON, M. K. (1953). *Rev. mod. Phys.* **25**, 100–107. [397, 405]
—— —— (1955). *Phys. Rev.* **97**, 304–10. [421, 433]
—— —— (1956). Unpublished. See Shull and Wollan (1956). [250]
—— WOLLAN, E. O. (1951). *Phys. Rev.* **81**, 527–35. [42, 45, 48, 97, 165, 594]
—— —— (1956). *Solid St. Phys.* **2**, 137–217. [481]
—— YAMADA, Y. (1962). *J. phys. Soc., Japan* **17**, Suppl. BIII, 1–6. [226]
—— STRAUSER, W. A., and WOLLAN, E. O. (1951). *Phys. Rev.* **83**, 333–45. [195, 211, 445, 452, 470]
—— WOLLAN, E. O., and KOEHLER, W. C. (1951). *Phys. Rev.* **84**, 912–21. [213, 397, 499, 528]
—— —— MORTON, G. A., and DAVIDSON, W. L. (1948). *Phys. Rev.* **73**, 842–7. [168, 324]
SIDHU, S. S., HEATON, LE ROY, ZAUBERIS, D. D., and CAMPOS, F. P. (1956). *J. appl. Phys.* **27**, 1040–2. [115]
SINCLAIR, R. N. and BROCKHOUSE, B. N. (1960). *Phys. Rev.* **120**, 1638–40. [308]
—— JOHNSON, D. A. G., DORE, J. C., CLARKE, J. H., and WRIGHT, A. C. (1974). *Nucl. Instrum. Meth.* **117**, 445–54. [154]
SLATER, J. C. (1930). *Phys. Rev.* **36**, 57. [202]
—— (1941). *J. chem. Phys.* **9**, 16–33. [340]

SMART, J. S. (1952). *Phys. Rev.* **86**, 968–74. [454]
SPEAKMAN, J. C. (1973). In *Molecular structure by diffraction methods*, Vol. 1, pp. 208–13 (eds G. A. Sim and L. E. Sutton), The Chemical Society, London. [326]
SQUIRE, C. F. (1939). *Phys. Rev.* **56**, 922. [444]
SQUIRES, G. L. (1952). *Proc. R. Soc.* A**212**, 192–206. [62, 166, 289]
—— (1954). *Proc. phys. Soc. Lond.* A**67**, 248–53. [314]
—— (1956). *Phys. Rev.* **103**, 304–12. [294]
STEINBERGER, J. and WICK, G. C. (1949). *Phys. Rev.* **76**, 994–5. [514]
STEINSVOLL, O., SHIRANE, G., NATHANS, R., BLUME, M., ALPERIN, H. A., and PICKART, S. J. (1967). *Phys. Rev.* **161**, 499–506. [248]
STEWART, R. F., DAVIDSON, E. R., and SIMPSON, W. T. (1965). *J. chem. Phys.* **42**, 3175–87. [355]
STILLER, H. H. and DANNER, H. R. (1961). In *Inelastic scattering of neutrons in solids and liquids*, p. 363, IAEA, Vienna. [297]
STIRLING, G. C. (1973). In *Chemical applications of thermal neutron scattering* (ed. B. T. M. Willis) p. 31–48. Clarendon Press, Oxford. [139, 298]
—— LUDMAN, C. J., and WADDINGTON, T. C. (1970). *J. chem. Phys.* **52**, 2730–5. [376]
STOLL, V. E. and HALG, W. (1965). *Z. angew. Maths. Phys.* **16**, 817. [109]
STONER, E. C. (1929). *Phil. Mag.* **8**, 250. [201]
STUHRMANN, H. B. (1974). *J. appl. Crystallogr.* **7**, 173–8. [599]
STURM, W. J. (1947). *Phys. Rev.* **71**, 757–76. [97]
SUTTON, R. B., HALL, T., ANDERSON, E. E., BRIDGE, H. S., DEWIRE, J. W., LAVATELLI, L. S., LONG, E. A., SNYDER, T., and WILLIAMS, R. W. (1947). *Phys. Rev.* **72**, 1147. [47, 168]
SYKES, C. (1935). *Proc. R. Soc.* A**148**, 422. [381]
SYRKIN, Y. K. and DYATKINA, M. E. (1950). *Structure of molecules and the chemical bond*, Butterworth, London. [200]
TOMBS, N. C. and ROOKSBY, H. P. (1950). *Nature, Lond.* **165**, 442. [449]
TRAMMELL, G. T. (1953). *Phys. Rev.* **92**, 1387–93. [224]
TURBERFIELD, K. C. (1970). In *Thermal neutron diffraction* (ed. B. T. M. Willis), pp. 34–50. Clarendon Press, Oxford. [149]
VAN HOVE, L. (1954*a*). *Phys. Rev.* **95**, 1374–84. [314, 546, 551]
—— (1954*b*). *Phys. Rev.* **93**, 268–9. [314, 546, 551]
—— (1954*c*). *Phys. Rev.* **95**, 249–62. [314, 546, 551]
VAN KRANENDONK, J. and VAN VLECK, J. H. (1958). *Rev. mod. Phys.* **30**, 1. [310]
VAN VLECK, J. H. (1932). *The theory of electric and magnetic susceptibilities*, Clarendon Press, Oxford. [201, 454]
—— (1939). *Phys. Rev.* **55**, 924–30. [305, 313]
—— (1941). *J. chem. Phys.* **9**, 85–90. [463]
VENKATARAMAN, G., USHA DENIZ, K., IYENGAR, P. K., VIJAYARAGHAVAN, P. R., and ROY, A. P. (1964). *Solid St. Commun.* **2**, 17–9. [378]
VERWEY, E. J. W., HAAYMANN, P. W., and ROMEIJN, F. C. (1947). *J. chem. Phys.* **15**, 181. [501]
—— HEILMANN, E. L. (1947). *J. chem. Phys.* **15**, 174. [387, 499]
VINEYARD, G. H. (1952). *Phys. Rev.* **85**, 633–6. [176]

VINEYARD, G. H. (1953). *Phys. Rev.* **91**, 239 and **96**, 93. [123]
DE VRIES, G. (1959). *J. Phys. Radiat.* **20**, 438. [402]
WALKER, C. B. and KEATING, D. T. (1963). *Phys. Rev.* **130**, 1726–34. [384]
WALLER, I. (1923). *Z. Phys.* **17**, 398. [55]
—— (1925). Dissertation. University of Uppsala. [55]
—— FROMAN, P. O. (1952). *Ark. Fys.* **4**, 183. [289]
WANG, S. P. and SHULL, C. G. (1962). *J. phys. Soc., Japan* **17**, Suppl. BIII, 340–1. [141]
WATSON, R. E. and FREEMAN, A. J. (1960). *Phys. Rev.* **120**, 1125–34. [237]
—— —— (1961). *Acta crystallogr.* **14**, 27–37. [237]
WEBB, F. J. and PEARCE, D. G. (1963). In *Inelastic scattering of neutrons in solids and liquids*, Vol. I, p. 83–94. IAEA, Vienna. [139]
WEBSTER, P. J. (1969). *Contemp. Phys.* **10**, 559–77. [427]
WEINSTOCK, R. (1944). *Phys. Rev.* **65**, 1–20. [51, 55]
WEISS, R. J. (1951). *Phys. Rev.* **83**, 379–89. [176, 592]
—— FREEMAN, A. J. (1959). *J. Phys. Chem. Solids* **10**, 147–61. [227]
WELLS, P. (1971). Thesis. University of Monash, Australia. [124]
WEST, J. (1930). *Z. Kristallogr.* **74**, 306. [335]
WHITAKER, M. D. (1937). *Phys. Rev.* **52**, 384. [193]
—— BEYER, H. G., and DUNNING, J. R. (1938). *Phys. Rev.* **54**, 771. [193]
—— BRIGHT, W. C. (1940). *Phys. Rev.* **57**, 1076. [193]
—— —— (1941). *Phys. Rev.* **60**, 280 [193]
WHITE, J. W. (1973). In *Chemical applications of thermal neutron scattering* (ed. B. T. M. Willis), pp. 49–77. Clarendon Press, Oxford. [372]
—— WRIGHT, C. J. (1970). *Chem. Commun.* 970–1. [369]
WIGNALL, G. D., SCHELTEN, J., and BALLARD, D. G. H. (1974). *J. appl. Crystallogr.* **7**, 190. [599]
WILD, S., GRIERSON, P., and JACK, K. H. (1972). *Special Ceramics* **5**, 385–95. [390]
WILKINSON, M. K., CABLE, J. W., WOLLAN, E. O., and KOEHLER, W. C. (1959). *Phys. Rev.* **113**, 497–507. [466]
—— SHULL, C. G. (1956). *Phys. Rev.* **103**, 516–24. [313, 596]
—— WOLLAN, E. O., KOEHLER, W. C., and CABLE, J. W. (1962). *Phys. Rev.* **127**, 2080–3. [198]
WILL, G., FRAZER, B. C., SHIRANE, G., COX, D. E., and BROWN, P. J. (1965). *Phys. Rev.* **140**A, 2139–42. [508]
—— PICKART, S. J., ALPERIN, H. A., and NATHANS, R. (1963). *J. Phys. Chem. Solids* **24**, 1679–81. [454]
WILLIAMS, J. M. and SCHNEEMEYER, L. F. (1973). *J. Am. chem. Soc.* **95**, 5780–1. [330]
WILLIAMS, W. G. (1973). Rutherford Lab. Rep. RL-73-034. H.M. Stationery Office. [541]
WILLIS, B. T. M. (1960). *Acta crystallogr.* **13**, 763–6. [103, 108]
—— (1962). *J. scient. Instrum.* **39**, 590. [129]
—— (1968). Private communication. [325]
—— (1970*b*). *Acta crystallogr.* **A26**, 396–401. [92]
WILSON, A. J. C. (1963). *Mathematical theory of X-ray powder diffractometry*. Philips' Technical Library, Amsterdam. [133]

WILSON, S. A. and COOPER, M. J. (1973). *Acta crystallogr.* **A29**, 90–1. [143]
WINDSOR, C. G. (1966). *Proc. phys. Soc. Lond.* **89**, 825–31. [313]
—— STEVENSON, R. W. H. (1966). *Proc. phys. Soc. Lond.* **87**, 501–4. [312]
WOLLAN, E. O., CHILD, H. R., KOEHLER, W. C., and WILKINSON, M. K. (1958). *Phys. Rev.* **112**, 1132–6. [496]
—— DAVIDSON, W. L. and SHULL, C. G. (1949). *Phys. Rev.* **75**, 1348–52. [326]
—— KOEHLER, W. C. (1955). *Phys. Rev.* **100**, 545–63. [135, 479]
—— KOEHLER, W. C., and WILKINSON, M. K. (1958). *Phys. Rev.* **110**, 638–46. [135, 466]
—— SHULL, C. G. (1948). *Phys. Rev.* **73**, 830–41. [9, 97]
—— —— MARNEY, M. C. (1948). *Phys. Rev.* **73**, 527–8. [141]
WOOD, J. H. and PRATT, G. W., JR. (1957). *Phys. Rev.* **107**, 995–1001. [226]
WOODS, A. D. B., BROCKHOUSE, B. N., SAKAMOTO, M., and SINCLAIR, R. N. (1961). In *Inelastic scattering of neutrons in solids and liquids*, IAEA, Vienna, p. 487–98. [297]
WORLTON, T. G., BRUGGER, R. M., and BENNION, R. B. (1968). *J. Phys. Chem. Solids* **29**, 435–8. [153, 476]
WYCKOFF, R. W. G. (1931). *The structure of crystals*. Reinhold, New York. [387]
YAFET, Y. and KITTEL, C. (1952). *Phys. Rev.* **87**, 290–4. [504]
YAMADA, T., KUNITOMI, N., NAKAI, Y., COX, D. E., and SHIRANE, G. (1970). *J. phys. Soc., Japan* **28**, 615–27. [407]
YOSHIMORI, A. (1959). *J. phys. Soc., Japan*, **14**, 807 [465]
ZACHARIASEN, W. H. (1945). *X-ray diffraction in crystals*. Wiley, New York. [69, 74, 101]
—— (1967). *Acta crystallogr.* **23**, 558–64. [84]
—— (1968*a*). *Acta crystallogr.* **A24**, 212–6. [85]
—— (1968*b*). *Acta crystallogr.* **A24**, 324–5. [85]
—— (1968*c*). *Acta crystallogr.* **A24**, 425–7. [85]
—— (1969). *Acta crystallogr.* **A25**, 102. [85]
ZEMLYANOV, M. G., KAGAN, YU., CHERNOPLEKOV, N. A., and CHICHERIN, A. G. (1963). In *Inelastic scattering of neutrons in solids and liquids*, Vol. II, pp. 125–44. IAEA, Vienna. [300]
ZENER, C. (1952). *Phys. Rev.* **85**, 324. [399]
ZINN, W. H. (1947). *Phys. Rev.* **71**, 752. [2]

AUTHOR INDEX

FOR JOINT PAPERS by four or more authors, a shortened form of reference in the text may result in the names of certain authors not appearing on the pages specified. In such cases their contribution can be recognized by consulting the full reference which is listed in the list of References.

Abeln, O., 140
Abragam, A., 541
Abrahams, S. C., 278
Adachi, K., 458
Adair, R. K., 44, 163
Adam, M. F., 138
Adams, M. J., 187
Ageef, N. V., 403
Ageron, P., 8
Ahmed, N., 541
Alcock, N. Z., 578
Aldred, B. K., 375
Allison, S. K., 101
Alvarez, L. W., 220
Alperin, H. A., 227, 241, 248, 253, 257, 454
Amer, J., 281
Andresen, A. F., 216, 220, 327, 458
Antal, J. J., 582
Antonini, B., 243
Argyle, B. E., 453
Arnold, G. P., 326
Ascarelli, P., 555
Atkinson, H. H., 581
Atoji, M., 197, 259, 281, 327, 345
Averbach, B. L., 123

Bacon, G. E., 3, 69, 74, 76, 88, 268, 333, 340, 344, 347, 349, 350, 351, 366, 386, 389, 390, 402, 403, 423, 424, 425, 428, 429, 503.
Baker, E. N., 187
Ballard, D. G. H., 599
Barrett, C. S., 99, 381
Barth, T. F. W., 387
Baur, W. H., 347
Beaumont, R. H., 402
Becka, L. N., 352
Becker, P. J., 87
Beckurts, K. H., 153
Begum, R. J., 221
Belov, N. V., 276, 281
Belson, J., 139
Bendt., P. J., 163
Bennion, R. B., 153, 476
Bergsma, J., 570
Bertaut, E. F., 450, 485, 486, 488
Bethe, H., 26, 168
Beyer, H. G., 193
Bizette, H., 203, 444, 465
Blatt, J. M., 168
Blech, I. A., 123
Blinowski, K., 316
Bloch, F., 191, 220, 512, 513, 524
Blume, M., 48, 206, 225, 245, 248, 250, 538
Blundell, T. L., 187
Borghini, M., 541
Bouchard, J. P., 485
Boutin, H., 372, 379
Bowman, A. L., 326
Bozorth, R. M., 328
Bradley, A. J., 112, 422
Bragg, W. H., 386
Brajovic, V., 372, 379
Breit, G., 25
Bright, W. C., 193
Brindley, G. W., 196
Brockhouse, B. N., 62, 183, 285, 288, 292, 297, 305, 308, 473, 568, 570
Brockman, F. G., 389, 503
Brown, N., 382
Brown, P. J., 90, 206, 252, 257, 403, 506, 508
Brugger, R. M., 153, 291, 476
Brun, T., 316
Buras, B., 140, 147
Burlet, P., 41
Burgy, M. T., 169, 519, 521, 524
Busing, W. R., 364
Bykov, V. N., 394, 403

Cable, J. W., 198, 415, 488, 538
Caglioti, G., 103, 107, 555

Campbell, S. J., 541
Carter, R. S., 295
Cassels, J. M., 33, 51, 55, 166, 289
Cheetham, A. K., 452
Chernoplekov, N. A., 300
Chicherin, A. G., 300
Chidambaran, R., 347
Chihara, H., 402
Child, H. R., 250, 496, 538
Christiansen, C., 175
Chumbley, L. C., Fig. 186
Claassen, A., 112
Clark, C. D., 139, 583, 591
Clarke, J. H., 154
Cocking, S. J., 14, 139, 572
Cohen, J. B., 453
Collins, M. F., 436, 441
Compton, A. H., 101
Condit, R. I., 512
Cooper, M. J., 85, 88, 90, 91, 113, 143
Coppens, P., 87, 325, 347, 356, 360
Corliss, L. M., 134, 271, 278, 281, 388, 389, 402, 449, 455, 458, 460, 503, 504
Coulson, C. A., 348, 354
Cowlam, N., 403, 425
Cox, D. E., 251, 271, 407, 474, 508
Crangle, J., 428
Cribier, D., 304, 305, 314, 315, 596
Cruickshank, D. W. J., 352
Curry, N. A., 251, 344, 345, 349, 350, 391

Dahlborg, U., 571
Dale, D. H., 187
Danner, H. R., 297
Darwin, C. G., 68
Dasannacharya, B. A., 570, 572
Davidson, E. R., 355
Davidson, W. L., 326
Davies, F., 139
Debye, P., 55
Delaplane, R. G., 356
Denkhaus, U., 599
DeWames, R. E., 209
Dienes, G. J., 582
Dixon, A. E., 285
Dodson, E. J., 187
Dodson, G. G., 187
Donnay, J. D. H., 278
Donnay, G., 278
Dore, J. C., 154
Drexel, W., 140
Duckworth, J. A. K., 91, 352
Dunlop, R. S., 342
Dunning, J. R., 163, 166, 193
Dwight, K., 505
Dyatkina, M. E., 200
Dyer, R. F., Fig. 186, 291

Ebdon, F. R., 541
Eden, R. C., 375
Egelstaff, P. A., 58, 288, 372, 553, 556, 557, 560, 563, 572
Eisenhauer, C. M., 300
Ekstein, H., 524
Elliott, N., 271, 278, 389, 391, 402, 449, 455, 458, 460, 503
Elliott, R. J., 314
Ellis, W. C., 381, 402
Ellison, R. D., 342
Elsasser, W. M., 1
Enderby, J. E., 556, 557, 560, 563
Erickson, R. A., 134, 212, 229, 257, 462, 463
Ericson, M., 305, 466
Ernst, M., 597
Ewald, P. P., 68

Farnoux, B., 596
Felcher, G. P., 243, 406
Feld, B. T., 163
Fender, B. E. F., 240, 254, 452, 589, 590
Fermi, E., 78, 156, 158
Ferrier, R. P., 24, 239
Ferraris, G., 347
Feshbach, H., 25
Fine, M. E., 402
Fischer, P., 41, 458
Fleeman, J., 515
Foldy, L. L., 23
Foote, H. L., 103
Forsyth, J. B., 403, 443
Fournet, G., 597
Fowler, I. L., 12
Franchini, A. M., 347
Frazer, B. C., 251, 333, 341, 506, 508
Freeman, A. J., 225, 227, 237, 245, 250, 538
Freier, G. D., 325
Froman, P. O., 289
Fruchart, D., 485, 486
Fruchart, R., 485, 486
Fujii, Y., 407

Gajewski, H., 579
Gans, R., 592
Garrett, B. S., 345
Gardner, W. E., 344
de Gennes, P. G., 305, 313, 314, 466
Glaser, W., 140, 153
Gobalakichena, S., 305
Golberg, D., 103
Goldberger, M. L., 162
Goldsmith, H. H., 163
Golovkin, V. S., 403
Gompf, F., 140, 153
Gorter, C. J., 543
Greiner, E. S., 381, 402
Grieveson, P., 390
Guinier, A., 597
Gurevich, I. I., 59, 61

Haaymann, P. W., 501
Halg, W., 109, 458
Halban, H., 1
Halliday, D., 47
Halpern, O., 32, 191, 205, 463, 512, 516, 521, 524, 531
Hamermesh, M., 24, 47, 512, 524
Hamilton, W. C., 326, 347, 355, 501
Harding, M. M., 187
Hastings, J. M., 134, 281, 388, 389, 391, 403, 449, 455, 458, 460, 503, 504
Havens, W. W. Jr., 163, 166
Haywood, B. C. G., 129, 150, 372
Heaton, Le Roy, 99, 115, 138
Heilmann, E. L., 387, 499
Heller, R. B., 519, 521
Helmholz, L., 328
Henfrey, A. W., 589, 590
Henson, R. W., 587
Hepworth, M. A., 496
Herpin, A., 424, 466
Hewat, A., 131, 132
Hibdon, C. T., 44
Hicks, T. J., 541
Hinks, D. G., 392
Hirone, T., 458
Hitterman, R. L., 138
Hodgkin, D. C., 187, 360
Holmryd, S., 571
Holstein, T., 516, 521
Holtzmann, R. H., 515, 521
van Hove, L., 314, 546, 551
Hubbard, C. R., 143
Hughes, D. J., 31, 169, 295, 300, 314, 515, 519, 521, 524
van de Hulst, H. C., 592
Hume-Rothery, W., 398
Hurst, D. G., 288, 578
Hvoslef, J., 391

Ibel, K., 596
Ibers, J. A., 347, 356
Ibser, H. W., 163
Irving, H. M., 398
Ishikawa, Y., 472
Iyengar, P. K., 297, 378
Izyumov, Yu. A., 278

Jack, K. H., 390, 496
Jackson, J. D., 168
Jacobson, A. J., 240, 254
Jacobson, R. A., 143
Jacrot, B., 139, 304, 305, 314, 315, 466, 596
James, R. W., 56, 66, 179, 196
Janik, J. A., 302
Janik, J. M., 302
Jauho, P., 48
Jay, A. H., 422
Johnson, C. K., 342
Johnson, D. A. G., 154
Johnson, M. H., 32, 191, 205, 463, 512, 531
Johnson, R. G., 568
Johnston, D. F., 251
Jones, F. W., 381
Jude, R. J., 88, 325, 351, 366

Kadar, G., 429, 431
Kagan, Yu., 300
Kaldis, E., 41
Kamb, B., 326
Kaplan, T. A., 505
Kasper, J. S., 2, 154, 393, 405, 432
Keating, D. T., 384
Khan, D. C., 229
Kimura, M., 2, 154
Kirste, R. G., 599
Kittel, C., 311, 504
Kleinman, D., 156
Knopf, K., 178
Kobla, J., 98
Koch, F., 453

Koehler, W. C., 41, 135, 165, 198, 213, 239, 246, 247, 250, 397, 411, 415, 416, 466, 477, 479, 481, 488, 499, 528, 530, 538
Koester, L., 173, 174, 175, 178
Konstantinovic, J., 315
Kouvel, J. S., 432
van Kranendonk, J., 310
Kren, E., 429, 431
Krieger, T. J., 302
Krohn, V. E., 24
Kruse, W. A., 599
Kunitomi, N., 103, 407
Kunumaura, W., 281
Kusch, P., 220
Kuznietz, M., 41

van Laar, B., 449, 450
Lampi, E. E., 325
Lander, G. H., 41
Landon, H. H., 103
Larsson, K. E., 571
Lax, M., 162
Leadbetter, A. J., 575, 576
Leake, J. W., 150
Leciejewicz, J., 327
Lehmann, M. S., 41
Lemaire, R., 450
Levy, H. A., 48, 82, 326, 328, 333, 342, 344, 364
Li, Y-Y., 447
Lindqvist, O., 41
Litchinsky, D., 575
Lomer, W. M., 59, 204
Lonsdale, K., 70
Loopstra, B. O., 110
Lorch, E., 574
Lorthioir, G., 486
Low, G. G. E., 59, 204, 291, 436, 441
Lovesey, S. W., 206, 253, 301, 531
Lowde, R. D., 69, 74, 141, 289, 291, 596
Ludman, C. J., 376
Lushchikov, V. I., 162
Lyons, D. H., 505

McGuire, T. R., 453
McReynolds, A. W., 168
Madar, R., 486
Madhav Rao, L., 596
Maier-Leibnitz, H., 161, 170
Marcinkowski, M. J., 382
Marcus, J. A., 406
Marney, M. C., 141
Marshall, L., 78, 158
Marshall, W., 206, 253, 301, 314, 531.
Martin, D. G., 584, 587
Mason, E. W., 268, 429
Massey, H. S. W., 25, 26
Mayer, A., 60
Mazzone, G., 243
Meardon, B. H., 591
Medrud, R. C., 327
Melkonian, E., 325
Mellor, J., 302
Meneghetti, D., 424
Menyuk, N., 505
Menzinger, F., 243
Mercier, M., 488
Meriel, P., 424, 466
Mika, K., 563
Millman, S., 220
Mitchell, E. W. J., 139, 583, 584, 588
Minakawa, N., 103
Modrzejewski, A., 98
Mook, H. A., 227, 234, 235, 236
Moon, R. M., 41, 227, 237, 239, 247, 250, 416, 530, 538
Moore, F. H., 187, 360
Moore, M. J., 2, 154
Morton, G. A., 168, 324
Morrison, J. A., 402
Motohashi, H., 103
Mott, N. F., 25, 26
Muehlhause, C. O., 44
Mueller, M. H., 41, 99, 138
Murthy, M. R. L. N., 221

Nakai, Y., 407
von Nardoff, R., 592
Nathans, R., 216, 219, 220, 227, 237, 241, 248, 252, 257, 282, 403, 422, 454, 472, 474
Néel, L., 446, 470, 499
Nelkin, M. S., 302
Neresen, N. G., 326
Neronova, N. N., 276, 281
Nicodemus, D. B., 515
Nilsson, N., 91
Nielsen, P., 316
Nishikawa, S., 386
Nix, F. C., 381
North, D. M., 556, 557, 563

Oberteuffer, J. A., 406
O'Connor, B. H., 187, 360
Olsen, C. E., 251
Oyamada, Y., 2, 154
Ozerov, R. P., 278

Padlo, I., 327
Page, D. I., 560, 561, 563
Pal, L., 429
Palevsky, H., 295, 300, 302, 314, 379
Palmberg, P. W., 209
Paoletti, A., 103, 219, 227, 237, 243
Parette, G., 314, 315
Passel, L., 316
Pauling, L., 195
Pauthenet, R., 488
Pawley, G. S., 91, 352
Peacock, R. D., 496
Pearce, D. G., 139
Pease, R. S., 333, 340
Peaslee, D. C., 25
Pelah, I., 300
Penney, W. G., 455
Pepinsky, R., 333, 341
Peterson, S. W., 48, 82, 326, 328, 333, 344
Pickart, S. J., 227, 241, 252, 257, 454, 472
Pirila, P., 48
La Placa, S. J., 281, 326, 355
Placzek, G., 555
Plant, J. S., 390, 424, 427
Pokotilovsky, Yu. N., 162
Pope, N. K., 568, 570, 580
Popovici, M., 103
Posnjak, E., 387
Powers, P. N., 1, 512
Prakash, A., 326
Pratt, G. W. Jr., 226
Preiswerk, P., 1
Price, D. L., 392
Prince, D. R., 392
Prince, E., 392, 504

Quicksall, C. O., 143

Rabi, I. I., 220
Rabidean, S. W., 326
Rainwater, L. J., 163, 166
Rao, K. R., 572
Raubenheimer, L. J., 250
Rayleigh, Lord, 592
Raynal, J. M., 591
Reddy, J. F., 41
Reichardt, W., 140, 153
Ricci, F. P., 103, 107
Rietveld, H. M., 93, 130
Rimmer, B., 187
Rimmer, D. E., 256
Ringo, G. R., 24, 169
Ripfel, H., 140
Riste, T., 220, 282, 474, 530
Roberts, B. W., 405
Roberts, F. F., 389, 503
Robertson, J. M., 350
Rodgers, A. L., 139
Rogers, M. T., 328
Romeijn, F. C., 501
Rooksby, H. P., 449
Rose, M. E., 47, 543
Ross, D. K., 139
Roth, M., 591
Roth, W. L., 281, 447, 452
Rouse, K. D., 85, 88, 91, 113
Rowe, J. M., 392
Roy, A. P., 378
Ruderman, I. W., 163, 193
Runciman, W. A., 391
Rundle, R. E., 327, 345
Rush, J. J., 302, 303, 392
Russell, H. N., 199

Sabine, T. M., 347, 356, 360
Safford, G. J., 372, 379
Sailor, V. L., 103
Sakamoto, M., 103, 297, 568
Samanathan, C. S., 221
Sanalan, Y., 139
Sass, R. L., 458
Satya Murthy, N. S., 221
Saunders, F. A., 199
Scatturin, V., 455
Schlecht, P., 601
Schelten, J., 596, 597, 599, 601
Schermer, R. I., 48
Schlapp, R., 455
Schmatz, W., 596, 597, 599, 601
Schneemeyer, L. F., 330
Schobinger-Papamentellos, P., 41
Schwartz, L. H., 406
Schweizer, J., 450
Schwinger, J., 23, 47, 168
Seitz, F., 162
Selove, W., 44
Sequeira, A., 347
Shafer, M. W., 453

Shapiro, F. L., 162
Sharrah, P. C., 544
Sheat, S., 187
Shirane, G., 212, 216, 220, 241, 248, 251, 271, 274, 282, 407, 472, 474, 508
Shockley, W., 381
Shull, C. G., 9, 23, 24, 34, 42, 45, 97, 103, 106, 141, 168, 178, 192, 211, 213, 216, 220, 226, 234, 235, 239, 250, 282, 313, 324, 326, 327, 397, 405, 421, 433, 444, 445, 452, 470, 474, 481, 499, 502, 528, 594, 596
Sidhu, S. S., 115, 424
Siegel, S., 381, 386
Sikka, S. K., 347
Simonsen, S. H., 333
Simpson, W. T., 355
Sinclair, R. N., 154, 297, 308
Slater, J. C., 202, 340
Smart, J. S., 444, 453, 454
Smirnova, T. S., 276, 281
Smith, G. P., 544
Snow, G., 156
Speakman, J. C., 326, 342
Springer, T., 161, 596
Squire, C. F., 203, 444
Squires, G. L., 62, 166, 289, 294, 314
Srinavasan, B. S., 221
Staub, H. H., 512, 515
Steinberger, J., 514
Steinsvoll, O., 248
Stevenson, R. W. H., 312
Stewart, R. F., 355
Stewart, R. J., 583, 588
Stiller, H. H., 297
Stirling, G. C., 139, 298, 376
Stoll, V. E., 109
Stoner, E. C., 201
Strauser, W. A., 192, 211, 445, 452, 470
Strȇlkov, A. V., 162
Stuhrmann, H. B., 599
Sugawara, M., 2, 154
Sturm, W. J., 97
Susman, S., 392
Sutton, R. B., 47, 168
Suzuki, T., 2, 154
Sykes, C., 381
Syrkin, Y. K., 200
Szabo, P., 429

Takeda, S., 2, 154
Takei, W. J., 271, 274, 474
Tamor, S., 463
Tarasov, L. V., 59, 61
Taylor, R. I., 452
Taylor, T. I., 302
Teller, E., 47, 168
Thewlis, J., 3
Titterton, D. H., 89, 347
Todd, M. C. J., 139
Tofield, B. C., 240
Tombs, N. C., 449
Tomiyoshi, S., 2, 154
Trammell, G. T., 224
Trego, A. L., 239
Treuting, R. G., 504
Tsai, B., 203, 444
Tuckey, G. S. G., 139
Tunnicliffe, P. R., 12
Turberfield, K. C., 149

Umakantha, N., 596
Ungerer, H., 175
Usha Deniz, K., 378

Van Vleck, J. H., 201, 305, 310, 313, 454, 463
Venkataraman, G., 378
Verwey, E. J. W., 387, 499, 501
Vijayaraghavan, P. R., 378
Vijayan, M., 187
Villain, J., 314, 424
Vineyard, G. H., 123, 176
Vogt, O., 41
Vos, A., 356
de Vries, G., 402

Waddington, T. C., 376
Walker, C. B., 384
Walker, J., 139
Wallace, J. R., 515, 519, 521
Wallace, J. W., 521
Waller, I., 55, 289
Wallis, D. E., Fig. 186
Wang, S. P., 141
Watanabe, N., 2, 154
Waterstrat, R. M., 393
Watson, R. E., 225, 237, 245, 250, 538
Wattenberg, A., 47
Webb, F. J., 14, 139, 372
Webster, P. J., 427
Wedepohl, P. T., 584

Wedgwood, F. A., 41, 254
Weinstock, R., 51, 55
Weiss, R. J., 168, 176, 227, 271, 402, 582, 592
Weisskopf, V. F., 25
Wells, P., 124
Wenzel, R. G., 326
West, J., 335
Westland, G. J., 496
Wheeler, D. A., 541
Whitaker, M. D., 193
White, J. W., 369, 372, 375
Wick, G. C., 514
Wignall, G. D., 599
Wigner, E., 25
Wild, S., 390
Wilkinson, C., 403
Wilkinson, M. K., 135, 198, 250, 313, 397, 405, 415, 421, 433, 466, 481, 496, 596
Will, G., 454, 508
Williams, H. J., 278
Williams, J. H., 325
Williams, J. M., 330
Williams, R. J. P., 398
Williams, W. G., 541
Willis, B. T. M., 85, 88, 91, 92, 103, 108, 187, 325, 352, 360
Wilson, A. J. C., 133
Wilson, S. A., 143, 349
Windsor, C. G., 312, 313
Wingfield, B. R., 560
Wollan, E. O., 9, 42, 45, 48, 97, 135, 141, 165, 168, 192, 198, 211, 213, 226, 246, 324, 326, 327, 397, 415, 445, 452, 466, 470, 477, 479, 481, 488, 496, 499, 528, 594
Wood, J. H., 226
Wood, R. E., 103
Woods, A. D. B., 285, 297
Woolf, W., 44, 519
Worlton, T. G., 153, 476
Wright, A. C., 154, 575, 576
Wright, C. J., 369
Wu, C. S., 166
Wyckoff, R. W. G., 387

Yafet, Y., 504
Yakel, H. L., 488
Yamada, T., 407
Yamada, Y., 226
Yoshimori, A., 465

Zacharias, J. R., 220
Zachariasen, W. H., 69, 74, 84, 85, 101
Zemlyanov, M. G., 300
Zener, C., 399
Zinn, W. H., 2, 158

SUBJECT INDEX

absorption coefficient, 8, 71, 163, 184, 325
  effect on gravity-mirror measurements, 173
  effect on single-crystal reflections, 74
  measurement of, 113
  powder diffraction correction for, 113, 147
  table of values, 72
abundances, 33, 37
acetic acid, 378
acoustic branch, 285
age-hardening alloys, 591
air-scattering of primary beam, 121
alkali metals, scattering amplitudes of, 178
alloys,
  liquid, 561
  magnetic structures of, 395, 418, 432
  order in, 380
aluminium,
  alloys, 591
  inelastic scattering, 294
  use of for sample holders, 114
ammonium
  chloride, 379
  perchlorate, 302
  sulphate, 301
amorphous solids, 154, 573
angular
  momentum, 199
  resolution, 10, 15, 105, 145, 448
anisotropic
  distribution of electrons, 230
  motion, 57, 351
anomalous scattering, 182, 362, 381
antiferromagnetic materials,
  dispersion relation, 311
  scattering by, 202, 208, 260
antiparallel position, 101, 106
antiphase domains, 271, 403
antisymmetry operations, 275
argon, liquid scattering, 559, 572
aromatic molecules, 349
atomic
  form factor, 22, 192, 205, 225, 226, 255
  motion in liquids, 567
  scattering factors, 22
    table of, 38
automatic data collection, 116, 126

background scattering, 9, 120, 160, 195
beam intensity, 96
benzene, C_6H_6, 349
beryllium,
  as monochromator, 98
  filter, 290, 581
    method of analysis, 297
beryllium fluoride, BeF_2, 576
biological materials, 358, 599
bismuth filter, 62
bond shortening, 363
borated paraffin, 10
boron,
  absorption of, 71
  complex amplitude, 24
  -trifluoride counter, 12, 150
bound-atom cross-section, 32, 164
  table of values, 38
Bragg
  equation, 5, 386
  reflection, polarization effects in, 527
Bravais lattice, magnetic, 276
Breit–Wigner formulae, 25, 31, 184
Brillouin function for antiferromagnetic intensities, 463

cadmium,
  absorption of, 71
  anomalous scattering, 182
  scattering amplitude of, 30, 182
caesium hydrogen chloride, 376
calcium
  carbide, 327
  hydroxide, $Ca(OH)_2$, 364
  sulphate, $CaSO_4 \cdot 2H_2O$, 345
carbides, 327
carbon,
  black, small-angle scattering, 594
  location of, 323
  scattering length of, 158, 165, 174
  tetrafluoride, gas scattering, 579
cerium carbide, 197, 259
chalcopyrite, $CuFeS_2$, 278
chlorine, scattering length of, 44, 174
Christiansen filter, 170, 175
chromites, 499
chromium,
  –antimony, CrSb, 210, 458
  diffraction pattern, 271

chromium—*cont.*
electron distribution, 239
fluoride, CrF_3, 496
magnetic structure of, 271, 401
oxide, Cr_2O_3, 473
paramagnetic scattering of, 198
potassium alum, 344
selenide, CrSe, 257
vanadate, $CrVO_4$, 506
clustering, 583
cobalt,
chloride, $CoCl_2$, 466
electron distribution, 237
fluoride, CoF_2, 462
fluoride, CoF_3, 496
iron, 215, 381, 530, 541
magnetic form factor of, 226
magnetic structure of, 395
mirror reflection, 526
oxide, CoO, 444
sulphate, $CoSO_4$, 506
'co-existing' structures, 420
coherent nuclear scattering cross-section, 32, 50
determination of, 155, 593
polarization analysis of, 533, 535
table of values, 38
cold source, 138, 161, 587
cold-worked metals, 404, 581
collimation, 3, 8, 10, 94, 159
optics of, 103
complex scattering length, 24, 182
compound nucleus, 25
conduction electrons, polarization of, 230
cone spiral, 474
configurational symmetry, 282
conical cycloid, 475
constant-Q method of analysis, 296
copper,
as monochromator, 98, 217
chloride, liquid, CuCl, 563
chloride, $CuCl_2 . 2H_2O$, 344
ferrite, $CuFe_2O_4$, 504
-gold alloys, 381, 386
manganese alloys, 424
sulphate, $CuSO_4 . 5H_2O$, 345
-tin, liquid, 563
-zinc alloys, 381, 385
correlation
functions, 546
techniques, 153
counters, 12
multiple technique, 117
covalency effects, 251
critical
angle for total reflection, 157
scattering, 313
cross-section,
absorption, 71
coherent, 25, 37, 51
disordered, 37
incoherent, 32
magnetic, 193
polarization, 512
cryostats, 133
crystal
monochromators, 5
vibrations, 283
crystalline field, 201
Curie
temperature, 202
-Weiss law, 191, 305, 454
cut-off wavelength, 52, 290
transmission beyond, 581
cyanides, 391
cycloidal spiral, 475, 509
cylindrical sample, 18, 112

data
analysis, 341
display, 117
Debye
factor, 55, 64
frequency distribution, 55, 284
defect structure, 452
defects,
in ferromagnetic alloys, 432
scattering by, 582, 587
deformed metals, 581
depolarization effects, 216, 515
detector, 12, 141, 150
position-sensitive, 149
deuteration, 324, 340, 376, 590, 599
deuterium,
gas scattering, 578
scattering amplitude of, 174, 326
difference projections in Fourier synthesis, 338, 354, 357
diffraction
broadening in small-angle scattering, 592
curve, angular width of, 105, 131
rings in Fourier projections, 332
diffractometer, 7, 16, 126, 219
four-circle, 126
programming of, 17, 116, 126

diffractometer—*cont.*
single-crystal, 126
time-of-flight, 149
diffuse scattering, 57, 191
diffusion in liquids, 375, 567
dilute alloys, 435
direct methods, 341
disordered scattering in ferromagnetic alloys, 432
dispersion
curve for phonons and magnons, 285, 303, 311
formula, Breit–Wigner, 25, 184
parameter in collimation, 105
display equipment, 117
distinct correlation function, 547
domain,
antiphase, 271, 403
magnetic, 210
domain size, measurement of, 519
double
Bragg reflection, 122, 581
refraction of iron, 512, 521
transmission effect, 517
doubling of unit cell, 209, 261
dynamical theory, 178
dysprosium, 409

Einstein model, 284
electronic
anisotropy, 230, 356
configuration, 189, 197
electron, scattering of neutrons by, 23, 63, 181, 189
electrons,
magnetic scattering of, 209
scattering of, 63
energy-analysis, 290
energy, measurement of change on scattering, 58
energy–momentum relation, 286
erbium, 409, 411
form factor, 245
manganite, $ErMnO_3$, 490
oxide, 246
europium,
anomalous scattering of, 185
oxide, 453
sulphide, 453
telluride, 453
exchange integral, 202
experimental methods, 94, 159, 219, 288, 530, 583, 587, 596
extinction, 65, 147
in magnetic scattering, 89

fast-neutron contamination, 160
ferric oxide (haematite), 470
ferrimagnetics, 387, 500
ferrites, 387, 499
ferroelectrics, 131, 340
ferromagnetic
alloys, 432
materials, 202, 207, 260, 308, 395
dispersion relation in, 308
ferrous
chloride, $FeCl_2$, 466
fluoride, FeF_2, 462
oxide, FeO, 444
sulphate, $FeSO_4$, 506
Fick's law, 567
filtered neutrons, 62, 290, 583, 587
filter, single-crystal, 62
fission chamber, 116
flux of neutrons, 8, 114, 117, 358
Foldy effect, 23, 181
form factor, 22, 192, 205, 225, 255, 438
Fourier
projection and synthesis, 330, 332
transform, 197, 229, 558
free-atom cross-section, 32, 38, 164
frequency-distribution function, 283
Friedel's law, 185
Fresnel zone, 66
furnaces, 137, 452

gadolinium,
anomalous scattering of, 185
form factor of, 250
oxide,
scattering by, 250
polarization analysis, 538
structure of, 409
–yttrium alloys, 418
gases,
coherent scattering amplitude, 168
diffraction patterns of, 578
refractive index of, 168
germanium,
defects in, 583
use as monochromator, 99, 117, 217
glasses, 573
glycollic acid, 342
gold–manganese,
AuMn, 420, 423
Au_2Mn, 261, 424

gold manganese—*cont.*
  aluminium, 268, 428
graphite,
  as monochromator, 98
  defects in, 582
  transmission of, 110
gravity-mirror refractometer, 170
guide tubes, 159, 596
gypsum, $CaSO_4 . 2H_2O$, 345

haematite, 470
  polarization analysis, 540
haemoglobin, 600
hard-sphere model, 559
heavy-element compounds, 326
helimagnetism, 261, 465
helium-3 counters, 14, 151
Heusler alloys, 217, 426
hexamethylene tetramine, 325
high-flux reactor, 8, 358, 596
high-temperature measurements, 133, 152
holmium, 409
  manganite, $HoMnO_3$, 490
  oxide, Ho_2O_3, 304
  –yttrium alloys, 418
'hot' source, 138
van Hove functions, 546
hydrates, 343
hydrides, 324, 590
hydrocarbon mirrors, 168
hydrogen bond, 328, 330, 343, 372
hydrogen,
  fluoride, HF, 372
  incoherent scattering, 46, 325, 371
    polarization analysis of, 535
  liquid, 372
  location of, 323, 343
  measurement of $\bar{b}$, 168, 174
  solid, 372
  total scattering cross-section, 46, 325

ice, structure of, 326, 345
ideally imperfect crystals, 75
ilmenite structures, 470
inclined holes, use of, 129
incoherent scattering, 32, 51, 299
  spectroscopy, 299, 369
  polarization analysis of, 535
indium foil, as detector, 141
inelastic
  incoherent scattering, 369
  scattering, 54, 166, 283, 369, 546, 550, 569
    magnetic, 303
infrared spectroscopy, 301, 369
insulin, 187, 362
integrated
  reflection, single-crystal, 69, 79
  scattering, 50
intensity,
  absolute measurement of, 115, 167
  magnetic, calculation, 262, 271, 454, 490
inversion, in spinel structures, 503
iron,
  –aluminium, 381, 422
  –cobalt, 381
    as polarizer, 215
  critical scattering, 314
  diffraction by magnetized samples, 527
  dilute alloys, 435
  fluorides, 496
  magnetic
    form factor, 226
    moment in metal, 234, 395
  mirror reflection, 526
  –nickel Ni_3Fe, 421
  oxides, 444, 470
  –platinum, 428
  polarization effects in, 512, 541
  –rhodium, 241
  scattering length of, 48
  small-angle scattering of, 314, 521, 595
  spin waves in, 596
isotopic
  incoherence, 36, 48
    polarization, analysis of, 533, 534
  substitution, 198, 375, 384, 562

Landé splitting factor, 199, 205, 224
lanthanum perovskites, 477
Larmor precession, 221
Laue pattern, 141, 144, 289
lead,
  liquid scattering, 544, 558
  monochromator, 98, 107, 217
  oxide, 327
least-squares analysis, 342
light elements, positions of, 323
linear accelerators, 2, 154

liquid
  air, immersion in, 81
  metals, 556
  mirrors, 170
liquids, scattering by, 544
lithium
  copper chloride, 278
  hydride, 324
lone-pair electrons, 347, 353
long-range order in alloys, 382
long-wavelength neutrons, 110, 139, 161, 439, 581
low-angle reflection in helimagnetics, 269
low-temperature measurements, 133, 152
luminosity, 105

macromolecules, 599
magnesium,
  –aluminium oxide, 386
  ferrite, 386, 503
  scattering by, 61
  sulphate, $MgSO_4 . 4H_2O$, 346
magnetic
  architecture, 189, 395
  coherent spectra, 201
  critical scattering, 313
  form factor, 192, 205, 225
  interaction vector, 203, 262
  refraction, 521
  satellites, 262
  scattering, 189, 204, 218, 529, 595
    calculation of, 259, 262, 490
    inelastic, 303
    polarization analysis of, 533, 539
    table of amplitudes, 211
  space groups, 276
  spin waves, 307
  structures,
    compilations of, 395
    identification of, 207, 259
  symmetry, 275
  unit cell, 260
magnetite,
  polarization of neutrons from, 215, 528
  structure of, 388, 499
  use of as monochromator, 99, 217
magnetized
  mirrors, 523
  samples, diffraction by, 212, 528
magnons, 307
manganese,
  arsenide, 458
  bismuthide, 458
manganese—*cont.*
  bromide, $MnBr_2$, 469
  chloride, $MnCl_2$, 466
  chromite, $MnCr_2O_4$, 504
  –copper, 424
  diffraction pattern of α-metal, 405
  dioxide, MnO_2, 464
  ferrite, $MnFe_2O_4$, 503
  fluoride, MnF_2,
    covalency in, 252
    magnetic structure, 462
    polarization analysis, 537
  fluoride, MnF_3, 498
  –gallium, Mn_3Ga, 431
  –gallium nitride, Mn_3GaN, 485
  γ-structure, 424
  –germanium, Mn_3Ge, 431
  –gold, MnAu, 420, 423
  –gold, $MnAu_2$, 261, 424
  magnetic form factor of, 226, 240
  –nickel nitride, Mn_3NiN, 487
  oxide, MnO, 210, 444
    Mn_2O_3, 305
  –platinum, Mn_3Pt, 429
  –rhodium, Mn_3Rh, 429
  scattering amplitude of, 44
  selenide, MnSe, 457
  selenide, $MnSe_2$, 460
  sulphate, $MnSO_4$, 305, 508
  sulphide, MnS, 455
  telluride, $MnTe_2$, 460
  –tin, Mn_3Sn, 431
  titanate, $MnTiO_3$, 473
  –zinc nitride, Mn_3ZnN, 486
Marx spectrometer, 298
Maxwell distribution, 3
mercury, scattering length of, 174
metals,
  liquid scattering, 556
  magnetic structures of, 395
methane, gas scattering, 578
methyl alcohol, 375
mirror reflection, 157
modulated-spin structure, 271, 403
molecular
  liquids, 563
  structure, 343
momentum,
  conservation of, 286
  transfer vector, 288
monitors of beam intensity, 116
monochromatic beam,
  analysis of, 101

monochromatic beam—*cont.*
  intensity of, 8, 18
  production of, 5, 94, 583
monochromator,
  adjustment of, 100
  choice of, 97
  choice of Bragg angle for, 108
  rocking curve of, 100
mosaic
  crystal, 63
    distribution of blocks in, 70
  spread, determination of, 79, 101
multi-axis spin structures, 450
multiple
  -counter technique, 117
  scattering, 122, 581
multiplicity of planes, 210, 281
myoglobin, 602

Néel
  structure, 499
  temperature, 202, 454
negative scattering length, 27, 43
neighbouring-atomic-number atoms, 380
neodymium
  ferrite, 485
  form factor, 245
  oxide, 246
neutron,
  crystallography, 82, 331
  Diffraction Commission, 42, 395
  diffractometer, 2, 7, 16, 126, 149
  electron interaction, 23
  inelastic incoherent spectroscopy, 299
  magnetic moment of, 189
  optics, 103, 155
  polarization, 512
  spectrum, 4, 138, 291
  wave nature of, 1
nickel
  -arsenide structures, 457
  coherent scattering amplitude, 48
  critical scattering, 314
  cross-section of, 48, 163
  disordered scattering, 48
    polarization analysis of, 534
  fluoride, NiF_2, 462
  form factor, 226
  magnetic electrons, 228, 235, 395
  –manganese, 382, 599
  oxide,
    covalency in, 253
    structure of, 444
nickel—*cont.*
  sulphate, $NiSO_4$, 506
  titanate, $NiTiO_3$, 473
niobium,
  carbide, NbC, 589
  diffraction pattern, 596
  hydride, 590
nitrogen, gas scattering, 168, 578
non-collinear spins, 206, 429, 487, 504
non-stoichiometric compounds, 588
nuclear
  polarization, 34, 531
  reactor, 2, 8, 114, 117, 358
  scattering, 22
    polarization analysis of, 539
  spin incoherence, 33, 45, 533
nul-matrix alloys, 115

optical branch, 285
orbital momentum, 199
  form factor for, 225, 245
ortho-hydrogen, 47
oxalic acid dihydrate, 325, 345, 357
oxides of transition metals, 444
oxygen,
  location of, 323, 339
  parameter in ferrites, 387
  scattering by gas, 578

pair-distribution function, 548
palladium, form-factor, 241
parabolic law, for single crystals, 78
para-hydrogen, 47
parallel position, 101, 106
paramagnetic scattering, 191, 259, 305
  polarization, analysis of, 535
partial structure factors, 562
Pauli spin matrices, 531
Pendellösung fringes, 178
penetration distance, 67
perfect crystal, 63
perovskite structures, 477
phase determination for structure amplitude, 185
phonon
  absorption and emission, 59, 286
  spectrum, 283
phosphonium iodide, PH_4I, 303
photographic recording, 141
Placzek correction, 555
platinum–iron alloys, Pt_3Fe, 428
plutonium, use of foil as filter, 99

polarization
analysis, 206, 512, 529
cross-section, 512
effects on Bragg intensities, 527
of neutrons by diffraction, 213, 512
by mirror-reflection, 512, 526
by transmission, 512
of nuclei, 34, 521
X-rays, 21
polarized neutrons, use of, 226, 442
polarizing
crystals, 215, 217
filters, 541
polymers, 599
position-sensitive detectors, 149
potassium
bifluoride, KHF_2, 328, 372
cobalticyanide, 391
cyanide, 391
dihydrogen phosphate, 333
tri-hydrogen succinate, 342
potential scattering, 25, 42
powder diffraction
method, 111, 130, 144
patterns,
angular resolution, 15, 105, 145, 154, 448
incoherent scattering in, 93
measurement of $\bar{b}$ from, 166
profile refinement of, 93, 130
powder samples,
containers for, 114
cylindrical form of, 112
size of, 114
transmission form of, 111
preferred orientation in profile refinement, 131
pressure, measurement at high, 152, 476
primary extinction, 65
profile-refinement of powder patterns, 20, 93, 130, 391
programming of diffractometer, 17, 116
protons, polarized, 541
pulsed reactors, 144, 152

Q,
–constant technique, 296
crystallographic measurement, 80
Q, momentum transfer vector, 288
quartz
as monochromator, 107
filter, 62
quasi-elastic scattering, 375, 546, 569
quenching of orbital momentum, 201, 454
radial distribution curve of electrons, 197, 227
radiation shield, 133
radium–beryllium source, 1, 8
radius,
electron, 204
nuclear, 26
of gyration, 597
rare-earth
alloys, 418
atoms,
magnetic moments of, 200, 221
scattering by, 221
carbides, 327
iron perovskites, 481
metals,
absorption of, 71
form factors of, 245
magnetic structures of, 407
oxides, 453
Raman spectroscopy, 301, 369
reactor, nuclear, flux of, 8, 114, 117, 358
reciprocal lattice, 149, 286
recording of diffraction patterns, 16, 117
refinement,
joint of neutron and X-ray data, 352
of structures, 130, 143, 391
reflection method, 75
reflectivity
of mirror surface, 157
of mosaic crystal, 75
refraction broadening in small-angle scattering, 592
refractive index, 155
of gases, 168
of magnetized iron, 521
refractometer, gravity mirror, 170
relaxation phenomena, 152
resolution, 105, 145, 154
resonance
parameters, 26, 182
scattering, 25, 42
resorcinol, $C_6H_4(OH)_2$, 349
rhodium,
form factor, 241
scattering length, 29
riding motion, 365
rigid-body motion, 351, 366
rocking curve of monochromator, 100
rotating-crystal spectrometer, 292

rotation of ions, 300, 361, 379
rubidium,
  liquid scattering, 559
  –manganese fluoride, $RbMnF_3$, 312
Russell–Saunders coupling, 199

salicylic acid, 366
samarium,
  anomalous scattering of, 185, 362
  form factor of, 246
  structure of, 416
sample, size of, 15, 83, 114
satellite reflections, 262, 403
scattering
  cross-section, 24, 32, 37
    measurement of, 163
  law for liquids, 551
  length, 24, 155, 204, 531
    comparison of for neutrons and X-rays, 38
    complex values of, 24, 182
    determination of, 155, 593
    sign of, 43, 155
    table of values, 38, 382
  vector, 203
Schwinger effect, 23
secondary extinction, 65, 329
    correction for, 88
second-order reflections, 6, 99, 391
selection rules in spectroscopy, 371
self-correlation function, 547
series-termination errors, 331
shielding material, 10, 14
short-range order, 385
  in glass, 573
  in liquids, 545
  magnetic, 195
short-wavelength neutrons, 140
Shubnikov groups, 276
sialons, 391
sign of scattering length, 43
σ-phases, 394
silica, diffraction pattern of, 574
silicon,
  nitrides, 391
  scattering length of, 178, 594
silver, scattering length of, 50
single-crystal
  analysis, 124, 149, 328
  diffractometer, 16, 149
  filter, 62
single crystals,
  experimental measurements with, 78
single crystals—*cont.*
  mounting of, 126
  permissible size of, 83, 351
  reflection by, 75
single-transmission effect, 514
sinusoidal variation of moments, 271, 403, 411
small-angle scattering, 176, 314, 521, 587, 591
sodium,
  cyanide, 391
  deuteride, 324
  frequency distribution, 286
  hydride, 324
  liquid scattering, 572
  sesquicarbonate, 343
Soller slits, 11, 105
space group, magnetic, 276
space–time correlation functions, 546
spectrometers, 2, 7, 16, 126, 219, 292, 298, 530
  beryllium-filter, 297
  constant-Q, 296
  Marx, 298
  polarized-neutrons, 219, 529
  rotating crystals, 292
  time-of-flight, 290
spectrum of neutrons, 4, 138, 291
spin
  -flip method of reversing polarization, 220, 530
  form factor, 192, 225, 245
  incoherence, 33, 45
    polarization analysis of, 533
  momentum, 199
  –orbital coupling, 198
  waves, 307
spinel structures, 386, 499
spiral-spin structures, 261, 403, 465
static approximation for liquids, 555
structure analysis, 19, 130, 143, 323
structure
  factor, 51, 206, 332
    comparison for powders and single crystals, 83
    for liquid, 553
  refinement, 130, 143
sulphur, scattering cross-section, 30
superconductivity, 596
superlattices in alloys, 380
susceptibility, 199, 202
symbols, list of, 603
symmetry, magnetic, 275, 282

techniques, experimental, 94, 159, 219, 288, 530, 583, 587, 596
terbium,
  carbide, 260
  form factor of, 248
  structure of, 409
  –yttrium alloys, 418
tetramethylammonium bromide, 372
thermal diffuse scattering, 90
thermal motion, 55, 335, 363
  effect on bond length, 363
thin-crystal criterion, 77, 81
third-order satellites, 274
thorium hydride, 326
thulium, 409, 414
time-of-flight technique,
  for crystallography, 140, 144, 406
  for inelastic scattering, 290
titanium, scattering length of, 48, 50
p-toluidinium bifluoride, 330
total reflection, 157, 170
transition metals,
  alloys of, 419
  electronic distributions of, 230
  form factors of, 226
  magnetic structures, 395
transmission, by magnetized materials, 512
  form of powder sample, 18
  method
    for measuring $\bar{b}$, 163
    for single crystals, 75
    for studying defects, 582
transmission of materials, 74
triangular spin arrangements, 485
triazine, 357
triple-axis spectrometer, 296

ultra-cold neutrons, 161
umbrella structure of CrSe, 459
uranium,
  depleted, 10
  hydride, 251, 326
  magnetic form factor of, 250
  nitride, 251
uranium—*cont.*
  oxides, 251, 326
  phosphide, 138, 251

vanadium,
  diffraction pattern of metal, 407
  incoherent scattering, 300
    polarization analysis of, 535
  phonon spectrum of, 300
  scattering by, 48
  use of for sample holders, 114, 136
velocity selector, 291
vertical diffractometer, 16
vitamin $\mathbf{B}_{12}$, 360

water,
  molecular shape in hydrates, 343
  scattering by, 569
wavelength of neutron, 2, 138
  change of in inelastic scattering, 58
  choice for diffraction, 6, 98
wave vector, 284
weighting of spin states, 33
white radiation, use of, 140
wood, shielding, 10

X-ray
  diffraction, comparison with neutrons, 2, 19, 63, 553
  scattering, 21
    table of values, 38
  –neutron joint refinement, 352
yttrium,
  alloys with rare earths, 418
  manganite, $YMnO_3$, 488

zinc, liquid scattering, 557
zinc ferrite, $ZnFe_2O_4$, 388, 503
zirconium,
  form factor of, 243
  hydride, 326
  isotopic incoherence, 50
  –zinc, $ZrZn_2$, 243